W9-BYJ-167

Guide to Univariate Process Monitoring and Control

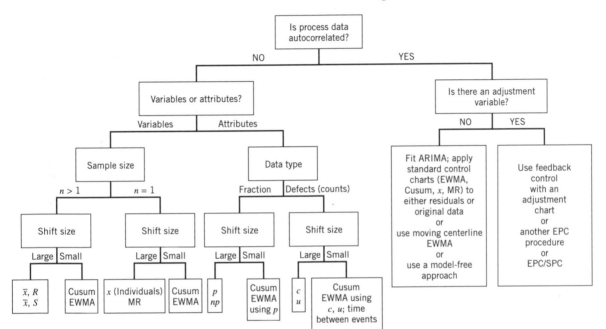

Is process data autocorrelated?

- NO → **Variables or attributes?**
  - Variables → **Sample size**
    - $n > 1$ → **Shift size**
      - Large → $\bar{x}, R$ / $\bar{x}, S$
      - Small → Cusum EWMA
    - $n = 1$ → **Shift size**
      - Large → x (Individuals) MR
      - Small → Cusum EWMA
  - Attributes → **Data type**
    - Fraction → **Shift size**
      - Large → p / np
      - Small → Cusum EWMA using p
    - Defects (counts) → **Shift size**
      - Large → c / u
      - Small → Cusum EWMA using c, u; time between events

- YES → **Is there an adjustment variable?**
  - NO → Fit ARIMA; apply standard control charts (EWMA, Cusum, x, MR) to either residuals or original data or use moving centerline EWMA or use a model-free approach
  - YES → Use feedback control with an adjustment chart or another EPC procedure or EPC/SPC

Introduction to
Statistical
Quality
Control

Fourth Edition

Introduction to
Statistical
Quality
Control

Fourth Edition

Douglas C. Montgomery
Arizona State University

John Wiley & Sons, Inc.

New York • Chichester • Weinheim • Brisbane • Singapore • Toronto

Editor	Wayne Anderson
Marketing Manager	Katherine Hepburn
Associate Production Director	Lucille Buonocore
Senior Production Editor	Monique Calello
Senior Designer	Kevin Murphy
Illustration Coordinator	Gene Aiello
Illustration Studio	Radiant/Precision Graphics

This book was set in Times Roman by Progressive Information Technologies and printed and bound by R.R. Donnelley & Sons, Crawfordsville. The cover was printed by Brady Palmer Printing Company.

This book is printed on acid-free paper.

The paper in this book was manufactured by a mill whose forest management programs include sustained yield harvesting of its timberlands. Sustained yield harvesting principles ensure that the number of trees cut each year does not exceed the amount of new growth.

Copyright © 2001, by John Wiley & Sons, Inc. All rights reserved.

No part of this publication may be reproduced, stored in a retrieval system or transmitted in any form or by any means, electronic, mechanical, photocopying, recording, scanning or otherwise, except as permitted under Sections 107 or 108 of the 1976 United States Copyright Act, without either the prior written permission of the Publisher, or authorization through payment of the appropriate per-copy fee to the Copyright Clearance Center, 222 Rosewood Drive, Danvers, MA 01923, (508) 750-8400, fax (508) 750-4470. Requests to the Publisher for permission should be addressed to the Permissions Department, John Wiley & Sons, Inc., 605 Third Avenue, New York, NY 10158-0012, (212) 850-6011, fax (212) 850-6008, E-Mail: PERMREQ@WILEY.COM.

To order books or for customer service call 1-800-CALL-WILEY (225-5945).

Library of Congress Cataloging-in-Publication Data
Montgomery, Douglas C.
  Introduction to statistical quality control / Douglas C. Montgomery.—4th ed.
    p. cm.
  Includes bibliographical references and index.
  ISBN 0-471-31648-2 (cloth : alk. paper)
  1. Quality control—Statistical methods. 2. Process control—Statistical methods. I. Title.

TS156.M64 2001   00-039237
658.5'62—dc21   CIP

Printed in the United States of America

10 9 8 7 6 5 4 3 2 1

About the Author

Douglas C. Montgomery, Professor of Industrial Engineering at Arizona State University, received his B.S., M.S., and Ph.D. degrees from Virginia Polytechnic Institute, all in engineering. From 1969 to 1984 he was a faculty member of the School of Industrial & Systems Engineering at the Georgia Institute of Technology; from 1984 to 1988 he was at the University of Washington, where he held the John M. Fluke Distinguished Chair of Manufacturing Engineering, was Professor of Mechanical Engineering, and Director of the Program in Industrial Engineering.

Dr. Montgomery has research and teaching interests in engineering statistics including statistical quality control techniques, design of experiments, regression analysis and empirical model building, and the application of operations research methodology to problems in manufacturing systems. He has authored and coauthored many technical papers in these fields and is an author of eleven other books. Dr. Montgomery is a Fellow of the American Society for Quality Control, a Fellow of the American Statistical Association, a Fellow of the Royal Statistical Society, and a Fellow of the Institute of Industrial Engineers. He is a Shewhart Medalist of the American Society for Quality Control, and has also received the Brombaugh Award, the William G. Hunter Award, and the Shewell Award from the ASQC. He is a recipient of the Ellis R. Ott Award. He is a former editor of the *Journal of Quality Technology*, the current editor of *Quality and Reliability Engineering International*, and serves on the editorial boards of several journals.

Preface

This book is about the use of modern statistical methods for quality control and improvement. It provides comprehensive coverage of the subject from basic principles to state-of-the-art concepts and applications. The objective is to give the reader a sound understanding of the principles and the basis for applying them in a variety of both product and nonproduct situations. Although statistical techniques are emphasized throughout, the book has a strong engineering and management orientation. Extensive knowledge of statistics is not a necessary prerequisite for using this book. Readers whose background includes a basic course in statistical methods will find much of the material easily accessible.

The book is an outgrowth of over 25 years of teaching, research, and consulting in the application of statistical methods in quality engineering and quality improvement. It is designed as a textbook for students enrolled in colleges and universities who are studying engineering, management, statistics, and related fields and are taking a first course in statistical quality control. This course is often taught at the junior or senior level. All of the basic topics for this course are covered in detail. There is also some advanced material in the book, which could be used with advanced undergraduates who have had some previous exposure to the basics, or in a course aimed at graduate students. I have also used the text materials extensively in programs for professional practitioners, including quality and reliability engineers, manufacturing and development engineers, managers, procurement specialists, marketing personnel, technicians and laboratory analysts, and some inspectors and operating personnel. Many professionals have also used the material for self-study and preparation for certification examinations.

Chapter 1 is an introduction to the philosophy and basic concepts of quality improvement. It notes that quality has become a major business strategy and that organizations that successfully improve quality can increase their productivity, enhance their market penetration, and achieve greater profitability and a strong competitive advantage. Some of the managerial and implementation aspects of quality improvement are included.

Following the introductory chapter, the book is divided into five parts. Part I is a description of statistical methods useful in quality improvement. Topics covered include sampling and descriptive statistics, the basic notions of probability and probability distributions, point and interval estimation of parameters, and statistical hypothesis testing. These topics are usually covered in a basic course in statistical methods; however, their presentation in this text is from the quality-engineering viewpoint. My experience has been that even readers with a strong statistical background will find the approach to this material useful and somewhat different from that of a standard statistics textbook.

Part II contains four chapters covering the basic methods of statistical process control (or SPC) and methods for process capability analysis. Although several SPC problem-solving tools are discussed (including Pareto charts and cause-and-effect diagrams, for example), the primary focus in this section is on the Shewhart control chart for variables and attributes. The Shewhart control chart is certainly not new, but its use in modern-day business and industry is of tremendous value. Process capability analysis and methods for evaluating measurement systems are discussed in detail in Chapter 7.

There are four chapters in Part III that present some more advanced SPC methods. Chapter 8 is devoted to the cumulative sum and exponentially weighted moving average control charts. This is not really very advanced material, but it is a nice transition between the traditional topics of Parts II and III. The other chapters in Part III cover other important univariate control charts such as procedures for short production runs, autocorrelated data, and multiple-stream processes (Chapter 9), multivariate process monitoring and control (Chapter 10), and feedback adjustment techniques (Chapter 11). Some of this material is at a higher level than Part II, but it is accessible by advanced undergraduates or first-year graduate students. Much of this material forms the basis of a second course in statistical quality control and improvement for this audience.

Part IV contains two chapters that show how statistically designed experiments can be used for process design, development, and improvement. Chapter 12 presents the fundamental concepts of designed experiments and introduces the reader to factorial and fractional factorial designs, with particular emphasis on the two-level system of designs. These designs are used extensively in industry for factor screening and process characterization. Although the treatment of the subject is not extensive and is no substitute for a formal course in experimental design, it will enable the reader to appreciate more sophisticated examples of experimental design. Chapter 13 introduces response surface methods and designs, illustrates evolutionary operation (EVOP) for process monitoring, and shows how statistically designed experiments can be used for process robustness studies. Chapters 12 and 13 emphasize the important interrelationship between statistical process control and experimental design for process improvement.

There are two chapters on acceptance sampling in Part V. The focus is on lot-by-lot acceptance sampling, although there is some discussion of continuous sampling and MIL STD 1235C in Chapter 15. Other sampling topics presented include various aspects of the design of acceptance-sampling plans, a discussion of MIL STD 105E, MIL STD 414 (and their civilian counterparts, ANSI/ASQC ZI.4 and ANSI/ASQC ZI.9), and other techniques such as chain sampling and skip-lot sampling.

Throughout the book guidelines are given for selecting the proper type of statistical technique to use in a wide variety of product and nonproduct situations. There are also extensive references to journal articles and other technical literature that should assist the reader in applying the methods described.

CHANGES IN THE FOURTH EDITION

Based on my own teaching experiences and extensive feedback from other users of the text, I have made several important changes in this edition of the book. I have grouped all the basic SPC and process capability techniques into a single section of the book (Part II) to facilitate topical flow in a first course. I have also expanded the material on process and measurement systems capability analysis. Much of the process monitoring and control material in Part III is either new or an expanded version of topics that were only briefly introduced in earlier editions. There are numerous examples that illustrate computer usage for construction and application of control charts and other techniques. Part IV on process improvement with designed experiments has been reorganized to move more quickly to the factorial design concept. I have also continued to illustrate how the computer is used in the analysis of data from designed experiments. I realize that it is possible to present this material without introducing the student to the analysis of variance (indeed, some books have done that) and give a manual analysis procedure for the two-level designs. I chose to use the analysis of variance because when the students plan and conduct experiments in the real world they will use a computer package that uses this method. It's a disservice to the students to teach them otherwise.

SUPPORTING TEXT MATERIALS

Computer Software

The computer plays an important role in a modern quality control course. This edition of the book uses Minitab as the primary illustrative software package. Instructors may order this book with Minitab included. Another option is a set of statistical programs supporting most of the control charting and experimental design methods in the book that are add-ins to the popular Microsoft Excel spreadsheet program. I strongly recommend that one of these alternatives be specified and that the course have a meaningful computing component.

Supplemental Text Material

I have written a set of supplemental material to augment the book. The supplemental material contains topics that I could not easily fit in the text without seriously disrupting the flow. Some of this material is proofs or derivations, new topics of a (sometimes) more advanced nature, supporting details concerning remarks or concepts presented in the text, and answers to frequently asked questions. There are many references to the supplemental material throughout the text. The supplemental material is an interesting set of accompanying readings for anyone curious about the field. It is available on the Instructor's CD-ROM and on the Web at www.wiley.com/college/montgomery.

Student Solutions Manual

The text contains answers to most of the odd-numbered exercises. There is also a Student Solutions Manual available on CD-ROM that presents comprehensive annotated solutions to these same odd-numbered problems. This is an excellent study aid that many text users will find extremely helpful.

Instructor's CD-ROM

The instructor's CD-ROM contains the following:

1. Solutions to the text problems
2. The supplemental text material described above
3. A set of Power Point slides for the basic SPC course
4. Data sets from the book in electronic form

Web Site

The Web Site, **www.wiley.com/college/montgomery**, contains the supplemental text material and the data sets in electronic form. It will also be used to post items of interest to text users.

ACKNOWLEDGMENTS

Many people have generously contributed their time and knowledge of statistics and quality improvement to this book. I would like to thank Dr. Bill Woodall, Dr. Joe Sullivan, Dr. George Runger, Dr. Brian Macpherson, Dr. Bob Hogg, Mr. Eric Ziegel, Dr. Joe Pignatiello, Dr. John Ramberg, Dr. Ernie Saniga, Dr. Enrique Del Castillo, and Dr. Jim Alloway for their thorough and insightful comments on the previous edition. They generously shared many of their ideas and teaching experiences with me, leading to substantial improvements in the book.

Over the years since the first edition was published I have received assistance and ideas from a great many other people. A complete list of colleagues with whom I have interacted would be impossible to enumerate. However, some of the major contributors and their professional affiliations are as follows: Dr. J. Bert Keats, Dr. George C. Runger, Dr. Mary R. Anderson, Dr. Dwayne A. Rollier, and Dr. Norma F. Hubele, Arizona State University; Mr. Seymour M. Selig, formerly of the Office of Naval Research; Dr. Lynwood A. Johnson, Dr. Russell G. Heikes, Dr. David E. Fyffe, and Dr. H. M. Wadsworth, Jr., Georgia Institute of Technology; Dr. Geoff Vining, Virginia Tech; Dr. Richard L. Storch, University of Washington; Dr. Cynthia A. Lowry, formerly of Texas Christian University; Dr. Christina M. Mastrangelo, the University of Virginia;

Dr. Smiley Cheng, Dr. John Brewster, and Dr. Fred Spiring, the University of Manitoba; Dr. Joseph D. Moder, University of Miami; Dr. Frank B. Alt, University of Maryland; Dr. Kenneth E. Case, Oklahoma State University; Dr. Daniel R. McCarville, Ms. Lisa Custer, Dr. Pat Spagon, and Mr. Robert Stuart, Motorola; Dr. Richard Post, Intel Corporation; Dr. Dale Sevier, Hybritech; Mr. John A. Butora, Mr. Leon V. Mason, Mr. Lloyd K. Collins, Mr. Dana D. Lesher, Mr. Roy E. Dent, Mr. Mark Fazey, Ms. Kathy Schuster, Mr. Dan Fritze, Dr. J. S. Gardiner, Mr. Ariel Rosentrater, Mr. Lolly Marwah, Mr. Ed Schleicher, Mr. Armin Weiner, and Ms. Elaine Baechtle, IBM; Mr. Thomas C. Bingham, Mr. K. Dick Vaughn, Mr. Robert LeDoux, Mr. John Black, Mr. Jack Wires, Dr. Julian Anderson, Mr. Richard Alkire, and Mr. Chase Nielsen, The Boeing Company; Ms. Karen Madison, Mr. Don Walton, and Mr. Mike Goza, Alcoa Corporation; Mr. Harry Peterson-Nedry, Ridgecrest Vineyards and The Chehalem Group; Dr. Russell A. Boyles, formerly of Precision Castparts Corporation; Dr. Sadre Khalessi and Mr. Franz Wagner, Signetics Corporation; Mr. Larry Newton and Mr. C. T. Howlett, Georgia Pacific Corporation; Mr. Robert V. Baxley, Monsanto Chemicals; Dr. Craig Fox, Dr. Thomas L. Sadosky, Mr. James F. Walker, and Mr. John Belvins, The Coca-Cola Company; Mr. Bill Wagner and Mr. Al Pariseau, Litton Industries; Mr. John M. Fluke, Jr., John Fluke Manufacturing Company; Dr. William DuMouchel, Bell Laboratories; Dr. Paul Tobias, Semitech; and Ms. Janet Olson, BBN Software Products Corporation. I would also like to acknowledge the contributions of my partner in Statistical Productivity Consultants, Mr. Sumner S. Averett. All of these individuals and many others have contributed to my knowledge of the quality-improvement field.

The editorial and production staff at Wiley, particularly Ms. Charity Robey (with whom I worked for many years) and Mr. Wayne Anderson have had much patience with me over the years, and have contributed greatly to the success of this book. Dr. Cheryl Jennings made many valuable contributions by her careful checking of the manuscript and proof materials, and in preparing the solutions. I also thank Dr. Gary Hogg, Chair of the Department of Industrial Engineering at Arizona State University, for his support and for providing a terrific environment in which to teach and conduct research.

I thank the various professional societies and publishers who have given permission to reproduce their materials in my text. Permission credit is acknowledged at appropriate places in the book.

I am also indebted to the member companies of the National Science Foundation/ Industry/University Cooperative Research Center in Quality and Reliability Engineering at Arizona State University. These companies, along with the Office of Naval Research, the National Science Foundation, the Aluminum Company of America, and the IBM Corporation have sponsored much of my research and many of my graduate students for a number of years. Finally, I would like to thank the many users of the previous editions of this book including students, practicing professionals, and my academic colleagues. Many of the changes and (hopefully) improvements in this edition of the book are the direct result of your feedback.

Douglas C. Montgomery
Tempe, Arizona

Contents

CHAPTER 1
Quality Improvement in the Modern Business
Environment   1

1-1 The Meaning of Quality and Quality
     Improvement   **2**
     1-1.1 Dimensions of Quality   **2**
     1-1.2 Quality Engineering Terminology   **6**
1-2 A Brief History of Quality Control and
     Improvement   **8**
1-3 Statistical Methods for Quality Control and
     Improvement   **12**
1-4 Other Aspects of Quality Control and
     Improvement   **17**
     1-4.1 Quality Philosophy and Management
            Strategies   **18**
     1-4.2 The Link between Quality and
            Productivity   **25**
     1-4.3 Quality Costs   **26**
     1-4.4 Legal Aspects of Quality   **32**
     1-4.5 Implementing Quality Improvement   **34**

PART I
Statistical Methods Useful in Quality
Improvement   37

CHAPTER 2
Modeling Process Quality   39

2-1 Describing Variation   **40**
     2-1.1 The Stem-and-Leaf Plot   **40**

2-1.2 The Frequency Distribution and
       Histogram   **43**
2-1.3 Numerical Summary of Data   **45**
2-1.4 The Box Plot   **48**
2-1.5 Sample Computer Output   **49**
2-1.6 Probability Distributions   **51**
2-2 Important Discrete Distributions   **56**
     2-2.1 The Hypergeometric Distribution   **56**
     2-2.2 The Binomial Distribution   **58**
     2-2.3 The Poisson Distribution   **60**
     2-2.4 The Pascal and Related
            Distributions   **61**
2-3 Important Continuous Distributions   **62**
     2-3.1 The Normal Distribution   **63**
     2-3.2 The Exponential Distribution   **69**
     2-3.3 The Gamma Distribution   **71**
     2-3.4 The Weibull Distribution   **73**
2-4 Some Useful Approximations   **74**
     2-4.1 The Binomial Approximation to the
            Hypergeometric   **75**
     2-4.2 The Poisson Approximation to the
            Binomial   **75**
     2-4.3 The Normal Approximation to the
            Binomial   **75**
     2-4.4   Comments on Approximations   **76**
2-5 Exercises   **77**

CHAPTER 3
Inferences about Process Quality   82

3-1 Statistics and Sampling Distributions   **83**
     3-1.1 Sampling from a Normal
            Distribution   **85**

3-1.2 Sampling from a Bernoulli Distribution **89**

3-1.3 Sampling from a Poisson Distribution **89**

3-2 Point Estimation of Process Parameters **90**

3-3 Statistical Inference for a Single Sample **93**

3-3.1 Inference on the Mean of a Population, Variance Known **94**

3-3.2 The Use of P-Values for Hypothesis Testing **98**

3-3.3 Inference on the Mean of a Normal Distribution, Variance Unknown **99**

3-3.4 Inference on the Variance of a Normal Distribution **103**

3-3.5 Inference on a Population Proportion **105**

3-3.6 The Probability of Type II Error **107**

3-3.7 Probability Plotting **110**

3-4 Statistical Inference for Two Samples **114**

3-4.1 Inference for a Difference in Means, Variances Known **114**

3-4.2 Inference for a Difference in Means of Two Normal Distributions, Variances Unknown **117**

3-4.3 Inference on the Variances of Two Normal Distributions **127**

3-4.4 Inference on Two Population Proportions **128**

3-5 What If There Are More Than Two Populations? The Analysis of Variance **130**

3-5.1 An Example **131**

3-5.2 The Analysis of Variance **133**

3-5.3 Checking Assumptions: Residual Analysis **140**

3-6 Exercises **142**

PART II
Basic Methods of Statistical Process Control and Capability Analysis **151**

CHAPTER 4
Methods and Philosophy of Statistical Process Control **153**

4-1 Introduction **154**

4-2 Chance and Assignable Causes of Quality Variation **154**

4-3 Statistical Basis of the Control Chart **156**

4-3.1 Basic Principles **156**

4-3.2 Choice of Control Limits **164**

4-3.3 Sample Size and Sampling Frequency **166**

4-3.4 Rational Subgroups **170**

4-3.5 Analysis of Patterns on Control Charts **172**

4-3.6 Discussion of Sensitizing Rules for Control Charts **175**

4-4 The Rest of the "Magnificent Seven" **177**

4-5 Implementing SPC **184**

4-6 An Application of SPC **186**

4-7 Nonmanufacturing Applications of Statistical Process Control **193**

4-8 Exercises **201**

CHAPTER 5
Control Charts for Variables **206**

5-1 Introduction **207**

5-2 Control Charts for $\bar{x}$ and R **207**

5-2.1 Statistical Basis of the Charts **207**

5-2.2 Development and Use of $\bar{x}$ and R Charts **212**

5-2.3 Charts Based on Standard Values **228**

5-2.4 Interpretation of $\bar{x}$ and R Charts **229**

5-2.5 The Effect of Nonnormality on $\bar{x}$ and R Charts **232**

5-2.6 The Operating-Characteristic Function **233**

5-2.7 The Average Run Length for the $\bar{x}$ Chart **236**

5-3 Control Charts for $\bar{x}$ and S **239**

    5-3.1 Construction and Operation of $\bar{x}$ and S Charts **239**

    5-3.2 The $\bar{x}$ and S Control Charts with Variable Sample Size **244**

    5-3.3 The S^2 Control Chart **248**

5-4 The Shewhart Control Chart for Individual Measurements **249**

5-5 Summary of Procedures for $\bar{x}$, R, and S Charts **260**

5-6 Applications of Variables Control Charts **260**

5-7 Exercises **265**

CHAPTER 6
Control Charts for Attributes 283

6-1 Introduction **284**

6-2 The Control Chart for Fraction Nonconforming **284**

    6-2.1 Development and Operation of the Control Chart **286**

    6-2.2 Variable Sample Size **298**

    6-2.3 Nonmanufacturing Applications **303**

    6-2.4 The Operating-Characteristic Function and Average Run Length Calculations **305**

6-3 Control Charts for Nonconformities (Defects) **308**

    6-3.1 Procedures with Constant Sample Size **308**

    6-3.2 Procedures with Variable Sample Size **319**

    6-3.3 Demerit Systems **322**

    6-3.4 The Operating-Characteristic Function **324**

    6-3.5 Dealing with Low Defect Levels **325**

    6-3.6 Nonmanufacturing Applications **328**

6-4 Choice between Attributes and Variables Control Charts **329**

6-5 Guidelines for Implementing Control Charts **333**

6-6 Exercises **339**

CHAPTER 7
Process and Measurement System Capability Analysis 349

7-1 Introduction **350**

7-2 Process Capability Analysis Using a Histogram or a Probability Plot **352**

    7-2.1 Using the Histogram **352**

    7-2.2 Probability Plotting **355**

7-3 Process Capability Ratios **357**

    7-3.1 Use and Interpretation of C_p **357**

    7-3.2 Process Capability Ratio for an Off-Center Process **362**

    7-3.3 Normality and the Process Capability Ratio **364**

    7-3.4 More about Process Centering **365**

    7-3.5 Confidence Intervals and Tests on Process Capability Ratios **367**

7-4 Process Capability Analysis Using a Control Chart **373**

7-5 Process Capability Analysis Using Designed Experiments **376**

7-6 Gage and Measurement System Capability Studies **377**

    7-6.1 Control Charts and Tabular Methods **377**

    7-6.2 Methods Based on Analysis of Variance **384**

7-7 Setting Specification Limits on Discrete Components **388**

    7-7.1 Linear Combinations **388**

    7-7.2 Nonlinear Combinations **392**

7-8 Estimating the Natural Tolerance Limits of a Process **394**

    7-8.1 Tolerance Limits Based on the Normal Distribution **395**

    7-8.2 Nonparametric Tolerance Limits **396**

7-9 Exercises **397**

PART III
Other Statistical Process Monitoring and
Control Techniques   403

CHAPTER 8
Cumulative Sum and Exponentially Weighted
Moving Average Control Charts   405

8-1 The Cumulative Sum Control Chart   **406**

   8-1.1 Basic Principles: The Cusum Control
   Chart for Monitoring the Process
   Mean   **406**

   8-1.2 The Tabular or Algorithmic
   Cusum for Monitoring the
   Process Mean   **410**

   8-1.3 Recommendations for Cusum
   Design   **415**

   8-1.4 The Standardized Cusum   **417**

   8-1.5 Rational Subgroups   **418**

   8-1.6 Improving Cusum Responsiveness for
   Large Shifts   **418**

   8-1.7 The Fast Initial Response or Headstart
   Feature   **419**

   8-1.8 One-Sided Cusums   **421**

   8-1.9 A Cusum for Monitoring Process
   Variability   **421**

   8-1.10 Cusums for Other Sample
   Statistics   **422**

   8-1.11 The V-Mask Procedure   **423**

8-2 The Exponentially Weighted Moving Average
Control Chart   **425**

   8-2.1 The Exponentially Weighted
   Moving Average Control Chart
   for Monitoring the Process
   Mean   **426**

   8-2.2 Design of an EWMA Control
   Chart   **431**

   8-2.3 Rational Subgroups   **432**

   8-2.4 Robustness of the EWMA to
   Nonnormality   **432**

   8-2.5 Extensions of the EWMA   **433**

8-3 The Moving Average Control
Chart   **437**

8-4 Exercises   **440**

CHAPTER 9
Other Univariate Statistical Process
Monitoring and Control Techniques   443

9-1 Statistical Process Control for Short Production
Runs   **444**

   9-1.1 $\bar{x}$ and R Charts for Short Production
   Runs   **444**

   9-1.2 Attributes Control Charts for Short
   Production Runs   **447**

   9-1.3 Other Methods   **448**

9-2 Modified and Acceptance Control
Charts   **449**

   9-2.1 Modified Control Limits for the $\bar{x}$
   Chart   **449**

   9-2.2 Acceptance Control Charts   **453**

9-3 Control Charts for Multiple-Stream
Processes   **454**

   9-3.1 Multiple-Stream Processes   **454**

   9-3.2 Group Control Charts   **455**

   9-3.3 Other Approaches   **457**

9-4 SPC with Autocorrelated Process
Data   **458**

   9-4.1 Sources and Effects of
   Autocorrelation in Process
   Data   **458**

   9-4.2 Model-Based Approaches   **462**

   9-4.3 A Model-Free Approach   **473**

9-5 Adaptive Sampling Procedures   **478**

9-6 Economic Design of Control
Charts   **479**

   9-6.1 Designing a Control Chart   **479**

   9-6.2 Process Characteristics   **480**

   9-6.3 Cost Parameters   **481**

   9-6.4 Early Work and Semieconomic
   Designs   **482**

   9-6.5 An Economic Model of the $\bar{x}$ Control
   Chart   **484**

   9-6.6 Other Work   **493**

9-7 Overview of Other Procedures   **494**

   9-7.1 Tool Wear   **494**

   9-7.2 Control Charts Based on Other Sample
   Statistics   **495**

   9-7.3 Fill Control Problems   **496**

9-7.4 Precontrol **497**

9-8 Exercises **499**

CHAPTER 10
Multivariate Process Monitoring and
Control **507**

10-1 The Multivariate Quality Control
Problem **508**
10-2 Description of Multivariate Data **510**
  10-2.1 The Multivariate Normal
  Distribution **510**
  10-2.2 The Sample Mean Vector and Covariance
  Matrix **511**
10-3 The Hotelling T^2 Control Chart **512**
  10-3.1 Subgrouped Data **512**
  10-3.2 Individual Observations **522**
10-4 The Multivariate EWMA Control
Chart **526**
10-5 Regression Adjustment **530**
10-6 Control Charts for Monitoring
Variability **532**
10-7 Latent Structure Methods **535**
  10-7.1 Principal Components **535**
  10-7.2 Partial Least Squares **542**
10-8 Exercises **542**

CHAPTER 11
Engineering Process Control and SPC **546**

11-1 Process Monitoring and Process
Regulation **547**
11-2 Process Control by Feedback
Adjustment **548**
  11-2.1 A Simple Adjustment Scheme: Integral
  Control **548**
  11-2.2 The Adjustment Chart **555**
  11-2.3 Variations of the Adjustment
  Chart **557**
  11-2.4 Other Types of Feedback
  Controllers **561**
11-3 Combining SPC and EPC **562**
11-4 Exercises **566**

PART IV
Process Design and Improvement with
Designed Experiments **569**

CHAPTER 12
Factorial and Fractional Factorial Experiments
for Process Design and Improvement **571**

12-1 What Is Experimental Design? **572**
12-2 Examples of Designed Experiments in Process
Improvement **573**
12-3 Guidelines for Designing Experiments **577**
12-4 Factorial Experiments **579**
  12-4.1 An Example **582**
  12-4.2 Statistical Analysis **583**
  12-4.3 Residual Analysis **588**
12-5 The 2^k Factorial Design **591**
  12-5.1 The 2^2 Design **591**
  12-5.2 The 2^k Design for $k \geq 3$ Factors **598**
  12-5.3 A Single Replicate of the 2^k
  Design **611**
  12-5.4 Addition of Center Points to the
  2^k Design **615**
  12-5.5 Blocking and Confounding in the
  2^k Design **620**
12-6 Fractional Replication of the 2^k Design **622**
  12-6.1 The One-Half Fraction of the 2^k
  Design **622**
  12-6.2 Smaller Fractions: The 2^{k-p} Fractional
  Factorial Design **628**
12-7 Exercises **635**

CHAPTER 13
Process Optimization with Designed
Experiments **639**

13-1 Response Surface Methods and Designs **640**
  13-1.1 The Method of Steepest Ascent **642**
  13-1.2 Analysis of a Second-Order Response
  Surface **645**
13-2 Process Robustness Studies **651**
  13-2.1 Background **651**

13-2.2 The Response Surface Approach to Process Robustness Studies **653**

13-3 Evolutionary Operation **662**

13-4 Exercises **669**

PART V
Acceptance Sampling   673

CHAPTER 14
Lot-by-Lot Acceptance Sampling for Attributes   675

14-1 The Acceptance-Sampling Problem **676**

14-1.1 Advantages and Disadvantages of Sampling **677**

14-1.2 Types of Sampling Plans **678**

14-1.3 Lot Formation **679**

14-1.4 Random Sampling **680**

14-1.5 Guidelines for Using Acceptance Sampling **680**

14-2 Single-Sampling Plans for Attributes **682**

14-2.1 Definition of a Single-Sampling Plan **682**

14-2.2 The OC Curve **682**

14-2.3 Designing a Single-Sampling Plan with a Specified OC Curve **689**

14-2.4 Rectifying Inspection **690**

14-3 Double, Multiple, and Sequential Sampling **694**

14-3.1 Double-Sampling Plans **694**

14-3.2 Multiple-Sampling Plans **701**

14-3.3 Sequential-Sampling Plans **701**

14-4 Military Standard 105E (ANSI/ASQC Z1.4, ISO 2859) **705**

14-4.1 Description of the Standard **705**

14-4.2 Procedure **707**

14-4.3 Discussion **712**

14-5 The Dodge–Romig Sampling Plans **714**

14-5.1 AOQL Plans **716**

14-5.2 LTPD Plans **716**

14-5.3 Estimation of Process Average **719**

14-6 Exercises **719**

CHAPTER 15
Other Acceptance-Sampling Techniques   722

15-1 Acceptance Sampling by Variables **723**

15-1.1 Advantages and Disadvantages of Variables Sampling **723**

15-1.2 Types of Sampling Plans Available **724**

15-1.3 Caution in the Use of Variables Sampling **725**

15-2 Designing a Variables Sampling Plan with a Specified OC Curve **726**

15-3 MIL STD 414 (ANSI/ASQC Z1.9) **729**

15-3.1 General Description of the Standard **729**

15-3.2 Use of the Tables **730**

15-3.3 Discussion of MIL STD 414 and ANSI/ASQC Z1.9 **733**

15-4 Other Variables Sampling Procedures **734**

15-4.1 Sampling by Variables to Give Assurance Regarding the Lot or Process Mean **734**

15-4.2 Sequential Sampling by Variables **735**

15-5 Chain Sampling **735**

15-6 Continuous Sampling **737**

15-6.1 CSP-1 **738**

15-6.2 Other Continuous-Sampling Plans **740**

15-7 Skip-Lot Sampling Plans **741**

15-8 Exercises **745**

Appendix   749

I. Summary of Common Probability Distributions Often Used in Statistical Quality Control **751**

II. Cumulative Standard Normal Distribution **752**

III. Percentage Points of the χ^2 Distribution **754**

IV. Percentage Points of the t Distribution **755**

V. Percentage Points of the F Distribution **756**

VI. Factors for Constructing Variables Control Charts **761**

VII. Factors for Two-Sided Normal Tolerance Limits **762**

VIII. Factors for One-Sided Normal Tolerance Limits **763**

IX. Random Numbers **764**

Bibliography **765**

Answers to Selected Exercises **777**

Index **791**

Quality Improvement in the Modern Business Environment

CHAPTER OUTLINE

1-1 THE MEANING OF QUALITY AND QUALITY IMPROVEMENT

   1-1.1 Dimensions of Quality

   1-1.2 Quality Engineering Terminology

1-2 A BRIEF HISTORY OF QUALITY CONTROL AND IMPROVEMENT

1-3 STATISTICAL METHODS FOR QUALITY CONTROL AND IMPROVEMENT

1-4 OTHER ASPECTS OF QUALITY CONTROL AND IMPROVEMENT

   1-4.1 Quality Philosophy and Management Strategies

   1-4.2 The Link Between Quality and Productivity

   1-4.3 Quality Costs

   1-4.4 Legal Aspects of Quality

   1-4.5 Implementing Quality Improvement

CHAPTER OVERVIEW

This book is about the use of statistical methods and other problem-solving techniques to improve the quality of the products used by our society. These products consist of **manufactured goods** such as automobiles, computers, and clothing, as well as **services** such as the generation and distribution of electrical energy, public transportation, banking, and health care. Quality-improvement methods can be applied to any area within a company or organization, including manufacturing, process development, engineering design, finance and accounting, marketing, and field service of products. This text presents the technical tools that are needed to achieve quality improvement in these organizations.

In this chapter we give the basic definitions of quality, quality improvement, and other quality engineering terminology. We also discuss the historical development of quality improvement methodology, and overview the statistical tools essential for modern professional practice. A brief discussion of some management and business aspects for implementing quality improvement is also given.

1-1 THE MEANING OF QUALITY AND QUALITY IMPROVEMENT

We may define **quality** in many ways. Most people have a conceptual understanding of quality as relating to one or more desirable characteristics that a product or service should possess. Although this conceptual understanding is certainly a useful starting point, we will give a more precise and useful definition.

Quality has become one of the most important consumer decision factors in the selection among competing products and services. The phenomenon is widespread, regardless of whether the consumer is an individual, an industrial organization, a retail store, or a military defense program. Consequently, understanding and improving quality is a key factor leading to business success, growth, and an enhanced competitive position. There is a substantial return on investment from improved quality and from successfully employing quality as an integral part of overall business strategy. In this section we give some operational definitions of quality and quality improvement. We begin with a brief discussion of the different dimensions of quality and some basic terminology.

1-1.1 Dimensions of Quality

The quality of a product can be evaluated in several ways. It is often very important to differentiate these different **dimensions of quality.** Garvin (1987) provides an excellent discussion of eight components or dimensions of quality. We summarize his key points concerning these dimensions of quality as follows:

1. **Performance** (will the product do the intended job?)
    Potential customers usually evaluate a product to determine if it will perform certain specific functions and determine how well it performs them. For example, you could evaluate spreadsheet software packages for a PC to determine which data manipulation operations they perform. You may discover that one outperforms another with respect to the execution speed.

2. **Reliability** (how often does the product fail?)
    Complex products, such as many appliances, automobiles, or airplanes, will usually require some repair over their service life. For example, you should expect that an automobile will require occasional repair, but if the car requires frequent repair, we say that it is unreliable. This is an industry in which the customer's view of quality is greatly impacted by the reliability dimension of quality.

3. **Durability** (how long does the product last?)
    This is the effective service life of the product. Customers obviously want products that perform satisfactorily over a long period of time. Again, the automobile and major appliance industries are examples of businesses where this dimension of quality is very important to most customers.

4. **Serviceability** (how easy is it to repair the product?)
    There are many industries in which the customer's view of quality is directly influenced by how quickly and economically a repair or routine maintenance activity can be accomplished. Examples include the appliance and automobile industries and many types of service industries (how long did it take a credit card company to correct an error in your bill?).

5. **Aesthetics** (what does the product look like?)
    This is the visual appeal of the product, often taking into account factors such as style, color, shape, packaging alternatives, tactile characteristics, and other sensory features. For example, soft-drink beverage manufacturers have relied on the visual appeal of their packaging to differentiate their product from other competitors.

6. **Features** (what does the product do?)
    Usually, customers associate high quality with products that have added features; that is, those that have features beyond the basic performance of the competition. For example, you might consider a spreadsheet software package to be of superior quality if it had built-in statistical analysis features while its competitors did not.

7. **Perceived Quality** (what is the reputation of the company or its product?)
    In many cases, customers rely on the past reputation of the company concerning quality of its products. This reputation is directly influenced by failures of the product that are highly visible to the public or that require product recalls, and by how the customer is treated when a quality-related problem with the product is reported. Perceived quality, customer loyalty, and repeated business are closely interconnected. For example, if you make regular business trips using a particular airline, and the flight almost always arrives on time and the airline company does not lose or damage your luggage, you will probably prefer to fly on that carrier instead of its competitors.

8. **Conformance to Standards** (is the product made exactly as the designer intended?)
    We usually think of a high-quality product as one that exactly meets the requirements placed on it. For example, how well does the hood fit on a new car? Is it perfectly flush with the fender height, and is the gap exactly the same on all sides? Manufactured parts that do not exactly meet the designer's requirements can cause significant quality problems when they are used as the components of a more complex assembly. An automobile consists of several thousand parts. If each one is just slightly too big or too small, many of the components will not fit together properly, and the vehicle (or its major subsystems) may not perform as the designer intended.

We see from the foregoing discussion that quality is indeed a multifaceted entity. Consequently, a simple answer to questions such as "What is quality?" or "What is quality improvement?" is not easy. The **traditional** definition of quality is based on the viewpoint that products and services must meet the requirements of those who use them.

Definition

Quality means fitness for use.

There are two general aspects of fitness for use: **quality of design** and **quality of conformance.** All goods and services are produced in various grades or levels of quality. These variations in grades or levels of quality are intentional, and, consequently, the appropriate technical term is quality of design. For example, all automobiles have as their basic objective providing safe transportation for the consumer. However, automobiles differ with respect to size, appointments, appearance, and performance. These differences are the result of intentional design differences between the types of automobiles. These design differences include the types of materials used in construction, specifications on the components, reliability obtained through engineering development of engines and drive trains, and other accessories or equipment.

The quality of conformance is how well the product conforms to the specifications required by the design. Quality of conformance is influenced by a number of factors, including the choice of manufacturing processes, the training and supervision of the workforce, the type of quality-assurance system used (process controls, tests, inspection activities, etc.), the extent to which these quality-assurance procedures are followed, and the motivation of the workforce to achieve quality.

Unfortunately, this definition has become associated more with the conformance aspect of quality than with design. This is in part due to the lack of formal education most designers and engineers receive in quality engineering methodology. This also leads to much less focus on the customer and more of a "conformance-to-specifications" approach to quality, regardless of whether the product, even when produced to standards, was actually "fit-for-use" by the customer. Also, there is still a widespread belief that quality is a problem that can be dealt with solely in manufacturing, or that the only way quality can be improved is by "gold-plating" the product.

We prefer a **modern** definition of quality:

Definition

Quality is inversely proportional to variability.

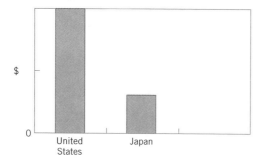

Figure 1-1   Warranty costs for transmissions.

Note that this definition implies that if variability[1] in the important characteristics of a product decreases, the quality of the product increases. As an example of the operational effectiveness of this definition, a few years ago, one of the automobile companies in the United States performed a comparative study of a transmission that was manufactured in a domestic plant and by a Japanese supplier. An analysis of warranty claims and repair costs indicated that there was a striking difference between the two sources of production, with the Japanese-produced transmission having much lower costs, as shown in Fig. 1-1. As part of the study to discover the cause of this difference in cost and performance, the company selected random samples of transmissions from each plant, disassembled them, and measured several critical quality characteristics.

Figure 1-2 is generally representative of the results of this study. Note that the distribution of the critical characteristics for the transmissions manufactured in the United States takes up about 75% of the width of the specifications, implying that very few nonconforming units would be produced. In fact, the plant was producing at a quality level that was quite good, based on the generally accepted view of quality within the company. However, the Japanese plant produced transmissions for which the same critical characteristics take up only about 25% of the specification band. As a result, there is considerably less variability in the critical quality characteristics of the Japanese-built transmissions in comparison to those built in the United States.

There are two obvious questions here: Why did the Japanese do this? How did they do this? The answer to the "why" question is obvious from examination of Fig. 1-1. Reduced variability has directly translated into lower costs. Furthermore, the Japanese-built transmissions shifted gears more smoothly, ran more quietly, and were generally perceived by the customer as superior to those built domestically. Fewer repairs and warranty claims means less **rework** and the reduction of wasted time, effort, and money. Thus, quality truly is inversely proportional to variability. Furthermore, it can be communicated very precisely in a language that everyone (particularly managers and executives) understands—namely, money.

---

[1] We are referring to unwanted or harmful variability. There are situations in which variability is actually good. As my good friend Bob Hogg has pointed out, "I really like Chinese food but I don't want to eat it every night."

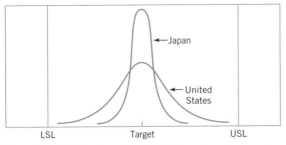

Figure 1-2   Distributions of critical dimensions for transmissions.

How did the Japanese do this? The answer lies in the systematic and effective use of the methods described in this book. It also leads to the following definition of **quality improvement.**

Definition

Quality improvement is the reduction of variability in processes and products.

Excessive variability in process performance often results in **waste.** For example, consider the wasted money, time, and effort that is associated with the repairs represented in Fig. 1-1. Therefore, an alternate and highly useful definition is that quality improvement is the **reduction of waste.** This definition is particularly effective in service industries, where there may not be as many things that can be directly measured (like the transmission critical dimensions in Fig. 1-2). In service industries, a quality problem may be an error or a mistake, the correction of which requires effort and expense. By improving the service process, this wasted effort and expense can be avoided.

We now present some quality engineering terminology that is used throughout the book.

1-1.2   Quality Engineering Terminology

Every product possesses a number of elements that jointly describe what the user or consumer thinks of as quality. These parameters are often called **quality characteristics.** Quality characteristics may be of several types:

1. **Physical:**   length, weight, voltage, viscosity
2. **Sensory:**   taste, appearance, color
3. **Time Orientation:**   reliability, durability, serviceability

Note that the different types of quality characteristics can relate directly or indirectly to the dimensions of quality discussed in the previous section.

Quality engineering is the set of operational, managerial, and engineering activities that a company uses to ensure that the quality characteristics of a product are at the nominal or required levels. The techniques discussed in the book form much of the basic methodology used by engineers and other technical professionals to achieve these goals.

Most organizations find it difficult (and expensive) to provide the customer with products that have quality characteristics that are always identical from unit to unit, or are at levels that match customer expectations. A major reason for this is **variability.** There is a certain amount of variability in every product; consequently, no two products are ever identical. For example, the thickness of the blades on a jet turbine engine impeller is not identical even on the same impeller. Blade thickness will also differ between impellers. If this variation in blade thickness is small, then it may have no impact on the customer. However, if the variation is large, then the customer may perceive the unit to be undesirable and unacceptable. Sources of this variability include differences in materials, differences in the performance and operation of the manufacturing equipment, and differences in the way the operators perform their tasks. This line of thinking led to the previous definition of quality improvement.

Since variability can only be described in statistical terms, **statistical methods** play a central role in quality improvement efforts. In the application of statistical methods to quality engineering, it is fairly typical to classify data on quality characteristics as either **attributes** or **variables** data. Variables data are usually continuous measurements, such as length, voltage, or viscosity. Attributes data, on the other hand, are usually discrete data, often taking the form of counts. We will describe statistical-based quality engineering tools for dealing with both types of data.

Quality characteristics are often evaluated relative to **specifications.** For a manufactured product, the specifications are the desired measurements for the quality characteristics on the components and subassemblies that make up the product, as well as the desired values for the quality characteristics in the final product. For example, the diameter of a shaft used in an automobile transmission cannot be too large or it will not fit into the mating bearing, nor can it be too small, resulting in a loose fit, causing vibration, wear, and early failure of the assembly. In the service industries, specifications are typically in terms of the maximum amount of time to process an order or to provide a particular service.

A value of a measurement that corresponds to the desired value for that quality characteristic is called the **nominal** or **target value** for that characteristic. These target values are usually bounded by a range of values that, most typically, we believe will be sufficiently close to the target so as to not impact the function or performance of the product if the quality characteristic is in that range. The largest allowable value for a quality characteristic is called the **upper specification limit (USL),** and the smallest allowable value for a quality characteristic is called the **lower specification limit (LSL).** Some quality characteristics have specification limits on only one side of the target. For example, the compressive strength of a component used in an automobile bumper likely has a target value and a lower specification limit, but not an upper specification limit.

Specifications are usually the result of the engineering design process for the product. Traditionally, design engineers have arrived at a product design configuration through the use of engineering science principles, which often results in the designer specifying the target values for the critical design parameters. Then prototype construction and testing follow. This testing is often done in a very unstructured manner, without the use of statistically based experimental design procedures, and without much interaction with or knowledge of the manufacturing processes that must produce the component parts and final product. However, through this general procedure, the specification limits are usually determined by the design engineer. Then the final product is released to manufacturing. We refer to this as the **over-the-wall** approach to design.

Problems in product quality usually are greater when the over-the-wall approach to design is used. In this approach, specifications are often set without regard to the inherent variability that exists in materials, processes, and other parts of the system, which results in components or products that are **nonconforming;** that is, that fail to meet one or more of its specifications. A specific type of failure is called a **nonconformity.** A nonconforming product is not necessarily unfit for use; for example, a detergent may have a concentration of active ingredient that is below the lower specification limit, but it may still perform acceptably if the customer uses a greater amount of the product. A nonconforming product is considered **defective** if it has one or more **defects,** which are nonconformities that are serious enough to significantly affect the safe or effective use of the product. Obviously, failure on the part of a company to improve its manufacturing processes can also cause nonconformities and defects.

The over-the-wall design process has been the subject of much attention in the last 20 years. CAD/CAM systems have done much to automate the design process and to more effectively translate specifications into manufacturing activities and processes. Design for manufacturability and assembly have emerged as an important part of overcoming the inherent problems with the over-the-wall approach to design, and most engineers receive some background on those areas today as part of their formal education. The recent emphasis on **concurrent engineering** has stressed a team approach to design, with specialists in manufacturing, quality engineering, and other disciplines working together with the product designer at the earliest stages of the product design process. Furthermore, the effective use of the quality improvement methodology on this book, at all levels of the process used in product design, development, and manufacturing, plays a crucial role in quality improvement.

1-2 A BRIEF HISTORY OF QUALITY CONTROL AND IMPROVEMENT

Quality always has been an integral part of virtually all products and services. However, our awareness of its importance and the introduction of formal methods for quality control and improvement have been an evolutionary development. Table 1-1 presents a timeline of some of the important milestones in this evolutionary process. We will briefly discuss some of the events on this timeline.

Table 1-1   A Timeline of Quality Methods

1700–1900	Quality is largely determined by the efforts of an individual craftsman. Eli Whitney introduces standardized, interchangeable parts to simplify assembly.
1875	Frederick W. Taylor introduces "Scientific Management" principles to divide work into smaller, more easily accomplished units—the first approach to dealing with more complex products and processes. The focus was on productivity. Later contributors were Gilbreth and Gantt.
1900–1930	Henry Ford—the assembly line—further refinement of work methods to improve productivity and quality; Ford developed mistake-proof assembly concepts, self-checking, and in-process inspection.
1901	First standards laboratories established in Great Britain.
1907–1908	AT&T begins systematic inspection and testing of products and materials.
1908	W. S. Gosset (writing as "Student") introduces the t-distribution—results from his work on quality control at Guiness Brewery.
1915–1919	WWI—British government begins a supplier certification program.
1919	Technical Inspection Association is formed in England; this later becomes the Institute of Quality Assurance.
1920s	AT&T Bell Laboratories forms a quality department—emphasizing quality, inspection and test, and product reliability. B. P. Dudding at General Electric in England uses statistical methods to control the quality of electric lamps.
1922–1923	R. A. Fisher publishes series of fundamental papers on designed experiments and their application to the agricultural sciences.
1924	W. A. Shewhart introduces the control chart concept in a Bell Laboratories technical memorandum.
1928	Acceptance sampling methodology is developed and refined by H. F. Dodge and H. G. Romig at Bell Labs.
1931	W. A. Shewhart publishes *Economic Control of Quality of Manufactured Product*—outlining statistical methods for use in production and control chart methods.
1932	W. A. Shewhart gives lectures on statistical methods in production and control charts at the University of London.
1932–1933	British textile and wooden industry and German chemical industry begin use of designed experiments for product/process development.
1933	The Royal Statistical Society forms the Industrial and Agricultural Research Section.
1938	W. E. Deming invites Shewhart to present seminars on control charts at the U.S. Department of Agriculture Graduate School.
1940	The U.S. War Department publishes a guide for using control charts to analyze process data.
1940–1943	Bell Labs develop the forerunners of the military standard sampling plans for the U.S. Army.
1942	In Great Britain, the Ministry of Supply Advising Service on Statistical Methods and Quality Control is formed.
1942–1946	Training courses on statistical quality control are given to industry; more than 15 quality societies are formed in North America.
1944	*Industrial Quality Control* begins publication.
1946	The American Society for Quality Control (ASQC) is formed as the merger of various quality societies. Deming is invited to Japan by the Economic and Scientific Services Section of the U.S. War Department to help occupation forces in rebuilding Japanese industry. The Japanese Union of Scientists and Engineers (JUSE) is formed.
1946–1949	Deming is invited to give statistical quality control seminars to Japanese industry.

(continued)

Table 1-1   (*Continued*)

1948	Professor G. Taguchi begins study and application of experimental design.
1950	Deming begins education of Japanese industrial managers; statistical quality control methods begin to be widely taught in Japan.
	Professor K. Ishikawa introduces the cause-and-effect diagram.
1950s	Classic texts on statistical quality control by Eugene Grant and A. J. Duncan appear.
1951	Dr. A. V. Feigenbaum publishes the first edition of his book, *Total Quality Control.*
	JUSE establishes the "Deming Prize" for significant achievement in quality control and quality methodology.
1951+	G. E. P. Box and K. B. Wilson publish fundamental work on using designed experiments and response surface methodology for process optimization; focus is on chemical industry. Applications of designed experiments in the chemical industry grow steadily after this.
1954	Dr. Joseph M. Juran is invited by the Japanese to give some lectures on quality management and improvement.
	British statistician E. S. Page introduces the cumulative sum (CUSUM) control chart.
1957	J. M. Juran and F. M. Gryna's *Quality Control Handbook* is first published.
1959	*Technometrics* (a journal of statistics for the physical, chemical, and engineering sciences) is established; J. Stuart Hunter is the founding editor.
	S. Roberts introduces the exponentially weighted moving average (EWMA) control chart.
	The U.S. manned spaceflight program makes industry aware of the need for reliable products; the field of reliability engineering grows from this starting point.
1960	G. E. P. Box and J. S. Hunter write fundamental papers on 2^{k-p} factorial designs.
	The quality control circle concept is introduced in Japan by K. Ishikawa.
1961	National Council for Quality and Productivity is formed in Great Britain as part of the British Productivity Council.
1960s	Courses in statistical quality control become widespread in Industrial Engineering academic programs.
	Zero defects (ZD) programs are introduced in certain U.S. industries.
1969	*Industrial Quality Control* ceases publication, replaced by *Quality Progress* and the *Journal of Quality Technology* (Dr. Lloyd S. Nelson is the founding editor of *JQT*).
1970s	In Great Britain the NCQP and the Institute of Quality Assurance merge to form the British Quality Association.
1975–1978	Books on designed experiments oriented toward engineers and scientists begin to appear.
	Interest in quality circles begins in North America—this grows into the total quality management (TQM) movement.
1980s	Experimental design methods are introduced to and adopted by a wider group of organizations, including electronics, aerospace, semiconductor, and the automotive industries.
	The works of Professor G. Taguchi on designed experiments appear in the United States for the first time.
1984	The American Statistical Association (ASA) establishes the Ad Hoc Committee on Quality and Productivity; this later becomes a full Section of the ASA.
1986	Box and others visit Japan, noting the extensive use of designed experiments and other statistical methods.
1988	The Malcolm Baldrige National Quality Award is established by the U.S. Congress.
1989	The journal *Quality Engineering* appears.
	Motorola's six-sigma initiative begins.
1990s	ISO 9000 certification activities increase in U.S. industry; applicants for the Baldrige award grow steadily; many states sponsor quality awards based on the Baldrige criteria.

1995	Many undergraduate engineering programs require formal courses in statistical techniques, focusing on basic methods for process characterization and improvement.
1997	Motorola's six-sigma approach spreads to other industries.
1998	The American Society for Quality Control becomes the American Society for Quality, attempting to indicate the broader aspects of the quality improvement field.

Frederick W. Taylor introduced some principles of scientific management as mass production industries began to develop prior to 1900. Taylor pioneered dividing work into tasks so that the product could be manufactured and assembled more easily. His work led to substantial improvements in productivity. Also, because of standardized production and assembly methods, the quality of manufactured goods was positively impacted as well. However, along with the standardization of work methods came the concept of work standards—a standard time to accomplish the work, or a specified number of units that must be produced per period. Frank Gilbreth and others extended this concept to the study of motion and work design. Much of this had a positive impact on productivity, but it often de-emphasized the quality aspect of work. Furthermore, if carried to extremes, work standards have the risk of halting innovation and continuous improvement, which we recognize today as being a vital aspect of all work activities.

Statistical methods and their application in quality improvement have had a long history. In 1924, Walter A. Shewhart of the Bell Telephone Laboratories developed the statistical control-chart concept, which is often considered the formal beginning of statistical quality control. Toward the end of the 1920s, Harold F. Dodge and Harry G. Romig, both of Bell Telephone Laboratories, developed statistically based acceptance sampling as an alternative to 100% inspection. By the middle of the 1930s, statistical quality-control methods were in wide use at Western Electric, the manufacturing arm of the Bell System. However, the value of statistical quality control was not widely recognized by industry.

World War II saw a greatly expanded use and acceptance of statistical quality-control concepts in manufacturing industries. Wartime experience made it apparent that statistical techniques were necessary to control and improve product quality. The American Society for Quality Control was formed in 1946. This organization promotes the use of quality improvement techniques for all types of products and services. It offers a number of conferences, technical publications, and training programs in quality assurance. The 1950s and 1960s saw the emergence of reliability engineering, the introduction of several important textbooks on statistical quality control, and the viewpoint that quality is a way of managing the organization.

In the 1950s, designed experiments for product and process improvement were first introduced in the United States. The initial applications were in the chemical industry. These methods were widely exploited in the chemical industry, and they are often cited as one of the primary reasons that the U.S. chemical industry is one of the most competitive in the world and has lost little business to foreign companies. The spread of these methods outside the chemical industry was relatively slow until the late 1970s or early 1980s, when many Western companies discovered that their Japanese competitors had been systematically using designed experiments since the 1960s for process troubleshooting, new process development, evaluation of new product designs, improvement of reliability and field performance of products, and many other aspects of product design, including selection of component and system tolerances. This discovery sparked further interest in statistically designed experiments and resulted in extensive efforts to introduce the methodology in engineering and development organizations in industry, as well as in academic engineering curricula.

Since 1980 there has been a profound growth in the use of statistical methods for quality improvement in the United States. This has been motivated, in part, by the widespread loss of business and markets suffered by many domestic companies that began during the 1970s. For example, the U.S. automobile industry was nearly destroyed by foreign competition during this period. One domestic automobile company estimated its operating losses at nearly $1 million *per hour* in 1980. The adoption and use of statistical methods have played a central role in the re-emergence of U.S. industry. Various management structures have also emerged as frameworks in which to implement quality improvement. In the next two sections we briefly discuss the statistical methods that are the central focus of this book and give an overview of some key aspects of TQM.

1-3   STATISTICAL METHODS FOR QUALITY CONTROL AND IMPROVEMENT

This textbook concentrates on statistical and engineering technology useful in quality improvement. Specifically, we focus on three major areas: **statistical process control, design of experiments,** and (to a lesser extent) **acceptance sampling.** In addition to these techniques, a number of other statistical tools are useful in analyzing quality problems and improving the performance of production processes. The role of some of these tools is illustrated in Fig. 1-3, which presents a production process as a system with a

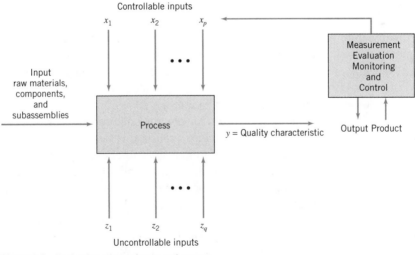

Figure 1-3   Production process inputs and outputs.

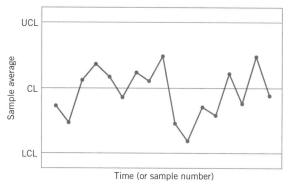

Figure 1-4   A typical control chart.

set of inputs and an output. The inputs $x_1, x_2, \ldots, x_p$ are controllable factors, such as temperatures, pressures, feed rates, and other process variables. The inputs $z_1, z_2, \ldots,$ z_q are uncontrollable (or difficult to control) inputs, such as environmental factors or properties of raw materials submitted by the vendor. The manufacturing process transforms these inputs into a finished product that has several quality characteristics. The output variable y is a measure of process quality.

A **control chart** is one of the primary techniques of **statistical process control** or **SPC.** A typical control chart is shown in Fig. 1-4. This chart plots the averages of measurements of a quality characteristic in samples taken from the process versus time (or the sample number). The chart has a center line (CL) and upper and lower control limits (UCL and LCL in Fig. 1-4). The center line represents where this process characteristic should fall if there are no unusual sources of variability present. The control limits are determined from some simple statistical considerations that we will discuss in Chapters 4, 5, and 6. Classically, control charts are applied to the output variable(s) in a system such as in Fig. 1-4. However, in some cases they can be usefully applied to the inputs as well.

The control chart is a very useful **process monitoring technique;** when unusual sources of variability are present, sample averages will plot outside the control limits. This is a signal that some investigation of the process should be made and corrective action to remove these unusual sources of variability taken. Systematic use of a control chart is an excellent way to reduce variability.

A **designed experiment** is extremely helpful in discovering the key variables influencing the quality characteristics of interest in the process. A designed experiment is an approach to systematically varying the controllable input factors in the process and determining the effect these factors have on the output product parameters. Statistically designed experiments are invaluable in reducing the variability in the quality characteristics and in determining the levels of the controllable variables that optimize process performance.

One major type of designed experiment is the **factorial design,** in which factors are varied together in such a way that all possible combinations of factor levels are tested. Figure 1-5 shows two possible factorial designs for the process in Fig. 1-3, for the cases of $p = 2$ and $p = 3$ controllable factors. In Fig. 1-5a the factors have two levels, low

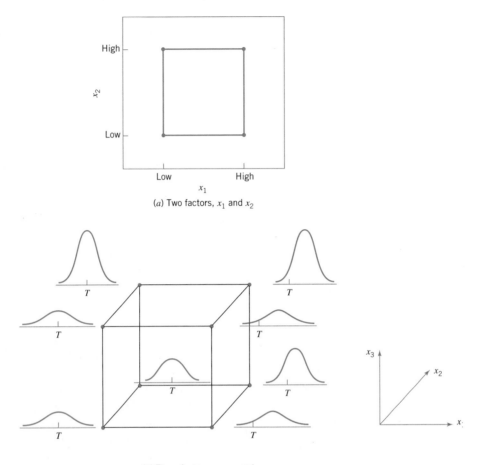

(a) Two factors, x_1 and x_2

(b) Three factors, x_1, x_2, and x_3

Figure 1-5   Factorial designs for the process in Fig. 1-3.

and high, and the four possible test combinations in this factorial experiment form the corners of a square. In Fig. 1-5b there are three factors each at two levels, giving an experiment with eight test combinations arranged at the corners of a cube. The distributions at the corners of the cube represent the process performance at each combination of the controllable factors x_1, x_2, and x_3. It is clear that some combinations of factor levels produce better results than others. For example, increasing x_1 from low to high increases the average level of the process output and could shift it off the target value (T). Furthermore, process variability seems to be substantially reduced when we operate the process along the back edge of the cube, where x_2 and x_3 are at their high levels.

Designed experiments are a major **off-line** quality-control tool, because they are often used during development activities and the early stages of manufacturing, rather than as a routine **on-line** or **in-process** procedure. They play a crucial role in reducing variability.

Once we have identified a list of important variables that affect the process output, it is usually necessary to model the relationship between the influential input variables and the output quality characteristics. Statistical techniques useful in constructing such models include regression analysis and time series analysis. Detailed discussions of designed experiments, regression analysis, and time series modeling are in Montgomery (1997), Montgomery and Peck (1992), and Box, Jenkins, and Reinsel (1994).

When the important variables have been identified and the nature of the relationship between the important variables and the process output has been quantified, then an on-line statistical process-control technique for monitoring and surveillance of the process can be employed with considerable effectiveness. Techniques such as control charts can be used to monitor the process output and detect when changes in the inputs are required to bring the process back to an in-control state. The models that relate the influential inputs to process outputs help determine the nature and magnitude of the adjustments required. In many processes, once the dynamic nature of the relationships between the inputs and the outputs are understood, it may be possible to routinely adjust the process so that future values of the product characteristics will be approximately on-target. This routine adjustment is often called **engineering control, automatic control,** or **feedback control.** We will briefly discuss these types of process control schemes in Chapter 11, and illustrate how SPC methods can be successfully integrated into a manufacturing system in which engineering control is in use.

The third area of quality control and improvement that we discuss is **acceptance sampling.** This is closely connected with inspection and testing of product, which is one of the earliest aspects of quality control, dating back to long before statistical methodology was developed for quality improvement. Inspection can occur at many points in a process. Acceptance sampling, defined as the inspection and classification of a sample of units selected at random from a larger batch or lot and the ultimate decision about disposition of the lot, usually occurs at two points: incoming raw materials or components, or final production.

Several different variations of acceptance sampling are shown in Fig. 1-6. In Fig. 1-6a, the inspection operation is performed immediately following production, before the product is shipped to the customer. This is usually called **outgoing inspection.** Figure 1-6b illustrates **incoming inspection;** that is, a situation in which lots of batches of product are sampled as they are received from the supplier. Various lot-dispositioning decisions are illustrated in Fig. 1-6c. Sampled lots may either be accepted or rejected. Items in a rejected lot are typically either scrapped or recycled, or they may be reworked or replaced with good units. This latter case is often called **rectifying inspection.**

Modern quality assurance systems usually place less emphasis on acceptance sampling and attempt to make statistical process control and designed experiments the focus of their efforts. Acceptance sampling tends to reinforce the "conformance to specification" view of quality and does not have any feedback into either the production process or engineering design or development that would necessarily lead to quality improvement.

Figure 1-7 shows the typical evolution in the use of these techniques in most organizations. At the lowest level of maturity, management may be completely unaware of quality issues, and there is likely to be no effective organized quality improvement effort. Frequently there will be some modest applications of acceptance-sampling methods, usually in receiving inspection. The first activity as maturity increases is to

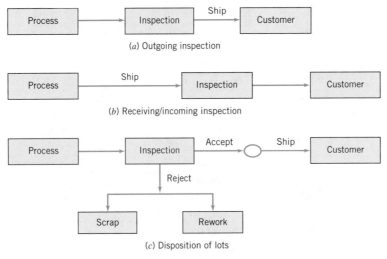

Figure 1-6   Variations of acceptance sampling.

intensify the use of sampling inspection. The use of sampling will increase until we realize that quality cannot be inspected or tested into the product.

At that point the organization usually begins to focus on process improvement. Statistical process control and experimental design potentially have major impact on manufacturing, product design activities, and process development. The systematic introduction of these methods usually marks the start of substantial quality, cost, and productivity improvements in the organization. At the highest levels of maturity, companies use designed experiments and statistical process control methods intensively, and make relatively modest use of acceptance sampling.

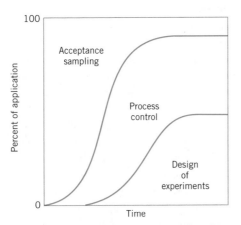

Figure 1-7   Phase diagram of the use of quality-engineering methods.

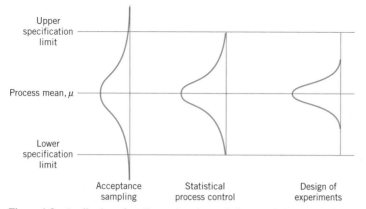

Figure 1-8   Application of quality-engineering techniques and the systematic reduction of process variability.

The primary **objective** of quality engineering efforts is the **systematic reduction of variability** in the key quality characteristics of the product. Figure 1-8 shows how this happens over time. In the early stages, when acceptance sampling is the major technique in use, process "fallout," or units that do not conform to the specifications, constitute a high percentage of the process output. The introduction of statistical process control will stabilize the process and reduce the variability. However, it is not satisfactory just to meet requirements—further reduction of variability usually leads to better product performance and enhanced competitive position, as was vividly demonstrated in the automobile transmission example discussed earlier. Statistically designed experiments can be employed in conjunction with statistical process control to minimize process variability in nearly all industrial settings.

1-4   OTHER ASPECTS OF QUALITY CONTROL AND IMPROVEMENT

Although statistical techniques are the critical technical tools for quality control and improvement, to be used most effectively they must be implemented within and be part of a management system that is quality driven. In effect, the management system must direct the quality improvement philosophy and ensure its implementation in all aspects of the business. One of the managerial frameworks used to accomplish this is **total quality management** (or **TQM**), although other widely used names include **company-wide quality control (CWQC), total quality assurance (TQA),** and **six-sigma.**

In this section we give a brief overview of some of the key elements of TQM. We discuss some of the important quality philosophies, the link between quality and productivity, economic and legal implication of quality, and some aspects of implementation.

1-4.1 Quality Philosophy and Management Strategies

Many people have contributed to the statistical methodology of quality improvement. However, in terms of implementation and management philosophy, three individuals emerge as the leaders: W. E. Deming, J. M. Juran, and A. V. Feigenbaum. We now review the approaches and philosophy of those leaders in quality management.

W. Edwards Deming

One reason that Japanese manufacturers exhibited broad capabilities in statistical quality control and improvement during the 1970s and 1980s is the work of Dr. W. Edwards Deming. During World War II, Dr. Deming worked for the War Department and the Census Bureau. Following the war, he became a consultant to Japanese industries and convinced their top management of the power of statistical methods and the importance of quality as a competitive weapon. This commitment to and use of statistical methods has been a key element in the expansion of Japan's industry and economy. The Japanese Union of Scientists and Engineers created the Deming Prize for quality improvement in his honor. Until his death in 1994, Dr. Deming was an active consultant and speaker; he was an inspirational force for quality improvement in this country and around the world. He firmly believed that the responsibility for quality rests with management; that is, most of the opportunities for quality improvement require management action, and very few opportunities lie at the workforce or operator level. Dr. Deming was a harsh critic of many American management practices.

The philosophy of Dr. Deming is an important framework for implementing quality and productivity improvement. This philosophy is summarized in his 14 points for management. We now give a brief statement and discussion of Dr. Deming's 14 points:

1. Create a constancy of purpose focused on the improvement of products and services. Constantly try to improve product design and performance. Investment in research, development, and innovation will have long-term payback to the organization.

2. Adopt a new philosophy of rejecting poor workmanship, defective products, or bad service. It costs as much to produce a defective unit as it does to produce a good one (and sometimes more). The cost of dealing with scrap, rework, and other losses created by defectives is an enormous drain on company resources.

3. Do not rely on mass inspection to "control" quality. All inspection can do is sort out defectives, and at this point it is too late because we have already paid to produce these defectives. Inspection typically occurs too late in the process, it is expensive, and it is often ineffective. Quality results from prevention of defectives through process improvement, not inspection.

4. Do not award business to suppliers on the basis of price alone, but also consider quality. Price is a meaningful measure of a supplier's product only if it is considered in relation to a measure of quality. In other words, the total cost of the item must be considered, not just the purchase price. When quality is

considered, the lowest bidder frequently is not the low-cost supplier. Preference should be given to suppliers who use modern methods of quality improvement in their business and who can demonstrate process control and capability.

5. Focus on continuous improvement. Constantly try to improve the production and service system. Involve the workforce in these activities and make use of statistical methods, particularly the statistically based problem-solving tools discussed in this book.

6. Practice modern training methods and invest in training for all employees. Everyone should be trained in the technical aspects of their job, and in modern quality- and productivity-improvement methods as well. The training should encourage all employees to practice these methods every day.

7. Practice modern supervision methods. Supervision should not consist merely of passive surveillance of workers but should be focused on helping the employees improve the system in which they work. The number one goal of supervision should be to improve the work system and the product.

8. Drive out fear. Many workers are afraid to ask questions, report problems, or point out conditions that are barriers to quality and effective production. In many organizations the economic loss associated with fear is large; only management can eliminate fear.

9. Break down the barriers between functional areas of the business. Teamwork among different organizational units is essential for effective quality and productivity improvement to take place.

10. Eliminate targets, slogans, and numerical goals for the workforce. A target such as "zero defects" is useless without a plan for the achievement of this objective. In fact, these slogans and "programs" are usually counterproductive. Work to improve the system and provide information on that.

11. Eliminate numerical quotas and work standards. These standards have historically been set without regard to quality. Work standards are often symptoms of management's inability to understand the work process and to provide an effective management system focused on improving this process.

12. Remove the barriers that discourage employees from doing their jobs. Management must listen to employee suggestions, comments, and complaints. The person who is doing the job knows the most about it and usually has valuable ideas about how to make the process work more effectively. The workforce is an important participant in the business, and not just an opponent in collective bargaining.

13. Institute an ongoing program of training and education for all employees. Education in simple, powerful statistical techniques should be mandatory for all employees. Use of the basic SPC problem-solving tools, particularly the control chart, should become widespread in the business. As these charts become widespread and as employees understand their uses, they will be more likely

to look for the causes of poor quality and to identify process improvements. Education is a way of making everyone partners in the quality-improvement process.

14. Create a structure in top management that will vigorously advocate the first 13 points.

As we read Dr. Deming's 14 points we notice two things: First, there is a strong emphasis on change. Second, the role of management in guiding this change process is of dominating importance. However, what should be changed, and how should this change process be started? For example, if we want to improve the yield of a semiconductor manufacturing process, what should we do? It is in this area that statistical methods come into play most frequently. To improve the semiconductor process, we must determine which controllable factors in the process influence the number of defective units produced. To answer this question, we must collect data on the process and see how the system reacts to change in the process variables. Statistical methods, such as designed experiments and control charts, can contribute to these activities.

Dr. Joseph M. Juran

Dr. Juran is one of the founding fathers of statistical quality control. He worked for Dr. Walter A. Shewhart at AT&T Bell Laboratories and has been at the leading edge of quality improvement ever since. He was invited to speak to Japanese industry leaders as they began their industrial transformation in the early 1950s. He is the co-author (with Frank M. Gryna) of the *Quality Control Handbook,* a standard reference for quality methods and improvement since its initial publication in 1957.

Dr. Juran is less focused than Dr. Deming on statistical methods. The Juran philosophy is based on organization for change and the implementation of improvement through what he calls "managerial breakthrough." The breakthrough sequence is really a structured problem-solving process. Dr. Juran is a firm believer that management action is required to improve quality. He shares with Dr. Deming the viewpoint that most of the opportunities (80%) for quality improvement can only be addressed by management and that a relatively small proportion of these opportunities (20%) can be dealt with at the workforce level.

Dr. Armand V. Feigenbaum

Dr. Feigenbaum first introduced the concept of company-wide quality control in his historic book *Total Quality Control* (the first edition was published in 1951). This book influenced much of the early philosophy of quality management in Japan in the early 1950s. In fact, many Japanese companies used the name "total quality control" to describe their efforts.

Dr. Feigenbaum is more concerned with organizational structure and a systems approach to improving quality than he is with statistical methods. He initially suggested that much of the technical capability be concentrated in a specialized department. This is in contrast to the more modern view that knowledge and use of statistical tools need to be widespread. However, the organizational aspects of Dr. Feigenbaum's work are important, as quality improvement does not usually spring forth as a "grass roots" activity; it requires a lot of management commitment to make it work.

The brief descriptions of the philosophies of Deming, Juran, and Feigenbaum have highlighted both the common aspects and differences of their viewpoints. In this author's opinion, there are more similarities than differences among them, and the similarities are what is important. All three of these pioneers stress the importance of quality as an essential competitive weapon, the important role that management must play in implementing quality improvement, and the importance of statistical methods and techniques in the "quality transformation" of an organization.

Total Quality Management
Total quality management (or **TQM**) is a strategy for implementing and managing quality improvement activities on an organization-wide basis. TQM began in the early 1980s, with the philosophies of Deming and Juran as the focal point. It evolved into a broader spectrum of concepts and ideas, involving participative organizations and work culture, customer focus, supplier quality improvement, integration of the quality system with business goals, and many other activities to focus all elements of the organization around the quality improvement goal. Typically, organizations that have implemented a TQM approach to quality improvement have quality councils or high-level teams that deal with strategic quality initiatives, workforce-level teams that focus on routine production or business activities, and cross-functional teams that address specific quality improvement issues.

TQM has only had moderate success for a variety of reasons, but frequently because there is insufficient effort devoted to widespread utilization of the technical tools of variability reduction. Many organizations saw the mission of TQM as one of training. Consequently, many TQM efforts engaged in widespread training of the workforce in the philosophy of quality improvement and a few basic methods. This training was usually placed in the hands of human resources departments, and much of it was ineffective. The trainers often had no real idea about *what* methods should be taught, and success was usually measured by the percentage of the workforce that had been "trained," not by whether any measurable impact on business results had been achieved. Some general reasons for the lack of conspicuous success of TQM include (1) lack of top-down, high-level management commitment and involvement; (2) inadequate use of statistical methods and insufficient recognition of variability reduction as a prime objective; (3) diffuse as opposed to focused, specific objectives; and (4) too much emphasis on widespread *training* as opposed to focused technical *education*.

Another reason for the erratic success of TQM is that many managers and executives have regarded it as just another "program" to improve quality. During the 1950s and 1960s, programs such as **zero defects** and **value engineering** abounded, but they had little real impact on quality and productivity improvement. During the heyday of TQM in the 1980s, another popular program was the **quality is free** initiative, in which management worked on identifying the cost of quality (or the cost of nonquality, as the "quality is free" devotees so cleverly put it). Indeed, Identification of quality costs can

be very useful (we discuss quality costs in Section 1-4.3), but the "quality is free" practitioners often had no idea about what to do to actually improve many types of complex
industrial processes. In fact, the leaders of this initiative had no knowledge about statistical methodology and completely failed to understand its role in quality improvement.
When TQM is wrapped around an ineffective program such as this, disaster is often the
result.

Quality Standards and Registration

The International Standards Organization (ISO) has developed a series of quality standards including the ISO 9000 series, which is also an American National Standards Institute and an ASQ standard as well. The focus of these standards is the quality system,
including components such as

1. Management responsibility for quality
2. Design control
3. Document and data control
4. Purchasing and contract management
5. Product identification and traceability
6. Inspection and testing, including control of measurement and inspection
   equipment
7. Process control
8. Handling of nonconforming product, corrective and preventive actions
9. Handling, storage, packaging, and delivery of product, including service
   activities
10. Control of quality records
11. Internal audits
12. Training
13. Statistical methodology

ISO 9000 sets standards for these and some other activities as well. Many organizations
have required their suppliers to become certified under ISO 9000 or one of the derivative
standards, such as those developed by the North American automobile industry, QS
9000. Consequently, there is currently a substantial effort in many organizations to become certified to the appropriate standard. Third-party registrars handle certification.

Much of the focus of ISO 9000 (and the derivative standards) is on formal documentation of the quality system. Organizations usually must make huge efforts to bring
their documentation into line with the requirements of the standards; this is the Achilles
heel of ISO 9000 and other standards. There is far too much effort devoted to paperwork
and bookkeeping and not nearly enough to actually reducing variability and improving
processes and products. Furthermore, many of the third-party registrars, auditors, and
consultants that work in this area are not sufficiently educated or experienced enough in
the **technical** tools of quality improvement. They are all too often unaware of what constitutes modern engineering and statistical practice, and usually are familiar with only

the most elementary techniques. Therefore, they concentrate largely on the bookkeeping, records, and paperwork aspects of certification.

It has been estimated that ISO registration activities are (approximately) a *40 billion dollar annual business,* worldwide. Much of this money flows to the registrars, auditors, and consultants. This amount does not include the internal costs incurred by organizations to achieve registration, such as the thousands of hours of engineering and management effort, travel, internal training, and internal auditing. It is not clear whether any significant fraction of this expenditure has made its way to the bottom line of the registered organizations. Many quality engineering authorities feel that ISO registration is largely a waste of effort. Often organizations would be far better off to "just say no to ISO" and spend a small fraction of that 40 billion dollars on their quality systems and another larger fraction on meaningful variability reduction efforts, develop their own internal (or perhaps industry-based) quality standards, rigorously enforce them, and pocket the difference. The focus of these standards should be principally on variability reduction and yield enhancement efforts, with systems and paperwork playing a supporting role.

Six-Sigma

High-technology products with many complex components typically have many opportunities for failure or defects to occur. Motorola developed the six-sigma program in the late 1980s as a response to the demand for these products. The focus of six-sigma is reducing variability in key product quality characteristics to the level at which failure or defects are extremely unlikely.

Figure 1-9*a* shows a normal probability distribution as a model for a quality characteristic with the specification limits at three standard deviations on either side of the mean. Now it turns out that in this situation the probability of producing a product within these specifications is 0.9973, which corresponds to 2700 parts per million (ppm) defective. This is referred to as **three-sigma quality performance,** and it actually sounds pretty good. However, suppose we have a product that consists of an assembly of 100 components or parts and all 100 of these parts must be nondefective for the product to function satisfactorily. The probability that any specific unit of product is nondefective is

$$0.9973 \times 0.9973 \times \cdots \times 0.9973 = (0.9973)^{100} = 0.7631$$

That is, about 23.7% of the products produced under three-sigma quality will be defective. This is not an acceptable situation, because many high-technology products are made up of thousands of components. An automobile has about 200,000 components and an airplane has several million!

The Motorola six-sigma concept is to reduce the variability in the process so that the specification limits are six standard deviations from the mean. Then, as shown in Fig. 1-9*a*, there will only be about 2 parts per *billion* defective. Under **six-sigma quality,** the probability that any specific unit of the hypothetical product above is nondefective is 0.9999998, or 0.2 ppm, a much better situation.

When the six-sigma concept was initially developed, an assumption was made that when the process reached the six-sigma quality level, the process mean was still subject

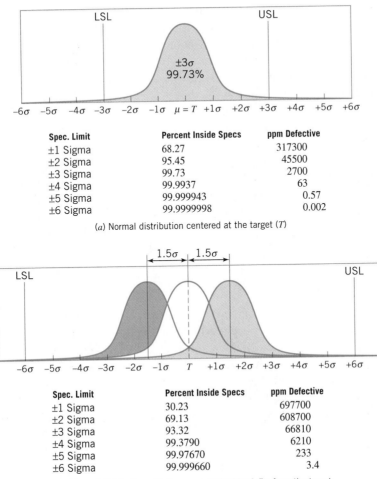

Spec. Limit	Percent Inside Specs	ppm Defective
±1 Sigma	68.27	317300
±2 Sigma	95.45	45500
±3 Sigma	99.73	2700
±4 Sigma	99.9937	63
±5 Sigma	99.999943	0.57
±6 Sigma	99.9999998	0.002

(*a*) Normal distribution centered at the target (*T*)

Spec. Limit	Percent Inside Specs	ppm Defective
±1 Sigma	30.23	697700
±2 Sigma	69.13	608700
±3 Sigma	93.32	66810
±4 Sigma	99.3790	6210
±5 Sigma	99.97670	233
±6 Sigma	99.999660	3.4

(*b*) Normal distribution with the mean shifted by 1.5σ from the target

Figure 1-9   The Motorola six-sigma concept.

to disturbances that could cause it to shift by as much as 1.5 standard deviations off target. This situation is shown in Fig. 1-9*b*. Under this scenario, a six-sigma process would produce about 3.4 ppm defective.

There is a logical inconsistency in this. As we will discuss in Chapter 7 on process capability, we can only make predictions about process performance when the process is **stable;** that is, when the mean (and standard deviation, too) is **constant.** If the mean is drifting around, and ends up as much as 1.5 standard deviations off target, a prediction of 3.4 ppm defective isn't very reliable, because the mean might shift by *more* than the "allowed" 1.5 standard deviations. Process **performance** isn't predictable unless the process **behavior** is stable.

Motorola established six-sigma as both an objective for the corporation and as a focal point for process and product quality improvement efforts. In recent years, six-sigma has spread beyond Motorola and has come to encompass much more. It has become a program for improving corporate *business* performance by both improving quality and paying attention to reducing costs. Companies involved in a six-sigma effort utilize teams to work on projects that have both quality *and* significant economic impact. The effort is better focused than in earlier TQM programs, and has been more successful in obtaining management commitment. However, remember Deming's point 10, which essentially says to eliminate slogans and programs to improve quality. Deming was right; we have seen many programs including zero defects, value engineering, quality is free, TQM, and so forth. Most of them have failed. A major component in successful quality improvement is driving the use of the proper statistical and engineering tools into the right places in the organization. Industry-based programs will never be completely successful in doing this—engineering, science, and management education at the university level must play a critical role.

Just-in-Time, Lean Manufacturing, Poka-Yoke, and Others

There have been many initiatives devoted to improving the production system. Some of these include the Just-in-Time approach emphasing in-process inventory reduction, rapid set-up, and a pull-type production system; Poka-Yoke or mistake-proofing of processes; the Toyota production system and other Japanese manufacturing techniques (with once popular management books by those names); reengineering; theory of constraints; agile manufacturing; lean manufacturing; and so on. Most of these "programs" devote far too little attention to variability reduction. It's virtually impossible to reduce the in-process inventory or operate a pull-type, agile, or lean production system when a large and unpredictable fraction of the process output is defective. Such efforts will not achieve their full potential without a major focus on statistical methods for process improvement and variability reduction to accompany them.

1-4.2   The Link between Quality and Productivity

Producing high-quality products in the modern industrial environment is not easy. A significant aspect of the problem is the rapid evolution of technology. The last 20 years have seen an explosion of technology in such diverse fields as electronics, metallurgy, ceramics, composite materials, biotechnology, and the chemical and pharmaceutical sciences that has resulted in many new products and services. For example, in the electronics field the development of the integrated circuit has revolutionized the design and manufacture of computers and many electronic office products. Basic integrated circuit technology has been supplanted by large-scale integration (LSI) and very large scale integration (VLSI) technology, with corresponding developments in semiconductor design and manufacturing. When technological advances occur rapidly and when the new technologies are used quickly to exploit competitive advantages, the problems of designing and manufacturing products of superior quality are greatly complicated.

Often, too little attention is paid to achieving all dimensions of an optimal process: economy, efficiency, productivity, and quality. Effective quality improvement can be instrumental in increasing productivity and reducing cost. To illustrate, consider the manufacture of a mechanical component used in a copier machine. The parts are manufactured in a machining process at a rate of approximately 100 parts per day. For various reasons, the process is operating at a first-pass yield of about 75%. (That is, about 75% of the process output conforms to specifications, and about 25% of the output is nonconforming.) About 60% of the fallout (the 25% nonconforming) can be reworked into an acceptable product, and the rest must be scrapped. The direct manufacturing cost through this stage of production per part is approximately $20. Parts that can be reworked incur an additional processing charge of $4. Therefore, the manufacturing cost per good part produced is

$$\text{Cost/good part} = \frac{\$20(100) + \$4(15)}{90} = \$22.89$$

Note that the total yield from this process, after reworking, is 90 good parts per day.

An engineering study of this process reveals that excessive process variability is responsible for the extremely high fallout. A new statistical process-control procedure is implemented that reduces variability, and consequently the process fallout decreases from 25% to 5%. Of the 5% fallout produced, about 60% can be reworked, and 40% are scrapped. After the process-control program is implemented, the manufacturing cost per good part produced is

$$\text{Cost/good part} = \frac{\$20(100) + \$4(3)}{98} = \$20.53$$

Note that the installation of statistical process control and the reduction of variability that follows result in a 10.3% reduction in manufacturing costs. Furthermore, productivity is up by almost 10%; 98 good parts are produced each day as opposed to 90 good parts previously. This amounts to an increase in production capacity of almost 10%, without any additional investment in equipment, workforce, or overhead. Efforts to improve this process by other methods (such as just-in-time, lean manufacturing, etc.) are likely to be completely ineffective until the basic problem of excessive variability is solved.

1-4.3 Quality Costs

All business organizations use financial controls. These financial controls involve a comparison of actual and budgeted costs, along with an associated analysis and action on the differences or *variances* between actual and budget. It is customary to apply these financial controls on a department or functional level. For many years, there was no direct effort to measure or account for the costs of the quality function. However, starting in the 1950s, many organizations began to formally evaluate the cost associated with quality. There are several reasons why the cost of quality should be explicitly considered in an organization. These include the following:

1. The increase in the cost of quality because of the increase in the complexity of manufactured products associated with advances in technology

2. Increasing awareness of life cycle costs, including maintenance, labor, spare parts, and the cost of field failures

3. The need for quality engineers and managers to effectively communicate the cost of quality in the language of general management—namely, money

As a result, quality costs have emerged as a financial control tool for management and as an aid in identifying opportunities for reducing quality costs.

Generally speaking, quality costs are those categories of costs that are associated with producing, identifying, avoiding, or repairing products that do not meet requirements. Many manufacturing and service organizations use four categories of quality costs: prevention costs, appraisal costs, internal failure costs, and external failure costs. These cost categories are shown in Table 1-2. We now discuss these categories in more detail.

Prevention Costs

Prevention costs are those costs associated with efforts in design and manufacturing that are directed toward the prevention of nonconformance. Broadly speaking, prevention costs are all costs incurred in an effort to "make it right the first time." The important subcategories of prevention costs follow.

Quality planning and engineering.   Costs associated with the creation of the overall quality plan, the inspection plan, the reliability plan, the data system, and all specialized plans and activities of the quality-assurance function; the preparation of manuals and procedures used to communicate the quality plan; and the costs of auditing the system.

Table 1-2   Quality Costs

Prevention Costs	Internal Failure Costs
Quality planning and engineering	Scrap
New products review	Rework
Product/process design	Retest
Process control	Failure analysis
Burn-in	Downtime
Training	Yield losses
Quality data acquisition and analysis	Downgrading (off-specing)
Appraisal Costs	External Failure Costs
Inspection and test of incoming material	Complaint adjustment
Product inspection and test	Returned product/material
Materials and services consumed	Warranty charges
Maintaining accuracy of test equipment	Liability costs
	Indirect costs

New products review.   Costs of the preparation of bid proposals, the evaluation of new designs from a quality viewpoint, the preparation of tests and experimental programs to evaluate the performance of new products, and other quality activities during the development and preproduction stages of new products or designs.

Product/process design.   Costs incurred during the design of the product or the selection of the production processes that are intended to improve the overall quality of the product. For example, an organization may decide to make a particular circuit component redundant because this will increase the reliability of the product by increasing the mean time between failures. Alternatively, it may decide to manufacture a component using process A rather than process B, because process A is capable of producing the product at tighter tolerances, which will result in fewer assembly and manufacturing problems. This may include a vendor's process, so the cost of dealing with other than the lowest bidder may also be a prevention cost.

Process control.   The cost of process-control techniques, such as control charts, that monitor the manufacturing process in an effort to reduce variation and build quality into the product.

Burn-in.   The cost of preshipment operation of the product to prevent early-life failures in the field.

Training.   The cost of developing, preparing, implementing, operating, and maintaining formal training programs for quality.

Quality data acquisition and analysis.   The cost of running the quality data system to acquire data on product and process performance; also the cost of analyzing these data to identify problems. It includes the work of summarizing and publishing quality information for management.

Appraisal Costs

Appraisal costs are those costs associated with measuring, evaluating, or auditing products, components, and purchased materials to ensure conformance to the standards that have been imposed. These costs are incurred to determine the condition of the product from a quality viewpoint and ensure that it conforms to specifications. The major subcategories follow.

Inspection and test of incoming material.   Costs associated with the inspection and testing of all vendor-supplied material. This subcategory includes receiving inspection and test; inspection, test, and evaluation at the vendor's facility; and a periodic audit of the vendor's quality-assurance system. This could also include intraplant vendors.

Product inspection and test.   The cost of checking the conformance of the product throughout its various stages of manufacturing, including final acceptance testing, packing and shipping checks, and any test done at the customer's facilities prior to turning the product over to the customer. This also includes life testing, environmental testing, and reliability testing.

Materials and services consumed. The cost of material and products consumed in a destructive test or devalued by reliability tests.

Maintaining accuracy of test equipment. The cost of operating a system that keeps the measuring instruments and equipment in calibration.

Internal Failure Costs

Internal failure costs are incurred when products, components, materials, and services fail to meet quality requirements, and this failure is discovered prior to delivery of the product to the customer. These costs would disappear if there were no defects in the product. The major subcategories of internal failure costs follow.

Scrap. The net loss of labor, material, and overhead resulting from defective product that cannot economically be repaired or used.

Rework. The cost of correcting nonconforming units so that they meet specifications. In some manufacturing operations rework costs include additional operations or steps in the manufacturing process that are created to solve either chronic defects or sporadic defects.

Retest. The cost of reinspection and retesting of products that have undergone rework or other modifications.

Failure analysis. The cost incurred to determine the causes of product failures.

Downtime. The cost of idle production facilities that results from nonconformance to requirements. The production line may be down because of nonconforming raw materials supplied by a vendor, which went undiscovered in receiving inspection.

Yield losses. The cost of process yields that are lower than might be attainable by improved controls (for example, soft-drink containers that are overfilled because of excessive variability in the filling equipment).

Downgrading/off-specing. The price differential between the normal selling price and any selling price that might be obtained for a product that does not meet the customer's requirements. Downgrading is a common practice in the textile, apparel goods, and electronics industries. The problem with downgrading is that products sold do not recover the full contribution margin to profit and overhead as do products that conform to the usual specifications.

External Failure Costs

External failure costs occur when the product does not perform satisfactorily after it is supplied to the customer. These costs would also disappear if every unit of product conformed to requirements. Subcategories of external failure costs follow.

Complaint adjustment. All costs of investigation and adjustment of justified complaints attributable to the nonconforming product.

Returned product/material. All costs associated with receipt, handling, and replacement of the nonconforming product or material that is returned from the field.

Warranty charges.   All costs involved in service to customers under warranty contracts.

Liability costs.   Costs or awards incurred as a result of product liability litigation.

Indirect costs.   In addition to direct operating costs of external failures, there are a significant number of indirect costs. These are incurred because of customer dissatisfaction with the level of quality of the delivered product. Indirect costs may reflect the customer's attitude toward the company. They include the costs of loss of business reputation, loss of future business, and loss of market share that inevitably results from delivering products and services that do not conform to the customer's expectations regarding fitness for use.

The Analysis and Use of Quality Costs

How large are quality costs? The answer, of course, depends on the type of organization and the success of their quality-improvement effort. In some organizations quality costs are 4% or 5% of sales, whereas in others they can be as high as 35% or 40% of sales. Obviously, the cost of quality will be very different for a high-technology computer manufacturer than for a typical service industry, such as a department store or hotel chain. In most organizations, however, quality costs are higher than necessary, and management should make continuing efforts to appraise, analyze, and reduce these costs.

The usefulness of quality costs stems from the **leverage effect;** that is, dollars invested in prevention and appraisal have a payoff in reducing dollars incurred in internal and external failures that exceeds the original investment. For example, a dollar invested in prevention may return $10 or $100 (or more) in savings from reduced internal and external failures.

Quality-cost analyses have as their principal objective cost reduction through identification of improvement opportunities. This is often done with a **Pareto analysis.** The Pareto analysis consists of identifying quality costs by category, or by product, or by type of defect or nonconformity. For example, inspection of the quality-cost information in Table 1-3 concerning defects or nonconformities in the assembly of electronic components onto printed circuit boards reveals that insufficient solder is the highest quality

Table 1-3   Monthly Quality-Costs Information for Assembly of Printed Circuit Boards

Type of Defect	Percent of Total Defects	Scrap and Rework Costs
Insufficient solder	42	$37,500.00 (52%)
Misaligned components	21	12,000.00
Defective components	15	8,000.00
Missing components	10	5,100.00
Cold solder joints	7	5,000.00
All other causes	5	4,600.00
Totals	100	$72,200.00

cost incurred in this operation. Insufficient solder accounts for 42% of the total defects in this particular type of board, and for almost 52% of the total scrap and rework costs. If the wave solder process can be improved, then there will be dramatic reductions in the cost of quality.

How much reduction in quality costs is possible? Although the cost of quality in many organizations can be significantly reduced, it is unrealistic to expect it can be reduced to zero. Before that level of performance is reached, the incremental costs of prevention and appraisal will rise more rapidly than the resulting cost reductions. However, paying attention to quality costs in conjunction with a focused effort on variability reduction has the capability of reducing quality costs by 50% or 60% provided that no organized effort has previously existed. This cost reduction also follows the Pareto principle; that is, most of the cost reductions will come from attacking the few problems that are responsible for the majority of quality costs.

In analyzing quality costs and in formulating plans for reducing the cost of quality, it is important to note the role of prevention and appraisal. Many organizations spend far too much of their quality-management budget on appraisal and not enough on prevention. This is an easy mistake for an organization to make, because appraisal costs are often budget line items in the quality-assurance or manufacturing areas. On the other hand, prevention costs may not be routinely budgeted items. It is not unusual to find in the early stages of a quality-cost program that appraisal costs are eight or ten times the magnitude of prevention costs. This is probably an unreasonable ratio, as dollars spent in prevention have a much greater payback than do dollars spent in appraisal.

Generating the quality-cost figures is not always easy, because most quality-cost categories are not directly reflected in the accounting records of the organization. Consequently, it may be difficult to obtain extremely accurate information on the costs incurred with respect to the various categories. The organization's accounting system can provide information on those quality-cost categories that coincide with the usual business accounts, such as, for example, product testing and evaluation. In addition, many companies will have detailed information on various categories of failure cost. The information for cost categories for which exact accounting information is not available should be generated by using estimates, or, in some cases, by creating special monitoring and surveillance procedures to accumulate those costs over the study period.

The reporting of quality costs is usually done on a basis that permits straightforward evaluation by management. Managers want quality costs expressed in an index that compares quality cost with the opportunity for quality cost. Consequently, the usual method of reporting quality costs is in the form of a ratio, where the numerator is quality-cost dollars and the denominator is some measure of activity, such as (1) hours of direct production labor, (2) dollars of direct production labor, (3) dollars of processing costs, (4) dollars of manufacturing cost, (5) dollars of sales, or (6) units of product.

Upper management may want a standard against which to compare the current quality-cost figures. It is difficult to obtain absolute standards and almost as difficult to obtain quality-cost levels of other companies in the same industry. Therefore, the usual approach is to compare current performance with past performance so that, in effect, quality-cost programs report variances from past performance. These variance analyses are primarily a device for detecting departures from standard and for bringing them to the attention of the appropriate managers. They are not necessarily in and of themselves a device for ensuring quality improvements.

This brings us to an interesting observation: Some quality-cost collection and analysis efforts fail. That is, a number of companies have started quality-cost analysis activities, used them for some time, and then abandoned the programs as ineffective. There are several reasons why this occurs. Chief among these is failure to use quality-cost information as a mechanism for generating improvement opportunities. If we use quality-cost information as a scorekeeping tool only, and do not make conscious efforts to identify problem areas and develop improved operating procedures and processes, then the programs will not be totally successful.

Another reason why quality-cost collection and analysis doesn't lead to useful results is that managers become preoccupied with perfection in the cost figures. Overemphasis in treating quality costs as part of the accounting systems rather than as a management control tool is a serious mistake. This approach greatly increases the amount of time required to develop the cost data, analyze them, and identify opportunities for quality improvements. As the time required to generate and analyze the data increases, management becomes more impatient and less convinced of the effectiveness of the activity. Any program that appears to management as going nowhere is likely to be abandoned.

A final reason for the failure of a quality-cost program is that management often underestimates the depth and extent of the commitment to prevention that must be made. The author has had numerous opportunities to examine quality cost data in many companies. In companies without effective quality-improvement programs, the dollars allocated to prevention rarely exceed 1% to 2% of revenue. This must be increased to a threshold of about 5% to 6% of revenue, and these additional prevention dollars must be spent largely on the technical methods of quality improvement, and not on establishing "programs," "TQM," "reengineering," or other similar activities. If management is persistent in this effort, then the cost of quality will decrease substantially. These cost savings will typically begin to occur in 1 to 2 years, although it could be longer in some companies.

1-4.4 Legal Aspects of Quality

Consumerism and product liability are important reasons for the recent re-emergence of quality assurance as an important business strategy. The rise of consumerism is in part due to the seemingly large number of failures in the field of consumer products. Highly visible field failures often prompt the questions of whether today's products are as good as their predecessors and whether manufacturers are really interested in quality. The answer to both of these questions is yes. Manufacturers are always vitally concerned about field failures because of heavy external failure costs and the related threat to their competitive position. Consequently, most producers have made product improvements directed toward reducing field failures. As examples, note that automobile tires now have over 10 times the life of many of their early predecessors and that solid-state and integrated-circuit technology has greatly reduced the failure of electronic equipment that once depended on the electron tube. Virtually every product line of today is superior to that of yesterday.

Consumer dissatisfaction and the general feeling that today's products are inferior to their predecessors arise from other phenomena. One of these is the explosion in the number of products. For example, a 1% field-failure rate for a consumer appliance with a production volume of 50,000 units per year means 500 field failures. However, if the production rate is 500,000 units per year and the field-failure rate remains the same, then 5000 units will fail in the field. This is equivalent, in the total number of dissatisfied customers, to a 10% failure rate at the lower production level. Increasing production volume increases the *liability exposure* of the manufacturer. Even in situations where the failure rate declines, if the production volume increases more rapidly than the decrease in failure rate, the total number of customers who experience failures will still increase.

A second aspect of the problem is that consumer tolerance for minor defects and aesthetic problems has decreased considerably, so that blemishes, surface-finish defects, noises, and appearance problems that were once tolerated now attract attention and result in adverse consumer reaction. Finally, the competitiveness of the marketplace forces many manufacturers to introduce new designs before they are fully evaluated and tested in order to remain competitive. These "early releases" of unproved designs are a major reason for new product quality failures. Eventually, these design problems are corrected, but the high failure rate connected with new products often supports the belief that today's quality is inferior to that of yesterday.

Product liability is a major social, market, and economic force. The legal obligation of manufacturers and sellers to compensate for injury or damage caused by defective products is not a recent phenomenon. The concept of product liability has been in existence for many years, but its emphasis has changed recently. The first major product liability case occurred in 1916 and was tried before the New York Court of Appeals. The court held that an automobile manufacturer had a product liability obligation to a car buyer, even though the sales contract was between the buyer and a third party—namely, a car dealer. The direction of the law has always been that manufacturers or sellers are likely to incur a liability when they have been unreasonably careless or negligent in what they have designed, or produced, or how they have produced it. In recent years, the courts have placed a more stringent rule in effect called *strict liability*. Two principles are characteristic of strict liability. The first is a strong responsibility for both manufacturer and merchandiser, requiring immediate responsiveness to unsatisfactory quality through product service, repair, or replacement of defective product. This extends into the period of actual use by the consumer. By producing a product, the manufacturer and seller must accept responsibility for the ultimate use of that product—not only for its performance, but also for its environmental effects, the safety aspects of its use, and so forth.

The second principle involves advertising and promotion of the product. Under strict product liability all advertising statements must be supportable by valid company quality or certification data, comparable to that now maintained for product identification under regulations for such products as automobiles.

These two strict product liability principles result in strong pressure on manufacturers, distributors, and merchants to develop and maintain a high degree of factually based evidence concerning the performance and safety of their products. This evidence must cover not only the quality of the product as it is delivered to the consumer, but also its durability or reliability, its protection from possible side effects or environmental hazards, and its safety aspects in actual use. A strong quality-assurance program can help management in ensuring that this information will be available, if needed.

1-4.5 Implementing Quality Improvement

In the last few sections we have discussed the philosophy of quality improvement, the link between quality and productivity, and both economic and legal implications of quality. These are important aspects of the management of quality within an organization. There are certain other aspects of the overall management of quality and the implementation of TQM that warrant some attention.

Management must recognize that quality is a multifaceted entity, incorporating the eight dimensions we discussed in Section 1-1.1. For convenient reference, Table 1-4 summarizes these quality dimensions.

A critical part of the **strategic management of quality** within any business is the recognition of these dimensions by management and the selection of dimensions along which the business will compete. It will be very difficult to compete against companies that can successfully accomplish this part of the strategy.

A good example is the Japanese dominance of the video cassette recorder (VCR) market. The Japanese did not invent the VCR; the first units for home use were designed and produced in Europe and North America. However, the early VCRs produced by these companies were very unreliable and frequently had high levels of manufacturing defects. When the Japanese entered the market, they elected to compete along the dimensions of reliability and conformance to standards (no defects). This strategy allowed them to quickly dominate the market. In subsequent years, they expanded the dimensions of quality to include added features, improved performance, easier serviceability, improved aesthetics, and so forth. They have used total quality as a competitive weapon to raise the entry barrier to this market so high that it is virtually impossible for a new competitor to enter.

Management must do this type of strategic thinking about quality. It is not necessary that the product be superior in all dimensions of quality, but management must

Table 1-4 The Eight Dimensions of Quality from Section 1-1.1

1. Performance	5. Aesthetics
2. Reliability	6. Features
3. Durability	7. Perceived quality
4. Serviceability	8. Conformance to standards

select and develop the "niches" of quality along which the company can successfully compete. Typically, these dimensions will be those that the competition has forgotten or ignored. The American automobile industry has been severely impacted by foreign competitors who expertly practiced this strategy.

The critical role of suppliers in quality management must not be forgotten. In fact, supplier selection and management may be the most critical aspects of successful quality management in industries such as automotive, aerospace, and electronics, where a very high percentage of the parts in the end item are manufactured by outside suppliers. Many companies have instituted formal supplier quality-improvement programs as part of their own **internal** quality-improvement efforts. Selection of suppliers based on **quality, schedule,** and **cost,** rather than on cost alone, is also a vital strategic management decision that can have a long-term significant impact on overall competitiveness.

It is also critical that management recognize that quality improvement must be a total, company-wide activity, and that every organizational unit *must* actively participate. Obtaining this participation is the responsibility of (and a significant challenge to) senior management. What is the role of the quality-assurance organization in this effect? The responsibility of quality assurance is to assist management in providing quality assurance for the companies' products. Specifically, the quality-assurance function is a technology warehouse that contains the skills and resources necessary to generate products of acceptable quality in the marketplace. Quality management also has the responsibility for evaluating and using quality-cost information for identifying improvement opportunities in the system, and for making these opportunities known to higher management. It is important to note, however, that the **quality function is not responsible for quality.** After all, the quality organization does not design, manufacture, distribute, or service the product. Thus, the responsibility for quality is distributed throughout the entire organization.

The philosophy of Deming, Juran, and Feigenbaum implies that responsibility for quality spans the entire organization. However, there is a danger that if we adopt the philosophy that "quality is everybody's job," then quality will become nobody's job. This is why quality planning and analysis are important. Because quality improvement activities are so broad, successful efforts require, as an initial step, top management commitment. This commitment involves emphasis on the importance of quality, identification of the respective quality responsibilities of the various organizational units, and explicit accountability for quality improvement of all managers and employees in the company.

Finally, strategic management of quality in an organization will always be most effective when *all* the individuals in the organization have an understanding of the basic tools of quality improvement. Central among these tools are the elementary statistical concepts that form the basis of process control and that are used for the analysis of process data. It is increasingly important that everyone in an organization, from top management to operating personnel, have an awareness of basic statistical methods and of how these methods are useful in manufacturing engineering design and development and in the general business environment. Certain individuals must have higher levels of skills; for example, those engineers and managers in the quality-assurance function would generally be experts in one or more areas of process control, reliability engineering, design of experiments, or engineering data analysis. However, the key point is the philosophy that statistical methodology is a language of communication about problems that enables management to mobilize resources rapidly and to efficiently develop solutions to such problems.

PART

I

Statistical Methods Useful in Quality Improvement

Statistics is a collection of techniques useful for making decisions about a process or population based on an analysis of the information contained in a sample from that population. Statistical methods play a vital role in quality improvement. They provide the principal means by which a product is sampled, tested, and evaluated, and the information in those data is used to control and improve the manufacturing process. Furthermore, statistics is the language in which development engineers, manufacturing, procurement, management, and other functional components of the business communicate about quality.

This part contains two chapters. Chapter 2 gives a brief introduction to **descriptive statistics,** showing how simple graphical and numerical techniques can be used to summarize the information in sample data. The use of **probability distributions** to model the behavior of product parameters in a process or lot is then discussed. Chapter 3 presents techniques of **statistical inference**—that is, how the information contained in a sample can be used to draw conclusions about the population from which the sample was drawn.

Modeling Process Quality

CHAPTER OUTLINE

2-1 DESCRIBING VARIATION
- 2-1.1 The Stem-and-Leaf Plot
- 2-1.2 The Frequency Distribution and Histogram
- 2-1.3 Numerical Summary of Data
- 2-1.4 The Box Plot
- 2-1.5 Sample Computer Output
- 2-1.6 Probability Distributions

2-2 IMPORTANT DISCRETE DISTRIBUTIONS
- 2-2.1 The Hypergeometric Distribution
- 2-2.2 The Binomial Distribution
- 2-2.3 The Poisson Distribution
- 2-2.4 The Pascal and Related Distributions

2-3 IMPORTANT CONTINUOUS DISTRIBUTIONS
- 2-3.1 The Normal Distribution
- 2-3.2 The Exponential Distribution
- 2-3.3 The Gamma Distribution
- 2-3.4 The Weibull Distribution

2-4 SOME USEFUL APPROXIMATIONS
- 2-4.1 The Binomial Approximation to the Hypergeometric
- 2-4.2 The Poisson Approximation to the Binomial
- 2-4.3 The Normal Approximation to the Binomial
- 2-4.4 Comments on Approximations

CHAPTER OVERVIEW

This textbook is about the use of statistical methodology in quality control and improvement. This chapter has two objectives. First, we show how simple tools of descriptive statistics can be used to express variation quantitatively in a quality characteristic when a **sample** of data on this characteristic is available. Generally, the sample is just a subset of data taken from some larger population or process. The second objective is to introduce **probability distributions** and show how they provide a tool for modeling or describing the quality characteristics of a process.

2-1 DESCRIBING VARIATION

2-1.1 The Stem-and-Leaf Plot

No two units of product produced by a manufacturing process are identical. Some **variation** is inevitable. As examples, the net content of a can of soft drink varies slightly from can to can, and the output voltage of a power supply is not exactly the same from one unit to the next. **Statistics** is the science of analyzing data and drawing conclusions, taking variation in the data into account.

There are several graphical methods that are very useful for summarizing and presenting data. One of the most useful graphical techniques is the **stem-and-leaf display.**

Suppose that the data are represented by $x_1, x_2, \ldots, x_n$ and that each number x_i consists of at least two digits. To construct a stem-and-leaf plot, we divide each number x_i into two parts: a stem, consisting of one or more of the leading digits; and a leaf, consisting of the remaining digits. For example, if the data consist of percent defective information between 0 and 100 on lots of semiconductor wafers, then we can divide the value 76 into the stem 7 and the leaf 6. In general, we should choose relatively few stems in comparison with the number of observations. It is usually best to choose between 5 and 20 stems. Once a set of stems has been chosen, they are listed along the left-hand margin of the display, and beside each stem all leaves corresponding to the observed data values are listed in the order in which they are encountered in the data set.

······· EXAMPLE 2-1 ···

To illustrate the construction of a stem-and-leaf plot, consider the data in Table 2-1, which represents weekly yield data from a semiconductor fabrication facility. To construct a stem-and-leaf plot, we select as stems the values 4, 5, 6, 7, 8, and 9. The resulting stem and leaf display is shown in Fig. 2-1. Inspection of the plot reveals that the

Stem	Leaf	Frequency
4	8 9 7 9 5	5
5	3 2 1 2 3 9 4	7
6	3 0 4 4 5 2 0 8 5 9	10
7	9 3 8 5 7 6 5 3	8
8	8 3 1 6 5 1 2	7
9	2 1 2	3

Figure 2-1   Stem-and-leaf display for the semiconductor yield data in Table 2-1.

Table 2-1 Weekly Yields from a Semiconductor Fabrication Facility

Week	Yield	Week	Yield
1	48	21	68
2	53	22	65
3	49	23	73
4	52	24	88
5	51	25	69
6	52	26	83
7	63	27	78
8	60	28	81
9	53	29	86
10	64	30	92
11	59	31	75
12	54	32	85
13	47	33	81
14	49	34	77
15	45	35	82
16	64	36	76
17	79	37	75
18	65	38	91
19	62	39	73
20	60	40	92

yield distribution has an approximately symmetric shape, with a single peak. The stem-and-leaf display allows us to quickly determine some important features of the data that are not obvious from the data table. For example, Fig. 2-1 gives a visual impression of shape, spread or variability, and the central tendency or middle of the data (which is close to 68).

· ·

Variations of the Stem-and-Leaf Display

An **ordered stem-and-leaf display** has the leaves arranged by magnitude, as shown in Fig. 2-2. This version of the display (which is produced automatically by some statistics

Stem	Leaf	Frequency
4	5 7 8 9 9	5
5	1 2 2 3 3 4 9	7
6	0 0 2 3 4 4 5 5 8 9	10
7	3 3 5 5 6 7 8 9	8
8	1 1 2 3 5 6 8	7
9	1 2 2	3

Figure 2-2 Ordered stem-and-leaf plot for the semiconductor yield data.

software packages) makes it very easy to find **percentiles** of the data. Generally, the 100 kth percentile is a value such that at least 100 k% of the data values are at or below this value and at least 100 $(1 - k)$% of the data values are at or above this value.

The **fiftieth percentile** of the data distribution is called the **sample median $\tilde{x}$.** The median can be thought of as the data value that exactly divides the sample in half, with half of the observations smaller than the median and half of them larger.

If n, the number of observations, is odd, finding the median is easy. First sort the observations in ascending order (or rank the data from smallest observation to largest observation). Then the median will be the observation in rank position $[(n - 1)/2 + 1]$ on this list. If n is even, the median is the average of the $(n/2)$st and $(n/2 + 1)$st ranked observations. Since in our example $n = 40$ is an even number, the median is the average of the two observations with rank 20 and 21, or

$$\tilde{x} = \frac{65 + 68}{2} = 66.5$$

The **tenth percentile** is the observation with rank $(0.1)(40) + 0.5 = 4.5$ (halfway between the fourth and fifth observation), or $(49 + 49)/2 = 49$. The **first quartile** is the observation with rank $(0.25)(40) + 0.5 = 10.5$ (halfway between the tenth and eleventh observation) or $(53 + 54)/2 = 53.5$, and the **third quartile** is the observation with rank $(0.75)(40) + 0.5 = 30.5$ (halfway between the thirtieth and thirty-first observation), or $(79 + 81)/2 = 80$. The first and third quartiles are occasionally denoted by the symbols Q1 and Q3, respectively, and the **interquartile range** IQR = Q3 − Q1 is occasionally used as a measure of variability. For the semiconductor yield data the interquartile range is IQR = Q3 − Q1 = 80 − 53.5 = 26.5.

In some stem-and-leaf displays, it may be desirable to provide more classes or stems. One way is to modify the original stems as follows: Divide the stem 5 (say) into two new stems, 5* and 5°. The stem 5* has leaves 0, 1, 2, 3, and 4, and the stem 5° has leaves 5, 6, 7, 8, and 9. This will double the number of original stems. We could increase the number of original stems by five by defining five new stems: 5* with leaves 0 and 1, 5t (for twos and threes) with leaves 2 and 3, 5f (for fours and fives) with leaves 4 and 5, 5s (for sixes and sevens) with leaves 6 and 7, and 5° with leaves 8 and 9.

Finally, although the stem-and-leaf display is an excellent way to visually show the variability in data, it does not take the **time order** of the observations into account. Time is often a very important factor that contributes to variability in quality improvement problems. We could, of course, simply plot the data values versus time; such a graph is called a **time series graph** or a **run chart.** However, a useful approach is to combine the time series graph with the stem-and-leaf display to produce a **digidot plot.**

Figure 2-3 shows the digidot plot for the semiconductor yield data. This display clearly indicates that time is an important source of variability in this production process. More specifically, yields in the first 20 weeks of production are substantially below the yields reported in the last 20 weeks. Something may have changed in the process (or have been deliberately changed by operating personnel or the process engineers) that is responsible for the yield improvement. Later in this book we formally

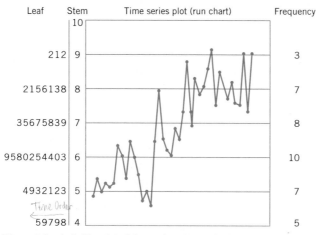

Leaf	Stem	Time series plot (run chart)	Frequency
	10		
212	9		3
2156138	8		7
35675839	7		8
9580254403	6		10
4932123	5		7
59798	4		5

Figure 2-3   A digidot plot of the semiconductor yield data.

introduce the control chart as a graphical technique for monitoring processes such as this one, and for producing a statistically based signal when a process change occurs.

2-1.2   The Frequency Distribution and Histogram

Table 2-2 presents 125 observations on the inside diameter of forged piston rings used in an automobile engine. The data were collected in 25 samples of five observations each. Note that there is some variability in piston-ring diameter. However, it is very difficult to see any *pattern* in the variability or *structure* in the data, with the observations arranged as they are in Table 2-2. A **frequency distribution** is an arrangement of the data by magnitude. It is a more compact summary of data than a stem-and-leaf display. For example, a frequency distribution of the piston-ring data is shown in Table 2-3. From this table we note that there was one ring that had a diameter between 73.965 mm and 73.970 mm, eight rings having diameters between 73.980 mm and 73.985 mm, and so forth.

A graph of the observed frequencies versus the ring diameter is shown in Fig. 2-4. This display is called a **histogram**. The height of each bar in Fig. 2-4 is equal to the frequency of occurrence of ring diameter. The histogram represents a visual display of the data in which one may more easily see three properties:

1. Shape
2. Location, or central tendency
3. Scatter, or spread

In the piston-ring diameter data, we see that the distribution of ring diameter is roughly symmetric and unimodal (mound shaped), with the central tendency very close to 74 mm. The variability in ring diameter is apparently relatively high, as some rings

Table 2-2   Forged Piston-Ring Inside Diameter (mm)

Sample Number	Observations				
1	74.030	74.002	74.019	73.992	74.008
2	73.995	73.992	74.001	74.011	74.004
3	73.988	74.024	74.021	74.005	74.002
4	74.002	73.996	73.993	74.015	74.009
5	73.992	74.007	74.015	73.989	74.014
6	74.009	73.994	73.997	73.985	73.993
7	73.995	74.006	73.994	74.000	74.005
8	73.985	74.003	73.993	74.015	73.988
9	74.008	73.995	74.009	74.005	74.004
10	73.998	74.000	73.990	74.007	73.995
11	73.994	73.998	73.994	73.995	73.990
12	74.004	74.000	74.007	74.000	73.996
13	73.983	74.002	73.998	73.997	74.012
14	74.006	73.967	73.994	74.000	73.984
15	74.012	74.014	73.998	73.999	74.007
16	74.000	73.984	74.005	73.998	73.996
17	73.994	74.012	73.986	74.005	74.007
18	74.006	74.010	74.018	74.003	74.000
19	73.984	74.002	74.003	74.005	73.997
20	74.000	74.010	74.013	74.020	74.003
21	73.988	74.001	74.009	74.005	73.996
22	74.004	73.999	73.990	74.006	74.009
23	74.010	73.989	73.990	74.009	74.014
24	74.015	74.008	73.993	74.000	74.010
25	73.982	73.984	73.995	74.017	74.013

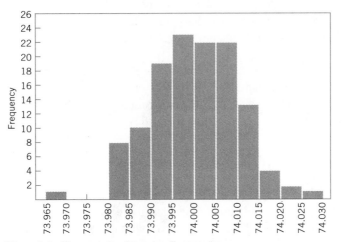

Figure 2-4   Histogram for piston-ring diameter data.

Table 2-3 Frequency Distribution for Piston-Ring Diameter

Ring Diameter, x (mm)	Tally	Frequency	Cumulative Frequency	Relative Frequency	Cumulative Relative Frequency
$73.965 \leq x < 73.970$	|	1	1	0.008	0.008
$73.970 \leq x < 73.975$		0	1	0.000	0.008
$73.975 \leq x < 73.980$		0	1	0.000	0.008
$73.980 \leq x < 73.985$	⫲⫲⫲ ⫲⫲⫲	8	9	0.064	0.072
$73.985 \leq x < 73.990$	⫲⫲⫲ ⫲⫲⫲	10	19	0.080	0.152
$73.990 \leq x < 73.995$	⫲⫲⫲ ⫲⫲⫲ ⫲⫲⫲ ⫲⫲⫲	19	38	0.152	0.304
$73.995 \leq x < 74.000$	⫲⫲⫲ ⫲⫲⫲ ⫲⫲⫲ ⫲⫲⫲ ⫲⫲⫲	23	61	0.184	0.488
$74.000 \leq x < 74.005$	⫲⫲⫲ ⫲⫲⫲ ⫲⫲⫲ ⫲⫲⫲ ⫲⫲⫲	22	83	0.176	0.664
$74.005 \leq x < 74.010$	⫲⫲⫲ ⫲⫲⫲ ⫲⫲⫲ ⫲⫲⫲ ⫲⫲⫲	22	105	0.176	0.840
$74.010 \leq x < 74.015$	⫲⫲⫲ ⫲⫲⫲ ⫲⫲⫲	13	118	0.104	0.944
$74.015 \leq x < 74.020$	⫲⫲⫲	4	122	0.032	0.976
$74.020 \leq x < 74.025$	||	2	124	0.016	0.992
$74.025 \leq x < 74.030$	|	1	125	0.008	1.000
Total		125		1.000	

are as small as 73.967 mm, while others are as large as 74.030 mm. Thus, the histogram gives some insight into the process that inspection of the raw data in Table 2-2 does not.

Several guidelines are helpful in constructing histograms. When the data are numerous, grouping them into bins or cells, as in the piston-ring example, is very useful. Generally,

1. Use between 4 and 20 bins—often choosing the number of bins approximately equal to the square root of the sample size works well.
2. Make the bins of uniform width.
3. Start the lower limit for the first bin just slightly below the smallest data value.

Grouping the data into bins condenses the original data, and as a result some detail is lost. Thus, when the number of observations is relatively small, or when the observations only take a few values, the histogram may be constructed from a frequency distribution of the ungrouped data. Alternatively, a stem-and-leaf display could be used. A primary advantage of the stem-and-leaf display is that the individual observations are preserved, whereas they are lost in a histogram.

2-1.3 Numerical Summary of Data

The stem-and-leaf display and the histogram provide a visual display of three properties of sample data: the shape of the distribution of the data, the central tendency in the data, and the scatter or variability in the data. It is also helpful to use numerical measures of central tendency and scatter.

Suppose that $x_1, x_2, \ldots, x_n$ are the observations in a sample. The most important measure of central tendency in the sample is the **sample average,**

$$\bar{x} = \frac{x_1 + x_2 + \cdots + x_n}{n}$$
$$= \frac{\sum\limits_{i=1}^{n} x_i}{n} \tag{2-1}$$

Note that the sample average $\bar{x}$ is simply the arithmetic mean of the n observations. The sample average for the piston-ring data is

$$\bar{x} = \frac{\sum\limits_{i=1}^{125} x_i}{125} = \frac{9250.147}{125} = 74.001 \text{ mm}$$

Refer to Fig. 2-4 and note that the sample average is the point at which the histogram exactly "balances." Thus, the sample average represents the center of mass of the sample data.

The variability in the sample data is measured by the **sample variance,**

$$S^2 = \frac{\sum\limits_{i=1}^{n} (x_i - \bar{x})^2}{n - 1} \tag{2-2}$$

Note that the sample variance is simply the sum of the squared deviations of each observation from the sample average $\bar{x}$, divided by the sample size minus one. If there is no variability in the sample, then each sample observation $x_i = \bar{x}$, and the sample variance $S^2 = 0$. Generally, the larger is the sample variance S^2, the greater is the variability in the sample data.

The units of the sample variance S^2 are the square of the original units of the data. This is often inconvenient and awkward to interpret, and so we usually prefer to use the square root of S^2, called the **sample standard deviation** S, as a measure of variability. It follows that

$$S = \sqrt{\frac{\sum_{i=1}^{n}(x_i - \bar{x})^2}{n-1}} \qquad (2\text{-}3)$$

The primary advantage of the sample standard deviation is that it is expressed in the original units of measurement. For the piston-ring data, we find that

$$S^2 = 0.000101 \text{ mm}^2$$

and

$$S = 0.010 \text{ mm}$$

To assist in understanding how the standard deviation describes variability, consider the two samples shown here:

Sample 1	Sample 2
$x_1 = 1$	$x_1 = 1$
$x_2 = 3$	$x_2 = 5$
$x_3 = 5$	$x_3 = 9$
$\bar{x} = 3$	$\bar{x} = 5$

Obviously, sample 2 has greater variability than sample 1. This is reflected in the standard deviation, which for sample 1 is

$$S = \sqrt{\frac{\sum_{i=1}^{3}(x_i - \bar{x})^2}{2}} = \sqrt{\frac{(1-3)^2 + (3-3)^2 + (5-3)^2}{2}} = \sqrt{4} = 2$$

and for sample 2 is

$$S = \sqrt{\frac{\sum_{i=1}^{3}(x_i - \bar{x})^2}{2}} = \sqrt{\frac{(1-5)^2 + (5-5)^2 + (9-5)^2}{2}} = \sqrt{16} = 4$$

Thus, the larger variability in sample 2 is reflected by its larger standard deviation. Now consider a third sample, say

Sample 3

$x_1 = 101$

$x_2 = 103$

$x_3 = 105$

$\bar{x} = 103$

The standard deviation for this third sample is $S = 2$, which is identical to the standard deviation of sample 1. Comparing the two samples, we see that both samples have identical variability or scatter about the average, and this is why they have the same standard deviations. This leads to an important point: **The standard deviation does not reflect the magnitude of the sample data, only the scatter about the average.**

Hand-held calculators are frequently used for calculating the sample average and standard deviation. Note that equations 2-2 and 2-3 are not very efficient computationally, because every number must be entered into the calculator twice. A more efficient formula is

$$ S = \sqrt{\frac{\sum\limits_{i=1}^{n} x_i^2 - \dfrac{\left(\sum\limits_{i=1}^{n} x_i\right)^2}{n}}{n - 1}} \qquad (2\text{-}4) $$

In using equation 2-4, each number would only have to be entered once, provided that $\sum_{i=1}^{n} x_i$ and $\sum_{i=1}^{n} x_i^2$ could be simultaneously accumulated in the calculator. Many inexpensive hand-held calculators perform this function and have automatic calculation of $\bar{x}$ and S.

2-1.4   The Box Plot

The stem-and-leaf display and the histogram provide a visual impression about a data set, whereas the sample average and standard deviation provide quantitative information about specific features of the data. The **box plot** is a graphical display that simultaneously displays several important features of the data, such as location or central tendency, spread or variability, departure from symmetry, and identification of observations that lie unusually far from the bulk of the data (these observations are often called "outliers").

A box plot displays the three quartiles, the minimum, and the maximum of the data on a rectangular box, aligned either horizontally or vertically. The box encloses the interquartile range with the left (or lower) line at the first quartile Q1 and the right (or

Table 2-4   Hole Diameters (in mm) in Wing Leading Edge Ribs

120.5	120.4	120.7
120.9	120.2	121.1
120.3	120.1	120.9
121.3	120.5	120.8

Figure 2-5 Box plot for the aircraft wing leading edge hole diameter data in Table 2-4.

upper) line at the third quartile Q3. A line is drawn through the box at the second quartile (which is the fiftieth percentile or the median) Q2 = $\tilde{x}$. A line at either end extends to the extreme values. These lines are usually called **whiskers.** Some authors refer to the box plot as the **box and whisker plot.** In some computer programs, the whiskers only extend a distance of 1.5 (Q3 − Q1) from the ends of the box, at most, and observations beyond these limits are flagged as potential outliers. This variation of the basic procedure is called a **modified box plot.**

······· EXAMPLE 2-2 ··

The data in Table 2-4 are diameters (in mm) of holes in a group of 12 wing leading edge ribs for a commercial transport airplane. Note that the median of the sample is halfway between the sixth and seventh rank-ordered observation, or (120.5 + 120.7)/2 = 120.6, and that the quartiles are Q1 = 120.35 and Q3 = 120.9. The box plot is shown in Fig. 2-5. This box plot indicates that the hole diameter distribution is not exactly symmetric around a central value, because the left and right whiskers and the left and right boxes around the median are not the same lengths.

···

Box plots are very useful in graphical comparisons among data sets, because they have visual impact and are easy to understand. For example, Fig. 2-6 shows the comparative box plots for a manufacturing quality index on products at three manufacturing plants. Inspection of this display reveals that there is too much variability at plant 2 and that plants 2 and 3 need to raise their quality index performance.

2-1.5   Sample Computer Output

Statistics software packages will produce most of the graphical and numerical data summaries that we have discussed. Figure 2-7 is a computer-generated histogram of the

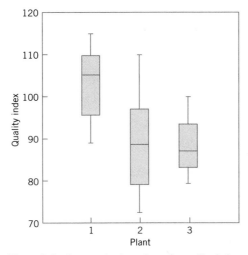

Figure 2-6   Comparative box plots of a quality index for products produced at three plants.

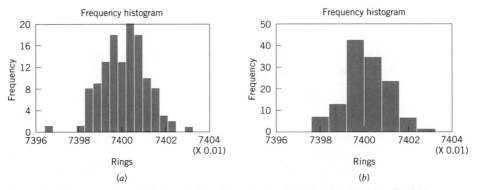

Figure 2-7   Computer-generated histogram of the piston-ring data in Table 2-2. (*a*) 20 bins, (*b*) 10 bins.

piston-ring data in Table 2-2. The left display, Fig. 2-7*a*, uses the "default" setting of 20 bins, whereas the histogram in Fig. 2-7*b* was constructed with 10 bins. Usually, with relatively large data sets, the shape of the histogram will not depend very much on the number of bins. For example, compare the histograms in Fig. 2-7 with the one in Fig. 2-4, which has 13 bins. All of these histograms give the same general information about shape. However, for small data sets, histograms may change dramatically in shape if the number and/or width of the bins change. Histograms are most suitable for larger data sets, preferably of size 75 to 100 observations or more.

Table 2-5 is a computer-generated numerical statistical summary for the piston-ring data. Note that the results agree with those reported previously, except that in some cases, more decimal places are reported. Computer programs often report four, five, or more decimal places in the output.

Table 2-5 Computer-Generated Summary Statistics for the Piston-Ring Data

Variable	QCDATA Rings
Sample size	125.
Average	74.001176
Median	74.001
Mode	74.
Variance	0.000101
Standard deviation	0.01007
Minimum	73.967
Maximum	74.03
Range	0.063
Lower quartile	73.994
Upper quartile	74.008
Interquartile range	0.014
Sum	9250.147

2-1.6   Probability Distributions

The histogram (or stem-and-leaf plot, or box plot) is used to describe *sample* data. A **sample** is a collection of measurements selected from some larger source or **population.** For example, the 125 piston-ring diameters in Table 2-1 are a sample of piston-ring diameters selected from the manufacturing process. The population in this example is the collection of all piston rings produced by that process. By using statistical methods, we may be able to analyze the sample piston-ring-diameter data and draw certain conclusions about the process that manufactures the rings.

A **probability distribution** is a mathematical model that relates the value of the variable with the probability of occurrence of that value in the population. In other words,

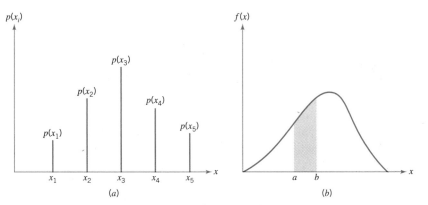

Figure 2-8   Probability distributions. (*a*) Discrete case. (*b*) Continuous case.

we might visualize piston-ring diameter as a **random variable,** because it takes on different values in the population according to some random mechanism, and then the probability distribution of ring diameter describes the probability of occurrence of any value of ring diameter in the population. There are two types of probability distributions.

Definition

1. **Continuous distributions.** When the variable being measured is expressed on a continuous scale, its probability distribution is called a *continuous distribution*. The probability distribution of piston-ring diameter is continuous.

2. **Discrete distributions.** When the parameter being measured can only take on certain values, such as the integers 0, 1, 2, . . . , the probability distribution is called a *discrete distribution*. For example, the distribution of the number of nonconformities or defects in printed circuit boards would be a discrete distribution.

Examples of discrete and continuous probability distributions are shown in Figs. 2-8*a* and 2-8*b*, respectively. The appearance of a discrete distribution is that of a series of vertical "spikes," with the height of each spike proportional to the probability. We write the probability that the random variable x takes on the specific value x_i as

$$P\{x = x_i\} = p(x_i)$$

The appearance of a continuous distribution is that of a smooth curve, with the area under the curve equal to probability, so that the probability that x lies in the interval from a to b is written as

$$P\{a \leq x \leq b\} = \int_a^b f(x)\, dx$$

········ **EXAMPLE 2-3** ···

A Discrete Distribution

A manufacturing process produces thousands of diodes per day. On the average, 1% of these diodes do not conform to specifications. Every hour, an inspector selects a random sample of 50 diodes and classifies each diode in the sample as conforming or nonconforming. If we let x be the random variable representing the number of nonconforming parts in the sample, then the probability distribution of x is

$$p(x) = \binom{50}{x}(0.01)^x(0.99)^{50-x} \qquad x = 0, 1, 2, \ldots, 50$$

where $\binom{50}{x} = 50!/[x!(50 - x)!]$. This is a *discrete* distribution, since the observed number of nonconformances is $x = 0, 1, 2, \ldots, 50$, and is called the **binomial distribution.** We may calculate the probability of finding one or fewer nonconforming parts in the sample as

$$
\begin{aligned}
P(x \leq 1) &= P(x = 0) + P(x = 1) \\
&= p(0) + p(1) \\
&= \sum_{x=0}^{1} \binom{50}{x} (0.01)^x (0.99)^{50-x} \\
&= \frac{50!}{0!50!} (0.99)^{50}(0.01)^0 + \frac{50!}{1!49!} (0.99)^{49}(0.01)^1 \\
&= 0.6050 + 0.3056 = 0.9106
\end{aligned}
$$

···

········ **EXAMPLE 2-4** ···

A Continuous Distribution

Suppose that x is a random variable that represents the actual contents in ounces of a 1-1b can of coffee. The probability distribution of x is assumed to be

$$
f(x) = \frac{1}{1.5} \qquad 15.5 \leq x \leq 17.0
$$

This is a *continuous* distribution, since the range of x is the interval [15.5, 17.0]. This distribution is called the **uniform distribution,** and it is shown graphically in Fig. 2-9. Note that the area under the function $f(x)$ corresponds to probability, so that the probability of a can containing less than 16.0 oz is

$$
\begin{aligned}
P\{x \leq 16.0\} &= \int_{15.5}^{16.0} f(x)\, dx \\
&= \int_{15.5}^{16.0} \frac{1}{1.5}\, dx \\
&= \left. \frac{x}{1.5} \right|_{15.5}^{16.0} = \frac{16.0 - 15.5}{1.5} = 0.3333
\end{aligned}
$$

This follows intuitively from inspection of Fig. 2-9.

···

In Sections 2-2 and 2-3 we present several useful discrete and continuous distributions.

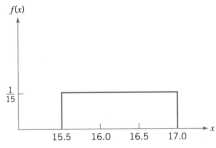

Figure 2-9   The uniform distribution for Example 2-4.

The **mean** μ of a probability distribution is a measure of the **central tendency** in the distribution, or its **location.** The mean is defined as

$$
\mu = \begin{cases} \displaystyle\int_{-\infty}^{\infty} x f(x)\, dx, & x \text{ continuous} \qquad (2\text{-}5a) \\[2ex] \displaystyle\sum_{i=1}^{\infty} x_i p(x_i), & x \text{ discrete} \qquad (2\text{-}5b) \end{cases}
$$

For the case of a discrete random variable with exactly N equally likely values [that is, $p(x_i) = 1/N$], then equation 2-5b reduces to

$$
\mu = \frac{\displaystyle\sum_{i=1}^{N} x_i}{N}
$$

Note the similarity of this last expression to the sample average $\bar{x}$ defined in equation 2-1. The mean is the point at which the distribution exactly "balances" (see Fig. 2-10). Thus, the mean is simply the center of mass of the probability distribution. Note from Fig. 2-10b that the mean is not necessarily the fiftieth percentile of the distribution (the **median**), and from Fig. 2-10c it is not necessarily the most likely value of the variable (which is called the **mode**). The mean simply determines the **location** of the distribution, as shown in Fig. 2-11.

The scatter, spread, or variability in a distribution is expressed by the **variance** σ^2. The definition of the variance is

$$
\sigma^2 = \begin{cases} \displaystyle\int_{-\infty}^{\infty} (x - \mu)^2 f(x)\, dx, & x \text{ continuous} \qquad (2\text{-}6a) \\[2ex] \displaystyle\sum_{i=1}^{\infty} (x_i - \mu)^2 p(x_i), & x \text{ discrete} \qquad (2\text{-}6b) \end{cases}
$$

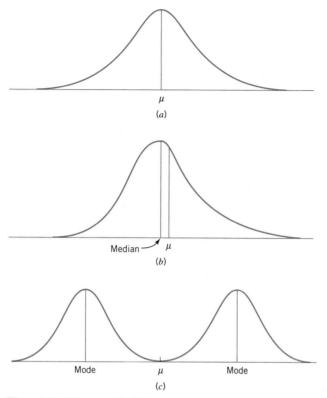

Figure 2-10   The mean of a distribution.

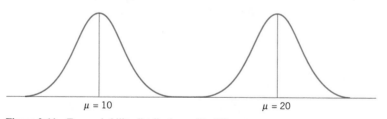

Figure 2-11   Two probability distributions with different means.

when the random variable is discrete with N equally likely values, then equation 2-6b becomes

$$\sigma^2 = \frac{\sum\limits_{i=1}^{N} (x_i - \mu)^2}{N}$$

and we observe that in this case the variance is the average squared distance of each element of the population from the mean. Note the similarity to the sample variance S^2,

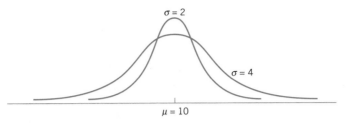

Figure 2-12 Two probability distributions with the same mean but different standard deviations.

defined in equation 2-2. If $\sigma^2 = 0$, there is no variability in the population. As the variability increases, the variance σ^2 increases. The variance is expressed in the square of the units of the original variable. For example, if we are measuring voltages, the units of the variance are (volts)^2. Thus, it is customary to work with the square root of the variance, called the **standard deviation** σ. It follows that

$$\sigma = \sqrt{\sigma^2} = \sqrt{\frac{\sum\limits_{i=1}^{N} (x_i - \mu)^2}{N}} \tag{2-7}$$

The standard deviation is a measure of spread or scatter in the population expressed in the original units. Two distributions with the same mean but different standard deviations are shown in Fig. 2-12.

2-2   IMPORTANT DISCRETE DISTRIBUTIONS

Several discrete probability distributions arise frequently in statistical quality control. In this section, we discuss the hypergeometric distribution, the binomial distribution, the Poisson distribution, and the Pascal or negative binomial distribution.

2-2.1   The Hypergeometric Distribution

Suppose that there is a finite population consisting of N items. Some number—say, $D(D \leq N)$—of these items fall into a class of interest. A random sample of n items is selected from the population *without replacement*, and the number of items in the sample that fall into the class of interest—say, x—is observed. Then x is a hypergeometric random variable with the probability distribution defined as follows.

Definition

The **hypergeometric probability distribution** is

$$p(x) = \frac{\binom{D}{x}\binom{N-D}{n-x}}{\binom{N}{n}} \qquad x = 0, 1, 2, \ldots, \min(n, D) \qquad (2\text{-}8)$$

The mean and variance of the distribution are

$$\mu = \frac{nD}{N} \qquad (2\text{-}9)$$

and

$$\sigma^2 = \frac{nD}{N}\left(1 - \frac{D}{N}\right)\left(\frac{N-n}{N-1}\right) \qquad (2\text{-}10)$$

In the above definition, the quantity

$$\binom{a}{b} = \frac{a!}{b!(a-b)!}$$

is the number of combinations of a items taken b at a time.

The hypergeometric distribution is the appropriate probability model for selecting a random sample of n items without replacement from a lot of N items of which D are nonconforming or defective. By a random sample, we mean a sample that has been selected in such a way that all possible samples have an equal chance of being chosen. In these applications, x usually represents the number of nonconforming items found in the sample. For example, suppose that a lot contains 100 items, 5 of which do not conform to requirements. If 10 items are selected at random without replacement, then the probability of finding one or fewer nonconforming items in the sample is

$$P\{x \le 1\} = P\{x = 0\} + P\{x = 1\}$$

$$= \frac{\binom{5}{0}\binom{95}{10}}{\binom{100}{10}} + \frac{\binom{5}{1}\binom{95}{9}}{\binom{100}{10}}$$

$$= 0.923$$

In Chapter 14 we show how probability models such as this can be used to design acceptance-sampling procedures.

2-2.2   The Binomial Distribution

Consider a process that consists of a sequence of n independent trials. By independent trials, we mean that the outcome of each trial does not depend in any way on the outcome of previous trials. When the outcome of each trial is either a "success" or a "failure," the trials are called **Bernoulli trials.** If the probability of "success" on any trial—say, p—is constant, then the number of "successes" x in n Bernoulli trials has the **binomial distribution** with parameters n and p, defined as follows.

Definition

The **binomial distribution** with parameters $n \geq 0$ and $0 < p < 1$ is

$$p(x) = \binom{n}{x} p^x (1 - p)^{n-x} \qquad x = 0, 1, \ldots, n \qquad (2\text{-}11)$$

The mean and variance of the binomial distribution are

$$\mu = np \qquad\qquad (2\text{-}12)$$

and

$$\sigma^2 = np(1 - p) \qquad\qquad (2\text{-}13)$$

The binomial distribution is used frequently in quality engineering. It is the appropriate probability model for sampling from an infinitely large population, where p represents the fraction of defective or nonconforming items in the population. In these applications, x usually represents the number of nonconforming items found in a random sample of n items. For example, if $p = 0.10$ and $n = 15$, then the probability of obtaining x nonconforming items is computed from equation 2-11 as

x	$p(x)$
0	0.2059
1	0.3432
2	0.2669
3	0.1285
4	0.0428
5	0.0105
6	0.0019
7	0.0003
8	0.0000
.	.
.	.
.	.
15	0.0000

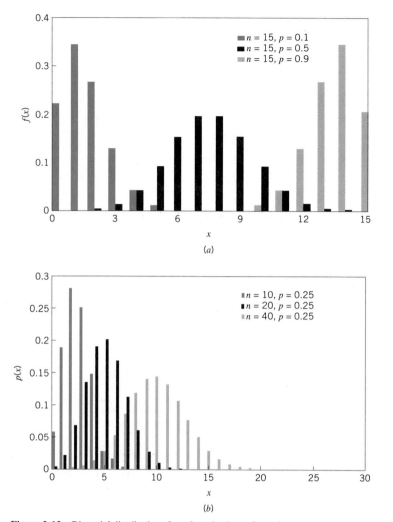

Figure 2-13   Binomial distributions for selected values of n and p.

Several binomial distributions are shown graphically in Fig. 2-13, where their shape is typical of all binomial distributions. For a fixed n, the distribution becomes more symmetric as p increases from 0 to 0.5 or decreases from 1 to 0.5. For a fixed p, the distribution becomes more symmetric as n increases.

A random variable that arises frequently in statistical quality control is

$$\hat{p} = \frac{x}{n} \qquad\qquad (2\text{-}14)$$

where x has a binomial distribution with parameters n and p. Often $\hat{p}$ is the ratio of the observed number of defective or nonconforming items in a sample (x) to the sample size (n) and this is usually called the **sample fraction defective** or **sample fraction**

nonconforming. The "^" symbol is used to indicate that $\hat{p}$ is an estimate of the true, unknown value of the binomial parameter p. The probability distribution of $\hat{p}$ is obtained from the binomial, since

$$P\{\hat{p} \le a\} = P\left\{\frac{x}{n} \le a\right\} = p\{x \le na\} = \sum_{x=0}^{[na]} \binom{n}{x} p^x (1-p)^{n-x}$$

where $[na]$ denotes the largest integer less than or equal to na. It is easy to show that the mean of $\hat{p}$ is p and that the variance of $\hat{p}$ is

$$\sigma_{\hat{p}}^2 = \frac{p(1-p)}{n}$$

2-2.3   The Poisson Distribution

A useful discrete distribution in statistical quality control is the Poisson distribution, defined as follows.

Definition

The **Poisson distribution** is

$$p(x) = \frac{e^{-\lambda} \lambda^x}{x!} \qquad x = 0, 1, \dots \qquad (2\text{-}15)$$

where the parameter $\lambda > 0$. The **mean** and **variance** of the Poisson distribution are

$$\mu = \lambda \qquad\qquad (2\text{-}16)$$

and

$$\sigma^2 = \lambda \qquad\qquad (2\text{-}17)$$

Note that the mean and variance of the Poisson distribution are both equal to the parameter λ.

A typical application of the Poisson distribution in quality control is as a model of the number of defects or nonconformities that occur in a unit of product. In fact, any random phenomenon that occurs on a per unit (or per unit area, per unit volume, per unit time, etc.) basis is often well approximated by the Poisson distribution. As an example, suppose that the number of wire-bonding defects per unit that occur in a semiconductor device is Poisson distributed with parameter $\lambda = 4$. Then the probability that a ran-

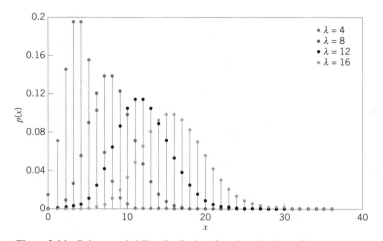

Figure 2-14 Poisson probability distributions for selected values of λ.

domly selected semiconductor device will contain two or fewer wire-bonding defects is

$$P\{x \le 2\} = \sum_{x=0}^{2} \frac{e^{-4}4^x}{x!}$$
$$= 0.0183 + 0.0733 + 0.1464 = 0.2380$$

Several Poisson distributions are shown in Fig. 2-14. Note that the distribution is *skewed;* that is, it has a long tail to the right. As the parameter λ becomes larger, the Poisson distribution becomes symmetric in appearance.

It is possible to derive the Poisson distribution as a limiting form of the binomial distribution. That is, in a binomial distribution with parameters n and p, if we let n approach infinity and p approach zero in such a way that $np = \lambda$ is a constant, then the Poisson distribution results. It is also possible to derive the Poisson distribution using a pure probability argument. For more information about the Poisson distribution, see Hines and Montgomery (1990), Montgomery and Runger (1999), and the supplemental text material.

2-2.4 The Pascal and Related Distributions

The Pascal distribution, like the binomial distribution, has its basis in Bernoulli trials. Consider a sequence of independent trials, each with probability of success p, and let x denote the trial on which the rth success occurs. Then x is a Pascal random variable with probability distribution defined as follows.

Definition

The **Pascal distribution** is

$$p(x) = \binom{x-1}{r-1} p^r (1-p)^{x-r} \qquad x = r, r+1, r+2, \ldots \qquad (2\text{-}18)$$

where $r \geq 1$ is an integer. The **mean** and **variance** of the Pascal distribution are

$$\mu = \frac{r}{p} \qquad (2\text{-}19)$$

and

$$\sigma^2 = \frac{r(1-p)}{p^2} \qquad (2\text{-}20)$$

respectively.

Two special cases of the Pascal distribution are of interest. The first of these is if $r > 0$ and not necessarily an integer. The resulting distribution is called the **negative binomial distribution.** It is relatively standard to refer to equation 2-18 as the negative binomial distribution, even when r is an integer. The negative binomial distribution, like the Poisson distribution, is sometimes useful as the underlying statistical model for various types of "count" data, such as the occurrence of nonconformities in a unit of product (see Section 6-3.1). There is an important duality between the binomial and negative binomial distributions. In the binomial distribution, we fix the sample size (number of Bernoulli trials) and observe the number of successes; in the negative binomial distribution, we fix the number of successes and observe the sample size (number of Bernoulli trials) required to achieve them. This concept is particularly important in various kinds of sampling problems.

The other special case of the Pascal distribution is if $r = 1$, in which case we have the **geometric distribution.** It is the distribution of the number of Bernoulli trials until the *first* success.

2-3 IMPORTANT CONTINUOUS DISTRIBUTIONS

In this section we discuss several continuous distributions that are important in statistical quality control. These include the normal distribution, the exponential distribution, the gamma distribution, and the Weibull distribution.

2-3.1   The Normal Distribution

The normal distribution is probably the most important distribution in both the theory and application of statistics. If x is a normal random variable, then the probability distribution of x is defined as follows.

Definition

The **normal distribution** is

$$f(x) = \frac{1}{\sigma\sqrt{2\pi}} e^{-\frac{1}{2}\left(\frac{x-\mu}{\sigma}\right)^2} \qquad -\infty < x < \infty \qquad (2\text{-}21)$$

The mean of the normal distribution is μ $(-\infty < \mu < \infty)$ and the variance is $\sigma^2 > 0$.

The normal distribution is used so much that we frequently employ a special notation, $x \sim N(\mu, \sigma^2)$, to imply that x is normally distributed with mean μ and variance σ^2. The visual appearance of the normal distribution is a symmetric, unimodal or **bell-shaped** curve, and is shown in Fig. 2-15.

There is a simple interpretation of the standard deviation σ of a normal distribution, which is illustrated in Fig. 2-16. Note that 68.26% of the population values fall between the limits defined by the mean plus and minus one standard deviation $(\mu \pm 1\sigma)$; 95.46% of the values fall between the limits defined by the mean plus and minus two standard deviations $(\mu \pm 2\sigma)$; and 99.73% of the population values fall within the limits defined by the mean plus and minus three standard deviations $(\mu \pm 3\sigma)$. Thus, the standard deviation measures the distance on the horizontal scale associated with the 68.26%, 95.46%, and 99.73% containment limits. It is common practice to round these percentages to 68%, 95%, and 99.7%.

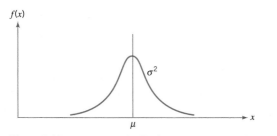

Figure 2-15   The normal distribution.

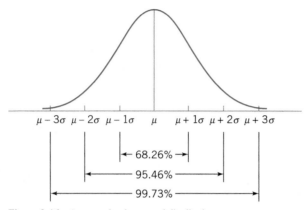

Figure 2-16   Areas under the normal distribution.

The cumulative normal distribution is defined as the probability that the normal random variable x is less than or equal to some value a, or

$$P\{x \le a\} = F(a) = \int_{-\infty}^{a} \frac{1}{\sigma\sqrt{2\pi}} e^{-\frac{1}{2}\left(\frac{x-\mu}{\sigma}\right)^{2}} dx \qquad (2\text{-}22)$$

This integral cannot be evaluated in closed form. However, by using the change of variable

$$z = \frac{x - \mu}{\sigma} \qquad (2\text{-}23)$$

the evaluation can be made independent of μ and σ^2. That is,

$$P\{x \le a\} = P\left\{z \le \frac{a - \mu}{\sigma}\right\} \equiv \Phi\left(\frac{a - \mu}{\sigma}\right)$$

where $\Phi(\cdot)$ is the cumulative distribution function of the **standard normal distribution** (mean = 0, standard deviation = 1). A table of the cumulative standard normal distribution is given in Appendix Table II. The transformation (2-23) is usually called **standardization,** because it converts a $N(\mu, \sigma^2)$ random variable into a $N(0, 1)$ random variable.

········ **EXAMPLE 2-5** ···

The tensile strength of paper used to make grocery bags is an important quality characteristic. It is known that the strength—say, x—is normally distributed with mean $\mu = $ 40 lb/in^2 and standard deviation $\sigma = 2$ lb/in^2, denoted $x \sim N(40, 2^2)$. The purchaser of the bags requires them to have a strength of at least 35 lb/in^2. The probability that a bag produced from this paper will meet or exceed this specification is $P\{x \ge 35\}$. Note that

$$P\{x \ge 35\} = 1 - P\{x \le 35\}$$

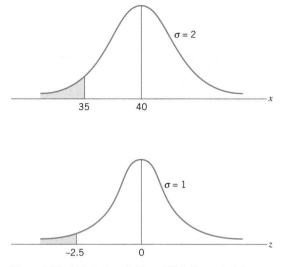

Figure 2-17   Calculation of $P\{x \leq 35\}$ in Example 2-5.

To evaluate this probability from the standard normal tables, we standardize the point 35 and find

$$
\begin{aligned}
P\{x \leq 35\} &= P\left\{z \leq \frac{35 - 40}{2}\right\} \\
&= P\{z \leq -2.5\} \\
&= \Phi(-2.5) \\
&= 0.0062
\end{aligned}
$$

Consequently, the desired probability is

$$
P\{x \geq 35\} = 1 - P\{x \leq 35\} = 1 - 0.0062 = 0.9938
$$

Figure 2-17 shows the tabulated probability for both the $N(40, 2^2)$ distribution and the standard normal distribution. Note that the shaded area to the left of 35 lb/in^2 in Fig. 2-17 represents the fraction nonconforming or "fallout" produced by the bag manufacturing process.

· ·

Appendix Table II gives only probabilities to the left of positive values of z. We will need to utilize the symmetry property of the normal distribution to evaluate probabilities. Specifically, note that

$$
P\{x \geq a\} = 1 - P\{x \leq a\} \tag{2-24a}
$$

$$
P\{x \leq -a\} = P\{x \geq a\} \tag{2-24b}
$$

and

$$P\{x \geq -a\} = P\{x \leq a\} \tag{2-24c}$$

It is helpful in problem solution to draw a graph of the distribution, as in Fig. 2-17.

······· **EXAMPLE 2-6** ···

The diameter of a metal shaft used in a disk-drive unit is normally distributed with mean 0.2508 in. and standard deviation 0.0005 in. The specifications on the shaft have been established as 0.2500 ± 0.0015 in. We wish to determine what fraction of the shafts produced conform to specifications. The appropriate normal distribution is shown in Fig. 2-18. Note that

$$P\{0.2485 \leq x \leq 0.2515\} = P\{x \leq 0.2515\} - P\{x \leq 0.2485\}$$

$$= \Phi\left(\frac{0.2515 - 0.2508}{0.0005}\right) - \Phi\left(\frac{0.2485 - 0.2508}{0.0005}\right)$$

$$= \Phi(1.40) - \Phi(-4.60)$$

$$= 0.9265 - 0.0000$$

$$= 0.9265$$

Thus, we would expect the process yield to be approximately 92.65%; that is, about 92.65% of the shafts produced conform to specifications.

Note that almost all of the nonconforming shafts are too large, because the process mean is located very near to the upper specification limit. Suppose we can recenter the manufacturing process, perhaps by adjusting the machine, so that the process mean is exactly equal to the nominal value of 0.2500. Then we have

$$P\{0.2485 \leq x \leq 0.2515\} = P\{x \leq 0.2515\} - P\{x \leq 0.2485\}$$

$$= \Phi\left(\frac{0.2515 - 0.2500}{0.0005}\right) - \Phi\left(\frac{0.2485 - 0.2500}{0.0005}\right)$$

$$= \Phi(3.00) - \Phi(-3.00)$$

$$= 0.99865 - 0.00135$$

$$= 0.9973$$

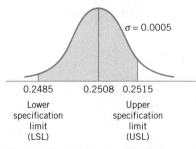

Figure 2-18 Distribution of shaft diameters, Example 2-6.

By recentering the process we have increased the yield of the process to approximately 99.73%.

· · · · · · · · **EXAMPLE 2-7** ·

Sometimes instead of finding the probability associated with a particular value of a normal random variable, we find it necessary to do the opposite—find a particular value of a normal random variable that results in a given probability. For example, suppose that $x \sim N(10, 9)$. We wish to find the value of x—say, a—such that $P\{x > a\} = 0.05$. Thus

$$P\{x > a\} = P\left\{z > \frac{a - 10}{3}\right\} = 0.05$$

or

$$P\left\{z \le \frac{a - 10}{3}\right\} = 0.95$$

From Appendix Table II, we have $P\{z \le 1.645\} = 0.95$, so

$$\frac{a - 10}{3} = 1.645$$

or

$$a = 10 + 3(1.645) = 14.935$$

· ·

The normal distribution has many useful properties. One of these is relative to linear combinations of normally and independently distributed random variables. If x_1, $x_2, \ldots, x_n$ are normally and independently distributed random variables with means $\mu_1, \mu_2, \ldots, \mu_n$ and variances $\sigma_1^2, \sigma_2^2, \ldots, \sigma_n^2$, respectively, then the distribution of

$$y = a_1 x_1 + a_2 x_2 + \cdots + a_n x_n$$

is normal with mean

$$\mu_y = a_1 \mu_1 + a_2 \mu_2 + \cdots + a_n \mu_n \tag{2-25}$$

and variance

$$\sigma_y^2 = a_1^2 \sigma_1^2 + a_2^2 \sigma_2^2 + \cdots + a_n^2 \sigma_n^2 \tag{2-26}$$

where $a_1, a_2, \ldots, a_n$ are constants.

The Central Limit Theorem

The normal distribution is often assumed as the appropriate probability model for a random variable. Later on, we will discuss how to check the validity of this assumption; however, the central limit theorem is often a justification of approximate normality.

Definition: The Central Limit Theorem

If $x_1, x_2, \ldots, x_n$ are independent random variables with mean μ_i and variance σ_i^2, and if $y = x_1 + x_2 + \cdots + x_n$, then the distribution of

$$\frac{y - \sum\limits_{i=1}^{n} \mu_i}{\sqrt{\sum\limits_{i=1}^{n} \sigma_i^2}}$$

approaches the $N(0, 1)$ distribution as n approaches infinity.

The central limit theorem implies that the sum of n independently distributed random variables is approximately normal, regardless of the distributions of the individual variables. The approximation improves as n increases. In many cases the approximation will be good for small n—say, $n < 10$—whereas in some cases we may require very large n—say, $n > 100$—for the approximation to be satisfactory. In general, if the x_i are identically distributed, and the distribution of each x_i does not depart radically from the normal, then the central limit theorem works quite well for $n \geq 3$ or 4. These conditions are met frequently in quality control problems.

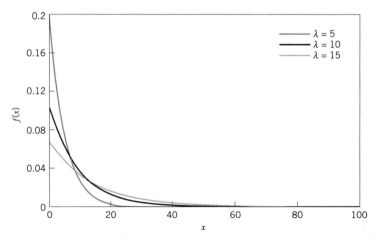

Figure 2-19   Exponential distributions for selected values of λ.

2-3.2   The Exponential Distribution

The probability distribution of the exponential random variable is defined as follows.

Definition

The **exponential distribution** is

$$f(x) = \lambda e^{-\lambda x} \qquad x \geq 0 \tag{2-27}$$

where $\lambda > 0$ is a constant. The **mean** and **variance** of the exponential distribution are

$$\mu = \frac{1}{\lambda} \tag{2-28}$$

and

$$\sigma^2 = \frac{1}{\lambda^2} \tag{2-29}$$

respectively.

Several exponential distributions are shown in Fig. 2-19.

The cumulative exponential distribution is

$$F(a) = P\{x \leq a\}$$

$$= \int_0^a \lambda e^{-\lambda t}\, dt$$

$$= 1 - e^{-\lambda a} \qquad a \geq 0 \tag{2-30}$$

Figure 2-20 depicts the exponential cumulative distribution function.

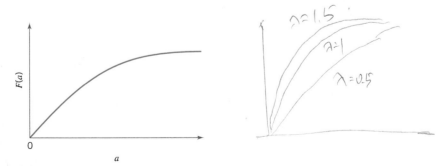

Figure 2-20 The cumulative exponential distribution function.

The exponential distribution is widely used in the field of **reliability engineering** as a model of the time to failure of a component or system. In these applications, the parameter λ is called the **failure rate** of the system, and the mean of the distribution $1/\lambda$ is called the **mean time to failure.**[1] For example, suppose that an electronic component in an airborne radar system has a useful life described by an exponential distribution with failure rate 10^{-4}/h; that is, $\lambda = 10^{-4}$. The mean time to failure for this component is $1/\lambda = 10^4 = 10,000$ h. If we wanted to determine the probability that this component would fail before its expected life, we would evaluate

$$P\left\{x \le \frac{1}{\lambda}\right\} = \int_0^{1/\lambda} \lambda e^{-\lambda t}\, dt = 1 - e^{-1} = 0.63212$$

This result holds regardless of the value of λ; that is, the probability that a value of an exponential random variable will be less than its mean is 0.63212. This happens, of course, because the distribution is not symmetric.

There is an important relationship between the exponential and Poisson distributions. If we consider the Poisson distribution as a model of the number of occurrences of some event in the interval $(0, t]$, then from equation 2-15 we have

$$p(x) = \frac{e^{-\lambda t}(\lambda t)^x}{x!}$$

Now $x = 0$ implies that there are no occurrences of the event in $(0, t]$, and $P\{x = 0\} = p(0) = e^{-\lambda t}$. We may think of $p(0)$ as the probability that the interval to the first occurrence is greater than t, or

$$P\{y > t\} = p(0) = e^{-\lambda t}$$

where y is the random variable denoting the interval to the first occurrence. Since

$$F(t) = P\{y \le t\} = 1 - e^{-\lambda t}$$

and using the fact that $f(y) = dF(y)/dy$, we have

$$f(y) = \lambda e^{-\lambda y} \tag{2-31}$$

as the distribution of the interval to the first occurrence. We recognize equation 2-31 as an exponential distribution with parameter λ. Therefore, we see that if the number of occurrences of an event has a Poisson distribution with parameter λ, then the distribution of the interval *between* occurrences is exponential with parameter λ.

---

[1] See the supplemental text material for more information.

2-3.3   The Gamma Distribution

The probability distribution of the gamma random variable is defined as follows.

Definition

The **gamma distribution** is

$$f(x) = \frac{\lambda}{\Gamma(r)} (\lambda x)^{r-1} e^{-\lambda x} \qquad x \geq 0 \qquad (2\text{-}32)$$

with **shape parameter** $r > 0$ and **scale parameter** $\lambda > 0$. The **mean** and **variance** of the gamma distribution are

$$\mu = \frac{r}{\lambda} \qquad (2\text{-}33)$$

and

$$\sigma^2 = \frac{r}{\lambda^2} \qquad (2\text{-}34)$$

respectively.[2]

Several gamma distributions are shown in Fig. 2-21. Note that if $r = 1$, the gamma distribution reduces to the exponential distribution with parameter λ (Section 2-3.2). The gamma distribution can assume many different shapes, depending on the values chosen for r and λ. This makes it useful as a model for a wide variety of continuous random variables.

If the parameter r is an integer, then the gamma distribution is the sum of r independently and identically distributed exponential distributions, each with parameter λ. That is, if $x_1, x_2, \ldots, x_r$ are exponential with parameter λ and independent, then

$$y = x_1 + x_2 + \cdots + x_r$$

is distributed as gamma with parameters r and λ. There are a number of important applications of this result.

······   **EXAMPLE 2-8**   ··

Consider the system shown in Fig. 2-22. This is called a **standby redundant system,** because while component 1 is on, component 2 is off, and when component 1 fails, the

---
[2]$\Gamma(r)$ in the denominator of equation 2-32 is the gamma function, defined as $\Gamma(r) = \int_0^\infty x^{r-1} e^{-x} \, dx, r > 0$. If r is a positive integer, then $\Gamma(r) = (r - 1)!$

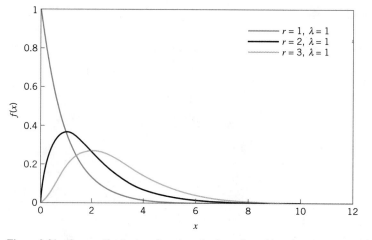

Figure 2-21   Gamma distributions for selected values of r and $\lambda = 1$.

switch automatically turns component 2 on. If each component has a life described by an exponential distribution with $\lambda = 10^{-4}$/h, say, then the system life is gamma distributed with parameters $r = 2$ and $\lambda = 10^{-4}$. Thus, the mean time to failure is $\mu = r/\lambda = 2/10^{-4} = 2 \times 10^4$ h.

The cumulative gamma distribution is

$$F(a) = 1 - \int_{a}^{\infty} \frac{\lambda}{\Gamma(r)} (\lambda t)^{r-1} e^{-\lambda t} \, dt \qquad (2\text{-}35)$$

If r is an integer, then equation 2-35 becomes

$$F(a) = 1 - \sum_{k=0}^{r-1} e^{-\lambda a} \frac{(\lambda a)^k}{k!} \qquad (2\text{-}36)$$

Consequently, the cumulative gamma distribution can be evaluated as the sum of r Poisson terms with parameter λa. This result is not too surprising, if we consider the Poisson distribution as a model of the number of occurrences of an event in a fixed interval, and the gamma distribution as the model of the portion of the interval required to obtain a specific number of occurrences.

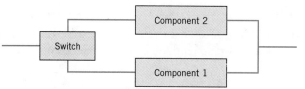

Figure 2-22   The standby redundant system for Example 2-8.

2-3.4 The Weibull Distribution

The Weibull distribution is defined as follows.

Definition

The **Weibull distribution** is

$$f(x) = \frac{\beta}{\theta} \left(\frac{x}{\theta} \right)^{\beta-1} \exp\left[-\left(\frac{x}{\theta} \right)^{\beta} \right] \qquad x \geq 0 \qquad (2\text{-}37)$$

where $\theta > 0$ is the **scale parameter,** and $\beta > 0$ is the **shape parameter.** The **mean** and **variance** of the Weibull distribution are

$$\mu = \theta \Gamma \left(1 + \frac{1}{\beta} \right) \qquad (2\text{-}38)$$

and

$$\sigma^2 = \theta^2 \left[\Gamma \left(1 + \frac{2}{\beta} \right) - \Gamma \left(1 + \frac{1}{\beta} \right)^2 \right] \qquad (2\text{-}39)$$

respectively.

The Weibull distribution is very flexible, and by appropriate selection of the parameters θ and β, the distribution can assume a wide variety of shapes. Several Weibull distributions are shown in Fig. 2-23 for $\theta = 1$ and $\beta = \frac{1}{2}$, 1, 2, and 4. Note that when $\beta = 1$,

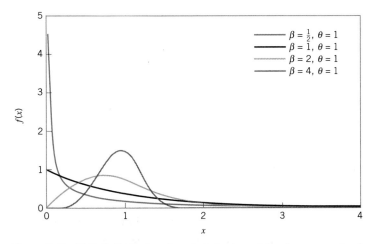

Figure 2-23 Weibull distributions for selected values of the shape parameter β and scale parameter $\theta = 1$.

the Weibull distribution reduces to the exponential distribution with mean $1/\theta$. The cumulative Weibull distribution is

$$F(a) = 1 - \exp\left[-\left(\frac{a}{\theta}\right)^{\beta}\right] \tag{2-40}$$

The Weibull distribution has been used extensively in reliability engineering as a model of time to failure for electrical and mechanical components and systems. Examples of situations in which the Weibull has been used include electronic devices such as memory elements, mechanical components such as bearings, and structural elements in aircraft and automobiles.[3]

······· **EXAMPLE 2-9** ··

The time to failure for an electronic subassembly used in a RISC workstation is satisfactorily modeled by a Weibull distribution with $\beta = \frac{1}{2}$ and $\theta = 1000$. The mean time to failure is

$$\mu = \theta\Gamma\left(1 + \frac{1}{\beta}\right)$$

$$= 1000\Gamma\left(1 + \frac{1}{1/2}\right)$$

$$= 1000\Gamma(3)$$

$$= 2000 \text{ h}$$

The fraction of subassemblies expected to survive $a = 4000$ h is

$$1 - F(a) = \exp\left[-\left(\frac{a}{\theta}\right)^{\beta}\right]$$

or

$$1 - F(4000) = \exp\left[-\left(\frac{4000}{1000}\right)^{1/2}\right]$$

$$= e^{-2}$$

$$= 0.1353$$

That is, all but about 13.53% of the subassemblies will fail by 4000 h.

···

2-4 SOME USEFUL APPROXIMATIONS

In certain quality control problems, it is sometimes useful to approximate one probability distribution with another. This is particularly helpful in situations where the original distribution is difficult to manipulate analytically. In this section, we present three such

---

[3] See the supplemental text material for more information.

approximations: (1) the binomial approximation to the hypergeometric, (2) the Poisson approximation to the binomial, and (3) the normal approximation to the binomial.

2-4.1   The Binomial Approximation to the Hypergeometric

Consider the hypergeometric distribution in equation 2-8. If the ratio n/N (often called the *sampling fraction*) is small—say, $n/N \leq 0.1$—then the binomial distribution with parameters $p = D/N$ and n is a good approximation to the hypergeometric. The approximation is better for small values of n/N.

This approximation is useful in the design of acceptance-sampling plans. Recall that the hypergeometric distribution is the appropriate model for the number of nonconforming items obtained in a random sample of n items from a lot of finite size N. Thus, if the sample size n is small relative to the lot size N, the binomial approximation may be employed, which usually simplifies the calculations considerably.

As an example, suppose that a production lot of 200 units contains 5 units that do not meet the specifications. The probability that a random sample of 10 units will contain no nonconforming items is, from equation 2-8,

$$p(0) = \frac{\binom{5}{0}\binom{195}{10}}{\binom{200}{10}} = 0.7717$$

Note that since $n/N = 10/200 = 0.05$ is relatively small, we could use the binomial approximation with $p = D/N = 5/200 = 0.025$ and $n = 10$ to calculate

$$p(0) = \binom{5}{0}(0.025)^0(0.975)^{10} = 0.7763$$

2-4.2   The Poisson Approximation to the Binomial

It was noted in Section 2-2.3 that the Poisson distribution could be obtained as a limiting form of the binomial distribution for the case where p approaches zero and n approaches infinity with $\lambda = np$ constant. This implies that, for small p and large n, the Poisson distribution with $\lambda = np$ may be used to approximate the binomial distribution. The approximation is usually good for large n and if $p < 0.1$. The larger is the value of n and the smaller is the value of p, the better is the approximation.

2-4.3   The Normal Approximation to the Binomial

In Section 2-2.2 we defined the binomial distribution as the sum of a sequence of n Bernoulli trials, each with probability of success p. If the number of trials n is large, then we may use the central limit theorem to justify the normal distribution with mean

np and variance $np(1 - p)$ as an approximation to the binomial. That is,

$$P\{x = a\} = \binom{n}{a} p^a (1 - p)^{n-a}$$

$$= \frac{1}{\sqrt{2\pi np(1 - p)}} e^{-\frac{1}{2}[(a - np)^2 / np(1 - p)]}$$

Since the binomial distribution is discrete and the normal distribution is continuous, it is common practice to use *continuity corrections* in the approximation, so that

$$P\{x = a\} \simeq \Phi\left(\frac{a + \frac{1}{2} - np}{\sqrt{np(1 - p)}}\right) - \Phi\left(\frac{a - \frac{1}{2} - np}{\sqrt{np(1 - p)}}\right)$$

where Φ denotes the standard normal cumulative distribution function. Other types of probability statements are evaluated similarly, such as

$$P\{a \le x \le b\} \simeq \Phi\left(\frac{b + \frac{1}{2} - np}{\sqrt{np(1 - p)}}\right) - \Phi\left(\frac{a - \frac{1}{2} - np}{\sqrt{np(1 - p)}}\right)$$

The normal approximation to the binomial is known to be satisfactory for p of approximately $\frac{1}{2}$ and $n > 10$. For other values of p, larger values of n are required. In general, the approximation is not adequate for $p < 1/(n + 1)$ or $p > n/(n + 1)$, or for values of the random variable outside an interval six standard deviations wide centered about the mean (i.e., the interval $np \pm 3\sqrt{np(1 - p)}$).

We may also use the normal approximation for the random variable $\hat{p} = x/n$—that is, the sample fraction defective of Section 2-2.2. The random variable $\hat{p}$ is approximately normally distributed with mean p and variance $p(1 - p)/n$, so that

$$P\{u \le \hat{p} \le v\} \simeq \Phi\left(\frac{v - p}{\sqrt{p(1 - p)/n}}\right) - \Phi\left(\frac{u - p}{\sqrt{p(1 - p)/n}}\right)$$

Since the normal will serve as an approximation to the binomial, and since the binomial and Poisson distributions are closely connected, it seems logical that the normal may serve to approximate the Poisson. This is indeed the case, and if the mean λ of the Poisson distribution exceeds 15 (or so), then the normal distribution with $\mu = \lambda$ and $\sigma^2 = \lambda$ is a satisfactory approximation.

2-4.4   Comments on Approximations

A summary of the approximations discussed above is presented in Fig. 2-24. In this figure, H, B, P, and N represent the hypergeometric, binomial, Poisson, and normal distributions, respectively. The widespread availability of modern microcomputers, good statistics software packages, and hand-held calculators has made reliance on these approximations largely unnecessary, but there are situations in which they are useful, particularly in the application of the popular three-sigma limit control charts.

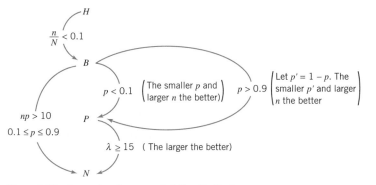

Figure 2-24   Approximations to probability distributions.

2-5 EXERCISES

2-1.   The fill volume of a soft-drink beverage is being analyzed for variability. Ten bottles, randomly selected
from the process, are measured, and
the results are as follows (in fluid
ounces): 10.05, 10.03, 10.02, 10.04,
10.05, 10.01, 10.02, 10.02, 10.03,
10.01.
(a) Calculate the sample average.
(b) Calculate the sample standard deviation.

2-2.   The bore diameters of eight randomly
selected bearings are shown here (in
mm):

50.001	50.002
49.998	50.006
50.005	49.996
50.003	50.004

(a) Calculate the sample average.
(b) Calculate the sample standard deviation.

2-3.   The nine measurements that follow
are furnace temperatures recorded on
successive batches in a semiconductor manufacturing process (units
are °F):

953	955	948
951	957	949
954	950	959

(a) Calculate the sample average.
(b) Calculate the sample standard deviation.

2-4.   Consider the furnace temperature data
in Exercise 2-3.
(a) Find the sample median of these
data.
(b) How much could the largest temperature measurement increase without
changing the sample median?

2-5.   Yield strengths of circular tubes with
end caps are measured. The first yields
(in kN) are as follows:

96	102	104	108
126	128	150	156

(a) Calculate the sample average.
(b) Calculate the sample standard deviation.

2-6.   The time to failure in hours of an
electronic component subjected to an
accelerated life test is shown here. To
accelerate the failure test, the units
were tested at an elevated temperature
(read down, then across).

127	124	121	118
125	123	136	131
131	120	140	125
124	119	137	133
129	128	125	141
121	133	124	125
142	137	128	140
151	124	129	131
160	142	130	129
125	123	122	126

(a) Calculate the sample average and standard deviation.

(b) Construct a histogram.

(c) Construct a stem-and-leaf plot.

(d) Find the sample median and the lower and upper quartiles.

2-7. The data shown here are chemical process yield readings on successive days (read down, then across). Construct a histogram for these data. Comment on the shape of the histogram. Does it resemble any of the distributions that we have discussed in this chapter?

94.1	87.3	94.1	92.4	84.6	85.4
93.2	84.1	92.1	90.6	83.6	86.6
90.6	90.1	96.4	89.1	85.4	91.7
91.4	95.2	88.2	88.8	89.7	87.5
88.2	86.1	86.4	86.4	87.6	84.2
86.1	94.3	85.0	85.1	85.1	85.1
95.1	93.2	84.9	84.0	89.6	90.5
90.0	86.7	87.3	93.7	90.0	95.6
92.4	83.0	89.6	87.7	90.1	88.3
87.3	95.3	90.3	90.6	94.3	84.1
86.6	94.1	93.1	89.4	97.3	83.7
91.2	97.8	94.6	88.6	96.8	82.9
86.1	93.1	96.3	84.1	94.4	87.3
90.4	86.4	94.7	82.6	96.1	86.4
89.1	87.6	91.1	83.1	98.0	84.5

2-8. An article in *Quality Engineering* (Vol. 4, 1992, pp. 487–495) presents viscosity data from a batch chemical process. A sample of these data is presented here (read down, then across).

13.3	14.9	15.8	16.0
14.5	13.7	13.7	14.9
15.3	15.2	15.1	13.6
15.3	14.5	13.4	15.3
14.3	15.3	14.1	14.3
14.8	15.6	14.8	15.6
15.2	15.8	14.3	16.1
14.5	13.3	14.3	13.9
14.6	14.1	16.4	15.2
14.1	15.4	16.9	14.4
14.3	15.2	14.2	14.0
16.1	15.2	16.9	14.4
13.1	15.9	14.9	13.7
15.5	16.5	15.2	13.8
12.6	14.8	14.4	15.6
14.6	15.1	15.2	14.5
14.3	17.0	14.6	12.8
15.4	14.9	16.4	16.1
15.2	14.8	14.2	16.6
16.8	14.0	15.7	15.6

(a) Construct a stem-and-leaf display for the viscosity data.

(b) Construct a frequency distribution and histogram.

(c) Convert the stem-and-leaf plot in part (a) into an ordered stem-and-leaf plot. Use this graph to assist in locating the median and the upper and lower quartiles of the viscosity data.

(d) What are the 90th and 10th percentiles of viscosity?

2-9. Consider the viscosity data in Exercise 2-8. Assume that reading down then across gives the data in time order. Construct and interpret a digidot plot.

2-10. Reconsider the yield data in Exercise 2-7. Construct a digidot plot for these data. Interpret the plot.

2-11. Consider the chemical process yield data in Exercise 2-7. Calculate the sample average and standard deviation.

2-12. Consider the chemical process yield data in Exercise 2-7. Construct a stem-and-leaf plot for the data and compare it with the histogram from Exercise 2-7. Which display provides more information about the process?

2-13. Construct a box plot for the data in Exercise 2-1.

2-14. Construct a box plot for the data in Exercise 2-2.

2-15. Suppose that two fair dice are tossed and the random variable observed — say, x — is the sum of the two up faces. Describe the sample space of this experiment, and determine the probability distribution of x.

2-16. Find the mean and variance of the random variable in Exercise 2-15.

2-17. A mechatronic assembly is subjected to a final functional test. Suppose that defects occur at random in these assemblies, and that defects occur according to a Poisson distribution with parameter $\lambda = 0.02$.
(a) What is the probability that an assembly will have exactly one defect?
(b) What is the probability that an assembly will have one or more defects?
(c) Suppose that you improve the process so that the occurrence rate of defects is cut in half to $\lambda = 0.01$. What effect does this have on the probability that an assembly will have one or more defects?

2-18. The probability distribution of x is $f(x) = ke^{-x}$, $0 \le x \le \infty$. Find the appropriate value of k. Find the mean and variance of x.

2-19. The random variable x takes on the values 1, 2, or 3 with probabilities $(1 + 3k)/3$, $(1 + 2k)/3$, and $(0.5 + 5k)/3$, respectively.
(a) Find the appropriate value of k.
(b) Find the mean and variance of x.
(c) Find the cumulative distribution function.

2-20. The probability distribution of the discrete random variable x is $p(x) = kr^x$, $0 < r < 1$. Find the appropriate value for k if $x = 0, 1, \ldots$.

2-21. A manufacturer of electronic calculators offers a 1-year warranty. If the calculator fails for any reason during this period, it is replaced. The time to failure is well modeled by the following probability distribution:

$$f(x) = 0.125e^{-0.125x} \qquad x > 0$$

(a) What percentage of the calculators will fail within the warranty period?
(b) The manufacturing cost of a calculator is $50, and the profit per sale is $25. What is the effect of warranty replacement on profit?

2-22. The net contents in ounces of a canned soft drink is a random variable with probability distribution

$$f(x) = \begin{cases} 4(x - 11.75) & 11.75 \le x \le 12.25 \\ 4(12.75 - x) & 12.25 \le x \le 12.75 \end{cases}$$

Find the probability that a can contains less than 12 oz of product.

2-23. A production process operates with 2% nonconforming output. Every hour a sample of 50 units of product is taken, and the number of nonconforming units counted. If one or more nonconforming units are found, the process is stopped and the quality control technician must search for the cause of nonconforming production. Evaluate the performance of this decision rule.

2-24. **Continuation of Exercise 2-23.** Consider the decision rule described in Exercise 2-23. Suppose that the process suddenly deteriorates to 4% nonconforming output. How many samples, on average, will be required to detect this?

2-25. A random sample of 100 units is drawn from a production process every half hour. The fraction of nonconforming product manufactured is 0.03. What is the probability that $\hat{p} \le 0.04$ if the fraction nonconforming really is 0.03?

2-26. A sample of 100 units is selected

from a production process that is 2% nonconforming. What is the probability that $\hat{p}$ will exceed the true fraction nonconforming by k standard deviations, where $k = 1, 2,$ and 3?

2-27. An electronic component for a laser range-finder is produced in lots of size $N = 25$. An acceptance testing procedure is used by the purchaser to protect against lots that contain too many nonconforming components. The procedure consists of selecting five components at random from the lot (without replacement) and testing them. If none of the components is nonconforming, the lot is accepted.

    (a) If the lot contains three nonconforming components, what is the probability of lot acceptance?

    (b) Calculate the desired probability in (a) using the binomial approximation. Is this approximation satisfactory? Why or why not?

    (c) Suppose the lot size was $N = 150$. Would the binomial approximation be satisfactory in this case?

    (d) Suppose that the purchaser will reject the lot with the decision rule of finding one or more nonconforming components in a sample of size n, and wants the lot to be rejected with probability at least 0.95 if the lot contains five or more nonconforming components. How large should the sample size n be?

2-28. A lot of size $N = 30$ contains three nonconforming units. What is the probability that a sample of five units selected at random contains exactly one nonconforming unit? What is the probability that it contains one or more nonconformances?

2-29. A textbook has 500 pages on which typographical errors could occur. Suppose that there are exactly 10 such errors randomly located on those pages. Find the probability that a ran-

dom selection of 50 pages will contain no errors. Find the probability that 50 randomly selected pages will contain at least two errors.

2-30. Surface-finish defects in a small electric appliance occur at random with a mean rate of 0.1 defects per unit. Find the probability that a randomly selected unit will contain at least one surface-finish defect.

2-31. Glass bottles are formed by pouring molten glass into a mold. The molten glass is prepared in a furnace lined with firebrick. As the firebrick wears, small pieces of brick are mixed into the molten glass and finally appear as defects (called "stones") in the bottle. If we can assume that stones occur randomly at the rate of 0.00001 per bottle, what is the probability that a bottle selected at random will contain at least one such defect?

2-32. The billing department of a major credit card company attempts to control errors (clerical, keypunch, etc.) on customers' bills. Suppose that errors occur according to a Poisson distribution with parameter $\lambda = 0.01$. What is the probability that a customer's bill selected at random will contain one error?

2-33. A production process operates in one of two states: the in-control state, in which most of the units produced conform to specifications, and an out-of-control state, in which most of the units produced are defective. The process will shift from the in-control to the out-of-control state at random. Every hour, a quality control technician checks the process, and if it is in the out-of-control state, the technician detects this with probability p. Assume that when the process shifts out of control it does so immediately following a check by the inspector, and once a shift has occurred, the process cannot automatically correct itself. If t denotes the number of periods the process remains out of control following a shift before detection, find the probability distribution of t. Find the mean number of periods the process will remain in the out-of-control state.

2-34. An inspector is looking for the nonconforming welds in a pipeline. The probability that any particular weld will be defective is 0.01. The inspector is determined to keep working until finding three defective welds. If the welds are located 100 ft apart, what is the probability that the inspector will have to walk 5000 ft? What is the probability that the inspector will have to walk more than 5000 ft?

2-35. The tensile strength of a metal part is normally distributed with mean 40 lb and standard deviation 8 lb. If 50,000 parts are produced, how many would fail to meet a minimum specification limit of 34-lb tensile strength? How many would have a tensile strength in excess of 48 lb?

2-36. The output voltage of a power supply is normally distributed with mean 12 V and standard deviation 0.05 V. If the lower and upper specifications for voltage are 11.90 V and 12.10 V, respectively, what is the probability that a power supply selected at random will conform to the specifications on voltage?

2-37. **Continuation of Exercise 2-36.** Reconsider the power supply manufacturing process in Exercise 2-36. Suppose we wanted to improve the process. Can shifting the mean reduce the number of nonconforming units produced? How much would the process variability need to be reduced in order to have all but one out of 1000 units conform to the specifications?

2-38. If x is normally distributed with mean μ and standard deviation four, and given that the probability that x is less than 32 is 0.0228, find the value of μ.

2-39. The life of an automotive battery is normally distributed with mean 900 days and standard deviation 35 days. What fraction of these batteries would be expected to survive beyond 1000 days?

2-40. A lightbulb has a normally distributed light output with mean 5000 end foot-candles and standard deviation of 50 end foot-candles. Find a lower specification limit such that only 0.5% of the bulbs will not exceed this limit.

2-41. The specifications on an electronic component in a target-acquisition system are that its life must be between 5000 and 10,000 h. The life is normally distributed with mean 7500 h. The manufacturer realizes a price of $10 per unit produced; however, defective units must be replaced at a cost of $5 to the manufacturer. Two different manufacturing processes can be used, both of which have the same mean life. However, the standard deviation of life for process 1 is 1000 h, whereas for process 2 it is only 500 h. Production costs for process 2 are twice those for process 1. What value of production costs will determine the selection between processes 1 and 2?

2-42. A quality characteristic of a product is normally distributed with mean μ and standard deviation one. Specifications on the characteristic are $6 \le x \le 8$. A unit that falls within specifications on this quality characteristic results in a profit of C_0. However, if $x < 6$, the profit is $-C_1$, whereas if $x > 8$, the profit is $-C_2$. Find the value of μ that maximizes the expected profit.

2-43. Derive the mean and variance of the binomial distribution.

2-44. Derive the mean and variance of the Poisson distribution.

2-45. Derive the mean and variance of the exponential distribution.

Inferences about Process Quality

CHAPTER OUTLINE

3-1 STATISTICS AND SAMPLING
DISTRIBUTIONS

    3-1.1 Sampling from a Normal
Distribution

    3-1.2 Sampling from a Bernoulli
Distribution

    3-1.3 Sampling from a Poisson
Distribution

3-2 POINT ESTIMATION OF PROCESS
PARAMETERS

3-3 STATISTICAL INFERENCE FOR A
SINGLE SAMPLE

    3-3.1 Inference on the Mean of a
Population, Variance Known

    3-3.2 The Use of P-Values for
Hypothesis Testing

    3-3.3 Inference on the Mean of a
Normal Distribution, Variance
Unknown

    3-3.4 Inference on the Variance of a
Normal Distribution

    3-3.5 Inference on a Population
Proportion

    3-3.6 The Probability of Type II Error

    3-3.7 Probability Plotting

3-4 STATISTICAL INFERENCE FOR
TWO SAMPLES

    3-4.1 Inference for a Difference in
Means, Variances Known

    3-4.2 Inference for a Difference in
Means of Two Normal
Distributions, Variances
Unknown

    3-4.3 Inference on the Variances of Two
Normal Distributions

    3-4.4 Inference on Two Population
Proportions

3-5 WHAT IF THERE ARE MORE
THAN TWO POPULATIONS? THE
ANALYSIS OF VARIANCE

    3-5.1 An Example

    3-5.2 The Analysis of Variance

    3-5.3 Checking Assumptions: Residual
Analysis

CHAPTER OVERVIEW

In the previous chapter we discussed the use of probability distributions in modeling or describing the output of a process. In all the examples presented we assumed that the parameters of the probability distribution, and hence, the parameters of the process, were known. This is usually a very unrealistic assumption. For example, in using the binomial distribution to model the number of nonconforming units found in sampling from a production process we assumed that the parameter p of the binomial distribution was known. The physical interpretation of p is that it is the true fraction of nonconforming units produced by the process. It is impossible to know this exactly in a real production process. Furthermore, if we did know the true value of p and it was relatively constant over time, we could argue that formal process monitoring and control procedures were unnecessary, provided p was "acceptably" small.

In general, the parameters of a process are unknown; furthermore, they can usually change over time. Therefore, we need to develop procedures to estimate the parameters of probability distributions and solve other inference or decision-oriented problems relative to them. The standard statistical techniques of parameter estimation and hypothesis testing are useful in this respect. These techniques are the underlying basis for much of the methodology of statistical quality control. In this chapter we present some of the elementary results of statistical inference, indicating its usefulness in quality-improvement problems. Key topics include point and confidence interval estimation of means, variances, and binomial parameters, hypothesis testing on means, variances, and binomial parameters, and the use of normal probability plots.

3-1 STATISTICS AND SAMPLING DISTRIBUTIONS

The objective of statistical inference is to draw conclusions or make decisions about a population based on a sample selected from the population. Frequently, we will assume that *random samples* are used in the analysis. The word "random" is often applied to any method or sample selection that lacks systematic direction. We will define a sample—say, $x_1, x_2, \ldots, x_n$—as a **random sample** of size n if it is selected so that the observations $\{x_i\}$ are independently and identically distributed. This definition is suitable for random samples drawn from infinite populations or from finite populations where sampling is performed with *replacement*. In sampling *without replacement* from a finite population of N items we say that a sample of n items is a random sample if each of the $\binom{N}{n}$ possible samples has an equal probability of being chosen. Figure 3-1 illustrates the relationship between the population and the sample.

Although most of the methods we will study assume that random sampling has been used, there are several other sampling strategies that are occasionally useful in quality control. Care must be exercised to use a method of analysis that is consistent with the sampling design; inference techniques intended for random samples can lead to serious errors when applied to data obtained from other sampling techniques.

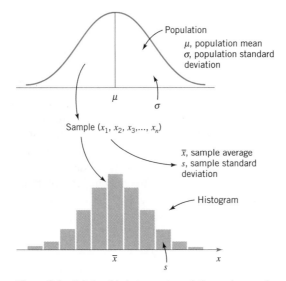

Figure 3-1   Relationship between a population and a sample.

Statistical inference uses quantities computed from the observations in the sample. A **statistic** is defined as any function of the sample data that does not contain unknown parameters. For example, let $x_1, x_2, \ldots, x_n$ represent the observations in a sample. Then the sample mean

$$\bar{x} = \frac{\sum\limits_{i=1}^{n} x_i}{n} \tag{3-1}$$

the sample variance

$$S^2 = \frac{\sum\limits_{i=1}^{n} (x_i - \bar{x})^2}{n-1} \tag{3-2}$$

and the sample standard deviation

$$S = \sqrt{\frac{\sum\limits_{i=1}^{n} (x_i - \bar{x})^2}{n-1}} \tag{3-3}$$

are statistics. The statistics $\bar{x}$ and S (or S^2) describe the central tendency and variability, respectively, of the sample.

If we know the probability distribution of the population from which the sample was taken, we can often determine the probability distribution of various statistics computed from the sample data. The probability distribution of a statistic is called a

sampling distribution. We now present the sampling distributions associated with three common sampling situations.

3-1.1 Sampling from a Normal Distribution

Suppose that x is a normally distributed random variable with mean μ and variance σ^2. If $x_1, x_2, \ldots, x_n$ is a random sample of size n from this process, then the distribution of the sample mean $\bar{x}$ is $N(\mu, \sigma^2/n)$. This follows directly from the results on the distribution of linear combinations of normal random variables in Section 2-3.1.

This property of the sample mean is not restricted exclusively to the case of sampling from normal populations. Note that we may write

$$\left(\frac{\bar{x} - \mu}{\sigma}\right)\sqrt{n} = \frac{\sum\limits_{i=1}^{n} x_i - n\mu}{\sigma\sqrt{n}}$$

From the central limit theorem we know that, regardless of the distribution of the population, the distribution of $\sum_{i=1}^{n} x_i$ is approximately normal with mean $n\mu$ and variance $n\sigma^2$. Therefore, regardless of the distribution of the population, the sampling distribution of the sample mean is approximately

$$\bar{x} \sim N\left(\mu, \frac{\sigma^2}{n}\right)$$

An important sampling distribution defined in terms of the normal distribution is the **chi-square** or χ^2 **distribution.** If $x_1, x_2, \ldots, x_n$ are normally and independently distributed random variables with mean zero and variance one, then the random variable

$$y = x_1^2 + x_2^2 + \cdots + x_n^2$$

is distributed as chi-square with n degrees of freedom. The chi-square probability distribution with n degrees of freedom is

$$f(y) = \frac{1}{2^{n/2}\Gamma\left(\dfrac{n}{2}\right)} y^{(n/2)-1} e^{-y/2} \qquad y > 0 \tag{3-4}$$

Several chi-square distributions are shown in Fig. 3-2. The distribution is skewed with mean $\mu = n$ and variance $\sigma^2 = 2n$. A table of the percentage points of the chi-square distribution is given in Appendix Table III.

To illustrate the use of the chi-square distribution, suppose that $x_1, x_2, \ldots, x_n$ is a random sample from an $N(\mu, \sigma^2)$ distribution. Then the random variable

$$y = \frac{\sum\limits_{i=1}^{n} (x_i - \bar{x})^2}{\sigma^2} \tag{3-5}$$

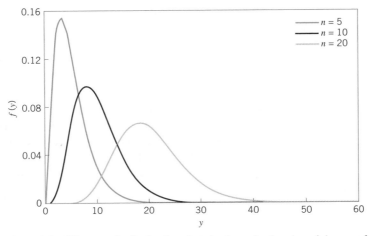

Figure 3-2 Chi-square distribution for selected values of n (number of degrees of freedom).

has a chi-square distribution with $n - 1$ degrees of freedom. However, using equation 3-2, which defines the sample variance, we may rewrite equation 3-5 as

$$y = \frac{(n-1)S^2}{\sigma^2}$$

That is, the sampling distribution of $(n - 1)S^2/\sigma^2$ is χ^2_{n-1} when sampling from a normal distribution.

Another useful sampling distribution is the t **distribution.** If x is a standard normal random variable and if y is a chi-square random variable with k degrees of freedom, and if x and y are independent, then the random variable

$$t = \frac{x}{\sqrt{y/k}} \qquad (3\text{-}6)$$

is distributed as t with k degrees of freedom. The probability distribution of t is

$$f(t) = \frac{\Gamma[(k+1)/2]}{\sqrt{k\pi}\,\Gamma(k/2)} \left(\frac{t^2}{k} + 1\right)^{-(k+1)/2} \qquad -\infty < t < \infty \qquad (3\text{-}7)$$

and the mean and variance of t are $\mu = 0$ and $\sigma^2 = k/(k - 2)$ for $k > 2$, respectively. The degrees of freedom for t are the degrees of freedom associated with the chi-square random variable in the denominator of equation 3-6. Several t distributions are shown in Fig. 3-3. Note that if $k = \infty$, the t distribution reduces to the standard normal distribution. A table of percentage points of the t distribution is given in Appendix Table IV.

As an example of a random variable that is distributed as t, suppose that x_1, $x_2, \ldots, x_n$ is a random sample from the $N(\mu, \sigma^2)$ distribution. If $\bar{x}$ and S^2 are

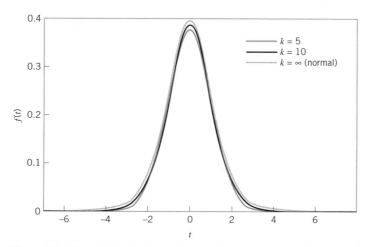

Figure 3-3 The t distribution for selected values of k (number of degrees of freedom).

computed from this sample, then

$$\frac{\bar{x} - \mu}{S/\sqrt{n}} = \frac{\dfrac{\bar{x} - \mu}{\sigma/\sqrt{n}}}{S/\sigma} \sim \frac{N(0, 1)}{\sqrt{\chi^2_{n-1}/(n-1)}}$$

using the fact that $(n - 1)S^2/\sigma^2 \sim \chi^2_{n-1}$. Now $\bar{x}$ and S^2 are independent, so the random variable

$$\frac{\bar{x} - \mu}{S/\sqrt{n}} \tag{3-8}$$

has a t distribution with $n - 1$ degrees of freedom.

The last sampling distribution based on the normal process that we will consider is the **F distribution.** If w and y are two independent chi-square random variables with u and v degrees of freedom, respectively, then the ratio

$$F_{u,v} = \frac{w/u}{y/v} \tag{3-9}$$

is distributed as F with u numerator degrees of freedom and v denominator degrees of freedom. If x is an F random variable with u numerator and v denominator degrees of freedom, then the distribution is

$$f(x) = \frac{\Gamma\left(\dfrac{u + v}{2}\right)\left(\dfrac{u}{v}\right)^{u/2}}{\Gamma\left(\dfrac{u}{2}\right)\Gamma\left(\dfrac{v}{2}\right)} \frac{x^{(u/2)-1}}{\left[\left(\dfrac{u}{2}\right)x + 1\right]^{(u+v)/2}} \qquad 0 < x < \infty \tag{3-10}$$

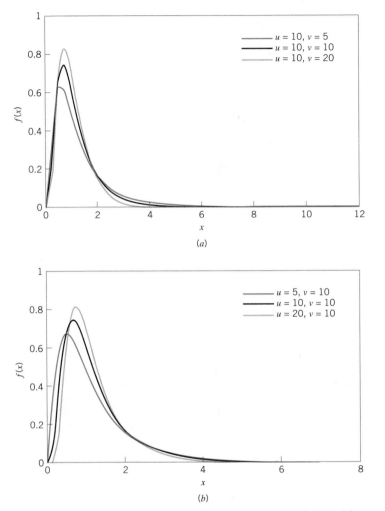

Figure 3-4   The F distribution for selected values of u (numerator degrees of freedom) and v (denominator degrees of freedom).

Several F distributions are shown in Fig. 3-4. A table of percentage points of the F distribution is given in Appendix Table V.

As an example of a random variable that is distributed as F, suppose we have two independent normal processes—say, $x_1 \sim N(\mu_1, \sigma_1^2)$, and $x_2 \sim N(\mu_2, \sigma_2^2)$. Let x_{11}, $x_{12}, \ldots, x_{1n_1}$ be a random sample of n_1 observations from the first normal process and $x_{21}, x_{22}, \ldots, x_{2n_2}$ be a random sample of size n_2 from the second. If S_1^2 and S_2^2 are the sample variances, then the ratio

$$\frac{S_1^2/\sigma_1^2}{S_2^2/\sigma_2^2} \sim F_{n_1-1,\,n_2-1}$$

This follows directly from the sampling distribution of S^2 discussed previously. The F distribution will be used in making inferences about the variances of two normal distributions.

3-1.2 Sampling from a Bernoulli Distribution

In this section, we discuss the sampling distributions of statistics associated with the Bernoulli distribution. The random variable x with probability function

$$p(x) = \begin{cases} p & x = 1 \\ (1 - p) = q & x = 0 \end{cases}$$

is called a Bernoulli random variable. That is, x takes on the value 1 with probability p and the value 0 with probability $1 - p = q$. A realization of this random variable is often called a Bernoulli trial. The sequence of Bernoulli trials $x_1, x_2, \ldots$, is a Bernoulli process. The outcome $x = 1$ is often called "success," and the outcome $x = 0$ is often called "failure."

Suppose that a random sample of n observations—say, $x_1, x_2, \ldots, x_n$—is taken from a Bernoulli process with constant probability of success p. Then the sum of the sample observations

$$x = x_1 + x_2 + \cdots + x_n \qquad (3\text{-}11)$$

has a binomial distribution with parameters n and p. Furthermore, since each x_i is either 0 or 1, the sample mean

$$\bar{x} = \frac{1}{n} \sum_{i=1}^{n} x_i \qquad (3\text{-}12)$$

is a discrete random variable with range space $\{0, 1/n, 2/n, \ldots, (n - 1)/n, 1\}$. The distribution of $\bar{x}$ can be obtained from the binomial since

$$P\{\bar{x} \leq a\} = P\{x \leq an\} = \sum_{k=0}^{[an]} \binom{n}{k} p^k (1 - p)^{n-k}$$

where $[an]$ is the largest integer less than or equal to an. The mean and variance of $\bar{x}$ are

$$\mu_{\bar{x}} = p$$

and

$$\sigma_{\bar{x}}^2 = \frac{p(1 - p)}{n} \qquad\qquad \hat{p} \quad \text{point estimator}$$

respectively. This same result was given previously in Section 2-2.2, where the random variable $\hat{p}$ (often called the sample fraction nonconforming) was introduced.

3-1.3 Sampling from a Poisson Distribution

The Poisson distribution was introduced in Section 2-2.3. Consider a random sample of size n from a Poisson distribution with parameter λ—say, $x_1, x_2, \ldots, x_n$. The distribution of the sample sum

$$x = x_1 + x_2 + \cdots + x_n \qquad (3\text{-}13)$$

is also Poisson with parameter $n\lambda$. More generally, the sum of n independent Poisson random variables is distributed Poisson with parameter equal to the sum of the individual Poisson parameters.

Now consider the distribution of the sample mean

$$\bar{x} = \frac{1}{n} \sum_{i=1}^{n} x_i \qquad (3\text{-}14)$$

This is a discrete random variable that takes on the values $\{0, 1/n, 2/n, \ldots\}$, and with probability distribution found from

$$P\{\bar{x} \le a\} = P\{x \le an\} = \sum_{k=0}^{[an]} \frac{e^{-n\lambda}(n\lambda)^k}{k!} \qquad (3\text{-}15)$$

where $[an]$ is the largest integer less than or equal to an. The mean and variance of $\bar{x}$ are

$$\bar{x} = \lambda$$

and

$$\sigma_{\bar{x}}^2 = \frac{\lambda}{n}$$

respectively.

Sometimes more general linear combinations of Poisson random variables are used in quality-engineering work. For example, consider the linear combination

$$L = a_1 x_1 + a_2 x_2 + \cdots + a_m x_m$$

$$= \sum_{i=1}^{m} a_i x_i \qquad (3\text{-}16)$$

where the $\{x_i\}$ are independent Poisson random variables each having parameter $\{\lambda_i\}$, respectively, and the $\{a_i\}$ are constants. This type of function occurs in situations where a unit of product can have m different types of defects or nonconformities (each modeled with a Poisson distribution with parameter λ_i) and the function used for quality monitoring purposes is a linear combination of the number of observed nonconformities of each type. The constants $\{a_i\}$ in equation 3-16 might be chosen to weight some types of nonconformities more heavily than others. For example, functional defects on a unit would receive heavier weight than appearance flaws. These schemes are sometimes called **demerit** procedures (see Section 6-3.3). In general, the distribution of L is not Poisson unless all $a_i = 1$ in equation 3-16; that is, sums of independent Poisson random variables are Poisson distributed, but more general linear combinations are not.

3-2 POINT ESTIMATION OF PROCESS PARAMETERS

A random variable is characterized or described by its probability distribution. This distribution is described by its **parameters.** For example, the mean μ and variance σ^2 of the normal distribution (equation 2-21) are its parameters, whereas λ is the parameter of the Poisson distribution (equation 2-15). In statistical quality control, the probability distribution is used to describe or model some quality characteristic, such as a critical dimension of a product or the fraction defective of the manufacturing process. Therefore, we are interested in making inferences about the parameters of probability distributions. Since the parameters are generally unknown, we require procedures to estimate them from sample data.

We may define an estimator of an unknown parameter as a statistic that corresponds to the parameter. A particular numerical value of an estimator, computed from sample data, is called an **estimate.** A **point estimator** is a statistic that produces a single numerical value as the estimate of the unknown parameter. To illustrate, consider the random variable x with probability distribution $f(x)$ shown in Fig. 3-1. Suppose that the mean μ and variance σ^2 of this distribution are both unknown. If a random sample of n observations is taken, then the **sample mean** $\bar{x}$ and **sample variance** S^2 are point estimators of the **population mean** μ and **population variance** σ^2, respectively. Suppose that this distribution represents a process producing bearings and the random variable x is the inside diameter. We want to obtain point estimates of the mean and variance of the inside diameter of bearings produced by this process. We could measure the inside diameters of a random sample of $n = 20$ bearings (say). Then the sample mean and sample variance could be computed. If this yields $\bar{x} = 1.495$ and $S^2 = 0.001$, then the point estimate of μ is $\hat{\mu} = \bar{x} = 1.495$ and the point estimate of σ^2 is $\hat{\sigma}^2 = S^2 = 0.001$. Recall that the "^" symbol is used to denote an estimate of a parameter.

The mean and variance of a distribution are not necessarily the parameters of the distribution. For example, the parameter of the Poisson distribution is λ, while its mean and variance are $\mu = \lambda$ and $\sigma^2 = \lambda$ (*both* the mean *and* variance are λ), and the parameters of the binomial distribution are n and p, while its mean and variance are $\mu = np$ and $\sigma^2 = np(1 - p)$, respectively. We may show that a good point estimator of the parameter λ of a Poisson distribution is

$$\hat{\lambda} = \frac{1}{n} \sum_{i=1}^{n} x_i = \bar{x}$$

and that a good point estimator of the parameter p of a binomial distribution is

$$\hat{p} = \frac{1}{n} \sum_{i=1}^{n} x_i = \bar{x}$$

for fixed n. In the binomial distribution the observations in the random sample $\{x_i\}$ are either 1 or 0, corresponding to "success" and "failure," respectively.

A number of important properties are required of good point estimators. Two of the most important of these properties are the following:

1. The point estimator should be **unbiased.** That is, the expected value of the point estimator should be the parameter being estimated.

2. The point estimator should have **minimum variance.** Any point estimator is a random variable. Thus, a minimum variance point estimator should have a variance that is smaller than the variance of any other point estimator of that parameter.

The *sample* mean and variance $\bar{x}$ and S^2 are unbiased estimators of the *population* mean and variance μ and σ^2, respectively. That is,

$$E(\bar{x}) = \mu \quad \text{and} \quad E(S^2) = \sigma^2$$

where the operator E is simply the expected value operator, a shorthand way of writing the process of finding the mean of a random variable.

The sample standard deviation S is *not* an unbiased estimator of the population standard deviation σ. It can be shown that

$$E(S) = \left(\frac{2}{n-1}\right)^{1/2} \frac{\Gamma(n/2)}{\Gamma[(n-1)/2]} \sigma$$

$$= c_4 \sigma \qquad (3\text{-}17)$$

Appendix Table VI gives values of c_4 for sample sizes $2 \le n \le 25$. We can obtain an unbiased estimate of the standard deviation from

$$\hat{\sigma} = \frac{S}{c_4} \qquad (3\text{-}18)$$

In many applications of statistics to quality-engineering problems, it is convenient to estimate the standard deviation by the **range method.** Let $x_1, x_2, \ldots, x_n$ be a random sample of n observations from a normal distribution with mean μ and variance σ^2. The **range** of the sample is

$$R = \max(x_i) - \min(x_i)$$

$$= x_{\max} - x_{\min} \qquad (3\text{-}19)$$

That is, the range R is simply the difference between the largest and smallest sample observations. The random variable $W = R/\sigma$ is called the **relative range.** The distribution of W has been well studied. The mean of W is a constant d_2 that depends on the size of the sample. That is, $E(W) = d_2$. Therefore, an unbiased estimator of the standard deviation σ of a normal distribution is

$$\hat{\sigma} = \frac{R}{d_2} \qquad (3\text{-}20)$$

Values of d_2 for sample sizes $2 \le n \le 25$ are given in Appendix Table VI.

Using the range to estimate σ dates from the earliest days of statistical quality control, and it was popular because it is very simple to calculate. With modern calculators and computers, this isn't a major consideration today. Generally, the "quadratic estimator" based on S is preferable. However, if the sample size n is relatively small, the range method actually works very well. The relative efficiency of the range method compared to S is shown here for various sample sizes:

Sample Size n	Relative Efficiency
2	1.000
3	0.992
4	0.975
5	0.955
6	0.930
10	0.850

For moderate values of n—say, $n \geq 10$—the range loses efficiency rapidly, as it ignores all of the information in the sample between the extremes. However, for small sample sizes—say, $n \leq 6$—it works very well and is entirely satisfactory. We will use the range method to estimate the standard deviation for certain types of control charts in Chapter 5. The supplemental text material contains more information about using the range to estimate variability.

3-3 STATISTICAL INFERENCE FOR A SINGLE SAMPLE

The techniques of statistical inference can be classified into two broad categories: **parameter estimation** and **hypothesis testing.** We have already briefly introduced the general idea of **point estimation** of process parameters.

A **statistical hypothesis** is a statement about the values of the parameters of a probability distribution. For example, suppose we think that the mean inside diameter of a bearing is 1.500 in. We may express this statement in a formal manner as

$$H_0: \quad \mu = 1.500$$
$$H_1: \quad \mu \neq 1.500 \tag{3-21}$$

The statement $H_0: \mu = 1.500$ in equation 3-21 is called the **null hypothesis,** and $H_1: \mu \neq 1.500$ is called the **alternative hypothesis.** In our example, H_1 specifies values of the mean diameter that are either greater than 1.500 or less than 1.500, and is called a **two-sided alternative hypothesis.** Depending on the problem, various one-sided alternative hypotheses may be appropriate.

Hypothesis testing procedures are quite useful in many types of statistical quality control problems. They also form the basis for most of the statistical process-control techniques to be described in Parts II and III of this textbook. An important part of any hypothesis testing problem is determining the parameter values specified in the null and alternative hypotheses. Generally, this is done in one of three ways. First, the values may result from past evidence or knowledge. This happens frequently in statistical quality control, where we use past information to specify values for a parameter corresponding to a state of control, and then periodically test the hypothesis that the parameter value has not changed. Second, the values may result from some theory or model of the process. Finally, the values chosen for the parameter may be the result of contractual or design specifications, a situation that occurs frequently. Statistical hypothesis testing procedures may be used to check the conformity of the process parameters to their specified values, or to assist in modifying the process until the desired values are obtained.

To test a hypothesis, we take a random sample from the population under study, compute an appropriate **test statistic,** and then either reject or fail to reject the null hypothesis H_0. The set of values of the test statistic leading to rejection of H_0 is called the **critical region** or **rejection region** for the test.

Two kinds of errors may be committed when testing hypotheses. If the null hypothesis is rejected when it is true, then a type I error has occurred. If the null hypothesis is not rejected when it is false, then a type II error has been made. The probabilities of these two types of errors are denoted as

$$\alpha = P\{\text{type I error}\} = P\{\text{reject } H_0 | H_0 \text{ is true}\}$$

$$\beta = P\{\text{type II error}\} = P\{\text{fail to reject } H_0 | H_0 \text{ is false}\}$$

Sometimes it is more convenient to work with the *power* of the test, where

$$\text{Power} = 1 - \beta = P\{\text{reject } H_0 | H_0 \text{ is false}\}$$

Thus, the power is the probability of *correctly* rejecting H_0. In quality control work, α is sometimes called the **producer's risk,** because it denotes the probability that a good lot will be rejected, or the probability that a process producing acceptable values of a particular quality characteristic will be rejected as performing unsatisfactorily. In addition, β is sometimes called the **consumer's risk,** because it denotes the probability of accepting a lot of poor quality, or allowing a process that is operating in an unsatisfactory manner relative to some quality characteristic to continue in operation.

The general procedure in hypothesis testing is to specify a value of the probability of type I error α, and then to design a test procedure so that a small value of the probability of type II error β is obtained. Thus, we speak of directly controlling or choosing the α risk. The β risk is generally a function of sample size and is controlled indirectly. The larger is the sample size(s) used in the test, the smaller is the β risk.

In this section we will review hypothesis testing procedures when a **single sample** of n observations has been taken from the process. We will also show how the information about the values of the process parameters that is in this sample can be expressed in terms of an interval estimate called a **confidence interval.** In Section 3-4 we will consider statistical inference for two samples from two possibly different processes.

3-3.1   Inference on the Mean of a Population, Variance Known

Hypothesis Testing

Suppose that x is a random variable with unknown mean μ and known variance σ^2. We wish to test the hypothesis that the mean is equal to a standard value—say, μ_0. The hypothesis may be formally stated as

$$H_0: \quad \mu = \mu_0$$
$$H_1: \quad \mu \neq \mu_0 \qquad \qquad (3\text{-}22)$$

The procedure for testing this hypothesis is to take a random sample of n observations on the random variable x, compute the test statistic

$$Z_0 = \frac{\bar{x} - \mu_0}{\sigma/\sqrt{n}} \qquad \qquad (3\text{-}23)$$

and reject H_0 if $|Z_0| > Z_{\alpha/2}$ where $Z_{\alpha/2}$ is the upper $\alpha/2$ percentage point of the standard normal distribution.

We may give an intuitive justification of this test procedure. From the central limit theorem, we know that the sample mean $\bar{x}$ is distributed approximately $N(\mu, \sigma^2/n)$. Now if $H_0: \mu = \mu_0$ is true, then the test statistic Z_0 is distributed approximately $N(0, 1)$; consequently, we would expect $100(1 - \alpha)\%$ of the values of Z_0 to fall between $-Z_{\alpha/2}$ and $Z_{\alpha/2}$. A sample producing a value of Z_0 outside of these limits would be unusual if the null hypothesis were true, and is evidence that $H_0: \mu = \mu_0$ should be rejected. Note that α is the probability of type I error for the test, and the intervals $(Z_{\alpha/2}, \infty)$ and $(-\infty, -Z_{\alpha/2})$ form the critical region for the test.

In some situations we may wish to reject H_0 only if the true mean is larger than μ_0. Thus, the **one-sided** alternative hypothesis is $H_1: \mu > \mu_0$, and we would reject $H_0: \mu = \mu_0$ only if $Z_0 > Z_\alpha$. If rejection is desired only when $\mu < \mu_0$, then the alternative hypothesis is $H_1: \mu < \mu_0$, and we reject H_0 only if $Z_0 < -Z_\alpha$.

.... EXAMPLE 3-1 ••

The internal pressure strength of glass bottles used to package a carbonated beverage is an important quality characteristic. The bottler wants to know whether the mean pressure strength exceeds 175 psi. From previous experience, he knows that the standard deviation of pressure strength is 10 psi. The glass manufacturer submits lots of these bottles to the bottler, who is interested in testing the hypothesis

$$H_0: \quad \mu = 175$$
$$H_1: \quad \mu > 175$$

Note that the lot will be accepted if the null hypothesis $H_0: \mu = 175$ is rejected. A random sample of 25 bottles is selected, and the bottles are placed on a hydrostatic pressure-testing machine that increases the pressure in the bottle until it fails. The sample average bursting strength is $\bar{x} = 182$ psi. The value of the test statistic is

$$Z_0 = \frac{\bar{x} - \mu_0}{\sigma/\sqrt{n}} = \frac{182 - 175}{10/\sqrt{25}} = 3.50$$

If we specify a type I error (or producer's risk) of $\alpha = 0.05$, then from Appendix Table II we find $Z_\alpha = Z_{0.05} = 1.645$. Therefore, we reject $H_0: \mu = 175$ and conclude that the lot mean pressure strength exceeds 175 psi.

- -

Confidence Intervals

An interval estimate of a parameter is the interval between two statistics that includes the true value of the parameter with some probability. For example, to construct an interval estimator of the mean μ, we must find two statistics L and U such that

$$P\{L \le \mu \le U\} = 1 - \alpha \qquad (3\text{-}24)$$

The resulting interval

$$L \le \mu \le U$$

is called a **100(1 $-$ α)% confidence interval** for the unknown mean μ. L and U are called the lower and upper confidence limits, respectively, and $1 - \alpha$ is called the confidence coefficient. Sometimes the half-interval width $U - \mu$ or $\mu - L$ is called the **accuracy** of the confidence interval. The interpretation of a confidence interval is that if a large number of such intervals are constructed, each resulting from a random sample, then $100(1 - \alpha)\%$ of these intervals will contain the true value of μ. Thus, confidence intervals have a frequency interpretation.

The confidence interval (3-24) might be more properly called a **two-sided** confidence interval, as it specifies both a lower and an upper limit on μ. Sometimes in quality control applications, a **one-sided** confidence interval might be more appropriate. A one-sided lower $100(1 - \alpha)\%$ confidence interval on μ would be given by the interval

$$L \le \mu \qquad (3\text{-}25)$$

where L, the lower confidence limit, is chosen so that

$$P\{L \le \mu\} = 1 - \alpha \qquad (3\text{-}26)$$

A one-sided upper $100(1 - \alpha)\%$ confidence interval on μ would be the interval

$$\mu \le U \qquad (3\text{-}27)$$

where U, the upper confidence limit, is chosen so that

$$P\{\mu \le U\} = 1 - \alpha \qquad (3\text{-}28)$$

Confidence Interval on the Mean with Variance Known

Consider the random variable x, with unknown mean μ and known variance σ^2. Suppose a random sample of n observations is taken—say, $x_1, x_2, \ldots, x_n$—and $\bar{x}$ is computed. Then the $100(1 - \alpha)\%$ two-sided confidence interval on μ is

$$\bar{x} - Z_{\alpha/2} \frac{\sigma}{\sqrt{n}} \leq \mu \leq \bar{x} + Z_{\alpha/2} \frac{\sigma}{\sqrt{n}} \qquad (3\text{-}29)$$

where $Z_{\alpha/2}$ is the percentage point of the $N(0, 1)$ distribution such that $P\{z \geq Z_{\alpha/2}\} = \alpha/2$.

Note that $\bar{x}$ is distributed approximately $N(\mu, \sigma^2/n)$ regardless of the distribution of x, per the central limit theorem. Consequently, equation 3-29 is an approximate $100(1 - \alpha)\%$ confidence interval for μ regardless of the distribution of x. If x is distributed $N(\mu, \sigma^2)$, then equation 3-29 is an exact $100(1 - \alpha)\%$ confidence interval. Furthermore, a $100(1 - \alpha)\%$ upper confidence interval on μ is

$$\mu \leq \bar{x} + Z_{\alpha} \frac{\sigma}{\sqrt{n}} \qquad (3\text{-}30)$$

whereas a $100(1 - \alpha)\%$ lower confidence interval on μ is

$$\bar{x} - Z_{\alpha} \frac{\sigma}{\sqrt{n}} \leq \mu \qquad (3\text{-}31)$$

········ **EXAMPLE 3-2** ···

Reconsider the bottle-testing scenario from Example 3-1. Since $\bar{x} = 182$ psi, we know that a reasonable point estimate of the mean bursting strength is $\hat{\mu} = \bar{x} = 182$ psi. We can also find a $100(1 - \alpha)\%$ confidence interval for μ. Suppose a 95% two-sided confidence interval is specified. Then from equation 3-29 we can compute

$$\bar{x} - Z_{\alpha/2} \frac{\sigma}{\sqrt{n}} \leq \mu \leq \bar{x} + Z_{\alpha/2} \frac{\sigma}{\sqrt{n}}$$

$$182 - 1.96 \frac{10}{\sqrt{25}} \leq \mu \leq 182 + 1.96 \frac{10}{\sqrt{25}}$$

$$178.08 \leq \mu \leq 185.92$$

Another way to express this result is that our estimate of mean bursting strength is 182 psi $\pm$ 3.92 psi with 95% confidence.

....

3-3.2   The Use of *P*-Values for Hypothesis Testing

The traditional way to report the results of a hypothesis test is to state that the null hypothesis was or was not rejected at a specified α-value or **level of significance.** For example, in the previous strength-testing problem, we can say that H_0: $\mu = 175$ was rejected at the 0.05 level of significance. This statement of conclusions is often inadequate, because it gives the analyst no idea about whether the computed value of the test statistic was just barely in the rejection region or very far into this region. Furthermore, stating the results this way imposes the predefined level of significance on other users of the information. This approach may be unsatisfactory, as some decision makers might be uncomfortable with the risks implied by $\alpha = 0.05$.

To avoid these difficulties the **P-value approach** has been adopted widely in practice. The *P*-value is the probability that the test statistic will take on a value that is at least as extreme as the observed value of the statistic when the null hypothesis H_0 is true. Thus, a *P*-value conveys much information about the weight of evidence against H_0, and so a decision maker can draw a conclusion at *any* specified level of significance. We now give a formal definition of a *P*-value.

Definition

The **P-value** is the smallest level of significance that would lead to rejection of the null hypothesis H_0.

It is customary to call the test statistic (and the data) significant when the null hypothesis H_0 is rejected; therefore, we may think of the *P*-value as the smallest level α at which the data are significant. Once the *P*-value is known, the decision maker can determine for himself or herself how significant the data are without the data analyst formally imposing a preselected level of significance.

For the normal distribution tests discussed above, it is relatively easy to compute the *P*-value. If z_0 is the computed value of the test statistic, then the *P*-value is

$$P = \begin{cases} 2[1 - \Phi|Z_0|] & \text{for a two-tailed test:} & H_0: \mu = \mu_0 \quad H_1: \mu \neq \mu_0 \\ 1 - \Phi(Z_0) & \text{for an upper-tailed test:} & H_0: \mu = \mu_0 \quad H_1: \mu > \mu_0 \\ \Phi(Z_0) & \text{for a lower-tailed test:} & H_0: \mu = \mu_0 \quad H_1: \mu < \mu_0 \end{cases}$$

Here, $\Phi(Z)$ is the standard normal cumulative distribution function defined in Chapter 2. To illustrate this, consider the bottle-testing problem in Example 3-1. The computed value of the test statistic is $Z_0 = 3.50$ and since the alternative hypothesis is one-tailed, the P-value is

$$P = 1 - \Phi(3.50) = 0.00023$$

Thus, H_0: $\mu = 175$ would be rejected at any level of significance $\alpha \geq P = 0.00023$. For example, H_0 would be rejected if $\alpha = 0.01$, but it would not be rejected if $\alpha = 0.0001$.

It is not always easy to compute the exact P-value for a test. However, most modern computer programs for statistical analysis report P-values, and they can be obtained using some hand-held calculators. It is also possible to use the statistical tables in the Appendix to approximate the P-value in some cases.

3-3.3   Inference on the Mean of a Normal Distribution, Variance Unknown

Hypothesis Testing

Suppose that x is a normal random variable with unknown mean μ and unknown variance σ^2. We wish to test the hypothesis that the mean equals a standard value μ_0; that is,

$$H_0: \mu = \mu_0$$
$$H_1: \mu \neq \mu_0 \tag{3-32}$$

Note that this problem is similar to that of Section 3-3.1, except that now the variance is unknown. Because the variance is unknown, we must make the additional assumption that the random variable is normally distributed. The normality assumption is needed to formally develop the statistical test, but moderate departures from normality will not seriously affect the results.

As σ^2 is unknown, it may be estimated by S^2. If we replace σ in equation 3-23 by S, we have the test statistic

$$t_0 = \frac{\bar{x} - \mu_0}{S/\sqrt{n}} \tag{3-33}$$

The null hypothesis H_0: $\mu = \mu_0$ will be rejected if $|t_0| > t_{\alpha/2, n-1}$, where $t_{\alpha/2, n-1}$ denotes the upper $\alpha/2$ percentage point of the t distribution with $n - 1$ degrees of freedom. The critical regions for the one-sided alternative hypotheses are as follows: if H_1: $\mu_1 > \mu_0$, reject H_0 if $t_0 > t_{\alpha, n-1}$, and if H_1: $\mu_1 < \mu_0$, reject H_0 if $t_0 < -t_{\alpha, n-1}$. One could also compute the P-value for a t-test. Most computer software packages report the P-value along with the computed value of t_0.

Table 3-1   Tensile Strength Measurements on Synthetic Fiber

Specimen	Strength (psi)
1	48.89
2	52.07
3	49.29
4	51.66
5	52.16
6	49.72
7	48.00
8	49.96
9	49.20
10	48.10
11	47.90
12	46.94
13	51.76
14	50.75
15	49.86
16	51.57

······· **EXAMPLE 3-3** ··

The mean tensile strength of a synthetic fiber is an important quality characteristic that is of interest to the manufacturer, who would like to test the hypothesis that the mean strength is 50 psi, using $\alpha = 0.05$. From past experience, the manufacturer is willing to assume that tensile strength is approximately normally distributed; however, both the mean tensile strength and standard deviation of tensile strength are unknown. A random sample of 16 fiber specimens is selected, and their tensile strengths are determined. The sample data are shown in Table 3-1.

We may calculate the sample mean and sample standard deviation of the tensile strength data as

$$\bar{x} = \frac{1}{n} \sum_{i=1}^{n} x_i = \frac{1}{16} (797.83) = 49.86 \text{ psi}$$

and

$$S = \sqrt{\frac{\sum_{i=1}^{n} x_i^2 - \frac{\left(\sum_{i=1}^{n} x_i\right)^2}{n}}{n-1}}$$

$$= \sqrt{\frac{39{,}824.69 - \frac{(797.83)^2}{16}}{15}}$$

$$= \sqrt{2.76} = 1.66 \text{ psi}$$

The hypotheses are

$$H_0: \quad \mu = 50$$
$$H_1: \quad \mu \neq 50$$

and the test statistic (equation 3-33) is

$$
\begin{aligned}
t_0 &= \frac{\bar{x} - \mu_0}{S/\sqrt{n}} \\
&= \frac{49.86 - 50}{1.66/\sqrt{16}} \\
&= -0.34
\end{aligned}
$$

Since the calculated value of t_0 does not exceed $t_{0.025,15} = 2.131$ or $-t_{0.025,15} = -2.131$, we cannot reject H_0. We conclude that there is no strong evidence to indicate that the mean strength differs from 50 psi.

···

Confidence Interval on the Mean of a Normal Distribution with Variance Unknown

Suppose that x is a normal random variable with unknown mean μ and unknown variance σ^2. From a random sample of n observations the sample mean $\bar{x}$ and sample variance S^2 are computed. Then a $100(1 - \alpha)\%$ two-sided confidence interval on the true mean is

$$\bar{x} - t_{\alpha/2,n-1}\frac{S}{\sqrt{n}} \leq \mu \leq \bar{x} + t_{\alpha/2,n-1}\frac{S}{\sqrt{n}} \qquad (3\text{-}34)$$

where $t_{\alpha/2,n-1}$ denotes the percentage point of the t distribution with $n - 1$ degrees of freedom such that $P\{t_{n-1} \geq t_{\alpha/2,n-1}\} = \alpha/2$. The corresponding upper and lower $100(1 - \alpha)\%$ confidence intervals are

$$\mu \leq \bar{x} + t_{\alpha,n-1}\frac{S}{\sqrt{n}} \qquad (3\text{-}35)$$

and

$$\bar{x} - t_{\alpha,n-1} \frac{S}{\sqrt{n}} \leq \mu \qquad\qquad (3\text{-}36)$$

respectively.

······ **EXAMPLE 3-4** ···

Reconsider the data from Example 3-3 concerning fiber tensile strength. Using equation 3-34, we can find a 95% confidence interval on the mean strength as follows:

$$\bar{x} - t_{\alpha/2,n-1} \frac{S}{\sqrt{n}} \leq \mu \leq \bar{x} + t_{\alpha/2,n-1} \frac{S}{\sqrt{n}}$$

$$49.86 - 2.132 \frac{1.66}{\sqrt{16}} \leq \mu \leq 49.86 + 2.132 \frac{1.66}{\sqrt{16}}$$

$$48.98 \leq \mu \leq 50.74$$

Another way to express this result is that our estimate of the mean tensile strength is 49.86 ± 0.88 psi with 95% confidence.

The manufacturer may only be concerned about tensile strength values that are too low and consequently may be interested in a one-sided confidence interval. The 95% lower confidence interval on mean tensile strength is found from equation 3-36, using $t_{0.05,15} = 1.753$, as

$$49.86 - 1.753 \frac{1.66}{\sqrt{16}} \leq \mu$$

or

$$49.13 \leq \mu$$

··

Computer Output

Table 3-2 presents the output from the one-sample t-test from Minitab (a widely-used PC statistics software package) for the fiber tensile strength data from Example 3-3. Note that the output includes the sample statistics $\bar{x}$ and S along with the "standard error of the mean," denoted SE mean, computed as $S/\sqrt{n} = 1.66/\sqrt{16} = 0.415$. Notice that the reported values differ slightly from those computed manually, due to rounding. The value of the t-statistic is also reported along with the P-value for the t-test. Note that $P = 0.749$, a very large value, clearly indicating that H_0: $\mu = 50$ cannot be rejected. The 95% confidence interval is also reported, and the numerical results are very similar to those reported in Example 3-4.

Table 3-2 Minitab Output for Example 3-3

```
Welcome  to  Minitab,  press  F1  for  help.
```

One-Sample T: Strength
```
Test  of  mu = 50  vs  mu  not = 50

Variable      N      Mean      StDev      SE  Mean
Strength     16    49.864     1.661         0.415

Variable         95.0% CI                 T        P
Strength     (48.979,  50.750)   -0.33   0.749
```

Figure 3-5 shows a box plot of the strength data from Example 3-3, also generated by Minitab. The box plot indicates that the data distribution is nearly symmetric. Another plot, the **normal probability plot,** introduced in Section 3-3.7, is useful in assessing the validity of the normality assumption underlying the *t*-test.

3-3.4 Inference on the Variance of a Normal Distribution

Hypothesis Testing
We now review hypothesis testing on the variance of a normal distribution. Whereas tests on means are relatively insensitive to the normality assumption, test procedures for variances are not.

Suppose we wish to test the hypothesis that the variance of a normal distribution equals a constant—say, σ_0^2. The hypotheses are

$$H_0: \quad \sigma^2 = \sigma_0^2$$
$$H_1: \quad \sigma^2 \neq \sigma_0^2 \tag{3-37}$$

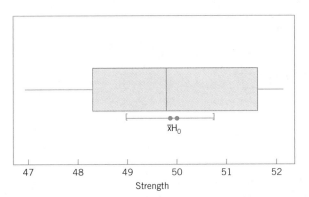

Figure 3-5 Minitab box plot of the fiber tensile strength data from Example 3-3.

The test statistic for this hypothesis is

$$\chi_0^2 = \frac{(n - 1)S^2}{\sigma_0^2} \tag{3-38}$$

where S^2 is the sample variance computed from a random sample of n observations. The null hypothesis is rejected if $\chi_0^2 > \chi_{\alpha/2,n-1}^2$ or if $\chi_0^2 < \chi_{1-\alpha/2,n-1}^2$, where $\chi_{\alpha/2,n-1}^2$ and $\chi_{1-\alpha/2,n-1}^2$ are the upper $\alpha/2$ and lower $1 - (\alpha/2)$ percentage points of the chi-square distribution with $n - 1$ degrees of freedom. If a one-sided alternative is specified—say, $H_1: \sigma^2 < \sigma_0^2$, then we would reject if $\chi_0^2 < \chi_{1-\alpha,n-1}^2$. For the other one-sided alternative $H_1: \sigma^2 > \sigma_0^2$, reject if $\chi_0^2 > \chi_{\alpha,n-1}^2$.

This test is very useful in many quality-improvement applications. For example, consider a normal random variable with mean μ and variance σ^2. If σ^2 is less than or equal to some value—say, σ_0^2—then the natural inherent scatter of the process will be well within the design requirements, and consequently, almost all of the production will conform to specifications. However, if σ^2 exceeds σ_0^2, then the natural scatter in the process will exceed the specification limits, resulting in a high percentage of nonconforming production or "fallout." In other words, **process capability** is directly related to **process variability.** Equations 3-37 and 3-38 may be used to analyze various other similar situations, and as we will see subsequently, they form the basis for a monitoring or control procedure for process variability.

Confidence Interval on the Variance of a Normal Distribution
Suppose that x is a normal random variable with unknown mean μ and unknown variance σ^2. Let the sample variance S^2 be computed from a random sample of n observations. Then a $100(1 - \alpha)\%$ two-sided confidence interval on the variance is

$$\frac{(n - 1)S^2}{\chi_{\alpha/2,n-1}^2} \leq \sigma^2 \leq \frac{(n - 1)S^2}{\chi_{1-\alpha/2,n-1}^2} \tag{3-39}$$

where $\chi_{\alpha/2,n-1}^2$ denotes the percentage point of the chi-square distribution such that $P\{\chi_{n-1}^2 \geq \chi_{\alpha/2,n-1}^2\} = \alpha/2$. If one-sided confidence intervals are desired, they may be obtained from equation 3-39 by using only the upper (or lower) limit with the probability level increased from $\alpha/2$ to α. That is, the upper and lower $100(1 - \alpha)\%$ confidence intervals are

$$\sigma^2 \leq \frac{(n-1)S^2}{\chi^2_{1-\alpha,n-1}} \quad \text{upper} \tag{3-40}$$

and

$$\frac{(n-1)S^2}{\chi^2_{\alpha,n-1}} \leq \sigma^2 \quad \text{lower} \tag{3-41}$$

respectively.

We may use the data from Example 3-3 to demonstrate the computation of a 95% (say) confidence interval on σ^2. Note that for the data in Table 3-1, we have $S^2 = 2.76$. From Appendix Table III, we find that $\chi^2_{0.025,15} = 27.49$ and $\chi^2_{0.975,15} = 6.27$. Therefore, from equation 3-39 we find the 95% two-sided confidence interval on σ^2 as

$$\frac{(15)2.76}{27.49} \leq \sigma^2 \leq \frac{(15)2.76}{6.27}$$

which reduces to

$$1.51 \leq \sigma^2 \leq 6.60$$

3-3.5 Inference on a Population Proportion

Hypothesis Testing
Suppose we wish to test the hypothesis that the proportion p of a population equals a standard value—say, p_0. The test we will describe is based on the normal approximation to the binomial. If a random sample of n items is taken from the population and x items in the sample belong to the class associated with p, then to test

$$H_0: \quad p = p_0$$
$$H_1: \quad p \neq p_0 \tag{3-42}$$

we use the statistic

$$
Z_0 = \begin{cases}
\dfrac{(x + 0.5) - np_0}{\sqrt{np_0(1 - p_0)}} & \text{if } x < np_0 \\[2ex]
\dfrac{(x - 0.5) - np_0}{\sqrt{np_0(1 - p_0)}} & \text{if } x > np_0
\end{cases}
\tag{3-43}
$$

The null hypothesis H_0: $p = p_0$ is rejected if $|Z_0| > Z_{\alpha/2}$. The one-sided alternative hypotheses are treated similarly.

······· **EXAMPLE 3-5** ··

A foundry produces steel castings used in the automotive industry. We wish to test the hypothesis that the fraction conforming or fallout from this process is 10%. In a random sample of 250 castings, 41 were found to be nonconforming. To test

$$
H_0: \quad p = 0.1
$$
$$
H_1: \quad p \neq 0.1
$$

we calculate the test statistic

$$
Z_0 = \frac{(x - 0.5) - np_0}{\sqrt{np_0(1 - p_0)}} = \frac{(41 - 0.5) - (250)(0.1)}{\sqrt{250(0.1)(1 - 0.1)}} = 3.27
$$

Using $\alpha = 0.05$ we find $Z_{0.025} = 1.96$, and therefore H_0: $p = 0.1$ is rejected (the P-value here is $P = 0.00108$). That is, the process fraction nonconforming or fallout is not equal to 10%.

···

Confidence Intervals on a Population Proportion

It is frequently necessary to construct $100(1 - \alpha)\%$ confidence intervals on a population proportion p. This parameter frequently corresponds to a lot or process fraction nonconforming. Now p is only one of the parameters of a binomial distribution, and we usually assume that the other binomial parameter n is known. If a random sample of n observations from the population has been taken, and x "nonconforming" observations have been found in this sample, then the unbiased point estimator of p is $\hat{p} = x/n$.

There are several approaches to constructing the confidence interval on p. If n is large and $p \geq 0.1$ (say), then the normal approximation to the binomial can be used, resulting in the $100(1 - \alpha)\%$ confidence interval.

$$\hat{p} - Z_{\alpha/2}\sqrt{\frac{\hat{p}(1-\hat{p})}{n}} \le p \le \hat{p} + Z_{\alpha/2}\sqrt{\frac{\hat{p}(1-\hat{p})}{n}} \qquad (3\text{-}44)$$

If n is small, then the binomial distribution should be used to establish the confidence interval on p. If n is large but p is small, then the Poisson approximation to the binomial is useful in constructing confidence intervals. Examples of these latter two procedures are given by Duncan (1986).

······· **EXAMPLE 3-6** ··

In a random sample of 80 automotive crankshaft bearings, 15 of the bearings have a surface finish that is rougher than the specifications will allow. The point estimate of the fraction nonconforming in the process is

$$\hat{p} = \frac{15}{80} = 0.1875$$

Assuming that the normal approximation to the binomial is appropriate, a 95% confidence interval on the process fraction nonconforming is found from equation 3-44 as

$$0.1875 - 1.96\sqrt{\frac{0.1875(0.8125)}{80}} \le p \le 0.1875 + 1.96\sqrt{\frac{0.1875(0.8125)}{80}}$$

which reduces to

$$0.1020 \le p \le 0.2730$$

3-3.6 The Probability of Type II Error

In most hypothesis testing situations, it is important to determine the probability of type II error associated with the test. Equivalently, we may elect to evaluate the power of the test. To illustrate how this may be done, we will find the probability of type II error associated with the test of

$$H_0: \quad \mu = \mu_0$$
$$H_1: \quad \mu \ne \mu_0$$

where the variance σ^2 is known. The test procedure was discussed in Section 3-3.1.

The test statistic for this hypothesis is

$$Z_0 = \frac{\bar{x} - \mu_0}{\sigma/\sqrt{n}}$$

and under the null hypothesis the distribution of Z_0 is $N(0, 1)$. To find the probability of type II error, we must assume that the null hypothesis $H_0: \mu = \mu_0$ is false and then find

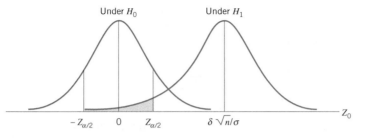

Figure 3-6 The distribution of Z_0 under H_0 and H_1.

the distribution of Z_0. Suppose that the mean of the distribution is really $\mu_1 = \mu_0 + \delta$, where $\delta > 0$. Thus, the alternative hypothesis H_1: $\mu \neq \mu_0$ is true, and under this assumption the distribution of the test statistic Z_0 is

$$Z_0 \sim N\left(\frac{\delta \sqrt{n}}{\sigma}, 1\right)\tag{3-45}$$

The distribution of the test statistic Z_0 under both hypotheses H_0 and H_1 is shown in Fig. 3-6. We note that the probability of type II error is the probability that Z_0 will fall between $-Z_{\alpha/2}$ and $Z_{\alpha/2}$ given that the alternative hypothesis H_1 is true. To evaluate this probability, we must find $F(Z_{\alpha/2}) - F(-Z_{\alpha/2})$, where F denotes the cumulative distribution function of the $N(\delta\sqrt{n}/\sigma, 1)$ distribution. In terms of the standard normal cumulative distribution, we then have

$$\beta = \Phi\left(Z_{\alpha/2} - \frac{\delta \sqrt{n}}{\sigma}\right) - \Phi\left(-Z_{\alpha/2} - \frac{\delta \sqrt{n}}{\sigma}\right)\tag{3-46}$$

as the probability of type II error. This equation will also work when $\delta < 0$.

······· **EXAMPLE 3-7** ···

The mean contents of coffee cans filled on a particular production line are being studied. Standards specify that the mean contents must be 16.0 oz, and from past experience it is known that the standard deviation of the can contents is 0.1 oz. The hypotheses are

$$H_0: \quad \mu = 16.0$$
$$H_1: \quad \mu \neq 16.0$$

A random sample of nine cans is to be used, and the type I error probability is specified as $\alpha = 0.05$. Therefore, the test statistic is

$$Z_0 = \frac{\bar{x} - 16.0}{0.1/\sqrt{9}}$$

and H_0 is rejected if $|Z_0| > Z_{0.025} = 1.96$. Suppose that we wish to find the probability of type II error if the true mean contents are $\mu_1 = 16.1$ oz. Since this implies that $\delta = \mu_1 - \mu_0 = 16.1 - 16.0 = 0.1$, we have

$$\beta = \Phi\left(Z_{\alpha/2} - \frac{\delta\sqrt{n}}{\sigma}\right) - \Phi\left(-Z_{\alpha/2} - \frac{\delta\sqrt{n}}{\sigma}\right)$$

$$= \Phi\left(1.96 - \frac{(0.1)(3)}{0.1}\right) - \Phi\left(-1.96 - \frac{(0.1)(3)}{0.1}\right)$$

$$= \Phi(-1.04) - \Phi(-4.96)$$

$$= 0.1492$$

That is, the probability that we will incorrectly fail to reject H_0 if the true mean contents are 16.1 oz is 0.1492. Equivalently, we can say that the power of the test is $1 - \beta = 1 - 0.1492 = 0.8508$.

We note from examining equation 3-46 and Fig. 3-6 that β is a function of n, δ, and α. It is customary to plot curves illustrating the relationship between these parameters. Such a set of curves is shown in Fig. 3-7 for $\alpha = 0.05$. Graphs such as these are usually called **operating-characteristic (OC)** curves. The parameter on the vertical axis of these curves is β, and the parameter on the horizontal axis is $d = |\delta|/\sigma$. From examining the operating-characteristic curves, we see that

1.   The further the true mean μ_1 is from the hypothesized value μ_0 (i.e., the larger the value of δ), the smaller is the probability of type II error for a given n and

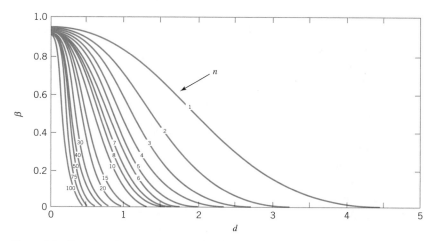

Figure 3-7   Operating-characteristic curves for the two-sided normal test with $\alpha = 0.05$. (Reproduced with permission from C. L. Ferris, F. E. Grubbs, and C. L. Weaver, "Operating Characteristic Curves for the Common Statistical Tests of Significance," *Annals of Mathematical Statistics*, June 1946.)

α. That is, for a specified sample size and α, the test will detect large differences more easily than small ones.

2. As the sample size n increases, the probability of type II error gets smaller for a specified δ and α. That is, to detect a specified difference we may make the test more powerful by increasing the sample size.

Operating-characteristic curves are useful in determining how large a sample is required to detect a specified difference with a particular probability. As an illustration, suppose that in Example 3-7 we wish to determine how large a sample will be necessary to have a 0.90 probability of rejecting H_0: $\mu = 16.0$ if the true mean is $\mu = 16.05$. Since $\delta = 16.05 - 16.0 = 0.05$, we have $d = |\delta|/\sigma = |0.05|/0.1 = 0.5$. From Fig. 3-7 with $\beta = 0.10$ and $d = 0.5$, we find $n = 45$, approximately. That is, 45 observations must be taken to ensure that the test has the desired probability of type II error.

Operating-characteristic curves are available for most of the standard statistical tests discussed in this chapter. For a detailed discussion of the use of operating-characteristic curves, refer to Hines and Montgomery (1990) or Montgomery and Runger (1999).

3-3.7 Probability Plotting

In many quality-engineering problems the analyst does not know the underlying distribution of the population, and it would be useful to have a procedure for testing the hypothesis that a particular distribution will be satisfactory as a model of the process or population. This general type of problem is often called **goodness-of-fit testing.** There are formal, analytical methods available for testing goodness of fit; for an introduction to these procedures, see Montgomery and Runger (1999).

Graphical methods are also useful when we are selecting a probability distribution to describe a population. **Probability plotting** is a graphical method for determining whether sample data conform to a hypothesized distribution based on a subjective visual examination of the data. The general procedure is very simple and can be performed quickly. Probability plotting typically uses special graph paper, known as **probability paper,** that has been designed for the hypothesized distribution. Probability paper is widely available for the normal, lognormal, Weibull, and various chi-square and gamma distributions.

To construct a probability plot, the observations in the sample are first ranked from smallest to largest. That is, the sample $x_1, x_2, \ldots, x_n$ is arranged as $x_{(1)}, x_{(2)}, \ldots, x_{(n)}$, where $x_{(1)}$ is the smallest observation, $x_{(2)}$ is the second smallest observation, and so forth, with $x_{(n)}$ the largest. The ordered observations $x_{(j)}$ are then plotted against their observed cumulative frequency $(j - 0.5)/n$ on the appropriate probability paper. If the hypothesized distribution adequately describes the data, the plotted points will fall approximately along a straight line; if the plotted points deviate significantly from a straight line, then the hypothesized model is not appropriate. Usually, the determination of whether or not the data plot as a straight line is subjective. The procedure is illustrated in the following example.

······   **EXAMPLE 3-8** ···

Ten observations on the effective life in minutes of a catalyst used in a chemical reaction are as follows: 1176, 1191, 1214, 1220, 1205, 1192, 1201, 1190, 1183, and 1185. We hypothesize that catalyst life is adequately modeled by a normal distribution. To use probability plotting to investigate this hypothesis, first arrange the observations in ascending order and calculate their cumulative frequencies $(j - 0.5)/10$ as follows.

j	$x_{(j)}$	$(j - 0.5)/10$
1	1176	0.05
2	1183	0.15
3	1185	0.25
4	1190	0.35
5	1191	0.45
6	1192	0.55
7	1201	0.65
8	1205	0.75
9	1214	0.85
10	1220	0.95

The pairs of values $x_{(j)}$ and $(j - 0.5)/10$ are now plotted on normal probability paper. This plot is shown in Fig. 3-8. Most normal probability paper plots $100(j - 0.5)/n$ on the left vertical scale and $100[1 - (j - 0.5)/n]$ on the right vertical scale, with the variable value plotted on the horizontal scale. A straight line, chosen subjectively, has been drawn through the plotted points. In drawing the straight line, you should be influenced more by the points near the middle of the plot than by the extreme points. A good rule of thumb is to draw the line approximately between the 25th and 75th percentile points,

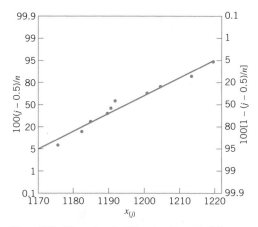

Figure 3-8   Normal probability plot, Example 3-8.

which is how the line in Fig. 3-8 was determined. In assessing the "closeness" of the points to the straight line, imagine a "fat pencil" lying along the line. If all the points are covered by this imaginary pencil, then a normal distribution adequately describes the data. Since the points in Fig. 3-8 would pass the "fat pencil" test, we conclude that the normal distribution is an appropriate model.

··

We can obtain an estimate of the mean and standard deviation directly from the normal probability plot. The mean is estimated as the 50th percentile on the probability plot, and the standard deviation is estimated as the difference between the 84th and 50th percentiles. To illustrate using the normal probability plot in Fig. 3-8, we observe that the 50th percentile is approximately 1196 minutes, so we would estimate mean life as $\hat{\mu} = 1196$ minutes. The 84th percentile is 1212 minutes (approximately), so we would estimate the standard deviation of life as $\hat{\sigma} = 1212 - 1196 = 16$ minutes.

A normal probability plot can also be constructed on ordinary graph paper by plotting the standardized normal scores z_j against $x_{(j)}$, where the standardized normal scores satisfy

$$\frac{j - 0.5}{n} = P(z \le z_j) = \Phi(z_j)$$

For example, if $(j - 0.5)/n = 0.05$, then $\Phi(z_j) = 0.05$ implies that $z_j = -1.64$. To illustrate, consider the data from Example 3-8. In the following table we have shown the standardized normal scores in the last column.

j	$x_{(j)}$	$(j - 0.5)/10$	z_j
1	1176	0.05	-1.64
2	1183	0.15	-1.04
3	1185	0.25	-0.67
4	1190	0.35	-0.39
5	1191	0.45	-0.13
6	1192	0.55	0.13
7	1201	0.65	0.39
8	1205	0.75	0.67
9	1214	0.85	1.04
10	1220	0.95	1.64

Figure 3-9 presents the plot of z_j versus $x_{(j)}$. This normal probability plot is equivalent to the one in Fig. 3-8.

A very important application of normal probability plotting is in **verification of assumptions** when using statistical inference procedures that require the normality assumption. For example, consider the fiber tensile strength data in Example 3-3. In that example, we used the t-test to investigate a hypothesis about the mean fiber tensile

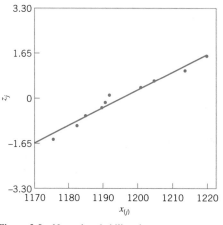

Figure 3-9 Normal probability plot.

strength. (We tested H_0: $\mu_0 = 50$ against H_1: $\mu_0 \neq 50$.) The t-test requires the assumption that the data come from normal distributions.

Figure 3-10 presents a normal probability plot (from Minitab) of the observations on tensile strength from Example 3-3. Note that the plot gives no indication that a normal distribution assumption is inappropriate for the population of tensile strength measurements. Of course, moderate departures from normality do not seriously affect the t-test, but a major violation of this assumption could point out the need to use other

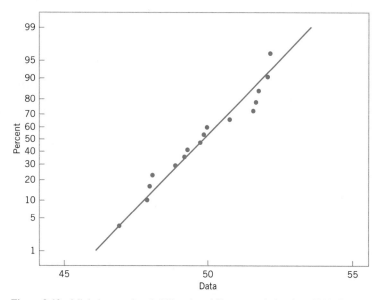

Figure 3-10   Minitab normal probability plot of fiber strength data from Table 3-1.

analysis methods, such as nonparametric statistical methods. See Montgomery and Runger (1999) for an introduction to these procedures.

3-4   STATISTICAL INFERENCE FOR TWO SAMPLES

The previous section presented hypothesis tests and confidence intervals for a single population parameter (the mean μ, the variance σ^2, or a proportion p). This section extends those results to the case of two independent populations.

The general situation is shown in Fig. 3-11. Population 1 has mean μ_1 and variance σ_1^2, whereas population 2 has mean μ_2 and variance σ_2^2. Inferences will be based on two random samples of sizes n_1 and n_2, respectively. That is, $x_{11}, x_{12}, \ldots, x_{1n_1}$ is a random sample of n_1 observations from population 1, and $x_{21}, x_{22}, \ldots, x_{2n_2}$ is a random sample of n_2 observations from population 2.

3-4.1   Inference for a Difference in Means, Variances Known

In this section we consider statistical inferences on the difference in means $\mu_1 - \mu_2$ of the populations shown in Fig. 3-11, where the variances σ_1^2 and σ_2^2 are known. The assumptions for this section are summarized here.

Assumptions

1. $x_{11}, x_{12}, \ldots, x_{1n_1}$ is a random sample from population 1.
2. $x_{21}, x_{22}, \ldots, x_{2n_2}$ is a random sample from population 2.
3. The two populations represented by x_1 and x_2 are independent.
4. Both populations are normal, or if they are not normal, the conditions of the central limit theorem apply.

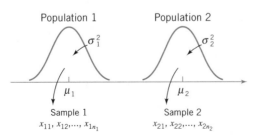

Figure 3-11   Two independent populations.

A logical point estimator of $\mu_1 - \mu_2$ is the difference in sample means $\bar{x}_1 - \bar{x}_2$. Based on the properties of expected values, we have

$$E(\bar{x}_1 - \bar{x}_2) = E(\bar{x}_1) - E(\bar{x}_2) = \mu_1 - \mu_2$$

and the variance of $\bar{x}_1 - \bar{x}_2$ is

$$V(\bar{x}_1 - \bar{x}_2) = V(\bar{x}_1) + V(\bar{x}_2) = \frac{\sigma_1^2}{n_1} + \frac{\sigma_2^2}{n_2}$$

Based on the assumptions and the preceding results, we may state the following.

The quantity

$$Z = \frac{\bar{x}_1 - \bar{x}_2 - (\mu_1 - \mu_2)}{\sqrt{\dfrac{\sigma_1^2}{n_1} + \dfrac{\sigma_2^2}{n_2}}} \qquad (3\text{-}47)$$

has a $N(0, 1)$ distribution.

This result will be used to form tests of hypotheses and confidence intervals on $\mu_1 - \mu_2$. Essentially, we may think of $\mu_1 - \mu_2$ as a parameter θ, and its estimator is $\hat{\theta} = \bar{x}_1 - \bar{x}_2$ with variance $\sigma_{\hat{\theta}}^2 = \sigma_1^2/n_1 + \sigma_2^2/n_2$. If θ_0 is the null hypothesis value specified for θ, then the test statistic will be $(\hat{\theta} - \theta_0)/\sigma_{\hat{\theta}}$. Note how similar this is to the test statistic for a single mean used in the previous section.

Hypothesis Tests for a Difference in Means, Variances Known
We now consider hypothesis testing on the difference in the means $\mu_1 - \mu_2$ of the two populations in Fig. 3-11. Suppose we are interested in testing that the difference in means $\mu_1 - \mu_2$ is equal to a specified value Δ_0. Thus, the null hypothesis will be stated as H_0: $\mu_1 - \mu_2 = \Delta_0$. Obviously, in many cases, we will specify $\Delta_0 = 0$ so that we are testing the equality of two means (i.e., H_0: $\mu_1 = \mu_2$). The appropriate test statistic would be found by replacing $\mu_1 - \mu_2$ in equation 3-47 by Δ_0, and this test statistic would have a standard normal distribution under H_0. Suppose that the alternative hypothesis is H_1: $\mu_1 - \mu_2 \neq \Delta_0$. Now, a sample value of $\bar{x}_1 - \bar{x}_2$ that is considerably different from Δ_0 is evidence that H_1 is true. Because Z_0 has the $N(0, 1)$ distribution when H_0 is true, we would take $-z_{\alpha/2}$ and $z_{\alpha/2}$ as the boundaries of the critical region just as we did in the single-sample hypothesis testing problem of Section 3-3.1. This would give a test with level of significance α. Critical regions for the one-sided alternatives would be located similarly. Formally, we summarize these results here.

Testing Hypotheses on $\mu_1 - \mu_2$, Variances Known

Null hypothesis: $H_0: \mu_1 - \mu_2 = \Delta_0$

Test statistic: $$Z_0 = \frac{\bar{x}_1 - \bar{x}_2 - \Delta_0}{\sqrt{\dfrac{\sigma_1^2}{n_1} + \dfrac{\sigma_2^2}{n_2}}} \tag{3-48}$$

Alternative Hypotheses	Rejection Criterion
$H_1: \mu_1 - \mu_2 \neq \Delta_0$	$z_0 > z_{\alpha/2}$ or $z_0 < -z_{\alpha/2}$
$H_1: \mu_1 - \mu_2 > \Delta_0$	$z_0 > z_\alpha$
$H_1: \mu_1 - \mu_2 < \Delta_0$	$z_0 < -z_\alpha$

······ **EXAMPLE 3-9** ···

A product developer is interested in reducing the drying time of a primer paint. Two formulations of the paint are tested; formulation 1 is the standard chemistry, and formulation 2 has a new drying ingredient that should reduce the drying time. From experience, it is known that the standard deviation of drying time is 8 minutes, and this inherent variability should be unaffected by the addition of the new ingredient. Ten specimens are painted with formulation 1, and another 10 specimens are painted with formulation 2; the 20 specimens are painted in random order. The two sample average drying times are $\bar{x}_1 = 121$ min and $\bar{x}_2 = 112$ min, respectively. What conclusions can the product developer draw about the effectiveness of the new ingredient, using $\alpha = 0.05$?

The hypotheses of interest here are

$$H_0: \quad \mu_1 - \mu_2 = 0$$
$$H_0: \quad \mu_1 - \mu_2 > 0$$

or equivalently,

$$H_0: \quad \mu_1 = \mu_2$$
$$H_1: \quad \mu_1 > \mu_2$$

Now since $\bar{x}_1 = 121$ min and $\bar{x}_2 = 112$ min, the test statistic is

$$Z_0 = \frac{121 - 112}{\sqrt{\dfrac{(8)^2}{10} + \dfrac{(8)^2}{10}}} = 2.52$$

Because the test statistic $Z_0 = 2.52 > Z_{0.05} = 1.645$, we reject $H_0: \mu_1 = \mu_2$ at the $\alpha = 0.05$ level and conclude that adding the new ingredient to the paint significantly reduces the drying time. Alternatively, we can find the P-value for this test as

$$P\text{-value} = 1 - \Phi(2.52) = 0.0059$$

Therefore, H_0: $\mu_1 = \mu_2$ would be rejected at any significance level $\alpha \geq 0.0059$.

Confidence Interval on a Difference in Means, Variances Known
The $100(1 - \alpha)\%$ confidence interval on the difference in two means $\mu_1 - \mu_2$ when the variances are known can be found directly from results given previously in this section. Recall that $x_{11}, x_{12}, \ldots, x_{1n_1}$ is a random sample of n_1 observations from the first population and $x_{21}, x_{22}, \ldots, x_{2n_2}$ is a random sample of n_2 observations from the second population. If $\bar{x}_1$ and $\bar{x}_2$ are the means of these two samples, then a $100(1 - \alpha)\%$ confidence interval on the difference in means $\mu_1 - \mu_2$ is given by the following.

$$\bar{x}_1 - \bar{x}_2 - z_{\alpha/2}\sqrt{\frac{\sigma_1^2}{n_1} + \frac{\sigma_2^2}{n_2}} \leq \mu_1 - \mu_2 \leq \bar{x}_1 - \bar{x}_2 + z_{\alpha/2}\sqrt{\frac{\sigma_1^2}{n_1} + \frac{\sigma_2^2}{n_2}} \tag{3-49}$$

This is a two-sided confidence interval. One-sided confidence intervals can be obtained by using the approach illustrated in Section 3-3 for the single-sample case.

3-4.2 Inference for a Difference in Means of Two Normal Distributions, Variances Unknown

We now extend the results of the previous section to the difference in means of the two distributions in Fig. 3-11 when the variances of both distributions σ_1^2 and σ_2^2 are unknown. If the sample sizes n_1 and n_2 exceed 30, then the normal distribution procedures in Section 3-4.1 could be used. However, when small samples are taken, we will assume that the populations are normally distributed and base our hypotheses tests and confidence intervals on the t distribution. This nicely parallels the case of inference on the mean of a single sample with unknown variance.

Hypotheses Tests for the Difference in Means
We now consider tests of hypotheses on the difference in means $\mu_1 - \mu_2$ of two normal distributions where the variances σ_1^2 and σ_2^2 are unknown. A t-statistic will be used to test these hypotheses. As noted above, the normality assumption is required to develop the test procedure, but moderate departures from normality do not adversely affect the

procedure. Two different situations must be treated. In the first case, we assume that the variances of the two normal distributions are unknown but equal; that is, $\sigma_1^2 = \sigma_2^2 = \sigma^2$. In the second, we assume that σ_1^2 and σ_2^2 are unknown and not necessarily equal.

Case 1: $\sigma_1^2 = \sigma_2^2 = \sigma^2$
Suppose we have two independent normal populations with unknown means μ_1 and μ_2, and unknown but equal variances, $\sigma_1^2 = \sigma_2^2 = \sigma^2$. We wish to test

$$H_0: \quad \mu_1 - \mu_2 = \Delta_0$$
$$H_1: \quad \mu_1 - \mu_2 \neq \Delta_0 \qquad (3\text{-}50)$$

Let $x_{11}, x_{12}, \ldots, x_{1n_1}$ be a random sample of n_1 observations from the first population and $x_{21}, x_{22}, \ldots, x_{2n_2}$ be a random sample of n_2 observations from the second population. Let $\bar{x}_1, \bar{x}_2, S_1^2, S_2^2$ be the sample means and sample variances, respectively. Now the expected value of the difference in sample means $\bar{x}_1 - \bar{x}_2$ is $E(\bar{x}_1 - \bar{x}_2) = \mu_1 - \mu_2$, so $\bar{x}_1 - \bar{x}_2$ is an unbiased estimator of the difference in means. The variance of $\bar{x}_1 - \bar{x}_2$ is

$$V(\bar{x}_1 - \bar{x}_2) = \frac{\sigma^2}{n_1} + \frac{\sigma^2}{n_2} = \sigma^2 \left(\frac{1}{n_1} + \frac{1}{n_2} \right)$$

It seems reasonable to combine the two sample variances S_1^2 and S_2^2 to form an estimator of σ^2. The **pooled estimator** of σ^2 is defined as follows.

The **pooled estimator** of σ^2, denoted by S_p^2, is defined by

$$S_p^2 = \frac{(n_1 - 1)S_1^2 + (n_2 - 1)S_2^2}{n_1 + n_2 - 2} \qquad (3\text{-}51)$$

It is easy to see that the pooled estimator S_p^2 can be written as

$$S_p^2 = \frac{n_1 - 1}{n_1 + n_2 - 2} S_1^2 + \frac{n_2 - 1}{n_1 + n_2 - 2} S_2^2$$
$$= wS_1^2 + (1 - w)S_2^2$$

where $0 < w \leq 1$. Thus S_p^2 is a **weighted average** of the two sample variances S_1^2 and S_2^2, where the weights w and $1 - w$ depend on the two sample sizes n_1 and n_2. Obviously, if $n_1 = n_2 = n$, then $w = 0.5$ and S_p^2 is simply the arithmetic average of S_1^2 and S_2^2. If $n_1 = 10$ and $n_2 = 20$ (say), then $w = 0.32$ and $1 - w = 0.68$. The first sample contributes $n_1 - 1$ degrees of freedom to S_p^2 and the second sample contributes $n_2 - 1$ degrees of freedom. Therefore, S_p^2 has $n_1 + n_2 - 2$ degrees of freedom.

Now we know that

$$Z = \frac{\bar{x}_1 - \bar{x}_2 - (\mu_1 - \mu_2)}{\sigma \sqrt{\dfrac{1}{n_1} + \dfrac{1}{n_2}}}$$

has a $N(0, 1)$ distribution. Replacing σ by S_p gives the following.

Given the assumptions of this section, the quantity

$$t = \frac{\bar{x}_1 - \bar{x}_2 - (\mu_1 - \mu_2)}{S_p \sqrt{\dfrac{1}{n_1} + \dfrac{1}{n_2}}} \qquad (3\text{-}52)$$

has a t distribution with $n_1 + n_2 - 2$ degrees of freedom.

The use of this information to test the hypotheses in equation 3-50 is now straight-forward: Simply replace $\mu_1 - \mu_2$ by Δ_0, and the resulting **test statistic** has a t distribution with $n_1 + n_2 - 2$ degrees of freedom under H_0: $\mu_1 - \mu_2 = \Delta_0$. The location of the critical region for both two-sided and one-sided alternatives parallels those in the one-sample case.

The Two-Sample Pooled t-Test[1]

Null hypothesis: H_0: $\mu_1 - \mu_2 = \Delta_0$

Test statistic: $$t_0 = \frac{\bar{x}_1 - \bar{x}_2 - \Delta_0}{S_p \sqrt{\dfrac{1}{n_1} + \dfrac{1}{n_2}}} \qquad (3\text{-}53)$$

Alternative Hypothesis	Rejection Criterion
H_1: $\mu_1 - \mu_2 \neq \Delta_0$	$t_0 > t_{\alpha/2, n_1 + n_2 - 2}$ or
	$t_0 < -t_{\alpha/2, n_1 + n_2 - 2}$
H_1: $\mu_1 - \mu_2 > \Delta_0$	$t_0 > t_{\alpha, n_1 + n_2 - 2}$
H_1: $\mu_1 - \mu_2 < \Delta_0$	$t_0 < -t_{\alpha, n_1 + n_2 - 2}$

---

[1] Although we have given the development of this procedure for the case where the sample sizes could be different, there is an advantage to using equal sample sizes $n_1 = n_2 = n$. When the sample sizes are the same from both populations, the t-test is very robust to the assumption of equal variances.

······   **EXAMPLE 3-10**   ···

Two catalysts are being analyzed to determine how they affect the mean yield of a chemical process. Specifically, catalyst 1 is currently in use, but catalyst 2 is acceptable. Since catalyst 2 is cheaper, it should be adopted, providing it does not change the process yield. A test is run in the pilot plant and results in the data shown in Table 3-3. Is there any difference between the mean yields? Use $\alpha = 0.05$. Assume equal variances.

The hypotheses are

$$H_0: \quad \mu_1 = \mu_2$$
$$H_1: \quad \mu_1 \neq \mu_2$$

From Table 3-3 we have $\bar{x}_1 = 92.255$, $S_1 = 2.39$, $n_1 = 8$, $\bar{x}_2 = 92.733$, $S_2 = 2.98$, and $n_2 = 8$. Therefore,

$$S_p^2 = \frac{(n_1 - 1)S_1^2 + (n_2 - 1)S_2^2}{n_1 + n_2 - 2} = \frac{(7)(2.39)^2 + 7(2.98)^2}{8 + 8 - 2} = 7.30$$

$$S_p = \sqrt{7.30} = 2.70$$

and

$$t_0 = \frac{\bar{x}_1 - \bar{x}_2}{2.70 \sqrt{\dfrac{1}{n_1} + \dfrac{1}{n_2}}} = \frac{92.255 - 92.733}{2.70 \sqrt{\dfrac{1}{8} + \dfrac{1}{8}}} = -0.35$$

Because $t_0 = -2.145 < -0.35 < 2.145$, the null hypothesis cannot be rejected. That is, at the 0.05 level of significance, we do not have strong evidence to conclude that

Table 3-3 Catalyst Yield Data, Example 3-10

Observation Number	Catalyst 1	Catalyst 2
1	91.50	89.19
2	94.18	90.95
3	92.18	90.46
4	95.39	93.21
5	91.79	97.19
6	89.07	97.04
7	94.72	91.07
8	89.21	92.75
	$\bar{x}_1 = 92.255$	$\bar{x}_2 = 92.733$
	$S_1 = 2.39$	$S_2 = 2.98$

catalyst 2 results in a mean yield that differs from the mean yield when catalyst 1 is used.

....

A *P*-value could also be used for decision making in this example. The actual value is $P = 0.7315$. (This value was obtained from a hand-held calculator.) Therefore, since the *P*-value exceeds $\alpha = 0.05$, the null hypothesis cannot be rejected.

Case 2: $\sigma_1^2 \neq \sigma_2^2$
In some situations, we cannot reasonably assume that the unknown variances σ_1^2 and σ_2^2 are equal. There is not an exact *t*-statistic available for testing $H_0: \mu_1 - \mu_2 = \Delta_0$ in this case. However, if $H_0: \mu_1 - \mu_2 = \Delta_0$ is true, then the statistic

$$t_0^* = \frac{\bar{x}_1 - \bar{x}_2 - \Delta_0}{\sqrt{\dfrac{S_1^2}{n_1} + \dfrac{S_2^2}{n_2}}} \tag{3-54}$$

is distributed approximately as *t* with degrees of freedom given by

$$v = \frac{\left(\dfrac{S_1^2}{n_1} + \dfrac{S_2^2}{n_2}\right)^2}{\dfrac{(S_1^2/n_1)^2}{n_1 + 1} + \dfrac{(S_2^2/n_2)^2}{n_2 + 1}} - 2 \tag{3-55}$$

Therefore, if $\sigma_1^2 \neq \sigma_2^2$, the hypotheses on differences in the means of two normal distributions are tested as in the equal variances case, except that t_0^* is used as the test statistic and $n_1 + n_2 - 2$ is replaced by v in determining the degrees of freedom for the test.

Confidence Interval on the Difference on Means, Variances Unknown

Case 1: $\sigma_1^2 = \sigma_2^2 = \sigma^2$
If $\bar{x}_1, \bar{x}_2, S_1^2$, and S_2^2 are the means and variances of two random samples of sizes n_1 and n_2, respectively, from two independent normal populations with unknown but

equal variances, then a $100(1 - \alpha)\%$ confidence interval on the difference in means $\mu_1 - \mu_2$ is

$$\bar{x}_1 - \bar{x}_2 - t_{\alpha/2,n_1+n_2-2}\, S_p \sqrt{\frac{1}{n_1} + \frac{1}{n_2}}$$

$$\leq \mu_1 - \mu_2 \leq \bar{x}_1 - \bar{x}_2 + t_{\alpha/2,n_1+n_2-2}\, S_p \sqrt{\frac{1}{n_1} + \frac{1}{n_2}} \quad (3\text{-}56)$$

where $S_p = \sqrt{[(n_1 - 1)S_1^2 + (n_2 - 1)S_2^2]/(n_1 + n_2 - 2)}$ is the pooled estimate of the common population standard deviation, and $t_{\alpha/2,n_1+n_2-2}$ is the upper $\alpha/2$ percentage point of the t distribution with $n_1 + n_2 - 2$ degrees of freedom.

Case 2: $\sigma_1^2 \neq \sigma_2^2$
If $\bar{x}_1, \bar{x}_2, S_1^2$, and S_2^2 are the means and variances of two random samples of sizes n_1 and n_2, respectively, from two independent normal populations with unknown and unequal variances, then an approximate $100(1 - \alpha)\%$ confidence interval on the difference in means $\mu_1 - \mu_2$ is

$$\bar{x}_1 - \bar{x}_2 - t_{\alpha/2,\nu}\sqrt{\frac{S_1^2}{n_1} + \frac{S_2^2}{n_2}} \leq \mu_1 - \mu_2 \leq \bar{x}_1 - \bar{x}_2 + t_{\alpha/2,\nu}\sqrt{\frac{S_1^2}{n_1} + \frac{S_2^2}{n_2}}$$

$$(3\text{-}57)$$

where ν is given by equation 3-55 and $t_{\alpha/2,\nu}$ is the upper $\alpha/2$ percentage point of the t distribution with ν degrees of freedom.

······· **EXAMPLE 3-11** ··

An article in the journal *Hazardous Waste and Hazardous Materials* (Vol. 6, 1989) reported the results of an analysis of the weight of calcium in standard cement and cement doped with lead. Reduced levels of calcium would indicate that the hydration mechanism in the cement is blocked and would allow water to attack various locations in the cement structure. Ten samples of standard cement had an average weight percent calcium of $\bar{x}_1 = 90.0$, with a sample standard deviation of $S_1 = 5.0$, and 15 samples of the lead-doped cement had an average weight percent calcium of $\bar{x}_2 = 87.0$, with a sample standard deviation of $S_2 = 4.0$.

We will assume that weight percent calcium is normally distributed and find a 95% confidence interval on the difference in means, $\mu_1 - \mu_2$, for the two types of cement. Furthermore, we will assume that both normal populations have the same standard deviation.

The pooled estimate of the common standard deviation is found using equation 3-51 as follows:

$$S_p^2 = \frac{(n_1 - 1)S_1^2 + (n_2 - 1)S_2^2}{n_1 + n_2 - 2}$$

$$= \frac{9(5.0)^2 + 14(4.0)^2}{10 + 15 - 2}$$

$$= 19.52$$

Therefore, the pooled standard deviation estimate is $S_p = \sqrt{19.75} = 4.4$. The 95% confidence interval is found using equation 3-56:

$$\bar{x}_1 - \bar{x}_2 - t_{0.025,23} S_p \sqrt{\frac{1}{n_1} + \frac{1}{n_2}} \leq \mu_1 - \mu_2 \leq \bar{x}_1 - \bar{x}_2 + t_{0.025,23} S_p \sqrt{\frac{1}{n_1} + \frac{1}{n_2}}$$

or upon substituting the sample values and using $t_{0.025,23} = 2.069$,

$$90.0 - 87.0 - 2.069(4.4) \sqrt{\frac{1}{10} + \frac{1}{15}} \leq \mu_1 - \mu_2$$

$$\leq 90.0 - 87.0 + 2.069(4.4) \sqrt{\frac{1}{10} + \frac{1}{15}}$$

which reduces to

$$-0.72 \leq \mu_1 - \mu_2 \leq 6.72$$

Note that the 95% confidence interval includes zero; therefore, at this level of confidence we cannot conclude that there is a difference in the means. Put another way, there is no evidence that doping the cement with lead affected the mean weight percent of calcium; therefore, we cannot claim that the presence of lead affects this aspect of the hydration mechanism at the 95% level of confidence.

....

Computer Solution

The two-sample t-test can be performed using most statistics software packages. Table 3-4 presents the output from the Minitab two-sample t-test routine for the catalyst yield data in Example 3-10. The output includes summary statistics for each sample, confidence intervals on the difference in means, and the hypothesis testing results. This analysis was performed assuming equal variances. Minitab has an option to perform the analysis assuming unequal variances. The confidence levels and α-value may be

Table 3-4   Minitab Two-Sample *t*-Test Output for Example 3-10

Two-Sample T-Test and CI: Catalyst 1, Catalyst 2

```
Two-sample T for Catalyst 1 vs Catalyst 2
              N    Mean    StDev    SE Mean
Catalyst 1    8    92.26   2.39     0.84
Catalyst 2    8    92.73   2.98     1.1
Difference = mu Catalyst 1 – mu Catalyst 2
Estimate for difference: –0.48
95% CI for difference: (–3.39, 2.44)
T–Test of difference = 0 (vs not = ): T–value = –0.35
   P–Value = 0.729   DF = 14
```

specified by the user. The hypothesis testing procedure indicates that we cannot reject the hypothesis that the mean yields are equal, which agrees with the conclusions we reached originally in Example 3-10.

Figure 3-12 shows comparative box plots for the yield data for the two types of catalysts in Example 3-10. These comparative box plots indicate that there is no obvious difference in the median of the two samples, although the second sample has a slightly larger sample dispersion or variance. There are no exact rules for comparing two samples with box plots; their primary value is in the visual impression they provide as a tool for explaining the results of a hypothesis test, as well as in verification of assumptions.

Figure 3-13 presents a Minitab normal probability plot of the two samples in Example 3-10. Note that both samples plot approximately along straight lines, and the straight lines for each sample have similar slopes. Therefore, we conclude that the normality and equal variances assumptions are reasonable.

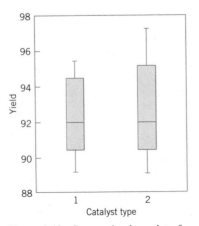

Figure 3-12 Comparative box plots for the catalyst yield data in Example 3-10.

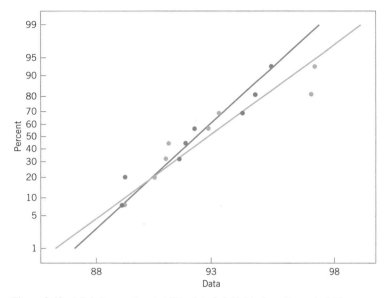

Figure 3-13   Minitab normal probability plot of yield data from Example 3-10.

Paired Data

It should be emphasized that we have assumed that the two samples used in the above tests are independent. In some applications, *paired* data are encountered. Observations in an experiment are often paired to prevent extraneous factors from inflating the estimate of the variance; hence, this method can be used to improve the precision of comparisons between means. For a further discussion of paired data, see Montgomery (1997) or Montgomery and Runger (1999). The analysis of such a situation is illustrated in the following example.

······ **EXAMPLE 3-12** ···

Two different types of machines are used to measure the tensile strength of synthetic fiber. We wish to determine whether or not the two machines yield the same average tensile strength values. Eight specimens of fiber are randomly selected, and one strength measurement is made using each machine on each specimen. The coded data are shown in Table 3-5.

The data in this experiment have been paired to prevent the difference between fiber specimens (which could be substantial) from affecting the test on the difference between machines. The test procedure consists of obtaining the differences of the pair of observations on each of the n specimens—say, $d_j = x_{1j} - x_{2j}$, $j = 1, 2, \ldots, n$—and then testing the hypothesis that the mean of the difference μ_d is zero. Note that testing H_0: $\mu_d = 0$ is equivalent to testing H_0: $\mu_1 = \mu_2$; furthermore, the test on μ_d is simply the one-sample t-test discussed in Section 3-3.3. The test statistic is

$$t_0 = \frac{\bar{d}}{S_d/\sqrt{n}}$$

Table 3-5   Paired Tensile Strength Data for Example 3-12

Specimen	Machine 1	Machine 2	Difference
1	74	78	-4
2	76	79	-3
3	74	75	-1
4	69	66	3
5	58	63	-5
6	71	70	1
7	66	66	0
8	65	67	-2

where

$$\bar{d} = \frac{1}{n} \sum_{j=1}^{n} d_j$$

and

$$S_d^2 = \frac{\sum_{j=1}^{n} (d_j - \bar{d})^2}{n-1} = \frac{\sum_{j=1}^{n} d_j^2 - \dfrac{\left(\sum_{j=1}^{n} d_j\right)^2}{n}}{n-1}$$

and $H_0: \mu_d = 0$ is rejected if $|t_0| > t_{\alpha/2, n-1}$.

In our example we find that

$$\bar{d} = \frac{1}{n} \sum_{j=1}^{n} d_j = \frac{1}{8} (-11) = -1.38$$

$$S_d^2 = \frac{\sum_{j=1}^{n} d_j^2 - \dfrac{\left(\sum_{j=1}^{n} d_j\right)^2}{n}}{n-1} = \frac{65 - \dfrac{(-11)^2}{8}}{7} = 7.13$$

Therefore, the test statistic is

$$t_0 = \frac{\bar{d}}{S_d/\sqrt{n}} = \frac{-1.38}{2.67/\sqrt{8}} = -1.46$$

Choosing $\alpha = 0.05$ results in $t_{0.025,7} = 2.365$, and we conclude that there is no strong evidence to indicate that the two machines differ in their mean tensile strength measurements (the P-value is $P = 0.1877$).

3-4.3   Inference on the Variances of Two Normal Distributions

Hypothesis Testing

Consider testing the hypothesis that the variances of two independent normal distributions are equal. If random samples of sizes n_1 and n_2 are taken from populations 1 and 2, respectively, then the test statistic for

$$H_0: \quad \sigma_1^2 = \sigma_2^2$$

$$H_1: \quad \sigma_1^2 \neq \sigma_2^2$$

is simply the ratio of the two sample variances,

$$F_0 = \frac{S_1^2}{S_2^2} \tag{3-58}$$

We would reject H_0 if $F_0 > F_{\alpha/2, n_1-1, n_2-1}$ or if $F_0 < F_{1-(\alpha/2), n_1-1, n_2-1}$, where $F_{(\alpha/2), n_1-1, n_2-1}$ and $F_{1-(\alpha/2), n_1-1, n_2-1}$ denote the upper $\alpha/2$ and lower $1 - (\alpha/2)$ percentage points of the F distribution with $n_1 - 1$ and $n_2 - 1$ degrees of freedom, respectively. The following display summarizes the test procedures for the one-sided alternative hypotheses.

Testing Hypotheses as $\sigma_1^2 = \sigma_2^2$ from Normal Distributions

Null hypothesis: $H_0: \sigma_1^2 = \sigma_2^2$

Alternative Hypothesis	Test Statistics	Rejection Criterion
$H_1: \sigma_1^2 < \sigma_2^2$	$F_0 = \dfrac{S_2^2}{S_1^2}$	$F_0 > F_{\alpha, n_2-1, n_1-1}$
$H_1: \sigma_1^2 > \sigma_2^2$	$F_0 = \dfrac{S_1^2}{S_2^2}$	$F_0 > F_{\alpha, n_1-1, n_2-1}$

Confidence Interval on the Ratio of the Variances
of Two Normal Distributions

Suppose that $x_1 \sim N(\mu_1, \sigma_1^2)$ and $x_2 \sim N(\mu_2, \sigma_2^2)$, where μ_1, σ_1^2, μ_2, and σ_2^2 are unknown, and we wish to construct a $100(1 - \alpha)\%$ confidence interval on σ_1^2/σ_2^2. If S_1^2

and S_2^2 are the sample variances, computed from random samples of n_1 and n_2 observations, respectively, then the $100(1 - \alpha)\%$ two-sided confidence interval is

$$\frac{S_1^2}{S_2^2} F_{1-\alpha/2,n_2-1,n_1-1} \leq \frac{\sigma_1^2}{\sigma_2^2} \leq \frac{S_1^2}{S_2^2} F_{\alpha/2,n_2-1,n_1-1} \tag{3-59}$$

where $F_{\alpha/2,u,v}$ is the percentage point of the F distribution with u and v degrees of freedom such that $P\{F_{u,v} \leq F_{\alpha/2,u,v}\} = \alpha/2$. The corresponding upper and lower confidence intervals are

$$\frac{\sigma_1^2}{\sigma_2^2} \leq \frac{S_1^2}{S_2^2} F_{\alpha,n_2-1,n_1-1} \tag{3-60}$$

and

$$\frac{S_1^2}{S_2^2} F_{1-\alpha,n_2-1,n_1-1} \leq \frac{\sigma_1^2}{\sigma_2^2} \tag{3-61}$$

respectively.[2]

3-4.4   Inference on Two Population Proportions

We now consider the case where there are two binomial parameters of interest—say, p_1 and p_2—and we wish to draw inferences about these proportions. We will present

---

[2] Appendix Table V gives only upper tail points of F; that is, $F_{\alpha,u,v}$. Lower tail points $F_{1-\alpha,u,v}$ may be found using the relationship $F_{1-\alpha,u,v} = 1/F_{\alpha,v,u}$.

large-sample hypothesis testing and confidence interval procedures based on the normal approximation to the binomial.

Large-Sample Test for H_0: $p_1 = p_2$

Suppose that the two independent random samples of sizes n_1 and n_2 are taken from two populations, and let x_1 and x_2 represent the number of observations that belong to the class of interest in samples 1 and 2, respectively. Furthermore, suppose that the normal approximation to the binomial is applied to each population, so that the estimators of the population proportions $\hat{p}_1 = x_1/n_1$ and $\hat{p}_2 = x_2/n_2$ and have approximate normal distributions. We are interested in testing the hypotheses

$$H_0: \quad p_1 = p_2$$
$$H_0: \quad p_1 \neq p_2$$

The statistic

$$Z = \frac{\hat{p}_1 - \hat{p}_2 - (p_1 - p_2)}{\sqrt{\dfrac{p_1(1 - p_1)}{n_1} + \dfrac{p_2(1 - p_2)}{n_2}}} \tag{3-62}$$

is distributed approximately as standard normal and is the basis of a test for H_0: $p_1 = p_2$. Specifically, if the null hypothesis H_0: $p_1 = p_2$ is true, then using the fact that $p_1 = p_2 = p$, the random variable

$$Z = \frac{\hat{p}_1 - \hat{p}_2}{\sqrt{p(1 - p)\left(\dfrac{1}{n_1} + \dfrac{1}{n_2}\right)}}$$

is distributed approximately $N(0, 1)$. An estimator of the common parameter p is

$$\hat{p} = \frac{x_1 + x_2}{n_1 + n_2}$$

The test statistic for H_0: $p_1 = p_2$ is then

$$Z_0 = \frac{\hat{p}_1 - \hat{p}_2}{\sqrt{\hat{p}(1 - \hat{p})\left(\dfrac{1}{n_1} + \dfrac{1}{n_2}\right)}}$$

This leads to the test procedures described here.

Null hypothesis: $H_0: p_1 = p_2$

Test statistic:
$$Z_0 = \frac{\hat{p}_1 - \hat{p}_2}{\sqrt{\hat{p}(1 - \hat{p})\left(\dfrac{1}{n_1} + \dfrac{1}{n_2}\right)}}$$ (3-63)

Alternative Hypotheses	**Rejection Criterion**
$H_1: p_1 \neq p_2$	$z_0 > z_{\alpha/2}$ or $z_0 < -z_{\alpha/2}$
$H_1: p_1 > p_2$	$z_0 > z_\alpha$
$H_1: p_1 < p_2$	$z_0 < -z_\alpha$

Confidence Interval on the Difference in Two Population Proportions
If there are two population proportions of interest—say, p_1 and p_2—it is possible to construct a $100(1 - \alpha)\%$ confidence interval on their difference. The confidence interval is as follows.

$$\hat{p}_1 - \hat{p}_2 - Z_{\alpha/2}\sqrt{\frac{\hat{p}_1(1 - \hat{p}_1)}{n_1} + \frac{\hat{p}_2(1 - \hat{p}_2)}{n_2}} \leq p_1 - p_2$$

$$\leq \hat{p}_1 - \hat{p}_2 + Z_{\alpha/2}\sqrt{\frac{\hat{p}_1(1 - \hat{p}_1)}{n_1} + \frac{\hat{p}_2(1 - \hat{p}_2)}{n_2}}$$ (3-64)

This result is based on the normal approximation to the binomial distribution.

3-5   WHAT IF THERE ARE MORE THAN TWO POPULATIONS? THE ANALYSIS OF VARIANCE

As this chapter has illustrated, testing and experimentation are a natural part of the engineering analysis process and arise often in quality control and engineering problems. Suppose, for example, that an engineer is investigating the effect of different heat-treating methods on the mean hardness of a steel alloy. The experiment would consist of testing several specimens of alloy using each of the proposed heat-treating methods and then measuring the hardness of each specimen. The data from this experiment could be used to determine which heat-treating method should be used to provide maximum mean hardness.

If there are only two heat-treating methods of interest, this experiment could be designed and analyzed using the two-sample *t*-test presented in this chapter. That is, the experimenter has a single **factor** of interest — heat-treating methods — and there are only two **levels** of the factor.

Many single-factor experiments require that more than two levels of the factor be considered. For example, the engineer may want to investigate five different heat-treating methods. In this chapter we show how the **analysis of variance** can be used for comparing means when there are more than two levels of a single factor. We will also discuss **randomization** of the experimental runs and the important role this concept plays in the overall experimentation strategy. In Part IV, we will discuss how to design and analyze experiments with several factors.

3-5.1 An Example

A manufacturer of paper used for making grocery bags is interested in improving the tensile strength of the product. Product engineering thinks that tensile strength is a function of the hardwood concentration in the pulp and that the range of hardwood concentrations of practical interest is between 5% and 20%. A team of engineers responsible for the study decides to investigate four levels of hardwood concentration: 5%, 10%, 15%, and 20%. They decide to make up six test specimens at each concentration level, using a pilot plant. All 24 specimens are tested on a laboratory tensile tester, in random order. The data from this experiment are shown in Table 3-6.

This is an example of a completely randomized single-factor experiment with four levels of the factor. The levels of the factor are sometimes called **treatments,** and each treatment has six observations or **replicates.** The role of **randomization** in this experiment is extremely important. By randomizing the order of the 24 runs, the effect of any nuisance variable that may influence the observed tensile strength is approximately balanced out. For example, suppose that there is a warm-up effect on the tensile testing machine; that is, the longer the machine is on, the greater the observed tensile strength.

Table 3-6 Tensile Strength of Paper (psi)

Hardwood Concentration (%)	Observations						Totals	Averages
	1	2	3	4	5	6		
5	7	8	15	11	9	10	60	10.00
10	12	17	13	18	19	15	94	15.67
15	14	18	19	17	16	18	102	17.00
20	19	25	22	23	18	20	127	21.17
							383	15.96

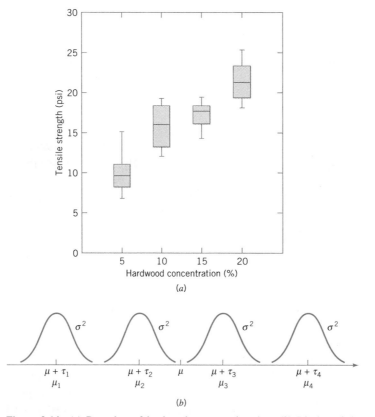

Figure 3-14 (*a*) Box plots of hardwood concentration data. (*b*) Display of the model in equation 3-65 for the completely randomized single-factor experiment.

If all 24 runs are made in order of increasing hardwood concentration (that is, all six 5% concentration specimens are tested first, followed by all six 10% concentration specimens, etc.), then any observed differences in tensile strength could also be due to the warm-up effect.

It is important to graphically analyze the data from a designed experiment. Figure 3-14*a* presents box plots of tensile strength at the four hardwood concentration levels. This figure indicates that changing the hardwood concentration has an effect on tensile strength; specifically, higher hardwood concentrations produce higher observed tensile strength. Furthermore, the distribution of tensile strength at a particular hardwood level is reasonably symmetric, and the variability in tensile strength does not change dramatically as the hardwood concentration changes.

Graphical interpretation of the data is always a good idea. Box plots show the variability of the observations *within* a treatment (factor level) and the variability *between* treatments. We now discuss how the data from a single-factor randomized experiment can be analyzed statistically.

Table 3-7   Typical Data for a Single-Factor Experiment

Treatment	Observations				Totals	Averages
1	y_{11}	y_{12}	$\cdots$	y_{1n}	$y_{1\cdot}$	$\bar{y}_{1\cdot}$
2	y_{21}	y_{22}	$\cdots$	y_{2n}	$y_{2\cdot}$	$\bar{y}_{2\cdot}$
.	.	.	$\cdots$	.	.	.
.	.	.	$\cdots$	.	.	.
.	.	.	$\cdots$	.	.	.
a	y_{a1}	y_{a2}	$\cdots$	y_{an}	$y_{a\cdot}$	$\bar{y}_{a\cdot}$
					$y_{\cdot\cdot}$	$\bar{y}_{\cdot\cdot}$

3-5.2   The Analysis of Variance

Suppose we have a different levels of a single factor that we wish to compare. Sometimes, each factor level is called a treatment, a very general term that can be traced to the early applications of experimental design methodology in the agricultural sciences. The response for each of the a treatments is a random variable. The observed data would appear as shown in Table 3-7. An entry in Table 3-7—say, y_{ij}—represents the jth observation taken under treatment i. We initially consider the case in which there are an equal number of observations, n, on each treatment.

We may describe the observations in Table 3-7 by the **linear statistical model**

$$Y_{ij} = \mu + \tau_i + \epsilon_{ij} \begin{cases} i = 1, 2, \ldots, a \\ j = 1, 2, \ldots, n \end{cases} \tag{3-65}$$

where Y_{ij} is a random variable denoting the (ij)th observation, μ is a parameter common to all treatments called the **overall mean**, τ_i is a parameter associated with the ith treatment called the ith **treatment effect,** and ϵ_{ij} is a random error component. Note that the model could have been written as

$$Y_{ij} = \mu_i + \epsilon_{ij} \begin{cases} i = 1, 2, \ldots, a \\ j = 1, 2, \ldots, n \end{cases}$$

where $\mu_i = \mu + \tau_i$ is the mean of the ith treatment. In this form of the model, we see that each treatment defines a population that has mean μ_i, consisting of the overall mean μ plus an effect τ_i that is due to that particular treatment. We will assume that the errors ϵ_{ij} are normally and independently distributed with mean zero and variance σ^2. Therefore, each treatment can be thought of as a normal population with mean μ_i and variance σ^2. See Fig. 3-14b.

Equation 3-65 is the underlying model for a single-factor experiment. Furthermore, since we require that the observations are taken in random order and that the environment (often called the experimental units) in which the treatments are used is as uniform as possible, this design is called a **experimental design.**

We now present the analysis of variance for testing the equality of a population means. This is called a **fixed effects model** analysis of variance. However, the analysis of variance is a far more useful and general technique; it will be used extensively in Chapters 12 and 13. In this section we show how it can be used to test for equality of treatment effects. The treatment effects τ_i are usually defined as deviations from the overall mean μ, so that

$$\sum_{i=1}^{a} \tau_i = 0 \tag{3-66}$$

Let $\bar{y}_i.$ represent the total of the observations under the ith treatment and $\bar{y}_i.$ represent the average of the observations under the ith treatment. Similarly, let $y..$ represent the grand total of all observations and $\bar{y}..$ represent the grand mean of all observations. Expressed mathematically,

$$y_i. = \sum_{j=1}^{n} y_{ij} \qquad \bar{y}_i. = y_i./n \qquad i = 1, 2, \ldots, a$$

$$y.. = \sum_{i=1}^{a} \sum_{j=1}^{n} y_{ij} \qquad \bar{y}.. = y../N \tag{3-67}$$

where $N = an$ is the total number of observations. Thus, the "dot" subscript notation implies summation over the subscript that it replaces.

We are interested in testing the equality of the a treatment means $\mu_1, \mu_2, \ldots, \mu_a$. Using equation 3-66, we find that this is equivalent to testing the hypotheses

$$H_0: \quad \tau_1 = \tau_2 = \cdots = \tau_a = 0$$
$$H_1: \quad \tau_i \neq 0 \quad \text{for at least one } i \tag{3-68}$$

Thus, if the null hypothesis is true, each observation consists of the overall mean μ plus a realization of the random error component ϵ_{ij}. This is equivalent to saying that all N observations are taken from a normal distribution with mean μ and variance σ^2. Therefore, if the null hypothesis is true, changing the levels of the factor has no effect on the mean response.

The analysis of variance partitions the total variability in the sample data into two component parts. Then, the test of the hypothesis in equation 3-68 is based on a comparison of two independent estimates of the population variance. The total variability in the data is described by the **total sum of squares**

$$SS_T = \sum_{i=1}^{a} \sum_{j=1}^{n} (y_{ij} - \bar{y}..)^2$$

The partition of the total sum of squares is given in the following definition.

The **sum of squares identity** is

$$\sum_{i=1}^{a} \sum_{j=1}^{n} (y_{ij} - \bar{y}..)^2 = n \sum_{i=1}^{a} (\bar{y}_i. - \bar{y}..)^2 + \sum_{i=1}^{a} \sum_{j=1}^{n} (y_{ij} - \bar{y}_i.)^2 \tag{3-69}$$

The proof of this identity is straightforward. Note that we may write

$$\sum_{i=1}^{a} \sum_{j=1}^{n} (y_{ij} - \bar{y}..)^2 = \sum_{i=1}^{a} \sum_{j=1}^{n} [(\bar{y}_{i.} - \bar{y}..) + (y_{ij} - \bar{y}_{i.})]^2$$

or

$$\sum_{i=1}^{a} \sum_{j=1}^{n} (y_{ij} - \bar{y}..)^2 = n \sum_{i=1}^{a} (\bar{y}_{i.} - \bar{y}..)^2 + \sum_{i=1}^{a} \sum_{j=1}^{n} (y_{ij} - \bar{y}_{i.})^2$$

$$+ 2 \sum_{i=1}^{a} \sum_{j=1}^{n} (\bar{y}_{i.} - \bar{y}..)(y_{ij} - \bar{y}_{i.}) \tag{3-70}$$

Note that the cross-product term in equation 3-70 is zero, since

$$\sum_{j=1}^{n} (y_{ij} - \bar{y}_{i.}) = y_{i.} - n\bar{y}_{i.} = y_{i.} - n(y_{i.}/n) = 0$$

Therefore, we have shown that equation 3-70 will reduce to equation 3-69.

The identity in equation 3-69 shows that the total variability in the data, measured by the total sum of squares, can be partitioned into a sum of squares of differences between treatment means and the grand mean and a sum of squares of differences of observations within a treatment from the treatment mean. Differences between observed treatment means and the grand mean measure the differences between treatments, whereas differences of observations within a treatment from the treatment mean can be due only to random error. Therefore, we write equation 3-69 symbolically as

$$SS_T = SS_{\text{Treatments}} + SS_E \tag{3-71}$$

where

$$SS_T = \sum_{i=1}^{a} \sum_{j=1}^{n} (y_{ij} - \bar{y}..)^2 = \text{total sum of squares}$$

$$SS_{\text{Treatments}} = n \sum_{i=1}^{a} (\bar{y}_{i.} - \bar{y}..)^2 = \text{treatment sum of squares}$$

and

$$SS_E = \sum_{i=1}^{a} \sum_{j=1}^{n} (y_{ij} - \bar{y}_{j.})^2 = \text{error sum of squares}$$

We can gain considerable insight into how the analysis of variance works by examining the expected values of $SS_{\text{Treatments}}$ and SS_E. This will lead us to an appropriate statistic for testing the hypothesis of no differences among treatment means (or $\tau_i = 0$).

The expected value of the treatment sum of squares is

$$E(SS_{\text{Treatments}}) = (a - 1)\sigma^2 + n \sum_{i=1}^{a} \tau_i^2$$

Now if the null hypothesis in equation 3-68 is true, each τ_i is equal to zero and

$$E\left(\frac{SS_{\text{Treatments}}}{a-1}\right) = \sigma^2$$

If the alternative hypothesis is true, then

$$E\left(\frac{SS_{\text{Treatments}}}{a-1}\right) = \sigma^2 + \frac{n\sum\limits_{i=1}^{a}\tau_i^2}{a-1}$$

The ratio $MS_{\text{Treatments}} = SS_{\text{Treatments}}/(a-1)$ is called the **mean square for treatments.** Thus, if H_0 is true, $MS_{\text{Treatments}}$ is an unbiased estimator of σ^2, whereas if H_1 is true, $MS_{\text{Treatments}}$ estimates σ^2 plus a positive term that incorporates variation due to the systematic difference in treatment means.

We can also show that the expected value of the error sum of squares is $E(SS_E) = a(n-1)\sigma^2$. Therefore, the **error mean square** $MS_E = SS_E/[a(n-1)]$ is an unbiased estimator of σ^2 regardless of whether or not H_0 is true.

The error mean square

$$MS_E = \frac{SS_E}{a(n-1)}$$

is an unbiased estimator of σ^2.

There is also a partition of the number of degrees of freedom that corresponds to the sum of squares identity in equation 3-69. That is, there are $an = N$ observations; thus, SS_T has $an-1$ degrees of freedom. There are a levels of the factor, so $SS_{\text{Treatments}}$ has $a-1$ degrees of freedom. Finally, within any treatment there are n replicates providing $n-1$ degrees of freedom with which to estimate the experimental error. Since there are a treatments, we have $a(n-1)$ degrees of freedom for error. Therefore, the degrees of freedom partition is

$$an - 1 = a - 1 + a(n-1)$$

Now assume that each of the a populations can be modeled as a normal distribution. Using this assumption we can show that if the null hypothesis H_0 is true, the ratio

$$F_0 = \frac{SS_{\text{Treatments}}/(a-1)}{SS_E/[a(n-1)]} = \frac{MS_{\text{Treatments}}}{MS_E} \tag{3-72}$$

has an F distribution with $a - 1$ and $a(n - 1)$ degrees of freedom. Furthermore, from the expected mean squares, we know that MS_E is an unbiased estimator of σ^2. Also, under the null hypothesis, $MS_{\text{Treatments}}$ is an unbiased estimator of σ^2. However, if the null hypothesis is false, then the expected value of $MS_{\text{Treatments}}$ is greater than σ^2. Therefore, under the alternative hypothesis, the expected value of the numerator of the test statistic (equation 3-72) is greater than the expected value of the denominator. Consequently, we should reject H_0 if the statistic is large. This implies an upper-tail, one-tail critical region. Therefore, we would reject H_0 if $F_0 > F_{\alpha, a-1, a(n-1)}$ where F_0 is computed from equation 3-72.

Efficient computational formulas for the sums of squares may be obtained by expanding and simplifying the definitions of $SS_{\text{Treatments}}$ and SS_T. This yields the following results.

Definition

The sums of squares computing formulas for the analysis of variance with equal sample sizes in each treatment are

$$SS_T = \sum_{i=1}^{a} \sum_{j=1}^{n} y_{ij}^2 - \frac{y_{..}^2}{N} \tag{3-73}$$

and

$$SS_{\text{Treatments}} = \sum_{i=1}^{a} \frac{y_{i.}^2}{n} - \frac{y_{..}^2}{N} \tag{3-74}$$

The error sum of squares is obtained by subtraction as

$$SS_E = SS_T - SS_{\text{Treatments}} \tag{3-75}$$

The computations for this test procedure are usually summarized in tabular form as shown in Table 3-8. This is called an **analysis of variance table.**

······· EXAMPLE 3-13 ··

Consider the paper tensile strength experiment described in Section 3-5.1. We can use the analysis of variance to test the hypothesis that different hardwood concentrations do not affect the mean tensile strength of the paper.

The hypotheses are

$$H_0: \quad \tau_1 = \tau_2 = \tau_3 = \tau_4 = 0$$
$$H_1: \quad \tau_i \neq 0 \text{ for at least one } i$$

Table 3-8   The Analysis of Variance for a Single-Factor Experiment

Source of Variation	Sum of Squares	Degrees of Freedom	Mean Square	F_0
Treatments	$SS_{\text{Treatments}}$	$a - 1$	$MS_{\text{Treatments}}$	$\dfrac{MS_{\text{Treatments}}}{MS_E}$
Error	SS_E	$a(n - 1)$	MS_E	
Total	SS_T	$an - 1$		

We will use $\alpha = 0.01$. The sums of squares for the analysis of variance are computed from equations 3-73, 3-74, and 3-75 as follows:

$$SS_T = \sum_{i=1}^{4} \sum_{j=1}^{6} y_{ij}^2 - \frac{y_{..}^2}{N}$$

$$= (7)^2 + (8)^2 + \cdots + (20)^2 - \frac{(383)^2}{24} = 512.96$$

$$SS_{\text{Treatments}} = \sum_{i=1}^{4} \frac{y_{i.}^2}{n} - \frac{y_{..}^2}{N}$$

$$= \frac{(60)^2 + (94)^2 + (102)^2 + (127)^2}{6} - \frac{(383)^2}{24} = 382.79$$

$$SS_E = SS_T - SS_{\text{Treatments}}$$

$$= 512.96 - 382.79 = 130.17$$

We usually do not perform these calculations by hand. The analysis of variance computed by Minitab is presented in Table 3-9. Since $F_{0.01,3,20} = 4.94$, we reject H_0 and conclude that hardwood concentration in the pulp significantly affects the strength of the paper. Note that the computer output reports a P-value for the test statistic $F = 19.61$ in Table 3-9 of 0. This is a truncated value; the actual P-value is $P = 3.59 \times 10^{-6}$. However, since the P-value is considerably smaller than $\alpha = 0.01$, we have strong evidence to conclude that H_0 is not true. Note that Minitab also provides some summary information about each level of hardwood concentration, including the confidence interval on each mean.

Finally, note that the analysis of variance tells us whether there is a difference among means. It does not tell us which means differ. If the analysis of variance indicates that there is a statistically significant difference among means, there is a simple graphical procedure that can be used to isolate the specific differences. Suppose that $\bar{y}_{1.}, \bar{y}_{2.}, \ldots, \bar{y}_{a.}$ are the observed averages for these factor levels. Each treatment average has standard deviation $\sigma/\sqrt{n}$, where σ is the standard deviation of an individual observation. If all treatment means are equal, the observed means $\bar{y}_{i.}$ would behave as if they were a set of observations drawn at random from a normal distribution with mean μ and standard deviation $\sigma/\sqrt{n}$.

Table 3-9 Minitab Analysis of Variance Output for the Paper Tensile Strength Experiment

```
                    One-Way  Analysis  of  Variance

Analysis  of  Variance
Source      DF        SS         MS        F         P
Factor       3     382.79     127.60    19.61     0.000
Error       20     130.17       6.51
Total       23     512.96
                                      Individual  95%  CIs  For  Mean
                                      Based  on  Pooled  StDev
Level       N       Mean      StDev   -----+---------+---------+---------+-
  5         6     10.000      2.828        (---*---)
 10         6     15.667      2.805                 (---*---)
 15         6     17.000      1.789                    (---*---)
 20         6     21.167      2.639                           (---*---)
                                      -----+---------+---------+---------+-
Pooled  StDev = 2.551                  10.0      15.0      20.0      25.0
```

Visualize this normal distribution capable of being slid along an axis below which the treatment means $\bar{y}_1., \bar{y}_2., \ldots, \bar{y}_a.$ are plotted. If all treatment means are equal, there should be some position for this distribution that makes it obvious that the $\bar{y}_i.$ values were drawn from the same distribution. If this is not the case, then the $\bar{y}_i.$ values that do not appear to have been drawn from this distribution are associated with treatments that produce different mean responses.

The only flaw in this logic is that σ is unknown. However, we can use $\sqrt{MS_E}$ from the analysis of variance to estimate σ. This implies that a t distribution should be used instead of the normal in making the plot, but since the t looks so much like the normal, sketching a normal curve that is approximately $6\sqrt{MS_E/n}$ units wide will usually work very well.

Figure 3-15 shows this arrangement for the hardwood concentration experiment in Section 3-5.1. The standard deviation of this normal distribution is

$$\sqrt{MS_E/n} = \sqrt{6.51/6} = 1.04$$

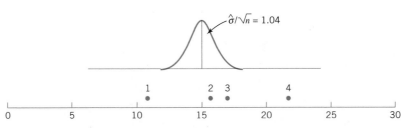

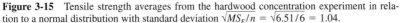

Figure 3-15 Tensile strength averages from the hardwood concentration experiment in relation to a normal distribution with standard deviation $\sqrt{MS_E/n} = \sqrt{6.51/6} = 1.04$.

If we visualize sliding this distribution along the horizontal axis, we note that there is no location for the distribution that would suggest that all four observations (the plotted means) are typical, randomly selected values from that distribution. This, of course, should be expected, because the analysis of variance has indicated that the means differ, and the display in Fig. 3-15 is simply a graphical representation of the analysis of variance results. The figure does indicate that treatment 4 (20% hardwood) produces paper with higher mean tensile strength than do the other treatments, and treatment 1 (5% hardwood) results in lower mean tensile strength than do the other treatments. The means of treatments 2 and 3 (10 and 15% hardwood, respectively) do not differ.

This simple procedure is a rough but very useful and effective technique for comparing means following an analysis of variance. For more details on these procedures, see Montgomery (1997).

3-5.3   Checking Assumptions: Residual Analysis

The analysis of variance assumes that the model errors (and as a result, the observations) are normally and independently distributed with the same variance in each factor level. These assumptions can be checked by examining the residuals. We define a residual as the difference between the actual observation y_{ij} and the value $\hat{y}_{ij}$ that would be obtained from a least squares fit of the underlying analysis of variance model to the sample data. For the type of experimental design in this situation, the value $\hat{y}_{ij}$ is the factor-level mean $\bar{y}_{i.}$. Therefore, the residual is $e_{ij} = y_{ij} - \bar{y}_{i.}$; that is, the difference between an observation and the corresponding factor-level mean. The residuals for the hardwood percentage experiment are shown in Table 3-10.

The normality assumption can be checked by constructing a normal probability plot of the residuals. To check the assumption of equal variances at each factor level, plot the residuals against the factor levels and compare the spread in the residuals. It is also useful to plot the residuals against $\bar{y}_{i.}$ (sometimes called the **fitted value**); the variability in the residuals should not depend in any way on the value of $\bar{y}_{i.}$. When a pattern appears in these plots, it usually suggests the need for data **transformation**—that is, analyzing the data in a different metric. For example, if the variability in the residuals increases with $\bar{y}_{i.}$, then a transformation such as $\log y$ or $\sqrt{y}$ should be considered. In some problems the dependency of residual scatter on $\bar{y}_{i.}$ is very important information. It may be

Table 3-10   Residuals for the Hardwood Experiment

Hardwood Concentration	Residuals					
5%	−3.00	−2.00	5.00	1.00	1.00	0.00
10%	−3.37	1.33	−2.67	2.33	−3.33	0.67
15%	−3.00	1.00	2.00	0.00	−1.00	1.00
20%	−2.17	3.83	0.83	1.83	−3.17	−1.17

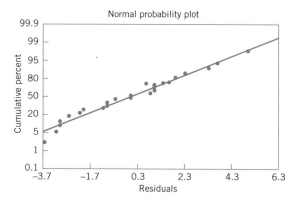

Figure 3-16 Normal probability plot of residuals from the hardwood concentration experiment.

desirable to select the factor level that results in maximum mean response; however, this level may also cause more variation in response from run to run.

The independence assumption can be checked by plotting the residuals against the run order in which the experiment was performed. A pattern in this plot, such as sequences of positive and negative residuals, may indicate that the observations are not independent. This suggests that run order is important or that variables that change over time are important and have not been included in the experimental design.

A normal probability plot of the residuals from the hardwood concentration experiment is shown in Fig. 3-16. Figures 3-17 and 3-18 present the residuals plotted against the factor levels and the fitted value $\bar{y}_i$. These plots do not reveal any model inadequacy or unusual problem with the assumptions.

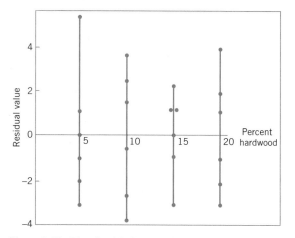

Figure 3-17 Plot of residuals versus factor levels.

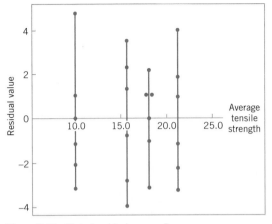

Figure 3-18   Plot of residuals versus $\bar{y}_{i\,\cdot\cdot}$.

3-6 EXERCISES

3-1.   The inside diameters of bearings used in an aircraft landing gear assembly are known to have a standard deviation of $\sigma = 0.002$ cm. A random sample of 15 bearings has an average inside diameter of 8.2535 cm.

(a) Test the hypothesis that the mean inside bearing diameter is 8.25 cm. Use a two-sided alternative and $\alpha = 0.05$.

(b) Find the P-value for this test.

(c) Construct a 95% two-sided confidence interval on mean bearing diameter.

3-2.   The tensile strength of a fiber used in manufacturing cloth is of interest to the purchaser. Previous experience indicates that the standard deviation of tensile strength is 2 psi. A random sample of eight fiber specimens is selected, and the average tensile strength is found to be 127 psi.

(a) Test the hypothesis that the mean tensile strength equals 125 psi versus the alternative that the mean exceeds 125 psi. Use $\alpha = 0.05$.

(b) What is the P-value for this test?

(c) Discuss why a one-sided alternative was chosen in part (a).

(d) Construct a 95% lower confidence interval on the mean tensile strength.

3-3.   The life of a battery used in a cardiac pacemaker is assumed to be normally distributed. A random sample of 10 batteries is subjected to an accelerated life test by running them continuously at an elevated temperature until failure, and the following lives are obtained.

25.5 h	26.1 h
26.8	23.2
24.2	28.4
25.0	27.8
27.3	25.7

(a) The manufacturer wants to be certain that the mean battery life exceeds 25 h. What conclusions can be drawn from these data (use $\alpha = 0.05$)?

(b) Construct a 90% two-sided confidence interval on mean life in the accelerated test.

(c) Construct a normal probability plot of the battery life data. What conclusions can you draw?

3-4. Using the data from Exercise 3-3, construct a 95% lower confidence interval on mean battery life. Why would the manufacturer be interested in a one-sided confidence interval?

3-5. A new process has been developed for applying photoresist to 125-mm silicon wafers used in manufacturing integrated circuits. Ten wafers were tested, and the photoresist thickness measurements shown here were observed:

13.3946 ($\times$ 1000 angstroms)	13.4002 ($\times$ 1000 angstroms)
13.3987	13.3957
13.3902	13.4015
13.4001	13.3918
13.3965	13.3925

(a) Test the hypothesis that mean thickness is 13.4×1000 Å. Use $\alpha = 0.05$ and assume a two-sided alternative.

(b) Find a 99% two-sided confidence interval on mean photoresist thickness. Assume that thickness is normally distributed.

(c) Does the normality assumption seem reasonable for these data?

3-6. A machine is used to fill containers with a liquid product. Fill volume can be assumed to be normally distributed. A random sample of 10 containers is selected, and the net contents are shown here.

12.03 oz	12.01 oz
12.04	12.02
12.05	11.98
11.96	12.02
12.05	11.99

(a) Suppose that the manufacturer wants to be sure that the mean net contents exceeds 12 oz. What con-

clusions can be drawn from the data (use $\alpha = 0.01$)?

(b) Construct a 95% two-sided confidence interval on the mean fill volume.

(c) Does the assumption of normality seem appropriate for the fill volume data?

3-7. Ferric chloride is used as a flux in some types of extraction metallurgy processes. This material is shipped in containers, and the container weight varies. It is important to obtain an accurate estimate of mean container weight. Suppose that from long experience a reliable value for the standard deviation of flux container weight is determined to be 4 lb. How large a sample would be required to construct a 95% two-sided confidence interval on the mean that has a total width of 1 lb?

3-8. The diameters of aluminum alloy rods produced on an extrusion machine are known to have a standard deviation of 0.0001 in. A random sample of 25 rods has an average diameter of 0.5046 in.

(a) Test the hypothesis that mean rod diameter is 0.5025 in. Assume a two-sided alternative and use $\alpha = 0.05$.

(b) Find the P-value for this test.

(c) Construct a 95% two-sided confidence interval on the mean rod diameter.

3-9. The output voltage of a power supply is assumed to be normally distributed. Sixteen observations taken at random on voltage are shown here.

10.35	9.30	10.00	9.96
11.65	12.00	11.25	9.58
11.54	9.95	10.28	8.37
10.44	9.25	9.38	10.85

(a) Test the hypothesis that the mean voltage equals 12 V against a two-sided alternative using $\alpha = 0.05$.

(b) Construct a 95% two-sided confidence interval on μ.

(c) Test the hypothesis that $\sigma^2 = 11$ using $\alpha = 0.05$.

(d) Construct a 95% two-sided confidence interval on σ.

(e) Construct a 95% upper confidence interval on σ.

(f) Does the assumption of normality seem reasonable for the output voltage?

3-10. Two machines are used for filling glass bottles with a soft-drink beverage. The filling processes have known standard deviations $\sigma_1 = 0.010$ liter and $\sigma_2 = 0.015$ liter, respectively. A random sample of $n_1 = 25$ bottles from machine 1 and $n_2 = 20$ bottles from machine 2 results in average net contents of $\bar{x}_1 = 2.04$ liters and $\bar{x}_2 = 2.07$ liters.

(a) Test the hypothesis that both machines fill to the same net contents, using $\alpha = 0.05$. What are your conclusions?

(b) Find the P-value for this test.

(c) Construct a 95% confidence interval on the difference in mean fill volume.

3-11. Two quality control technicians measured the surface finish of a metal part, obtaining the data shown. Assume that the measurements are normally distributed.

Technician 1	Technician 2
1.45	1.54
1.37	1.41
1.21	1.56
1.54	1.37
1.48	1.20
1.29	1.31
1.34	1.27
	1.35

(a) Test the hypothesis that the mean surface finish measurements made by the two technicians are equal. Use $\alpha = 0.05$, and assume equal variances.

(b) What are the practical implications of the test in part (a)? Discuss what practical conclusions you would draw if the null hypothesis were rejected.

(c) Assuming that the variances are equal, construct a 95% confidence interval on the mean difference in surface-finish measurements.

(d) Test the hypothesis that the variances of the measurements made by the two technicians are equal. Use $\alpha = 0.05$. What are the practical implications if the null hypothesis is rejected?

(e) Construct a 95% confidence interval estimate of the ratio of the variances of technician measurement error.

(f) Construct a 95% confidence interval on the variance of measurement error for technician 2.

(g) Does the normality assumption seem reasonable for the data?

3-12. Suppose that $x_1 \sim N(\mu_1, \sigma_1^2)$ and $x_2 \sim N(\mu_2, \sigma_2^2)$, and that x_1 and x_2 are independent. Develop a procedure for constructing a $100(1 - \alpha)\%$ confidence interval on $\mu_1 - \mu_2$, assuming that σ_1^2 and σ_2^2 are unknown and cannot be assumed equal.

3-13. Two different hardening processes, (1) saltwater quenching and (2) oil quenching, are used on samples of a particular type of metal alloy. The results are shown here. Assume that hardness is normally distributed.

Saltwater Quench	Oil Quench
145	152
150	150
153	147
148	155
141	140

Saltwater Quench	Oil Quench
152	146
146	158
154	152
139	151
148	143

(a) Test the hypothesis that the mean hardness for the saltwater quenching process equals the mean hardness for the oil quenching process. Use $\alpha = 0.05$ and assume equal variances.

(b) Assuming that the variances σ_1^2 and σ_2^2 are equal, construct a 95% confidence interval on the difference in mean hardness.

(c) Construct a 95% confidence interval on the ratio σ_1^2/σ_2^2. Does the assumption made earlier of equal variances seem reasonable?

(d) Does the assumption of normality seem appropriate for these data?

3-14. A random sample of 200 printed circuit boards contains 18 defective or nonconforming units. Estimate the process fraction nonconforming.

(a) Test the hypothesis that the true fraction nonconforming in this process is 0.10. Use $\alpha = 0.05$. Find the P-value.

(b) Construct a 90% two-sided confidence interval on the true fraction nonconforming in the production process.

3-15. A random sample of 500 connecting rod pins contains 65 nonconforming units. Estimate the process fraction nonconforming.

(a) Test the hypothesis that the true fraction defective in this process is 0.08. Use $\alpha = 0.05$.

(b) Find the P-value for this test.

(c) Construct a 95% upper confidence interval on the true process fraction nonconforming.

3-16. Two processes are used to produce forgings used in an aircraft wing assembly. Of 200 forgings selected from process 1, 10 do not conform to the strength specifications, whereas of 300 forgings selected from process 2, 20 are nonconforming.

(a) Estimate the fraction nonconforming for each process.

(b) Test the hypothesis that the two processes have identical fractions nonconforming. Use $\alpha = 0.05$.

(c) Construct a 90% confidence interval on the difference in fraction nonconforming between the two processes.

3-17. A new purification unit is installed in a chemical process. Before its installation, a random sample yielded the following data about the percentage of impurity: $\bar{x}_1 = 9.85$, $S_1^2 = 81.73$, and $n_1 = 10$. After installation, a random sample resulted in $\bar{x}_2 = 8.08$, $S_2^2 = 78.46$, and $n_2 = 8$.

(a) Can you conclude that the two variances are equal? Use $\alpha = 0.05$.

(b) Can you conclude that the new purification device has reduced the mean percentage of impurity? Use $\alpha = 0.05$.

3-18. Two different types of glass bottles are suitable for use by a soft-drink beverage bottler. The internal pressure strength of the bottle is an important quality characteristic. It is known that $\sigma_1 = \sigma_2 = 3.0$ psi. From a random sample of $n_1 = n_2 = 16$ bottles, the mean pressure strengths are observed to be $\bar{x}_1 = 175.8$ psi and $\bar{x}_2 = 181.3$ psi. The company will not use bottle design 2 unless its pressure strength exceeds that of bottle design 1 by at least 5 psi. Based on the sample data, should they use bottle design 2 if we use $\alpha = 0.05$? What is the P-value for this test?

3-19. The diameter of a metal rod is measured by 12 inspectors, each using both a micrometer caliper and a vernier caliper. The results are shown here. Is there a difference between the mean measurements produced by the two types of caliper? Use $\alpha = 0.01$.

Inspector	Micrometer Caliper	Vernier Caliper
1	0.150	0.151
2	0.151	0.150
3	0.151	0.151
4	0.152	0.150
5	0.151	0.151
6	0.150	0.151
7	0.151	0.153
8	0.153	0.155
9	0.152	0.154
10	0.151	0.151
11	0.151	0.150
12	0.151	0.152

3-20. The cooling system in a nuclear submarine consists of an assembly pipe through which a coolant is circulated. Specifications require that weld strength must meet or exceed 150 psi.

(a) Suppose the designers decide to test the hypothesis H_0: $\mu = 150$ versus H_1: $\mu > 150$. Explain why this choice of alternative is preferable to H_1: $\mu < 150$.

(b) A random sample of 20 welds results in $\bar{x} = 153.7$ psi and $S = 11.5$ psi. What conclusions can you draw about the hypothesis in part (a)? Use $\alpha = 0.05$.

3-21. An experiment was conducted to investigate the filling capability of packaging equipment at a winery in Newberg, Oregon. Twenty bottles of Pinot Gris were randomly selected and the fill volume (in ml) measured. Assume that fill volume has a normal distribution. The data are as follows:

753	751	752	753	753
753	752	753	754	754
752	751	752	750	753
755	753	756	751	750

(a) Do the data support the claim that the standard deviation of fill volume is less than 1 ml? Use $\alpha = 0.05$.

(b) Find a 95% two-sided confidence interval on the standard deviation of fill volume.

(c) Does it seem reasonable to assume that fill volume has a normal distribution?

3-22. Suppose we wish to test the hypotheses

$$H_0: \quad \mu = 15$$
$$H_1: \quad \mu \neq 15$$

where we know that $\sigma^2 = 9.0$. If the true mean is really 20, what sample size must be used to ensure that the probability of type II error is no greater than 0.10? Assume that $\alpha = 0.05$.

3-23. Consider the hypotheses

$$H_0: \quad \mu = \mu_0$$
$$H_1: \quad \mu \neq \mu_0$$

where σ^2 is known. Derive a general expression for determining the sample size for detecting a true mean of $\mu_1 \neq \mu_0$ with probability $1 - \beta$ if the type I error is α.

3-24. **Sample size allocation.**   Suppose we are testing the hypotheses

$$H_0: \quad \mu_1 = \mu_2$$
$$H_1: \quad \mu_1 \neq \mu_2$$

where σ_1^2 and σ_2^2 are known. Resources are limited, and consequently the total sample size $n_1 + n_2 = N$. How should we allocate the N observations between the two populations to obtain the most powerful test?

3-25. Develop a test for the hypotheses

$$H_0: \quad \mu_1 = 2\mu_2$$
$$H_1: \quad \mu_1 \neq 2\mu_2$$

where σ_1^2 and σ_2^2 are known.

3-26. Nonconformities occur in glass bottles according to a Poisson distribution. A random sample of 100 bottles contains a total of 11 nonconformities.
  (a) Develop a procedure for testing the hypothesis that the mean of a Poisson distribution λ equals a specified value λ_0. Hint: Use the normal approximation to the Poisson.
  (b) Use the results of part (a) to test the hypothesis that the mean occurrence rate of nonconformities is $\lambda = 0.15$. Use $\alpha = 0.01$.

3-27. An inspector counts the surface-finish defects in dishwashers. A random sample of five dishwashers contains three such defects. Is there reason to conclude that the mean occurrence rate of surface-finish defects per dishwasher exceeds 0.5? Use the results of part (a) of Exercise 3-26 and assume that $\alpha = 0.05$.

3-28. An in-line tester is used to evaluate the electrical function of printed circuit boards. This machine counts the number of defects observed on each board. A random sample of 1000 boards contains a total of 688 defects. Is it reasonable to conclude that the mean occurrence rate of defects is $\lambda = 1$? Use the results of part (a) of Exercise 3-26 and assume that $\alpha = 0.05$.

3-29. An article in *Solid State Technology* (May 1987) describes an experiment to determine the effect of C_2F_6 flow rate on etch uniformity on a silicon wafer used in integrated-circuit manufacturing. Three flow rates are tested, and the resulting uniformity (in percent) is observed for six test units at each flow rate. The data are shown in the following table.

C_2F_6 Flow (SCCM)	Observations					
	1	2	3	4	5	6
125	2.7	2.6	4.6	3.2	3.0	3.8
160	4.6	4.9	5.0	4.2	3.6	4.2
200	4.6	2.9	3.4	3.5	4.1	5.1

  (a) Does C_2F_6 flow rate affect etch uniformity? Answer this question by using an analysis of variance with $\alpha = 0.05$.
  (b) Construct a box plot of the etch uniformity data. Use this plot, together with the analysis of variance results, to determine which gas flow rate would be best in terms of etch uniformity (a small percentage is best).
  (c) Plot the residuals versus predicted C_2F_6 flow. Interpret this plot.
  (d) Does the normality assumption seem reasonable in this problem?

3-30. Compare the mean etch uniformity values at each of the C_2F_6 flow rates from Exercise 3-29 with a scaled t distribution. Does this analysis indicate that there are differences in mean etch uniformity at the different flow rates? Which flows produce different results?

3-31. An article in the *ACI Materials Journal* (Vol. 84, 1987, pp. 213–216) describes several experiments investigating the rodding of concrete to remove entrapped air. A 3-in. diameter cylinder was used, and the number of times this rod was used is the design variable. The resulting compressive strength of the concrete specimen is the response. The data are shown in the following table.

Rodding Level	Compressive Strength		
10	1530	1530	1440
15	1610	1650	1500
20	1560	1730	1530
25	1500	1490	1510

(a) Is there any difference in compressive strength due to the rodding level? Answer this question by using the analysis of variance with $\alpha = 0.05$.

(b) Construct box plots of compressive strength by rodding level. Provide a practical interpretation of these plots.

(c) Construct a normal probability plot of the residuals from this experiment. Does the assumption of a normal distribution for compressive strength seem reasonable?

3-32.   Compare the mean compressive strength at each rodding level from Exercise 3-31 with a scaled t distribution. What conclusions would you draw from this plot?

3-33.   An aluminum producer manufactures carbon anodes and bakes them in a ring furnace prior to use in the smelting operation. The baked density of the anode is an important quality characteristic, as it may affect anode life. One of the process engineers suspects that firing temperature in the ring furnace affects baked anode density. An experiment was run at four different temperature levels, and six anodes were baked at each temperature level. The data from the experiment follow.

Temperature (°C)	Density					
500	41.8	41.9	41.7	41.6	41.5	41.7
525	41.4	41.3	41.7	41.6	41.7	41.8
550	41.2	41.0	41.6	41.9	41.7	41.3
575	41.0	40.6	41.8	41.2	41.9	41.5

(a) Does firing temperature in the ring furnace affect mean baked anode density?

(b) Find the residuals for this experiment and plot them on a normal probability scale. Comment on the plot.

(c) What firing temperature would you recommend using?

3-34.   Plot the residuals from Exercise 3-33 against the firing temperatures. Is there any indication that variability in baked anode density depends on the firing temperature? What firing temperature would you recommend using?

3-35.   An article in *Environmental International* (Vol. 18, No. 4, 1992) describes an experiment in which the amount of radon released in showers was investigated. Radon-enriched water was used in the experiment, and six different orifice diameters were tested in shower heads. The data from the experiment are shown in the following table.

Orifice Diameter	Radon Released (%)			
0.37	80	83	83	85
0.51	75	75	79	79
0.71	74	73	76	77
1.02	67	72	74	74
1.40	62	62	67	69
1.99	60	61	64	66

(a) Does the size of the orifice affect the mean percentage of radon released? Use the analysis of variance and $\alpha = 0.05$.

(b) Analyze the results from this experiment.

3-36. An article in the *Journal of the Electrochemical Society* (Vol. 139, No. 2, 1992, pp. 524–532) describes an experiment to investigate the low-pressure vapor deposition of polysilicon. The experiment was carried out in a large-capacity reactor at Sematech in Austin, Texas. The reactor has several wafer positions, and four of these positions are selected at random. The response variable is film thickness uniformity. Three replicates of the experiment were run, and the data are as follows.

Wafer Position	Uniformity		
1	2.76	5.67	4.49
2	1.43	1.70	2.19
3	2.34	1.97	1.47
4	0.94	1.36	1.65

(a) Is there a difference in the wafer positions? Use the analysis of variance and $\alpha = 0.05$.

(b) Estimate the variability due to wafer positions.

(c) Estimate the random error component.

(d) Analyze the residuals from this experiment and comment on model adequacy.

PART

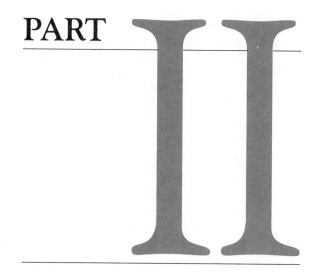

II

Basic Methods of
Statistical Process
Control and
Capability Analysis

It is impossible to inspect or test quality into a product; the product must be built right the first time. This implies that the manufacturing processing must be stable and that all individuals involved with the process (including operators, engineers, quality-assurance personnel, and management) must continuously seek to improve process performance and reduce variability in key parameters. On-line **statistical process control** (SPC) is a primary tool for achieving this objective. Control charts are the simplest type of on-line statistical process-control procedure. Chapters 4 through 7 present many of the basic SPC techniques, concentrating primarily on the type of control chart proposed by Dr. Walter A. Shewhart and called the **Shewhart control chart.**

Chapter 4 is an introduction to the general methodology of statistical process control. This chapter describes several fundamental SPC problem-solving tools, including an introduction to the Shewhart control chart. A discussion of how to implement SPC is given, along with some comments on SPC in the nonmanufacturing environment. Chapter 5 introduces Shewhart control charts for measurement data, sometimes called **variables control charts.** The $\bar{x}$ and R control charts are discussed in detail, along with several important variations of these charts. Chapter 6 presents Shewhart control charts for **attribute data,** such as a fraction defective or nonconforming, nonconformities (defects), or nonconformities per unit of product. Chapter 7 explores **process-capability analysis;** that is, how control charts and other statistical techniques can be used to estimate the natural capability of a process and to determine how it will perform relative to specifications on the product. Some aspects of setting specifications and tolerances, including the tolerance "stack-up" problem, are also presented.

Throughout this section we stress the three fundamental uses of a control chart:

1. Reduction of process variability
2. Monitoring and surveillance of a process
3. Estimation of product or process parameters

Methods and
Philosophy of
Statistical Process
Control

CHAPTER OUTLINE

4-1 INTRODUCTION

4-2 CHANCE AND ASSIGNABLE
    CAUSES OF QUALITY
    VARIATION

4-3 STATISTICAL BASIS OF THE
    CONTROL CHART

    4-3.1 Basic Principles

    4-3.2 Choice of Control Limits

    4-3.3 Sample Size and Sampling
          Frequency

    4-3.4 Rational Subgroups

    4-3.5 Analysis of Patterns on Control
          Charts

    4-3.6 Discussion of Sensitizing Rules for
          Control Charts

4-4 THE REST OF THE
    "MAGNIFICENT SEVEN"

4-5 IMPLEMENTING SPC

4-6 AN APPLICATION OF SPC

4-7 NONMANUFACTURING
    APPLICATIONS OF STATISTICAL
    PROCESS CONTROL

CHAPTER OVERVIEW

This chapter has three objectives. The first is to present the basic SPC problem-solving tools, called the "magnificent seven," and to illustrate how these tools form a cohesive, practical framework for quality improvement. The second objective is to describe the statistical basis of the Shewhart control chart. The reader will see how decisions about sample size, sampling interval, and placement of control limits affect the performance of a control chart. Other key concepts include the idea of rational subgroups, interpretation of control chart signals and patterns, and the average run length as a measure of control chart performance. The third objective is to discuss and illustrate some practical issues in the implementation of SPC.

4-1   INTRODUCTION

If a product is to meet customer requirements, generally it should be produced by a process that is stable or repeatable. More precisely, the process must be capable of operating with little variability around the target or nominal dimensions of the product's quality characteristics. **Statistical process control (SPC)** is a powerful collection of problem-solving tools useful in achieving process stability and improving capability through the reduction of variability.

SPC can be applied to *any* process. Its seven major tools are

1. Histogram or stem-and-leaf display
2. Check sheet
3. Pareto chart
4. Cause-and-effect diagram
5. Defect concentration diagram
6. Scatter diagram
7. Control chart

Although these tools, often called **"the magnificent seven,"** are an important part of SPC, they comprise only its technical aspects. SPC builds an environment in which all individuals in an organization desire continuous improvement in quality and productivity. This environment is best developed when management becomes involved in an ongoing quality-improvement process. Once this environment is established, routine application of the magnificent seven becomes part of the usual manner of doing business, and the organization is well on its way to achieving its quality-improvement objectives.

In this chapter we will present an overview of the magnificent seven. Of these tools, the Shewhart control chart is probably the most technically sophisticated. It was developed in the 1920s by Dr. Walter A. Shewhart of the Bell Telephone Laboratories. To understand the statistical concepts that form the basis of SPC, we must first describe Shewhart's theory of variability.

4-2   CHANCE AND ASSIGNABLE CAUSES OF QUALITY VARIATION

In any production process, regardless of how well designed or carefully maintained it is, a certain amount of inherent or natural variability will always exist. This natural variability or "background noise" is the cumulative effect of many small, essentially unavoidable causes. In the framework of statistical quality control, this natural variability is often called a "stable system of chance causes." A process that is operating with only **chance causes of variation** present is said to be **in statistical control.** In other words, the chance causes are an inherent part of the process.

Other kinds of variability may occasionally be present in the output of a process. This variability in key quality characteristics usually arises from three sources: improp-

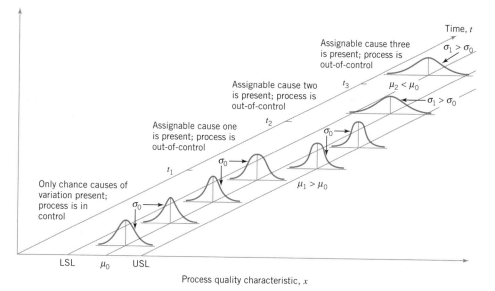

Figure 4-1   Chance and assignable causes of variation.

erly adjusted or controlled machines, operator errors, or defective raw material. Such variability is generally large when compared to the background noise, and it usually represents an unacceptable level of process performance. We refer to these sources of variability that are not part of the chance cause pattern as "assignable causes." A process that is operating in the presence of **assignable causes** is said to be **out of control.**[1]

These chance and assignable causes of variation are illustrated in Fig. 4-1. Until time t_1 the process shown in this figure is in control; that is, only chance causes of variation are present. As a result, both the mean and standard deviation of the process are at their in-control values (say, μ_0 and σ_0). At time t_1 an assignable cause occurs. As shown in Fig. 4-1, the effect of this assignable cause is to shift the process mean to a new value $\mu_1 > \mu_0$. At time t_2 another assignable cause occurs, resulting in $\mu = \mu_0$, but now the process standard deviation has shifted to a larger value $\sigma_1 > \sigma_0$. At time t_3 there is another assignable cause present, resulting in both the process mean and standard deviation taking on out-of-control values. From time t_1 forward, the presence of assignable causes has resulted in an out-of-control process.

Often production processes will operate in the in-control state, producing acceptable product for relatively long periods of time. Eventually, however, assignable causes will occur, seemingly at random, resulting in a "shift" to an out-of-control state where a larger proportion of the process output does not conform to requirements. For example, note from Fig. 4-1 that when the process is in control, most of the production will fall between the lower and upper specification limits (LSL and USL, respectively). When

---

[1] The terminology **chance** and **assignable causes** was developed by Dr. Walter A. Shewhart. Today, some writers use the terminology **common cause** instead of **chance cause** and **special cause** instead of **assignable cause.**

the process is out of control, a higher proportion of the process lies outside of these specifications.

A major objective of statistical process control is to quickly detect the occurrence of assignable causes of process shifts so that investigation of the process and corrective action may be undertaken before many nonconforming units are manufactured. The control chart is an on-line process-monitoring technique widely used for this purpose. Control charts may also be used to estimate the parameters of a production process, and, through this information, to determine process capability. The control chart may also provide information useful in improving the process. Finally, remember that the eventual goal of statistical process control is the **elimination of variability in the process.** It may not be possible to completely eliminate variability, but the control chart is an effective tool in reducing variability as much as possible.

We now present the statistical concepts that form the basis of control charts. Chapters 5 and 6 develop the details of construction and use of the standard types of control charts.

4-3 STATISTICAL BASIS OF THE CONTROL CHART

4-3.1 Basic Principles

A typical control chart is shown in Fig. 4-2, which is a graphical display of a quality characteristic that has been measured or computed from a sample versus the sample number or time. The chart contains a **center line** that represents the average value of the quality characteristic corresponding to the in-control state. (That is, only chance causes are present.) Two other horizontal lines, called the **upper control limit** (UCL) and the **lower control limit** (LCL), are also shown on the chart. These control limits are chosen so that if the process is in control, nearly all of the sample points will fall between them. As long as the points plot within the control limits, the process is assumed to be in

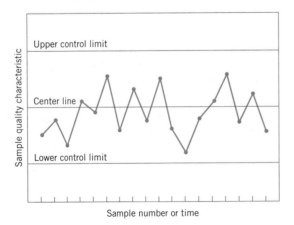

Figure 4-2   A typical control chart.

control, and no action is necessary. However, a point that plots outside of the control limits is interpreted as evidence that the process is out of control, and investigation and corrective action are required to find and eliminate the assignable cause or causes responsible for this behavior. It is customary to connect the sample points on the control chart with straight-line segments, so that it is easier to visualize how the sequence of points has evolved over time.

Even if all the points plot inside the control limits, if they behave in a systematic or nonrandom manner, then this could be an indication that the process is out of control. For example, if 18 of the last 20 points plotted above the center line but below the upper control limit and only two of these points plotted below the center line but above the lower control limit, we would be very suspicious that something was wrong. If the process is in control, all the plotted points should have an essentially random pattern. Methods for looking for sequences or nonrandom patterns can be applied to control charts as an aid in detecting out-of-control conditions. Usually, there is a reason why a particular nonrandom pattern appears on a control chart, and if it can be found and eliminated, process performance can be improved. This topic is discussed further in Sections 4-3.5 and 5-2.4.

There is a close connection between control charts and hypothesis testing. To illustrate this connection, suppose that the vertical axis in Fig. 4-2 is the sample average $\bar{x}$. Now, if the current value of $\bar{x}$ plots between the control limits, we conclude that the process mean is in control; that is, it is equal to some value μ_0. On the other hand, if $\bar{x}$ exceeds either control limit, we conclude that the process mean is out of control; that is, it is equal to some value $\mu_1 \neq \mu_0$. In a sense, then, the control chart is a test of the hypothesis that the process is in a state of statistical control. A point plotting within the control limits is equivalent to failing to reject the hypothesis of statistical control, and a point plotting outside the control limits is equivalent to rejecting the hypothesis of statistical control. This hypothesis testing framework is useful in many ways, but there are some differences in viewpoint between control charts and hypothesis tests. For example, when testing statistical hypotheses, we usually check the validity of assumptions, whereas control charts are used to detect departures from an assumed state of statistical control. Furthermore, the assignable cause can result in many different types of shifts in the process parameters. For example, the mean could shift instantaneously to a new value and remain there (this is sometimes called a *sustained* shift); or it could shift abruptly but the assignable cause could be short lived and the mean could then return to its nominal or in-control value; or the assignable cause could result in a steady drift or trend in the value of the mean. Only the sustained shift fits nicely within the usual statistical hypothesis testing model.

One place where the hypothesis testing framework is useful is in analyzing the **performance** of a control chart. For example, we may think of the probability of type I error of the control chart (concluding the process is out of control when it is really in control) and the probability of type II error of the control chart (concluding the process is in control when it is really out of control). It is occasionally helpful to use the operating-characteristic curve of a control chart to display its probability of type II error. This would be an indication of the ability of the control chart to detect process shifts of different magnitudes.

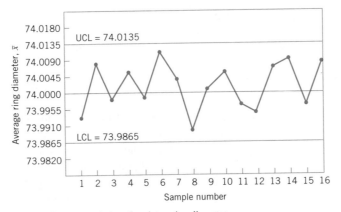

Figure 4-3 $\bar{x}$ control chart for piston-ring diameter.

To illustrate the preceding ideas, we give an example of a control chart. In the manufacture of automotive engine piston rings, a critical quality characteristic is the inside diameter of the ring. The process can be controlled at a mean inside ring diameter of 74 mm, and it is known that the standard deviation of ring diameter is 0.01 mm. A control chart for the average ring diameter is shown in Fig. 4-3. Every hour a random sample of five rings is taken, the average ring diameter of the sample (say, $\bar{x}$) computed, and $\bar{x}$ plotted on the chart. Because this control chart utilizes the sample average $\bar{x}$ to monitor the process mean, it is usually called an $\bar{x}$ control chart. Note that all the points fall within the control limits, so the chart indicates that the process is in statistical control.

To assist in understanding the statistical basis of this control chart, consider how the control limits were determined. The process mean is 74 mm, and the process standard deviation is $\sigma = 0.01$ mm. Now if samples of size $n = 5$ are taken, the standard deviation of the sample average $\bar{x}$ is

$$\sigma_{\bar{x}} = \frac{\sigma}{\sqrt{n}} = \frac{0.01}{\sqrt{5}} = 0.0045$$

Therefore, if the process is in control with a mean diameter of 74 mm, then by using the central limit theorem to assume that $\bar{x}$ is approximately normally distributed, we would expect $100(1 - \alpha)\%$ of the sample mean diameters $\bar{x}$ to fall between $74 + Z_{\alpha/2}(0.0045)$ and $74 - Z_{\alpha/2}(0.0045)$. We will arbitrarily choose the constant $Z_{\alpha/2}$ to be 3, so that the upper and lower control limits become

$$UCL = 74 + 3(0.0045) = 74.0135$$

and

$$LCL = 74 - 3(0.0045) = 73.9865$$

as shown on the control chart. These are typically called **"three-sigma"**[2] **control limits.**

---

[2]Note that "sigma" refers to the standard deviation of the statistic plotted on the chart (i.e., $\sigma_{\bar{x}}$), *not* the standard deviation of the quality characteristic.

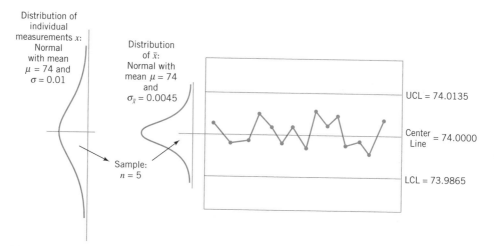

Figure 4-4   How the control chart works.

The width of the control limits is inversely proportional to the sample size n for a given multiple of sigma. Note that choosing the control limits is equivalent to setting up the critical region for testing the hypothesis

$$H_0: \quad \mu = 74$$
$$H_1: \quad \mu \neq 74$$

where $\sigma = 0.01$ is known. Essentially, the control chart tests this hypothesis repeatedly at different points in time. The situation is illustrated graphically in Fig. 4-4.

We may give a general **model** for a control chart. Let w be a sample statistic that measures some quality characteristic of interest, and suppose that the mean of w is μ_w and the standard deviation of w is σ_w. Then the center line, the upper control limit, and the lower control limit become

$$\text{UCL} = \mu_w + L\sigma_w$$
$$\text{Center line} = \mu_w \qquad\qquad (4\text{-}1)$$
$$\text{LCL} = \mu_w - L\sigma_w$$

where L is the "distance" of the control limits from the center line, expressed in standard deviation units. This general theory of control charts was first proposed by Dr. Walter S. Shewhart, and control charts developed according to these principles are often called **Shewhart control charts.**

The control chart is a device for describing in a precise manner exactly what is meant by statistical control; as such, it may be used in a variety of ways. In many applications, it is used for on-line process surveillance. That is, sample data are collected and

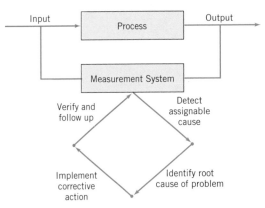

Figure 4-5   Process improvement using the control chart.

used to construct the control chart, and if the sample values of $\bar{x}$ (say) fall within the control limits and do not exhibit any systematic pattern, we say the process is in control at the level indicated by the chart. Note that we may be interested here in determining *both* whether the past data came from a process that was in control and whether future samples from this process indicate statistical control.

The most important use of a control chart is to **improve** the process. We have found that, generally,

1. Most processes do not operate in a state of statistical control.
2. Consequently, the routine and attentive use of control charts will identify assignable causes. If these causes can be eliminated from the process, variability will be reduced and the process will be improved.

This process improvement activity using the control chart is illustrated in Fig. 4-5. Note that

3. The control chart will only **detect** assignable causes. Management, operator, and engineering **action** will usually be necessary to eliminate the assignable causes.

In identifying and eliminating assignable causes, it is important to find the underlying **root cause** of the problem and to attack it. A cosmetic solution will not result in any real, long-term process improvement. Developing an effective system for corrective action is an essential component of an effective SPC implementation.

A very important part of the corrective action process associated with control chart usage is the **Out-Of-Control-Action-Plan** or **OCAP**. An OCAP is a flow chart or text-based description of the sequence of activities that must take place following the occurrence of an *activating event*. These are usually out-of-control signals from the control chart. The OCAP consists of *checkpoints*, which are potential assignable causes, and *terminators*, which are actions taken to resolve the out-of-control condition hopefully by eliminating the assignable cause. It is very important that the OCAP specify as complete a set as possible of checkpoints and terminators, and that these be arranged in an order that facilitates process diagnostic activities. Often analysis of prior failure modes of the process and/or product can be helpful in designing this aspect of the OCAP. Furthermore, an OCAP is a *living document* in the sense that it will be modified over time as more knowledge and understanding of the process is gained. Consequently, when a control chart is introduced, an initial OCAP should accompany it. Control charts without an OCAP are not likely to be very useful as a process improvement tool.

We may also use the control chart as an **estimating device.** That is, from a control chart that exhibits statistical control, we may estimate certain process parameters, such as the mean, standard deviation, fraction nonconforming or fallout, and so forth. These estimates may then be used to determine the **capability** of the process to produce acceptable products. Such **process-capability studies** have considerable impact on many management decision problems that occur over the product cycle, including make or buy decisions, plant and process improvements that reduce process variability, and contractual agreements with customers or vendors regarding product quality.

Control charts may be classified into two general types. If the quality characteristic can be measured and expressed as a number on some continuous scale of measurement, it is usually called a **variable.** In such cases, it is convenient to describe the quality characteristic with a measure of central tendency and a measure of variability. Control charts for central tendency and variability are collectively called **variables control charts.** The $\bar{x}$ chart is the most widely used chart for controlling central tendency, whereas charts based on either the sample range or the sample standard deviation are used to control process variability. Control charts for variables are discussed in Chapter 5. Many quality characteristics are not measured on a continuous scale or even a quantitative scale. In these cases, we may judge each unit of product as either conforming or nonconforming on the basis of whether or not it possesses certain attributes, or we may count the number of nonconformities (defects) appearing on a unit of product. Control charts for such quality characteristics are called **attributes control charts** and are discussed in Chapter 6.

An important factor in control chart usage is the **design of the control chart.** By this we mean the selection of the sample size, control limits, and frequency of sampling. For example, in the $\bar{x}$ chart of Fig. 4-3, we specified a sample size of five measurements, three-sigma control limits, and the sampling frequency to be every hour. In most quality control problems, it is customary to design the control chart using primarily statistical considerations. For example, we know that increasing the sample size will decrease the probability of type II error, thus enhancing the chart's ability to detect an out-of-control state, and so forth. The use of statistical criteria such as these along with industrial experience has led to general guidelines and procedures for designing control charts. These procedures usually consider cost factors only in an implicit manner. Recently, however, we have begun to examine control chart design from an **economic** point of view, considering explicitly the cost of sampling, losses from allowing defective product to be produced, and the costs of investigating out-of-control signals that are really "false alarms."

Another important consideration in control chart usage is the **type of variability** exhibited by the process. Figure 4-6 presents data from three different processes.

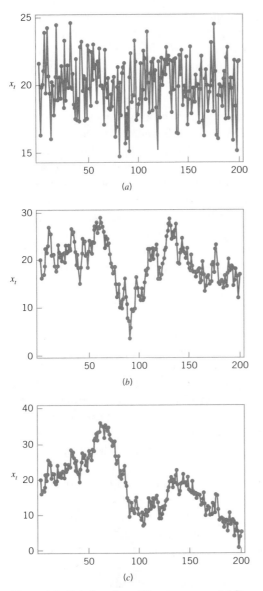

Figure 4-6   Data from three different processes. (*a*) Stationary and uncorrelated (white noise). (*b*) Stationary and autocorrelated. (*c*) Nonstationary.

Figures 4-6*a* and 4-6*b* illustrate **stationary behavior.** By this we mean that the process data vary around a fixed mean in a stable or predictable manner. This is the type of behavior that Shewhart implied was produced by an **in-control process.**

Even a cursory examination of Figs. 4-6*a* and 4-6*b* reveals some important differences. The data in Fig. 4-6*a* are **uncorrelated;** that is, the observations give the appearance of having been drawn at random from a stable population, perhaps a normal distribution. This type of data is referred to by time series analysts as **white noise.** (Time series analysis is a field of statistics devoted exclusively to studying and modeling time-oriented data.) In this type of process, the order in which the data occur does not tell us much that is useful to analyze the process. In other words, the past values of the data are of no help in predicting any of the future values.

Figure 4-6b illustrates stationary but **autocorrelated** process data. Notice that successive observations in these data are **dependent;** that is, a value above the mean tends to be followed by another value above the mean, whereas a value below the mean is usually followed by another such value. This produces a data series that has a tendency to move in moderately long "runs" on either side of the mean.

Figure 4-6c illustrates **nonstationary** variation. This type of process data occurs frequently in the chemical and process industries. Note that the process is very unstable in that it drifts or "wanders" about without any sense of a stable or fixed mean. In many industrial settings, we stabilize this type of behavior by using **engineering process control** (such as **feedback control**). This approach to process control is required when there are factors that affect the process that cannot be stabilized, such as environmental variables or properties of raw materials. When the control scheme is effective, the process output will *not* look like Fig. 4-6c but will hopefully resemble either Fig. 4-6a or 4-6b.

Shewhart control charts are most effective when the in-control process data look like Fig. 4-6a. By this we mean that the charts can be designed so that their performance is predictable and reasonable to the user, and that they are effective in reliably detecting out-of-control conditions. Most of our discussion of control charts in this chapter and in chapters 5 and 6 will assume that the in-control process data are stationary and uncorrelated.

With some modifications, Shewhart control charts and other types of control charts can be applied to autocorrelated data. We discuss this in more detail in Part III of the book. We also discuss feedback control and the use of SPC in systems where feedback control is employed in Part III.

Control charts have had a long history of use in U.S. industries and in many offshore industries as well. There are at least five reasons for their popularity.

1. **Control charts are a proven technique for improving productivity.**   A successful control chart program will reduce scrap and rework, which are the primary productivity killers in *any* operation. If you reduce scrap and rework, then productivity increases, cost decreases, and production capacity (measured in the number of *good* parts per hour) increases.

2. **Control charts are effective in defect prevention.**   The control chart helps keep the process in control, which is consistent with the "do it right the first time" philosophy. It is never cheaper to sort out "good" units from "bad" units later on than it is to build it right initially. If you do not have effective process control, you are paying someone to make a nonconforming product.

3. **Control charts prevent unnecessary process adjustment.**   A control chart can distinguish between background noise and abnormal variation; no other device including a human operator is as effective in making this distinction. If

process operators adjust the process based on periodic tests unrelated to a control chart program, they will often overreact to the background noise and make unneeded adjustments. These unnecessary adjustments can actually result in a deterioration of process performance. In other words, the control chart is consistent with the "if it isn't broken, don't fix it" philosophy.

4. **Control charts provide diagnostic information.** Frequently, the pattern of points on the control chart will contain information of diagnostic value to an experienced operator or engineer. This information allows the implementation of a change in the process that improves its performance.

5. **Control charts provide information about process capability.** The control chart provides information about the value of important process parameters and their stability over time. This allows an estimate of process capability to be made. This information is of tremendous use to product and process designers.

Control charts are among the most important management control tools; they are as important as cost controls and material controls. Modern computer technology has made it easy to implement control charts in *any* type of process, as data collection and analysis can be performed on a microcomputer or a local area network terminal in real-time, on-line at the work center. Some additional guidelines for implementing a control chart program are given at the end of Chapter 6.

4-3.2   Choice of Control Limits

Specifying the control limits is one of the critical decisions that must be made in designing a control chart. By moving the control limits farther from the center line, we decrease the risk of a type I error—that is, the risk of a point falling beyond the control limits, indicating an out-of-control condition when no assignable cause is present. However, widening the control limits will also increase the risk of a type II error—that is, the risk of a point falling between the control limits when the process is really out of control. If we move the control limits closer to the center line, the opposite effect is obtained: The risk of type I error is increased, while the risk of type II error is decreased.

For the $\bar{x}$ chart shown in Fig. 4-3, where three-sigma control limits were used, if we assume that the piston-ring diameter is normally distributed, we find from the standard normal table that the probability of type I error is 0.0027. That is, an incorrect out-of-control signal or false alarm will be generated in only 27 out of 10,000 points. Furthermore, the probability that a point taken when the process is in control will exceed the three-sigma limits in one direction only is 0.00135. Instead of specifying the control limit as a multiple of the standard deviation of $\bar{x}$, we could have directly chosen the type I error probability and calculated the corresponding control limit. For example, if we specified a 0.001 type I error probability in one direction, then the appropriate multiple of the standard deviation would be 3.09. The control limits for the $\bar{x}$ chart would then be

$$UCL = 74 + 3.09(0.0045) = 74.0139$$
$$LCL = 74 - 3.09(0.0045) = 73.9861$$

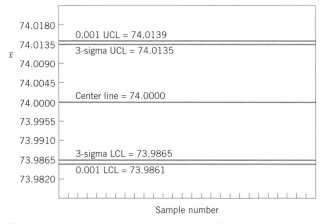

Figure 4-7 Comparison of three-sigma and 0.001 probability limits for the $\bar{x}$ chart.

These control limits are called 0.001 **probability limits.** The $\bar{x}$ chart with both three-sigma limits and 0.001 limits is shown in Fig. 4-7. There is only a slight difference between the two limits.

Regardless of the distribution of the quality characteristic, it is standard practice in the United States to determine the control limits as a multiple of the standard deviation of the statistic plotted on the chart. The multiple usually chosen is 3; hence, three-sigma limits are customarily employed on control charts, regardless of the type of chart employed. In the United Kingdom and parts of Western Europe, probability limits are used, with the standard probability level being 0.001.

We typically justify the use of three-sigma control limits on the basis that they give good results in practice. Moreover, in many cases, the true distribution of the quality characteristic is not known well enough to compute exact probability limits. If the distribution of the quality characteristic is reasonably approximated by the normal distribution, then there will be little difference between three-sigma and 0.001 probability limits.

Warning Limits on Control Charts

Some analysts suggest using two sets of limits on control charts, such as those shown in Fig. 4-8. The outer limits—say, at three-sigma—are the usual **action limits;** that is, when a point plots outside of this limit, a search for an assignable cause is made and corrective action is taken if necessary. The inner limits, usually at two-sigma, are called **warning limits.** In Fig. 4-8, we have shown the three-sigma upper and lower control limits for the $\bar{x}$ chart for the piston-ring diameter. The upper and lower warning limits are located at

$$\text{UWL} = 74 + 2(0.0045) = 74.0090$$
$$\text{LWL} = 74 - 2(0.0045) = 73.9910$$

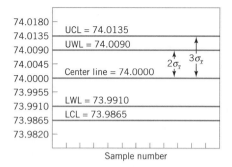

Figure 4-8   An $\bar{x}$ chart with two-sigma warning limits.

When probability limits are used, the action limits are generally 0.001 limits and the warning limits are 0.025 limits.

If one or more points fall between the warning limits and the control limits, or very close to the warning limit, we should be suspicious that the process may not be operating properly. One possible action to take when this occurs is to increase the sampling frequency and/or the sample size so that more information about the process can be obtained quickly. Process control schemes that change the sample size and/or the sampling frequency depending on the position of the current sample value are called **adaptive** or **variable sampling interval** (or **variable sample size,** etc.) schemes. These techniques have been used in practice for many years and have recently been studied extensively by researchers in the field. We will discuss this technique again in Part III of this book.

The use of warning limits can increase the **sensitivity** of the control chart; that is, it can allow the control chart to signal a shift in the process more quickly. One of their disadvantages is that they may be confusing to operating personnel. This is not usually a serious objection, however, and many practitioners use warning limits routinely on control charts. A more serious objection is that although the use of warning limits can improve the sensitivity of the chart, they also result in an **increased risk of false alarms.** We will discuss the use of sensitizing rules (such as warning limits) more thoroughly in Section 4-3.6.

4-3.3   Sample Size and Sampling Frequency

In designing a control chart, we must specify both the **sample size** to use and the **frequency of sampling.** In general, larger samples will make it easier to detect small shifts in the process. This is demonstrated in Fig. 4-9, where we have plotted the operating characteristic curve for the $\bar{x}$ chart in Fig. 4-3 for various sample sizes. Note that the probability of detecting a shift from 74.0000 mm to 74.0100 mm (for example) increases as the sample size n increases. When choosing the sample size, we must keep in mind the size of the shift that we are trying to detect. If the process shift is relatively

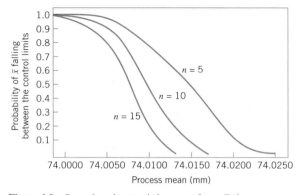

Figure 4-9 Operating-characteristic curves for an $\bar{x}$ chart.

large, then we use smaller sample sizes than those that would be employed if the shift of interest were relatively small.

We must also determine the frequency of sampling. The most desirable situation from the point of view of detecting shifts would be to take large samples very frequently; however, this is usually not economically feasible. The general problem is one of **allocating sampling effort.** That is, either we take small samples at short intervals or larger samples at longer intervals. Current industry practice tends to favor smaller, more frequent samples, particularly in high-volume manufacturing processes, or where a great many types of assignable causes can occur. Furthermore, as automatic sensing and measurement technology develops, it is becoming possible to greatly (increase) sampling frequencies. Ultimately, every unit can be tested as it is manufactured. Automatic measurement systems and microcomputers with statistical process-control software applied at the work center for real-time on-line process control is an increasingly effective way to apply statistical process control.

Another way to evaluate the decisions regarding sample size and sampling frequency is through the **average run length (ARL)** of the control chart. Essentially, the ARL is the average number of points that must be plotted before a point indicates an out-of-control condition. If the process observations are uncorrelated, then for any Shewhart control chart, the ARL can be calculated easily from

$$\text{ARL} = \frac{1}{p} \qquad (4\text{-}2)$$

where p is the probability that any point exceeds the control limits. This equation can be used to evaluate the performance of the control chart.

To illustrate, for the $\bar{x}$ chart with three-sigma limits, $p = 0.0027$ is the probability that a single point falls outside the limits when the process is in control. Therefore, the average run length of the $\bar{x}$ chart when the process is in control (called ARL_0) is

$$ARL_0 = \frac{1}{p} = \frac{1}{0.0027} = 370$$

That is, even if the process remains in control, an out-of-control signal will be generated every 370 samples, on the average.

The use of average run lengths to describe the performance of control charts has been subjected to criticism in recent years. The reasons for this arise because the distribution of run length for a Shewhart control chart is a geometric distribution (refer to Section 2-2.4). Consequently, there are two concerns with ARL: (1) the standard deviation of the run length is very large, and (2) the geometric distribution is very skewed, so the mean of the distribution (the ARL) is not necessarily a very "typical" value of the run length.

For example, consider the Shewhart $\bar{x}$ control chart with three-sigma limits. When the process is in control, we have noted that $p = 0.0027$ and the in-control ARL_0 is $ARL_0 = 1/p = 1/0.0027$. This is the mean of the geometric distribution. Now the standard deviation of the geometric distribution is

$$\sqrt{(1-p)}/p = \sqrt{(1-0.0027)}/0.0027 \approx 370$$

That is, the standard deviation of the geometric distribution in this case is approximately equal to its mean. As a result, the actual ARL_0 observed in practice for the Shewhart $\bar{x}$ control chart will likely vary considerably. Furthermore, for the geometric distribution with $p = 0.0027$, the 10th and 50th percentiles of the distribution are 38 and 256, respectively. This means that approximately 10% of the time the in-control run length will be less than or equal to 10 samples and 50% of the time it will be less than or equal to 256 samples. This occurs because the geometric distribution with $p = 0.0027$ is quite skewed to the right.

It is also occasionally convenient to express the performance of the control chart in terms of its **average time to signal (ATS).** If samples are taken at fixed intervals of time that are h hours apart, then

$$ATS = ARL\, h \qquad (4\text{-}3)$$

Consider the piston-ring process discussed earlier, and suppose we are sampling every hour. Equation 4-3 indicates that we will have a **false alarm** about every 370 hours on the average.

Now consider how the control chart performs in detecting shifts in the mean. Suppose we are using a sample size of $n = 5$ and that when the process goes out of control the mean shifts to 74.015 mm. From the operating characteristic curve in Fig. 4-9 we find that if the process mean is 74.015 mm, the probability of $\bar{x}$ falling between the control limits is approximately 0.50. Therefore, p in equation 4-2 is 0.50, and the out-of-control ARL (called ARL_1) is

$$ARL_1 = \frac{1}{p} = \frac{1}{0.5} = 2$$

That is, the control chart will require two samples to detect the process shift, on the average, and since the time interval between samples is $h = 1$ hour, the average time required to detect this shift is

$$\begin{aligned} ATS &= ARL_1 \, h \\ &= 2\,(1) \\ &= 2 \text{ hours} \end{aligned}$$

Suppose that this is unacceptable, because production of piston rings with a mean diameter of 74.015 mm results in excessive scrap costs and delays final engine assembly. How can we reduce the time needed to detect the out-of-control condition? One method is to sample more frequently. For example, if we sample every half hour, then the average time to signal for this scheme is $ATS = ARL_1 \, h = 2(\frac{1}{2}) = 1$; that is, only one hour will elapse (on the average) between the shift and its detection. The second possibility is to increase the sample size. For example, if we use $n = 10$, then Fig. 4-9 shows that the probability of $\bar{x}$ falling between the control limits when the process mean is 74.015 mm is approximately 0.1, so that $p = 0.9$, and from equation 4-2 the out-of-control ARL or ARL_1 is

$$ARL_1 = \frac{1}{p} = \frac{1}{0.9} = 1.11$$

and, if we sample every hour, the average time to signal is

$$\begin{aligned} ATS &= ARL_1 \, h \\ &= 1.11\,(1) \\ &= 1.11 \text{ hours} \end{aligned}$$

Thus, the larger sample size would allow the shift to be detected about twice as quickly as the old one. If it became important to detect the shift in the (approximately) first hour after it occurred, two control chart designs would work:

Design 1	**Design 2**
Sample Size: $n = 5$	Sample Size: $n = 10$
Sampling Frequency: every half hour	Sampling Frequency: every hour

To answer the question of sampling frequency more precisely, we must take several factors into account, including the cost of sampling, the losses associated with allowing the process to operate out of control, the rate of production, and the probabilities with which various types of process shifts occur. We discuss various methods for selecting an appropriate sample size and sampling frequency for a control chart in the next four chapters.

4-3.4   Rational Subgroups

A fundamental idea in the use of control charts is the collection of sample data according to what Shewhart called the **rational subgroup** concept. To illustrate this concept, suppose that we are using an $\bar{x}$ control chart to detect changes in the process mean. Then the rational subgroup concept means that subgroups or samples should be selected so that if assignable causes are present, the chance for differences *between* subgroups will be maximized, while the chance for differences due to these assignable causes *within* a subgroup will be minimized.

When control charts are applied to production processes, the time order of production is a logical basis for rational subgrouping. Even though time order is preserved, it is still possible to form subgroups erroneously. If some of the observations in the sample are taken at the end of one shift and the remaining observations are taken at the start of the next shift, then any differences between shifts might not be detected. Time order is frequently a good basis for forming subgroups because it allows us to detect assignable causes that occur over time.

Two general approaches to constructing rational subgroups are used. In the first approach, each sample consists of units that were produced at the same time (or as closely together as possible). Ideally, we would like to take **consecutive** units of production. This approach is used when the primary purpose of the control chart is to detect process shifts. It minimizes the chance of variability due to assignable causes *within* a sample, and it maximizes the chance of variability *between* samples if assignable causes are present. It also provides a better estimate of the standard deviation of the process in the case of variables control charts. This approach to rational subgrouping essentially gives a "snapshot" of the process at each point in time where a sample is collected.

Figure 4-10 illustrates this type of sampling strategy. In Fig. 4-10*a* we show a process for which the mean experiences a series of sustained shifts, and the corresponding observations obtained from this process at the points in time along the horizontal axis, assuming that five consecutive units are selected. Figure 4-10*b* shows the $\bar{x}$ **control chart** and an **R chart** (or **range chart**) for these data. The center line and control limits on the R chart are constructed using the range of each sample in the upper part of the figure (details will be given in Chapter 5). Note that although the process mean is shifting, the process variability is stable. Furthermore, the within-sample measure of variability is used to construct the control limits on the $\bar{x}$ chart. Note that the $\bar{x}$ chart in Fig. 4-10*b* has points out of control corresponding to the shifts in the process mean.

In the second approach, each sample consists of units of product that are representative of *all* units that have been produced since the last sample was taken. Essentially, each subgroup is a **random sample of all process output over the sampling interval.** This method of rational subgrouping is often used when the control chart is employed to make decisions about the acceptance of all units of product that have been produced since the last sample. In fact, if the process shifts to an out-of-control state and then back in control again *between* samples, it is sometimes argued that the first method of rational subgrouping defined above will be ineffective against these types of shifts, and so the second method must be used.

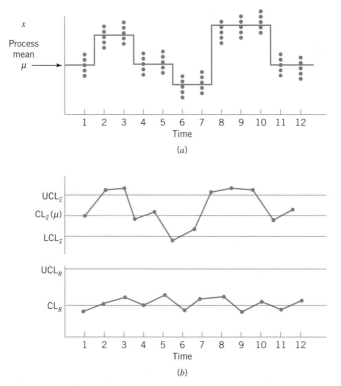

Figure 4-10   The "snapshot" approach to rational subgroups. (*a*) Behavior of the process mean. (*b*) Corresponding $\bar{x}$ and R control charts.

When the rational subgroup is a random sample of all units produced over the sampling interval, considerable care must be taken in interpreting the control charts. If the process mean drifts between several levels during the interval between samples, this may cause the range of the observations within the sample to be relatively large, resulting in wider limits on the $\bar{x}$ chart. This scenario is illustrated in Fig. 4-11. In fact, **we can often make any process appear to be in statistical control just by stretching out the interval between observations in the sample.** It is also possible for shifts in the process average to cause points on a control chart for the range or standard deviation to plot out of control, even though there has been no shift in process variability.

There are other bases for forming rational subgroups. For example, suppose a process consists of several machines that pool their output into a common stream. If we sample from this common stream of output, it will be very difficult to detect whether or not some of the machines are out of control. A logical approach to rational subgrouping here is to apply control chart techniques to the output for each individual machine. Sometimes this concept needs to be applied to different heads on the same machine, different work stations, different operators, and so forth. In many situations the rational subgroup will consist of a single observation. This situation occurs frequently in the

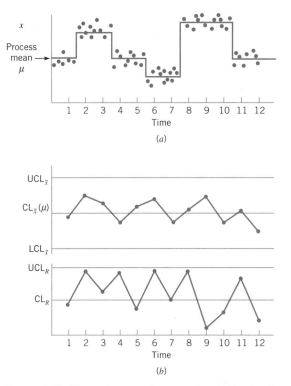

Figure 4-11 The random sample approach to rational subgroups. (*a*) Behavior of the process mean. (*b*) Corresponding $\bar{x}$ and R control charts.

chemical and process industries where the quality characteristic of the product changes relatively slowly and samples taken very close together in time are virtually identical, apart from measurement or analytical error.

The rational subgroup concept is very important. The proper selection of samples requires careful consideration of the process, with the objective of obtaining as much useful information as possible from the control chart analysis.

4-3.5 Analysis of Patterns on Control Charts

A control chart may indicate an out-of-control condition either when one or more points fall beyond the control limits or when the plotted points exhibit some nonrandom pattern of behavior. For example, consider the $\bar{x}$ chart shown in Fig. 4-12. Although all 25 points fall within the control limits, the points do not indicate statistical control because their pattern is very nonrandom in appearance. Specifically, we note that 19 of 25 points plot below the center line, while only 6 of them plot above. If the points are truly random, we should expect a more even distribution of them above and below the center

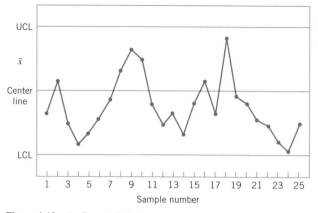

Figure 4-12 An $\bar{x}$ control chart.

line. We also observe that following the fourth point, five points in a row increase in magnitude. This arrangement of points is called a **run.** Since the observations are increasing, we could call this a run up. Similarly, a sequence of decreasing points is called a run down. This control chart has an unusually long run up (beginning with the fourth point) and an unusually long run down (beginning with the eighteenth point.)

In general, we define a run as a sequence of observations of the same type. In addition to runs up and runs down, we could define the types of observations as those above and below the center line, respectively, so that two points in a row above the center line would be a run of length 2.

A run of length 8 or more points has a very low probability of occurrence in a random sample of points. Consequently, any type of run of length 8 or more is often taken as a signal of an out-of-control condition. For example, eight consecutive points on one side of the center line will indicate that the process is out of control.

Although runs are an important measure of nonrandom behavior on a control chart, other types of patterns may also indicate an out-of-control condition. For example, consider the $\bar{x}$ chart in Fig. 4-13. Note that the plotted sample averages exhibit a cyclic

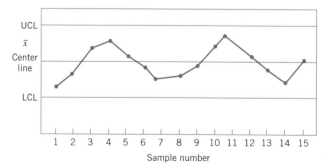

Figure 4-13 An $\bar{x}$ chart with a cyclic pattern.

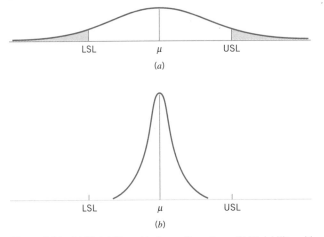

Figure 4-14   (*a*) Variability with the cyclic pattern. (*b*) Variability with the cyclic pattern eliminated.

behavior, yet they all fall within the control limits. Such a pattern may indicate a problem with the process such as operator fatigue, raw material deliveries, heat or stress buildup, and so forth. Although the process is not really out of control, the yield may be improved by elimination or reduction of the sources of variability causing this cyclic behavior (see Fig. 4-14).

The problem is one of **pattern recognition**—that is, recognizing systematic or nonrandom patterns on the control chart and identifying the reason for this behavior. The ability to interpret a particular pattern in terms of assignable causes requires experience and knowledge of the process. That is, we must not only know the statistical principles of control charts, but we must also have a good understanding of the process. We discuss the interpretation of patterns on control charts in more detail in Chapter 5.

The Western Electric Handbook (1956) suggests a set of decision rules for detecting nonrandom patterns on control charts. Specifically, it suggests concluding that the process is out of control if either

1.   One point plots outside the three-sigma control limits;
2.   Two out of three consecutive points plot beyond the two-sigma warning limits;
3.   Four out of five consecutive points plot at a distance of one-sigma or beyond from the center line;

or

4.   Eight consecutive points plot on one side of the center line.

Those rules apply to one side of the center line at a time. Therefore, a point above the *upper* warning limit followed immediately by a point below the *lower* warning limit would not signal an out-of-control alarm. These are often used in practice for enhancing the sensitivity of control charts. That is, the use of these rules can allow smaller process

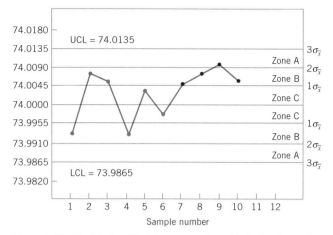

Figure 4-15   The Western Electric or zone rules, with the last four points showing a violation of rule 3.

shifts to be detected more quickly than would be the case if our only criterion was the usual three-sigma control limit violation.

Figure 4-15 shows an $\bar{x}$ control chart for the piston-ring process with the one-sigma, two-sigma, and three-sigma limits used in the Western Electric procedure. Note that these limits partition the control chart into three zones A, B, and C on each side of the center line. Consequently, the Western Electric rules are sometimes called the **zone rules** for control charts. Note that the last four points fall in zone B or beyond. Thus, since four of five consecutive points exceed the one-sigma limit, the Western Electric procedure will conclude that the pattern is nonrandom and the process is out of control.

4-3.6   Discussion of Sensitizing Rules for Control Charts

As may be gathered from earlier sections, several criteria may be applied simultaneously to a control chart to determine whether the process is out of control. The basic criterion is one or more points outside of the control limits. The supplementary criteria are sometimes used to increase the **sensitivity** of the control charts to a small process shift so that we may respond more quickly to the assignable cause. Some of the **sensitizing rules** that are widely used in practice are shown in Table 4-1. For a good discussion of some of these rules, see Nelson (1984). Frequently we will inspect the control chart and conclude that the process is out of control if any one or more of the criteria in Table 4-1 are met.

When several of these sensitizing rules are applied simultaneously, we often use a **graduated response** to out-of-control signals. For example, if a point exceeded a control limit, we would immediately begin to search for the assignable cause, but if one or two consecutive points exceeded only the two-sigma warning limit, we might increase the frequency of sampling from every hour—say, to every 10 minutes. This **adaptive**

Table 4-1 Some Sensitizing Rules for Shewhart Control Charts

Standard Action Signal →	1. One or more points outside of the control limits.
	2. Two of three consecutive points outside the two-sigma warning limits but still inside the control limits.
	3. Four of five consecutive points beyond the one-sigma limits.
	4. A run of eight consecutive points on one side of the center line.
	5. Six points in a row steadily increasing or decreasing.
	6. Fifteen points in a row in zone C (both above and below the center line).
	7. Fourteen points in a row alternating up and down.
	8. Eight points in a row on both sides of the center line with none in zone C.
	9. An unusual or nonrandom pattern in the data.
	10. One or more points near a warning or control limit.

Rules 1–4 are bracketed as "Western Electric Rules"

sampling response might not be as severe as a complete search for an assignable cause, but if the process were really out of control, it would give us a high probability of detecting this situation more quickly than we would by maintaining the longer sampling interval.

In general, care should be exercised when using several decision rules simultaneously. Suppose that the analyst uses k decision rules and that criterion i has type I error probability α_i. Then the overall type I error or false-alarm probability for the decision based on all k tests is

$$\alpha = 1 - \prod_{i=1}^{k} (1 - \alpha_i) \qquad (4\text{-}4)$$

provided that all k decision rules are independent. However, the independence assumption is not valid with the usual sensitizing rules. Furthermore, the value of α_i is not always clearly defined for the sensitizing rules, because these rules involve several observations.

Champ and Woodall (1987) investigated the average run length performance for the Shewhart control chart with various sensitizing rules. They found that the use of these rules does improve the ability of the control chart to detect smaller shifts, but the in-control average run length can be substantially degraded. For example, assuming independent process data and using a Shewhart control chart with the Western Electric rules results in an in-control ARL of 91.25, in contrast to 370 for the Shewhart control chart alone.

Thus, the sensitizing rules need to be used with **considerable caution,** as an excessive number of false alarms can be harmful to an effective SPC program. Furthermore, as more supplemental rules are applied to the chart, the decision process becomes more complicated, and the inherent simplicity of the Shewhart control chart is lost. Many of the sensitizing rules that we have discussed have been implemented in modern SPC computer programs. However, the analyst should be extremely careful about their routine use. Runs rules and other sensitizing rules can be helpful when the control chart is first applied and the focus is on stabilizing an out-of-control process. However, once the process is reasonably stable, the routine use of these sensitizing rules to detect small shifts, or to try to react more quickly to assignable causes, should be discouraged. The cumulative sum and exponentially weighted moving average control charts (see Chapter 8 in Part III) should be used when smaller process shifts are of interest.

4-4 THE REST OF THE "MAGNIFICENT SEVEN"

Although the control chart is a very powerful problem-solving and process improvement tool, it is most effective when its use is fully integrated into a comprehensive SPC program. The seven major SPC problem-solving tools should be widely taught throughout the organization and used routinely to identify improvement opportunities and to assist in reducing variability and eliminating waste. These "magnificent seven," introduced in Section 4-1, are listed again here for convenience:

1. Histogram or stem-and-leaf display
2. Check sheet
3. Pareto chart
4. Cause-and-effect diagram
5. Defect concentration diagram
6. Scatter diagram
7. Control chart

We have already introduced the histogram and the stem-and-leaf display (Chapter 2), and control chart. In this section we will briefly illustrate the rest of the tools.

Check Sheet
In the early stages of an SPC implementation, it will often become necessary to collect either historical or current operating data about the process under investigation. A **check sheet** can be very useful in this data collection activity. The check sheet shown in Fig. 4-16 was developed by an engineer at an aerospace firm who was investigating the various types of defects that occurred on a tank used in one of their products with a view toward improving the process. The engineer designed this check sheet to facilitate summarizing all the historical defect data available concerning the tanks. Because only a few tanks were manufactured each month, it seemed appropriate to summarize the data monthly and to identify as many different types of defects as possible. The **time-oriented summary** is particularly

CHECK SHEET
DEFECT DATA FOR 1988–1989 YTD

Part No.: TAX-41
Location: Bellevue
Study Date: 6/5/89
Analyst: TCB

Defect	1988 1	2	3	4	5	6	7	8	9	10	11	12	1989 1	2	3	4	5	Total
Parts damaged		1		3	1	2			1	10	3		2	2	7	2		34
Machining problems			3	3				1	8		3		8	3				29
Supplied parts rusted			1	1		2	9											13
Masking insufficient		3	6	4	3	1												17
Misaligned weld	2																	2
Processing out of order	2															2		4
Wrong part issued		1					2											3
Unfinished fairing			3															3
Adhesive failure			1								1		2		1	1		6
Powdery alodine				1														1
Paint out of limits						1								1				2
Paint damaged by etching		1																1
Film on parts						3		1	1									5
Primer cans damaged								1										1
Voids in casting									1	1								2
Delaminated composite									2									2
Incorrect dimensions											13	7	13	1		1	1	36
Improper test procedure									1									1
Salt-spray failure													4					4
TOTAL	4	5	14	12	5	9	9	6	10	14	20	7	29	7	7	6	2	166

Figure 4-16 A check sheet to record defects on a tank used in an aerospace application.

valuable in looking for **trends** or other meaningful patterns. For example, if many defects occur during the summer, one possible cause that should be investigated is the use of temporary workers during a heavy vacation period.

When designing a check sheet, it is important to clearly specify the type of data to be collected, the part or operation number, the date, the analyst, and any other information useful in diagnosing the cause of poor performance. If the check sheet is the basis for performing further calculations or is used as a worksheet for data entry into a computer, then it is important to be sure that the check sheet will be adequate for this purpose before considerable effort is expended in actually collecting data. In some cases, a "trial-run" to validate the check sheet layout and design may be helpful.

Pareto Chart

The **Pareto chart** is simply a frequency distribution (or histogram) of attribute data arranged by category. To illustrate a Pareto chart, consider the tank defect data presented

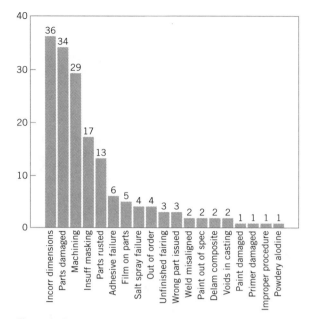

Figure 4-17 Pareto chart of the tank defect data.

in Fig. 4-16. Plot the total frequency of occurrence of each defect type (the last column of the table in Fig. 4-16) against the various defect types to produce Fig. 4-17, which is called a Pareto chart.[3] Through this chart the user can quickly and visually identify the most frequently occurring types of defects. For example, Fig. 4-17 indicates that incorrect dimensions, parts damaged, and machining are the most commonly encountered defects. Thus the causes of these defect types should probably be identified and attacked first.

Note that the Pareto chart does not automatically identify the most *important* defects, but rather only those that occur most frequently. For example, in Fig. 4-17 casting voids occur very infrequently (2 of 166 defects, or 1.2%). However, voids could result in scrapping the tank, a potentially large cost exposure—perhaps so large that casting voids should be elevated to a major defect category. When the list of defects contains a mixture of those that might have extremely serious consequences and others of much less importance, one of two methods can be used:

1. Use a weighting scheme to modify the frequency counts. Weighting schemes for defects are discussed in Chapter 6.

2. Accompany the **frequency Pareto chart** analysis with a **cost** or **exposure Pareto chart**.

---

[3]The name Pareto chart is derived from Italian economist Vilfredo Pareto (1848–1923), who theorized that in certain economies the majority of the wealth was held by a disproportionately small segment of the population. Quality engineers have observed that defects usually follow a similar Pareto distribution.

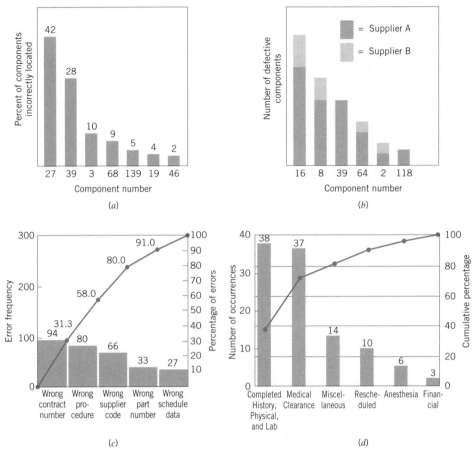

Figure 4-18   Various examples of Pareto charts.

There are many variations of the basic Pareto chart. Figure 4-18*a* shows a Pareto chart applied to an electronics assembly process using surface-mount components. The vertical axis is the percentage of components incorrectly located, and the horizontal axis is the component number, a code that locates the device on the printed circuit board. Note that locations 27 and 39 account for 70% of the errors. This may be the result of the *type* or *size* of components at these locations, or of *where* these locations are on the board layout. Figure 4-18*b* presents another Pareto chart from the electronics industry. The vertical axis is the number of defective components, and the horizontal axis is the component number. Note that each vertical bar has been broken down by supplier to produce a *stacked Pareto chart*. This analysis clearly indicates that supplier A provides a disproportionally large share of the defective components.

Pareto charts are widely used in **nonmanufacturing applications** of quality-improvement methods. A Pareto chart used by a quality-improvement team in a procurement organization is shown in Fig. 4-18*c*. The team was investigating errors on purchase orders in an effort to reduce the number of purchase order changes issued by the organization. (Each change typically costs between $100 and $500, and this organization issued several hundred purchase order changes each month.) This Pareto chart has two scales—one for the actual error frequency and another for the percentage of errors. Figure 4-18d presents a Pareto chart constructed by a quality-improvement team in a hospital to reflect the reasons for cancellation of scheduled outpatient surgery.

In general, the Pareto chart is one of the most useful of the "magnificent seven." Its applications to quality improvement are limited only by the ingenuity of the analyst.

Cause-and Effect-Diagram

Once a defect, error, or problem has been identified and isolated for further study, we must begin to analyze potential **causes** of this undesirable **effect.** In situations where causes are not obvious (sometimes they are), the cause-and-effect diagram is a formal tool frequently useful in unlayering potential causes. The cause-and-effect diagram constructed by a quality-improvement team assigned to identify potential problem areas in the tank manufacturing process mentioned earlier is shown in Fig. 4-19. The steps in constructing the cause-and-effect diagram are as follows:

How to Construct a Cause-and-Effect Diagram

1. Define the problem or effect to be analyzed.
2. Form the team to perform the analysis. Often the team will uncover potential causes through brainstorming.
3. Draw the effect box and the center line.
4. Specify the major potential cause categories and join them as boxes connected to the center line.
5. Identify the possible causes and classify them into the categories in step 4. Create new categories, if necessary.
6. Rank order the causes to identify those that seem most likely to impact the problem.
7. Take corrective action.

In analyzing the tank defect problem, the team elected to lay out the major categories of tank defects as machines, materials, methods, personnel, measurement, and environment. A brainstorming session ensued to identify the various subcauses in each of these major categories and to prepare the diagram in Fig. 4-19. Then through discussion and the process of elimination, the group decided that materials and methods contained the most likely cause categories.

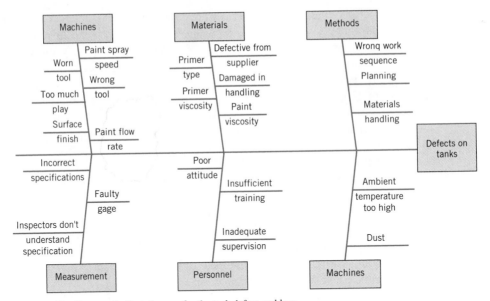

Figure 4-19   Cause-and-effect diagram for the tank defect problem.

Cause-and-effect analysis is an extremely powerful tool. A highly detailed cause-and-effect diagram can serve as an effective troubleshooting aid. Furthermore, the construction of a cause-and-effect diagram as a **team experience** tends to get people involved in attacking a problem rather than in affixing blame.

Defect Concentration Diagram

A **defect concentration diagram** is a picture of the unit, showing all relevant views. Then the various types of defects are drawn on the picture, and the diagram is analyzed to determine whether the **location** of the defects on the unit conveys any useful information about the potential causes of the defects.

Figure 4-20 presents a defect concentration diagram for the final assembly stage of a refrigerator manufacturing process. Surface-finish defects are identified by the dark shaded areas on the refrigerator. From inspection of the diagram it seems clear that materials handling is responsible for the majority of these defects. The unit is being moved by securing a belt around the middle, and this belt is either too loose (tight), worn out, made of abrasive material, or too narrow. Furthermore, when the unit is moved the corners are being damaged. It is possible that worker fatigue is a factor in this process. In any event, proper work methods and improved materials handling will likely improve this process dramatically.

Figure 4-21 shows the defect concentration diagram for the tank problem mentioned earlier. Note that this diagram shows several different broad categories of defects, each identified with a specific code. Often different colors are used to indicate different types of defects.

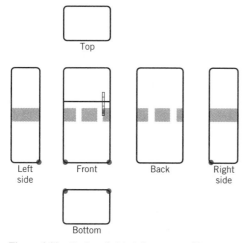

Figure 4-20   Surface-finish defects on a refrigerator.

When defect data are portrayed on a defect concentration diagram over a sufficient number of units, patterns frequently emerge, and the location of these patterns often contains much information about the causes of the defects. We have found defect concentration diagrams to be important problem-solving tools in many industries, including plating, painting and coating, casting and foundry operations, machining, and electronics assembly.

Scatter Diagram
The scatter diagram is a useful plot for identifying a potential relationship between two variables. Data are collected in pairs on the two variables—say, (y_i, x_i)—for $i = 1,$

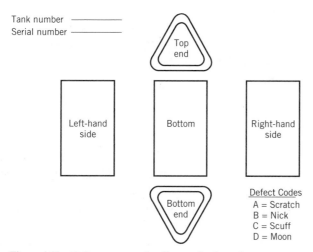

Figure 4-21   Defect concentration diagram for the tank.

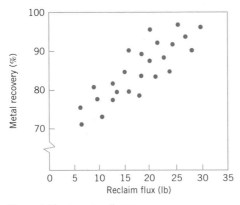

Figure 4-22   A scatter diagram.

2, . . . , n. Then y_i is plotted against the corresponding x_i. The shape of the scatter diagram often indicates what type of relationship may exist between the two variables.

Figure 4-22 shows a scatter diagram relating metal recovery (in percent) from a magnathermic smelting process for magnesium against corresponding values of the amount of reclaim flux added to the crucible. The scatter diagram indicates a strong **positive correlation** between metal recovery and flux amount; that is, as the amount of flux added is increased, the metal recovery also increases. It is tempting to conclude that the relationship is one based on cause and effect: By increasing the amount of reclaim flux used, we can always ensure high metal recovery. This thinking is potentially dangerous, because **correlation** does not necessarily imply **causality.** This apparent relationship could be caused by something quite different. For example, both variables could be related to a third one, such as the temperature of the metal prior to the reclaim pouring operation, and this relationship could be responsible for what we see in Fig. 4-22. If higher temperatures lead to higher metal recovery and the practice is to add reclaim flux in proportion to temperature, adding more flux when the process is running at low temperature will do nothing to enhance yield. The scatter diagram is useful for identifying **potential relationships. Designed experiments** [see Montgomery (1997)] **must be used to verify causality.**

4-5   IMPLEMENTING SPC

The methods of statistical process control can provide significant payback to those companies that can successfully implement them. Although SPC seems to be a collection of statistically based problem-solving tools, there is more to the successful use of SPC than learning and using these tools. **Management involvement and commitment** to the quality-improvement process is the most vital component of SPC's potential success. Management is a role model, and others in the organization will look to management for guidance and as an example. A team approach is also important, as it is usually difficult for one person alone to introduce process improvements. Many of the "magnificent seven" are helpful in building an improvement team, including cause-and-effect diagrams, Pareto charts, and defect concentration diagrams. The basic SPC problem-solving tools must become widely known and widely used throughout the organization. Ongoing education of personnel about SPC and other methods for reducing variability are necessary to achieve this widespread knowledge of the tools.

The objective of an SPC-based variability reduction program is continuous improvement on a weekly, quarterly, and annual basis. SPC is not a one-time program to be applied when the business is in trouble and later abandoned. Quality improvement that is focused on reduction of variability must become part of the culture of the organization.

The control chart is an important tool for process improvement. Processes do not naturally operate in an in-control state, and the use of control charts is an important step that must be taken early in an SPC program to eliminate assignable causes, reduce process variability, and stabilize process performance. To improve quality and productivity, we must begin to manage with facts and data, and not simply rely on judgment. Control charts are an important part of this change in management approach.

In implementing a company-wide effort to reduce variability and improve quality, we have found that several elements are usually present in all successful efforts. These elements are as follows:

Elements of a Successful SPC Program

1. Management leadership
2. A team approach
3. Education of employees at all levels
4. Emphasis on reducing variability
5. Measuring success in quantitative (economic) terms
6. A mechanism for communicating successful results throughout the organization

---

We cannot overemphasize the importance of **management leadership** and the **team approach.** Successful quality improvement is a "top-down" management-driven activity. It is also important to measure progress and success in quantitative (economic) terms, and to spread knowledge of this success throughout the organization. When successful improvements are communicated throughout the company, this can provide motivation and incentive to improve other processes and to make continuous improvement a normal part of the way of doing business.

4-6   AN APPLICATION OF SPC

In this section, we give an account of applying SPC methods to improve quality and productivity in a copper plating operation at a printed circuit board fabrication facility. This process was characterized by high levels of defects such as brittle copper and copper voids and by long flow time. The long flow time was particularly troublesome, as it had led to an extensive work backlog and was a major contributor to poor conformance to the factory production schedule.

Management chose this process area for an initial implementation of SPC. An improvement team was formed, consisting of the plating tank operator, the manufacturing engineer responsible for the process, and a quality engineer. All members of the team had been exposed to the "magnificent seven" in a company-sponsored SPC seminar. During the first team meeting, it was decided to concentrate on reducing the flow time through the process, as the missed delivery targets were considered to be the most serious obstacle to improving productivity. The team quickly determined (based on operator experience) that excessive downtime on the controller that regulated the copper concentration in the plating tank was a major factor in the excessive flow time, as controller downtime translated directly into lost production.

The team decided to use a cause-and-effect analysis to begin to isolate the potential causes of controller downtime. Figure 4-23 shows the cause-and-effect diagram that was produced during a brainstorming session focused on controller downtime. The team was able to quickly identify 11 major potential causes of controller downtime. However, when

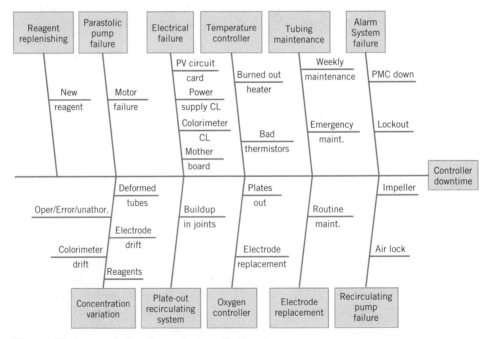

Figure 4-23   Cause-and-effect diagram for controller downtime.

WEEKLY TALLY	OPERATOR_____		
WEEK ENDING_____	ERRORS	DESCRIPTION	ACTION

1. CONCENTRATION VARIATION
   a. Colorimeter drift
   b. Electrode failure
   c. Reagents
   d. Deformed tubes
   e. Oper/error/unauthorized
	_____
	_____
	_____
	_____
	_____		

2. ALARM SYSTEM FAILURE
   a. PMC down
   b. Lockout
	_____
	_____		

3. RECIRCULATING PUMP FAILURE
   a. Air lock
   b. Impeller
	_____
	_____		

4. REAGENT REPLENISHING **a.** New reagent	_____		

5. TUBING MAINTENANCE
   a. Weekly maintenance
   b. Emergency maintenance
	_____
	_____		

6. ELECTRODE REPLACEMENT **a.** Routine maintenance	_____		

7. TEMPERATURE CONTROLLER
   a. Burned out heater
   b. Bad thermistors
	_____
	_____		

8. OXYGEN CONTROLLER
   a. Plates out
   b. Electrode replacement
	_____
	_____		

9. PARASTOLIC PUMP FAILURE **a.** Motor failure	_____		

10. ELECTRICAL FAILURE
   a. PV circuit card
   b. Power supply CL
   c. Colorimeter CL
   d. Mother board
	_____
	_____
	_____
	_____		

11. PLATE-OUT RECIRCULATING **a.** Buildup at joints	_____		
TOTAL COUNT			

Figure 4-24 Check sheet for logbook.

they examined the equipment logbook to make a more definitive diagnosis of the causes of downtime based on actual process performance, the results were disappointing. The logbook contained little useful information about causes of downtime; instead, it contained only a chronological record of when the machine was up and when it was down.

The team then decided that it would be necessary to collect valid data about the causes of controller downtime. They designed the check sheet shown in Fig. 4-24 as a supplemental page for the logbook. The team agreed that whenever the equipment was down one team member would assume responsibility for filling out the check sheet. Note that the major causes of controller downtime identified on the cause-and-effect diagram have been used to structure the headings and subheadings on the check sheet. The team agreed that data would be collected over a 4- to 6-week period.

As more reliable data concerning the causes of controller downtime became available, the team was able to analyze it using other SPC techniques. Figure 4-25 presents the Pareto analysis of the controller failure data produced during the 6-week study of the process. Note that concentration variation is a major cause of downtime. Actually, the situation is probably more complex than it appears. The third largest category of downtime causes is reagent replenishment. Frequently, the reagent in the colorimeter on the controller is replenished because concentration has varied so far outside the process specifications that reagent replenishment and colorimeter recalibration is the only step that can be used to bring the process back on line. Therefore, it is possible that up to 50% of the downtime associated with controller failures can be attributed to concentration variation. Figure 4-26 presents a Pareto analysis of only the concentration variation data. From this diagram we know that colorimeter drift and problems with reagents are major causes of concentration variation. This information led the manufacturing engineer on the team to conclude that rebuilding the colorimeter would be an important step in improving the process.

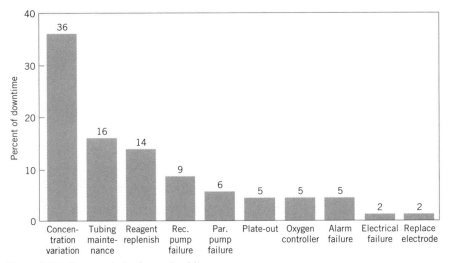

Figure 4-25   Pareto analysis of controller failures.

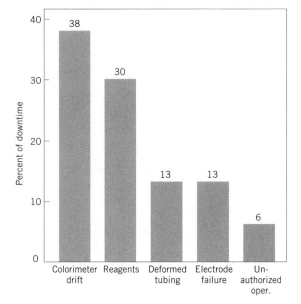

Figure 4-26   Pareto analysis of concentration variation.

During the time that these process data were collected, the team decided to run statistical control charts on the process. The information collected to this point about process performance was the basis for constructing the initial OCAPs (out-of-control-action-plans) for these control charts. Copper concentration is measured in this process manually three times per day. Figure 4-27 presents the $\bar{x}$ control chart for average daily copper concentration; that is, each point plotted in the figure is a daily average. The

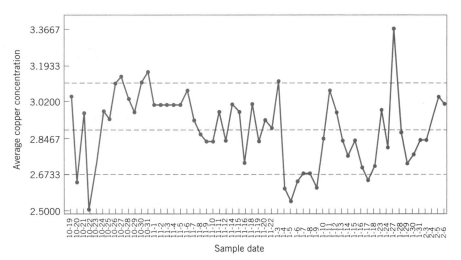

Figure 4-27 $\bar{x}$ chart for the average daily copper concentration.

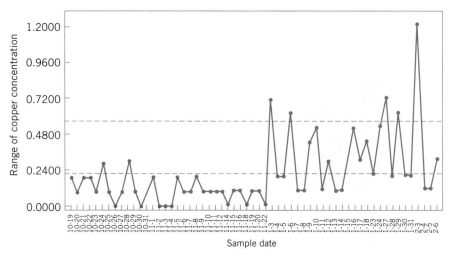

Figure 4-28   *R* chart for daily copper concentration.

chart shows the center line and three-sigma statistical control limits. (We will discuss the construction of these limits in more detail in the next few chapters.) Note that there are a number of points outside the control limits, indicating that assignable causes are present in the process. Figure 4-28 presents the range or *R* chart for daily copper concentration. On this chart *R* represents the difference between the maximum and minimum copper concentration readings in a day. Note that the *R* chart also exhibits a lack of statistical control. In particular, the second half of the *R* chart appears much more unstable than the first half. Examining the dates along the horizontal axis, the team noted that severe variation in average daily copper concentration only appeared after January 3. The last observations on copper concentration had been taken on November 22. From November 23 until January 3 the process had been in a shutdown mode because of holidays. Apparently, when the process was restarted, substantial deterioration in controller/colorimeter performance had occurred. This hastened engineering's decision to rebuild the colorimeter.

Figure 4-29 presents a **tolerance diagram** of daily copper concentration readings. In this figure, each day's copper concentration readings are plotted, and the extremes connected with a vertical line. In some cases more than one observation is plotted at a single position, so a numeral is used to indicate the number of observations plotted at each particular point. The center line on this chart is the process average over the time period studied, and the upper and lower limits are the specification limits on copper concentration. Every instance in which a point is outside the specification limits would correspond to nonscheduled downtime on the process. Several things are evident from examining the tolerance diagram. First, the process average is significantly different from the nominal specification on copper concentration (the midpoint of the upper and lower tolerance band). This implies that the calibration of the colorimeter may be inadequate. That is, we are literally aiming at the wrong target. Second, we note that there is considerably more

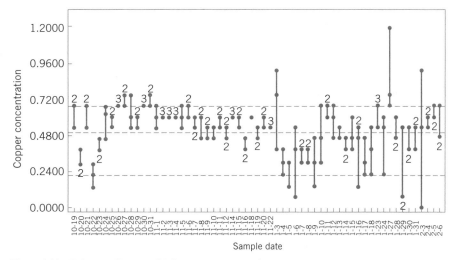

Figure 4-29 Tolerance diagram of daily copper concentration.

variation in the daily copper concentration readings after January 3 than there was prior to shutdown. Finally, if we could reduce variation in the process to a level roughly consistent with that observed prior to shutdown and correct the process centering, many of the points outside specifications would not have occurred, and downtime on the process should be reduced.

In early February, the colorimeter and controller were rebuilt by manufacturing engineering. The result of this maintenance activity was to restore the variability in daily copper concentration readings to the pre-shutdown level. The rebuilt colorimeter was recalibrated and was subsequently able to hold the correct target. This recentering and recalibration of the process reduced the downtime on the controller from approximately 60% to less than 20%. At this point the process was capable of meeting the required production rate.

Once this aspect of process performance was improved, the team directed its efforts to reducing the number of defective units produced by the process. Generally, as noted earlier, defects fell into two major categories: brittle copper and copper voids. The team decided that, although control charts and statistical process control techniques could be applied to this problem, the use of **experimental design methods** might lead to a more rapid solution. As noted in Chapter 1, the objective of a designed experiment is to generate information that will allow us to understand and model the relationship between these process variables and measures of the process performance.

The designed experiment for the plating process is shown in Table 4-2 and Fig. 4-30. The objective of this experiment was to provide information that would be useful in minimizing plating defects. The process variables considered in the experiment were copper concentration, sodium hydroxide concentration, formaldehyde concentration, temperature, and oxygen. A low and high level, represented symbolically by the minus and plus signs in Table 4-2, were chosen for each process variable. The team

Table 4-2   A Designed Experiment for the Plating Process

Objective: Minimize Plating Defects		
Process Variables	Low Level	High Level
A = Copper concentration	−	+
B = Sodium hydroxide concentration	−	+
C = Formaldehyde concentration	−	+
D = Temperature	−	+
E = Oxygen	−	+

	Experimental Design					
	Variables					Response
Run	A	B	C	D	E	(Defects)
1	−	−	−	−	+	
2	+	−	−	−	−	
3	−	+	−	−	−	
4	+	+	−	−	+	
5	−	−	+	−	−	
6	+	−	+	−	+	
7	−	+	+	−	+	
8	+	+	+	−	−	
9	−	−	−	+	−	
10	+	−	−	+	+	
11	−	+	−	+	+	
12	+	+	−	+	−	
13	−	−	+	+	+	
14	+	−	+	+	−	
15	−	+	+	+	−	
16	+	+	+	+	+	

initially considered a **factorial experiment**—that is, an experimental design in which all possible combinations of these factor levels would be run. This design would have required 32 runs—that is, a run at each of the 32 corners of the cubes in Fig. 4-30. Since this is too many runs, a **fractional factorial design** that used only 16 runs was ultimately selected. This fractional factorial design is shown in the bottom half of Table 4-2 and geometrically in Fig. 4-30. In this experimental design, each row of the table is a run on the process. The combination of minus and plus signs in each column of that row determines the low and high levels of the five process variables to be used during that run. For example, in run 1 copper concentration, sodium hydroxide concentration, formaldehyde concentration, and temperature are run at the low level and oxygen is run at the high level. The process would be run at each of the 16 sets of conditions described by the design (for reasons to be discussed later, the runs would not be made in the order shown in Table 4-2), and a response variable—an observed number of plating defects—would be recorded for each run. Then these data could be analyzed

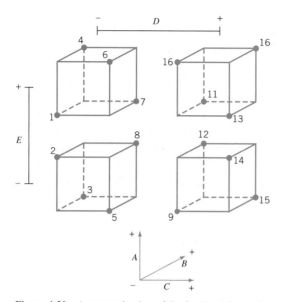

Figure 4-30 A geometric view of the fractional factorial design for the plating process experiment.

using simple statistical techniques to determine which factors have a significant influence on plating defects, whether or not any of the factors jointly influence the occurrence of defects, and whether it is possible to adjust these variables to new levels that will reduce plating defects below their current level. Although a complete discussion of design of experiments is beyond the scope of this text, we will present examples of designed experiments for improving process performance in Part IV.

After the team conducted the experiment shown in Table 4-2 and analyzed the resulting process data, they determined that several of the process variables that they identified for the study were important and had significant impact on the occurrence of plating defects. They were able to adjust these factors to new levels, and as a result, plating defects were reduced by approximately a factor of 10. Therefore, at the conclusion of the team's initial effort at applying SPC to the plating process, it had made substantial improvements in product flow time through the process and had taken a major step in improving the process capability.

4-7 NONMANUFACTURING APPLICATIONS OF STATISTICAL PROCESS CONTROL

This book presents the underlying principles of statistical process control. Most of the examples used to reinforce these principles are in an industrial, product-oriented framework. There have been many successful applications of statistical process-control methods in the manufacturing environment. However, the principles themselves are general;

consequently, there are many nonindustrial or service industry applications of statistical process-control and quality-improvement methodology.

These nonmanufacturing applications do not differ substantially from the more usual industrial applications. As an example, the control chart for fraction nonconforming (which is discussed in Chapter 6) could be applied to reducing billing errors in a bank credit card operation as easily as it could be used to reduce the fraction of nonconforming printed circuit boards produced in an electronics plant. The $\bar{x}$ and R charts discussed in this chapter and applied to the piston-ring manufacturing process could be used to monitor and control the flow time of accounts payable through a finance function. Nonmanufacturing or nonproduct applications of statistical process-control and quality-improvement methodology sometimes require ingenuity beyond that normally required for the more typical manufacturing applications. There seems to be two primary reasons for this difference:

1. Most nonmanufacturing operations do not have a natural measurement system that allows the analyst to easily define quality.

2. The system that is to be improved is usually fairly obvious in a manufacturing setting, whereas the observability of the process in a nonmanufacturing setting may be fairly low.

For example, if we are trying to improve the performance of a personal computer assembly line, then it is likely that the line will be contained within one facility and the activities of the system will be readily observable. However, if we are trying to improve the operation of a finance organization, then the observability of the process may be low. The actual activities of the process may be performed by a group of people who work in different locations, and the operation steps or workflow sequence may be difficult to observe. Furthermore, the lack of a quantitative and objective measurement system in most nonmanufacturing processes complicates the problem.

The key to applying statistical process-control and other quality-improvement methods in a nonmanufacturing environment is to focus initial efforts on resolving these two issues. We have found that once the system is adequately defined and a valid measurement system has been developed, most of the SPC tools discussed in this chapter can be easily applied to a wide variety of nonmanufacturing operations including finance, marketing, material and procurement, customer support, field service, engineering development and design, and software development and programming.

Flow charts and **operation process charts** are particularly useful in developing process definition and process understanding. A flow chart is simply a chronological sequence of process steps or workflow. Sometimes flow charting is called **process mapping.** Flow charts or process maps must be constructed in sufficient detail to identify **value-added** versus **nonvalue-added** work activity in the process.

Most nonmanufacturing processes have scrap, rework, and other **nonvalue-add operations,** such as unnecessary work steps and choke-points or bottlenecks. A systematic analysis of these processes can often eliminate many of these nonvalue-add activities. The flow chart is very helpful in visualizing and defining the process so that nonvalue-add activities can be identified. Some ways to remove nonvalue-add activities and simplify the process are summarized in the following box:

Ways to Eliminate Nonvalue-Add Activities

1. Rearranging the sequence of worksteps
2. Rearranging the physical location of the operator in the system
3. Changing work methods
4. Changing the type of equipment used in the process
5. Redesigning forms and documents for more efficient use
6. Improving operator training
7. Improving supervision
8. Identifying more clearly the function of the process to all employees
9. Trying to eliminate unnecessary steps
10. Trying to consolidate process steps

Figure 4-31 is an example of a flow chart for a process in a service industry. It was constructed by a process improvement team in an accounting firm that was studying the process of preparing Form 1040 income tax returns, and this particular flow chart documents only one particular subprocess, that of assembling final tax documents. Note the high level of detail in the flow chart, which will assist the team in finding waste or nonvalue-add activities. In this example, the team used special symbols in their flow chart. Specifically, they used the operation process chart symbols shown as follows:

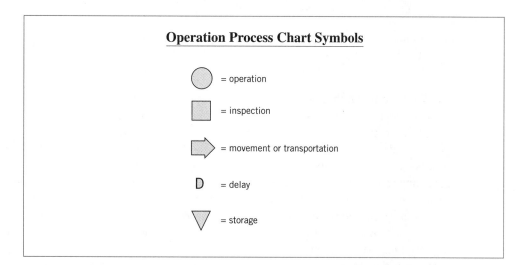

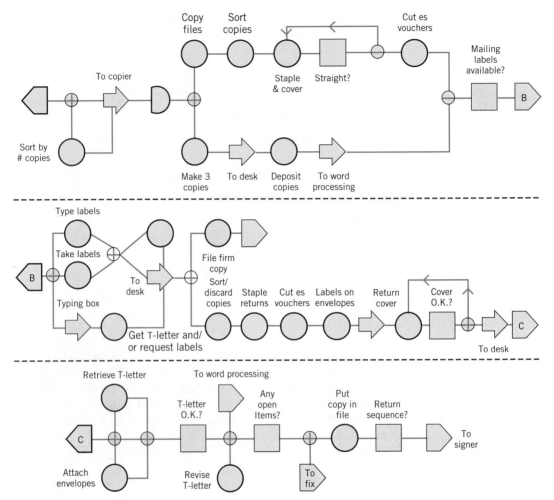

Figure 4-31   Flow chart of the assembly portion of the Form 1040 tax return process.

We have found that these symbols are very useful in helping team members identify improvement opportunities. For example, delays, most inspection, and many movements usually represent nonvalue-add activities. The accounting firm that supplied this example was able to use quality-improvement methods very successfully in their Form 1040 process, reducing the tax document preparation flow time (and work content) by about 25%, and reducing the cycle time for preparing the client bill from over 60 days to zero (that's right, zero!!!). The client's bill is now included with his or her tax return.

As a more complete illustration of some of these ideas, we present an example of applying quality-improvement methods in a planning organization. This planning organization, part of a large aerospace manufacturing concern, produces the plans and documents that accompany each job to the factory floor. These plans are quite extensive,

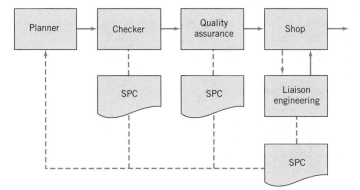

Figure 4-32   A high-level flow chart of the planning process.

often several hundred pages long. Errors in the planning process can have a major impact on the factory floor, contributing to scrap and rework, lost production time, overtime, missed delivery schedules, and many other problems.

Figure 4-32 presents a high-level flow chart of this planning process. After plans are produced, they are sent to a checker who tries to identify obvious errors and defects in the plans. The plans are also reviewed by a quality-assurance organization to ensure that process specifications are being met and that the final product will conform to engineering standards. Then the plans are sent to the shop, where a liaison engineering organization deals with any errors in the plan encountered by manufacturing. This flow chart is useful in presenting an overall picture of the planning system. It is not particularly helpful in uncovering nonvalue-add activities as there is insufficient detail in each of the major blocks. However, each block, such as the planner, checker, and quality-assurance block, could be broken down into a more detailed sequence of work activities and steps. The step-down approach is frequently helpful in constructing flow charts for complex processes. However, even at the relatively high level shown, it is possible to identify at least three areas in which SPC methods could be usefully applied in the planning process.

The management of the planning organization decided to use the reduction of planning errors as a quality-improvement project for their organization. A team of managers, planners, and checkers was chosen to begin this implementation. The team decided that each week three plans would be selected at random from the week's output of plans and that these plans would be analyzed extensively to record all planning errors that could be found. The check sheet shown in Fig. 4-33 was used to record the errors found in each plan. These weekly data were summarized monthly, using the summary check sheet presented in Fig. 4-34. After several weeks, the team was able to summarize the planning error data obtained using the Pareto analysis in Fig. 4-35. This Pareto chart implies that errors in the operations section of the plan are predominant. In fact, 65% of the planning errors are in the operations section. Figure 4-36 presents a further Pareto analysis of the operations section errors. We see that omitted operations and process specifications are the major contributors to the problem.

Data Sheet P/N_____	ERROR	DESCRIPTION	ACTION TAKEN
1. HEADER SECT.			
a. PART NO.			
b. ITEM			
c. MODEL			
2. DWG/DOC SECT.			
3. COMPONENT PART SECT.			
a. PROCUREMENT CODES			
b. STAGING			
c. MOA (# SIGNS)			
4. MOTE SECT.			
5. MATERIAL SECT.			
a. MCC CODE (NON MP&R)			
6. OPERATION SECT.			
a. ISSUE STORE(S)			
b. EQUIPMENT USAGE			
c. OPC FWC MNEMONICS			
d. SEQUENCING			
e. OPER'S OMITTED			
f. PROCESS SPECS			
g. END ROUTE STORE			
h. WELD GRID			
7. TOOL/SHOP AIDS ORDERS			
8. CAR/SHOP STOCK PREP.			
REMARKS:			

CHECKER _____

NO. OF OPERATIONS _____   DATE   _____

Figure 4-33   The check sheet for the planning example.

The team decided that many of the operations errors were occurring because planners were not sufficiently familiar with the manufacturing operations and the process specifications that were currently in place. Consequently, a program was undertaken to refamiliarize planners with the details of factory floor operations and to provide more feedback on the type of planning errors actually experienced. Figure 4-37 presents a run chart of the planning errors per operation for 25 consecutive weeks. Note that there is a general tendency for the planning errors per operation to decline over the first half of the study period. This decline may be due partly to the increased training and supervision activities for the planners and partly to the additional feedback they were given regarding the types of planning errors that were occurring. The team also recommended that substantial changes be made in the work methods used to prepare plans. Rather than having an individual planner with overall responsibility for the operations section, it rec-

Monthly Data Summary					
1. HEADER SECT.					
a. PART NO.					
b. ITEM					
c. MODEL					
2. DWG/DOC SECT.					
3. COMPONENT PART SECT.					
a. PROCUREMENT CODES					
b. STAGING					
c. MOA (# SIGNS)					
4. MOTE SECT.					
5. MATERIAL SECT.					
a. MCC CODE (NON MP&R)					
6. OPERATION SECT.					
a. ISSUE STORE(S)					
b. EQUIPMENT USAGE					
c. OPC FWC MNEMONICS					
d. SEQUENCING					
e. OPER'S OMITTED					
f. PROCESS SPECS					
g. END ROUTE STORE					
h. WELD GRID					
7. TOOL/SHOP AIDS ORDERS					
TOTAL NUMBER ERRORS					
TOTAL OPERATIONS CHECKED					
WEEK ENDING					

Figure 4-34   The summary check sheet.

ommended that this task become a team activity so that knowledge and experience regarding the interface between factor and planning operations could be shared in an effort to further improve the process.

This planning organization began to use other statistical process-control tools as part of their quality-improvement effort. For example, note that the run chart in Fig. 4-37 could be converted to a Shewhart control chart with the addition of a center line and appropriate control limits. Once the planners were exposed to the concepts of SPC, these control charts came into use in the organization. The control charts proved effective in identifying assignable causes; that is, periods of time in which the error rates produced by the system were higher than those that could be justified by chance cause alone. It is its ability to differentiate between assignable and chance causes that makes the control chart so indispensable. Management must react differently to an assignable

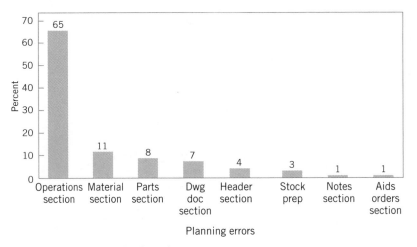

Figure 4-35   Pareto analysis of planning errors.

cause than it does to a chance or random cause. Assignable causes are due to phenomena external to the system, and they must be tracked down and their underlying root causes eliminated. Chance or random causes are part of the system itself. They can only be reduced or eliminated by making changes in how the system operates. This may mean changes in work methods and procedures, improved levels of operator training, different types of equipment and facilities, or improved input materials, all of which are the responsibility of management. In the planning process, many of the common causes identified were related to the experience, training, and supervision of the individual planners, as well as poor input information from design and development engineering.

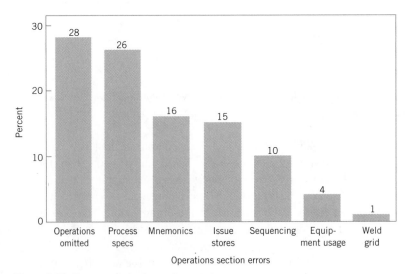

Figure 4-36   Pareto analysis of operations section errors.

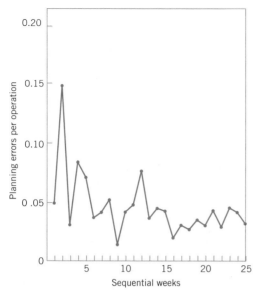

Figure 4-37   A run chart of planning errors.

These common causes were systematically removed from the process, and the long-term impact of the SPC implementation in this organization was to reduce planning errors to a level of less than one planning error per 1000 operations.

4-8 EXERCISES

4-1.   What are chance and assignable causes of variability? What part do they play in the operation and interpretation of a Shewhart control chart?

4-2.   Discuss the relationship between a control chart and statistical hypothesis testing.

4-3.   Discuss type I and type II errors relative to the control chart. What practical implication in terms of process operation do these two types of errors have?

4-4.   What is meant by the statement that a process is in a state of statistical control?

4-5.   If a process is in a state of statistical control, does it necessarily follow that all or nearly all of the units of product produced will be within the specification limits?

4-6.   Discuss the logic underlying the use

of three-sigma limits on Shewhart control charts. How will the chart respond if narrower limits are chosen? How will it respond if wider limits are chosen?

4-7.   What are warning limits on a control chart? How can they be used?

4-8.   Discuss the rational subgroup concept. What part does it play in control chart analysis?

4-9.   When taking samples or subgroups from a process, do you want assignable causes occurring within the subgroups or between them? Fully explain your answer.

4-10.   A molding process uses a five-cavity mold for a part used in an automotive assembly. The wall thickness of the part is the critical quality characteristic. It has been suggested to use $\bar{x}$ and

R charts to monitor this process, and to use as the subgroup or sample all five parts that result from a single "shot" of the machine. What do you think of this sampling strategy? What impact does it have on the ability of the charts to detect assignable causes?

4-11. A manufacturing process produces 500 parts per hour. A sample part is selected about every half-hour, and after five parts are obtained, the average of these five measurements is plotted on an $\bar{x}$ control chart.

(a) Is this an appropriate sampling scheme if the assignable cause in the process results in an instantaneous upward shift in the mean that is of very short duration?

(b) If your answer is no, propose an alternative procedure.

4-12. Consider the sampling scheme proposed in Exercise 4-11. Is this scheme appropriate if the assignable cause results in a slow, prolonged upward drift in the mean? If your answer is no, propose an alternative procedure.

4-13. If the time order of production has not been recorded in a set of data from a process, is it possible to detect the presence of assignable causes?

4-14. What information is provided by the operating-characteristic curve of a control chart?

4-15. How do the costs of sampling, the costs of producing an excessive number of defective units, and the costs of searching for assignable causes impact on the choice of parameters of a control chart?

4-16. Is the average run length performance of a control chart a more meaningful measure of performance than the type I and type II error probabilities? What information does ARL convey that the statistical error probabilities do not?

4-17. Consider the control chart shown here. Does the pattern appear random?

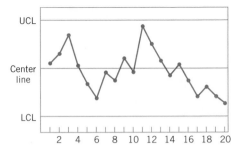

4-18. Consider the control chart shown here. Does the pattern appear random?

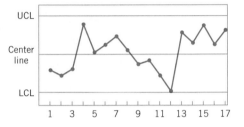

4-19. Consider the control chart shown here. Does the pattern appear random?

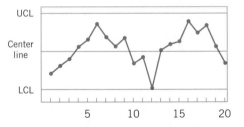

4-20. Consider the control chart shown in Exercise 4-17. Would the use of warning limits reveal any potential out-of-control conditions?

4-21. Apply the Western Electric rules to the control chart in Exercise 4-17. Are any of the criteria for declaring the process out of control satisfied?

4-22. Sketch warning limits on the control chart in Exercise 4-19. Do these limits indicate any potential out-of-control conditions?

4-23. Apply the Western Electric rules to the control chart presented in Exercise 4-19. Would these rules result in any out-of-control signals?

4-24. Consider the time-varying process behavior shown in Figs. *a–e*. Match each of these several patterns of process performance to the corresponding $\bar{x}$ and R charts shown in Figs. 1–5.

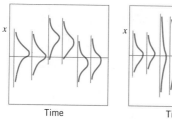

(a)

(b)

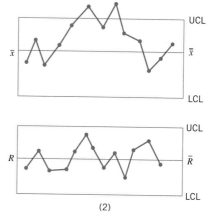

(2)

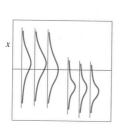

(c)

(d)

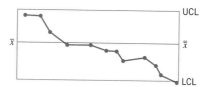

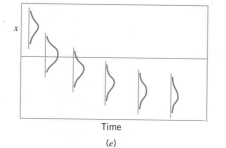

(e)

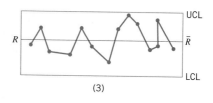

(3)

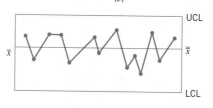

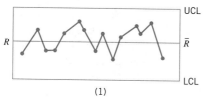

(1)

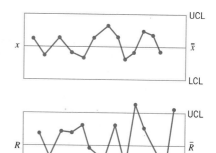

(4)

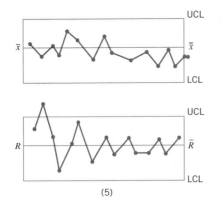

(5)

4-25. Consider the $\bar{x}$ chart for the piston-ring example in this chapter, Fig. 4-3. Let ring diameter be normally distributed, and the sample size is $n = 5$.
(a) Find the two-sigma control limits for this chart.
(b) Suppose it was suggested that the two-sigma limits be used instead of the typical three-sigma limits. What effect would this have on the occurrence of false alarms?
(c) What effect would the use of two-sigma limits have on the in-control ARL of the chart?
(d) Discuss the meaning of your findings in parts (b) and (c).

4-26. Two decision rules are given here. Assume they apply to a normally distributed quality characteristic, the control chart has three-sigma control limits, and the sample size is $n = 5$.

Rule 1: If *one or more* of the next seven samples yield values of the sample average that fall outside the control limits, conclude that the process is out of control.

Rule 2: If *all* of the next seven sample averages fall on the same side of the center line, conclude that the process is out of control.

What is the type I error probability for each of these rules?

4-27. **Continuation of Exercise 4-26.** Consider the situation in Exercise 4-26. If the mean of the quality characteristic shifts one standard deviation—that is, goes out of control by one-sigma—and remains there during the collection of the next seven samples, what is the β-risk associated with each decision rule?

4-28. A normally distributed quality characteristic is monitored by a control chart with L-sigma control limits. Develop a general expression for the probability that a point will plot outside the control limits when the process is really in control.

4-29. A quality characteristic is monitored by a control chart designed so the probability that a certain out-of-control condition will be detected on the first sample following the shift to that state is $1 - \beta$. Find the following:
(a) The probability that the out-of-control condition will be detected on the second sample following the shift.
(b) The probability that the out-of-control condition will be detected on the mth sample following the shift.
(c) The expected number of subgroups analyzed before the shift is detected.
(d) The probability that the first sample following the shift produces a statistic that plots inside the control limits.
(e) The probability that the shift is not detected on the mth subsequent sample.
(f) The probability that two consecutive samples produce statistics that plot outside of the control limits.
(g) The probability that at least one of the next m samples yields values of the test statistics that plot outside the control limits.

4-30. You consistently arrive at your office about one-half hour later than you would like. Develop a cause-and-effect diagram that identifies and outlines the possible causes of this event.

4-31. A car has gone out of control during a snowstorm and struck a tree. Construct a cause-and-effect diagram that identifies and outlines the possible causes of the accident.

4-32. Laboratory glassware has been shipped from the manufacturer to your plant via an overnight package service, and it has arrived damaged. Develop a cause-and-effect diagram that identifies and outlines the possible causes of this event.

Control Charts
for Variables

CHAPTER OUTLINE

5-1 INTRODUCTION

5-2 CONTROL CHARTS FOR $\bar{x}$ AND R

    5-2.1 Statistical Basis of the Charts

    5-2.2 Development and Use of $\bar{x}$ and R Charts

    5-2.3 Charts Based on Standard Values

    5-2.4 Interpretation of $\bar{x}$ and R Charts

    5-2.5 The Effect of Nonnormality on $\bar{x}$ and R Charts

    5-2.6 The Operating-Characteristic Function

    5-2.7 The Average Run Length for the $\bar{x}$ Chart

5-3 CONTROL CHARTS FOR $\bar{x}$ AND S

    5-3.1 Construction and Operation of $\bar{x}$ and S Charts

    5-3.2 The $\bar{x}$ and S Control Charts with Variable Sample Size

    5-3.3 The S^2 Control Chart

5-4 THE SHEWHART CONTROL CHART FOR INDIVIDUAL MEASUREMENTS

5-5 SUMMARY OF PROCEDURES FOR $\bar{x}$, R, AND S CHARTS

5-6 APPLICATIONS OF VARIABLES CONTROL CHARTS

CHAPTER OVERVIEW

A quality characteristic that is measured on a numerical scale is called a **variable.** Examples include dimensions such as length or width, temperature, and volume. This chapter presents Shewhart control charts for these types of quality characteristics. The $\bar{x}$ and R control charts are widely used to monitor the mean and variability of variables. Several variations of the $\bar{x}$ and R charts are also given, including a procedure to adapt them to individual measurements. The chapter concludes with typical applications of variables control charts.

5-1 INTRODUCTION

Many quality characteristics can be expressed in terms of a **numerical measurement.** For example, the diameter of a bearing could be measured with a micrometer and expressed in millimeters. A single measurable quality characteristic, such as a dimension, weight, or volume, is called a **variable.** Control charts for variables are used extensively.

When dealing with a quality characteristic that is a variable, it is usually necessary to monitor both the mean value of the quality characteristic and its variability. Control of the process average or mean quality level is usually done with the control chart for means, or the $\bar{x}$ chart. Process variability can be monitored with either a control chart for the standard deviation, called the S chart, or a control chart for the range, called an R chart. The R chart is more widely used. Usually, separate $\bar{x}$ and R charts are maintained for each quality characteristic of interest. (However, if the quality characteristics are closely related, this can sometimes cause misleading results; refer to Chapter 11 of Part III.) The $\bar{x}$ and R (or S) charts are among the most important and useful on-line statistical process monitoring and control techniques.

Note that it is important to maintain control over both the process mean and process variability. Figure 5-1 illustrates the output of a production process. In Fig. 5-1a both the mean μ and standard deviation σ are in control at their nominal values (say, μ_0 and σ_0); consequently, most of the process output falls within the specification limits. However, in Fig. 5-1b the mean has shifted to a value $\mu_1 > \mu_0$, resulting in a higher fraction of nonconforming product. In Fig. 5-1c the process standard deviation has shifted to a value $\sigma_1 > \sigma_0$. This also results in higher process fallout, even though the process mean is still at the nominal value.

5-2 CONTROL CHARTS FOR $\bar{x}$ AND R

5-2.1 Statistical Basis of the Charts

Suppose that a quality characteristic is normally distributed with mean μ and standard deviation σ, where both μ and σ are known. If $x_1, x_2, \ldots, x_n$ is a sample of size n, then the average of this sample is

$$\bar{x} = \frac{x_1 + x_2 + \cdots + x_n}{n}$$

and we know that $\bar{x}$ is normally distributed with mean μ and standard deviation $\sigma_{\bar{x}} = \sigma/\sqrt{n}$. Furthermore, the probability is $1 - \alpha$ that any sample mean will fall between

$$\mu + Z_{\alpha/2}\sigma_{\bar{x}} = \mu + Z_{\alpha/2}\frac{\sigma}{\sqrt{n}} \qquad (5\text{-}1a)$$

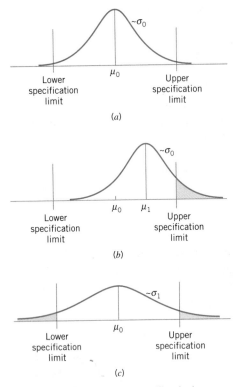

Figure 5-1 The need for controlling both process mean and process variability. (*a*) Mean and standard deviation at nominal levels. (*b*) Process mean $\mu_1 > \mu_0$. (*c*) Process standard deviation $\sigma_1 > \sigma_0$.

and

$$\mu - Z_{\alpha/2}\sigma_{\bar{x}} = \mu - Z_{\alpha/2}\frac{\sigma}{\sqrt{n}} \tag{5-1b}$$

Therefore, if μ and σ are known, equations 5-1a and 5-1b could be used as upper and lower control limits on a control chart for sample means. As noted in Chapter 4, it is customary to replace $Z_{\alpha/2}$ by 3, so that three-sigma limits are employed. If a sample mean falls outside of these limits, it is an indication that the process mean is no longer equal to μ.

We have assumed that the distribution of the quality characteristic is normal. However, the above results are still approximately correct even if the underlying distribution is nonnormal, because of the central limit theorem. We discuss the effect of the normality assumption on variables control charts in Section 5-2.5.

In practice, we usually will not know μ and σ. Therefore, they must be estimated from preliminary samples or subgroups taken when the process is thought to be in control. These estimates should usually be based on at least 20 to 25 samples. Suppose that

m samples are available, each containing n observations on the quality characteristic. Typically, n will be small, often either 4, 5, or 6. These small sample sizes usually result from the construction of rational subgroups and from the fact that the sampling and inspection costs associated with variables measurements are usually relatively large. Let $\bar{x}_1, \bar{x}_2, \ldots, \bar{x}_m$ be the average of each sample. Then the best estimator of μ, the process average, is the grand average—say,

$$\bar{\bar{x}} = \frac{\bar{x}_1 + \bar{x}_2 + \cdots + \bar{x}_m}{m} \tag{5-2}$$

Thus, $\bar{\bar{x}}$ would be used as the center line on the $\bar{x}$ chart.

To construct the control limits, we need an estimate of the standard deviation σ. Recall from Chapter 3 (Section 3-2) that we may estimate σ from either the standard deviations or the ranges of the m samples. For the present, we will concentrate on the range method. If $x_1, x_2, \ldots, x_n$ is a sample of size n, then the range of the sample is the difference between the largest and smallest observations; that is,

$$R = x_{max} - x_{min}$$

Let $R_1, R_2, \ldots, R_m$ be the ranges of the m samples. The average range is

$$\bar{R} = \frac{R_1 + R_2 + \cdots + R_m}{m} \tag{5-3}$$

We may now give the formulas for constructing the control limits on the $\bar{x}$ chart. They are as follows:

Control Limits for the $\bar{x}$ Chart

$$\text{UCL} = \bar{\bar{x}} + A_2\bar{R}$$
$$\text{Center line} = \bar{\bar{x}} \tag{5-4}$$
$$\text{LCL} = \bar{\bar{x}} - A_2\bar{R}$$

The constant A_2 is tabulated for various sample sizes in Appendix Table VI.

Process variability may be monitored by plotting values of the sample range R on a control chart. The center line and control limits of the R chart are as follows:

Control Limits for the R Chart

$$\text{UCL} = D_4 \overline{R}$$

$$\text{Center line} = \overline{R} \tag{5-5}$$

$$\text{LCL} = D_3 \overline{R}$$

The constants D_3 and D_4 are tabulated for various values of n in Appendix Table VI.

Development of Equations 5-4 and 5-5

The development of the equations for computing the control limits on the $\overline{x}$ and R control charts is relatively easy. In Chapter 3 (Section 3-2) we observed that there is a well-known relationship between the range of a sample from a normal distribution and the standard deviation of that distribution. The random variable $W = R/\sigma$ is called the *relative range*. The parameters of the distribution of W are a function of the sample size n. The mean of W is d_2. Consequently, an estimator of σ is $\hat{\sigma} = R/d_2$. Values of d_2 for various sample sizes are given in Appendix Table VI. Therefore, if $\overline{R}$ is the average range of the m preliminary samples, we may use

---

$$\hat{\sigma} = \frac{\overline{R}}{d_2} \tag{5-6}$$

---

to estimate σ. This is an unbiased estimator of σ.

If we use $\overline{\overline{x}}$ as an estimator of μ and $\overline{R}/d_2$ as an estimator of σ, then the parameters of the $\overline{x}$ chart are

$$\text{UCL} = \overline{\overline{x}} + \frac{3}{d_2\sqrt{n}}\overline{R}$$

$$\text{Center line} = \overline{\overline{x}} \tag{5-7}$$

$$\text{LCL} = \overline{\overline{x}} - \frac{3}{d_2\sqrt{n}}\overline{R}$$

If we define

$$A_2 = \frac{3}{d_2\sqrt{n}} \qquad (5\text{-}8)$$

then equation 5-7 reduces to equation 5-4.

Now consider the R chart. The center line will be $\bar{R}$. To determine the control limits, we need an estimate of σ_R. Assuming that the quality characteristic is normally distributed, $\hat{\sigma}_R$ can be found from the distribution of the relative range $W = R/\sigma$. The standard deviation of W, say d_3, is a known function of n. Thus, since

$$R = W\sigma$$

the standard deviation of R is

$$\sigma_R = d_3\sigma$$

Since σ is unknown, we may estimate σ_R by

$$\hat{\sigma}_R = d_3 \frac{\bar{R}}{d_2} \qquad (5\text{-}9)$$

Consequently, the parameters of the R chart with the usual three-sigma control limits are

$$\text{UCL} = \bar{R} + 3\hat{\sigma}_R = \bar{R} + 3d_3\frac{\bar{R}}{d_2}$$

$$\text{Center line} = \bar{R} \qquad (5\text{-}10)$$

$$\text{LCL} = \bar{R} - 3\hat{\sigma}_R = \bar{R} - 3d_3\frac{\bar{R}}{d_2}$$

If we let

$$D_3 = 1 - 3\frac{d_3}{d_2}$$

and

$$D_4 = 1 + 3\frac{d_3}{d_2}$$

equation 5-10 reduces to equation 5-5.

Trial Control Limits

When preliminary samples are used to construct $\bar{x}$ and R control charts, it is customary to treat the control limits obtained from equations 5-4 and 5-5 as **trial control limits.** They allow us to determine whether the process was in control when the m initial samples were selected. To test the hypothesis of past control, plot the values of $\bar{x}$ and R from each sample on the charts and analyze the resulting display. If all points plot inside the control limits and no systematic behavior is evident, then we conclude that the process was in control in the past, and the trial control limits are suitable for controlling current or future production. This analysis of past data is sometimes referred to as a phase 1 analysis. It is highly desirable to have 20–25 samples or subgroups of size n (typically n is between 3 and 5) to compute the trial control limits. We can, of course, work with less data but the control limits are not as reliable.

Suppose that one or more of the values of either $\bar{x}$ or R plot out of control when compared to the trial control limits. Clearly, if control limits for current or future production are to be meaningful, they must be based on data from a process that is in control. Therefore, when the hypothesis of past control is rejected, it is necessary to **revise** the trial control limits. This is done by examining each of the out-of-control points, looking for an assignable cause. If an assignable cause is found, the point is discarded and the trial control limits are recalculated, using only the remaining points. Then these remaining points are reexamined for control. (Note that points that were in control initially may now be out of control, because the new trial control limits will generally be tighter than the old ones.) This process is continued until all points plot in control, at which point the trial control limits are adopted for current use.

In some cases, it may not be possible to find an assignable cause for a point that plots out of control. There are two courses of action open to us. The first of these is to eliminate the point, just as if an assignable cause had been found. There is no analytical justification for choosing this action, other than that points that are outside of the control limits are likely to have been drawn from a probability distribution characteristic of an out-of-control state. The alternative is to retain the point (or points) considering the trial control limits as appropriate for current control. Of course, if the point really does represent an out-of-control condition, the resulting control limits will be too wide. However, if there are only one or two such points, this will not distort the control chart significantly. If future samples still indicate control, then the unexplained points can probably be safely dropped.

Occasionally, when the initial sample values of $\bar{x}$ and R are plotted against the trial control limits, many points will plot out of control. Clearly, if we arbitrarily drop the out-of-control points, we will have an unsatisfactory situation, as few data will remain with which we can recompute reliable control limits. We also suspect that this approach would ignore much useful information in the data. On the other hand, searching for an assignable cause for *each* out-of-control point is unlikely to be successful. We have found that when many of the initial samples plot out of control against the trial limits, it is better to concentrate on the *pattern* formed by these points. Such a pattern will almost always exist. Usually, the assignable cause associated with the pattern of out-of-control points is fairly easy to identify. Removal of this process problem usually results in a major process improvement.

5-2.2   Development and Use of $\bar{x}$ and R Charts

In the previous section we presented the statistical background for $\bar{x}$ and R control charts. We now illustrate the construction and application of these charts. We also discuss some guidelines for using these charts in practice.

....   **EXAMPLE 5-1**   ....

Piston rings for an automotive engine are produced by a forging process. We wish to establish statistical control of the inside diameter of the rings manufactured by this process using $\bar{x}$ and R charts. Twenty-five samples, each of size five, have been taken when we think the process is in control. The inside diameter measurement data from these samples are shown in Table 5-1.

When setting up $\bar{x}$ and R control charts, it is best to begin with the R chart. Because the control limits on the $\bar{x}$ chart depend on the process variability, unless process variability is in control, these limits will not have much meaning. Using the data in Table 5-1, we find that the center line for the R chart is

$$\bar{R} = \frac{\sum_{i=1}^{25} R_i}{25} = \frac{0.581}{25} = 0.023$$

Table 5-1   Inside Diameter Measurements (mm) on Forged Piston Rings, Example 5-1

Sample Number	Observations					$\bar{x}_i$	R_i
1	74.030	74.002	74.019	73.992	74.008	74.010	0.038
2	73.995	73.992	74.001	74.011	74.004	74.001	0.019
3	73.988	74.024	74.021	74.005	74.002	74.008	0.036
4	74.002	73.996	73.993	74.015	74.009	74.003	0.022
5	73.992	74.007	74.015	73.989	74.014	74.003	0.026
6	74.009	73.994	73.997	73.985	73.993	73.996	0.024
7	73.995	74.006	73.994	74.000	74.005	74.000	0.012
8	73.985	74.003	73.993	74.015	73.988	73.997	0.030
9	74.008	73.995	74.009	74.005	74.004	74.004	0.014
10	73.998	74.000	73.990	74.007	73.995	73.998	0.017
11	73.994	73.998	73.994	73.995	73.990	73.994	0.008
12	74.004	74.000	74.007	74.000	73.996	74.001	0.011
13	73.983	74.002	73.998	73.997	74.012	73.998	0.029
14	74.006	73.967	73.994	74.000	73.984	73.990	0.039
15	74.012	74.014	73.998	73.999	74.007	74.006	0.016
16	74.000	73.984	74.005	73.998	73.996	73.997	0.021
17	73.994	74.012	73.986	74.005	74.007	74.001	0.026
18	74.006	74.010	74.018	74.003	74.000	74.007	0.018
19	73.984	74.002	74.003	74.005	73.997	73.998	0.021
20	74.000	74.010	74.013	74.020	74.003	74.009	0.020
21	73.982	74.001	74.015	74.005	73.996	74.000	0.033
22	74.004	73.999	73.990	74.006	74.009	74.002	0.019
23	74.010	73.989	73.990	74.009	74.014	74.002	0.025
24	74.015	74.008	73.993	74.000	74.010	74.005	0.022
25	73.982	73.984	73.995	74.017	74.013	73.998	0.035
						$\Sigma = 1850.028$	0.581
						$\bar{\bar{x}} = 74.001$	$\bar{R} = 0.023$

For samples of $n = 5$, we find from Appendix Table VI that $D_3 = 0$ and $D_4 = 2.115$. Therefore, the control limits for the R chart are, using equation 5-5,

$$\text{LCL} = \bar{R}D_3 = 0.023(0) = 0$$

$$\text{UCL} = \bar{R}D_4 = 0.023(2.115) = 0.049$$

The R chart is shown in Fig. 5-2. When the 25 sample ranges are plotted on this chart, there is no indication of an out-of-control condition.

Since the R chart indicates that process variability is in control, we may now construct the $\bar{x}$ chart. The center line is

$$\bar{\bar{x}} = \frac{\sum_{i=1}^{25} \bar{x}_i}{25} = \frac{1850.028}{25} = 74.001$$

To find the control limits on the $\bar{x}$ chart, we use $A_2 = 0.577$ from Appendix Table VI for samples of size $n = 5$ and equation 5-4 to find

$$\text{UCL} = \bar{\bar{x}} + A_2\bar{R} = 74.001 + (0.577)(0.023) = 74.014$$

and

$$\text{LCL} = \bar{\bar{x}} - A_2\bar{R} = 74.001 - (0.577)(0.023) = 73.988$$

The $\bar{x}$ chart is shown in Fig. 5-3. When the preliminary sample averages are plotted on this chart, no indication of an out-of-control condition is observed. Therefore, since both the $\bar{x}$ and R charts exhibit control, we would conclude that the process is in control at

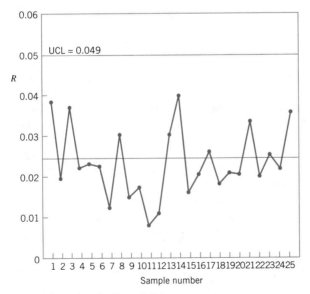

Figure 5-2 R chart for Example 5-1.

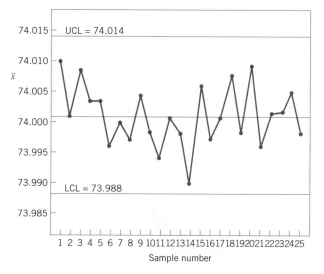

Figure 5-3 $\bar{x}$ chart for Example 5-1.

the stated levels and adopt the trial control limits for use in on-line statistical process control.

In practice, it is common to use computer software packages in control chart construction. Figure 5-4 shows the computer-generated $\bar{x}$ and R charts for this example from Minitab, one popular PC-based statistics package.

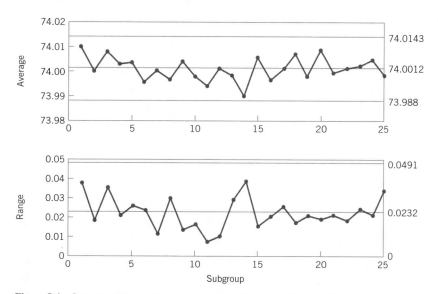

Figure 5-4   Computer-generated control charts from Minitab for Example 5-1.

Estimating Process Capability

The $\bar{x}$ and R charts provide information about the performance or **capability** of the process. From the $\bar{x}$ chart, we may estimate the mean diameter of the piston ring as $\bar{\bar{x}} = 74.001$ mm. The process standard deviation may be estimated using equation 5-6; that is,

$$\hat{\sigma} = \frac{\bar{R}}{d_2} = \frac{0.023}{2.326} = 0.0099$$

where the value of d_2 for samples of size five is found in Appendix Table VI. The specification limits on this piston ring are 74.000 ± 0.05 mm. The control chart data may be used to describe the capability of the process to produce piston rings relative to these specifications. Assuming that piston-ring diameter is a normally distributed random variable, with mean 74.001 and standard deviation 0.0099, we may estimate the fraction of nonconforming piston rings produced as

$$p = P\{x < 73.950\} + P\{x > 74.050\}$$

$$= \Phi\left(\frac{73.950 - 74.001}{0.0099}\right) + 1 - \Phi\left(\frac{74.050 - 74.001}{0.0099}\right)$$

$$= \Phi(-5.15) + 1 - \Phi(4.04)$$

$$\approx 0 + 1 - 0.99998$$

$$\approx 0.00002$$

That is, about 0.002% [20 parts per million (ppm)] of the piston rings produced will be outside of the specifications.

Another way to express process capability is in terms of the **process capability ratio (PCR)** C_p, which for a quality characteristic with both upper and lower specification limits (USL and LSL, respectively) is

$$C_p = \frac{\text{USL} - \text{LSL}}{6\sigma} \tag{5-11}$$

Note that the 6σ spread of the process is the basic definition of process capability. Since σ is usually unknown, we must replace it with an estimate. We frequently use $\hat{\sigma} = \bar{R}/d_2$ as an estimate of σ, resulting in an estimate $\hat{C}_p$ of C_p. For the piston-ring process, since $\bar{R}/d_2 = \hat{\sigma} = 0.0099$, we find that

$$\hat{C}_p = \frac{74.05 - 73.95}{6(0.0099)}$$

$$= \frac{0.10}{0.0594}$$

$$= 1.68$$

This implies that the "natural" tolerance limits in the process (three-sigma above and below the mean) are well inside the lower and upper specification limits. Consequently, a relatively low number of nonconforming piston rings will be produced. The PCR C_p

may be interpreted another way. The quantity

$$P = \left(\frac{1}{C_p}\right) 100\%$$

is simply the percentage of the specification band that the process uses up. For the piston-ring process an estimate of P is

$$\hat{P} = \left(\frac{1}{\hat{C}_p}\right) 100\%$$

$$= \left(\frac{1}{1.68}\right) 100\%$$

$$= 59.5\%$$

That is, the process uses up about 60% of the specification band.

Figure 5-5 illustrates three cases of interest relative to the PCR C_p and process specifications. In Fig. 5-5a the PCR C_p is greater than unity. This means that the process uses up much less than 100% of the tolerance band. Consequently, relatively few non-conforming units will be produced by this process. Figure 5-5b shows a process for which the PCR $C_p = 1$; that is, the process uses up all the tolerance band. For a normal

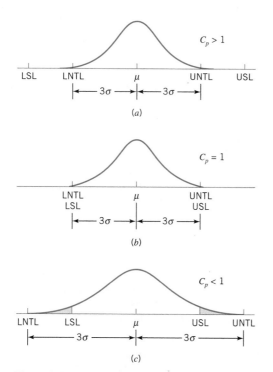

Figure 5-5 Process fallout and the process capability ratio C_p.

distribution this would imply about 0.27% (or 2700 ppm) nonconforming units. Finally, Fig. 5-5c presents a process for which the PCR $C_p < 1$; that is, the process uses up more than 100% of the tolerance band. In this case the process is very yield-sensitive, and a large number of nonconforming units will be produced.

Note that all the cases in Fig. 5-5 assume that the process is centered at the midpoint of the specification band. In many situations this will not be the case, and as we will see in Chapter 7 (which is devoted to a more extensive treatment of process capability analysis), some modification of the PCR C_p is necessary to describe this situation adequately.

Revision of Control Limits and Center Lines

We should always treat the initial set of control limits as *trial* limits, subject to subsequent revision. Generally, the effective use of any control chart will require **periodic revision** of the control limits and center lines. Some practitioners establish regular periods for review and revision of control chart limits, such as every week, every month, or every 25, 50, or 100 samples. When revising control limits, remember that it is highly desirable to use at least 25 samples or subgroups (some authorities recommend 200–300 individual observations) in computing control limits.

Sometimes the user will replace the center line of the $\bar{x}$ chart with a **target value,** say $\bar{\bar{x}}_0$. If the R chart exhibits control, this can be helpful in shifting the process average to the desired value, particularly in processes where the mean may be changed by a fairly simple adjustment of a manipulatable variable in the process. If the mean is not easily influenced by a simple process adjustment, then it is likely to be a complex and unknown function of several process variables and a target value $\bar{\bar{x}}_0$ may not be helpful, as use of that value could result in many points outside the control limits. In such cases we would not necessarily know whether the point was really associated with an assignable cause or whether it plotted outside the limits because of a poor choice for the center line.

When the R chart is out of control, we often eliminate the out-of-control points and recompute a revised value of $\bar{R}$. This value is then used to determine new limits and center line on the R chart and new limits on the $\bar{x}$ chart. This will tighten the limits on both charts, making them consistent with a process standard deviation σ consistent with use of the *revised* $\bar{R}$ in the relationship $\bar{R}/d_2$. This estimate of σ could be used as the basis of a preliminary analysis of process capability.

Continuation of the $\bar{x}$ and R Charts

Once a set of reliable control limits is established, we use the control chart for monitoring future production. This is sometimes called phase 2 of control chart usage.

Fifteen additional samples from the piston-ring manufacturing process were collected after the control charts were established. The data from these new samples are shown in Table 5-2, and the continuations of the $\bar{x}$ and R charts are shown in Fig. 5-6. The control charts indicate that the process is in control, until the $\bar{x}$-value from the thirty-seventh sample is plotted. Since this point and the three subsequent ones plot above the upper control limit, we would suspect that an assignable cause has occurred

Table 5-2   Additional Samples for Example 5-1

Sample Number, i		Observations			$\bar{x}_i$	R_i
26	74.012	74.015	74.030	73.986	74.000	74.009	0.044
27	73.995	74.010	73.990	74.015	74.001	74.002	0.025
28	73.987	73.999	73.985	74.000	73.990	73.992	0.015
29	74.008	74.010	74.003	73.991	74.006	74.004	0.019
30	74.003	74.000	74.001	73.986	73.997	73.997	0.017
31	73.994	74.003	74.015	74.020	74.004	74.007	0.026
32	74.008	74.002	74.018	73.995	74.005	74.006	0.023
33	74.001	74.004	73.990	73.996	73.998	73.998	0.014
34	74.015	74.000	74.016	74.025	74.000	74.011	0.025
35	74.030	74.005	74.000	74.016	74.012	74.013	0.030
36	74.001	73.990	73.995	74.010	74.024	74.004	0.034
37	74.015	74.020	74.024	74.005	74.019	74.017	0.019
38	74.035	74.010	74.012	74.015	74.026	74.020	0.025
39	74.017	74.013	74.036	74.025	74.026	74.023	0.023
40	74.010	74.005	74.029	74.000	74.020	74.015	0.029

around that time. The general pattern of points on the $\bar{x}$ chart from about subgroup 34 and 35 onward is indicative of a shift in the process mean.

Once the control chart is established and is being used in on-line process monitoring, one is often tempted to use the sensitizing rules (or Western Electric rules) discussed in Chapter 4 (Section 4-3.6) to speed up shift detection. Here, for example, the use of such rules would likely result in the shift being detected around sample 35 or 36. However, recall the discussion from Section 4-3.3 in which we discouraged the routine use of these sensitizing rules for on-line monitoring of a stable process because they greatly increase the occurrence of false alarms.

In examining control chart data, it is sometimes helpful to construct a run chart of the individual observations in each sample. This chart is sometimes called a **tolerance chart** or **tier diagram.** This may reveal some pattern in the data, or it may show that a particular value of $\bar{x}$ or R was produced by one or two unusual observations in the sample. When the sample size is larger than 7 or 8, the box plot will usually be a good alternative to the tier diagram. A tier chart of the individual observations in the last 15 samples is shown in Fig. 5-7. This chart does not indicate that the out-of-control signals were generated by unusual individual observations, but instead, they probably resulted from a shift in the mean around the time that sample 34 or 35 was taken. The average of the averages of samples 34 through 40 is 74.015 mm. The specification limits of 74.000 $\pm$ 0.05 mm are plotted in Fig. 5-7, along with a sketch of the normal distribution that represents process output when the process mean equals the in-control value 74.001 mm. A sketch of the normal distribution representing process output at the new apparent mean diameter of 74.015 mm is also shown in Fig. 5-7. It is obvious that a much higher percentage of nonconforming piston rings will be produced at this new mean diameter; in

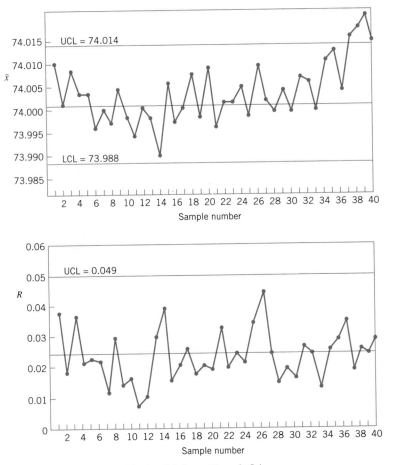

Figure 5-6   Continuation of the $\bar{x}$ and R charts, Example 5-1.

fact, about 0.021% (or 210 ppm) of the process output is now nonconforming. A search for the cause of this shift in the mean must be conducted. The out-of-control-action-plan or **OCAP** for this control chart would play a key role in these activities by directing operating personnel through a series of sequential activities to find the assignable cause. Often additional input and support from engineers, management, and the quality engineering staff is necessary to find and eliminate assignable causes.

Control Limits, Specification Limits, and Natural Tolerance Limits

A point that should be emphasized is that there is no connection or relationship between the *control limits* on the $\bar{x}$ and R charts and the *specification limits* on the process. The control limits are driven by the natural variability of the process (measured by the process standard deviation σ), that is, by the **natural tolerance limits** of the process. It is customary to define the upper and lower natural tolerance limits, say UNTL and LNTL, as 3σ above and below the process mean. The specification limits, on the other

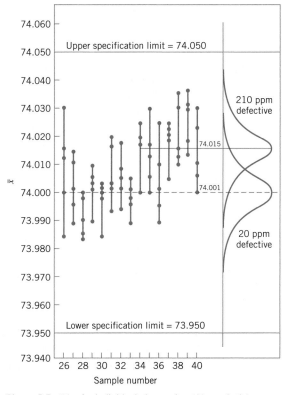

Figure 5-7 Plot for individual observations, Example 5-1.

hand, are determined externally. They may be set by management, the manufacturing engineers, the customer, or by product developers/designers. One should have knowledge of inherent process variability when setting specifications, but **remember that there is no mathematical or statistical relationship between the control limits and specification limits.** The situation is summarized in Fig. 5-8. We have encountered practitioners who have plotted specification limits on the $\bar{x}$ control chart. This practice is completely incorrect and should not be done. When dealing with plots of *individual* observations (not averages), as in Fig. 5-7, it is helpful to plot the specification limits on that chart.

Rational Subgroups

The rational subgroup concept plays an important role in the use of $\bar{x}$ and R control charts. Defining a rational subgroup in practice may be easier if we have a clear understanding of the functions of the two types of control charts. The $\bar{x}$ chart monitors the average quality level in the process. Therefore, samples should be selected in such a way that maximizes the chances for shifts in the process average to occur between samples, and thus to show up as out-of-control points on the $\bar{x}$ chart. The R chart, on the other hand, measures the variability *within* a sample. Therefore, samples should be selected so

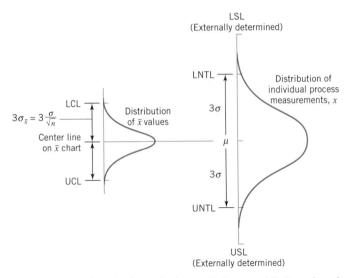

Figure 5-8   Relationship of natural tolerance limits, control limits, and specification limits.

that variability within samples measures only chance or random causes. Another way of saying this is that the $\bar{x}$ chart monitors **between-sample variability** (variability in the process over time), and the R chart measures **within-sample variability** (the instantaneous process variability at a given time).

An important aspect of this is evident from carefully examining how the control limits for the $\bar{x}$ and R charts are determined from past data. The estimate of the process standard deviation σ used in constructing the control limits is calculated from the variability **within** each sample (i.e., from the individual sample ranges). Consequently, the estimate of σ reflects **within-sample variability** only. It is not correct to estimate σ based on the usual quadratic estimation—say,

$$S = \sqrt{\frac{\sum_{i=1}^{m} \sum_{j=1}^{n} (x_{ij} - \bar{\bar{x}})^2}{mn - 1}}$$

where x_{ij} is the jth observation in the ith sample, because if the sample *means* differ, then this will cause S to be too large. Consequently, σ will be overestimated. Pooling all of the preliminary data in this manner to estimate σ is not a good practice because it potentially combines *both between*-sample and *within*-sample variability. The control limits must be based on **within-sample variability** only. Refer to the supplemental text material for more details.

Guidelines for the Design of the Control Chart
To design the $\bar{x}$ and R charts, we must specify the sample size, control limit width, and frequency of sampling to be used. It is not possible to give an exact solution to the prob-

lem of control chart design, unless the analyst has detailed information about both the statistical characteristics of the control chart tests and the economic factors that affect the problem. A complete solution of the problem would require knowledge of the cost of sampling, the costs of investigating and possibly correcting the process in response to out-of-control signals, and the costs associated with producing a product that does not meet specifications. Given this kind of information, an economic decision model could be constructed to allow economically optimum control chart design. In Chapter 9 (Section 9-6) we briefly discuss this approach to the problem. However, it is possible to give some general guidelines now that will aid in control chart design.

If the $\bar{x}$ chart is being used primarily to detect moderate to large process shifts — say, on the order of 2σ or larger — then relatively small samples of size $n = 4, 5,$ or 6 are reasonably effective. On the other hand, if we are trying to detect small shifts, then larger sample sizes of possibly $n = 15$ to $n = 25$ are needed. When smaller samples are used, there is less risk of a process shift occurring while a sample is taken. If a shift does occur while a sample is taken, the sample average can obscure this effect. Consequently, this is an argument for using as small a sample size as is consistent with the magnitude of the process shift that one is trying to detect. An alternative to increasing the sample size is to use warning limits and other sensitizing procedures to enhance the ability of the control chart to detect small process shifts. However, as we discussed in Chapter 4, we do not favor the routine use of these sensitizing rules. If you are interested in small shifts, use the CUSUM or EWMA charts in Chapter 8.

The R chart is relatively insensitive to shifts in the process standard deviation for small samples. For example, samples of size $n = 5$ have only about a 40% chance of detecting on the first sample a shift in the process standard deviation from σ to 2σ. Larger samples would seem to be more effective, but we also know that the range method for estimating the standard deviation drops dramatically in efficiency as n increases. Consequently, for large n — say, $n > 10$ or 12 — it is probably best to use a control chart for S or S^2 instead of the R chart. Details of the construction of these charts are shown in Sections 5-3.1 and 5-3.2.

From a statistical point of view, the operating-characteristic curves of the $\bar{x}$ and R charts can be helpful in choosing the sample size. They will provide the analyst with a feel for the magnitude of process shift that will be detected with a stated probability for any sample size n. These operating-characteristic curves are discussed in Section 5-2.6.

The problem of choosing the sample size and the frequency of sampling is one of **allocating sampling effort.** Generally, the decision maker will have only a limited number of resources to allocate to the inspection process. The available strategies will usually be either to take small, frequent samples or to take larger samples less frequently. For example, the choice may be between samples of size 5 every half hour or samples of size 20 every 2 hours. It is impossible to say which strategy is best in all cases, but current industry practice favors small, frequent samples. The general feeling is that if the interval between samples is too great, too much defective product will be produced before another opportunity to detect the process shift occurs. From economic considerations, if the cost associated with producing defective items is high, smaller, more frequent samples are better than larger, less frequent ones. Variable sample interval and variable sample size schemes could, of course, be used. Refer to Chapter 9.

The rate of production also influences the choice of sample size and sampling frequency. If the rate of production is high—say, 50,000 units per hour—then more frequent sampling is called for than if the production rate is extremely slow. At high rates of production, many nonconforming units of product will be produced in a very short time when process shifts occur. Furthermore, at high production rates, it is sometimes possible to obtain fairly large samples economically. For example, if we produce 50,000 units per hour, it does not take an appreciable difference in time to collect a sample of size 20 compared to a sample of size 5. If per unit inspection and testing costs are not excessive, high-speed production processes are often monitored with moderately large sample sizes.

The use of three-sigma control limits on the $\bar{x}$ and R control charts is a widespread practice. There are situations, however, when departures from this customary choice of control limits are helpful. For example, if false alarms or type I errors (an out-of-control signal is generated when the process is really in control) are very expensive to investigate, then it may be best to use wider control limits than three-sigma—perhaps as wide as 3.5-sigma. However, if the process is such that out-of-control signals are quickly and easily investigated with a minimum of lost time and cost, then narrower control limits, perhaps at 2.5-sigma, are appropriate.

Changing Sample Size on the $\bar{x}$ and R Charts

We have presented the development of $\bar{x}$ and R charts assuming that the sample size n is constant from sample to sample. However, there are situations in which the sample size n is not constant. One situation is that of *variable* sample size; that is, each sample may consist of a different number of observations. The $\bar{x}$ and R charts are generally not used in this case because they lead to a changing center line on the R chart, which is difficult to interpret for many users. The $\bar{x}$ and S charts in Section 5-3.2 would be preferable in this case.

Another situation is that of making a **permanent** (or **semipermanent**) **change** in the sample size because of cost or because the process has exhibited good stability and fewer resources are being allocated for process monitoring. In this case it is easy to recompute the new control limits directly from the old ones without collecting additional samples based on the new sample size. Let

$$\bar{R}_{old} = \text{average range for the old sample size}$$

$$\bar{R}_{new} = \text{average range for the new sample size}$$

$$n_{old} = \text{old sample size}$$

$$n_{new} = \text{new sample size}$$

$$d_2(old) = \text{factor } d_2 \text{ for the old sample size}$$

$$d_2(new) = \text{factor } d_2 \text{ for the new sample size}$$

For the $\bar{x}$ chart the new control limits are

$$UCL = \bar{\bar{x}} + A_2 \left[\frac{d_2(\text{new})}{d_2(\text{old})} \right] \bar{R}_{\text{old}}$$

$$LCL = \bar{\bar{x}} - A_2 \left[\frac{d_2(\text{new})}{d_2(\text{old})} \right] \bar{R}_{\text{old}}$$

(5-12)

where the center line $\bar{\bar{x}}$ is unchanged and the factor A_2 is selected for the *new* sample size. For the R chart, the new parameters are

$$UCL = D_4 \left[\frac{d_2(\text{new})}{d_2(\text{old})} \right] \bar{R}_{\text{old}}$$

$$CL = \bar{R}_{\text{new}} = \left[\frac{d_2(\text{new})}{d_2(\text{old})} \right] \bar{R}_{\text{old}}$$

$$LCL = \max \left\{ 0, D_3 \left[\frac{d_2(\text{new})}{d_2(\text{old})} \right] \bar{R}_{\text{old}} \right\}$$

(5-13)

where D_3 and D_4 are selected for the *new* sample size.

······· **EXAMPLE 5-2** ··

To illustrate the above procedure, consider the $\bar{x}$ and R charts developed for the piston-ring data in Example 5-1. These charts were based on a sample size of five rings. Suppose that since the process exhibits good control, manufacturing engineering personnel want to reduce the sample size to three rings. From Example 5-1, we know that

$$n_{\text{old}} = 5 \qquad \bar{R}_{\text{old}} = 0.023$$

and from Appendix Table VI we have

$$d_2(\text{old}) = 2.326 \qquad d_2(\text{new}) = 1.693$$

Therefore, the new control limits on the $\bar{x}$ chart are found from equation 5-12 as

$$UCL = \bar{\bar{x}} + A_2 \left[\frac{d_2(\text{new})}{d_2(\text{old})} \right] \bar{R}_{\text{old}}$$

$$= 74.001 + (1.023) \left[\frac{1.693}{2.326} \right] (0.023)$$

$$= 74.018$$

and

$$\text{LCL} = \bar{\bar{x}} - A_2 \left[\frac{d_2(\text{new})}{d_2(\text{old})} \right] \bar{R}_{\text{old}}$$

$$= 74.001 - (1.023) \left[\frac{1.693}{2.326} \right] (0.023)$$

$$= 73.984$$

For the R chart, the new parameters are given by equation 5-13:

$$\text{UCL} = D_4 \left[\frac{d_2(\text{new})}{d_2(\text{old})} \right] \bar{R}_{\text{old}}$$

$$= (2.578) \left[\frac{1.693}{2.326} \right] (0.023)$$

$$= 0.043$$

$$\text{CL} = \bar{R}_{\text{new}} = \left[\frac{d_2(\text{new})}{d_2(\text{old})} \right] \bar{R}_{\text{old}}$$

$$= \left[\frac{1.693}{2.326} \right] (0.023)$$

$$= 0.017$$

$$\text{LCL} = \max \left\{ 0, D_3 \left[\frac{d_2(\text{new})}{d_2(\text{old})} \right] \bar{R}_{\text{old}} \right\}$$

$$= 0$$

Figure 5-9 shows the new control limits. Note that the effect of reducing the sample size is to *increase* the width of the limits on the $\bar{x}$ chart (because $\sigma/\sqrt{n}$ is smaller when $n = 5$ than when $n = 3$) and to *lower* the center line and the upper control limit on the R chart (because the expected range from a sample of $n = 3$ is smaller than the expected range from a sample of $n = 5$).

· ·

Probability Limits on the $\bar{x}$ and R Charts

It is customary to express the control limits on the $\bar{x}$ and R charts as a multiple of the standard deviation of the statistic plotted on the charts. If the multiple chosen is k, then the limits are referred to as k-sigma limits, the usual choice being $k = 3$. As mentioned in Chapter 4, however, it is also possible to define the control limits by specifying the type I error level for the test. Such control limits are called probability limits, and are used extensively in the United Kingdom and some parts of Western Europe.

It is easy to choose probability limits for the $\bar{x}$ chart. Since $\bar{x}$ is approximately normally distributed, we may obtain a desired type I error of α by choosing the multiple of sigma for the control limit as $k = Z_{\alpha/2}$, where $Z_{\alpha/2}$ is the upper $\alpha/2$ percentage point

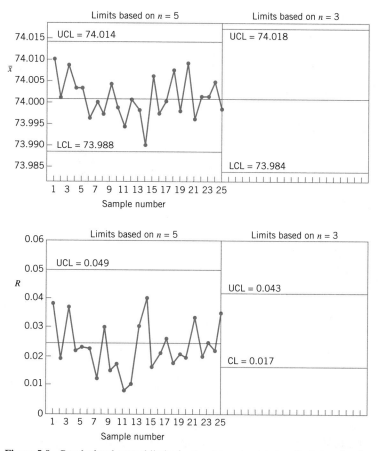

Figure 5-9 Recalculated control limits for the piston-ring data to reflect changing the sample size from $n = 5$ to $n = 3$.

of the standard normal distribution. Note that the usual three-sigma limits imply that the type I error probability is $\alpha = 0.0027$. If we choose $\alpha = 0.002$, for example, as most writers who recommend probability limits do, then $Z_{\alpha/2} = Z_{0.001} = 3.09$. Consequently, there is very little difference between such control limits and three-sigma control limits.

We may also construct R charts using probability limits. If $\alpha = 0.002$, the 0.001 and 0.999 percentage points of the distribution of the relative range $W = R/\sigma$ are required. These points obviously depend on the subgroup size n. Denoting these points by $W_{0.001}(n)$ and $W_{0.999}(n)$, and estimating σ by $\bar{R}/d_2$, we would have the 0.001 and 0.999 limits for R as $W_{0.001}(n)(\bar{R}/d_2)$ and $W_{0.999}(n)(\bar{R}/d_2)$. If we let $D_{0.001} = W_{0.001}(n)/d_2$ and $D_{0.999} = W_{0.999}(n)/d_2$, then the probability limits for the R chart are

$$\text{UCL} = D_{0.999}\bar{R}$$
$$\text{LCL} = D_{0.001}\bar{R}$$

Tables of pairs of values $(D_{0.001}, D_{0.999})$, $(D_{0.005}, D_{0.995})$, and $(D_{0.025}, D_{0.975})$ for $2 \leq n \leq 10$ are in Grant and Leavenworth (1980). These control limits will not differ substantially from the customary three-sigma limits. However, for sample sizes $3 \leq n \leq 6$, they will produce a positive lower control limit for the R chart whereas the conventional three-sigma limits do not.

5-2.3   Charts Based on Standard Values

When it is possible to specify standard values for the process mean and standard deviation, we may use these standards to establish the control charts for $\bar{x}$ and R without analysis of past data. Suppose that the standards given are μ and σ. Then the parameters of the $\bar{x}$ chart are

$$
\begin{aligned}
\text{UCL} &= \mu + 3 \frac{\sigma}{\sqrt{n}} \\
\text{Center line} &= \mu \\
\text{LCL} &= \mu - 3 \frac{\sigma}{\sqrt{n}}
\end{aligned}
\tag{5-14}
$$

The quantity $3/\sqrt{n} = A$, say, is a constant that depends on n, which has been tabulated in Appendix Table VI. Consequently, we could write the parameters of the $\bar{x}$ chart as

$$
\begin{aligned}
\text{UCL} &= \mu + A\sigma \\
\text{Center line} &= \mu \\
\text{LCL} &= \mu - A\sigma
\end{aligned}
\tag{5-15}
$$

To construct the R chart with a standard value of σ, recall that $\sigma = R/d_2$, where d_2 is the mean of the distribution of the relative range. Furthermore, the standard deviation of R is $\sigma_R = d_3\sigma$, where d_3 is the standard deviation of the distribution of the relative range. Therefore, the parameters of the control chart are

$$
\begin{aligned}
\text{UCL} &= d_2\sigma + 3d_3\sigma \\
\text{Center line} &= d_2\sigma \\
\text{LCL} &= d_2\sigma - 3d_3\sigma
\end{aligned}
\tag{5-16}
$$

It is customary to define the constants

$$D_1 = d_2 - 3d_3$$
$$D_2 = d_2 + 3d_3$$

These constants are tabulated in Appendix Table VI. Thus, the parameters of the R chart with standard σ given are

$$\text{UCL} = D_2\sigma$$
$$\text{Center line} = d_2\sigma \qquad (5\text{-}17)$$
$$\text{LCL} = D_1\sigma$$

One must exercise care when standard values of μ and σ are given. It may be that these standards are not really applicable to the process, and as a result, the $\bar{x}$ and R charts will produce many out-of-control signals relative to the specified standards. If the process is really in control at some *other* mean and standard deviation, then the analyst may spend considerable effort looking for assignable causes that do not exist. Standard values of σ seem to give more trouble than standard values of μ. In processes where the mean of the quality characteristic is controlled by adjustments to the machine, standard or target values of μ are sometimes helpful in achieving management goals with respect to process performance.

5-2.4 Interpretation of $\bar{x}$ and R Charts

We have noted previously that a control chart can indicate an out-of-control condition even though no single point plots outside the control limits, if the pattern of the plotted points exhibits nonrandom or systematic behavior. In many cases, the pattern of the plotted points will provide useful diagnostic information on the process, and this information can be used to make process modifications that reduce variability (the goal of statistical process control). Furthermore, these patterns occur fairly often in phase 1 (retrospective study of past data), and their elimination is frequently crucial in bringing a process into control.

   In this section, we briefly discuss some of the more common patterns that appear on $\bar{x}$ and R charts, and indicate some of the process characteristics that may produce these patterns. To effectively interpret $\bar{x}$ and R charts, the analyst must be familiar with both the statistical principles underlying the control chart and the process itself. Additional information on the interpretation of patterns on control charts is in the Western Electric *Statistical Quality Control Handbook* (1956, pp. 149–183).

   In interpreting patterns on the $\bar{x}$ chart, we must first determine whether or not the R chart is in control. Some assignable causes show up on *both* the $\bar{x}$ and R charts. If both

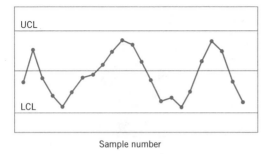

Figure 5-10 Cycles on a control chart.

the $\bar{x}$ and R charts exhibit a nonrandom pattern, the best strategy is to eliminate the R chart assignable causes first. In many cases, this will automatically eliminate the nonrandom pattern on the $\bar{x}$ chart. Never attempt to interpret the $\bar{x}$ chart when the R chart indicates an out-of-control condition.

Cyclic patterns occasionally appear on the control chart. A typical example is shown in Fig. 5-10. Such a pattern on the $\bar{x}$ chart may result from systematic environmental changes such as temperature, operator fatigue, regular rotation of operators and/or machines, or fluctuation in voltage or pressure or some other variable in the production equipment. R charts will sometimes reveal cycles because of maintenance schedules, operator fatigue, or tool wear resulting in excessive variability. In one study in which this author was involved, systematic variability in the fill volume of a metal container was caused by the on–off cycle of a compressor in the filling machine.

A **mixture** is indicated when the plotted points tend to fall near or slightly outside the control limits, with relatively few points near the center line, as shown in Fig. 5-11. A mixture pattern is generated by two (or more) overlapping distributions generating the process output. The probability distributions that could be associated with the mixture pattern in Fig. 5-11 are shown on the right-hand side of that figure. The severity of the mixture pattern depends on the extent to which the distributions overlap. Sometimes mixtures result

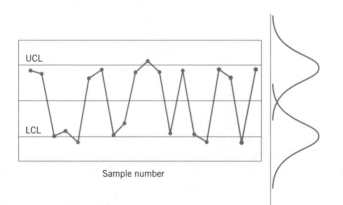

Figure 5-11 A mixture pattern.

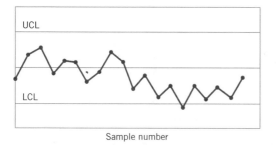

Figure 5-12   A shift in process level.

from "overcontrol," where the operators make process adjustments too often, responding to random variation in the output rather than systematic causes. A mixture pattern can also occur when output product from several sources (such as parallel machines) are fed into a common stream which is then sampled for process monitoring purposes.

A **shift in process level** is illustrated in Fig. 5-12. These shifts may result from the introduction of new workers, methods, raw materials, or machines; a change in the inspection method or standards; or a change in either the skill, attentiveness, or motivation of the operators. Sometimes an improvement in process performance is noted following introduction of a control chart program, simply because of motivational factors influencing the workers.

A **trend,** or continuous movement in one direction, is shown on the control chart in Fig. 5-13. Trends are usually due to a gradual wearing out or deterioration of a tool or some other critical process component. In chemical processes they often occur because of settling or separation of the components of a mixture. They can also result from human causes, such as operator fatigue or the presence of supervision. Finally, trends can result from seasonal influences, such as temperature. When trends are due to tool wear or other systematic causes of deterioration, this may be directly incorporated into the control chart model. A device useful for monitoring and analyzing processes with trends is the **regression control chart** [see Mandel (1969)]. The **modified control chart,** discussed in Chapter 9, is also used when the process exhibits tool wear.

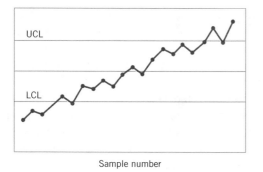

Figure 5-13   A trend in process level.

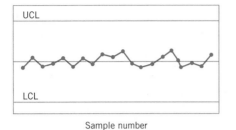

Figure 5-14   A stratification pattern.

Stratification, or a tendency for the points to cluster artificially around the center line, is illustrated in Fig. 5-14. We note that there is a marked lack of natural variability in the observed pattern. One potential cause of stratification is incorrect calculation of control limits. This pattern may also result when the sampling process collects one or more units from several different underlying distributions within each subgroup. For example, suppose that a sample of size five is obtained by taking one observation from each of five parallel processes. If the largest and smallest units in each sample are relatively far apart because they come from two different distributions, then $\bar{R}$ will be incorrectly inflated, causing the limits on the $\bar{x}$ chart to be too wide. In this case $\bar{R}$ incorrectly measures the variability *between* the different underlying distributions, in addition to the chance cause variation that it is intended to measure.

In interpreting patterns on the $\bar{x}$ and R charts, one should consider the two charts jointly. If the underlying distribution is normal, then the random variables $\bar{x}$ and R computed from the same sample are statistically independent. Therefore, $\bar{x}$ and R should behave independently on the control chart. If there is correlation between the $\bar{x}$ and R values—that is, if the points on the two charts "follow" each other—then this indicates that the underlying distribution is skewed. If specifications have been determined assuming normality, then those analyses may be in error.

5-2.5   The Effect of Nonnormality on $\bar{x}$ and R Charts

A fundamental assumption in the development of $\bar{x}$ and R control charts is that the underlying distribution of the quality characteristic is **normal.** In many situations we may have reason to doubt the validity of this assumption. For example, we may know that the underlying distribution is not normal, because we have collected extensive data that indicate the normality assumption is inappropriate. Now if we know the form of the underlying distribution, it is possible to derive the sampling distributions of $\bar{x}$ and R (or some other measure of process variability) and to obtain exact probability limits for the control charts. This approach could be difficult in some cases, and most analysts would probably prefer to use the standard approach based on the normality assumption if they felt that the effect of departure from this assumption was not serious. However, we may know nothing about the form of the underlying distribution, and then our only choice may be to use the normal theory results. Obviously, in either case, we would be interested in knowing the effect of departures from normality on the usual control charts for $\bar{x}$ and R.

Several authors have investigated the effect of departures from normality on control charts. Burr (1967) notes that the usual normal theory control limit constants are very robust to the normality assumption and can be employed unless the population is extremely nonnormal. Schilling and Nelson (1976), Chan, Hapuarachchi, and Macpherson (1988), and Yourstone and Zimmer (1992) have also studied the effect of nonnormality on the control limits of the $\bar{x}$ chart. Schilling and Nelson investigated the uniform, right triangular, gamma (with $\lambda = 1$ and $r = \frac{1}{2}$, 1, 2, 3, and 4) and two bimodal distributions formed as mixtures of two normal distributions. Their study indicates that, in most cases, samples of size four or five are sufficient to ensure reasonable robustness to the normality assumption. The worst cases observed were for small values of r in the gamma distribution [$r = \frac{1}{2}$ and $r = 1$ (the exponential distribution)]. For example, they report the actual α-risk to be 0.014 or less if $n \geq 4$ for the gamma distribution with $r = \frac{1}{2}$, as opposed to a theoretical value of 0.0027 for the normal distribution.

While the use of three-sigma control limits on the $\bar{x}$ chart will produce an α-risk of 0.0027 if the underlying distribution is normal, the same is not true for the R chart. The sampling distribution of R is not symmetric, even when sampling from the normal distribution, and the long tail of the distribution is on the high or positive side. Thus, symmetric three-sigma limits are only an approximation, and the α-risk on such an R chart is *not* 0.0027. (In fact, for $n = 4$, it is $\alpha = 0.00461$.) Furthermore, the R chart is more sensitive to departures from normality than the $\bar{x}$ chart.

5-2.6   The Operating-Characteristic Function

The ability of the $\bar{x}$ and R charts to detect shifts in process quality is described by their operating-characteristic (OC) curves. In this section, we present these OC curves for charts used for on-line control of a process.

Consider the OC curve for an $\bar{x}$ chart with the standard deviation σ known and constant. If the mean shifts from the in-control value—say, μ_0—to another value $\mu_1 = \mu_0 + k\sigma$, the probability of *not* detecting this shift on the first subsequent sample or the β-risk is

$$\beta = P\{\text{LCL} \leq \bar{x} \leq \text{UCL} \mid \mu = \mu_1 = \mu_0 + k\sigma\} \tag{5-18}$$

Since $\bar{x} \sim N(\mu, \sigma^2/n)$, and the upper and lower control limits are $\text{UCL} = \mu_0 + L\sigma/\sqrt{n}$ and $\text{LCL} = \mu_0 - L\sigma/\sqrt{n}$, we may write equation 5-18 as

$$\beta = \Phi\left[\frac{\text{UCL} - (\mu_0 + k\sigma)}{\sigma/\sqrt{n}}\right] - \Phi\left[\frac{\text{LCL} - (\mu_0 + k\sigma)}{\sigma/\sqrt{n}}\right]$$

$$= \Phi\left[\frac{\mu_0 + L\sigma/\sqrt{n} - (\mu_0 + k\sigma)}{\sigma/\sqrt{n}}\right] - \Phi\left[\frac{\mu_0 - L\sigma/\sqrt{n} - (\mu_0 + k\sigma)}{\sigma/\sqrt{n}}\right]$$

where Φ denotes the standard normal cumulative distribution function. This reduces to

$$\beta = \Phi(L - k\sqrt{n}) - \Phi(-L - k\sqrt{n}) \qquad (5\text{-}19)$$

To illustrate the use of equation 5-19, suppose that we are using an $\bar{x}$ chart with $L = 3$ (the usual three-sigma limits), the sample size $n = 5$, and we wish to determine the probability of detecting a shift to $\mu_1 = \mu_0 + 2\sigma$ on the first sample following the shift. Then, since $L = 3$, $k = 2$, and $n = 5$, we have

$$\beta = \Phi[3 - 2\sqrt{5}] - \Phi[-3 - 2\sqrt{5}]$$
$$= \Phi(-1.47) - \Phi(-7.37)$$
$$\simeq 0.0708$$

This is the β-risk, or the probability of not detecting such a shift. The probability that such a shift *will* be detected on the first subsequent sample is $1 - \beta = 1 - 0.0708 = 0.9292$.

To construct the OC curve for the $\bar{x}$ chart, plot the β-risk against the magnitude of the shift we wish to detect expressed in standard deviation units for various sample sizes n. These probabilities may be evaluated directly from equation 5-19. This OC curve is shown in Fig. 5-15 for the case of three-sigma limits ($L = 3$).

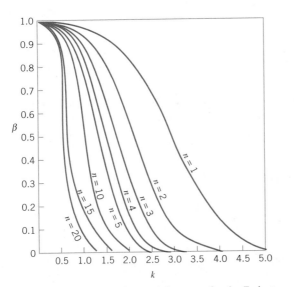

Figure 5-15 Operating-characteristic curves for the $\bar{x}$ chart with three-sigma limits. $\beta = P$ (not detecting a shift of $k\sigma$ in the mean on the first sample following the shift).

Figure 5-15 indicates that for the typical sample sizes of four, five, and six, the $\bar{x}$ chart is not particularly effective in detecting a small shift—say, those on the order of 1.5σ or less—on the first sample following the shift. For example, if the shift is 1.0σ and $n = 5$, then from Fig. 5-15, we have $\beta = 0.75$, approximately. Thus, the probability that the shift will be detected on the first sample is only $1 - \beta = 0.25$. However, the probability that the shift is detected on the second sample is $\beta(1 - \beta) = 0.75(0.25) = 0.19$, whereas the probability that it is detected on the third sample is $\beta^2(1 - \beta) = (0.75^2)0.25 = 0.14$. Thus, the probability that the shift will be detected on the rth subsequent sample is simply $1 - \beta$ times the probability of not detecting the shift on each of the initial $r - 1$ samples, or

$$\beta^{r-1}(1 - \beta)$$

In general, the expected number of samples taken before the shift is detected is simply the **average run length,** or

$$\text{ARL} = \sum_{r=1}^{\infty} r\beta^{r-1}(1 - \beta) = \frac{1}{1 - \beta}$$

Therefore, in our example, we have

$$\text{ARL} = \frac{1}{1 - \beta} = \frac{1}{0.25} = 4$$

In other words, the expected number of samples taken to detect a shift of 1.0σ with $n = 5$ is 4.

The above discussion provides a supportive argument for the use of small sample sizes on the $\bar{x}$ chart. Even though small sample sizes often result in a relatively large β-risk, because samples are collected and tested periodically, there is a very good chance that the shift will be detected reasonably quickly, although perhaps not on the first sample following the shift.

To construct the OC curve for the R chart, the distribution of the relative range $W = R/\sigma$ is employed. Suppose that the in-control value of the standard deviation is σ_0. Then the OC curve plots the probability of not detecting a shift to a new value of σ—say, $\sigma_1 > \sigma_0$—on the first sample following the shift. Figure 5-16 presents the OC curve, in which β is plotted against $\lambda = \sigma_1/\sigma_0$ (the ratio of new to old process standard deviation) for various values of n.

From examining Fig. 5-16, we observe that the R chart is not very effective in detecting process shifts for small sample sizes. For example, if the process standard deviation doubles (i.e., $\lambda = \sigma_1/\sigma_0 = 2$), which is a fairly large shift, then samples of size five have only about a 40% chance of detecting this shift on each subsequent sample. Most quality engineers feel that the R chart is insensitive to small or moderate shifts for the usual subgroup sizes of $n = 4, 5,$ or 6. If $n > 10$ or 12, the S chart discussed in Section 5-3.1 should generally be used instead of the R chart.

The OC curves in Figs. 5-15 and 5-16 assume that the $\bar{x}$ and R charts are used for on-line process control. It is occasionally useful to study the OC curve for the chart used to analyze past data. This can give some indication of how the number of

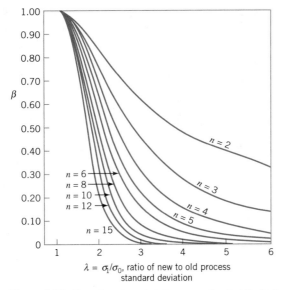

Figure 5-16 Operating-characteristic curves for the R chart with three-sigma limits. (Adapted from A. J. Duncan, "Operating Characteristics of R Charts," *Industrial Quality Control*, vol. 7, no. 5, pp. 40–41, 1951, with permission of the American Society for Quality Control.)

preliminary subgroups used to establish the control chart affects the ability of the chart to detect out-of-control conditions that existed when the data were collected. It is from such analytical studies, as well as practical experience, that the recommendation to use about 20 to 25 preliminary subgroups in establishing $\bar{x}$ and R charts has evolved.

5-2.7 The Average Run Length for the $\bar{x}$ Chart

For any Shewhart control chart, we have noted previously that the ARL can be expressed as

$$\text{ARL} = \frac{1}{P(\text{one point plots out of control})}$$

or

$$\text{ARL}_0 = \frac{1}{\alpha} \qquad (5\text{-}20)$$

for the in-control ARL and

$$\text{ARL}_1 = \frac{1}{1 - \beta} \tag{5-21}$$

for the out-of-control ARL. These results are actually rather intuitive. If the observations plotted on the control chart are independent, then the number of points that must be plotted until the first point exceeds a control limit is a geometric random variable with parameter p (see Chapter 2). The mean of this geometric distribution is simply $1/p$, the average run length.

Since it is relatively easy to develop a general expression for β for the $\bar{x}$ control chart to detect a shift in the mean of $k\sigma$ (see equation 5-19), then it is not difficult to construct a set of ARL curves for the $\bar{x}$ chart. Figure 5-17 presents these ARL curves for sample sizes of $n = 1, 2, 3, 4, 5, 7, 9$, and 16 for the $\bar{x}$ control chart, where the ARL is in terms of the expected number of *samples* taken in order to detect the shift. To illustrate the use of Fig. 5-17, note that if we wish to detect a shift of 1.5σ using a sample size of $n = 3$, then the average number of samples required will be $\text{ARL}_1 = 3$. Note also that we could reduce the ARL_1 to approximately 1 if we increased the sample size to $n = 16$.

Recall our discussion in Chapter 4 (Section 4-3.3) indicating that ARLs are subject to some criticism as performance measures for control charts. We noted there (and

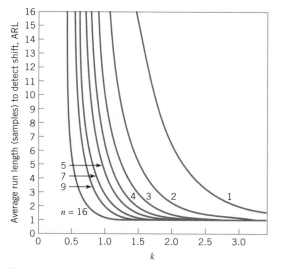

Figure 5-17 Average run length (samples) for the $\bar{x}$ chart with 3-sigma limits, where the process mean shifts by $k\sigma$. (Adapted from *Modern Methods for Quality Control and Improvement*, by H. M. Wadsworth, K. S. Stephens, and A. B. Godfrey, John Wiley & Sons, 1986.)

above as well) that the distribution of run length for a Shewhart control chart is geometric and that this can be a very skewed distribution, so the average (that is, the ARL) may not be the best measure of a "typical" run length. There is another issue concerning ARL related to the fact that the computations for a specific control chart are usually based on **estimates** of the process parameters. This results in inflation of both ARL_0 and ARL_1. For example, suppose that the center line of the chart is estimated perfectly, but the process standard deviation is overestimated by 10%. This would result in $ARL_0 = 517$, considerably longer than the nominal or "theoretical" value of 370. Now with a normally distributed process, we are just as likely to underestimate the process standard deviation by 10%, and this would result in $ARL_0 = 268$, a value considerably shorter than 370. The average is $(268 + 517)/2 = 392$, suggesting that errors in estimating the process standard deviation result in ARLs that are overestimated.

Two other performance measures based on ARL are sometimes of interest. The average time to signal is the number of time periods that occur until a signal is generated on the control chart. If samples are taken at equally spaced intervals of time h, then the **average time to signal** or the ATS is

$$ATS = ARL\ h \qquad\qquad (5\text{-}22)$$

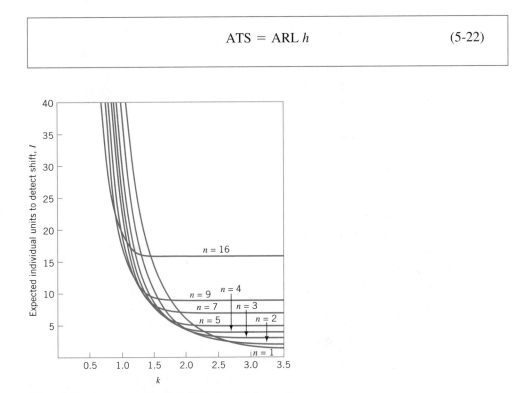

Figure 5-18   Average run length (individual units) for the $\bar{x}$ chart with three-sigma limits, where the process mean shifts by $k\sigma$. (Adapted from *Modern Methods for Quality Control and Improvement*, by H. M. Wadsworth, K. S. Stephens, and A. B. Godfrey, John Wiley & Sons, 1986.)

It may also be useful to express the ARL in terms of the expected number of individual *units* sampled—say, I—rather than the number of samples taken to detect a shift. If the sample size is n, the relationship between I and ARL is

$$I = n \text{ ARL} \qquad (5\text{-}23)$$

Figure 5-18 presents a set of curves that plot the expected number of individual units I that must be sampled for the $\bar{x}$ chart to detect a shift of $k\sigma$. To illustrate the use of the curve, note that to detect a shift of 1.5σ, an $\bar{x}$ chart with $n = 16$ will require that approximately 16 units be sampled, whereas if the sample size is $n = 3$, only about 9 units will be required, on the average.

5-3 CONTROL CHARTS FOR $\bar{x}$ AND S

Although $\bar{x}$ and R charts are widely used, it is occasionally desirable to estimate the process standard deviation directly instead of indirectly through the use of the range R. This leads to control charts for $\bar{x}$ and S, where S is the sample standard deviation.[1] Generally, $\bar{x}$ and S charts are preferable to their more familiar counterparts, $\bar{x}$ and R charts, when either

1. The sample size n is moderately large—say, $n > 10$ or 12. (Recall that the range method for estimating σ loses statistical efficiency for moderate to large samples.)
2. The sample size n is variable.

In this section, we illustrate the construction and operation of $\bar{x}$ and S control charts. We also show how to deal with variable sample size and discuss an alternative to the S chart.

5-3.1 Construction and Operation of $\bar{x}$ and S Charts

Setting up and operating control charts for $\bar{x}$ and S requires about the same sequence of steps as those for $\bar{x}$ and R charts, except that for each sample we must calculate the sample average $\bar{x}$ and the sample standard deviation S. For example, Table 5-3 for Example 5-3 presents the inside diameter measurements on the piston rings used in Example 5-1. Note that we have calculated the sample average and sample standard deviation for each of the 25 samples. We will use these data to illustrate the construction and operation of $\bar{x}$ and S charts.

---

[1]Some authors refer to the S chart as the σ chart.

Table 5-3   Inside Diameter Measurements (mm) for Automobile Engine Piston Rings

Sample Number			Observations			$\bar{x}_i$	S_i
1	74.030	74.002	74.019	73.992	74.008	74.010	0.0148
2	73.995	73.992	74.001	74.011	74.004	74.001	0.0075
3	73.988	74.024	74.021	74.005	74.002	74.008	0.0147
4	74.002	73.996	73.993	74.015	74.009	74.003	0.0091
5	73.992	74.007	74.015	73.989	74.014	74.003	0.0122
6	74.009	73.994	73.997	73.985	73.993	73.996	0.0087
7	73.995	74.006	73.994	74.000	74.005	74.000	0.0055
8	73.985	74.003	73.993	74.015	73.988	73.997	0.0123
9	74.008	73.995	74.009	74.005	74.004	74.004	0.0055
10	73.998	74.000	73.990	74.007	73.995	73.998	0.0063
11	73.994	73.998	73.994	73.995	73.990	73.994	0.0029
12	74.004	74.000	74.007	74.000	73.996	74.001	0.0042
13	73.983	74.002	73.998	73.997	74.012	73.998	0.0105
14	74.006	73.967	73.994	74.000	73.984	73.990	0.0153
15	74.012	74.014	73.998	73.999	74.007	74.006	0.0073
16	74.000	73.984	74.005	73.998	73.996	73.997	0.0078
17	73.994	74.012	73.986	74.005	74.007	74.001	0.0106
18	74.006	74.010	74.018	74.003	74.000	74.007	0.0070
19	73.984	74.002	74.003	74.005	73.997	73.998	0.0085
20	74.000	74.010	74.013	74.020	74.003	74.009	0.0080
21	74.982	74.001	74.015	74.005	73.996	74.000	0.0122
22	74.004	73.999	73.990	74.006	74.009	74.002	0.0074
23	74.010	73.989	73.990	74.009	74.014	74.002	0.0119
24	74.015	74.008	73.993	74.000	74.010	74.005	0.0087
25	73.982	73.984	73.995	74.017	74.013	73.998	0.0162

$$\Sigma = 1850.028 \qquad 0.2350$$
$$\bar{\bar{x}} = 74.001 \qquad \bar{S} = 0.0094$$

If σ^2 is the unknown variance of a probability distribution, then an unbiased estimator of σ^2 is the sample variance

$$S^2 = \frac{\sum\limits_{i=1}^{n} (x_i - \bar{x})^2}{n - 1}$$

However, the sample standard deviation S is *not* an unbiased estimator of σ. In Chapter 3 (Section 3-2) we observed that if the underlying distribution is normal, then S actually estimates $c_4\sigma$, where c_4 is a constant that depends on the sample size n. Furthermore, the standard deviation of S is $\sigma\sqrt{1 - c_4^2}$. This information can be used to establish control charts on $\bar{x}$ and S.

Consider the case where a standard value is given for σ. Since $E(S) = c_4\sigma$, the center line for the chart is $c_4\sigma$. The three-sigma control limits for S are then

$$UCL = c_4\sigma + 3\sigma\sqrt{1 - c_4^2}$$
$$LCL = c_4\sigma - 3\sigma\sqrt{1 - c_4^2}$$

It is customary to define the two constants

$$B_5 = c_4 - 3\sqrt{1 - c_4^2} \tag{5-24a}$$

and

$$B_6 = c_4 + 3\sqrt{1 - c_4^2} \tag{5-24b}$$

Consequently, the parameters of the S chart with a standard value for σ given become

$$UCL = B_6\sigma$$
$$\text{Center line} = c_4\sigma \tag{5-25}$$
$$LCL = B_5\sigma$$

Values of B_5 and B_6 are tabulated for various sample sizes in Appendix Table VI. The parameters of the corresponding $\bar{x}$ chart are given in equation 5-15, Section 5-2.3.

If no standard is given for σ, then it must be estimated by analyzing past data. Suppose that m preliminary samples are available, each of size n, and let S_i be the standard deviation of the ith sample. The average of the m standard deviations is

$$\bar{S} = \frac{1}{m}\sum_{i=1}^{m} S_i$$

The statistic $\bar{S}/c_4$ is an unbiased estimator of σ. Therefore, the parameters of the S chart would be

$$UCL = \bar{S} + 3\frac{\bar{S}}{c_4}\sqrt{1 - c_4^2}$$

$$\text{Center Line} = \bar{S}$$

$$LCL = \bar{S} - 3\frac{\bar{S}}{c_4}\sqrt{1 - c_4^2}$$

We usually define the constants

$$B_3 = 1 - \frac{3}{c_4}\sqrt{1 - c_4^2} \tag{5-26a}$$

and

$$B_4 = 1 + \frac{3}{c_4} \sqrt{1 - c_4^2} \qquad (5\text{-}26b)$$

Consequently, we may write the parameters of the S chart as

$$
\begin{aligned}
\text{UCL} &= B_4 \overline{S} \\
\text{Center line} &= \overline{S} \qquad\qquad (5\text{-}27)\\
\text{LCL} &= B_3 \overline{S}
\end{aligned}
$$

Note that $B_4 = B_6/c_4$ and $B_3 = B_5/c_4$.

When $\overline{S}/c_4$ is used to estimate σ, we may define the control limits on the corresponding $\overline{x}$ chart as

$$\text{UCL} = \overline{\overline{x}} + \frac{3\overline{S}}{c_4 \sqrt{n}}$$

$$\text{Center Line} = \overline{\overline{x}}$$

$$\text{LCL} = \overline{\overline{x}} - \frac{3\overline{S}}{c_4 \sqrt{n}}$$

Let the constant $A_3 = 3/(c_4 \sqrt{n})$. Then the $\overline{x}$ chart parameters become

$$
\begin{aligned}
\text{UCL} &= \overline{\overline{x}} + A_3 \overline{S} \\
\text{Center line} &= \overline{\overline{x}} \qquad\qquad (5\text{-}28)\\
\text{LCL} &= \overline{\overline{x}} - A_3 \overline{S}
\end{aligned}
$$

The constants B_3, B_4, and A_3 for construction of $\overline{x}$ and S charts from past data are listed in Appendix Table VI for various sample sizes.

Note that we have assumed that the sample standard deviation is defined as

$$S = \sqrt{\frac{\sum_{i=1}^{n} (x_i - \overline{x})^2}{n - 1}} \qquad (5\text{-}29)$$

Some authors define S with n in the denominator of equation 5-29 instead of $n - 1$. When this is the case, the definitions of the constants c_4, B_3, B_4, and A_3 are altered. The corresponding constants based on the use of n in calculating S are called c_2, B_1, B_2, and A_1, respectively. See Bowker and Lieberman (1972) for their definitions.

Traditionally, quality engineers have preferred the R chart to the S chart because of the simplicity of calculating R from each sample. The current availability of hand-held calculators with automatic calculation of S and the increased availability of microcomputers for on-line implementation of control charts directly at the workstation have eliminated any computational difficulty.

••••••• **EXAMPLE 5-3** ••

We will illustrate the construction of $\bar{x}$ and S charts using the piston-ring inside diameter measurements in Table 5-3. The grand average and the average standard deviation are

$$\bar{\bar{x}} = \frac{1}{25} \sum_{i=1}^{25} \bar{x}_i = \frac{1}{25}(1850.028) = 74.001$$

and

$$\bar{S} = \frac{1}{25} \sum_{i=1}^{25} S_i = \frac{1}{25}(0.2350) = 0.0094$$

respectively. Consequently, the parameters for the $\bar{x}$ chart are

$$\text{UCL} = \bar{\bar{x}} + A_3\bar{S} = 74.001 + (1.427)(0.0094) = 74.014$$
$$\text{CL} = \bar{\bar{x}} = 74.001$$
$$\text{LCL} = \bar{\bar{x}} - A_3\bar{S} = 74.001 - (1.427)(0.0094) = 73.988$$

and for the S chart

$$\text{UCL} = B_4\bar{S} = (2.089)(0.0094) = 0.0196$$
$$\text{CL} = \bar{S} = 0.0094$$
$$\text{LCL} = B_3\bar{S} = (0)(0.0094) = 0$$

The control charts are shown in Fig. 5-19.

Note that the control limits for the $\bar{x}$ chart based on $\bar{S}$ are identical to the $\bar{x}$ chart control limits in Example 5-1 where the limits were based on $\bar{R}$. They will not always be the same, and in general, the $\bar{x}$ chart control limits based on $\bar{S}$ will be slightly different than limits based on $\bar{R}$.

•••

Estimation of σ

We can estimate the process standard deviation using the fact that S/c_4 is an unbiased estimate of σ. Therefore, since $c_4 = 0.9400$ for samples of size five, our estimate of the process standard deviation is

$$\hat{\sigma} = \frac{\bar{S}}{c_4} = \frac{0.0094}{0.9400} = 0.01$$

This estimate is very similar to that of σ obtained via the range method in Example 5-1.

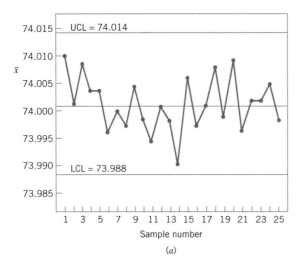

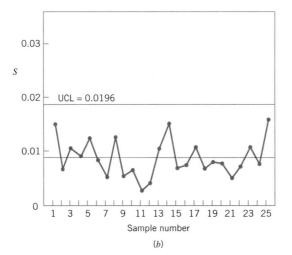

Figure 5-19   The $\bar{x}$ and S control charts for Example 5-3. (*a*) The $\bar{x}$ chart with control limits based on $\bar{S}$. (*b*) The S control chart.

5-3.2   The $\bar{x}$ and S Control Charts with Variable Sample Size

The $\bar{x}$ and S control charts are relatively easy to apply in cases where the sample sizes are variable. In this case, we should use a weighted average approach in calculating $\bar{\bar{x}}$ and $\bar{S}$. If n_i is the number of observations in the ith sample, then use

$$\bar{\bar{x}} = \frac{\sum\limits_{i=1}^{m} n_i \bar{x}_i}{\sum\limits_{i=1}^{m} n_i} \tag{5-30}$$

and

$$\bar{S} = \left[\frac{\sum\limits_{i=1}^{m} (n_i - 1)S_i^2}{\sum\limits_{i=1}^{m} n_i - m} \right]^{1/2} \qquad (5\text{-}31)$$

as the center lines on the $\bar{x}$ and S control charts, respectively. The control limits would be calculated from equations 5-27 and 5-28, respectively, but the constants A_3, B_3, and B_4 will depend on the sample size used in each individual subgroup.

······· **EXAMPLE 5-4** ···

Consider the data in Table 5-4, which is a modification of the piston-ring data used previously. Note that the sample sizes vary from $n = 3$ to $n = 5$. We may use the procedure described above to set up the $\bar{x}$ and S control charts. The weighted grand

Table 5-4 Inside Diameter Measurements (mm) on Automobile Engine Piston Rings

Sample Number	Observations					$\bar{x}_i$	S_i
1	74.030	74.002	74.019	73.992	74.008	74.010	0.0148
2	73.995	73.992	74.001			73.996	0.0046
3	73.988	74.024	74.021	74.005	74.002	74.008	0.0147
4	74.002	73.996	73.993	74.015	74.009	74.003	0.0091
5	73.992	74.007	74.015	73.989	74.014	74.003	0.0122
6	74.009	73.994	73.997	73.985		73.996	0.0099
7	73.995	74.006	73.994	74.000		73.999	0.0055
8	73.985	74.003	73.993	74.015	73.988	73.997	0.0123
9	74.008	73.995	74.009	74.005		74.004	0.0064
10	73.998	74.000	73.990	74.007	73.995	73.998	0.0063
11	73.994	73.998	73.994	73.995	73.990	73.994	0.0029
12	74.004	74.000	74.007	74.000	73.996	74.001	0.0042
13	73.983	74.002	73.998			73.994	0.0100
14	74.006	73.967	73.994	74.000	73.984	73.990	0.0153
15	74.012	74.014	73.998			74.008	0.0087
16	74.000	73.984	74.005	73.998	73.996	73.997	0.0078
17	73.994	74.012	73.986	74.005		73.999	0.0115
18	74.006	74.010	74.018	74.003	74.000	74.007	0.0070
19	73.984	74.002	74.003	74.005	73.997	73.998	0.0085
20	74.000	74.010	74.013			74.008	0.0068
21	73.982	74.001	74.015	74.005	73.996	74.000	0.0122
22	74.004	73.999	73.990	74.006	74.009	74.002	0.0074
23	74.010	73.989	73.990	74.009	74.014	74.002	0.0119
24	74.015	74.008	73.993	74.000	74.010	74.005	0.0087
25	73.982	73.984	73.995	74.017	74.013	73.998	0.0162

mean and weighted average standard deviation are computed from equations 5-30 and 5-31 as follows:

$$\bar{\bar{x}} = \frac{\sum\limits_{i=1}^{25} n_i \bar{x}_i}{\sum\limits_{i=1}^{25} n_i} = \frac{5(74.010) + 3(73.996) + \cdots + 5(73.998)}{5 + 3 + \cdots + 5}$$

$$= \frac{8362.075}{113}$$

$$= 74.001$$

and

$$\bar{S} = \left[\frac{\sum\limits_{i=1}^{25} (n_i - 1)S_i^2}{\sum\limits_{i=1}^{25} n_i - 25} \right]^{1/2} = \left[\frac{4(0.0148)^2 + 2(0.0046)^2 + \cdots + 4(0.0162)^2}{5 + 3 + \cdots + 5 - 25} \right]^{1/2}$$

$$= \left[\frac{0.008426}{88} \right]^{1/2}$$

$$= 0.0098$$

Therefore, the center line of the $\bar{x}$ chart is $\bar{\bar{x}} = 74.001$, and the center line of the S chart is $\bar{S} = 0.0098$. The control limits may now be easily calculated. To illustrate, consider the first sample. The limits for the $\bar{x}$ chart are

$$\text{UCL} = 74.001 + (1.427)(0.0098) = 74.015$$
$$\text{CL} = 74.001$$
$$\text{LCL} = 74.001 - (1.427)(0.0098) = 73.987$$

The control limits for the S chart are

$$\text{UCL} = (2.089)(0.0098) = 0.020$$
$$\text{CL} = 0.0098$$
$$\text{LCL} = 0(0.0098) = 0$$

Note that we have used the values of A_3, B_3, and B_4 for $n_1 = 5$. The limits for the second sample would use the values of these constants for $n_2 = 3$. The control limit calculations for all 25 samples are summarized in Table 5-5. The control charts are plotted in Fig. 5-20.

··

An alternative to using variable-width control limits on the $\bar{x}$ and S control charts is to base the control limit calculations on an average sample size $\bar{n}$. If the n_i are not very

Table 5-5 Computation of Control Limits for $\bar{x}$ and S Charts with Variable Sample Size

Sample	n	$\bar{x}$	S	A_3	$\bar{x}$ Chart LCL	$\bar{x}$ Chart UCL	B_3	B_4	S Chart LCL	S Chart UCL
1	5	74.010	0.0148	1.427	73.987	74.015	0	2.089	0	0.020
2	3	73.996	0.0046	1.954	73.982	74.020	0	2.568	0	0.025
3	5	74.008	0.0147	1.427	73.987	74.015	0	2.089	0	0.020
4	5	74.003	0.0091	1.427	73.987	74.015	0	2.089	0	0.020
5	5	74.003	0.0122	1.427	73.987	74.015	0	2.089	0	0.020
6	4	73.996	0.0099	1.628	73.985	74.017	0	2.266	0	0.022
7	4	73.999	0.0055	1.628	73.985	74.017	0	2.266	0	0.022
8	5	73.997	0.0123	1.427	73.987	74.015	0	2.089	0	0.020
9	4	74.004	0.0064	1.628	73.985	74.017	0	2.266	0	0.022
10	5	73.998	0.0063	1.427	73.987	74.015	0	2.089	0	0.020
11	5	73.994	0.0029	1.427	73.987	74.015	0	2.089	0	0.020
12	5	74.001	0.0042	1.427	73.987	74.015	0	2.089	0	0.020
13	3	73.994	0.0100	1.954	73.982	74.020	0	2.568	0	0.025
14	5	73.990	0.0153	1.427	73.987	74.015	0	2.089	0	0.020
15	3	74.008	0.0087	1.954	73.982	74.020	0	2.568	0	0.025
16	5	73.997	0.0078	1.427	73.987	74.015	0	2.089	0	0.020
17	4	73.999	0.0115	1.628	73.985	74.017	0	2.226	0	0.022
18	5	74.007	0.0070	1.427	73.987	74.015	0	2.089	0	0.020
19	5	73.998	0.0085	1.427	73.987	74.015	0	2.089	0	0.020
20	3	74.008	0.0068	1.954	73.982	74.020	0	2.568	0	0.025
21	5	74.000	0.0122	1.427	73.987	74.015	0	2.089	0	0.020
22	5	74.002	0.0074	1.427	73.987	74.015	0	2.089	0	0.020
23	5	74.002	0.0119	1.427	73.987	74.015	0	2.089	0	0.020
24	5	74.005	0.0087	1.427	73.987	74.015	0	2.089	0	0.020
25	5	73.998	0.0162	1.427	73.987	74.015	0	2.089	0	0.020

different, this approach may be satisfactory in some situations; it is particularly helpful if the charts are to be used in a presentation to management. Since the average sample size n_i may not be an integer, a useful alternative is to base these approximate control limits on a modal (most common) sample size.

Estimation of σ

We may estimate the process standard deviation, σ, from the individual sample values S_i. First, average all the values of S_i for which $n_i = 5$ (the most frequently occurring value of n_i). This gives

$$\bar{S} = \frac{0.1605}{17}$$
$$= 0.0094$$

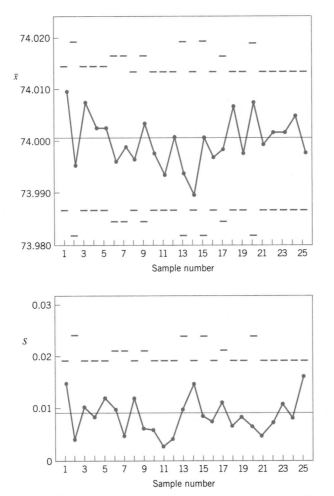

Figure 5-20 The $\bar{x}$ and S control charts for piston-ring data with variable sample size, Example 5-4.

The estimate of the process σ is then

$$\hat{\sigma} = \frac{\overline{S}}{c_4} = \frac{0.0094}{0.9400} = 0.01$$

where the value of c_4 used is for samples of size $n = 5$.

5-3.3 The S^2 Control Chart

Most quality engineers use either the R chart or the S chart to monitor process variability, with S preferable to R for moderate to large sample sizes. Some practitioners recommend a control chart based directly on the sample variance S^2. The parameters for the S^2 control chart are

$$UCL = \frac{\overline{S}^2}{n-1} \chi^2_{\alpha/2, n-1}$$

$$\text{Center line} = \overline{S}^2 \qquad\qquad (5\text{-}32)$$

$$LCL = \frac{\overline{S}^2}{n-1} \chi^2_{1-(\alpha/2), n-1}$$

where $\chi^2_{\alpha/2, n-1}$ and $\chi^2_{1-(\alpha/2), n-1}$ denote the upper and lower $\alpha/2$ percentage points of the chi-square distribution with $n-1$ degrees of freedom, and $\overline{S}^2$ is an average sample variance obtained from the analysis of preliminary data. A standard value σ^2 could be used in equation 5-32 instead of $\overline{S}^2$ if one were available. Note that this control chart is defined with probability limits.

5-4 THE SHEWHART CONTROL CHART FOR INDIVIDUAL MEASUREMENTS

There are many situations in which the sample size used for process monitoring is $n = 1$; that is, the sample consists of an individual unit. Some examples of these situations are as follows:

1. Automated inspection and measurement technology is used, and every unit manufactured is analyzed so there is no basis for rational subgrouping.
2. The production rate is very slow, and it is inconvenient to allow samples sizes of $n > 1$ to accumulate before analysis. The long interval between observations will cause problems with rational subgrouping.
3. Repeat measurements on the process differ only because of laboratory or analysis error, as in many chemical processes.
4. Multiple measurements are taken on the same unit of product, such as measuring oxide thickness at several different locations on a wafer in semiconductor manufacturing.
5. In process plants, such as papermaking, measurements on some parameter such as coating thickness *across* the roll will differ very little and produce a standard deviation that is much too small if the objective is to control coating thickness *along* the roll.

In such situations, the control chart for individual units is useful. (The cumulative sum and exponentially weighted moving-average control charts discussed in Chapter 8 will be a better alternative when the magnitude of the shift in process mean that is of interest is small.) In many applications of the individuals control chart we use the moving range of two successive observations as the basis of estimating the process variability. The moving range is defined as

$$MR_i = |x_i - x_{i-1}|$$

It is also possible to establish a control chart on the moving range. The procedure is illustrated in the following example.

······· **EXAMPLE 5-5** ···

The viscosity of an aircraft primer paint is an important quality characteristic. The product is produced in batches, and because each batch takes several hours to produce, the production rate is too slow to allow sample sizes greater than one. The viscosity of the previous 15 batches is shown in Table 5-6.

To set up the control chart for individual observations, note that the sample average of the 15 viscosity readings is $\bar{x} = 33.52$ and that the average of the moving ranges of two observations is $\overline{MR} = 0.48$. To set up the moving-range chart, we note that $D_3 = 0$ and $D_4 = 3.267$ for $n = 2$. Therefore, the moving-range chart has center line

Table 5-6   Viscosity of Aircraft Primer Paint

Batch Number	Viscosity x	Moving Range MR
1	33.75	
2	33.05	0.70
3	34.00	0.95
4	33.81	0.19
5	33.46	0.35
6	34.02	0.56
7	33.68	0.34
8	33.27	0.41
9	33.49	0.22
10	33.20	0.29
11	33.62	0.42
12	33.00	0.62
13	33.54	0.54
14	33.12	0.42
15	33.84	0.72
	$\bar{x} = 33.52$	$\overline{MR} = 0.48$

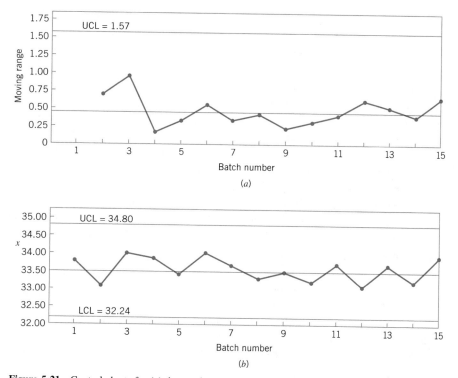

Figure 5-21   Control charts for (*a*) the moving range and for (*b*) individual observations on viscosity.

$\overline{MR}$ = 0.48, LCL = 0, and UCL = $D_4\overline{MR}$ = (3.267)0.48 = 1.57. The control chart is shown in Fig. 5-21*a*. Since no points exceed the upper control limit, we may now set up the control chart for individual viscosity measurements.

For the control chart for individual measurements, the parameters are

$$UCL = \bar{x} + 3\,\frac{\overline{MR}}{d_2}$$

$$\text{Center line} = \bar{x} \qquad\qquad (5\text{-}33)$$

$$LCL = \bar{x} - 3\,\frac{\overline{MR}}{d_2}$$

If a moving range of $n = 2$ observations is used, then $d_2 = 1.128$. For the data in Table 5-6, we have

$$UCL = \bar{x} + 3\,\frac{\overline{MR}}{d_2} = 33.52 + 3\,\frac{0.48}{1.128} = 34.80$$

$$\text{Center line} = \bar{x} = 33.52$$

$$LCL = \bar{x} - 3\,\frac{\overline{MR}}{d_2} = 33.52 - 3\,\frac{0.48}{1.128} = 32.24$$

The control chart for individual batch viscosity measurements is shown in Fig. 5-21b. There is no indication of an out-of-control condition.

• •

Interpretation of the Charts

The chart for individuals can be interpreted much like an ordinary $\bar{x}$ control chart. A shift in the process average will result in either a point (or points) outside the control limits, or a pattern consisting of a run on one side of the center line.

Table 5-7 contains data on aircraft primer paint viscosity for batches 16–30. These data are plotted in Fig. 5-22 on the continuation of the control chart for individuals and the moving-range control chart developed in Example 5-5. As this figure makes clear, an upward shift in mean viscosity has occurred around batch 20 or 21, since there is an obvious "shift in process level" pattern on the chart for individuals. Note that the moving-range chart also reacts to this level shift with a single large spike at sample 20. This spike on the moving-range chart is sometimes helpful in identifying exactly where a process shift in the mean has occurred.

Some care should be exercised in interpreting patterns on the moving-range chart. The moving ranges are correlated, and this correlation may often induce a pattern of runs or cycles on the chart. Such a pattern is evident on the moving-range chart in Fig. 5-22. The individual measurements on the x chart are assumed to be uncorrelated, however, and any apparent pattern on this chart should be carefully investigated.

Table 5-7   Viscosity for Aircraft Primer Paint, Batches 16–30

Batch Number	Viscosity x	Moving Range MR
16[a]	33.50	0.34
17	33.25	0.25
18	33.40	0.15
19	33.27	0.13
20	34.65	1.38
21	34.80	0.15
22	34.55	0.25
23	35.00	0.45
24	34.75	0.25
25	34.50	0.25
26	34.70	0.20
27	34.29	0.41
28	34.61	0.32
29	34.49	0.12
30	35.03	0.54

[a] The moving range for batch 16 was computed as the difference in viscosity readings between batches 16 and 15; that is, $MR_{16} = |x_{16} - x_{15}| = |33.50 - 33.84| = 0.34$.

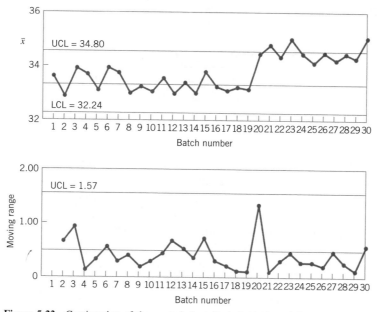

Figure 5-22 Continuation of the control chart for individuals and the moving range using the additional data in Table 5-7.

Some authorities have recommended not constructing and plotting the moving range chart. They point out (correctly, we add) that the moving range chart cannot really provide useful information about a shift in process variability. In effect, shifts in the process mean also show up on the moving range chart. Our feeling is that as long as the analyst is careful in interpretation and relies primarily on the individuals chart, little trouble will ensue from plotting both charts.

Average Run Lengths

Crowder (1987b) has studied the average run length of the combined control chart for individuals and moving-range chart. He produces ARLs for various settings of the control limits and shifts in the process mean and standard deviation. In general, his work shows that the ARL_0 of the combined procedure will generally be much less than the ARL_0 of a standard Shewhart control chart when the process is in control (recall that ARL_0 for a Shewhart chart is 370 samples), if we use the conventional three-sigma limits on the charts. In general, results closer to the Shewhart in-control ARL are obtained if we use three-sigma limits on the chart for individuals and compute the upper control limit on the moving-range chart from

$$\text{UCL} = D\overline{\text{MR}}$$

where the constant D should be chosen such that $4 \leq D \leq 5$.

One can get a very good idea about the ability of the individuals control chart to detect process shifts by looking at the OC curves in Fig. 5-15 or the ARL curves in

Fig. 5-17. For an individuals control chart with three-sigma limits, we can compute the following:

Size of Shift	β	ARL_1
1σ	0.9772	43.96
2σ	0.8413	6.30
3σ	0.5000	2.00

Note that the ability of the individuals control chart to detect small shifts is very poor. For instance, consider a continuous chemical process in which samples are taken every hour. If a shift in the process mean of about one standard deviation occurs, the information above tells us that it will take about 44 samples, on the average, to detect the shift. This is nearly two full days of continuous production in the out-of-control state, a situation that has potentially devastating economic consequences.

Some individuals have suggested that control limits narrower than three-sigma be used on the chart for individuals to enhance its ability to detect small process shifts. This is a dangerous suggestion, as narrower limits will dramatically reduce the value of ARL_0 and increase the occurrence of false alarms to the point where the charts are ignored and hence become useless. If we are interested in detecting small shifts, then the correct approach is to use either the cumulative-sum control chart or the exponentially weighted moving-average control chart (see Chapter 8).

Normality

Our discussion in this section has made an assumption that the observations follow a normal distribution. Borror, Montgomery, and Runger (1999) have studied the behavior of the Shewhart control chart for individuals when the process data are not normal. They investigated various gamma distributions to represent skewed process data and t distributions to represent symmetric normal-like data. They found that the in-control ARL is dramatically affected by nonnormal data. For example, if the individuals chart has three-sigma limits so that for normal data $ARL_0 = 370$, the actual ARL_0 for gamma-distributed data is between 45 and 97, depending on the shape of the gamma distribution (more highly skewed distributions yield poorer performance). For the t distribution, the ARL_0 values range from 76 to 283 as the degrees of freedom increase from 4 to 50 (that is, as the t becomes more like the normal distribution).

In the face of these results we conclude that if the process shows evidence of even moderate departure from normality, the control limits given here may be entirely inappropriate. One approach to dealing with the problem of nonnormality would be to determine the control limits for the individuals control chart based on the percentiles of the correct underlying distribution. These percentiles could be obtained from a histogram if a large sample (at least 100 but preferably 200 observations) were available, or from a probability distribution fit to the data. See Willemain and Runger (1996) for details on designing control charts from empirical reference distributions. Another approach would be to transform the original variable to a new variable that is approximately normally distributed, and then apply control charts to the new variable. Borror,

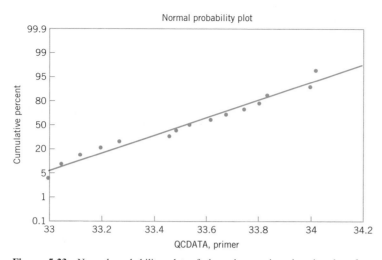

Figure 5-23 Normal probability plot of the primer paint viscosity data from Table 5-6, Example 5-5.

Montgomery, and Runger (1999) show how a properly designed EWMA control chart is very insensitive to the normality assumption. This approach will be discussed in Chapter 8.

It is important to check the normality assumption when using the control chart for individuals. A simple way to do this is with the normal probability plot. Figure 5-23 is the normal probability plot for the primer paint viscosity data in Table 5-6. There is no obvious problem with the normality assumption in these data. However, remember that the normal probability plot is but a **crude check** of the normality assumption and the individuals control chart is very sensitive to nonnormality. We suggest that the Shewhart individuals chart be used with extreme caution.

······· **EXAMPLE 5-6** ···

Table 5-8 presents consecutive measurements on the resistivity of 25 silicon wafers after an epitaxial layer is deposited in a single-wafer deposition process. We would like to apply an individuals control chart to this process. A normal probability plot of the resistivity measurements is shown in Fig. 5-24. This plot was constructed by Minitab, which fits the line to the points by least squares (not the best method). It is clear from inspection of the normal probability plot that the normality assumption for resistivity is at best questionable, so it would be dangerous to apply an individuals control chart to the original process data.

Figure 5-24 indicates that the distribution of resistivity has a long tail to the right, and consequently we would expect the log transformation (or a similar transformation) to produce a distribution that is closer to normal. The natural log of resistivity is shown in column three of Table 5-8, and the normal probability plot of the natural log of resistivity is shown in Fig. 5-25. Clearly the log transformation has resulted in a new variable that is more nearly approximated by a normal distribution than were the original resistivity measurements.

Table 5-8 Resistivity Data for Example 5-6

Sample, i	Resistivity (x_i)	ln (x_i)	MR
1	216	5.37528	
2	290	5.66988	0.294603
3	236	5.46383	0.206049
4	228	5.42935	0.0344862
5	244	5.49717	0.0678226
6	210	5.34711	0.150061
7	139	4.93447	0.412634
8	310	5.73657	0.802098
9	240	5.48064	0.255933
10	211	5.35186	0.128781
11	175	5.16479	0.187072
12	447	6.10256	0.937773
13	307	5.72685	0.375711
14	242	5.48894	0.23791
15	168	5.12396	0.364974
16	360	5.8861	0.76214
17	226	5.42053	0.465569
18	253	5.53339	0.112854
19	380	5.94017	0.406782
20	131	4.8752	1.06497
21	173	5.15329	0.278094
22	224	5.41165	0.258354
23	195	5.273	0.138646
24	199	5.2933	0.0203053
25	226	5.42053	0.12723
		$\overline{\ln (x_i)} = 5.44402$	$\overline{\text{MR}} = 0.337119$

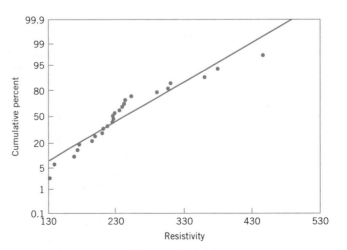

Figure 5-24 Normal probability plot of resistivity.

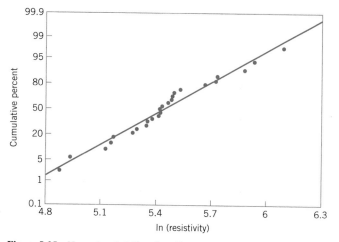

Figure 5-25   Normal probability plot of ln(resistivity).

The last column of Table 5-8 shows the moving ranges of the natural log of resistivity. Figure 5-26 presents the individuals and moving range control charts for the natural log of resistivity. Note that there is no indication of an out-of-control process.

More about Estimating σ

Very often in practice we use moving ranges in estimating σ for the individuals control chart. Recall that moving ranges are defined as $MR_i = |x_i - x_{i-1}|, i = 2, 3, \ldots , m.$ More properly, this statistic should be called a moving range of **span two** since the number of observations used to calculate the range in the moving window could be

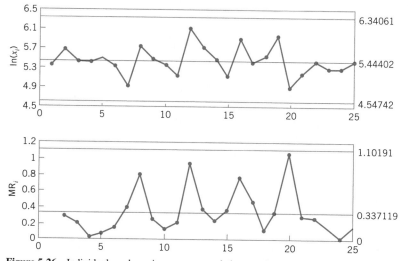

Figure 5-26   Individuals and moving range control charts on ln(resistivity), Example 5-6.

increased. The most common estimator is the one we used in Example 5-5, based on the **average moving range** $\overline{MR} = \sum_{i=2}^{m} MR_i/(m-1)$ and can be written as

$$\hat{\sigma}_1 = 0.8865 \, \overline{MR}$$

where the constant 0.8865 is the reciprocal of d_2 for samples of size two. For in-control processes, Cryer and Ryan (1990), among others, have pointed out that a more efficient estimator is one based on the sample standard deviation

$$\hat{\sigma}_2 = \frac{S}{c_4}$$

Both of these estimators are unbiased, assuming that no assignable causes are present in the sequence of m individual observations.

   If assignable causes are present, then both $\hat{\sigma}_1$ and $\hat{\sigma}_2$ result in biased estimates of the process standard deviation. To illustrate, suppose that in the sequence of individual observations

$$x_1, x_2, \ldots, x_t, x_{t+1}, \ldots, x_m$$

the process is in control with mean μ_0 and standard deviation σ for the first t observations, but between x_t and x_{t+1} an assignable cause occurs that results in a sustained shift in the process mean to a new level $\mu = \mu_0 + \delta\sigma$ and the mean remains at this new level for the remaining sample observations $x_{t+1}, \ldots, x_m$. Under these conditions, Woodall and Montgomery (2001) show that

$$E(S^2) = \sigma^2 + \frac{t(m-t)}{m(m-1)}(\delta\sigma)^2$$

In fact, this result holds for any case in which the mean of t of the observations is μ_0 and the mean of the remaining observations is $\mu_0 + \delta\sigma$, since the order of the observations is not relevant in computing S^2. Therefore, S^2 is biased upwards, and consequently $\hat{\sigma}_2 = S/c_4$ will tend to overestimate σ. Note that the extent of the bias in $\hat{\sigma}_2$ depends on the magnitude of the shift in the mean $(\delta\sigma)$, the time period following which the shift occurs (t), and the number of available observations (m). Now the moving range is only impacted by the shift in the mean during one period $(t + 1)$, so the bias in $\hat{\sigma}_1$ depends only on the shift magnitude and m. If $1 < t < m - 1$, the bias in $\hat{\sigma}_1$ will always be smaller than the bias in $\hat{\sigma}_2$. Cruthis and Rigdon (1992–93) show how the ratio

$$F^* = \left(\frac{\hat{\sigma}_1}{\hat{\sigma}_2}\right)^2$$

can be used to determine whether the process was in control when both estimates were calculated. They use simulation to obtain the approximate 90th, 95th, 99th, and 99.9th percentiles of the distribution of F^* for sample sizes $m = 10(5)100$, assuming that the process is in control. Since this is an empirical reference distribution, observed values of F^* that exceed one of these percentiles is an indication that the process was not in control over the time period during which the m observations were collected.

One way to reduce or possibly eliminate the bias in estimating σ when a sustained shift in the mean is present is to base the estimator on the **median** of the moving ranges of span two, as suggested by Clifford (1959) and Bryce, Gaudard, and Joiner (1998). This estimator is

$$\hat{\sigma}_3 = 1.047 \widetilde{\text{MR}}$$

where $\widetilde{\text{MR}}$ is the median of the span-two moving ranges, and 1.047 is the reciprocal of the control chart constant d_4 for subgroups of size two defined such that $E(\tilde{R}) = d_4\sigma$ and $\tilde{R}$ is the median range. A table of d_4 values is in Wadsworth, Stephens, and Godfrey (1986). Essentially, only one of the span-two moving ranges should be affected by the sustained shift, and this single large moving range will have little impact on the value of the median moving range, certainly much less impact than it will have on the average moving range. Constructing an individuals control chart using the median moving range to estimate σ is an option in Minitab.

Now suppose that the assignable cause affects a single observation rather than causing a sustained shift in the mean. If there is a single observation that has mean $\mu_0 + \delta\sigma$, then

$$E(S^2) = \sigma^2 + \frac{1}{m}(\delta\sigma)^2$$

and this observation will affect two of the span-two moving ranges. If there are two adjacent such observations, then

$$E(S^2) = \sigma^2 + \frac{2(m-2)}{m(m-1)}(\delta\sigma)^2$$

and two of the span-two moving ranges will be affected by the out-of-control observations. Thus, when the assignable cause affects one or only a few adjacent observations, we expect the bias in S^2 to be smaller than when a sustained shift occurs. However, if an assignable cause producing a sustained shift in the mean occurs either very early in the sequence of observations or very late, it will produce much the same effect as an assignable cause affecting only one or a few adjacent points.

Some authors have suggested basing the estimate of σ on moving ranges of span greater than two, and some computer programs for control charts offer this as an option. It is easy to show that this will always lead to potentially increased bias in the estimate of σ when assignable causes are present. Note that if one uses a span-three moving range and there is a single observation whose mean is affected by the assignable cause, then this single observation will affect up to three of the moving ranges. Thus, a span-three moving range will result in more bias in the estimate of σ than will the moving range of span two. Furthermore, two span-three moving ranges will be affected by a sustained shift. In general, if one uses a span-w moving range and there is a single observation whose mean is affected by the assignable cause, up to w of these moving ranges will be impacted by this observation. Furthermore, if there is a sustained shift in the mean, up to $w - 1$ of the moving ranges will be affected by the shift in the mean. Consequently, increasing the span of a moving

Table 5-9 Formulas for Control Charts, Standards Given

Chart	Center Line	Control Limits
$\bar{x}$ (μ and σ given)	μ	$\mu \pm A\sigma$
R (σ given)	$d_2\sigma$	UCL $= D_2\sigma$, LCL $= D_1\sigma$
S (σ given)	$c_4\sigma$	UCL $= B_6\sigma$, LCL $= B_5\sigma$

average beyond two results in increasing the bias in the estimate of σ if assignable causes that either produce sustained shifts in the process mean or that affect the mean of a single observation (or a few adjacent ones) are present. In fact, Wetherill and Brown (1991) advise plotting the estimate of σ versus the span of the moving average used to obtain the estimate. A sharply rising curve indicates the presence of assignable causes.

5-5 SUMMARY OF PROCEDURES FOR $\bar{x}$, R, AND S CHARTS

It is convenient to summarize in one place the various computational formulas for the major types of variables control charts discussed in this chapter. Table 5-9 summarizes the formulas for $\bar{x}$, R, and S charts when standard values for μ and σ are given. Table 5-10 provides the corresponding summary when no standard values are given and trial control limits must be established from analysis of past data. The constants given for the S chart assume that $n - 1$ is used in the denominator of S. All constants are tabulated for various sample sizes in Appendix Table VI.

5-6 APPLICATIONS OF VARIABLES CONTROL CHARTS

There are many interesting applications of variables control charts. In this section, a few of them will be described to give the reader additional insights into how the control chart works, as well as ideas for further applications.

Table 5-10 Formulas for Control Charts, Control Limits Based on Past Data (No Standards Given)

Chart	Center Line	Control Limits
$\bar{x}$ (using R)	$\bar{\bar{x}}$	$\bar{\bar{x}} \pm A_2\bar{R}$
$\bar{x}$ (using S)	$\bar{\bar{x}}$	$\bar{\bar{x}} \pm A_3\bar{S}$
R	$\bar{R}$	UCL $= D_4\bar{R}$, LCL $= D_3\bar{R}$
S	$\bar{S}$	UCL $= B_4\bar{S}$, LCL $= B_3\bar{S}$

········   **EXAMPLE 5-7**   ·· ·

Using Control Charts to Improve Suppliers' Processes

A large aerospace manufacturer purchased an aircraft component from two suppliers. These components frequently exhibited excessive variability on a key dimension that made it impossible to assemble them into the final product. This problem always resulted in expensive rework costs and occasionally caused delays in finishing the assembly of an airplane.

The materials-receiving group performed 100% inspection of these parts in an effort to improve the situation. They maintained $\bar{x}$ and R charts on the dimension of interest for both suppliers. They found that the fraction of nonconforming units was about the same for both suppliers, but for very different reasons. Supplier A could produce parts with mean dimension equal to the required value, but the process was out of statistical control. Supplier B could maintain good statistical control and, in general, produced a part that exhibited considerably less variability than parts from supplier A, but his process was centered so far off the nominal required dimension that many parts were out of specification.

This situation convinced the procurement organization to work with both suppliers, persuading supplier A to install an SPC activity and to begin working at continuous improvement, and assisting supplier B to find out why his process was consistently centered incorrectly. Supplier B's problem was ultimately tracked to some incorrect code in an NC (numerical-controlled) machine, and the use of SPC at supplier A resulted in considerable reduction in variability over a 6-month period. As a result of these actions, the problem with these parts was essentially eliminated.

········   **EXAMPLE 5-8**   ·· ·

Using SPC to Purchase a Machine Tool

An article in *Manufacturing Engineering* ("Picking a Marvel at Deere," January 1989, pp. 74–77) describes how the John Deere Company uses SPC methods to help choose production equipment. When a machine tool is purchased, it must go through the company capability demonstration prior to shipment to demonstrate that the tool has the ability to meet or exceed the established performance criteria. The procedure was applied to a programmable controlled bandsaw. The bandsaw supplier cut 45 pieces that were analyzed using $\bar{x}$ and R charts to demonstrate statistical control and to provide the basis for process capability analysis. The saw proved capable, and the supplier learned many useful things about the performance of his equipment. Control and capability tests such as this one are becoming a basic part of the equipment selection and acquisition process in many companies.

········   **EXAMPLE 5-9**   ·· ·

SPC Implementation in a Short-Run Job-Shop

One of the more interesting aspects of SPC is the successful implementation of control charts in a job-shop manufacturing environment. Most job-shops are characterized by short production runs, and many of these shops produce parts on production runs of fewer than 50 units. This situation can make the routine use of control charts appear to be somewhat of a challenge, as not enough units are produced in any one batch to establish the control limits.

This problem can usually be easily solved. Since statistical process-control methods are most frequently applied to a characteristic of a product, we can extend SPC to the job-shop environment by focusing on the **process characteristic** in each unit of product. To illustrate, consider a drilling operation in a job-shop. The operator drills holes of various sizes in each part passing through the machine center. Some parts require one hole, and others several holes of different sizes. It is almost impossible to construct an $\bar{x}$ and R chart on hole diameter, since each part is potentially different. The correct approach is to focus on the characteristic of interest in the *process*. In this case, the manufacturer is interested in drilling holes that have the correct diameter, and therefore wants to reduce the variability in hole diameter as much as possible. This may be accomplished by control charting the *deviation* of the actual hole diameter from the nominal diameter. Depending on the process production rate and the mix of parts produced, either a control chart for individuals with a moving-range control chart or a conventional $\bar{x}$ and R chart can be used. In these applications it is usually very important to mark the start of each lot or to batch carefully on the control chart, so that if changing the size, position, or number of holes drilled on each part affects the process the resulting pattern on the control charts will be easy to interpret.

·········· **EXAMPLE 5-10** ···

Use of $\bar{x}$ and R Charts in a Nonmanufacturing Environment
Variables control charts have found frequent application in both manufacturing and non-manufacturing settings. A fairly widespread but erroneous notion about these charts is that they do not apply to the nonmanufacturing environment because the "product is different." Actually, if we can make measurements on the product that are reflective of quality, function, or performance, then the *nature* of the product has no bearing on the general applicability of control charts. There are, however, two commonly encountered differences between manufacturing and nonmanufacturing situations: (1) In the nonmanufacturing environment specification limits rarely apply to the product, so the notion of process capability is often undefined; and (2) more imagination may be required to select the proper variable or variables for measurement.

One application of $\bar{x}$ and R control charts in a nonmanufacturing environment involved the efforts of a finance group to reduce the time required to process its accounts payable. The division of the company in which the problem occurred had recently experienced a considerable increase in business volume, and along with this expansion came a gradual lengthening of the time the finance department needed to process check requests. As a result, many suppliers were being paid beyond the normal 30-day period, and the company was failing to capture the discounts available from its suppliers for prompt payment. The quality-improvement team assigned to this

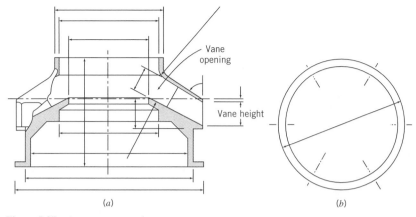

Figure 5-27   An aerospace casting.

project used the flow time through the finance department as the variable for control chart analysis. Five completed check requests were selected each day, and the average and range of flow time were plotted on $\bar{x}$ and R charts. Although management and operating personnel had addressed this problem before, the use of $\bar{x}$ and R charts was responsible for substantial improvements. Within 9 months, the finance department had reduced the percentage of invoices paid late from over 90% to under 3%, resulting in an annual savings of several hundred thousand dollars in realized discounts to the company.

········ **EXAMPLE 5-11** ···

The Need for Care in Selecting Rational Subgroups
Figure 5-27a shows a casting used in a gas turbine jet aircraft engine. This part is typical of those produced by both casting and machining processes for use in gas turbine engines and auxiliary power units in the aerospace industry—cylindrical parts created by rotating the cross section around a central axis. The vane height on this part is a critical quality characteristic.

Data on vane heights are collected by randomly selecting five vanes on each casting produced. Initially, the company constructed $\bar{x}$ and S control charts on these data to control and improve the process. This usually produced many out-of-control points on the $\bar{x}$ chart, with an occasional out-of-control point on the S chart. Figure 5-28 shows typical $\bar{x}$ and S charts for 20 castings. A more careful analysis of the control-charting procedure revealed that the chief problem was the use of the five measurements on a single part as a rational subgroup, and that the out-of-control conditions on the $\bar{x}$ chart did not provide a valid basis for corrective action.

Remember that the control chart for $\bar{x}$ deals with the issue of whether or not the between-sample variability is consistent with the within-sample variability. In this case it is not: The vanes on a single casting are formed together in a common wax mold assembly. It is likely that the vane heights on a specific casting will be very similar, and it is

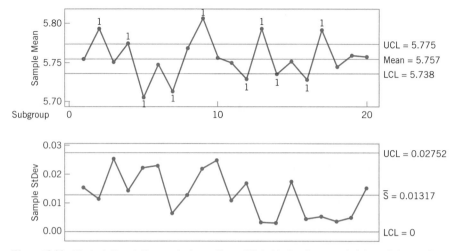

Figure 5-28 Typical $\bar{x}$ and S control charts (from Minitab) for the vane heights of the castings in Fig. 5-27.

reasonable to believe that there will be more variation in average vane height between the castings.

This situation was handled by using the S chart in the ordinary way to measure variation in vane height. However, as this standard deviation is clearly too small to provide a valid basis for control of $\bar{x}$, the quality engineer at the company decided to treat the average vane height on each casting as an *individual measurement* and to control average vane height by using a control chart for individuals with a moving-range chart. This solution worked extremely well in practice, and the group of three control charts provided an excellent basis for process improvement.

Figure 5-29 shows this set of three control charts as generated by Minitab. The Minitab package generates these charts automatically, referring to them as "between/within" control charts. Note that the individuals chart exhibits control, whereas the $\bar{x}$ chart in Fig. 5-28 did not. Essentially the moving range of the average vane heights provides a much more reasonable estimate of the variability in height **between parts.** The S chart can be thought of as a measure of the variability in vane height on a single casting. We want this variability to be as small as possible, so that all vanes on the same part will be nearly identical.

Situations such as the one described in Example 5-11 occur frequently in the application of control charts. For example, there are many similar problems in the semiconductor industry. In such cases, it is important to study carefully the behavior of the variables being measured and to have a clear understanding of the purpose of the control charts. For instance, if the variation in vane height on a specific casting were completely unrelated, using the average height as an individual measurement could be very inappropriate. It would be necessary to (1) use a control chart on each individual vane included in the sample, (2) investigate the use of a control chart technique for multistream

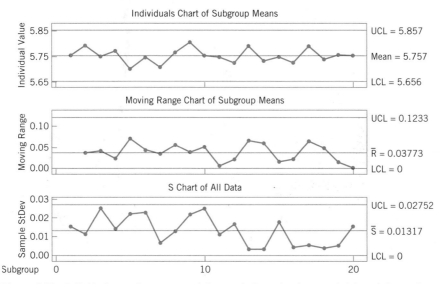

Figure 5-29 Individuals, moving range, and S control charts for the vane heights of the castings in Fig. 5-27.

processes, or (3) use some multivariate control chart technique. Some of these possibilities are discussed in the chapters in Part III of the text.

5-7 EXERCISES

5-1. The data shown here are $\bar{x}$ and R values for 24 samples of size $n = 5$ taken from a process producing bearings. The measurements are made on the inside diameter of the bearing, with only the last three decimals recorded (i.e., 34.5 should be 0.50345).

Sample Number	$\bar{x}$	R	Sample Number	$\bar{x}$	R
1	34.5	3	13	35.4	8
2	34.2	4	14	34.0	6
3	31.6	4	15	37.1	5
4	31.5	4	16	34.9	7
5	35.0	5	17	33.5	4
6	34.1	6	18	31.7	3
7	32.6	4	19	34.0	8
8	33.8	3	20	35.1	4
9	34.8	7	21	33.7	2
10	33.6	8	22	32.8	1
11	31.9	3	23	33.5	3
12	38.6	9	24	34.2	2

(a) Set up $\bar{x}$ and R charts on this process. Does the process seem to be in statistical control? If necessary, revise the trial control limits.

(b) If specifications on this diameter are 0.5030 ± 0.0010, find the percentage of nonconforming bearings produced by this process. Assume that diameter is normally distributed.

5-2. A high-voltage power supply should have a nominal output voltage of 350 V. A sample of four units is selected each day and tested for process-control purposes. The data shown give the difference between the observed reading on each unit and the nominal voltage times ten; that is,

$$x_i = (\text{observed voltage on unit } i - 350)10$$

Sample Number	x_1	x_2	x_3	x_4
1	6	9	10	15
2	10	4	6	11
3	7	8	10	5
4	8	9	6	13
5	9	10	7	13
6	12	11	10	10
7	16	10	8	9
8	7	5	10	4
9	9	7	8	12
10	15	16	10	13
11	8	12	14	16
12	6	13	9	11
13	16	9	13	15
14	7	13	10	12
15	11	7	10	16
16	15	10	11	14
17	9	8	12	10
18	15	7	10	11
19	8	6	9	12
20	14	15	12	16

(a) Set up $\bar{x}$ and R charts on this process. Is the process in statistical control?

(b) If specifications are at 350 V ± 5 V, what can you say about process capability?

(c) Is there evidence to support the claim that voltage is normally distributed?

5-3. The data shown here are the deviations from nominal diameter for holes drilled in a carbon-fiber composite material used in aerospace manufacturing. The values reported are deviations from nominal in ten-thousandths of an inch.

Sample Number	x_1	x_2	x_3	x_4	x_5
1	−30	+50	−20	+10	+30
2	0	+50	−60	−20	+30
3	−50	+10	+20	+30	+20
4	−10	−10	+30	−20	+50
5	+20	−40	+50	+20	+10
6	0	0	+40	−40	+20
7	0	0	+20	−20	−10
8	+70	−30	+30	−10	0
9	0	0	+20	−20	+10
10	+10	+20	+30	+10	+50
11	+40	0	+20	0	+20
12	+30	+20	+30	+10	+40
13	+30	−30	0	+10	+10
14	+30	−10	+50	−10	−30
15	+10	−10	+50	+40	0
16	0	0	+30	−10	0
17	+20	+20	+30	+30	−20
18	+10	−20	+50	+30	+10
19	+50	−10	+40	+20	0
20	+50	0	0	+30	+10

(a) Set up $\bar{x}$ and R charts on the process. Is the process in statistical control?

(b) Estimate the process standard deviation using the range method.

(c) If specifications are at nominal ±100, what can you say about the capability of this process? Calculate the PCR C_p.

5-4. The thickness of a printed circuit board is an important quality parameter. Data on board thickness (in inches) are given here for 25 samples of three boards each.

Sample Number	x_1	x_2	x_3
1	0.0629	0.0636	0.0640
2	0.0630	0.0631	0.0622
3	0.0628	0.0631	0.0633
4	0.0634	0.0630	0.0631
5	0.0619	0.0628	0.0630
6	0.0613	0.0629	0.0634
7	0.0630	0.0639	0.0625
8	0.0628	0.0627	0.0622
9	0.0623	0.0626	0.0633
10	0.0631	0.0631	0.0633
11	0.0635	0.0630	0.0638
12	0.0623	0.0630	0.0630
13	0.0635	0.0631	0.0630
14	0.0645	0.0640	0.0631
15	0.0619	0.0644	0.0632
16	0.0631	0.0627	0.0630
17	0.0616	0.0623	0.0631
18	0.0630	0.0630	0.0626
19	0.0636	0.0631	0.0629
20	0.0640	0.0635	0.0629
21	0.0628	0.0625	0.0616
22	0.0615	0.0625	0.0619
23	0.0630	0.0632	0.0630
24	0.0635	0.0629	0.0635
25	0.0623	0.0629	0.0630

(a) Set up $\bar{x}$ and R control charts. Is the process in statistical control?

(b) Estimate the process standard deviation.

(c) What are the limits that you would expect to contain nearly all the process measurements?

(d) If the specifications are at 0.0630 in. $\pm$ 0.0015 in., what is the value of the PCR C_p?

5-5. The fill volume of soft-drink beverage bottles is an important quality characteristic. The volume is measured (approximately) by placing a gauge over the crown and comparing the height of the liquid in the neck of the bottle against a coded scale. On this scale, a reading of zero corresponds to the correct fill height. Fifteen samples of size $n = 10$ have been analyzed, and the fill heights are shown next.

Sample Number	x_1	x_2	x_3	x_4	x_5	x_6	x_7	x_8	x_9	x_{10}
1	2.5	0.5	2.0	-1.0	1.0	-1.0	0.5	1.5	0.5	-1.5
2	0.0	0.0	0.5	1.0	1.5	1.0	-1.0	1.0	1.5	-1.0
3	1.5	1.0	1.0	-1.0	0.0	-1.5	-1.0	-1.0	1.0	-1.0
4	0.0	0.5	-2.0	0.0	-1.0	1.5	-1.5	0.0	-2.0	-1.5
5	0.0	0.0	0.0	-0.5	0.5	1.0	-0.5	-0.5	0.0	0.0
6	1.0	-0.5	0.0	0.0	0.0	0.5	-1.0	1.0	-2.0	1.0
7	1.0	-1.0	-1.0	-1.0	0.0	1.5	0.0	1.0	0.0	0.0
8	0.0	-1.5	-0.5	1.5	0.0	0.0	0.0	-1.0	0.5	-0.5
9	-2.0	-1.5	1.5	1.5	0.0	0.0	0.5	1.0	0.0	1.0
10	-0.5	3.5	0.0	-1.0	-1.5	-1.5	-1.0	-1.0	1.0	0.5
11	0.0	1.5	0.0	0.0	2.0	-1.5	0.5	-0.5	2.0	-1.0
12	0.0	-2.0	-0.5	0.0	-0.5	2.0	1.5	0.0	0.5	-1.0
13	-1.0	-0.5	-0.5	-1.0	0.0	0.5	0.5	-1.5	-1.0	-1.0
14	0.5	1.0	-1.0	-0.5	-2.0	-1.0	-1.5	0.0	1.5	1.5
15	1.0	0.0	1.5	1.5	1.0	-1.0	0.0	1.0	-2.0	-1.5

(a) Set up $\bar{x}$ and S control charts on this process. Does the process exhibit statistical control? If necessary, construct revised control limits.

(b) Set up an R chart, and compare with the S chart in part (a).

(c) Set up an S^2 chart, and compare with the S chart in part (a).

5-6. The net weight (in oz) of a dry bleach product is to be monitored by $\bar{x}$ and R control charts using a sample size of $n = 5$. Data for 20 preliminary samples are as follows.

Sample Number	x_1	x_2	x_3	x_4	x_5
1	15.8	16.3	16.2	16.1	16.6
2	16.3	15.9	15.9	16.2	16.4
3	16.1	16.2	16.5	16.4	16.3

Sample Number	x_1	x_2	x_3	x_4	x_5
4	16.3	16.2	15.9	16.4	16.2
5	16.1	16.1	16.4	16.5	16.0
6	16.1	15.8	16.7	16.6	16.4
7	16.1	16.3	16.5	16.1	16.5
8	16.2	16.1	16.2	16.1	16.3
9	16.3	16.2	16.4	16.3	16.5
10	16.6	16.3	16.4	16.1	16.5
11	16.2	16.4	15.9	16.3	16.4
12	15.9	16.6	16.7	16.2	16.5
13	16.4	16.1	16.6	16.4	16.1
14	16.5	16.3	16.2	16.3	16.4
15	16.4	16.1	16.3	16.2	16.2
16	16.0	16.2	16.3	16.3	16.2
17	16.4	16.2	16.4	16.3	16.2
18	16.0	16.2	16.4	16.5	16.1
19	16.4	16.0	16.3	16.4	16.4
20	16.4	16.4	16.5	16.0	15.8

(a) Set up $\bar{x}$ and R control charts using these data. Does the process exhibit statistical control?

(b) Estimate the process mean and standard deviation.

(c) Does fill weight seem to follow a normal distribution?

(d) If the specifications are at 16.2 ± 0.5, what conclusions would you draw about process capability?

(e) What fraction of containers produced by this process is likely to be below the lower specification limit of 15.7 oz?

5-7.   Rework Exercise 5-2 using the S chart.

5-8.   Rework Exercise 5-3 using the S chart.

5-9.   Control charts on $\bar{x}$ and S are to be maintained on the torque readings of a bearing used in a wingflap actuator assembly. Samples of size $n = 10$ are to be used, and we know from past experience that when the process is in control, bearing torque has a normal distribution with mean $\mu = 80$ inch-pounds and standard deviation $\sigma = 10$ inch-pounds. Find the center line and control limits for these control charts.

5-10. Samples of $n = 8$ items each are taken from a manufacturing process at regular intervals. A quality characteristic is measured, and $\bar{x}$ and R values are calculated for each sample. After 50 samples, we have

$$\sum_{i=1}^{50} \bar{x}_i = 2000 \quad \text{and} \quad \sum_{i=1}^{50} R_i = 250$$

Assume that the quality characteristic is normally distributed.

(a) Compute control limits for the $\bar{x}$ and R control charts.

(b) All points on both control charts fall between the control limits computed in part (a). What are the natural tolerance limits of the process?

(c) If the specifications limits are 41 ± 5.0, what are your conclusions regarding the ability of the process to produce items within these specifications?

(d) Assuming that if an item exceeds the upper specification limit it can be reworked, and if it is below the lower specification limit it must be scrapped, what percent scrap and rework is the process producing?

(e) Make suggestions as to how the process performance could be improved.

5-11. Samples of $n = 6$ items are taken from a manufacturing process at regular intervals. A normally distributed quality characteristic is measured and $\bar{x}$ and S values are calculated for each sample. After 50 subgroups have been analyzed, we have

$$\sum_{i=1}^{50} \bar{x}_i = 1000 \quad \text{and} \quad \sum_{i=1}^{50} S_i = 75$$

(a) Compute the control limit for the $\bar{x}$ and S control charts.

(b) Assume that all points on both charts plot within the control limits. What are the natural tolerance limits of the process?

(c) If the specification limits are 19 ± 4.0, what are your conclusions regarding the ability of the process to produce items conforming to specifications?

(d) Assuming that if an item exceeds the upper specification limit it can be reworked, and if it is below the lower specification limit it must be scrapped, what percent scrap and rework is the process now producing?

(e) If the process were centered at $\mu = 19.0$, what would be the effect on percent scrap and rework?

5-12. The table shown presents 20 subgroups of five measurements on the critical dimension of a part produced by a machining process.

Sample Number	x_1	x_2	x_3	x_4	x_5	$\bar{x}$	R
1	138.1	110.8	138.7	137.4	125.4	130.1	27.9
2	149.3	142.1	105.0	134.0	92.3	124.5	57.0
3	115.9	135.6	124.2	155.0	117.4	129.6	39.1
4	118.5	116.5	130.2	122.6	100.2	117.6	30.0
5	108.2	123.8	117.1	142.4	150.9	128.5	42.7
6	102.8	112.0	135.0	135.0	145.8	126.1	43.0
7	120.4	84.3	112.8	118.5	119.3	111.0	36.1
8	132.7	151.1	124.0	123.9	105.1	127.4	46.0
9	136.4	126.2	154.7	127.1	173.2	143.5	46.9
10	135.0	115.4	149.1	138.3	130.4	133.6	33.7
11	139.6	127.9	151.1	143.7	110.5	134.6	40.6
12	125.3	160.2	130.4	152.4	165.1	146.7	39.8
13	145.7	101.8	149.5	113.3	151.8	132.4	50.0
14	138.6	139.0	131.9	140.2	141.1	138.1	9.2
15	110.1	114.6	165.1	113.8	139.6	128.7	54.8
16	145.2	101.0	154.6	120.2	117.3	127.6	53.3
17	125.9	135.3	121.5	147.9	105.0	127.1	42.9
18	129.7	97.3	130.5	109.0	150.5	123.4	53.2
19	123.4	150.0	161.6	148.4	154.2	147.5	38.3
20	144.8	138.3	119.6	151.8	142.7	139.4	32.2

(a) Set up $\bar{x}$ and R control charts on this process. Verify that the process is in statistical control.

(b) Following the establishment of control charts in part (a) above, 10 new samples were collected. Plot the $\bar{x}$ and R values on the control chart you established in part (a) and draw conclusions.

Sample Number	x_1	x_2	x_3	x_4	x_5	$\bar{x}$	R
1	131.0	184.8	182.2	143.3	212.8	170.8	81.8
2	181.3	193.2	180.7	169.1	174.3	179.7	24.0
3	154.8	170.2	168.4	202.7	174.4	174.1	48.0
4	157.5	154.2	169.1	142.2	161.9	157.0	26.9
5	216.3	174.3	166.2	155.5	184.3	179.3	60.8
6	186.9	180.2	149.2	175.2	185.0	175.3	37.8
7	167.8	143.9	157.5	171.8	194.9	167.2	51.0
8	178.2	186.7	142.4	159.4	167.6	166.9	44.2
9	162.6	143.6	132.8	168.9	177.2	157.0	44.5
10	172.1	191.7	203.4	150.4	196.3	182.8	53.0

(c) Suppose that the assignable cause responsible for the action signals generated in part (b) has been identified and adjustments made to the process to correct its performance. Plot the $\bar{x}$ and R values from the following new subgroups, which were taken following the adjustment, against the control chart limits established in part (a). What are your conclusions?

Sample Number	x_1	x_2	x_3	x_4	x_5	$\bar{x}$	R
1	131.5	143.1	118.5	103.2	121.6	123.6	39.8
2	111.0	127.3	110.4	91.0	143.9	116.7	52.8
3	129.8	98.3	134.0	105.1	133.1	120.1	35.7
4	145.2	132.8	106.1	131.0	99.2	122.8	46.0
5	114.6	111.0	108.8	177.5	121.6	126.7	68.7
6	125.2	86.4	64.4	137.1	117.5	106.1	72.6
7	145.9	109.5	84.9	129.8	110.6	116.1	61.0
8	123.6	114.0	135.4	83.2	107.6	112.8	52.2
9	85.8	156.3	119.7	96.2	153.0	122.2	70.6
10	107.4	148.7	127.4	125.0	127.5	127.2	41.3

5-13. Parts manufactured by an injection molding process are subjected to a compressive strength test. Twenty samples of five parts each are collected, and the compressive strengths (in psi) are shown in the following table.

Sample Number	x_1	x_2	x_3	x_4	x_5	$\bar{x}$	R
1	83.0	81.2	78.7	75.7	77.0	79.1	7.3
2	88.6	78.3	78.8	71.0	84.2	80.2	17.6
3	85.7	75.8	84.3	75.2	81.0	80.4	10.4
4	80.8	74.4	82.5	74.1	75.7	77.5	8.4
5	83.4	78.4	82.6	78.2	78.9	80.3	5.2
6	75.3	79.9	87.3	89.7	81.8	82.8	14.5
7	74.5	78.0	80.8	73.4	79.7	77.3	7.4
8	79.2	84.4	81.5	86.0	74.5	81.1	11.4
9	80.5	86.2	76.2	64.1	80.2	81.4	9.9
10	75.7	75.2	71.1	82.1	74.3	75.7	10.9
11	80.0	81.5	78.4	73.8	78.1	78.4	7.7
12	80.6	81.8	79.3	73.8	81.7	79.4	8.0
13	82.7	81.3	79.1	82.0	79.5	80.9	3.6
14	79.2	74.9	78.6	77.7	75.3	77.1	4.3
15	85.5	82.1	82.8	73.4	71.7	79.1	13.8
16	78.8	79.6	80.2	79.1	80.8	79.7	2.0
17	82.1	78.2	75.5	78.2	82.1	79.2	6.6
18	84.5	76.9	83.5	81.2	79.2	81.1	7.6
19	79.0	77.8	81.2	84.4	81.6	80.8	6.6
20	84.5	73.1	78.6	78.7	80.6	79.1	11.4

(a) Establish $\bar{x}$ and R control charts for compressive strength using these data. Is the process in statistical control?

(b) After establishing the control charts in part (a), 15 new subgroups were collected and the compressive strengths are shown next. Plot the $\bar{x}$ and R values against the control units from part (a) and draw conclusions.

Sample Number	x_1	x_2	x_3	x_4	x_5	$\bar{x}$	R
1	68.9	81.5	78.2	80.8	81.5	78.2	12.6
2	69.8	68.6	80.4	84.3	83.9	77.4	15.7
3	78.5	85.2	78.4	80.3	81.7	80.8	6.8

Sample Number	x_1	x_2	x_3	x_4	x_5	$\bar{x}$	R
4	76.9	86.1	86.9	94.4	83.9	85.6	17.5
5	93.6	81.6	87.8	79.6	71.0	82.7	22.5
6	65.5	86.8	72.4	82.6	71.4	75.9	21.3
7	78.1	65.7	83.7	93.7	93.4	82.9	27.9
8	74.9	72.6	81.6	87.2	72.7	77.8	14.6
9	78.1	77.1	67.0	75.7	76.8	74.9	11.0
10	78.7	85.4	77.7	90.7	76.7	81.9	14.0
11	85.0	60.2	68.5	71.1	82.4	73.4	24.9
12	86.4	79.2	79.8	96.0	75.4	81.3	10.9
13	78.5	99.0	78.3	71.4	81.8	81.7	27.6
14	68.8	62.0	82.0	77.5	76.1	73.3	19.9
15	83.0	83.7	73.1	82.2	95.3	83.5	22.2

5-14. Reconsider the data presented in Exercise 5-13.

(a) Rework both parts (a) and (b) of Exercise 5-13 using the $\bar{x}$ and S charts.

(b) Does the S chart detect the shift in process variability more quickly than the R chart did originally in part (b) of Exercise 5-13?

5-15. Consider the $\bar{x}$ and R charts you established in Exercise 5-1 using $n = 5$.

(a) Suppose that you wished to continue charting this quality characteristic using $\bar{x}$ and R charts based on a sample size of $n = 3$. What limits would be used on the $\bar{x}$ and R charts?

(b) What would be the impact of the decision you made in part (a) on the ability of the $\bar{x}$ chart to detect a 2σ shift in the mean?

(c) Suppose you wished to continue charting this quality characteristic using $\bar{x}$ and R charts based on a sample size of $n = 8$. What limits would be used on the $\bar{x}$ and R charts?

(d) What is the impact of using $n = 8$ on the ability of the $\bar{x}$ chart to detect a 2σ shift in the mean?

5-16. Control charts for $\bar{x}$ and R are maintained for an important quality characteristic. The sample size is $n = 7$; $\bar{x}$ and R are computed for each sample. After 35 samples, we have found that

$$\sum_{i=1}^{35} \bar{x}_i = 7805 \quad \text{and} \quad \sum_{i=1}^{35} R_i = 1200$$

(a) Set up $\bar{x}$ and R charts using these data.
(b) Assuming that both charts exhibit control, estimate the process mean and standard deviation.
(c) If the quality characteristic is normally distributed and if the specifications are 220 ± 35, can the process meet the specifications? Estimate the fraction nonconforming.
(d) Assuming the variance to remain constant, state where the process mean should be located to minimize the fraction nonconforming. What would be the value of the fraction nonconforming under these conditions?

5-17. Samples of size $n = 5$ are taken from a manufacturing process every hour. A quality characteristic is measured, and $\bar{x}$ and R are computed for each sample. After 25 samples have been analyzed, we have

$$\sum_{i=1}^{25} \bar{x}_i = 662.50 \quad \text{and} \quad \sum_{i=1}^{25} R_i = 9.00$$

The quality characteristic is normally distributed.
(a) Find the control limits for the $\bar{x}$ and R charts.
(b) Assume that both charts exhibit control. If the specifications are 26.40 ± 0.50, estimate the fraction nonconforming.
(c) If the mean of the process were 26.40, what fraction nonconforming would result?

5-18. Samples of size $n = 5$ are collected from a process every half hour. After 50 samples have been collected, we calculate $\bar{\bar{x}} = 20.0$ and $\bar{S} = 1.5$. As-

sume that both charts exhibit control and that the quality characteristic is normally distributed.
(a) Estimate the process standard deviation.
(b) Find the control limits on the $\bar{x}$ and S charts.
(c) If the process mean shifts to 22, what is the probability of concluding that the process is still in control?

5-19. An $\bar{x}$ chart is used to control the mean of a normally distributed quality characteristic. It is known that $\sigma = 6.0$ and $n = 4$. The center line = 200, UCL = 209, and LCL = 191. If the process mean shifts to 188, find the probability that this shift is detected on the first subsequent sample.

5-20. A parameter of a part being produced on a lathe has specifications 100 ± 10. Control chart analysis indicates that the process is in control with $\bar{\bar{x}} = 104$ and $\bar{R} = 9.30$. The control charts use samples of size $n = 5$. If we assume that the characteristic is normally distributed, can the mean be located (by adjusting the tool position) so that all output meets specifications? What is the present capability of the process?

5-21. A process is to be monitored with standard values $\mu = 10$ and $\sigma = 2.5$. The sample size is three.
(a) Find the center line and control limits for the $\bar{x}$ chart.
(b) Find the center line and control limits for the R chart.
(c) Find the center line and the control limits for the S chart.

5-22. Samples of $n = 5$ units are taken from a process every hour. The $\bar{x}$ and R values for a particular quality characteristic are determined. After 25 samples have been collected, we calculate $\bar{\bar{x}} = 20$ and $\bar{R} = 4.56$.
(a) What are the three-sigma control limits for $\bar{x}$ and R?

(b) Both charts exhibit control. Estimate the process standard deviation.

(c) Assume that the process output is normally distributed. If the specifications are 19 ± 5, what are your conclusions regarding the process capability?

(d) If the process mean shifts to 24, what is the probability of not detecting this shift on the first subsequent sample?

5-23. Control charts for $\bar{x}$ and R are to be established to control the tensile strength of a metal part. Assume that tensile strength is normally distributed. Thirty samples of size $n = 6$ parts are collected over a period of time with the following results:

$$\sum_{i=1}^{30} \bar{x}_i = 6000 \quad \text{and} \quad \sum_{i=1}^{30} R_i = 150$$

(a) Calculate control limits for $\bar{x}$ and R.

(b) Both charts exhibit control. The specifications on tensile strength are 200 ± 5. What are your conclusions regarding process capability?

(c) For the above $\bar{x}$ chart, find the β-risk when the true process mean is 199.

5-24. An $\bar{x}$ chart has a center line of 100, uses three-sigma control limits, and is based on a sample size of four. The process standard deviation is known to be six. If the process mean shifts from 100 to 92, what is the probability of detecting this shift on the first sample following the shift?

5-25. The following data were collected from a process manufacturing power supplies. The variable of interest is output voltage, and $n = 5$.

Sample Number	$\bar{x}$	R
1	103	4
2	102	5

Sample Number	$\bar{x}$	R
3	104	2
4	105	11
5	104	4
6	106	3
7	102	7
8	105	2
9	106	4
10	104	3
11	105	4
12	103	2
13	102	3
14	105	4
15	104	5
16	105	3
17	106	5
18	102	2
19	105	4
20	103	2

(a) Compute center lines and control limits suitable for controlling future production.

(b) Assume that the quality characteristic is normally distributed. Estimate the process standard deviation.

(c) What are the apparent three-sigma natural tolerance limits of the process?

(d) What would be your estimate of the process fraction nonconforming if the specifications on the characteristic were 103 ± 4?

(e) What approaches to reducing the fraction nonconforming can you suggest?

5-26. Control charts on $\bar{x}$ and R for samples of size $n = 5$ are to be maintained on the tensile strength in pounds of a yarn. To start the charts, 30 samples were selected, and the mean and range of each computed. This yields

$$\sum_{i=1}^{30} \bar{x}_i = 607.8 \quad \text{and} \quad \sum_{i=1}^{30} R_i = 144$$

(a) Compute the center line and control limits for the $\bar{x}$ and R control charts.

(b) Suppose both charts exhibit control. There is a single lower specification limit of 16 lb. If strength is normally distributed, what fraction of yarn would fail to meet specifications?

5-27. Specifications on a cigar lighter detent are 0.3220 and 0.3200 in. Samples of size five are taken every 45 min with the following results (measured as deviations from 0.3210 in 0.0001 in.).

Sample Number	x_1	x_2	x_3	x_4	x_5
1	1	9	6	9	6
2	9	4	3	0	3
3	0	9	0	3	2
4	1	1	0	2	1
5	−3	0	−1	0	−4
6	−7	2	0	0	2
7	−3	−1	−1	0	−2
8	0	−2	−3	−3	−2
9	2	0	−1	−3	−1
10	0	2	−1	−1	2
11	−3	−2	−1	−1	2
12	−16	2	0	−4	−1
13	−6	−3	0	0	−8
14	−3	−5	5	0	5
15	−1	−1	−1	−2	−1

(a) Set up an R chart and examine the process for statistical control.

(b) What parameters would you recommend for an R chart for on-line control?

(c) Estimate the standard deviation of the process.

(d) What is the process capability?

5-28. **Continuation of Exercise 5-27.** Reconsider the data from Exercise 5-27 and establish $\bar{x}$ and R charts with appropriate trial control limits. Revise these trial limits as necessary to produce a set of control charts for monitoring future production. Suppose that the following new data are observed.

Sample Number	x_1	x_2	x_3	x_4	x_5
16	2	10	9	6	5
17	1	9	5	9	4
18	0	9	8	2	5
19	−3	0	5	1	4
20	2	10	9	3	1
21	−5	4	0	6	−1
22	0	2	−5	4	6
23	10	0	3	1	5
24	−1	2	5	6	−3
25	0	−1	2	5	−2

(a) Plot these new observations on the control chart. What conclusions can you draw about process stability?

(b) Use all 25 observations to revise the control limits for the $\bar{x}$ and R charts. What conclusions can you draw now about the process?

5-29. Two parts are assembled as shown in the figure. Assume that the dimensions x and y are normally distributed with means μ_x and μ_y and standard deviations σ_x and σ_y, respectively. The parts are produced on different machines and are assembled at random. Control charts are maintained on each dimension for the range of each sample ($n = 5$). Both range charts are in control.

(a) Given that for 20 samples on the range chart controlling x and 10 samples on the range chart controlling y, we have

$$\sum_{i=1}^{20} R_{x_i} = 18.608 \quad \text{and} \quad \sum_{i=1}^{10} R_{y_i} = 6.978$$

Estimate σ_x and σ_y.

(b) If it is desired that the probability of a smaller clearance (i.e., $x - y$) than 0.09 should be 0.006, what distance between the average dimensions (i.e., $\mu_x - \mu_y$) should be specified?

5-30. Control charts for $\bar{x}$ and R are maintained on the tensile strength of a metal fastener. After 30 samples of size $n = 6$ are analyzed, we find that

$$\sum_{i=1}^{30} \bar{x}_i = 12{,}870 \quad \text{and} \quad \sum_{i=1}^{30} R_i = 1350$$

(a) Compute control limits on the R chart.

(b) Assuming that the R chart exhibits control, estimate the parameters μ and σ.

(c) If the process output is normally distributed, and if the specifications are 440 ± 40, can the process meet the specifications? Estimate the fraction nonconforming.

(d) If the variance remains constant, where should the mean be located to minimize the fraction nonconforming?

5-31. Control charts for $\bar{x}$ and S are maintained on a quality characteristic. The sample size is $n = 4$. After 30 samples, we obtain

$$\sum_{i=1}^{30} \bar{x}_i = 12{,}870 \quad \text{and} \quad \sum_{i=1}^{30} S_i = 410$$

(a) Find the three-sigma limits for the S chart.

(b) Assuming that both charts exhibit control, estimate the parameters μ and σ.

5-32. An $\bar{x}$ chart on a normally distributed quality characteristic is to be established with the standard values $\mu = 100$, $\sigma = 8$, and $n = 4$. Find the following:

(a) The two-sigma control limits.

(b) The 0.005 probability limits.

5-33. An $\bar{x}$ chart with three-sigma limits has parameters as follows:

$$UCL = 104$$
$$\text{Center line} = 100$$
$$LCL = 96$$
$$n = 5$$

Suppose the process quality characteristic being controlled is normally distributed with a true mean of 98 and a standard deviation of 8. What is the probability that the control chart would exhibit lack of control by at least the third point plotted?

5-34. Consider the $\bar{x}$ chart defined in Exercise 5-33. Find the ARL_1 for the chart.

5-35. Control charts for $\bar{x}$ and S with $n = 4$ are maintained on a quality characteristic. The parameters of these charts are as follows:

$\bar{x}$ Chart	S Chart
UCL = 201.88	UCL = 2.266
Center line = 200.00	Center line = 1.000
LCL = 198.12	LCL = 0

Both charts exhibit control. Specifications on the quality characteristic are 197.50 and 202.50. What can be said about the ability of the process to produce product that conforms to specifications?

5-36. Statistical monitoring of a quality characteristic uses both an $\bar{x}$ and an S chart. The charts are to be based on the standard values $\mu = 200$ and $\sigma = 10$, with $n = 4$.

(a) Find three-sigma control limits for the S chart.

(b) Find a center line and control limits for the $\bar{x}$ chart such that the probability of a type I error is 0.05.

5-37. Specifications on a normally distributed dimension are 600 ± 20. $\bar{x}$ and R charts are maintained on this dimension and have been in control over a long period of time. The parameters of these control charts are as follows ($n = 9$).

$\bar{x}$ Chart	S Chart
UCL = 616	UCL = 32.36
Center line = 610	Center line = 17.82
LCL = 604	LCL = 3.28

(a) What are your conclusions regarding the capability of the process to produce items within specifications?

(b) Construct an OC curve for the $\bar{x}$ chart assuming that σ is constant.

5-38. Thirty samples each of size seven have been collected to establish control over a process. The following data were collected:

$$\sum_{i=1}^{30} \bar{x}_i = 2700 \quad \text{and} \quad \sum_{i=1}^{30} R_i = 120$$

(a) Calculate trial control limits for the two charts.

(b) On the assumption that the R chart is in control, estimate the process standard deviation.

(c) Suppose an S chart were desired. What would be the appropriate control limits and center line?

5-39. An $\bar{x}$ chart is to be established based on the standard values $\mu = 600$ and $\sigma = 12$, with $n = 9$. The control limits are to be based on an α-risk of 0.01. What are the appropriate control limits?

5-40. $\bar{x}$ and R charts with $n = 4$ are used to monitor a normally distributed quality characteristic. The control chart parameters are

$\bar{x}$ Chart	R Chart
UCL = 815	UCL = 46.98
Center line = 800	Center line = 20.59
LCL = 785	LCL = 0

Both charts exhibit control. What is the probability that a shift in the process mean to 790 will be detected on the first sample following the shift?

5-41. Consider the $\bar{x}$ chart in Exercise 5-40. Find the average run length for the chart.

5-42. Control charts for $\bar{x}$ and R are in use with the following parameters:

$\bar{x}$ Chart	R Chart
UCL = 363.0	UCL = 16.18
Center line = 360.0	Center line = 8.91
LCL = 357.0	LCL = 1.64

The sample size is $n = 9$. Both charts exhibit control. The quality characteristic is normally distributed.

(a) What is the α-risk associated with the $\bar{x}$ chart?

(b) Specifications on this quality characteristic are 358 ± 6. What are your conclusions regarding the ability of the process to produce items within specifications?

(c) Suppose the mean shifts to 357. What is the probability that the shift will not be detected on the first sample following the shift?

(d) What would be the appropriate control limits for the $\bar{x}$ chart if the type I error probability were to be 0.01?

5-43. A normally distributed quality characteristic is monitored through use of an $\bar{x}$ and an R chart. These charts have the following parameters ($n = 4$):

$\bar{x}$ Chart	R Chart
UCL = 626.0	UCL = 18.795
Center line = 620.0	Center line = 8.236
LCL = 614.0	LCL = 0

Both charts exhibit control.
(a) What is the estimated standard deviation of the process?
(b) Suppose an S chart were to be substituted for the R chart. What would be the appropriate parameters of the S chart?
(c) If specifications on the product were 610 ± 15, what would be your estimate of the process fraction nonconforming?
(d) What could be done to reduce this fraction nonconforming?
(e) What is the probability of detecting a shift in the process mean to 610 on the first sample following the shift (σ remains constant)?
(f) What is the probability of detecting the shift in part (e) by at least the third sample after the shift occurs?

5-44. Control charts for $\bar{x}$ and S have been maintained on a process and have exhibited statistical control. The sample size is $n = 6$. The control chart parameters are as follows:

$\bar{x}$ Chart	S Chart
UCL = 708.20	UCL = 3.420
Center line = 706.00	Center line = 1.738
LCL = 703.80	LCL = 0.052

(a) Estimate the mean and standard deviation of the process.
(b) Estimate the natural tolerance limits for the process.
(c) Assume that the process output is well modeled by a normal distribution. If specifications are 703 and 709, estimate the fraction nonconforming.
(d) Suppose the process mean shifts to 702.00 while the standard deviation remains constant. What is the probability of an out-of-control

signal occurring on the first sample following the shift?
(e) For the shift in part (d), what is the probability of detecting the shift by at least the third subsequent sample?

5-45. The following $\bar{x}$ and S charts based on $n = 4$ have shown statistical control:

$\bar{x}$ Chart	S Chart
UCL = 710	UCL = 18.08
Center line = 700	Center line = 7.979
LCL = 690	LCL = 0

(a) Estimate the process parameters μ and σ.
(b) If the specifications are at 705 ± 15, and the process output is normally distributed, estimate the fraction nonconforming.
(c) For the $\bar{x}$ chart, find the probability of a type I error, assuming σ is constant.
(d) Suppose the process mean shifts to 693 and the standard deviation simultaneously shifts to 12. Find the probability of detecting this shift on the $\bar{x}$ chart on the first subsequent sample.
(e) For the shift of part (d), find the average run length.

5-46. One-pound coffee cans are filled by a machine, sealed, and then weighed automatically. After adjusting for the weight of the can, any package that weighs less than 16 oz is cut out of the conveyor. The weights of 25 successive cans are shown here. Set up a moving-range control chart and a control chart for individuals. Estimate the mean and standard deviation of the amount of coffee packed in each can. Is it reasonable to assume that can weight is normally distributed? If the process remains in control at this level, what percentage of cans will be underfilled?

Can Number	Weight	Can Number	Weight
1	16.11	14	16.12
2	16.08	15	16.10
3	16.12	16	16.08
4	16.10	17	16.13
5	16.10	18	16.15
6	16.11	19	16.12
7	16.12	20	16.10
8	16.09	21	16.08
9	16.12	22	16.07
10	16.10	23	16.11
11	16.09	24	16.13
12	16.07	25	16.10
13	16.13		

5-47. Fifteen successive heats of a steel alloy are tested for hardness. The resulting data are shown here. Set up a control chart for the moving range and a control chart for individual hardness measurements. Is it reasonable to assume that hardness is normally distributed?

Heat	Hardness (coded)	Heat	Hardness (coded)
1	52	9	58
2	51	10	51
3	54	11	54
4	55	12	59
5	50	13	53
6	52	14	54
7	50	15	55
8	51		

5-48. The viscosity of a polymer is measured hourly. Measurements for the last 20 hours are shown as follows:

Test	Viscosity	Test	Viscosity
1	2838	4	3064
2	2785	5	2996
3	3058	6	2882
7	2878	14	2975
8	2920	15	2719
9	3050	16	2861
10	2870	17	2797
11	3174	18	3078
12	3102	19	2964
13	2762	20	2805

(a) Does viscosity follow a normal distribution?

(b) Set up a control chart on viscosity and a moving range chart. Does the process exhibit statistical control?

(c) Estimate the process mean and standard deviation.

5-49. **Continuation of Exercise 5-48.** The next five measurements on viscosity are 3163, 3199, 3054, 3147, and 3156. Do these measurements indicate that the process is in statistical control?

5-50. (a) Thirty observations on the oxide thickness of individual silicon wafers are shown here. Use these data to set up a control chart on oxide thickness and a moving range chart. Does the process exhibit statistical control? Does oxide thickness follow a normal distribution?

Wafer	Oxide Thickness	Wafer	Oxide Thickness
1	45.4	10	46.3
2	48.6	11	53.9
3	49.5	12	49.8
4	44.0	13	46.9
5	50.9	14	49.8
6	55.2	15	45.1
7	45.5	16	58.4
8	52.8	17	51.0
9	45.3	18	41.2

Wafer	Oxide Thickness	Wafer	Oxide Thickness	Wafer	Oxide Thickness	Wafer	Oxide Thickness
19	47.1	25	47.2	1	43.4	11	50.0
20	45.7	26	48.0	2	46.7	12	61.2
21	60.6	27	55.9	3	44.8	13	46.9
22	51.0	28	50.0	4	51.3	14	44.9
23	53.0	29	47.9	5	49.2	15	46.2
24	56.0	30	53.4	6	46.5	16	53.3
				7	48.4	17	44.1
				8	50.1	18	47.4
				9	53.7	19	51.3
				10	45.6	20	42.5

(b) Following the establishment of the control charts in part (a), ten new wafers were observed. The oxide thickness measurements are as follows:

Wafer	Oxide Thickness
1	54.3
2	57.5
3	64.8
4	62.1
5	59.6
6	51.5
7	58.4
8	67.5
9	61.1
10	63.3

Plot these observations against the control limits determined in part (a). Is the process in control?

(c) Suppose the assignable cause responsible for the out-of-control signal in part (b) is discovered and removed from the process. Twenty additional wafers are subsequently sampled. Plot the oxide thickness against the part (a) control limits. What conclusions can you draw? The new data are shown here.

5-51. Thirty observations on concentration (in g/l) of the active ingredient in a liquid cleaner produced in a continuous chemical process are shown here.

Observation	Concentration	Observation	Concentration
1	60.4	16	99.9
2	69.5	17	59.3
3	78.4	18	60.0
4	72.8	19	74.7
5	78.2	20	75.8
6	78.7	21	76.6
7	56.9	22	68.4
8	78.4	23	83.1
9	79.6	24	61.1
10	100.8	25	54.9
11	99.6	26	69.1
12	64.9	27	67.5
13	75.5	28	69.2
14	70.4	29	87.2
15	68.1	30	73.0

(a) A normal probability plot of the concentration data is shown next. The straight line was fit by eye to pass approximately through the twentieth and eightieth percentiles.

Does the normality assumption seem reasonable here?

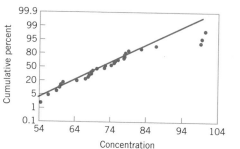

(b) Set up an individuals and moving range control chart for the concentration data. Interpret the charts.

(c) Construct a normal probability plot for the natural log of concentration. Is the transformed variable normally distributed?

(d) Repeat part (b), using the natural log of concentration as the charted variable. Comment on any differences in the charts you note in comparison to those constructed in part (b).

5-52. The purity of a chemical product is measured on each batch. Purity determinations for 20 successive batches are shown here.

Batch	Purity	Batch	Purity
1	0.81	11	0.81
2	0.82	12	0.83
3	0.81	13	0.81
4	0.82	14	0.82
5	0.82	15	0.81
6	0.83	16	0.85
7	0.81	17	0.83
8	0.80	18	0.87
9	0.81	19	0.86
10	0.82	20	0.84

(a) Is purity normally distributed?
(b) Is the process in statistical control?

(c) Estimate the process mean and standard deviation.

5-53. Reconsider the situation in Exercise 5-46. Construct an individuals control chart using the median of the span-two moving ranges to estimate variability. Compare this control chart to the one constructed in Exercise 5-46 and discuss.

5-54. Reconsider the hardness measurements in Exercise 5-47. Construct an individuals control chart using the median of the span-two moving ranges to estimate variability. Compare this control chart to the one constructed in Exercise 5-47 and discuss.

5-55. Reconsider the polymer viscosity data in Exercise 5-48. Use the median of the span-two moving ranges to estimate σ and set up the individuals control chart. Compare this chart to the one originally constructed using the average moving range method to estimate σ.

5-56. **Continuation of Exercises 5-50 and 5-51.** Use all 50 observations on oxide thickness in both exercises.

(a) Set up an individuals control chart with σ estimated by the average moving range method.

(b) Set up an individuals control chart with σ estimated by the median moving range method.

(c) Compare and discuss the two control charts.

5-57. Consider the individuals measurement data shown next.

(a) Estimate σ using the average of the moving ranges of span two.

(b) Estimate σ using S/c_4.

(c) Estimate σ using the median of the span-two moving ranges.

(d) Estimate σ using the average of the moving ranges of span 3, 4, . . . , 20.

(e) Discuss the results you have obtained.

Observation	x
1	10.07
2	10.47
3	9.45
4	9.44
5	8.99
6	7.74
7	10.63
8	9.78
9	9.37
10	9.95
11	12.04
12	10.93
13	11.54
14	9.58
15	8.80
16	12.94
17	10.78
18	11.26
19	9.48
20	11.28
21	12.54
22	11.48
23	13.26
24	11.10
25	10.82

Casting	Vane 1	Vane 2	Vane 3	Vane 4	Vane 5
3	5.77314	5.71216	5.74810	5.77292	5.75591
4	5.77030	5.75903	5.77157	5.79687	5.78063
5	5.72047	5.68587	5.73302	5.70472	5.68116
6	5.77265	5.76426	5.74373	5.71338	5.74765
7	5.70581	5.70835	5.71866	5.71252	5.72089
8	5.76466	5.78766	5.76115	5.77523	5.75590
9	5.79397	5.83308	5.77902	5.81122	5.82335
10	5.78671	5.76411	5.75941	5.75619	5.71787
11	5.75352	5.74144	5.74109	5.76817	5.75019
12	5.72787	5.70716	5.75349	5.72389	5.73488
13	5.79707	5.79231	5.79022	5.79694	5.79805
14	5.73765	5.73615	5.73249	5.74006	5.73265
15	5.72477	5.76565	5.76963	5.74993	5.75196
16	5.73199	5.72926	5.72963	5.72259	5.73513
17	5.79166	5.79516	5.79903	5.78548	5.79826
18	5.74973	5.74863	5.73994	5.74405	5.74682
19	5.76449	5.75632	5.76197	5.76684	5.75474
20	5.75168	5.75579	5.73979	5.77963	5.76933

5-59. The diameter of the casting in Fig. 5-27 is also an important quality characteristic. A coordinate measuring machine is used to measure the diameter of each casting at five different locations. Data for 20 castings are shown in the following table.

(a) Set up $\bar{x}$ and R charts for this process, assuming the measurements on each casting form a rational subgroup.

(b) Discuss the charts you have constructed in part (a).

(c) Construct "between/within" charts for this process.

(d) Do you believe that the charts in part (c) are more informative than those in part (a)? Discuss why.

(e) Provide a practical interpretation of the "within" chart.

5-58. The vane heights for 20 of the castings from Fig. 5-27 are shown here. Construct the "between/within" control charts for these process data using a range chart to monitor the within-castings vane height. Compare these to the control charts shown in Fig. 5-29.

Casting	Vane 1	Vane 2	Vane 3	Vane 4	Vane 5
1	5.77799	5.74907	5.76672	5.74836	5.74122
2	5.79090	5.78043	5.79163	5.79393	5.81158

	Diameter				
Casting	1	2	3	4	5
1	11.7629	11.7403	11.7511	11.7474	11.7374
2	11.8122	11.7506	11.7787	11.7736	11.8412

Casting	1	2	3	4	5
3	11.7742	11.7114	11.7530	11.7532	11.7773
4	11.7833	11.7311	11.7777	11.8108	11.7804
5	11.7134	11.6870	11.7305	11.7419	11.6642
6	11.7925	11.7611	11.7588	11.7012	11.7611
7	11.6916	11.7205	11.6958	11.7440	11.7062
8	11.7109	11.7832	11.7496	11.7496	11.7318
9	11.7984	11.8887	11.7729	11.8485	11.8416
10	11.7914	11.7613	11.7356	11.7628	11.7070
11	11.7260	11.7329	11.7424	11.7645	11.7571
12	11.7202	11.7537	11.7328	11.7582	11.7265
13	11.8356	11.7971	11.8023	11.7802	11.7903
14	11.7069	11.7112	11.7492	11.7329	11.7289
15	11.7116	11.7978	11.7982	11.7429	11.7154
16	11.7165	11.7284	11.7571	11.7597	11.7317
17	11.8022	11.8127	11.7864	11.7917	11.8167
18	11.7775	11.7372	11.7241	11.7773	11.7543
19	11.7753	11.7870	11.7574	11.7620	11.7673
20	11.7572	11.7626	11.7523	11.7395	11.7884

5-60. In the semiconductor industry, the production of microcircuits involves many steps. The wafer fabrication process typically builds these microcircuits on silicon wafers, and there are many microcircuits per wafer. Each production lot consists of between 16 and 48 wafers. Some processing steps treat each wafer separately, so that the batch size for that step is one wafer. It is usually necessary to estimate several components of variation: within-wafer, between-wafer, between-lot, and the total variation.

(a) Suppose that one wafer is randomly selected from each lot and that a single measurement on a critical dimension of interest is taken. Which components of variation could be estimated with these data? What type of control charts would you recommend?

(b) Suppose that each wafer is tested at five fixed locations (say, the center and four points at the circumference). The average and range of these within-wafer measurements are $\bar{x}_{ww}$ and R_{ww}, respec-

tively. What components of variability are estimated using control charts based on these data?

(c) Suppose that one measurement point on each wafer is selected and that this measurement is recorded for five consecutive wafers. The average and range of these between-wafer measurements are $\bar{x}_{BW}$ and R_{BW}, respectively. What components of variability are estimated using control charts based on these data? Would it be necessary to run separate $\bar{x}$ and R charts for all five locations on the wafer?

(d) Consider the question in part (c). How would your answer change if the test sites on each wafer were randomly selected and varied from wafer to wafer?

(e) What type of control charts and rational subgroup scheme would you recommend to control the batch-to-batch variability?

5-61. Consider the situation described in Exercise 5-60. A critical dimension (measured in μm) is of interest to the process engineer. Suppose that five fixed positions are used on each wafer (position 1 is the center) and that two consecutive wafers are selected from each batch. The data that result from several batches are shown here.

(a) What can you say about overall process capability?

(b) Can you construct control charts that allow within-wafer variability to be evaluated?

(c) What control charts would you establish to evaluate variability between wafers? Set up these charts and use them to draw conclusions about the process.

(d) What control charts would you use to evaluate lot-to-lot variability? Set up these charts and use them to draw conclusions about lot-to-lot variability.

Lot Number	Wafer Number	Position 1	2	3	4	5	Lot Number	Wafer Number	Position 1	2	3	4	5
1	1	2.15	2.13	2.08	2.12	2.10	11	1	2.15	2.13	2.14	2.09	2.08
	2	2.13	2.10	2.04	2.08	2.05		2	2.11	2.13	2.10	2.14	2.10
2	1	2.02	2.01	2.06	2.05	2.08	12	1	2.03	2.06	2.05	2.01	2.00
	2	2.03	2.09	2.07	2.06	2.04		2	2.04	2.08	2.03	2.10	2.07
3	1	2.13	2.12	2.10	2.11	2.08	13	1	2.05	2.03	2.05	2.09	2.08
	2	2.03	2.08	2.03	2.09	2.07		2	2.08	2.01	2.03	2.04	2.10
4	1	2.04	2.01	2.10	2.11	2.09	14	1	2.08	2.04	2.05	2.01	2.08
	2	2.07	2.14	2.12	2.08	2.09		2	2.09	2.11	2.06	2.04	2.05
5	1	2.16	2.17	2.13	2.18	2.10	15	1	2.14	2.13	2.10	2.10	2.08
	2	2.17	2.13	2.10	2.09	2.13		2	2.13	2.10	2.09	2.13	2.15
6	1	2.04	2.06	2.00	2.10	2.08	16	1	2.06	2.08	2.05	2.03	2.09
	2	2.03	2.10	2.05	2.07	2.04		2	2.03	2.01	2.00	2.06	2.05
7	1	2.04	2.02	2.01	2.00	2.05	17	1	2.05	2.03	2.08	2.01	2.04
	2	2.06	2.04	2.03	2.08	2.10		2	2.06	2.05	2.03	2.05	2.00
8	1	2.13	2.10	2.10	2.15	2.13	18	1	2.03	2.08	2.04	2.00	2.03
	2	2.10	2.09	2.13	2.14	2.11		2	2.04	2.03	2.05	2.01	2.04
9	1	2.00	2.03	2.08	2.07	2.08	19	1	2.16	2.13	2.10	2.13	2.12
	2	2.01	2.03	2.06	2.05	2.04		2	2.13	2.15	2.18	2.19	2.13
10	1	2.04	2.08	2.09	2.10	2.01	20	1	2.06	2.03	2.04	2.09	2.10
	2	2.06	2.04	2.07	2.04	2.01		2	2.01	2.00	2.05	2.08	2.06

Control Charts
for Attributes

CHAPTER OUTLINE

6-1 INTRODUCTION

6-2 THE CONTROL CHART FOR FRACTION NONCONFORMING

  6-2.1 Development and Operation of the Control Chart

  6-2.2 Variable Sample Size

  6-2.3 Nonmanufacturing Applications

  6-2.4 The Operating-Characteristic Function and Average Run Length Calculations

6-3 CONTROL CHARTS FOR NONCONFORMITIES (DEFECTS)

  6-3.1 Procedures with Constant Sample Size

  6-3.2 Procedures with Variable Sample Size

  6-3.3 Demerit Systems

  6-3.4 The Operating-Characteristic Function

  6-3.5 Dealing with Low Defect Levels

  6-3.6 Nonmanufacturing Applications

6-4 CHOICE BETWEEN ATTRIBUTES AND VARIABLES CONTROL CHARTS

6-5 GUIDELINES FOR IMPLEMENTING CONTROL CHARTS

CHAPTER OVERVIEW

Many quality characteristics cannot be conveniently represented numerically. In such cases, we usually classify each item inspected as either conforming or nonconforming to the specifications on that quality characteristic. The terminology **"defective"** or **"non-defective"** is often used to identify these two classifications of product. More recently, the terminology **"conforming"** and **"nonconforming"** has become popular. Quality characteristics of this type are called **attributes.** Some examples of quality characteristics that are attributes are the occurrence of warped automobile engine connecting rods in a day's production and the proportion of nonfunctional semiconductor chips in a production run.

In this chapter, we present three widely used attributes control charts. The first of these relates to the fraction of nonconforming or defective product produced by a manufacturing process, and is called the **control chart for fraction nonconforming,** or *p* chart. In some situations it is more convenient to deal with the number of **defects** or **nonconformities** observed rather than the fraction nonconforming. The second type of control chart that we study, called the **control chart for nonconformities,** or the *c* chart, is designed to deal with this case. Finally, we present a **control chart for nonconformities per unit,** or the *u* chart, which is useful in situations where the average number of nonconformities per unit is a more convenient basis for process control. An excellent supplement to this chapter is the paper by Woodall (1997) that summarizes over 250 papers on attributes control charts and provides a comprehensive bibliography. The chapter concludes with some guidelines for implementing control charts.

6-1   INTRODUCTION

In Chapter 5, we introduced control charts for quality characteristics that are expressed as variables. Although these control charts enjoy widespread application they are not universally applicable because not all quality characteristics can be expressed with variables data. For example, consider a glass container for a liquid product. Suppose we examine a container and classify it into one of the two categories called "conforming" or "nonconforming," depending on whether the container meets the requirements on one or more quality characteristics. This is an example of **attributes** data, and a control chart for the fraction of nonconforming containers could be established (we show how to do this in Section 6-2). Alternatively, in some processes we may examine a unit of product and count defects or nonconformities on the unit. This type of data is widely encountered in the semiconductor industry, for example. In Section 6-3 we show how to establish control charts for counts, or for the average number of counts per unit.

Attributes charts are generally not as informative as variables charts because there is typically more information in a numerical measurement than in merely classifying a unit as conforming or nonconforming. However, attribute charts do have important applications. They are particularly useful in service industries and in nonmanufacturing quality-improvement efforts because so many of the quality characteristics found in these environments are not easily measured on a numerical scale.

6-2   THE CONTROL CHART FOR FRACTION NONCONFORMING

The **fraction nonconforming** is defined as the ratio of the number of nonconforming items in a population to the total number of items in that population. The items may have *several* quality characteristics that are examined simultaneously by the inspector. If the item does not conform to standard on one or more of these characteristics, it is classified as nonconforming. We usually express the fraction nonconforming as a decimal, although occasionally the percent nonconforming (which is simply 100% times the fraction nonconforming) is used. When demonstrating or displaying the control chart to production personnel or presenting results to management, the percent nonconforming is often used, as it has more intuitive appeal. Although it is customary to work with fraction nonconforming, we could also analyze the fraction conforming just as easily, resulting in a control chart on *process yield*. For example, many manufacturing organizations operate a yield-management system at each stage of their manufacturing process, with the first-pass yield tracked on a control chart.

The statistical principles underlying the control chart for fraction nonconforming are based on the binomial distribution. Suppose the production process is operating in a stable manner, such that the probability that any unit will not conform to specifications is p, and that successive units produced are independent. Then each unit produced is a realization of a Bernoulli random variable with parameter p. If a random sample of n units of product is selected, and if D is the number of units of product that are nonconforming, then D has a binomial distribution with parameters n and p; that is,

$$P\{D = x\} = \binom{n}{x} p^x (1 - p)^{n-x} \qquad x = 0, 1, \ldots, n \qquad (6\text{-}1)$$

From Section 2-2.2 we know that the mean and variance of the random variable D are np and $np(1 - p)$, respectively.

The **sample fraction nonconforming** is defined as the ratio of the number of nonconforming units in the sample D to the sample size n; that is,

$$\hat{p} = \frac{D}{n} \qquad\qquad (6\text{-}2)$$

As noted in Section 2-2.2, the distribution of the random variable $\hat{p}$ can be obtained from the binomial. Furthermore, the mean and variance of $\hat{p}$ are

$$\mu = p \qquad\qquad (6\text{-}3)$$

and

$$\sigma_{\hat{p}}^2 = \frac{p(1 - p)}{n} \qquad\qquad (6\text{-}4)$$

respectively. We will now see how this theory can be applied to the development of a control chart for fraction nonconforming. Because the chart monitors the process fraction nonconforming p, it is also called the p chart.

6-2.1   Development and Operation of the Control Chart

In Chapter 4, we discussed the general statistical principles on which the Shewhart control chart is based. If w is a statistic that measures a quality characteristic, and if the mean of w is μ_w and the variance of w is σ_w^2, then the general model for the Shewhart control chart is as follows:

$$\text{UCL} = \mu_w + L\sigma_w$$
$$\text{Center line} = \mu_w \tag{6-5}$$
$$\text{LCL} = \mu_w - L\sigma_w$$

where L is the distance of the control limits from the center line, in multiples of the standard deviation of w. It is customary to choose $L = 3$.

Suppose that the true fraction nonconforming p in the production process is known or is a **standard value** specified by management. Then from equation 6-5, the center line and control limits of the fraction nonconforming control chart would be as defined here.

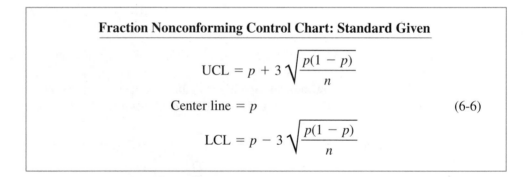

Fraction Nonconforming Control Chart: Standard Given

$$\text{UCL} = p + 3\sqrt{\frac{p(1-p)}{n}}$$
$$\text{Center line} = p \tag{6-6}$$
$$\text{LCL} = p - 3\sqrt{\frac{p(1-p)}{n}}$$

The actual operation of this chart would consist of taking subsequent samples of n units, computing the sample fraction nonconforming $\hat{p}$, and plotting the statistic $\hat{p}$ on the chart. As long as $\hat{p}$ remains within the control limits and the sequence of plotted points does not exhibit any systematic nonrandom pattern, we can conclude that the process is in control at the level p. If a point plots outside of the control limits, or if a nonrandom pattern in the plotted points is observed, we can conclude that the process fraction nonconforming has most likely shifted to a new level and the process is out of control.

When the process fraction nonconforming p is not known, then it must be estimated from observed data. The usual procedure is to select m preliminary samples, each of size n. As a general rule, m should be 20 or 25. Then if there are D_i nonconforming units in sample i, we compute the fraction nonconforming in the ith sample as

$$\hat{p}_i = \frac{D_i}{n} \qquad i = 1, 2, \ldots, m$$

and the average of these individual sample fractions nonconforming is

$$\bar{p} = \frac{\sum_{i=1}^{m} D_i}{mn} = \frac{\sum_{i=1}^{m} \hat{p}_i}{m} \tag{6-7}$$

The statistic $\bar{p}$ estimates the unknown fraction nonconforming p. The center line and control limits of the control chart for fraction nonconforming are computed as defined here.

Fraction Nonconforming Control Chart: No Standard Given

$$\text{UCL} = \bar{p} + 3\sqrt{\frac{\bar{p}(1-\bar{p})}{n}}$$

$$\text{Center line} = \bar{p} \tag{6-8}$$

$$\text{LCL} = \bar{p} - 3\sqrt{\frac{\bar{p}(1-\bar{p})}{n}}$$

The control limits defined in equation 6-8 should be regarded as **trial control limits.** The sample values of $\hat{p}_i$ from the preliminary subgroups should be plotted against the trial limits to test whether the process was in control when the preliminary data were collected. Any points that exceed the trial control limits should be investigated. If assignable causes for these points are discovered, they should be discarded and new trial control limits determined. Refer to the discussion of trial control limits for the $\bar{x}$ and R charts in Chapter 5.

If the control chart is based on a known or standard value for the fraction nonconforming p, then the calculation of trial control limits is generally unnecessary. However, one should be cautious when working with a standard value for p. Since in practice the true value of p would rarely be known with certainty, we would usually be given a standard value of p that represents a desired or **target value** for the process fraction nonconforming. If this is the case, and future samples indicate an out-of-control condition, we must determine whether the process is out of control at the *target* p but in control at some *other* value of p. For example, suppose management specifies a target value of $p = 0.01$, but the process is really in control at a larger value of fraction nonconforming—say, $p = 0.05$. Using the control chart based on $p = 0.01$, we see that many of the points will plot above the upper control limit, indicating an out-of-control condition. However, the process is really out of control only with respect to the target $p = 0.01$. Sometimes it may be possible to "improve" the level of quality by using target values, or to bring a process into control at a particular level of quality performance. In processes where the fraction nonconforming can be controlled by relatively simple process adjustments, target values of p may be useful.

······   **EXAMPLE 6-1**   ··

Frozen orange juice concentrate is packed in 6-oz cardboard cans. These cans are formed on a machine by spinning them from cardboard stock and attaching a metal bottom panel. By inspection of a can, we may determine whether, when filled, it could possibly leak either on the side seam or around the bottom joint. Such a nonconforming can has an improper seal on either the side seam or the bottom panel. We wish to set up a control chart to improve the fraction of nonconforming cans produced by this machine.

To establish the control chart, 30 samples of $n = 50$ cans each were selected at half-hour intervals over a three-shift period in which the machine was in continuous operation. The data are shown in Table 6-1.

We construct a preliminary control chart to see whether the process was in control when these data were collected. Since the 30 samples contain $\sum_{i=1}^{30} D_i = 347$ nonconforming cans, we find from equation 6-7,

$$\bar{p} = \frac{\sum\limits_{i=1}^{m} D_i}{mn} = \frac{347}{(30)(50)} = 0.2313$$

Using $\bar{p}$ as an estimate of the true process fraction nonconforming, we can now calculate the upper and lower control limits as

$$\bar{p} \pm 3 \sqrt{\frac{\bar{p}(1-\bar{p})}{n}} = 0.2313 \pm 3 \sqrt{\frac{0.2313(0.7687)}{50}}$$
$$= 0.2313 \pm 3(0.0596)$$
$$= 0.2313 \pm 0.1789$$

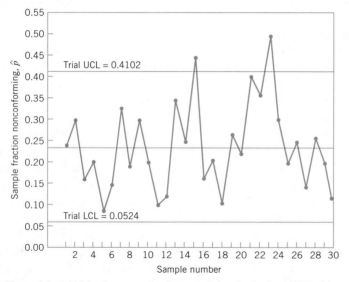

Figure 6-1   Initial fraction nonconforming control chart for the data in Table 6-1.

Table 6-1 Data for Trial Control Limits, Example 6-1 Sample Size $n = 50$

Sample Number	Number of Nonconforming Cans, D_i	Sample Fraction Nonconforming, $\hat{p}_i$	Sample Number	Number of Nonconforming Cans, D_i	Sample Fraction Nonconforming, $\hat{p}_i$
1	12	0.24	17	10	0.20
2	15	0.30	18	5	0.10
3	8	0.16	19	13	0.26
4	10	0.20	20	11	0.22
5	4	0.08	21	20	0.40
6	7	0.14	22	18	0.36
7	16	0.32	23	24	0.48
8	9	0.18	24	15	0.30
9	14	0.28	25	9	0.18
10	10	0.20	26	12	0.24
11	5	0.10	27	7	0.14
12	6	0.12	28	13	0.26
13	17	0.34	29	9	0.18
14	12	0.24	30	6	0.12
15	22	0.44		347	$\bar{p} = 0.2313$
16	8	0.16			

Therefore,

$$\text{UCL} = \bar{p} + 3\sqrt{\frac{\bar{p}(1 - \bar{p})}{n}} = 0.2313 + 0.1789 = 0.4102$$

and

$$\text{LCL} = \bar{p} - 3\sqrt{\frac{\bar{p}(1 - \bar{p})}{n}} = 0.2313 - 0.1789 = 0.0524$$

The control chart with center line at $\bar{p} = 0.2313$ and the above upper and lower control limits is shown in Fig. 6-1. The sample fraction nonconforming from each preliminary sample is plotted on this chart. We note that two points, those from samples 15 and 23, plot above the upper control limit, so the process is not in control. These points must be investigated to see whether an assignable cause can be determined.

Analysis of the data from sample 15 indicates that a new batch of cardboard stock was put into production during that half-hour period. The introduction of new batches of raw material sometimes causes irregular production performance, and it is reasonable to believe that this has occurred here. Furthermore, during the half-hour period in which sample 23 was obtained, a relatively inexperienced operator had been temporarily assigned to the machine, and this could account for the high fraction nonconforming obtained from that sample. Consequently, samples 15 and 23 are eliminated, and the new center line and revised control limits are calculated as

$$\bar{p} = \frac{301}{(28)(50)} = 0.2150$$

$$\text{UCL} = \bar{p} + 3\sqrt{\frac{\bar{p}(1 - \bar{p})}{n}} = 0.2150 + 3\sqrt{\frac{0.2150(0.7850)}{50}} = 0.3893$$

$$\text{LCL} = \bar{p} - 3\sqrt{\frac{\bar{p}(1 - \bar{p})}{n}} = 0.2150 - 3\sqrt{\frac{0.2150(0.7850)}{50}} = 0.0407$$

The revised center line and control limits are shown on the control chart in Fig. 6-2. Note that we have not dropped samples 15 and 23 from the chart, but they have been excluded from the control limit calculations, and we have noted this directly on the control chart. This annotation of the control chart to indicate unusual points, process adjustments, or the type of investigation made at a particular point in time forms a useful record for future process analysis and should become a standard practice in control chart usage.

Note also that the fraction nonconforming from sample 21 now exceeds the upper control limit. However, analysis of the data does not produce any reasonable or logical assignable cause for this, and we decide to retain the point. Therefore, we conclude that the new control limits in Fig. 6-2 can be used for future samples. Thus, we have concluded the control limit estimation phase of control chart usage.

Sometimes examination of control chart data reveals information that affects other points that are not necessarily outside the control limits. For example, if we had found that the temporary operator working when sample 23 was obtained was actually working

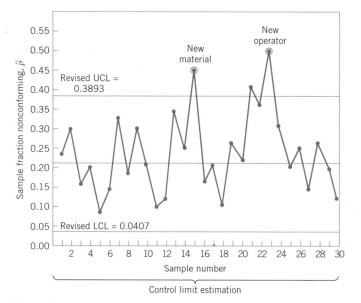

Figure 6-2   Revised control limits for the data in Table 6-1.

during the entire 2-hour period in which samples 21–24 were obtained, then we should discard all four samples, even if only sample 21 exceeded the control limits, on the grounds that this inexperienced operator probably had some adverse influence on the fraction nonconforming during the entire period.

Before we conclude that the process is in control at this level, we could examine the remaining 28 samples for runs and other nonrandom patterns. The largest run is one of length 5 above the center line, and there are no obvious patterns present in the data. There is no strong evidence of anything other than a random pattern of variation about the center line.

We conclude that the process is in control at the level $p = 0.2150$ and that the revised control limits should be adopted for monitoring current production. However, we note that although the process is in control, the fraction nonconforming is much too high. That is, the process is operating in a stable manner, and no unusual **operator-controllable** problems are present. It is unlikely that the process quality can be improved by action at the work-force level. The nonconforming cans produced are **management controllable** because an intervention by management in the process will be required to improve performance. Plant management agrees with this observation and directs that, in addition to implementing the control chart program, the engineering staff should analyze the process in an effort to improve the process yield. This study indicates that several adjustments can be made on the machine that should improve its performance.

During the next three shifts following the machine adjustments and the introduction of the control chart, an additional 24 samples of $n = 50$ observations each are collected. These data are shown in Table 6-2, and the sample fractions nonconforming are plotted on the control chart in Fig. 6-3.

Table 6-2   Orange Juice Concentrate Can Data in Samples of Size $n = 50$

Sample Number	Number of Nonconforming Cans, D_i	Sample Fraction Nonconforming, $\hat{p}_i$	Sample Number	Number of Nonconforming Cans, D_i	Sample Fraction Nonconforming, $\hat{p}_i$
31	9	0.18	44	6	0.12
32	6	0.12	45	5	0.10
33	12	0.24	46	4	0.08
34	5	0.10	47	8	0.16
35	6	0.12	48	5	0.10
36	4	0.08	49	6	0.12
37	6	0.12	50	7	0.14
38	3	0.06	51	5	0.10
39	7	0.14	52	6	0.12
40	6	0.12	53	3	0.10
41	2	0.04	54	5	0.10
42	4	0.08		133	$\bar{p} = 0.1108$
43	3	0.06			

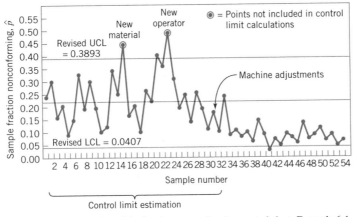

Figure 6-3   Continuation of the fraction nonconforming control chart, Example 6-1.

From an examination of Fig. 6-3, our immediate impression is that the process is now operating at a new quality level that is substantially better than the center line level of $\bar{p} = 0.2150$. One point, that from sample 41, is below the lower control limit. No assignable cause for this out-of-control signal can be determined. The only logical reasons for this ostensible change in process performance are the machine adjustments made by the engineering staff and possibly, the operators themselves. It is not unusual to find that process performance improves following the introduction of formal statistical process-control procedures, often because the operators are more aware of process quality and because the control chart provides a continuing visual display of process performance.

We may test the hypothesis that the process fraction nonconforming in this current three-shift period differs from the process fraction nonconforming in the preliminary data, using the procedure given in Section 3-3.4. The hypotheses are

$$H_0: \quad p_1 = p_2$$
$$H_1: \quad p_1 > p_2$$

where p_1 is the process fraction nonconforming from the preliminary data and p_2 is the process fraction nonconforming in the current period. We may estimate p_1 by $\hat{p}_1 = \bar{p} = 0.2150$, and p_2 by

$$\hat{p}_2 = \frac{\sum_{i=31}^{54} D_i}{(50)(24)} = \frac{133}{1200} = 0.1108$$

The (approximate) test statistic for the above hypothesis is, from equations 3-62 and 3-63,

$$Z_0 = \frac{\hat{p}_1 - \hat{p}_2}{\sqrt{\hat{p}(1 - \hat{p})\left(\dfrac{1}{n_1} + \dfrac{1}{n_2}\right)}}$$

where

$$\hat{p} = \frac{n_1 \hat{p}_1 + n_2 \hat{p}_2}{n_1 + n_2}$$

In our example, we have

$$\hat{p} = \frac{(1400)(0.2150) + (1200)(0.1108)}{1400 + 1200} = 0.1669$$

and

$$Z_0 = \frac{0.2150 - 0.1108}{\sqrt{(0.1669)(0.8331)\left(\dfrac{1}{1400} + \dfrac{1}{1200}\right)}} = 7.10$$

Comparing this to the upper 0.05 point of the standard normal distribution, we find that $Z_0 = 7.10 > Z_{0.05} = 1.645$. Consequently, we reject H_0 and conclude that there has been a significant decrease in the process fallout.

Based on the apparently successful process adjustments, it seems logical to revise the control limits again, using only the most recent samples (numbers 31–54). This results in the new control chart parameters:

$$\text{Center line} = \bar{p} = 0.1108$$

$$\text{UCL} = \bar{p} + 3\sqrt{\frac{\bar{p}(1 - \bar{p})}{n}} = 0.1108 + 3\sqrt{\frac{(0.1108)(0.8892)}{50}} = 0.2440$$

$$\text{LCL} = \bar{p} - 3\sqrt{\frac{\bar{p}(1 - \bar{p})}{n}} = 0.1108 - 3\sqrt{\frac{(0.1108)(0.8892)}{50}} = -0.0224 = 0$$

Figure 6-4 shows the control chart with these new parameters. Note that since the calculated lower control limit is less than zero, we have set LCL = 0. Therefore, the new

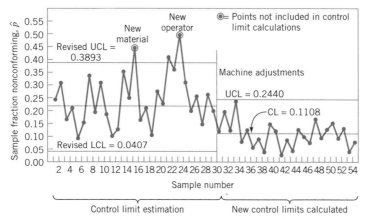

Figure 6-4 New control limits on the fraction nonconforming control chart, Example 6-1.

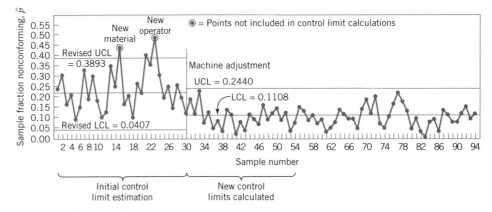

Figure 6-5   Completed fraction nonconforming control chart, Example 6-1.

control chart will have only an upper control limit. From inspection of Fig. 6-4, we see that all the points would fall inside the revised upper control limit; therefore, we conclude that the process is in control at this new level.

The continued operation of this control chart for the next five shifts is shown in Fig. 6-5. Data for the process during this period are shown in Table 6-3. The control

Table 6-3   New Data for the Fraction Nonconforming Control Chart in Fig. 6-5, $n = 50$

Sample Number	Number of Nonconforming Cans, D_i	Sample Fraction Nonconforming, $\hat{p}_i$	Sample Number	Number of Nonconforming Cans, D_i	Sample Fraction Nonconforming, $\hat{p}_i$
55	8	0.16	75	5	0.10
56	7	0.14	76	8	0.16
57	5	0.10	77	11	0.22
58	6	0.12	78	9	0.18
59	4	0.08	79	7	0.14
60	5	0.10	80	3	0.06
61	2	0.04	81	5	0.10
62	3	0.06	82	2	0.04
63	4	0.08	83	1	0.02
64	7	0.14	84	4	0.08
65	6	0.12	85	5	0.10
66	5	0.10	86	3	0.06
67	5	0.10	87	7	0.14
68	3	0.06	88	6	0.12
69	7	0.14	89	4	0.08
70	9	0.18	90	4	0.08
71	6	0.12	91	6	0.12
72	10	0.20	92	8	0.16
73	4	0.08	93	5	0.10
74	3	0.06	94	6	0.12

chart does not indicate lack of control. Despite the improvement in yield following the engineering changes in the process and the introduction of the control chart, the process fallout of $\bar{p} = 0.1108$ is still too high. Further management action will be required to improve the yield. These management interventions may be further adjustments to the machine. Statistically designed experiments (see Part III) are an appropriate way to determine which machine adjustments are critical to defect-free manufacturing, and the appropriate magnitude and direction of these adjustments. The control chart should be continued during the period in which the adjustments are made. By marking the time scale of the control chart *when* a process change is made, the control chart becomes a **logbook** in which the timing of process interventions and their subsequent effect on process performance are easily seen. This logbook aspect of control chart usage is extremely important.

Design of the Fraction Nonconforming Control Chart

The fraction nonconforming control chart has three parameters that must be specified: the sample size, the frequency of sampling, and the width of the control limits. Ideally, we should have some general guidelines for selecting those parameters.

It is relatively common to base a control chart for fraction nonconforming on 100% inspection of *all* process output over some convenient period of time, such as a shift or a day. In this case, both sample size and sampling frequency are interrelated. We would generally select a sampling frequency appropriate for the production rate, and this fixes the sample size. Rational subgrouping may also play a role in determining the sampling frequency. For example, if there are three shifts, and we suspect that shifts differ in their general quality level, then we should use the output of each shift as a subgroup rather than pooling the output of all three shifts together to obtain a daily fraction defective.

If we are to select a sample of process output, then we must choose the sample size n. Various rules have been suggested for the choice of n. If p is very small, we should choose n sufficiently large so that we have a high probability of finding at least one nonconforming unit in the sample. Otherwise, we might find that the control limits are such that the presence of only one nonconforming unit in the sample would indicate an out-of-control condition. For example, if $p = 0.01$ and $n = 8$, we find that the upper control limit is

$$\text{UCL} = p + 3\sqrt{\frac{p(1-p)}{n}} = 0.01 + 3\sqrt{\frac{(0.01)(0.99)}{8}} = 0.1155$$

If there is one nonconforming unit in the sample, then $\hat{p} = \frac{1}{8} = 0.1250$, and we can conclude that the process is out of control. Since for any $p > 0$ there is a positive probability of producing *some* defectives, it is unreasonable in many cases to conclude that the process is out of control on observing a single nonconforming item.

To avoid this pitfall, we can choose the sample size n so that the probability of finding at least one nonconforming unit per sample is at least γ. For example, suppose that $p = 0.01$, and we want the probability of at least one nonconforming unit in the sample to be at least 0.95. If D denotes the number of nonconforming units, then we want to find n such that $P\{D \geq 1\} \geq 0.95$. Using the Poisson approximation to the binomial, we find from the cumulative Poisson table that $\lambda = np$ must exceed 3.00. Consequently, since $p = 0.01$, this implies that the sample size should be 300.

Duncan (1986) has suggested that the sample size should be large enough that we have approximately a 50% chance of detecting a process shift of some specified amount. For example, suppose that $p = 0.01$, and we want the probability of detecting a shift to $p = 0.05$ to be 0.50. Assuming that the normal approximation to the binomial applies, we should choose n so that the upper control limit exactly coincides with the fraction nonconforming in the out-of-control state.[1] If δ is the magnitude of the process shift, then n must satisfy

$$\delta = L \sqrt{\frac{p(1 - p)}{n}} \tag{6-9}$$

Therefore,

$$n = \left(\frac{L}{\delta}\right)^2 p(1 - p) \tag{6-10}$$

In our example, $p = 0.01$, $\delta = 0.05 - 0.01 = 0.04$, and if three-sigma limits are used, then from equation 6-10,

$$n = \left(\frac{3}{0.04}\right)^2 (0.01)(0.99) = 56$$

If the in-control value of the fraction nonconforming is small, another useful criterion is to choose n large enough so that the control chart will have a positive lower control limit. This ensures that we will have a mechanism to force us to investigate one or more samples that contain an unusually small number of nonconforming items. Since we wish to have

$$\text{LCL} = p - L \sqrt{\frac{p(1 - p)}{n}} > 0 \tag{6-11}$$

this implies that

$$n > \frac{(1 - p)}{p} L^2 \tag{6-12}$$

For example, if $p = 0.05$ and three-sigma limits are used, the sample size must be

$$n > \frac{0.95}{0.05} (3)^2 = 171$$

Thus, if $n \geq 172$ units, the control chart will have a positive lower control limit.

---

[1] If $\hat{p}$ is approximately normal, then the probability that $\hat{p}$ exceeds the UCL is 0.50 if the UCL equals the out-of-control fraction nonconforming p, due to the symmetry of the normal distribution. See Section 2-4.3 for a discussion of the normal approximation to the binomial.

Three-sigma control limits are usually employed on the control chart for fraction nonconforming on the grounds that they have worked well in practice. As discussed in Section 4-3.2, narrower control limits would make the control chart more sensitive to small shifts in p but at the expense of more frequent "false alarms." Occasionally, we have seen narrower limits used in an effort to *force* improvement in process quality. Care must be exercised in this, however, as too many false alarms will destroy the operating personnel's confidence in the control chart program.

We should note that the fraction nonconforming control chart is not a universal model for *all* data on fraction nonconforming. It is based on the binomial probability model; that is, the probability of occurrence of a nonconforming unit is *constant,* and successive units of production are *independent.* In processes where nonconforming · units are clustered together, or where the probability of a unit being nonconforming depends on whether or not previous units were nonconforming, the fraction nonconforming control chart is often of little use. In such cases, it is necessary to develop a control chart based on the correct probability model.

Interpretation of Points on the Control Chart for Fraction Nonconforming

Example 6-1 illustrated how points that plot beyond the control limits are treated, both in establishing the control chart and during its routine operation. Care must be exercised in interpreting points that plot *below* the lower control limit. These points often do not represent a real improvement in process quality. Frequently, they are caused by errors in the inspection process resulting from inadequately trained or inexperienced inspectors or from improperly calibrated test and inspection equipment. We have also seen cases in which inspectors deliberately passed nonconforming units or reported fictitious data. The analyst must keep these warnings in mind when looking for assignable causes if points plot below the lower control limits. Not all "downward shifts" in p are attributable to improved quality.

The np Control Chart

It is also possible to base a control chart on the number nonconforming rather than the fraction nonconforming. This is often called an np control chart. The parameters of this chart are as follows.

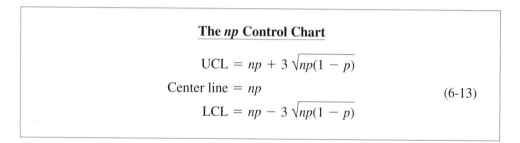

The np Control Chart

$$\text{UCL} = np + 3\sqrt{np(1-p)}$$
$$\text{Center line} = np$$
$$\text{LCL} = np - 3\sqrt{np(1-p)}$$

(6-13)

If a standard value for p is unavailable, then $\bar{p}$ can be used to estimate p. Many nonstatistically trained personnel find the np chart easier to interpret than the usual fraction nonconforming control chart.

······   **EXAMPLE 6-2**   ···

To illustrate the construction of an np control chart, consider the data in Example 6-1 for the fraction nonconforming orange juice concentrate cans. Using the data in Table 6-1, we found that

$$\bar{p} = 0.2313 \qquad n = 50$$

Therefore, the parameters of the np control chart would be

$$\text{UCL} = n\bar{p} + 3\sqrt{n\bar{p}(1 - \bar{p})}$$

$$= 50(0.2313) + 3\sqrt{(50)(0.2313)(0.7687)}$$

$$= 20.510$$

$$\text{Center line} = n\bar{p} = 50(0.2313) = 11.565$$

$$\text{LCL} = n\bar{p} - 3\sqrt{n\bar{p}(1 - \bar{p})}$$

$$= 50(0.2313) - 3\sqrt{(50)(0.2313)(0.7687)}$$

$$= 2.620$$

Now in practice, the number of nonconforming units in each sample is plotted on the np control chart, and the number of nonconforming units is an integer. Thus, if 20 units are nonconforming the process is in control, but if 21 occur the process is out of control. Similarly, there are three nonconforming units in the sample and the process is in control, but two nonconforming units would imply an out-of-control process. Some practitioners prefer to use integer values for control limits on the np chart instead of their decimal fraction counterparts. In this example we could choose 2 and 21 as the LCL and UCL, respectively, and the process would be considered out of control if a sample value of np plotted at or beyond the control limits.

··

6-2.2   Variable Sample Size

In some applications of the control chart for fraction nonconforming, the sample is a 100% inspection of process output over some period of time. Since different numbers of units could be produced in each period, the control chart would then have a variable sample size. There are three approaches to constructing and operating a control chart with a variable sample size.

Variable-Width Control Limits

The first and perhaps the most simple approach is to determine control limits for each individual sample that are based on the specific sample size. That is, if the ith sample is of size n_i, then the upper and lower control limits are $p \pm 3\sqrt{p(1 - p)/n_i}$. Note that the width of the control limits is inversely proportional to the square root of the sample size. To illustrate this approach, consider the data in Table 6-4. For the 25 samples, we calculate

$$\bar{p} = \frac{\sum_{i=1}^{25} D_i}{\sum_{i=1}^{25} n_i} = \frac{234}{2450} = 0.096$$

Consequently, the center line is at 0.096, and the control limits are

$$\mathrm{UCL} = \bar{p} + 3\hat{\sigma}_{\hat{p}} = 0.096 + 3\sqrt{\frac{(0.096)(0.904)}{n_i}}$$

Table 6-4 Data for a Control Chart for Fraction Nonconforming with Variable Sample Size

Sample Number, i	Sample Size, n_i	Number of Nonconforming Units, D_i	Sample Fraction Nonconforming, $\hat{p}_i = D_i/n_i$	Standard Deviation $\hat{\sigma}_{\hat{p}} = \sqrt{\dfrac{(0.096)(0.904)}{n_i}}$	Control Limits LCL	UCL
1	100	12	0.120	0.029	0.009	0.183
2	80	8	0.100	0.033	0	0.195
3	80	6	0.075	0.033	0	0.195
4	100	9	0.090	0.029	0.009	0.183
5	110	10	0.091	0.028	0.012	0.180
6	110	12	0.109	0.028	0.012	0.180
7	100	11	0.110	0.029	0.009	0.183
8	100	16	0.160	0.029	0.009	0.183
9	90	10	0.110	0.031	0.003	0.189
10	90	6	0.067	0.031	0.003	0.189
11	110	20	0.182	0.028	0.012	0.180
12	120	15	0.125	0.027	0.015	0.177
13	120	9	0.075	0.027	0.015	0.177
14	120	8	0.067	0.027	0.015	0.177
15	110	6	0.055	0.028	0.012	0.180
16	80	8	0.100	0.033	0	0.195
17	80	10	0.125	0.033	0	0.195
18	80	7	0.088	0.033	0	0.195
19	90	5	0.056	0.031	0.003	0.189
20	100	8	0.080	0.029	0.009	0.183
21	100	5	0.050	0.029	0.009	0.183
22	100	8	0.080	0.029	0.009	0.183
23	100	10	0.100	0.029	0.009	0.183
24	90	6	0.067	0.031	0.003	0.189
25	90	9	0.100	0.031	0.003	0.189
	2450	234	0.096			

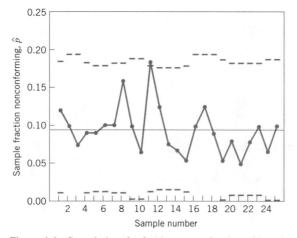

Figure 6-6 Control chart for fraction nonconforming with variable sample size.

and

$$\text{LCL} = \bar{p} - 3\hat{\sigma}_{\hat{p}} = 0.096 - 3\sqrt{\frac{(0.096)(0.904)}{n_i}}$$

where $\hat{\sigma}_{\hat{p}}$ is the estimate of the standard deviation of the sample fraction nonconforming $\hat{p}$. The calculations to determine the control limits are displayed in the last three columns of Table 6-4. The manually constructed control chart is plotted in Fig. 6-6.

Many popular quality control computer programs will handle the variable sample size case. Figure 6-7 presents the computer-generated control chart corresponding to Fig. 6-6. This control chart was obtained using Minitab.

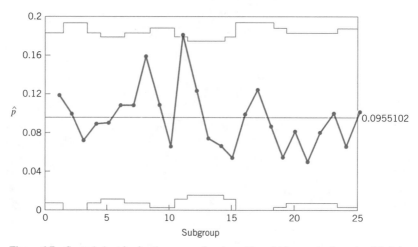

Figure 6-7 Control chart for fraction nonconforming with variable sample size using Minitab.

Control Limits Based on an Average Sample Size
The second approach is to base the control chart on an ***average* sample size,** resulting in an **approximate set of control limits.** This assumes that future sample sizes will not differ greatly from those previously observed. If this approach is used, the control limits will be *constant*, and the resulting control chart will not look as formidable to operating personnel as the control chart with variable limits. However, if there is an unusually large variation in the size of a particular sample or if a point plots near the approximate control limits, then the *exact* control limits for that point should be determined and the point examined relative to that value. For the data in Table 6-4, we find that the average sample size is

$$\bar{n} = \frac{\sum_{i=1}^{25} n_i}{25} = \frac{2450}{25} = 98$$

Therefore, the approximate control limits are

$$\text{UCL} = \bar{p} + 3\sqrt{\frac{\bar{p}(1-\bar{p})}{\bar{n}}} = 0.096 + 3\sqrt{\frac{(0.096)(0.904)}{98}} = 0.185$$

and

$$\text{LCL} = \bar{p} - 3\sqrt{\frac{\bar{p}(1-\bar{p})}{\bar{n}}} = 0.096 - 3\sqrt{\frac{(0.096)(0.904)}{98}} = 0.007$$

The resulting control chart is shown in Fig. 6-8. Note that $\hat{p}$ for sample 11 plots close to the approximate upper control limit, yet it appears to be in control. However, when compared to its exact upper control limit (0.180, from Table 6-4), the point indicates an out-of-control condition. Similarly, points that are outside the approximate control limits

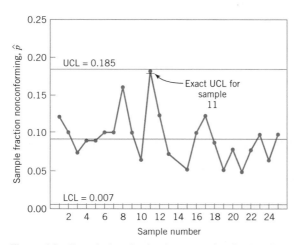

Figure 6-8 Control chart for fraction nonconforming based on average sample size.

may be inside their exact control limits. In general, care must be taken in the interpretation of points near the approximate control limits.

We must also be careful in analyzing runs or other apparently abnormal patterns on control charts with variable sample sizes. The problem is that a change in the sample fraction nonconforming $\hat{p}$ must be interpreted relative to the sample size. For example, suppose that $p = 0.20$ and that two successive sample fractions nonconforming are $\hat{p}_i = 0.28$ and $\hat{p}_{i+1} = 0.24$. The first observation seems to indicate poorer quality than the second, since $\hat{p}_i > \hat{p}_{i+1}$. However, suppose that the sample sizes are $n_i = 50$ and $n_{i+1} = 250$. In *standard deviation units*, the first point is 1.41 units above average and the second point is 1.58 units above average. That is, the second point actually represents a *greater* deviation from the standard of $p = 0.20$ than does the first, even though the second point is the smaller of the two. Clearly, looking for runs or other nonrandom patterns is virtually meaningless here.

The Standardized Control Chart

The third approach to dealing with variable sample size is to use a "standardized" control chart, where the points are plotted in standard deviation units. Such a control chart has the center line at zero, and upper and lower control limits of $+3$ and -3, respectively. The variable plotted on the chart is

$$Z_i = \frac{\hat{p}_i - p}{\sqrt{\dfrac{p(1 - p)}{n_i}}} \qquad (6\text{-}14)$$

where p (or $\bar{p}$ if no standard is given) is the process fraction nonconforming in the in-control state. The standardized control chart for the data in Table 6-4 is shown in Fig. 6-9. The calculations associated with this control chart are shown in Table 6-5. Tests for runs and pattern-recognition methods could safely be applied to this chart, because the relative changes from one point to another are all expressed in terms of the same units of measurement.

The standardized control chart is no more difficult to construct or maintain than either of the other two procedures discussed in the section. In fact, many quality control software packages either automatically execute this as a standard feature or can be programmed to plot a standardized control chart. For example, the version of Fig. 6-9 shown in Fig. 6-10 was created using Minitab. Conceptually, however, it may be more difficult for operating personnel to understand and interpret, because reference to the actual process fraction defective has been "lost." However, if there is large variation in sample size, then runs and pattern-recognition methods can only be safely applied to the standardized control chart. In such a case, it might be advisable to maintain a control

Table 6-5 Calculations for the Standardized Control Chart in Fig. 6-9, $\bar{p} = 0.096$

Sample Number, i	Sample Size, n_i	Number of Noncon-forming Units, D_i	Sample Fraction Noncon-forming, $\hat{p}_i = D_i/n_i$	Standard Deviation $\hat{\sigma}_{\hat{p}} = \sqrt{\dfrac{(0.096)(0.904)}{n_i}}$	$Z_i = \dfrac{\hat{p}_i - \bar{p}}{\sqrt{\dfrac{(0.096)(0.904)}{n_i}}}$
1	100	12	0.120	0.029	0.83
2	80	8	0.100	0.033	0.12
3	80	6	0.075	0.033	−0.64
4	100	9	0.090	0.029	−0.21
5	110	10	0.091	0.028	−0.18
6	110	12	0.109	0.028	0.46
7	100	11	0.110	0.029	0.48
8	100	16	0.160	0.029	2.21
9	90	10	0.110	0.031	0.45
10	90	6	0.067	0.031	−0.94
11	110	20	0.182	0.028	3.07
12	120	15	0.125	0.027	1.07
13	120	9	0.075	0.027	−0.78
14	120	8	0.067	0.027	−1.07
15	110	6	0.055	0.028	−1.46
16	80	8	0.100	0.033	0.12
17	80	10	0.125	0.033	0.88
18	80	7	0.088	0.033	−0.24
19	90	5	0.056	0.031	−1.29
20	100	8	0.080	0.029	−0.55
21	100	5	0.050	0.029	−1.59
22	100	8	0.080	0.029	−0.55
23	100	10	0.100	0.029	0.14
24	90	6	0.067	0.031	−0.94
25	90	9	0.100	0.031	0.13

chart with individual control limits for each sample (as in Fig. 6-6) for the operating personnel, while simultaneously maintaining a standardized control chart for the quality engineer's use.

The standardized control chart is also recommended when the length of the production run is short, as in many job-shop settings. Control charts for short production runs are discussed in Chapter 8.

6-2.3 Nonmanufacturing Applications

The control chart for fraction nonconforming is widely used in nonmanufacturing applications of statistical process control. In the nonmanufacturing environment, many quality

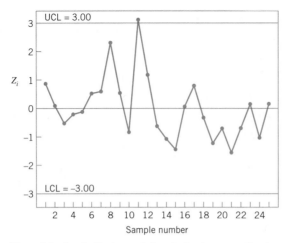

Figure 6-9   Standardized control chart for fraction nonconforming.

characteristics can be observed on a conforming or nonconforming basis. Examples would include the number of employee paychecks that are in error or distributed late during a pay period, the number of check requests that are not paid within the standard accounting cycle, and the number of deliveries made by a supplier that are not on time.

Many nonmanufacturing applications of the fraction nonconforming control chart will involve the variable sample size case. For example, the total number of check requests during an accounting cycle is most likely not constant, and since information about the timeliness of processing for all check requests is generally available, we would calculate $\hat{p}$ as the ratio of all late checks to the total number of checks processed during the period.

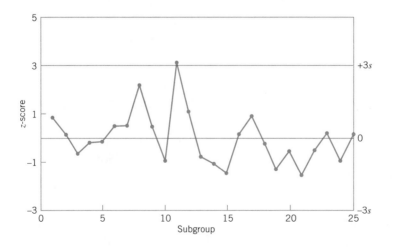

Figure 6-10   Computer-generated standardized control chart for fraction nonconforming, Table 6-4.

As an illustration consider the data in Table 6-4. These data originally came from the purchasing group of a large aerospace company. This group issues purchase orders weekly to the company's suppliers. The sample sizes in Table 6-4 are the actual number of purchase orders issued each week. Note that not all purchase orders are correct. Among the most common errors are specifying incorrect part numbers, incorrect delivery dates, incorrect prices or terms, and wrong supplier number. Any of these errors can result in a purchase order change, which is costly and may delay delivery of material. The purchasing group's quality-improvement team decided to investigate how many purchase orders issued each week resulted in a corresponding purchase order change document owing to errors in the original work. This quantity is the number nonconforming in Table 6-4, and the fraction conforming in this example can be viewed as the fraction of purchase orders in error for the week. The use of this control chart was a key initial step in identifying the underlying root cause of the errors on purchase orders and in developing the corrective actions necessary to improve the process.

6-2.4   The Operating-Characteristic Function and Average Run Length Calculations

The operating-characteristic (or OC) function of the fraction nonconforming control chart is a graphical display of the probability of incorrectly accepting the hypothesis of statistical control (i.e., a type II or β-error) against the process fraction nonconforming. The OC curve provides a measure of the *sensitivity* of the control chart—that is, its ability to detect a shift in the process fraction nonconforming from the nominal value $\bar{p}$ to some other value p. The probability of type II error for the fraction nonconforming control chart may be computed from

$$\beta = p\{\hat{p} < \text{UCL}\,|\,p\} - P\{\hat{p} \le \text{LCL}\,|\,p\}$$
$$= p\{D < n\text{UCL}\,|\,p\} - P\{D \le n\text{LCL}\,|\,p\} \qquad (6\text{-}15)$$

Since D is a binomial random variable with parameters n and p, the β-error defined in equation 6-15 can be obtained from the cumulative binomial distribution.

Table 6-6 illustrates the calculations required to generate the OC curve for a control chart for fraction nonconforming with parameters $n = 50$, LCL = 0.0303, and UCL = 0.3697. Using these parameters, equation 6-15 becomes

$$\beta = P\{D < (50)(0.3697)\,|\,p\} - P\{D \le (50)(0.0303)\,|\,p\}$$
$$= P\{D < 18.49\,|\,p\} - P\{D \le 1.52\,|\,p\}$$

Table 6-6[a]   Calculations for Constructing the OC Curve for a Control Chart for Fraction Nonconforming with $n = 50$, LCL = 0.0303, and UCL = 0.3697

p	$P\{D \leq 18 \mid p\}$	$P\{D \leq 1 \mid p\}$	$\beta = P\{D \leq 18 \mid p\} - P\{D \leq 1 \mid p\}$
0.01	1.0000	0.9106	0.0894
0.03	1.0000	0.5553	0.4447
0.05	1.0000	0.2794	0.7206
0.10	1.0000	0.0338	0.9662
0.15	0.9999	0.0291	0.9708
0.20	0.9975	0.0002	0.9973
0.25	0.9713	0.0001	0.9712
0.30	0.8594	0.0000	0.8594
0.35	0.6216	0.0000	0.6216
0.40	0.3356	0.0000	0.3356
0.45	0.1273	0.0000	0.1273
0.50	0.0325	0.0000	0.0325
0.55	0.0053	0.0000	0.0053

[a]The probabilities in this table were found by evaluating the cumulative binomial distribution. For small p ($p < 0.1$, say) the Poisson approximation could be used, and for larger values of p the normal approximation could be used.

However, since D must be an integer, we find that

$$\beta = P\{D \leq 18 \mid p\} - P\{D \leq 1 \mid p\}$$

The OC curve is plotted in Fig. 6-11.

We may also calculate average run lengths (ARLs) for the fraction nonconforming control chart. Recall from Chapter 4 that for uncorrelated process data the ARL for any Shewhart control chart can be written as

$$ARL = \frac{1}{P(\text{sample point plots out of control})}$$

Thus, if the process is in control, ARL_0 is

$$ARL_0 = \frac{1}{\alpha}$$

and if it is out of control, then

$$ARL_1 = \frac{1}{1 - \beta}$$

These probabilities (α, β) can be calculated directly from the binomial distribution or read from an OC curve.

To illustrate, consider the control chart for fraction nonconforming used in the OC curve calculations in Table 6-6. This chart has parameters $n = 50$, UCL = 0.3697, LCL = 0.0303, and the center line is $\bar{p} = 0.20$. From Table 6-6 (or the OC curve in Fig. 6-11) we find that if the process is in control with $p = \bar{p}$, the probability of a point plotting in control is 0.9973. Thus, in this case $\alpha = 1 - \beta = 0.0027$, and the value of ARL_0 is

$$ARL_0 = \frac{1}{\alpha} = \frac{1}{0.0027} = 370$$

Therefore, if the process is really in control, we will experience a "false-alarm" out-of-control signal about every 370 samples. (This will be approximately true, in general, for any Shewhart control chart with three-sigma limits.) This in-control ARL_0 is generally considered to be satisfactorily large. Now suppose that the process shifts out of control to $p = 0.3$. Table 6-6 indicates that if $p = 0.3$, then $\beta = 0.8594$. Therefore, the value of ARL_1 is

$$ARL_1 = \frac{1}{1 - \beta} = \frac{1}{1 - 0.8594} = 7$$

and it will take about seven samples, on the average, to detect this shift with a point outside of the control limits. If this is unsatisfactory, then action must be taken to reduce the out-of-control ARL_1. Increasing the sample size would result in a smaller value of β and a shorter out-of-control ARL_1. Another approach would be to reduce the interval *between* samples. That is, if we are currently sampling every hour, it will take about 7 hours, on

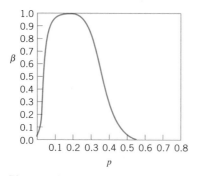

Figure 6-11   Operating-characteristic curve for the fraction nonconforming control chart with $\bar{p} = 0.20$, LCL = 0.0303, and UCL = 0.3697.

the average, to detect the shift. If we take the sample every half hour, it will require only 3.5 hours, on the average, to detect the shift. Another approach is to use a control chart that is more responsive to small shifts, such as the cumulative sum charts in Chapter 7.

6-3 CONTROL CHARTS FOR NONCONFORMITIES (DEFECTS)

A nonconforming item is a unit of product that does not satisfy one or more of the specifications for that product. Each specific point at which a specification is not satisfied results in a **defect** or **nonconformity.** Consequently, a nonconforming item will contain at least one nonconformity. However, depending on their nature and severity, it is quite possible for a unit to contain several nonconformities and *not* be classified as nonconforming. As an example, suppose we are manufacturing personal computers. Each unit could have one or more very minor flaws in the cabinet finish, and since these flaws do not seriously affect the unit's functional operation, it could be classified as conforming. However, if there are too many of these flaws, the personal computer should be classified as nonconforming, since the flaws would be very noticeable to the customer and might affect the sale of the unit. There are many practical situations in which we prefer to work directly with the number of defects or nonconformities rather than the fraction nonconforming. These include the number of defective welds in 100 m of oil pipeline, the number of broken rivets in an aircraft wing, the number of functional defects in an electronic logic device, and so forth.

It is possible to develop control charts for either the **total number of nonconformities** in a unit or the **average number of nonconformities per unit.** These control charts usually assume that the occurrence of nonconformities in samples of constant size is well modeled by the Poisson distribution. Essentially, this requires that the number of opportunities or potential locations for nonconformities be infinitely large and that the probability of occurrence of a nonconformity at any location be small and constant. Furthermore, the **inspection unit** must be the same for each sample. That is, each inspection unit must always represent an identical "area of opportunity" for the occurrence of nonconformities. In addition, we can count nonconformities of several different types on one unit, as long as the above conditions are satisfied for each class of nonconformity.

In most practical situations, these conditions will not be satisfied exactly. The number of opportunities for the occurrence of nonconformities may be finite, or the probability of occurrence of nonconformities may not be constant. As long as these departures from the assumptions are not severe, the Poisson model will usually work reasonably well. There are cases, however, in which the Poisson model is completely inappropriate. These situations are discussed in more detail at the end of Section 6-3.1.

6-3.1 Procedures with Constant Sample Size

Consider the occurrence of nonconformities in an inspection unit of product. In most cases, the inspection unit will be a single unit of product, although this is not necessarily always so. The inspection unit is simply an entity for which it is convenient to keep records. It could be a group of 5 units of product, 10 units of product, and so on. Suppose that defects or nonconformities occur in this inspection unit according to the Poisson distribution; that is,

$$p(x) = \frac{e^{-c}c^x}{x!} \qquad x = 0, 1, 2, \ldots$$

where x is the number of nonconformities and $c > 0$ is the parameter of the Poisson distribution. From Section 2-2.3 we recall that both the mean and variance of the Poisson distribution are the parameter c. Therefore, a control chart for nonconformities with three-sigma limits[2] would be defined as follows.

Control Chart for Nonconformities: Standard Given

$$\text{UCL} = c + 3\sqrt{c}$$
$$\text{Center line} = c \qquad\qquad (6\text{-}16)$$
$$\text{LCL} = c - 3\sqrt{c}$$

---

assuming that a standard value for c is available. Should these calculations yield a negative value for the LCL, set LCL $= 0$.

If no standard is given, then c may be estimated as the observed average number of nonconformities in a preliminary sample of inspection units—say, $\bar{c}$. In this case, the control chart has parameters defined as follows.

Control Chart for Nonconformities: No Standard Given

$$\text{UCL} = \bar{c} + 3\sqrt{\bar{c}}$$
$$\text{Center line} = \bar{c} \qquad\qquad (6\text{-}17)$$
$$\text{LCL} = \bar{c} - 3\sqrt{\bar{c}}$$

When no standard is given, the control limits in equation 6-17 should be regarded as *trial* control limits, and the preliminary samples examined for lack of control. The control chart for nonconformities is also sometimes called the c chart.

---

[2]The α-risk for three-sigma limits is not equally allocated above the UCL and below the LCL, because the Poisson distribution is asymmetric. Some authors recommend the use of probability limits for this chart, particularly when c is small.

······· **EXAMPLE 6-3** ···

Table 6-7 presents the number of nonconformities observed in 26 successive samples of 100 printed circuit boards. Note that, for reasons of convenience, the inspection unit is defined as 100 boards. Since the 26 samples contain 516 total nonconformities, we estimate c by

$$\bar{c} = \frac{516}{26} = 19.85$$

Therefore, the trial control limits are given by

$$\text{UCL} = \bar{c} + 3\sqrt{\bar{c}} = 19.85 + 3\sqrt{19.85} = 33.22$$
$$\text{Center line} = \bar{c} = 19.85$$
$$\text{LCL} = \bar{c} - 3\sqrt{\bar{c}} = 19.85 - 3\sqrt{19.85} = 6.48$$

The control chart is shown in Fig. 6-12. The number of observed nonconformities from the preliminary samples is plotted on this chart. Two points plot outside the control limits, samples 6 and 20. Investigation of sample 6 revealed that a new inspector had examined the boards in this sample and that he did not recognize several of the types of nonconformities that could have been present. Furthermore, the unusually large number of nonconformities in sample 20 resulted from a temperature control problem in the wave soldering machine, which was subsequently repaired. Therefore, it seems reasonable to exclude these two samples and revise the trial control limits. The estimate of c is now computed as

$$\bar{c} = \frac{472}{24} = 19.67$$

and the revised control limits are

$$\text{UCL} = \bar{c} + 3\sqrt{\bar{c}} = 19.67 + 3\sqrt{19.67} = 32.97$$
$$\text{Center line} = \bar{c} = 19.67$$
$$\text{LCL} = \bar{c} - 3\sqrt{\bar{c}} = 19.67 - 3\sqrt{19.67} = 6.37$$

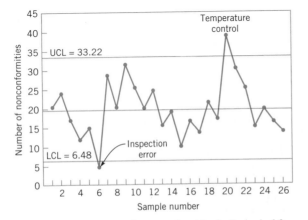

Figure 6-12   Control chart for nonconformities for Example 6-3.

Table 6-7 Data on the Number of Nonconformities in Samples of 100 Printed Circuit Boards

Sample Number	Number of Nonconformities	Sample Number	Number of Nonconformities
1	21	14	19
2	24	15	10
3	16	16	17
4	12	17	13
5	15	18	22
6	5	19	18
7	28	20	39
8	20	21	30
9	31	22	24
10	25	23	16
11	20	24	19
12	24	25	17
13	16	26	15

These become the standard values against which production in the next period can be compared.

Twenty new samples, each consisting of one inspection unit (i.e., 100 boards), are subsequently collected. The number of nonconformities in each sample is noted and recorded in Table 6-8. These points are plotted on the control chart in Fig. 6-13. No lack of control is indicated; however, the number of nonconformities per board is still unacceptably high. Management action is necessary to improve the process.

Table 6-8 Additional Data for the Control Chart for Nonconformities, Example 6-3

Sample Number	Number of Nonconformities	Sample Number	Number of Nonconformities
27	16	37	18
28	18	38	21
29	12	39	16
30	15	40	22
31	24	41	19
32	21	42	12
33	28	43	14
34	20	44	9
35	25	45	16
36	19	46	21

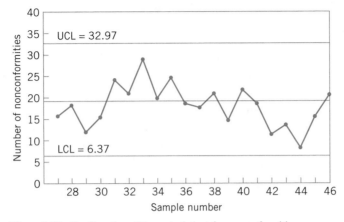

Figure 6-13   Continuation of the control chart for nonconformities, Example 6-3.

Further Analysis of Nonconformities

Defect or nonconformity data are always more informative than fraction nonconforming, because there will usually be several different *types* of nonconformities. By analyzing the nonconformities by type, we can often gain considerable insight into their cause. This can be of considerable assistance in developing the **out-of-control-action-plans** (OCAPs) that must accompany control charts.

For example, in the printed circuit board process, there are 16 different types of defects. Defect data for 500 boards are plotted on a Pareto chart in Fig. 6-14. Note that over 60% of the total number of defects is due to *two defect types:* solder insufficiency

Defect code		Freq	Cum. freq.	Percent	Cum. percent
Sold. insufficie	*************************************	40	40	40.82	40.82
Sold. cold joint	*******************	20	60	20.41	61.23
Sold. opens/dewe	*******	7	67	7.14	68.37
Comp. improper 1	******	6	73	6.12	74.49
Sold. splatter/w	*****	5	78	5.10	79.59
Tst. mark ec mark	***	3	81	3.06	82.65
Tst. mark white m	***	3	84	3.06	85.71
Raw cd shroud re	***	3	87	3.06	88.78
Comp. extra part	**	2	89	2.04	90.82
Comp. damaged	**	2	91	2.04	92.86
Comp. missing	**	2	93	2.04	94.90
Wire incorrect s	*	1	94	1.02	95.92
Stamping oper id	*	1	95	1.02	96.94
Stamping missing	*	1	96	1.02	97.96
Sold. short	*	1	97	1.02	98.98
Raw cd damaged	*	1	98	1.02	100.00

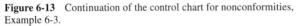

2 4 6 8 10 12 14 16 18 20 22 24 26 28 30 32 34 36 38 40
Percentage

Figure 6-14   Pareto analysis of nonconformities for the printed circuit board process.

and solder cold joints. This points to further problems with the wave soldering process. If these problems can be isolated and eliminated, there will be a dramatic increase in process yield. Notice that the nonconformities follow the Pareto distribution; that is, most of the defects are attributable to a few (in this case, two) defect types.

This process manufactures several different types of printed circuit boards. Therefore, it may be helpful to examine the occurrence of defect type by type of printed circuit board (part number). Table 6-9 presents this information. Note that all 40 solder insufficiencies and all 20 solder cold joints occurred on the same part number, 0001285. This implies that this particular type of board is very susceptible to problems in wave soldering, and special attention must be directed toward improving this step of the process for this part number.

Another useful technique for further analysis of nonconformities is the *cause-and-effect diagram* discussed in Chapter 4. The cause-and-effect diagram is used to illustrate the various sources of nonconformities in products and their interrelationships. It is useful in focusing the attention of operators, manufacturing engineers, and managers on quality problems. Developing a good cause-and-effect diagram usually advances the level of technological understanding of the process.

A cause-and-effect diagram for the printed circuit board assembly process is shown in Fig. 6-15. Since most of the defects in this example were solder related, the cause-and-effect diagram could be used here to help choose the variables for a designed experiment to optimize the wave soldering process. There are several ways to draw the diagram. This one focuses on the three main generic sources of nonconformities: materials, operators, and equipment. Another useful approach is to organize the diagram according to the flow of material through the process.

Choice of Sample Size: The u Chart

Example 6-3 illustrates a control chart for nonconformities with the sample size exactly equal to one inspection unit. The inspection unit is chosen for operational or data-collection simplicity. However, there is no reason why the sample size must be restricted to

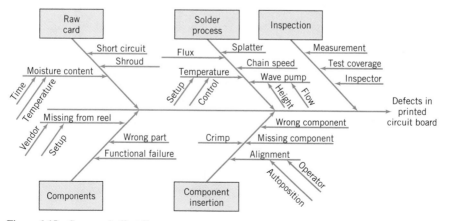

Figure 6-15   Cause-and-effect diagram.

Table 6-9 Table of Defects Classified by Part Number and Defect Code

Part Number / Frequency / Percent / Row Percent / Column Percent	Component Missing	Damaged (NO)	Component Extra Part	Component Improper I	Raw Card Shroud RE	Card Damaged	Solder Short	Solder Opens/DEWE	Cold Joint
0001285	1	0	0	0	0	1	0	5	20
	1.02	0.00	0.00	0.00	0.00	1.02	0.00	5.10	20.41
	1.41	0.00	0.00	0.00	0.00	1.41	0.00	7.04	28.17
	50.00	0.00	0.00	0.00	0.00	100.00	0.00	71.43	100.00
0001481	1	2	2	6	3	0	1	2	0
	1.02	2.04	6.12	3.06	0.00	0.00	1.02	2.04	0.00
	3.70	7.41	22.22	11.11	0.00	0.00	3.70	7.41	0.00
	50.00	100.00	100.00	100.00	100.00	0.00	100.00	28.57	0.00
0006429	0	0	0	0	0	0	0	0	0
	0.00	0.00	0.00	0.00	0.00	0.00	0.00	0.00	0.00
	0.00	0.00	0.00	0.00	0.00	0.00	0.00	0.00	0.00
	0.00	0.00	0.00	0.00	0.00	0.00	0.00	0.00	0.00
Total	2	2	2	6	3	1	1	7	20
	2.04	2.04	2.04	6.12	3.06	1.02	1.02	7.14	20.41

Part Number Frequency Percent Row Percent Column Percent	Solder Insufficiencies	Solder Splatter	Code Stamping Missing	Defect Test Stamping Operator ID	Test Mark White *M*	Wire Mark EC Mark	Incorrect 5	Good Unit(s)	Total
0001285	40	0	0	0	2	1	1	0	71
	40.82	0.00	0.00	0.00	2.04	1.02	1.02	0.00	72.45
	56.32	0.00	0.00	0.00	2.82	1.41	1.41	0.00	
	100.00	0.00	0.00	0.00	66.67	33.33	100.00	0.00	
0001481	0	5	1	1	1	2	0	0	27
	0.00	5.10	1.02	1.02	1.02	2.04	0.00	0.00	27.55
	0.00	18.52	3.70	3.70	3.70	7.41	0.00	0.00	
	0.00	100.00	100.00	100.00	33.33	66.67	0.00	0.00	
0006429	0	0	0	0	0	0	0	0	0
	0.00	0.00	0.00	0.00	0.00	0.00	0.00	0.00	0.00
	0.00	0.00	0.00	0.00	0.00	0.00	0.00	0.00	
	0.00	0.00	0.00	0.00	0.00	0.00	0.00	0.00	
Total	40	5	1	1	3	3	1	0	98
	40.82	5.10	1.02	1.02	3.06	3.06	1.02	0.00	100.00

one inspection unit. In fact, we would often prefer to use *several* inspection units in the sample, thereby increasing the area of opportunity for the occurrence of nonconformities. The sample size should be chosen according to statistical considerations, such as specifying a sample size large enough to ensure a positive lower control limit or to obtain a particular probability of detecting a process shift. Alternatively, economic factors could enter into sample-size determination.

Suppose we decide to base the control chart on a sample size of n inspection units. Note that n does not have to be an integer. To illustrate this, suppose that in Example 6-3 we were to specify a subgroup size of $n = 2.5$ inspection units. Then the sample size becomes $(2.5)(100) = 250$ boards. There are two general approaches to constructing the revised chart once a new sample size has been selected. One approach is simply to redefine a new inspection unit that is equal to n times the old inspection unit. In this case, the center line on the new control chart is $n\bar{c}$ and the control limits are located at $n\bar{c} \pm 3\sqrt{n\bar{c}}$, where $\bar{c}$ is the observed mean number of nonconformities in the *original* inspection unit. Suppose that in Example 6-2, after revising the trial control limits, we decided to use a sample size of $n = 2.5$ inspection units. Then the center line would have been located at $n\bar{c} = (2.5)(19.67) = 49.18$ and the control limits would have been $49.18 \pm 3\sqrt{49.18}$ or LCL $= 28.14$ and UCL $= 70.22$.

The second approach involves setting up a control chart based on the average number of nonconformities per inspection unit. If we find x *total* nonconformities in a sample of n inspection units, then the *average* number of nonconformities per inspection unit is

$$u = \frac{x}{n} \qquad (6\text{-}18)$$

Note that x is a Poisson random variable; consequently, the parameters of the control chart for the average number of nonconformities per unit are as follows.

Control Chart for Average Number of Nonconformities per Unit

$$\text{UCL} = \bar{u} + 3\sqrt{\frac{\bar{u}}{n}}$$

$$\text{Center line} = \bar{u} \qquad (6\text{-}19)$$

$$\text{LCL} = \bar{u} - 3\sqrt{\frac{\bar{u}}{n}}$$

where $\bar{u}$ represents the observed average number of nonconformities per unit in a preliminary set of data. Control limits found from equation 6-19 would be regarded as trial control limits. The control chart for nonconformities per unit is often called the **u chart.**

······· **EXAMPLE 6-4** ···

A personal computer manufacturer wishes to establish a control chart for nonconformities per unit on the final assembly line. The sample size is selected as five computers. Data on the number of nonconformities in 20 samples of 5 computers each are shown in Table 6-10. From these data, we would estimate the average number of nonconformities per unit to be

$$\bar{u} = \frac{\sum_{i=1}^{20} u_i}{20} = \frac{38.60}{20} = 1.93$$

Table 6-10   Data on Number of Nonconformities in Personal Computers

Sample Number, i	Sample Size n	Total Number of Nonconformities, x_i	Average Number of Nonconformities per Unit, $u_i = x_i/n$
1	5	10	2.0
2	5	12	2.4
3	5	8	1.6
4	5	14	2.8
5	5	10	2.0
6	5	16	3.2
7	5	11	2.2
8	5	7	1.4
9	5	10	2.0
10	5	15	3.0
11	5	9	1.8
12	5	5	1.0
13	5	7	1.4
14	5	11	2.2
15	5	12	2.4
16	5	6	1.2
17	5	8	1.6
18	5	10	2.0
19	5	7	1.4
20	5	5	1.0
		193	38.6

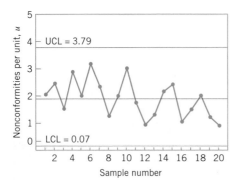

Figure 6-16 A control chart for nonconformities per unit.

Therefore, the parameters of the control chart are

$$\text{UCL} = \bar{u} + 3\sqrt{\frac{\bar{u}}{n}} = 1.93 + 3\sqrt{\frac{1.93}{5}} = 3.79$$

$$\text{Center line} = \bar{u} = 1.93$$

$$\text{LCL} = \bar{u} - 3\sqrt{\frac{\bar{u}}{n}} = 1.93 - 3\sqrt{\frac{1.93}{5}} = 0.07$$

The control chart is shown in Fig. 6-16. The preliminary data do not exhibit lack of statistical control; therefore, the trial control limits given here would be adopted for current control purposes. Once again, note that, although the process is in control, the average number of nonconformities per unit is unacceptably high. Even if these are nonfunctional or appearance nonconformities, there are too many of them. Management must take action to improve the process.

Alternative Probability Models for Count Data

Most applications of the c chart assume that the Poisson distribution is the correct probability model underlying the process. However, it is not the only distribution that could be utilized as a model of "count" or nonconformities per unit-type data. Various types of phenomena can produce distributions of defects that are not well modeled by the Poisson distribution. For example, suppose that nonconformities tend to occur in **clusters;** that is, if there is one nonconformity in some part of a product, then it is likely that there will be others. Note that there are at least two random processes at work here: one generating the number and location of clusters, and the second generating the number of nonconformities within each cluster. If the number of clusters has a Poisson distribution and the number of nonconformities within each cluster has a common distribution (say, f), then the total number of nonconformities has a **compound Poisson distribution.** Many types of compound or generalized distributions could be used as a model for count-type data. As an illustration, if the number of clusters has a Poisson distribution and the number of nonconformities within each cluster is also Poisson, then Neyman's type-A distribution models the total number of nonconformities. Alternatively, if the cluster distribution is gamma and the number of nonconformities within each cluster is Poisson, the negative binomial distribution results. Johnson and Kotz (1969) give a good summary of these and other discrete distributions that could be useful in modeling count-type data.

Mixtures of various types of nonconformities can lead to situations in which the total number of nonconformities is not adequately modeled by the Poisson distribution. Similar situations occur when the count data have either too many or too few zeros. A good discussion of this general problem is the paper by Jackson (1972). The use of the negative binomial distribution to model count data in inspection units of varying size has been studied by Sheaffer and Leavenworth (1976). The dissertation by Gardiner (1987) describes the use of various discrete distributions to model the occurrence of defects in integrated circuits.

6-3.2 Procedures with Variable Sample Size

Control charts for nonconformities are occasionally formed using 100% inspection of the product. When this method of sampling is used, the number of inspection units in a sample will usually not be constant. For example, the inspection of rolls of cloth or paper often leads to a situation in which the size of the sample varies, because not all rolls are exactly the same length or width. If a control chart for nonconformities (c chart) is used in this situation, both the center line and the control limits will vary with the sample size. Such a control chart would be very difficult to interpret. The correct procedure is to use a control chart for nonconformities per unit (u chart). This chart will have a constant center line; however, the control limits will vary inversely with the square root of the sample size n.

······· **EXAMPLE 6-5** ···

In a textile finishing plant, dyed cloth is inspected for the occurrence of defects per 50 square meters. The data on ten rolls of cloth are shown in Table 6-11. We will use these data to set up a control chart for nonconformities per unit.

The center line of the chart should be the average number of nonconformities per inspection unit—that is, the average number of nonconformities per 50 square meters, computed as

$$\bar{u} = \frac{153}{107.5} = 1.42$$

Note that $\bar{u}$ is the ratio of the total number of observed nonconformities to the total number of inspection units.

The control limits on this chart are computed from equation 6-19 with n replaced by n_i. The width of the control limits will vary inversely with n_i, the number of inspection

Table 6-11   Occurrence of Nonconformities in Dyed Cloth

Roll Number	Number of Square Meters	Total Number of Nonconformities	Number of Inspection Units in Roll, n	Number of Nonconformities per Inspection Unit
1	500	14	10.0	1.40
2	400	12	8.0	1.50
3	650	20	13.0	1.54
4	500	11	10.0	1.10
5	475	7	9.5	0.74
6	500	10	10.0	1.00
7	600	21	12.0	1.75
8	525	16	10.5	1.52
9	600	19	12.0	1.58
10	625	23	12.5	1.84
		153	107.50	

units in the roll. The calculations for the control limits are displayed in Table 6-12. Figure 6-17 plots the manually constructed control chart, and Fig. 6-18 shows a corresponding computer-generated control chart.

· ·

As noted previously, the u chart should always be used when the sample size is variable. The most common implementation involves variable control limits, as illustrated in Example 6-5. There are, however, two other possible approaches:

1. Use control limits based on an average sample size

$$\bar{n} = \sum_{i=1}^{m} n_i / m$$

Table 6-12   Calculation of Control Limits, Example 6-5

Roll Number, i	n_i	UCL $= \bar{u} + 3\sqrt{\bar{u}/n_i}$	LCL $= \bar{u} - 3\sqrt{\bar{u}/n_i}$
1	10.0	2.55	0.29
2	8.0	2.68	0.16
3	13.0	2.41	0.43
4	10.0	2.55	0.29
5	9.5	2.58	0.26
6	10.0	2.55	0.29
7	12.0	2.45	0.39
8	10.5	2.52	0.32
9	12.0	2.45	0.39
10	12.5	2.43	0.41

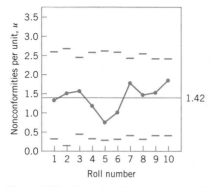

Figure 6-17 Nonconformities per unit control chart with variable sample size, Example 6-5.

2. Use a standardized control chart (this is the preferred option).

This second alternative would involve plotting a standardized statistic

$$Z_i = \frac{u_i - \overline{u}}{\sqrt{\dfrac{\overline{u}}{n_i}}} \tag{6-20}$$

on a control chart with LCL = -3 and UCL = $+3$ and the center line at zero. This chart is appropriate if tests for runs and other pattern-recognition methods are to be used in conjunction with the chart. Figure 6-19 shows the standardized version of the control chart in Example 6-5. This standardized control chart could also be useful in the short production run situation (see Chapter 9, Section 9-1).

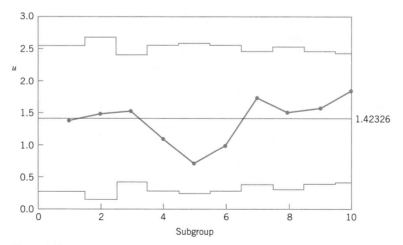

Figure 6-18 Computer-generated (Minitab) version of Fig. 6-17.

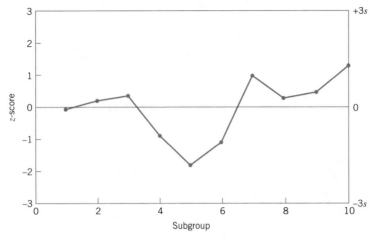

Figure 6-19   Standardized control chart for nonconformities per unit, Example 6-5.

6-3.3   Demerit Systems

With complex products such as automobiles, computers, or major appliances, we usually find that many different types of nonconformities or defects can occur. Not all of these types of defects are equally important. A unit of product having one very serious defect would probably be classified as nonconforming to requirements, but a unit having several minor defects might not necessarily be nonconforming. In such situations, we need a method to classify nonconformities or defects according to severity and to weight the various types of defects in a reasonable manner.

One possible demerit scheme is defined as follows.

Class A Defects—Very Serious. The unit is either completely unfit for service, or will fail in service in such a manner that cannot be easily corrected in the field, or will cause personal injury or property damage.

Class B Defects—Serious. The unit will possibly suffer a Class A operating failure, or will certainly cause somewhat less serious operating problems, or will certainly have reduced life or increased maintenance cost.

Class C Defects—Moderately Serious. The unit will possibly fail in service, or cause trouble that is less serious than operating failure, or possibly have reduced life or increased maintenance costs, or have a major defect in finish, appearance, or quality of work.

Class D Defects—Minor. The unit will not fail in service but has minor defects in finish, appearance, or quality of work.

Let c_{iA}, c_{iB}, C_{iC}, and C_{iD} represent the number of Class A, Class B, Class C, and Class D defects, respectively, in the ith inspection unit. We assume that each class of defect is independent, and the occurrence of defects in each class is well modeled by a Poisson distribution. Then we define the number of **demerits** in the inspection unit as

$$d_i = 100c_{iA} + 50c_{iB} + 10c_{iC} + c_{iD} \tag{6-21}$$

The demerit weights of Class A—100, Class B—50, Class C—10, and Class D—1 are used fairly widely in practice. However, any reasonable set of weights appropriate for a specific problem may also be used.

Suppose that a sample of n inspection units is used. Then the number of demerits per unit is

$$u_i = \frac{D}{n} \tag{6-22}$$

where $D = \sum_{i=1}^{n} d_i$ is the total number of demerits in all n inspection units. Since u is a linear combination of independent Poisson random variables, the statistics u_i could be plotted on a control chart with the following parameters:

where

$$
\begin{aligned}
\text{UCL} &= \bar{u} + 3\hat{\sigma}_u \\
\text{Center line} &= \bar{u} \\
\text{LCL} &= \bar{u} - 3\hat{\sigma}_u
\end{aligned}
\tag{6-23}
$$

where

$$\bar{u} = 100\bar{u}_A + 50\bar{u}_B + 10\bar{u}_C + \bar{u}_D \tag{6-24}$$

and

$$\hat{\sigma}_u = \left[\frac{(100)^2\bar{u}_A + (50)^2\bar{u}_B + (10)^2\bar{u}_C + \bar{u}_D}{n} \right]^{1/2} \tag{6-25}$$

In the preceding equations, $\bar{u}_A$, $\bar{u}_B$, $\bar{u}_C$, and $\bar{u}_D$ represent the average number of Class A, Class B, Class C, and Class D defects per unit. The values of $\bar{u}_A$, $\bar{u}_B$, $\bar{u}_C$, and $\bar{u}_D$ are obtained from the analysis of preliminary data, taken when the process is supposedly operating in control. Standard values for u_A, u_B, u_C, and u_D may also be used, if they are available.

Jones, Woodall, and Conerly (1999) provide a very thorough discussion of demerit-based control charts. They show how probability-based limits can be computed as alternatives to the traditional three-sigma limits used above. They also show that in general the probability limits give superior performance. They are, however, more complicated to compute.

Many variations of this idea are possible. For example, we can classify nonconformities as either **functional defects** or **appearance defects** if a two-class system is preferred. It is also fairly common practice to maintain separate control charts on each defect class rather than combining them into one chart.

6-3.4   The Operating-Characteristic Function

The operating-characteristic (OC) curves for both the c chart and the u chart can be obtained from the Poisson distribution. For the c chart, the OC curve plots the probability of type II error β against the true mean number of defects c. The expression for β is

$$\beta = P\{x < \text{UCL}|c\} - P\{x \leq \text{LCL}|c\} \tag{6-26}$$

where x is a Poisson random variable with parameter c.

We will generate the OC curve for the c chart in Example 6-3. For this example, since the LCL = 6.48 and the UCL = 33.22, equation 6-26 becomes

$$\beta = P\{x < 33.22|c\} - P\{x \leq 6.48|c\}$$

Since the number of nonconformities must be integer, this is equivalent to

$$\beta = P\{x \leq 33|c\} - P\{x \leq 6|c\}$$

These probabilities are evaluated in Table 6-13. The OC curve is shown in Fig. 6-20. For the u chart, we may generate the OC curve from

$$\begin{aligned}
\beta &= P\{x < \text{UCL}|u\} - P\{x \leq \text{LCL}|u\} \\
&= P\{c < n\text{UCL}|u\} - P\{c \leq n\text{LCL}|u\} \\
&= P\{n\text{LCL} < x \leq n\text{UCL}|u\} \\
&= \sum_{x=\langle n\text{LCL}\rangle}^{[n\text{UCL}]} \frac{e^{-nu}(nu)^x}{x!}
\end{aligned} \tag{6-27}$$

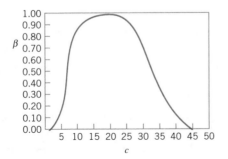

Figure 6-20   OC curve of a c chart with LCL = 6.48 and UCL = 33.22.

Table 6-13   Calculation of the OC Curve for a c Chart with UCL = 33.22 and LCL = 6.48

c	$P\{x \leq 33 \mid c\}$	$P\{x \leq 6 \mid c\}$	$\beta = P\{x \leq 33 \mid c\} - P\{x \leq 6 \mid c\}$
1	1.000	0.999	0.001
3	1.000	0.966	0.034
5	1.000	0.762	0.238
7	1.000	0.450	0.550
10	1.000	0.220	0.780
15	0.999	0.008	0.991
20	0.997	0.000	0.997
25	0.950	0.000	0.950
30	0.744	0.000	0.744
33	0.546	0.000	0.546
35	0.410	0.000	0.410
40	0.151	0.000	0.151
45	0.038	0.000	0.038

where $\langle nLCL \rangle$ denotes the smallest integer greater than or equal to $nLCL$ and $[nUCL]$ denotes the largest integer less than or equal to $nUCL$. The limits on the summation in equation 6-26 follow from the fact that the total number of nonconformities observed in a sample of n inspection units must be an integer. Note that n need not be an integer.

6-3.5   Dealing with Low Defect Levels

When defect levels or in general, **count rates,** in a process become very low—say, under 1000 occurrences per million—there will be very long periods of time between the occurrence of a nonconforming unit. In these situations many samples will have zero defects, and a control chart with the statistic consistently plotting at zero will be relatively uninformative. Thus, conventional c and u charts become ineffective as count rates are driven into the low parts per million (ppm) range.

One way to deal with this problem is to control chart a new variable, the **time between** successive occurrences of the counts. The time-between-events control chart has been very effective as a process-control procedure for processes with low defect levels.

Suppose that defects or counts or "events" of interest occur according to a Poisson distribution. Then the probability distribution of the time between events is the exponential distribution. Therefore, constructing a time-between-events control chart is essentially equivalent to control charting an exponentially distributed variable. However, the exponential distribution is highly skewed, and as a result, the corresponding control chart would be very asymmetric. Such a control chart would certainly look unusual, and might present some difficulties in interpretation for operating personnel.

Nelson (1994) has suggested solving this problem by transforming the exponential random variable to a Weibull random variable such that the resulting Weibull distribution is well approximated by the normal distribution. If y represents the original exponential random variable, the appropriate transformation is

$$x = y^{1/3.6} = y^{0.2777} \qquad\qquad (6\text{-}28)$$

One would now construct a control chart on x, assuming that x follows a normal distribution.

EXAMPLE 6-6

A chemical engineer wants to set up a control chart for monitoring the occurrence of failures of an important valve. She has decided to use the number of hours between failures as the variable to monitor. Table 6-14 shows the number of hours between failures for the last 20 failures of this valve. Figure 6-21 is a normal probability plot of the time between failures. Clearly, time between failures is not normally distributed.

Table 6-14   Time between Failure Data, Example 6-6

Failure	Time between Failures, y (hr)	Transformed Value of Time between Failures, $x = y^{0.2777}$
1	286	4.80986
2	948	6.70903
3	536	5.72650
4	124	3.81367
5	816	6.43541
6	729	6.23705
7	4	1.46958
8	143	3.96768
9	431	5.39007
10	8	1.78151
11	2837	9.09619
12	596	5.89774
13	81	3.38833
14	227	4.51095
15	603	5.91690
16	492	5.59189
17	1199	7.16124
18	1214	7.18601
19	2831	9.09083
20	96	3.55203

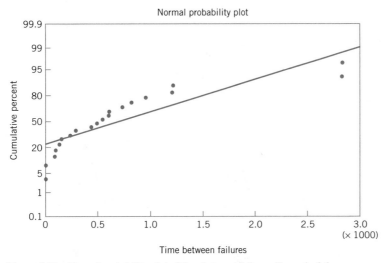

Figure 6-21   Normal probability plot of time between failures, Example 6-6.

Table 6-14 also shows the values of the transformed time between events, computed from equation 6-27. Figure 6-22 is a normal probability plot of the transformed time between failures. Note that the plot indicates that the distribution of this transformed variable is well approximated by the normal.

Figure 6-23 is a control chart for individuals and a moving range control chart for the transformed time between failures. Note that the control charts indicate a state of control, implying that the failure mechanism for this valve is constant. If a process change is made that improves the failure rate (such as a different type of maintenance action), then we would expect to see the mean time between failures get longer. This

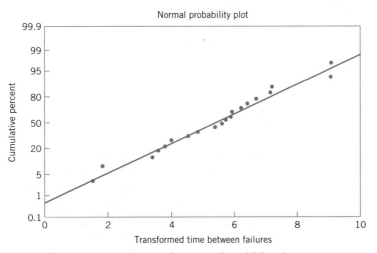

Figure 6-22   Normal probability plot for the transformed failure data.

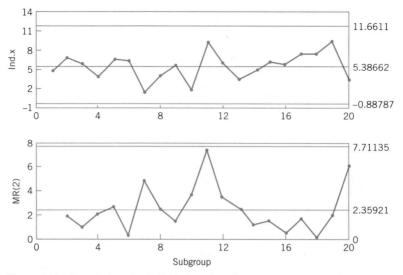

Figure 6-23   Control charts for individuals and moving range control chart for the transformed time between failures, Example 6-6.

would result in points plotting above the upper control limit on the individuals control chart in Fig. 6-23.

The previous example illustrated the use of the individuals control chart with time-between-events data. In many cases, the cusum and EWMA control charts in Chapter 8 would be better alternatives, because these charts are more effective in detecting small shifts in the mean.

Kittlitz (1999) has also investigated transforming the exponential for control charting purposes. He notes that a log transformation will stabilize the variance of the exponential distribution, but produces a rather negatively skewed distribution. Kittlitz suggests using the transformation $x = y^{0.25}$, noting that it is very similar to Nelson's recommendation and it is also very easy to compute (simply push the square-root key on the calculator twice!).

6-3.6   Nonmanufacturing Applications

The c chart and u chart are widely used in nonmanufacturing applications of statistical process control. In effect, we can treat **errors** in the nonmanufacturing environment the same as we treat defects or nonconformities in the manufacturing world. To give just a few examples, we can plot errors on engineering drawings, errors on plans and documents, and errors in computer software as c or u charts. An example using u charts to monitor errors in computer software during product development is given in Gardiner and Montgomery (1987).

6-4   CHOICE BETWEEN ATTRIBUTES AND VARIABLES CONTROL CHARTS

In many applications, the analyst will have to choose between using a variables control chart, such as the $\bar{x}$ and R charts, and an attributes control chart, such as the p chart. In some cases, the choice will be clear-cut. For example, if the quality characteristic is the color of the item, such as might be the case in carpet or cloth production, then attributes inspection would often be preferred over an attempt to quantify the quality characteristic "color." In other cases, the choice will not be obvious, and the analyst must take several factors into account in choosing between attributes and variables control charts.

Attributes control charts have the advantage that several quality characteristics can be considered jointly and the unit classified as nonconforming if it fails to meet the specification on any one characteristic. On the other hand, if the several quality characteristics are treated as variables, then each one must be measured, and either a separate $\bar{x}$ and R chart must be maintained on each or some multivariate control technique that considers all the characteristics must simultaneously be employed. There is an obvious simplicity associated with the attributes chart in this case. Furthermore, expensive and time-consuming measurements may be avoided by attributes inspection.

Variables control charts, in contrast, provide much more useful information about process performance than does an attributes control chart. Specific information about the process mean and variability is obtained directly. In addition, when points plot out of control on variables control charts, usually much more information is provided relative to the potential *cause* of that out-of-control signal. For a process capability study, variables control charts are almost always preferable to attributes control charts. The exceptions to this are studies relative to nonconformities produced by machines or operators in which there are a very limited number of sources of nonconformities, or studies directly concerned with process yields and fallouts.

Perhaps the most important advantage of the $\bar{x}$ and R control charts is that they often provide an indication of impending trouble and allow operating personnel to take corrective action *before* any defectives are actually produced. Thus, $\bar{x}$ and R charts are *leading indicators* of trouble, whereas p charts (or c and u charts) will not react unless the process has already changed so that *more* nonconforming units are being produced.

To illustrate, consider the production process depicted in Fig. 6-24. When the process mean is at μ_1, few nonconforming units are produced. Suppose the process mean begins to shift upward. By the time it has reached μ_2, the $\bar{x}$ and R charts will have reacted to the change in the mean by generating a strong nonrandom pattern and possibly several out-

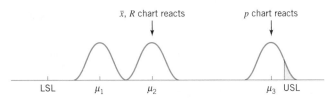

Figure 6-24   Why the $\bar{x}$ and R charts can warn of impending trouble.

of-control points. However, a p chart would not react until the mean had shifted all the way to μ_3, or until the actual number of nonconforming units produced had increased. Thus, the $\bar{x}$ and R charts are more powerful control tools than the p chart.

For a specified level of protection against process shifts, variables control charts usually require a much smaller sample size than does the corresponding attributes control chart. Thus, although variables-type inspection is usually more expensive and time consuming on a per unit basis than attributes inspection, many fewer units must be examined. This is an important consideration, particularly in cases where inspection is destructive (such as opening a can to measure the volume of product within or to test chemical properties of the product). The following example demonstrates the economic advantage of variables control charts.

······· **EXAMPLE 6-7** ···

The nominal value of the mean of a quality characteristic is 50, and the standard deviation is 2. The process is controlled by an $\bar{x}$ chart. Specification limits on the process are established at ± 3-sigma, such that the lower specification limit is 44 and the upper specification limit is 56. When the process is in control at the nominal level of 50, the fraction of nonconforming product produced, assuming that the quality characteristic is normally distributed, is 0.0027. Suppose that the process mean were to shift to 52. The fraction of nonconforming product produced following the shift is approximately 0.0228. If we want the probability of detecting this shift on the first subsequent sample to be 0.50 (say), then the sample size on the $\bar{x}$ chart must be large enough for the upper three-sigma control limit to be 52. This implies that

$$50 + \frac{3(2)}{\sqrt{n}} = 52$$

or $n = 9$. If a p chart is used, then we may find the required sample size to give the same probability of detecting the shift from equation 6-10, that is,

$$n = \left(\frac{L}{\delta}\right)^2 p(1 - p)$$

where $L = 3$ is the width of the control limits, $p = 0.0027$ is the in-control fraction nonconforming, and $\delta = 0.0228 - 0.0027 = 0.0201$ is the magnitude of the shift. Consequently, we find

$$n = (3/0.0201)^2 (0.0027)(0.9973) = 59.98$$

or $n \simeq 60$ would be required for the p chart. Unless the cost of measurements inspection is more than seven times as costly as attributes inspection, the $\bar{x}$ chart is less expensive to operate.

··

Generally speaking, variables control charts are preferable to attributes. However, this logic can be carried to an illogical extreme, as shown in the next example.

EXAMPLE 6-8 ·· ·······

A Misapplication of $\bar{x}$ and R Charts

This example illustrates a misapplication of $\bar{x}$ and R charts that the author encountered in the electronics industry. A company manufacturing a box-level product inspected a sample of the production units several times each shift using attributes inspection. The output of each sample inspection was an estimate of the process fraction nonconforming $\hat{p}_i$. The company personnel were well aware that attributes data did not contain as much information about the process as variables data, and were exploring ways to get more useful information about their process. A consultant to the company (*not* the author) had suggested that they could achieve this objective by converting their fraction noncon-forming data into $\bar{x}$ and R charts. To do so, each group of five successive values of $\hat{p}_i$ was treated as if it were a sample of five *variables* measurements; then the average and

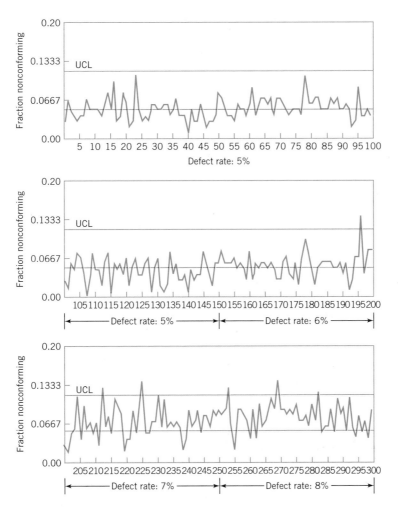

Figure 6-25   Fraction nonconforming control chart for Example 6-8.

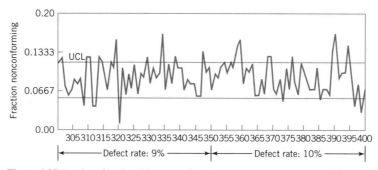

Figure 6-25 *(continued)*   Fraction nonconforming control chart for Example 6-8.

range were computed as

$$\bar{x} = \frac{1}{5} \sum_{i=1}^{5} \hat{p}_i$$

and

$$R = \max(\hat{p}_i) - \min(\hat{p}_i)$$

and these values were plotted on $\bar{x}$ and R charts. The consultant claimed that this procedure would provide more information than the fraction nonconforming control chart.

This suggestion was incorrect. If the inspection process actually produces attributes data governed by the binomial distribution with fixed n, then the sample fraction nonconforming contains *all* the information in the sample (this is an application of the concept of minimal sufficient statistics) and forming two new functions of $\hat{p}_i$ will not provide any additional information.

To illustrate this idea, consider the control chart for fraction nonconforming in Fig. 6-25. This chart was produced by drawing 100 samples (each of size 200) from a process for which $p = 0.05$ and by using these data to compute the control limits. Then the sample draws were continued until sample 150, where the population fraction nonconforming was increased to $p = 0.06$. At each subsequent 50-sample interval, the value of p was increased by 0.01. Note that the control chart reacts to the shift in p at sample number 196. Figures 6-26 and 6-27 present the $\bar{x}$ and R charts obtained by

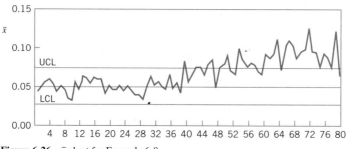

Figure 6-26 $\bar{x}$ chart for Example 6-8.

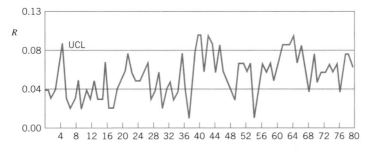

Figure 6-27   *R* chart for Example 6-8.

subgrouping the sample values of $\hat{p}_i$ as suggested above. The first 20 of those subgroups were used to compute the center line and control limits on the $\bar{x}$ and R charts. Note that the $\bar{x}$ chart reacts to the shift in $\hat{p}$ at about subgroup number 40. (This would correspond to *original* samples 196–200.) This result is to be expected, as the $\bar{x}$ chart is really monitoring the fraction nonconforming p. The R chart in Fig. 6-27 is misleading, however. One subgroup within the original set used to construct the control limits is out of control. (This is a false alarm, since $p = 0.05$ for all 100 original samples.) Furthermore, the out-of-control points beginning at about subgroup 40 do not contribute any additional useful information about the process because when $\hat{p}$ shifts from 0.05 to 0.06 (say), the standard deviation of p will *automatically* increase. Therefore, in this case there is no added benefit to the user from $\bar{x}$ and R charts.

This is not to say that the conventional fraction nonconforming control chart based on the binomial probability distribution is the right control chart for all fraction nonconforming data, just as the c chart (based on the Poisson distribution) is not always the right control chart for defect data. If the variability in $\hat{p}_i$ from sample to sample is greater than that which could plausibly be explained by the binomial model, then the analyst should determine the *correct* underlying probability model and base the control chart on that distribution.

··

6-5   GUIDELINES FOR IMPLEMENTING CONTROL CHARTS

Almost any process will benefit from statistical process-control methods, including the use of control charts. In this section, we present some general guidelines helpful in implementing control charts. Specifically, we deal with the following:

1. Determining *which* process characteristics to control
2. Determining *where* the charts should be implemented in the process
3. Choosing the proper *type* of control charts
4. Taking actions to improve processes as the result of SPC/control chart analysis
5. Selecting data-collection systems and computer software

The guidelines are applicable to both variables and attributes control charts. Remember, control charts are not only for process surveillance; they should be used as an active, on-line method for reduction of process variability.

Determining Which Characteristics to Control and Where to Put the Control Charts

At the start of a control chart program, it is usually difficult to determine which product or process characteristics should be controlled and at which points in the process to apply control charts. Some useful guidelines follow.

1. At the beginning of a control chart program, control charts should be applied to any product characteristics or manufacturing operations believed to be important. The charts will provide immediate feedback as to whether they are actually needed.

2. The control charts found to be unnecessary should be removed, and others that engineering and operator judgment indicates may be required should be added. More control charts will usually be employed at the beginning than after the process has stabilized.

3. Information on the number and types of control charts on the process should be kept current. It is best to keep separate records on the variables and attributes charts. In general, after the control charts are first installed, we often find that the number of control charts tends to increase rather steadily. After that, it will usually decrease. When the process stabilizes, we typically find that it has the same number of charts from one year to the next. However, they are not necessarily the same charts.

4. If control charts are being used effectively and if new knowledge is being gained about the key process variables, we should find that the number of $\bar{x}$ and R charts increases and the number of attributes control charts decreases.

5. At the beginning of a control chart program there will usually be more attributes control charts, applied to semifinished or finished units near the *end* of the manufacturing process. As we learn more about the process, these charts will be replaced with $\bar{x}$ and R charts applied *earlier* in the process to the critical parameters and operations that result in nonconformities in the finished product. Generally, **the earlier that process control can be established, the better.** In a complex assembly process, this may imply that process controls need to be implemented at the vendor or supplier level.

6. Control charts are an on-line, process-monitoring procedure. They should be implemented and maintained as close to the work center as possible, so that feedback will be rapid. Furthermore, the process operators and manufacturing engineering should have direct responsibility for collecting the process data, maintaining the charts, and interpreting the results. The operators and engineers have the detailed knowledge of the process required to correct process upsets and use the control chart as a device to improve process performance.

Microcomputers can speed up the feedback and should be an integral part of any modern, on-line, process-control procedure.

Choosing the Proper Type of Control Chart

A. **$\bar{x}$ and R (or $\bar{x}$ and S) charts.** Consider using measurements control charts in these situations:

1. A new process is coming on stream, or a new product is being manufactured by an existing process.

2. The process has been in operation for some time, but it is chronically in trouble or unable to hold the specified tolerances.

3. The process is in trouble, and the control chart can be useful for diagnostic purposes (troubleshooting).

4. Destructive testing (or other expensive testing procedures) is required.

5. It is desirable to reduce acceptance-sampling or other downstream testing to a minimum when the process can be operated in control.

6. Attributes control charts have been used, but the process is either out of control or in control but the yield is unacceptable.

7. There are very tight specifications, overlapping assembly tolerances, or other difficult manufacturing problems.

8. The operator must decide whether or not to adjust the process, or when a set-up must be evaluated.

9. A change in product specifications is desired.

10. Process stability and capability must be continually demonstrated, such as in regulated industries.

B. **Attributes Charts (p charts, c charts, and u charts).** Consider using attributes control charts in these situations:

1. Operators control the assignable causes, and it is necessary to reduce process fallout.

2. The process is a complex assembly operation and product quality is measured in terms of the occurrence of nonconformities, successful or unsuccessful product function, and so forth. (Examples include computers, office automation equipment, automobiles, and the major subsystems of these products.)

3. Process control is necessary, but measurement data cannot be obtained.

4. A historical summary of process performance is necessary. Attributes control charts, such as p charts, c charts, and u charts, are very effective for summarizing information about the process for management review.

5. Remember that attributes charts are generally inferior to charts for measurements. Always use $\bar{x}$ and R or $\bar{x}$ and S charts whenever possible.

C. Control Charts for Individuals. Consider using the control chart for individuals in conjunction with a moving-range chart in these situations:

1. It is inconvenient or impossible to obtain more than one measurement per sample, or repeat measurements will only differ by laboratory or analysis error. Examples often occur in chemical processes.

2. Automated testing and inspection technology allow measurement of every unit produced. In these cases, also consider the cumulative sum control chart and the exponentially weighted moving average control chart discussed in Chapter 7.

3. The data become available very slowly, and waiting for a larger sample will be impractical or make the control procedure too slow to react to problems. This often happens in nonproduct situations; for example, accounting data may become available only monthly.

4. Generally, individuals charts have poor performance in shift detection and can be very sensitive to departures from normality. Always use the EWMA and cusum charts of Chapter 8 instead of individuals charts whenever possible.

Actions Taken to Improve the Process

Process improvement is the primary objective of statistical process control. The application of control charts will give information on two key aspects of the process: statistical control and capability. Figure 6-28 shows the possible states in which the process may exist with respect to these two issues. Technically speaking, the capability of a process cannot be adequately assessed until statistical control has been established, but we will use a less precise definition of capability that is just a qualitative assessment of whether or not the level of nonconforming units produced is low enough to warrant no immediate *additional* effort to further improve the process.

Figure 6-28 gives the answers to two questions: "Is the process in control?" and "Is the process capable (in the sense of the previous paragraph)?" Each of the four cells

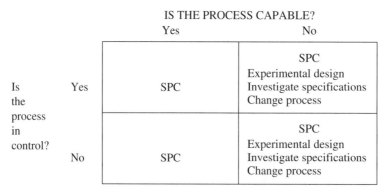

		IS THE PROCESS CAPABLE?	
		Yes	No
Is the process in control?	Yes	SPC	SPC Experimental design Investigate specifications Change process
	No	SPC	SPC Experimental design Investigate specifications Change process

Figure 6-28 Actions taken to improve a process.

in the figure contains some recommended courses of action that depend on the answers to these two questions. The box in the northwest corner is the "ideal state"; the process is in statistical control and exhibits adequate capability for present business objectives. In this case, SPC methods are valuable for process monitoring and for warning against the occurrence of any new assignable causes that could cause slippage in performance. The northeast corner implies that the process exhibits statistical control but has poor capability. Perhaps the PCR is lower than the value required by the customer, or there is sufficient variability remaining to result in excessive scrap or rework. In this case SPC methods may be useful for process diagnosis and improvement, primarily through the recognition of *patterns* on the control chart, but the control charts will not produce very many out-of-control signals. It will usually be necessary to intervene actively in the process to improve it. Experimental design methods are helpful in this regard [see Montgomery (1991)]. Usually, it is also helpful to reconsider the specifications: They may have been set at levels tighter than necessary to achieve function or performance from the part. As a last resort we may have to consider changing the process—that is, investigating or developing new technology that has less variability with respect to this quality characteristic than the existing process.

The lower two boxes in Fig. 6-28 deal with the case of an out-of-control process. The southeast corner presents the case of a process that is out of control and not capable. (Remember our nontechnical use of the term *capability*.) The actions recommended here are identical to those for the box in the northeast corner, except that SPC would be expected to yield fairly rapid results now, because the control charts should be identifying the presence of assignable causes. The other methods of attack will warrant consideration and use in many cases, however. Finally, the southwest corner treats the case of a process that exhibits lack of statistical control but does not produce a meaningful number of defectives because the specifications are very wide. SPC methods should still be used to establish control and reduce variability in this case, for the following reasons:

1. Specifications can change without notice.
2. The customer may require both *control* and capability.
3. The fact that the process experiences assignable causes implies that unknown forces are at work; these unknown forces could result in poor capability in the near future.

Selection of Data-Collection Systems and Computer Software
The last few years have witnessed an explosion of quality control software and electronic data-collection devices. For example, each year, one issue of *Quality Progress* contains a directory of software packages and suppliers, arranged in several applications categories. (Some software packages appear in more than one category). Some SPC consultants have recommended against using the computer, noting that it is unnecessary, since most applications of SPC in Japan emphasized the *manual* use of control charts. If the Japanese were successful in the 1960s and 1970s using manual control charting methods, then does the computer truly have a useful role in SPC?

The answer to this question is yes, for several reasons:

1. Although it is usually helpful to begin with manual methods of control charting at the start of an SPC implementation, it is necessary to move successful applications to the computer very soon. The computer is a great productivity-improvement device. We don't drive cars with the passenger saftey systems of the 1960s and we don't fly airplanes with 1960s avionics technology. We shouldn't use 1960s technology with control charts either.

2. The computer will enable the SPC data to become part of the company-wide manufacturing database, and in that form, the data will be useful (and hence more likely to be used) to everyone—management, engineering, marketing, and so on, *not* only manufacturing and quality.

3. A computer-based SPC system can provide more information than any manual system. It permits the user to monitor many quality characteristics and to provide automatic signaling of assignable causes.

What type of software should be used? That is a difficult question to answer, because all applications have unique requirements and the capability of the software is constantly changing. However, several features are necessary for successful results:

1. The software should be capable of stand-alone operation on a personal computer, on a multiterminal local area network, or on a multiterminal minicomputer-host system. SPC packages that are exclusively tied to a large mainframe system are frequently not very useful because they usually cannot produce control charts and other routine reports in a timely manner.

2. The system must be "user friendly." If operating personnel are to use the system, it must have limited options, be easy to use, provide adequate error-correction opportunities, and contain many on-line help features. It should ideally be possible to tailor or customize the system for each application, although this installation activity may have to be carried out by engineering/technical personnel.

3. The system should provide video display of control charts for at least the last 25 samples. Ideally, the length of record displayed should be controlled by the user. Printed output should be immediately available on either a line printer or a plotter.

4. File storage should be sufficient to accommodate a reasonable amount of process history. Editing and updating files should be straightforward. Provision to transfer data to other storage media or to transfer the data to a master manufacturing database is critical.

5. The system should be able to handle multiple files simultaneously. Only rarely does a process have only one quality characteristic that needs to be examined.

6. The user should be able to calculate control limits from any subset of the data on the file. The user should have the capability to input center lines and control limits directly.

7. The system should be able to accept a variety of inputs, including manual data entry, RS-232 input from an electronic gage, or input from another computer or instrument controller. It is becoming increasingly important to do real-time process monitoring, or to transfer data from a real-time data acquisition system.

8. The system should support other statistical applications, including as a minimum histograms and computation of process capability indices.

9. Service and support from the software supplier after purchase is always an important factor in deciding which software package to use.

The purchase price of commercially available software ranges from $50 to over $150,000. Obviously, the total cost of software is very different from the purchase price. In many cases a $500 SPC package is really a $10,000 package when we take into account the total costs of making the package work correctly in the intended application. It is also relatively easy to establish control charts with most of the popular spreadsheet software packages.

There are several sources of free software. In addition to the packages available on various personal computer bulletin boards, the *Journal of Quality Technology* has published computer programs in either BASIC or FORTRAN since 1969.

6-6 EXERCISES

6-1. The data that follow give the number of nonconforming bearing and seal assemblies in samples of size 100. Construct a fraction nonconforming control chart for these data. If any points plot out of control, assume that assignable causes can be found and determine the revised control limits.

Construct a fraction nonconforming control chart for these data. Does the process appear to be in control? If not, assume that assignable causes can be found for all points outside the control limits and calculate the revised control limits.

Sample Number	Number of Nonconforming Assemblies	Sample Number	Number of Nonconforming Assemblies
1	7	11	6
2	4	12	15
3	1	13	0
4	3	14	9
5	6	15	5
6	8	16	1
7	10	17	4
8	5	18	5
9	2	19	7
10	7	20	12

Sample Number	Number of Nonconforming Switches	Sample Number	Number of Nonconforming Switches
1	8	11	6
2	1	12	0
3	3	13	4
4	0	14	0
5	2	15	3
6	4	16	1
7	0	17	15
8	1	18	2
9	10	19	3
10	6	20	0

6-2. The number of nonconforming switches in samples of size 150 are shown here.

6-3. The following data represent the results of inspecting all units of a personal computer produced for the last

10 days. Does the process appear to be in control?

Day	Units Inspected	Nonconforming Units	Fraction Nonconforming
1	80	4	0.050
2	110	7	0.064
3	90	5	0.056
4	75	8	0.107
5	130	6	0.038
6	120	6	0.050
7	70	4	0.057
8	125	5	0.040
9	105	8	0.076
10	95	7	0.074

6-4. A process that produces titanium forgings for automobile turbocharger wheels is to be controlled through use of a fraction nonconforming chart. Initially, one sample of size 150 is taken each day for 20 days, and the results shown here are observed.

Day	Nonconforming Units	Day	Nonconforming Units
1	3	11	2
2	2	12	4
3	4	13	1
4	2	14	3
5	5	15	6
6	2	16	0
7	1	17	1
8	2	18	2
9	0	19	3
10	5	20	2

(a) Establish a control chart to monitor future production.

(b) What is the smallest sample size that could be used for this process and still give a positive lower control limit on the chart?

6-5. A process produces rubber belts in lots of size 2500. Inspection records on the last 20 lots reveal the following data.

Lot Number	Number of Nonconforming Belts	Lot Number	Number of Nonconforming Belts
1	230	11	456
2	435	12	394
3	221	13	285
4	346	14	331
5	230	15	198
6	327	16	414
7	285	17	131
8	311	18	269
9	342	19	221
10	308	20	407

(a) Compute trial control limits for a fraction nonconforming control chart.

(b) If you wanted to set up a control chart for controlling future production, how would you use these data to obtain the center line and control limits for the chart?

6-6. Based on the following data, if an np chart is to be established, what would you recommend as the center line and control limits? Assume that $n = 500$.

Day	Number of Nonconforming Units
1	3
2	4
3	3
4	2
5	6
6	12
7	5
8	1
9	2
10	2

6-7. A control chart indicates that the current process fraction nonconforming is 0.02. If 50 items are inspected each day, what is the probability of detecting a shift in the fraction nonconforming to 0.04 on the first day after the shift? By the end of the third day following the shift?

6-8. A company purchases a small metal bracket in containers of 5000 each. Ten containers have arrived at the unloading facility, and 250 brackets are selected at random from each container. The fraction nonconforming in each sample are 0, 0, 0, 0.004, 0.008, 0.020, 0.004, 0, 0, and 0.008. Do the data from this shipment indicate statistical control?

6-9. Diodes used on printed circuit boards are produced in lots of size 1000. We wish to control the process producing these diodes by taking samples of size 64 from each lot. If the nominal value of the fraction nonconforming is $p = 0.10$, determine the parameters of the appropriate control chart. To what level must the fraction nonconforming increase to make the β-risk equal to 0.50? What is the minimum sample size that would give a positive lower control limit for this chart?

6-10. A control chart for the number of nonconforming piston rings is maintained on a forging process with $np = 16.0$. A sample of size 100 is taken each day and analyzed.

(a) What is the probability that a shift in the process average to $np = 20.0$ will be detected on the first day following the shift? What is the probability that the shift will be detected by at least the end of the third day?

(b) Find the smallest sample size that will give a positive lower control limit.

6-11. A control chart for the fraction nonconforming is to be established using a center line of $p = 0.10$. What sample size is required if we wish to detect a shift in the process fraction nonconforming to 0.20 with probability 0.50?

6-12. A process is controlled with a fraction nonconforming control chart with three-sigma limits, $n = 100$, UCL = 0.161, center line = 0.080, and LCL = 0.

(a) Find the equivalent control chart for the number nonconforming.

(b) Use the Poisson approximation to the binomial to find the probability of a type I error.

(c) Use the correct approximation to find the probability of a type II error if the process fraction nonconforming shifts to 0.2.

(d) What is the probability of detecting the shift in part (c) by at most the fourth sample after the shift?

6-13. A process is being controlled with a fraction nonconforming control chart. The process average has been shown to be 0.07. Three-sigma control limits are used, and the procedure calls for taking daily samples of 400 items.

(a) Calculate the upper and lower control limits.

(b) If the process average should suddenly shift to 0.10, what is the probability that the shift would be detected on the first subsequent sample?

(c) What is the probability that the shift in part (b) would be detected on the first or second sample taken after the shift?

6-14. In designing a fraction nonconforming chart with center line at $p = 0.20$ and three-sigma control limits, what is the sample size required to yield a positive lower control limit? What is the value of n necessary to give a probability of 0.50 of detecting a shift in the process to 0.26?

6-15. A control chart is used to control the fraction nonconforming for a plastic part manufactured in an injection molding process. Ten subgroups yield the following data:

Sample Number	Sample Size	Number Nonconforming
1	100	10
2	100	15
3	100	31
4	100	18
5	100	24
6	100	12
7	100	23
8	100	15
9	100	8
10	100	8

(a) Set up a control chart for the number nonconforming in samples of $n = 100$.

(b) For the chart established in part (a), what is the probability of detecting a shift in the process fraction nonconforming to 0.30 on the first sample after the shift has occurred?

6-16. A control chart for fraction nonconforming indicates that the current process average is 0.03. The sample size is constant at 200 units.

(a) Find the three-sigma control limits for the control chart.

(b) What is the probability that a shift in the process average to 0.08 will be detected on the first subsequent sample? What is the probability that this shift will be detected at least by the fourth sample following the shift?

6-17. (a) A control chart for the number nonconforming is to be established, based on samples of size 400. To start the control chart, 30 samples were selected and the number nonconforming in each sample determined, yielding $\sum_{i=1}^{30} D_i = 1200$. What are the parameters of the np chart?

(b) Suppose the process average fraction nonconforming shifted to

0.15. What is the probability that the shift would be detected on the first subsequent sample?

6-18. A fraction nonconforming control chart with center line 0.10, UCL = 0.19, and LCL = 0.01 is used to control a process.

(a) If three-sigma limits are used, find the sample size for the control chart.

(b) Use the Poisson approximation to the binomial to find the probability of type I error.

(c) Use the Poisson approximation to the binomial to find the probability of type II error if the process fraction defective is actually $p = 0.20$.

6-19. Consider the control chart designed in Exercise 6-18. Find the average run length to detect a shift to a fraction nonconforming of 0.15.

6-20. Consider the control chart in Exercise 6-18. Find the average run length if the process fraction nonconforming shifts to 0.20.

6-21. A maintenance group improves the effectiveness of its repair work by monitoring the number of maintenance requests that require a second call to complete the repair. Twenty weeks of data are available.

Week	Total Requests	Second Visit Required	Week	Total Requests	Second Visit Required
1	200	6	11	100	1
2	250	8	12	100	0
3	250	9	13	100	1
4	250	7	14	200	4
5	200	3	15	200	5
6	200	4	16	200	3
7	150	2	17	200	10
8	150	1	18	200	4
9	150	0	19	250	7
10	150	2	20	250	6

(a) Find trial control limits for this process.

(b) Design a control chart for controlling future production.

6-22. Analyze the data in Exercise 6-21 using an average sample size.

6-23. Construct a standardized control chart for the data in Exercise 6-21.

6-24. **Continuation of Exercise 6-21.** Note that in Exercise 6-21 there are only four different sample sizes; $n = 100$, 150, 200, and 250. Prepare a control chart that has a set of limits for each possible sample size and show how it could be used as an alternative to the variable-width control limit method used in Exercise 6-21. How easy would this method be to use in practice?

6-25. A fraction nonconforming control chart has center line 0.01, UCL = 0.0399, LCL = 0, and $n = 100$. If three-sigma limits are used, find the smallest sample size that would yield a positive lower control limit.

6-26. Why is the np chart not appropriate with variable sample size?

6-27. A fraction nonconforming control chart with $n = 400$ has the following parameters:

$$UCL = 0.0809$$
$$\text{Center line} = 0.0500$$
$$LCL = 0.0191$$

(a) Find the width of the control limits in standard deviation units.

(b) What would be the corresponding parameters for an equivalent control chart based on the number nonconforming?

(c) What is the probability that a shift in the process fraction nonconforming to 0.0300 will be detected on the first sample following the shift?

6-28. A fraction nonconforming control chart with $n = 400$ has the following parameters:

$$UCL = 0.0962$$
$$\text{Center line} = 0.0500$$
$$LCL = 0.0038$$

(a) Find the width of the control limits in standard deviation units.

(b) Suppose the process fraction nonconforming shifts to 0.15. What is the probability of detecting the shift on the first subsequent sample?

6-29. A fraction nonconforming control chart is to be established with a center line of 0.01 and two-sigma control limits.

(a) How large should the sample size be if the lower control limit is to be nonzero?

(b) How large should the sample size be if we wish the probability of detecting a shift to 0.04 to be 0.50?

6-30. The following fraction nonconforming control chart with $n = 100$ is used to control a process:

$$UCL = 0.0750$$
$$\text{Center line} = 0.0400$$
$$LCL = 0.0050$$

(a) Use the Poisson approximation to the binomial to find the probability of a type I error.

(b) Use the Poisson approximation to the binomial to find the probability of a type II error, if the true process fraction nonconforming is 0.0600.

(c) Draw the OC curve for this control chart.

(d) Find the ARL when the process is in control and the ARL when the process fraction nonconforming is 0.0600.

6-31. A process that produces bearing housings is controlled with a fraction nonconforming control chart, using sample size $n = 100$ and a center line $\bar{p} = 0.02$.

(a) Find the three-sigma limits for this chart.

(b) Analyze the ten new samples ($n = 100$) shown here for statistical control. What conclusions can you draw about the process now?

Sample Number	Number Nonconforming
1	5
2	2
3	3
4	8
5	4
6	1
7	2
8	6
9	3
10	4

6-32. Consider an np chart with k-sigma control limits. Derive a general formula for determining the minimum sample size to ensure that the chart has a positive lower control limit.

6-33. Consider the fraction nonconforming control chart in Exercise 6-4. Find the equivalent np chart.

6-34. Consider the fraction nonconforming control chart in Exercise 6-5. Find the equivalent np chart.

6-35. Construct a standardized control chart for the data in Exercise 6-3.

6-36. Surface defects have been counted on 25 rectangular steel plates, and the data are shown here. Set up a control chart for nonconformities using these data. Does the process producing the plates appear to be in statistical control?

C-chart

Plate Number	Number of Nonconformities	Plate Number	Number of Nonconformities
1	1	14	0
2	0	15	2
3	4	16	1
4	3	17	3
5	1	18	5
6	2	19	4
7	5	20	6
8	0	21	3
9	2	22	1
10	1	23	0
11	1	24	2
12	0	25	4
13	8		

6-37. A paper mill uses a control chart to monitor the imperfection in finished rolls of paper. Production output is inspected for 20 days, and the resulting data are shown here. Use these data to set up a control chart for nonconformities per roll of paper. Does the process appear to be in statistical control? What center line and control limits would you recommend for controlling current production?

Day	Number of Rolls Produced	Total Number of Imperfections
1	18	12
2	18	14
3	24	20
4	22	18
5	22	15
6	22	12
7	20	11
8	20	15
9	20	12
10	20	10
11	18	18
12	18	14
13	18	9
14	20	10
15	20	14
16	20	13
17	24	16
18	24	18
19	22	20
20	21	17

6-38. **Continuation of Exercise 6-37.** Consider the papermaking process in Exercise 6-37. Set up a u chart based on an average sample size to control this process.

6-39. **Continuation of Exercise 6-37.** Consider the papermaking process in

Exercise 6-37. Set up a standardized u chart for this process.

(6-40) The number of nonconformities found on final inspection of a cassette deck is shown here. Can you conclude that the process is in statistical control? What center line and control limits would you recommend for controlling future production?

Deck Number	Number of Nonconformities	Deck Number	Number of Nonconformities
2412	0	2421	1
2413	1	2422	0
2414	1	2423	3
2415	0	2424	2
2416	2	2425	5
2417	1	2426	1
2418	1	2427	2
2419	3	2428	1
2420	2	2429	1

6-41. The following data represent the number of nonconformities per 1000 meters in telephone cable. From analysis of these data, would you conclude that the process is in statistical control? What control procedure would you recommend for future production?

Sample Number	Number of Nonconformities	Sample Number	Number of Nonconformities
1	1	12	6
2	1	13	9
3	3	14	11
4	7	15	15
5	8	16	8
6	10	17	3
7	5	18	6
8	13	19	7
9	0	20	4
10	19	21	9
11	24	22	20

6-42. Consider the data in Exercise 6-40. Suppose we wish to define a new inspection unit of four cassette decks.
(a) What are the center line and control limits for a control chart for moni-

toring future production based on the total number of defects in the new inspection unit?
(b) What are the center line and control limits for a control chart for nonconformities per unit used to monitor future production?

6-43. Consider the data in Exercise 6-41. Suppose a new inspection unit is defined as 2500 m of wire.
(a) What are the center line and control limits for a control chart for monitoring future production based on the total number of nonconformities in the new inspection unit?
(b) What are the center line and control limits for a control chart for average nonconformities per unit used to monitor future production?

6-44. An automobile manufacturer wishes to control the number of nonconformities in a subassembly area producing manual transmissions. The inspection unit is defined as four transmissions, and data from 16 samples (each of size 4) are shown here.

Sample Number	Number of Nonconformities	Sample Number	Number of Nonconformities
1	1	9	2
2	3	10	1
3	2	11	0
4	1	12	2
5	0	13	1
6	2	14	1
7	1	15	2
8	5	16	3

(a) Set up a control chart for nonconformities per unit.
(b) Do these data come from a controlled process? If not, assume that assignable causes can be found for all out-of-control points and calculate the revised control chart parameters.
(c) Suppose the inspection unit is redefined as eight transmissions.

Design an appropriate control chart for monitoring future production.

6-45. Find the three-sigma control limits for
(a) a c chart with process average equal to four nonconformities.
(b) a u chart with $c = 4$ and $n = 4$.

6-46. Find 0.900 and 0.100 probability limits for a c chart when the process average is equal to 16 nonconformities.

6-47. Find the three-sigma control limits for
(a) a c chart with process average equal to nine nonconformities.
(b) a u chart with $c = 16$ and $n = 4$.

6-48. Find 0.980 and 0.020 probability limits for a control chart for nonconformities per unit when $u = 6.0$ and $n = 3$.

6-49. Find 0.975 and 0.025 probability limits for a control chart for nonconformities when $c = 7.6$.

6-50. A control chart for nonconformities per unit uses 0.95 and 0.05 probability limits. The center line is at $u = 1.4$. Determine the control limits if the sample size is $n = 10$.

6-51. The number of workmanship nonconformities observed in the final inspection of disk-drive assemblies has been tabulated as shown here. Does the process appear to be in control?

Day	Number of Assemblies Inspected	Total Number of Nonconformities
1	2	10
2	4	30
3	2	18
4	1	10
5	3	20
6	4	24
7	2	15
8	4	26
9	3	21
10	1	8

6-52. A control chart for nonconformities is to be constructed with $c = 2.0$, LCL $= 0$, and UCL such that the probability of a point plotting outside control limits when $c = 2.0$ is only 0.005.
(a) Find the UCL.
(b) What is the type I error probability if the process is assumed to be out of control only when two consecutive points fall outside the control limits?

6-53. A textile mill wishes to establish a control procedure on flaws in towels it manufactures. Using an inspection unit of 50 units, past inspection data show that 100 previous inspection units had 850 total flaws. What type of control chart is appropriate? Design the control chart such that it has two-sided probability control limits of $\alpha = 0.06$, approximately. Give the center line and control limits.

6-54. The manufacturer wishes to set up a control chart at the final inspection station for a gas water heater. Defects in workmanship and visual quality features are checked in this inspection. For the last 22 working days, 176 water heaters were inspected and a total of 924 nonconformities reported.
(a) What type of control chart would you recommend here and how would you use it?
(b) Using two water heaters as the inspection unit, calculate the center line and control limits that are consistent with the past 22 days of inspection data.
(c) What is the probability of type I error for the control chart in part (b)?

6-55. Assembled portable television sets are subjected to a final inspection for surface defects. A total procedure is established based on the requirement that if the average number of nonconformities per unit is 4.0, the probability of concluding that the process is in control will be 0.99. There is to be no lower control limit. What is the appropriate type of control chart and what is the required upper control limit?

6-56. A control chart is to be established on a process producing refrigerators. The inspection unit is one refrigerator, and a common chart for nonconformities is to be used. As preliminary data, 16 nonconformities were counted in inspecting 30 refrigerators.

(a) What are the three-sigma control limits?

(b) What is the α-risk for this control chart?

(c) What is the β-risk if the average number of defects is actually two (i.e., if $c = 2.0$)?

(d) Find the average run length if the average number of defects is actually two.

6-57. Consider the situation described in Exercise 6-56.

(a) Find two-sigma control limits and compare these with the control limits found in part (a) of Exercise 6-56.

(b) Find the α-risk for the control chart with two-sigma control limits and compare with the results of part (b) of Exercise 6-56.

(c) Find the β-risk for $c = 2.0$ for the chart with two-sigma control limits and compare with the results of part (c) of Exercise 6-56.

(d) Find the ARL if $c = 2.0$ and compare with the ARL found in part (d) of Exercise 6-56.

6-58. A control chart for nonconformities is to be established in conjunction with final inspection of a radio. The inspection unit is to be a group of ten radios. The average number of nonconformities per radio has, in the past, been 0.5. Find three-sigma control limits for a c chart based on this size inspection unit.

6-59. A control chart for nonconformities is maintained on a process producing desk calculators. The inspection unit is

defined as two calculators. The average number of nonconformities per machine when the process is in control is estimated to be two.

(a) Find the appropriate three-sigma control limits for this size inspection unit.

(b) What is the probability of type I error for this control chart?

6-60. A production line assembles electric clocks. The average number of nonconformities per clock is estimated to be 0.75. The quality engineer wishes to establish a c chart for this operation, using an inspection unit of six clocks. Find the three-sigma limits for this chart.

6-61. Suppose that we wish to design a control chart for nonconformities per unit with L-sigma limits. Find the minimum sample size that would result in a positive lower control limit for this chart.

6-62. Kittlitz (1999) presents data on homicides in Waco, Texas, for the years 1980–1989 (data taken from the *Waco Tribune-Herald*, 29 December 1989). There were 29 homicides in 1989. The following table gives the dates of the 1989 homicides and the number of days between each homicide.

Month	Date	Days Between
Jan.	20	
Feb.	23	34
Feb.	25	2
March	5	8
March	10	5
April	4	25
May	7	33
May	24	17
May	28	4
June	7	10
June	16[*]	9.25
June	16[*]	0.50

Month	Date	Days Between
June	22*	5.25
June	25	3
July	6	11
July	8	2
July	9	1
July	26	17
Sep.	9	45
Sep.	22	13
Sep.	24	2
Oct.	1	7
Oct.	4	3
Oct.	8	4
Oct.	19	11
Nov.	2	14
Nov.	25	23
Dec.	28	33
Dec.	29	1

The * refers to the fact that two homicides occurred on 16 June and were determined to be 12 hours apart.

(a) Plot the days-between-homicides data on a normal probability plot. Does the assumption of a normal distribution seem reasonable for these data?

(b) Transform the data using the 0.2777 root of the data. Plot the transformed data on a normal probability plot. Does this plot indicate that the transformation has been successful in making the new data more closely resemble data from a normal distribution?

(c) Transform the data using the fourth root (0.25) of the data. Plot the transformed data on a normal probability plot. Does this plot indicate that the transformation has been successful in making the new data more closely resemble data from a normal distribution? Is the plot very different from the one in part (b)?

(d) Construct an individuals control chart using the transformed data from part (b).

(e) Construct an individuals control chart using the transformed data from part (c). How similar is it to the one you constructed in part (d)?

(f) Is the process "stable"? Provide a practical interpretation of the control chart.

6-63. Suggest at least two nonmanufacturing scenarios in which attributes control charts could be useful for process monitoring.

6-64. What practical difficulties could be encountered in monitoring time-between-events data?

Process and Measurement System Capability Analysis

CHAPTER OUTLINE

7-1 INTRODUCTION

7-2 PROCESS CAPABILITY ANALYSIS USING A HISTOGRAM OR A PROBABILITY PLOT

    7-2.1 Using the Histogram

    7-2.2 Probability Plotting

7-3 PROCESS CAPABILITY RATIOS

    7-3.1 Use and Interpretation of C_p

    7-3.2 Process Capability Ratio for an Off-Center Process

    7-3.3 Normality and the Process Capability Ratio

    7-3.4 More about Process Centering

    7-3.5 Confidence Intervals and Tests on Process Capability Ratios

7-4 PROCESS CAPABILITY ANALYSIS USING A CONTROL CHART

7-5 PROCESS CAPABILITY ANALYSIS USING DESIGNED EXPERIMENTS

7-6 GAGE AND MEASUREMENT SYSTEM CAPABILITY STUDIES

    7-6.1 Control Charts and Tabular Methods

    7-6.2 Methods Based on Analysis of Variance

7-7 SETTING SPECIFICATION LIMITS ON DISCRETE COMPONENTS

    7-7.1 Linear Combinations

    7-7.2 Nonlinear Combinations

7-8 ESTIMATING THE NATURAL TOLERANCE LIMITS OF A PROCESS

    7-8.1 Tolerance Limits Based on the Normal Distribution

    7-8.2 Nonparametric Tolerance Limits

CHAPTER OVERVIEW

In Chapter 5, we formally introduced the concept of process capability, or how the inherent variability in a process compares with the specifications or requirements for the product. This chapter provides more extensive discussion of process capability, including several ways to study or analyze the capability of a process. We believe that the control chart is a simple and effective process capability analysis technique. We also extend the presentation of process capability ratios that we began in Chapter 5, showing how to interpret these ratios and discussing their dangers. The chapter also contains information on evaluating measurement system performance, illustrating graphical and tabular methods, as well as techniques based on the analysis of variance. We also discuss setting specifications on individual discrete parts or components and estimating the natural tolerance limits of a process.

7-1   INTRODUCTION

Statistical techniques can be helpful throughout the product cycle, including development activities prior to manufacturing, in quantifying process variability, in analyzing this variability relative to product requirements or specifications, and in assisting development and manufacturing in eliminating or greatly reducing this variability. This general activity is called **process capability analysis.**

Process capability refers to the **uniformity** of the process. Obviously, the variability in the process is a measure of the uniformity of output. There are two ways to think of this variability:

1. The natural or inherent variability at a specified time; that is, "instantaneous" variability
2. The variability over time

We present methods for investigating and assessing both aspects of process capability.

It is customary to take the six-sigma spread in the distribution of the product quality characteristic as a measure of process capability. Figure 7-1 shows a process for which the quality characteristic has a normal distribution with mean μ and standard deviation σ. The upper and lower "natural tolerance limits" of the process fall at $\mu + 3\sigma$ and $\mu - 3\sigma$, respectively; that is,

$$\text{UNTL} = \mu + 3\sigma$$

$$\text{LNTL} = \mu - 3\sigma$$

For a normal distribution, the natural tolerance limits include 99.73% of the variable, or put another way, only 0.27% of the process output will fall outside the natural tolerance limits. Two points should be remembered:

1. 0.27% outside the natural tolerances sounds small, but this corresponds to 2700 nonconforming parts per million.

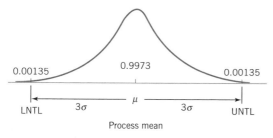

0.00135   0.9973   0.00135

LNTL   3σ μ 3σ UNTL

Process mean

Figure 7-1   Upper and lower natural tolerance limits in the normal distribution.

2. If the distribution of process output is nonnormal, then the percentage of output falling outside $\mu \pm 3\sigma$ may differ considerably from 0.27%.

We define **process capability analysis** as an engineering study to estimate process capability. The estimate of process capability may be in the form of a probability distribution having a specified shape, center (mean), and spread (standard deviation). For example, we may determine that the process output is normally distributed with mean $\mu = 1.0$ cm and standard deviation $\sigma = 0.001$ cm. In this sense, a process capability analysis may be performed **without regard to specifications on the quality characteristic.** Alternatively, we may express process capability as a percentage outside of specifications. However, specifications are not *necessary* to perform a process capability analysis.

A process capability study usually measures functional parameters on the product, not the process itself. When the analyst can directly observe the process and can control or monitor the data-collection activity, the study is a true process capability study, because by controlling the data collection and knowing the time sequence of the data, inferences can be made about the stability of the process over time. However, when we have available only sample units of product, perhaps supplied by the vendor or obtained via receiving inspection, and there is no direct observation of the process or time history of production, then the study is more properly called **product characterization.** In a product characterization study we can only estimate the distribution of the product quality characteristic or the process yield (fraction conforming to specifications); we can say nothing about the dynamic behavior of the process or its state of statistical control.

Process capability analysis is a vital part of an overall quality-improvement program. Among the major uses of data from a process capability analysis are the following:

1. Predicting how well the process will hold the tolerances
2. Assisting product developers/designers in selecting or modifying a process
3. Assisting in establishing an interval between sampling for process monitoring
4. Specifying performance requirements for new equipment
5. Selecting between competing vendors

6. Planning the sequence of production processes when there is an interactive effect of processes on tolerances

7. Reducing the variability in a manufacturing process

Thus, process capability analysis is a technique that has application in many segments of the product cycle, including product and process design, vendor sourcing, production or manufacturing planning, and manufacturing.

Three primary techniques are used in process capability analysis: **histograms** or **probability plots, control charts,** and **designed experiments.** We will discuss and illustrate each of these methods in the next three sections. We will also discuss the process capability ratio (PCR) introduced in Chapter 5 and some useful variations of this ratio.

7-2   PROCESS CAPABILITY ANALYSIS USING A HISTOGRAM OR A PROBABILITY PLOT

7-2.1   Using the Histogram

The histogram can be helpful in estimating process capability. Alternatively, a stem-and-leaf diagram may be substituted for the histogram. At least 100 or more observations should be available for the histogram (or the stem-and-leaf diagram) to be moderately stable so that a reasonably reliable estimate of process capability may be obtained. If the quality engineer has access to the process and can control the data-collection effort, the following steps should be followed prior to data collection:

1. Choose the machine or machines to be used. If the results based on one (or a few) machines are to be extended to a larger population of machines, the machine selected should be representative of those in the population. Furthermore, if the machine has multiple workstations or heads, it may be important to collect the data so that head-to-head variability can be isolated. This may imply that designed experiments should be used.

2. Select the process operating conditions. Carefully define conditions, such as cutting speeds, feed rates, and temperatures, for future reference. It may be important to study the effects of varying these factors on process capability.

3. Select a representative operator. In some studies, it may be important to estimate *operator* variability. In these cases, the operators should be selected at random from the population of operators.

4. Carefully monitor the data-collection process, and record the time order in which each unit is produced.

The histogram, along with the sample average $\bar{x}$ and sample standard deviation S, provides information about process capability. You may wish to review the guidelines for constructing histograms in Chapter 2.

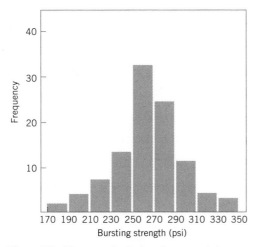

Figure 7-2 Histogram for the bursting-strength data.

•••••• **EXAMPLE 7-1** ••

To illustrate the use of a histogram to estimate process capability, consider Fig. 7-2, which presents a histogram of the bursting strength of 100 glass 1-liter soft-drink bottles. The data are shown in Table 7-1, and the frequency distribution is shown in Table 7-2. Analysis of the 100 observations gives

$$\bar{x} = 264.06 \qquad S = 32.02$$

Consequently, the process capability would be estimated as

$$\bar{x} \pm 3S$$

or

$$264.06 \pm 3(32.02) \simeq 264 \pm 96 \text{ psi}$$

Table 7-1 Bursting Strengths for 100 One-Liter Glass Soft-Drink Bottles

265	197	346	280	265	200	221	265	261	278
205	286	317	242	254	235	176	262	248	250
263	274	242	260	281	246	248	271	260	265
307	243	258	321	294	328	263	245	274	270
220	231	276	228	223	296	231	301	337	298
268	267	300	250	260	276	334	280	250	257
260	281	208	299	308	264	280	274	278	210
234	265	187	258	235	269	265	253	254	280
299	214	264	267	283	235	272	287	274	269
215	318	271	293	277	290	283	258	275	251

Table 7-2 Frequency Distribution for the Bursting-Strength Data

Class Interval (psi)	Frequency	Relative Frequency	Cumulative Relative Frequency
$170 \leq x < 190$	2	0.02	0.02
$190 \leq x < 210$	4	0.04	0.06
$210 \leq x < 230$	7	0.07	0.13
$230 \leq x < 250$	13	0.13	0.26
$250 \leq x < 270$	32	0.32	0.58
$270 \leq x < 290$	24	0.24	0.82
$290 \leq x < 310$	11	0.11	0.93
$310 \leq x < 330$	4	0.04	0.97
$330 \leq x < 350$	3	0.03	1.00
	100	1.00	

Furthermore, the shape of the histogram implies that the distribution of bursting strength is approximately normal. Thus, we can estimate that approximately 99.73% of the bottles manufactured by this process will burst between 168 and 360 psi. Note that we can estimate process capability *independent of the specifications on bursting strength.*

An advantage of using the histogram to estimate process capability is that it gives an immediate, visual impression of process performance. It may also immediately show the reason for poor process performance. For example, Fig. 7-3*a* shows a process with adequate potential capability, but the process target is poorly located, whereas Fig. 7-3*b* shows a process with poor capability resulting from excess variability.

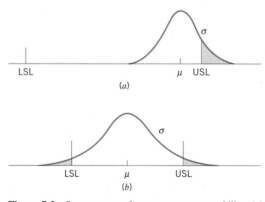

Figure 7-3 Some reasons for poor process capability. (*a*) Poor process centering. (*b*) Excess process variability.

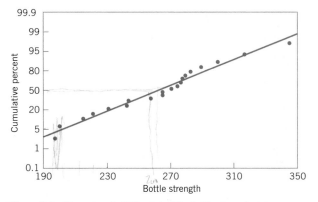

Figure 7-4   Normal probability plot of the bottle-strength data.

7-2.2   Probability Plotting

Probability plotting is an alternative to the histogram that can be used to determine the shape, center, and spread of the distribution. It has the advantage that it is unnecessary to divide the range of the variable into class intervals, and it often produces reasonable results for moderately small samples (which the histogram will not). Generally, a probability plot is a graph of the ranked data versus the sample cumulative frequency on special paper with a vertical scale chosen so that the cumulative distribution of the assumed type is a straight line. In Chapter 2 we discussed and illustrated **normal probability plots.** These plots are very useful in process capability studies.

   To illustrate the use of a normal probability plot in a process capability study, consider the following 20 observations on bottle bursting strength: 197, 200, 215, 221, 231, 242, 245, 258, 265, 265, 271, 275, 277, 278, 280, 283, 290, 301, 318, and 346. Figure 7-4 is the normal probability plot of strength. Note that the data lie nearly along a straight line, implying that the distribution of bursting strength is normal. Recall from Chapter 3 that the mean of the normal distribution is the fiftieth percentile, which we may estimate from Fig. 7-4 as approximately 260 psi. The standard deviation of the distribution is the *slope* of the straight line. It is convenient to estimate the standard deviation as the difference between the eighty-fourth and the fiftieth percentiles. For the strength data, this yields

$$\hat{\sigma} = 84\text{th percentile} - 50\text{th percentile}$$
$$= 298 - 260 \text{ psi}$$
$$= 38 \text{ psi}$$

Note that $\mu = 260$ psi and $\sigma = 38$ psi are not far from the sample average $\bar{x} = 264.06$ and standard deviation $S = 32.02$.

   The normal probability plot can also be used to estimate process yields and fallouts. For example, the specification on bottle strength is LSL = 200 psi. From Fig. 7-4, we

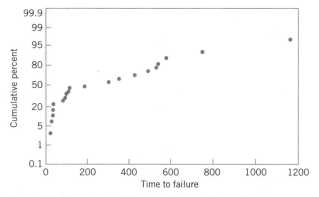

Figure 7-5   Normal probability plot of time to failure of a valve.

would estimate that about 5% of the bottles manufactured by this process would burst below this limit.

Care should be exercised in using probability plots. If the data do not come from the assumed distribution, inferences about process capability drawn from the plot may be seriously in error. Figure 7-5 presents a normal probability plot of times to failure (in hours) of a valve in a chemical plant. From examining this plot, we can see that the distribution of failure time is not normal.

An obvious disadvantage of probability plotting is that it is not an objective procedure. It is possible for two analysts to arrive at different conclusions using the same data. For this reason, it is often desirable to supplement probability plots with more formal statistically based goodness-of-fit tests. A good introduction to these tests is in Shapiro (1980). Choosing the distribution to fit the data is also an important step in probability plotting. Sometimes we can use our knowledge of the physical phenomena or past experience to suggest the choice of distribution. In other situations, the display in Fig. 7-6 may be useful in selecting a distribution that describes the data. This figure shows the regions in the β_1, β_2 plane for several standard probability distributions, where β_1 and β_2 are the measures of *skewness* and *kurtosis,* respectively. To use Fig. 7-6, calculate estimates of skewness and kurtosis from the sample — say,

$$\sqrt{\hat{\beta}_1} = \frac{M_3}{(M_2)^{3/2}}$$

and

$$\hat{\beta}_2 = \frac{M_4}{M_2^2}$$

where

$$M_j = \frac{\sum_{i=1}^{n}(x_i - \bar{x})^j}{n} \qquad j = 1, 2, 3, 4$$

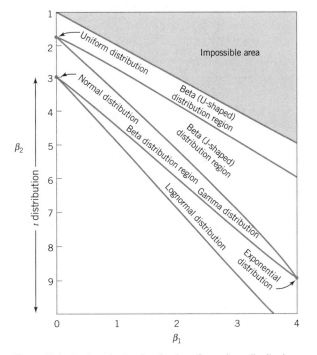

Figure 7-6 Regions in the β_1, β_2 plane for various distributions. (From G. J. Hahn and S. S. Shapiro, *Statistical Models in Engineering,* John Wiley, New York, 1967.)

and plot the point $(\hat{\beta}_1, \hat{\beta}_2)$ on the graph. If the plotted point falls close to a point, line, or region that corresponds to one of the distributions in the figure, then this distribution is a logical choice to use as a model for the data. If the point falls in regions of the β_1, β_2 plane where *none* of the distributions seems appropriate, other, more general probability distributions, such as the Johnson or Pearson families of distributions, may be required. Procedures similar to that in Fig. 7-6 for fitting these distributions and graphs are in Hahn and Shapiro (1967).

7-3   PROCESS CAPABILITY RATIOS

7-3.1   Use and Interpretation of C_p

It is frequently convenient to have a simple, quantitative way to express process capability. One way to do so is through the process capability ratio (PCR) C_p first introduced in Chapter 5. Recall that

$$C_p = \frac{\text{USL} - \text{LSL}}{6\sigma} \qquad (7\text{-}1)$$

where USL and LSL are the upper and lower specification limits, respectively. C_p and other process capability ratios are used extensively in industry. They are also widely *misused.* We will point out some of the more common abuses of process capability ratios. An excellent recent book on process capability ratios that is highly recommended is Kotz and Lovelace (1998).

In a practical application, the process standard deviation σ is almost always unknown and must be replaced by an estimate σ. To estimate σ we typically use either the *sample standard deviation S* or $\overline{R}/d_2$ (when variables control charts are used in the capability study). This results in an estimate of C_p—say,

$$\hat{C}_p = \frac{\text{USL} - \text{LSL}}{6\hat{\sigma}} \qquad (7\text{-}2)$$

To illustrate the calculation of C_p, recall the diameter of automobile engine piston rings, first analyzed in Example 5-1 using $\bar{x}$ and R charts. The specifications on piston-ring diameter are USL = 74.05 mm and LSL = 73.95 mm, and from the R chart we estimated $\sigma = \overline{R}/d_2 = 0.0099$. Thus, our estimate of the PCR C_p is

$$\hat{C}_p = \frac{\text{USL} - \text{LSL}}{6\hat{\sigma}}$$

$$= \frac{74.05 - 73.95}{6(0.0099)}$$

$$= 1.68$$

In Chapter 5, we assumed that piston-ring diameter is approximately normally distributed (a reasonable assumption, based on the histogram in Fig. 2-1) and the cumulative normal distribution table in the Appendix was used to estimate that the process produces approximately 20 ppm (parts per million) defective.

The PCR C_p in equation 7-1 has a useful practical interpretation—namely,

$$P = \left(\frac{1}{C_p}\right) 100 \qquad (7\text{-}3)$$

is the percentage of the specification band used up by the process. The piston-ring process uses

$$P = \left(\frac{1}{1.68} \right) 100$$
$$= 59.5$$

percent of the specification band.

Equations 7-1 and 7-2 assume that the process has both upper and lower specification limits. For one-sided specifications, we define one-sided PCRs as follows.

$$C_{pu} = \frac{\text{USL} - \mu}{3\sigma} \qquad \text{(upper specification only)} \qquad (7\text{-}4)$$

$$C_{pl} = \frac{\mu - \text{LSL}}{3\sigma} \qquad \text{(lower specification only)} \qquad (7\text{-}5)$$

Estimates $\hat{C}_{pu}$ and $\hat{C}_{pl}$ would be obtained by replacing μ and σ in equations 7-4 and 7-5 by estimates $\hat{\mu}$ and $\hat{\sigma}$, respectively.

······ **EXAMPLE 7-2** ···

To illustrate the use of the one-sided process-capability ratios, consider the bottle bursting-strength data in Example 7-1. Suppose that the lower specification limit on bursting strength is 200 psi. We will use $\bar{x} = 264$ and $S = 32$ as estimates of μ and σ, respectively, and the resulting estimate of the one-sided lower process-capability ratio is

$$\hat{C}_{pl} = \frac{\hat{\mu} - \text{LSL}}{3\hat{\sigma}}$$
$$= \frac{264 - 200}{3(32)}$$
$$= \frac{64}{96}$$
$$= 0.67$$

The fraction of defective bottles produced by this process is estimated by finding the area to the left of $Z = (\text{LSL} - \mu)/\sigma = (200 - 264)/32 = -2$ under the standard normal distribution. The estimated fallout is about 2.28% defective, or about 22,800 nonconforming bottles per million. Note that if the normal distribution were an inappropriate model for strength, then this last calculation would have to be performed using the appropriate probability distribution.

···

Table 7-3   Values of the Process Capability Ratio (C_p) and Associated Process Fallout for a Normally Distributed Process (in defective ppm) That Is in Statistical Control

PCR	Process Fallout (in defective ppm)	
	One-Sided Specifications	Two-Sided Specifications
0.25	226,628	453,255
0.50	66,807	133,614
0.60	35,931	71,861
0.70	17,865	35,729
0.80	8,198	16,395
0.90	3,467	6,934
1.00	1,350	2,700
1.10	484	967
1.20	159	318
1.30	48	96
1.40	14	27
1.50	4	7
1.60	1	2
1.70	0.17	0.34
1.80	0.03	0.06
2.00	0.0009	0.0018

The process capability ratio is a measure of the ability of the process to manufacture product that meets the specifications. Table 7-3 presents several values of the PCR C_p along with the associated values of process fallout, expressed in defective parts or non-conforming units of product per million (ppm). These ppm quantities were calculated using the following **very important assumptions:**

1. The quality characteristic has a normal distribution.
2. The process is in statistical control.
3. In the case of two-sided specifications, the process mean is centered between the lower and upper specification limits.

These assumptions are absolutely **critical** to the **accuracy** and **validity** of the reported numbers, and if they are not valid, then the reported quantities may be seriously in error. For example, Somerville and Montgomery (1996) report an extensive investigation of the errors in using the normality assumption to make inferences about the ppm level of a process when in fact the underlying distribution is nonnormal. They investigated various nonnormal distributions and observed that errors of several orders of magnitude can result in predicting ppm by erroneously making the normality assumption. Even when using a t distribution with as many as 30 degrees of freedom, substantial errors result. Thus even though a t distribution with 30 degrees of freedom is symmetrical and almost visually indistinguishable from the normal, the longer and heavier tails of the t distribution make a significant difference when estimating the ppm. Consequently, symmetry in the distribution of process output alone is insufficient to ensure that any PCR will provide a reliable prediction of process ppm. We will discuss the nonnormality issue in more detail in Section 7-3.3.

Stability or statistical control of the process is also essential to the correct interpretation of any PCR. Unfortunately, it is fairly common practice to compute a PCR from a sample of historical process data without any consideration of whether or not the process is in statistical control. If the process is not in control, then of course its parameters are unstable, and the value of these parameters in the future is uncertain. Thus the predictive aspects of the PCR regarding process ppm performance are lost.

Finally, remember that what we actually observe in practice is an *estimate* of the PCR. This estimate is subject to error in estimation, since it depends on sample statistics. English and Taylor (1993) report that large errors in estimating PCRs from sample data can occur, so the estimate one actually has at hand may not be very reliable. It is always a good idea to report the estimate of any PCR in terms of a **confidence interval.** We will show how to do this for some of the commonly used PCRs in Section 7-3.5.

To illustrate the use of Table 7-3, notice that a PCR for a normally distributed stable process of $C_p = 1.00$ implies a fallout rate of 2700 ppm for two-sided specifications, whereas a PCR of $C_p = 1.50$ for this process implies a fallout rate of 4 ppm for one-sided specifications.

Table 7-4 presents some recommended guidelines for minimum values of the PCR. The bottle-strength characteristic is a parameter closely related to the safety of the product; bottles with inadequate pressure strength may fail and injure consumers. This implies that the PCR should be at least 1.45. Perhaps one way the PCR could be improved

Table 7-4 Recommended Minimum Values of the Process Capability Ratio

	Two-Sided Specifications	One-Sided Specifications
Existing processes	1.33	1.25
New processes	1.50	1.45
Safety, strength, or critical parameter, existing process	1.50	1.45
Safety, strength, or critical parameter, new process	1.67	1.60

would be by increasing the mean strength of the bottles—say, by pouring more glass in the mold.

We point out that the values in Table 7-4 are only **recommended minimums.** In recent years, many companies have adopted criteria for evaluating their processes that include process capability objectives that are more stringent than those of Table 7-4. For example, the "six-sigma" program pioneered by Motorola essentially requires that when the process mean is in control, it will not be closer than six standard deviations from the nearest specification limit. This, in effect, requires that the process capability ratio will be at least 2.0.

7-3.2 Process Capability Ratio for an Off-Center Process

The process capability ratio C_p does not take into account *where* the process mean is located relative to the specifications. C_p simply measures the spread of the specifications relative to the six-sigma spread in the process. For example, the top two normal distributions in Fig. 7-7 both have $C_p = 2.0$, but the process in panel (b) of the figure clearly has lower capability than the process in panel (a) because it is not operating at the midpoint of the interval between the specifications.

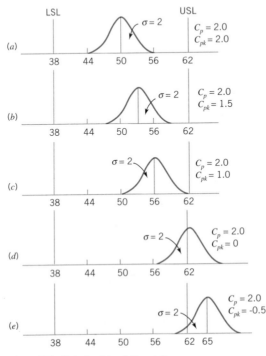

Figure 7-7 Relationship of C_p and C_{pk}.

This situation may be more accurately reflected by defining a new process capability ratio that takes process centering into account. This quantity is

$$C_{pk} = \min (C_{pu}, C_{pl})$$ (7-6)

Note that C_{pk} is simply the one-sided PCR for the specification limit nearest to the process average. For the process shown in Fig. 7-7b, we would have

$$C_{pk} = \min (C_{pu}, C_{pl})$$
$$= \min \left(C_{pu} = \frac{\text{USL} - \mu}{3\sigma}, C_{pl} = \frac{\mu - \text{LSL}}{3\sigma} \right)$$
$$= \min \left(C_{pu} = \frac{62 - 53}{3(2)} = 1.5, C_{pl} = \frac{53 - 38}{3(2)} = 2.5 \right)$$
$$= 1.5$$

Generally, if $C_p = C_{pk}$, the process is centered at the midpoint of the specifications, and when $C_{pk} < C_p$ the process is off-center.

The magnitude of C_{pk} relative to C_p is a direct measure of how off-center the process is operating. Several commonly encountered cases are illustrated in Fig. 7-7. Note in panel (c) of Fig. 7-7 that $C_{pk} = 1.0$ while $C_p = 2.0$. One can use Table 7-3 to get a quick estimate of potential improvement that would be possible by centering the process. If we take $C_p = 1.0$ in Table 7-3 and read the fallout from the one-sided specifications column, we can estimate the *actual* fallout as 1350 ppm. However, if we can center the process, then $C_p = 2.0$ can be achieved, and Table 7-3 (using $C_p = 2.0$ and two-sided specifications) suggests that the *potential* fallout is 0.0018 ppm, an improvement of several orders of magnitude in process performance. Thus, we usually say that C_p measures **potential capability** in the process, whereas C_{pk} measures **actual capability.**

Panel (d) of Fig. 7-7 illustrates the case in which the process mean is exactly equal to one of the specification limits, leading to $C_{pk} = 0$. As panel (e) illustrates, when $C_{pk} < 0$ the implication is that the process mean lies outside the specifications. Clearly, if $C_{pk} < -1$, the entire process lies outside the specification limits. Some authors define C_{pk} to be nonnegative, so that values less than zero are defined as zero.

Many quality-engineering authorities have advised against the routine use of process capability ratios such as C_p and C_{pk} (or the others discussed later in this section) on the grounds that they are an oversimplification of a complex phenomenon. Certainly, any statistic that combines information about both location (the mean and process centering) and variability and that requires the assumption of normality for its meaningful interpolation is likely to be misused (or abused). Furthermore, as we will see, point estimates of process capability ratios are virtually useless if they are computed from small samples. Clearly, these ratios need to be used and interpreted very carefully.

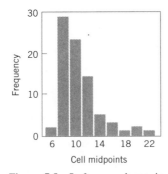

Figure 7-8 Surface roughness in microinches for a machined part.

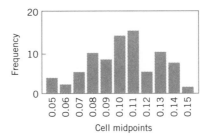

Figure 7-9 Reciprocals of surface roughness. (Adapted from data in the "Statistics Corner" Column in *Quality Progress,* March 1989, with permission of the American Society for Quality Control.)

7-3.3   Normality and the Process Capability Ratio

An important assumption underlying our discussion of process capability and the ratios C_p and C_{pk} is that their usual interpretation is based on a **normal distribution** of process output. If the underlying distribution is nonnormal, then as we previously cautioned, the statements about expected process fallout attributed to a particular value of C_p or C_{pk} may be in error.

To illustrate this point, consider the data in Fig. 7-8, which is a histogram of 80 measurements of surface roughness on a machined part (measured in microinches). The upper specification limit is at USL = 32 microinches. The sample average and standard deviation are $\bar{x} = 10.44$ and $S = 3.053$, implying that $\hat{C}_{pu} = 2.35$, and Table 7-3 would suggest that the fallout is less than one part per billion. However, since the histogram is highly skewed, we are fairly certain that the distribution is nonnormal. Thus, this estimate of capability is unlikely to be correct.

One approach to dealing with this situation is to **transform the data** so that in the new, transformed metric the data have a normal distribution appearance. There are various graphical and analytical approaches to selecting a transformation. In this example, a reciprocal transformation was used. Figure 7-9 presents a histogram of the reciprocal values $x^* = 1/x$. In the transformed scale, $\bar{x}^* = 0.1025$ and $S^* = 0.0244$, and the upper specification limit is USL = $1/32 = 0.03125$. This gives $\hat{C}_{pu} = 0.97$, which implies that about 1350 ppm are outside of specifications. This estimate of process performance is clearly much more realistic than the one resulting from the usual "normal theory" assumption.

Other approaches have been considered in dealing with nonnormal data. There have been various attempts to extend the definitions of the standard capability indices to the case of nonnormal distributions. Luceño (1996) introduced the index C_{pc}, defined as

$$C_{pc} = \frac{\text{USL} - \text{LSL}}{6\sqrt{\dfrac{\pi}{2} E|X - T|}} \qquad (7\text{-}7)$$

where the process target value $T = \frac{1}{2}(\text{USL} + \text{LSL})$. Luceño uses the second subscript in C_{pc} to stand for *confidence,* and he stresses that the confidence intervals based on C_{pc} are reliable; of course, this statement should be interpreted cautiously. The author has also used the constant $6\sqrt{\pi/2} = 7.52$ in the denominator, to make it equal to 6σ when the underlying distribution is normal. We will give the confidence interval for C_{pc} in Section 7-3.5.

There have also been attempts to modify the usual capability indices so that they are appropriate for two **general families** of distributions, the Pearson and Johnson families. This would make PCRs broadly applicable for both normal and nonnormal distributions. Good discussions of these approaches are in Rodriguez (1992) and Kotz and Lovelace (1998).

The general idea is to use appropriate **quantiles** of the process distribution—say, $x_{0.00135}$ and $x_{0.99865}$—to define a quantile-based PCR—say,

$$C_p(q) = \frac{\text{USL} - \text{LSL}}{x_{0.99865} - x_{0.00135}} \tag{7-8}$$

Now since in the normal distribution $x_{0.00135} = \mu - 3\sigma$ and $x_{0.99865} = \mu + 3\sigma$, we see that in the case of a normal distribution $C_p(q)$ reduces to C_p. Clements (1989) proposed a method for determining the quantiles based on the Pearson family of distributions. In general, however, we could fit *any* distribution to the process data, determine its quantiles $x_{0.99865}$ and $x_{0.00135}$, and apply equation 7-8. Refer to Kotz and Lovelace (1998) for more information.

7-3.4   More about Process Centering

The process capability ratio C_{pk} was initially developed because C_p does not adequately deal with the case of a process with mean μ that is not centered between the specification limits. However, C_{pk} *alone* is still an inadequate measure of process centering. For example, consider the two processes shown in Fig. 7-10. Both processes A and B have $C_{pk} = 1.0$, yet their centering is clearly different. To characterize process centering satisfactorily, C_{pk} must be compared to C_p. For Process A, $C_{pk} = C_p = 1.0$, implying that the process is centered, whereas for process B, $C_p = 2.0 > C_{pk} = 1.0$, implying that the process is off-center. For any fixed value of μ in the interval from LSL to USL, C_{pk} depends inversely on σ and becomes large as σ approaches zero. This characteristic can make C_{pk} unsuitable as a measure of centering. That is, a large value of C_{pk} does not really tell us anything about the location of the mean in the interval from LSL to USL.

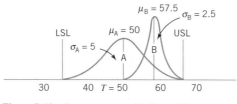

Figure 7-10   Two processes with $C_{pk} = 1.0$.

One way to address this difficulty is to use a process capability ratio that is a better indicator of centering. One such ratio is

$$C_{pm} = \frac{\text{USL} - \text{LSL}}{6\tau} \tag{7-9}$$

where τ is the square root of expected squared deviation from target $T = \frac{1}{2}(\text{USL} + \text{LSL})$,

$$\begin{aligned} \tau^2 &= E[(x - T)^2] \\ &= E[(x - \mu)^2] + (\mu - T)^2 \\ &= \sigma^2 + (\mu - T)^2 \end{aligned}$$

Thus, equation 7-9 can be written as

$$\begin{aligned} C_{pm} &= \frac{\text{USL} - \text{LSL}}{6\sqrt{\sigma^2 + (\mu - T)^2}} \\ &= \frac{C_p}{\sqrt{1 + \xi^2}} \end{aligned} \tag{7-10}$$

where

$$\xi = \frac{\mu - T}{\sigma} \tag{7-11}$$

A logical way to estimate C_{pm} is by

$$\hat{C}_{pm} = \frac{\hat{C}_p}{\sqrt{1 + V^2}} \tag{7-12}$$

where

$$V = \frac{\bar{x} - T}{S} \tag{7-13}$$

Chan et al. (1988) discussed this ratio, various estimators of C_{pm}, and their sampling properties. Boyles (1991) has provided a definitive analysis of C_{pm} and its usefulness in measuring process centering. He notes that both C_{pk} and C_{pm} coincide with C_p when $\mu = T$ and decrease as μ moves away from T. However, $C_{pk} < 0$ for $\mu > \text{USL}$ or $\mu < \text{LSL}$, whereas C_{pm} approaches zero asymptotically as $|\mu - T| \to \infty$. Boyles also shows that the C_{pm} of a process with $|\mu - T| = \Delta > 0$ is strictly bounded above by the C_p value of a process with $\sigma = \Delta$. That is,

$$C_{pm} < \frac{\text{USL} - \text{LSL}}{6|\mu - T|} \tag{7-14}$$

Thus, a necessary condition for $C_{pm} \geq 1$ is

$$|\mu - T| < \frac{1}{6}(\text{USL} - \text{LSL})$$

This statistic says that if the target value T is the midpoint of the specifications, a C_{pm} of one or greater implies that the mean μ lies within the middle third of the specification band. A similar statement can be made for any value of C_{pm}. For instance, $C_{pm} \geq \frac{4}{3}$ implies that $|\mu - T| < \frac{1}{8}(\text{USL} - \text{LSL})$. Thus, a given value of C_{pm} places a constraint on the difference between μ and the target value T.

······· **EXAMPLE 7-3** ···

To illustrate the use of C_{pm}, consider the two processes A and B in Fig. 7-10. For process A we find that

$$C_{pm} = \frac{C_p}{\sqrt{1 + \xi^2}} = \frac{1.0}{\sqrt{1 + 0}} = 1.0$$

since process A is centered at the target value $T = 50$. Note that $C_{pm} = C_{pk}$ for process A. Now consider process B:

$$C_{pm} = \frac{C_p}{\sqrt{1 + \xi^2}} = \frac{2.0}{\sqrt{1 + (-3)^2}} = 0.63$$

If we use equation 7-14, this is equivalent to saying that the process mean lies approximately within the middle half of the specification range. Visual examination of Fig. 7-10 reveals this to be the case.

···

Pearn et al. (1992) proposed the process capability ratio

$$C_{pkm} = \frac{C_{pk}}{\sqrt{1 + \left(\dfrac{\mu - T}{\sigma}\right)^2}}$$

$$= \frac{C_{pk}}{\sqrt{1 + \xi^2}} \tag{7-15}$$

This is sometimes called a "third generation" process capability ratio, since it is constructed from the "second generation" ratios C_{pk} and C_{pm} in the same way that they were generated from the "first generation" ratio C_p. The motivation of this new ratio is increased sensitivity to departures of the process mean μ from the desired target T. For more details, see Kotz and Lovelace (1998).

7-3.5 Confidence Intervals and Tests on Process Capability Ratios

Confidence Intervals on Process Capability Ratios

Much of the industrial use of process capability ratios focuses on computing and interpreting the **point estimate** of the desired quantity. Practitioners often forget that $\hat{C}_p$ or $\hat{C}_{pk}$ (for examples) are simply point estimates, and, as such, are subject to statistical fluctuation. An alternative that should become standard practice is to report **confidence intervals** for process capability ratios.

It is easy to find a confidence interval for the "first generation" ratio C_p. If we replace σ by S in the equation for C_p, we produce the usual point estimator $\hat{C}_p$. If the quality characteristic follows a normal distribution, then a $100(1 - \alpha)\%$ confidence interval on C_p is obtained from

$$\frac{\text{USL} - \text{LSL}}{6S} \sqrt{\frac{\chi^2_{1-\alpha/2, n-1}}{n-1}} \leq C_p \leq \frac{\text{USL} - \text{LSL}}{6S} \sqrt{\frac{\chi^2_{\alpha/2, n-1}}{n-1}} \qquad (7\text{-}16)$$

or

$$\hat{C}_p \sqrt{\frac{\chi^2_{1-\alpha/2, n-1}}{n-1}} \leq C_p \leq \hat{C}_p \sqrt{\frac{\chi^2_{\alpha/2, n-1}}{n-1}} \qquad (7\text{-}17)$$

where $\chi^2_{1-\alpha/2, n-1}$ and $\chi^2_{\alpha/2, n-1}$ are the lower $\alpha/2$ and upper $\alpha/2$ percentage points of the chi-square distribution with $n-1$ degrees of freedom. These percentage points are tabulated in Appendix Table III.

EXAMPLE 7-4

Suppose that a stable process has upper and lower specifications at USL = 62 and LSL = 38. A sample of size $n = 20$ from this process reveals that the process mean is centered approximately at the midpoint of the tolerance interval and that the sample standard deviation $S = 1.75$. Therefore, a point estimate of C_p is

$$\hat{C}_p = \frac{\text{USL} - \text{LSL}}{6S}$$

$$= \frac{62 - 38}{6(1.75)}$$

$$= 2.29$$

The 95% confidence interval on C_p is found from equation 7-15 as follows:

$$\hat{C}_p = \sqrt{\frac{\chi^2_{1-\alpha/2, n-1}}{n-1}} \leq C_p \leq \hat{C}_p \sqrt{\frac{\chi^2_{\alpha/2, n-1}}{n-1}}$$

$$2.29 \sqrt{\frac{8.91}{19}} \leq C_p \leq 2.29 \sqrt{\frac{32.85}{19}}$$

$$1.57 \leq C_p \leq 3.01$$

where $\chi^2_{0.975, 19} = 8.91$ and $\chi^2_{0.025, 19} = 32.85$ were taken from Appendix Table III.

The confidence interval on C_p in Example 7-4 is relatively wide because the sample standard deviation S exhibits considerable fluctuation in small to moderately large samples. This means, in effect, that confidence intervals on C_p based on small samples will be wide.

Note also that the confidence interval uses S rather than $\overline{R}/d_2$ to estimate σ. This further emphasizes that the process must be **in statistical control** for PCRs to have any real meaning. If the process is not in control, S and $\overline{R}/d_2$ could be very different, leading to very different values of the PCR.

For more complicated ratios such as C_{pk} and C_{pm}, various authors have developed approximate confidence intervals; for example, see Zhang et al. (1990), Bissell (1990), Kushler and Hurley (1992), and Pearn et al. (1992). If the quality characteristic is normally distributed, then an approximate $100(1 - \alpha)\%$ confidence interval on C_{pk} is given as follows.

$$\hat{C}_{pk}\left[1 - Z_{\alpha/2}\sqrt{\frac{1}{9n\hat{C}_{pk}^2} + \frac{1}{2(n-1)}}\right] \le C_{pk}$$

$$\le \hat{C}_{pk}\left[1 + Z_{\alpha/2}\sqrt{\frac{1}{9n\hat{C}_{pk}^2} + \frac{1}{2(n-1)}}\right] \quad (7\text{-}18)$$

Kotz and Lovelace (1998) give an extensive summary of confidence intervals for various PCRs.

········ **EXAMPLE 7-5** ···

A sample of size $n = 20$ from a stable process is used to estimate C_{pk}, with the result that $\hat{C}_{pk} = 1.33$. Using equation 7-18, an approximate 95% confidence interval on C_{pk} is

$$\hat{C}_{pk}\left[1 - Z_{\alpha/2}\sqrt{\frac{1}{9n\hat{C}_{pk}^2} + \frac{1}{2(n-1)}}\right]$$

$$\le C_{pk} \le \hat{C}_{pk}\left[1 + Z_{\alpha/2}\sqrt{\frac{1}{9n\hat{C}_{pk}^2} + \frac{1}{2(n-1)}}\right]$$

$$1.33\left[1 - 1.96\sqrt{\frac{1}{9(20)(1.33)^2} + \frac{1}{2(19)}}\right]$$

$$\le C_{pk} \le 1.33\left[1 + 1.96\sqrt{\frac{1}{9(20)(1.33)^2} + \frac{1}{2(19)}}\right]$$

or

$$0.99 \le C_{pk} \le 1.67$$

This is an extremely wide confidence interval. Based on the sample data, the ratio C_{pk} could be less than one (a very bad situation), or it could be as large as 1.67 (a very good situation). Thus, we have learned very little about actual process capability, because C_{pk} is very imprecisely estimated. The reason for this, of course, is that a very small sample ($n = 20$) has been used.

....

For nonnormal data, the PCR C_{pc} developed by Luceño (1996) can be employed. Recall that C_{pc} was defined in equation 7-7. Luceño developed the confidence interval for C_{pc} as follows: First, evaluate $|x - T|$, whose expected value is estimated by

$$\bar{c} = \frac{1}{n} \sum_{i=1}^{n} |x_i - T|$$

leading to the estimator

$$\hat{C}_{pc} = \frac{\text{USL} - \text{LSL}}{6\sqrt{\dfrac{\pi}{2}}\, c}$$

A $100(1 - \alpha)\%$ confidence interval for $E|x - T|$ is given as

$$\bar{c} \pm t_{\alpha/2,\, n-1} \frac{s_c}{\sqrt{n}}$$

where

$$s_c^2 = \frac{1}{n-1} \sum_{i=1}^{n} (|x_i - T| - \bar{c})^2 = \frac{1}{n-1} \left(\sum_{i=1}^{n} |x_i - T|^2 - n\bar{c}^2 \right)$$

Therefore, a $100(1 - \alpha)\%$ confidence interval for C_{pc} is given by

$$\frac{\hat{C}_{pc}}{1 + t_{\alpha/2,\, n-1}[s_c/(\bar{c}\sqrt{n})]} \leq C_{pc} \leq \frac{\hat{C}_{pc}}{1 - t_{\alpha/2,\, n-1}[s_c/(\bar{c}\sqrt{n})]} \qquad (7\text{-}19)$$

Testing Hypotheses about PCRs
A practice that is becoming increasingly common in industry is to require a supplier to **demonstrate** process capability as part of the contractual agreement. Thus, it is frequently necessary to demonstrate that the process capability ratio C_p meets or exceeds some particular target value—say, C_{p0}. This problem may be formulated as a hypothesis testing problem

$$H_0: \quad C_p = C_{p0} \text{ (or the process is not capable)}$$
$$H_1: \quad C_p \geq C_{p0} \text{ (or the process is capable)}$$

Table 7-5 Sample Size and Critical Value Determination for Testing
H_0: $C_p = C_{p0}$

Sample Size, n	(a) $\alpha = \beta = 0.10$		(b) $\alpha = \beta = 0.05$	
	C_p(High)/ C_p(Low)	C/C_p(Low)	C_p(High)/ C_p(Low)	C/C_p(Low)
10	1.88	1.27	2.26	1.37
20	1.53	1.20	1.73	1.26
30	1.41	1.16	1.55	1.21
40	1.34	1.14	1.46	1.18
50	1.30	1.13	1.40	1.16
60	1.27	1.11	1.36	1.15
70	1.25	1.10	1.33	1.14
80	1.23	1.10	1.30	1.13
90	1.21	1.10	1.28	1.12
100	1.20	1.09	1.26	1.11

Source: Adapted from Kane (1986), with permission of the American Society for Quality Control.

We would like to reject H_0 (recall that in statistical hypothesis testing rejection of H_0 is always a strong conclusion), thereby demonstrating that the process is capable. We can formulate the statistical test in terms of $\hat{C}_p$, so that we will reject H_0 if $\hat{C}_p$ exceeds a critical value C.

Kane (1986) has investigated this test, and provides a table of sample sizes and critical values C to assist in testing process capability. We may define C_p(High) as a process capability that we would like to accept with probability $1 - \alpha$ and C_p(Low) as a process capability that we would like to reject with probability $1 - \beta$. Table 7-5 gives values of C_p(High)/C_p(Low) and C/C_p(Low) for varying sample sizes and $\alpha = \beta = 0.05$ or $\alpha = \beta = 0.10$. The following example illustrates the use of this table.

······· **EXAMPLE 7-6** ···

A customer has told his supplier that, in order to qualify for business with his company, the supplier must demonstrate that his process capability exceeds $C_p = 1.33$. Thus, the supplier is interested in establishing a procedure to test the hypotheses

$$H_0: \quad C_p = 1.33$$
$$H_1: \quad C_p > 1.33$$

The supplier wants to be sure that if the process capability is below 1.33 there will be a high probability of detecting this (say, 0.90), whereas if the process capability exceeds 1.66 there will be a high probability of judging the process capable (again, say, 0.90). This would imply that C_p(Low) = 1.33, C_p(High) = 1.66, and $\alpha = \beta = 0.10$. To find the sample size and critical value C from Table 7-5, compute

$$\frac{C_p(\text{High})}{C_p(\text{Low})} = \frac{1.66}{1.33} = 1.25$$

and enter the table value in panel (a) where $\alpha = \beta = 0.10$. This yields

$$n = 70$$

and

$$C/C_p(\text{Low}) = 1.10$$

from which we calculate

$$C = C_p(\text{Low})1.10$$

$$= 1.33(1.10)$$

$$= 1.46$$

Thus, to demonstrate capability, the supplier must take a sample of $n = 70$ parts, and the sample process capability ratio $\hat{C}_p$ must exceed $C = 1.46$.

....

This example shows that, in order to demonstrate that process capability is at least equal to 1.33, the *observed sample* $\hat{C}_p$ will have to exceed 1.33 by a considerable amount. This illustrates that some common industrial practices may be questionable statistically. For example, it is fairly common practice to accept the process as capable at the level $C_p \geq 1.33$ if the *sample* $\hat{C}_p \geq 1.33$ based on a sample size of $30 \leq n \leq 50$ parts. Clearly, this procedure does not account for sampling variation in the estimate of σ, and larger values of n and/or higher acceptable values of $\hat{C}_p$ may be necessary in practice.

Process Performance Indices

In 1991 the Automotive Industry Action Group, or the **AIAG**, was formed. This group consists of representatives of the "big three" (Ford, General Motors, and Daimler/ Chrysler) and the American Society for Quality Control (now the American Society for Quality). One of their objectives was to standardize the reporting requirements from suppliers and in general for their industry. The AIAG recommends using the process capability indices C_p and C_{pk} when the process is in control, with the process standard deviation estimated by $\sigma = \overline{R}/d_2$. When the process is *not* in control, the AIAG recommends using **process performance indices** P_p and P_{pk}, where, for example,

$$\hat{P}_p = \frac{\text{USL} - \text{LSL}}{6S}$$

and S is the usual sample standard deviation $S = \sqrt{\Sigma_{i=1}^{n}(x_i - \overline{x})^2/(n - 1)}$. Even the American National Standards Institute in ANSI Standard Z1 on Process Capability Analysis (1996) states that P_p and P_{pk} should be used when the process is not in control.

Now it is clear that when the process is normally distributed and in control, $\hat{P}_p$ is essentially $\hat{C}_p$ and $\hat{P}_{pk}$ is essentially $\hat{C}_{pk}$ because for a stable process the difference between S and $\sigma = \overline{R}/d_2$ is minimal. However, please note that if the process is not in control, the indices P_p and P_{pk} have no meaningful interpretation relative to process capability, because they cannot predict process performance. Furthermore, their statistical properties are not determinable, and so no valid inference can be made regarding their true (or population) values. Also, P_p and P_{pk} provide no motivation or incentive to the companies that use them to bring their processes into control.

Kotz and Lovelace (1998) strongly recommend **against** the use of P_p and P_{pk}, indicating that these indices are actually a step **backwards** in quantifying process capability. They refer to the mandated use of P_p and P_{pk} through quality standards or industry guidelines as undiluted **statistical terrorism** (i.e., the use or misuse of statistical methods along with threats and/or intimidation to achieve a business objective).

This author agrees completely with Kotz and Lovelace. The process performance indices P_p and P_{pk} are actually more than a step backwards. **They are a waste of engineering and management effort—they tell you nothing.** Instead of using meaningless indices, organizations should devote effort to developing and implementing a process characterization and control plan. The U.S. semiconductor industry did this in the late 1980s (at Semitech) with great success. This is a much more reasonable and effective approach to process improvement.

7-4 PROCESS CAPABILITY ANALYSIS USING A CONTROL CHART

Histograms, probability plots, and process capability ratios summarize the performance of the process. They do not necessarily display the **potential capability** of the process because they do not address the issue of **statistical control,** or show systematic patterns in process output that, if eliminated, would reduce the variability in the quality characteristic. Control charts are very effective in this regard. The control chart should be regarded as the primary technique of process capability analysis.

Both attributes and variables control charts can be used in process capability analysis. The $\bar{x}$ and R charts should be used whenever possible, because of the greater power and better information they provide relative to attributes charts. However, both p charts and c (or u) charts are useful in analyzing process capability. Techniques for constructing and using these charts are given in Chapters 5 and 6. Remember that to use the p chart there must be specifications on the product characteristics. The $\bar{x}$ and R charts allow us to study processes without regard to specifications.

The $\bar{x}$ and R control charts allow both the instantaneous variability (short-term process capability) and variability across time (long-term process capability) to be analyzed. It is particularly helpful if the data for a process capability study are collected in two to three different time periods (such as different shifts, different days, etc.).

Table 7-6 presents the soft-drink bottle bursting-strength data in 20 samples of five observations each. The calculations for the $\bar{x}$ and R charts are summarized here:

Table 7-6 Bottle-Strength Data

Sample	Data					$\bar{x}$	R
1	265	205	263	307	220	252.0	102
2	268	260	234	299	215	255.2	84
3	197	286	274	243	231	246.2	89
4	267	281	265	214	318	269.0	104
5	346	317	242	258	276	287.8	104
6	300	208	187	264	271	246.0	113
7	280	242	260	321	228	266.2	93
8	250	299	258	267	293	273.4	49
9	265	254	281	294	223	263.4	71
10	260	308	235	283	277	272.6	73
11	200	235	246	328	296	261.0	128
12	276	264	269	235	290	266.8	55
13	221	176	248	263	231	227.8	87
14	334	280	265	272	283	286.8	69
15	265	262	271	245	301	268.8	56
16	280	274	253	287	258	270.4	34
17	261	248	260	274	337	276.0	89
18	250	278	254	274	275	266.2	28
19	278	250	265	270	298	272.2	48
20	257	210	280	269	251	253.4	70
						$\bar{\bar{x}} = 264.06$	$\bar{R} = 77.3$

R Chart

Center line $= \bar{R} = 77.3$

$$\text{UCL} = D_4\bar{R} = (2.115)(77.3) = 163.49$$
$$\text{LCL} = D_3\bar{R} = (0)(77.3) = 0$$

$\bar{x}$ Chart

Center line $= \bar{\bar{x}} = 264.06$

$$\text{UCL} = \bar{\bar{x}} + A_2\bar{R} = 264.06 + (0.577)(77.3) = 308.66$$
$$\text{LCL} = \bar{\bar{x}} - A_2\bar{R} = 264.06 - (0.577)(77.3) = 219.46$$

Figure 7-11 presents the $\bar{x}$ and R charts for the 20 samples in Table 7-6. Both charts exhibit statistical control. The process parameters may be estimated from the control chart as

$$\hat{\mu} = \bar{\bar{x}} = 264.06$$

$$\hat{\sigma} = \frac{\bar{R}}{d_2} = \frac{77.3}{2.326} = 33.23$$

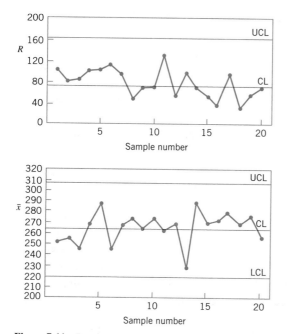

Figure 7-11 $\bar{x}$ and R charts for the bottle-strength data.

Thus, the one-sided lower process capability ratio is estimated by

$$\hat{C}_{pl} = \frac{\mu - \text{LSL}}{3\hat{\sigma}}$$

$$= \frac{264.06 - 200}{3(33.23)}$$

$$= 0.64$$

Clearly, since strength is a safety-related parameter, the process capability is inadequate.

This example illustrates a process that is in control but operating at an unacceptable level. There is no evidence to indicate that the production of nonconforming units is **operator-controllable.** Management intervention will be required either to improve the process or to change the requirements if the quality problems with the bottles are to be solved. The objective of these interventions is to increase the process capability ratio to at least a minimum acceptable level. The control chart can be used as a monitoring device or logbook to show the effect of changes in the process on process performance.

Sometimes the process capability analysis indicates an out-of-control process. It is **unsafe** to estimate process capability in such cases. The process must be stable in order to produce a reliable estimate of process capability. When the process is out of control in the early stages of process capability analysis, the first objective is finding and eliminating the assignable causes in order to bring the process into an in-control state.

7-5   PROCESS CAPABILITY ANALYSIS USING DESIGNED EXPERIMENTS

The design of experiments is a systematic approach to varying the input **controllable** variables in the process and analyzing the effects of these process variables on the output. Designed experiments are also useful in discovering *which* set of process variables are influential on the output, and at what levels these variables should be held to optimize process performance. Thus, design of experiments is useful in more general manufacturing and development problems than merely estimating process capability. For an introduction to design of experiments, see Montgomery (1997). Part IV of this textbook provides more information on experimental design methods and on their use in process improvement.

One of the major uses of designed experiments is in isolating and estimating the **sources of variability** in a process. For example, consider a machine that fills bottles with a soft-drink beverage. Each machine has a large number of filling heads that must be independently adjusted. The quality characteristic measured is the syrup content (in degrees brix) of the finished product. There can be variation in the observed brix (σ_B^2) because of machine variability (σ_M^2), head variability (σ_H^2), and analytical test variability (σ_A^2). The variability in the observed brix value is

$$\sigma_B^2 = \sigma_M^2 + \sigma_H^2 + \sigma_A^2$$

An experiment can be designed, involving sampling from several machines and several heads on each machine, and making several analyses on each bottle, which would allow estimation of the variances σ_M^2, σ_H^2, and σ_A^2. Suppose that the results appear as in Fig. 7-12. Since a substantial portion of the total variability in observed brix is due to

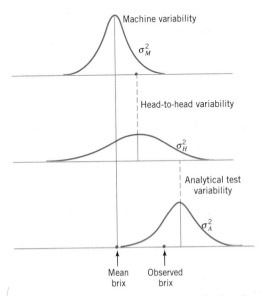

Figure 7-12 Sources of variability in the bottling line example.

variability between heads, this indicates that the process can perhaps best be improved by reducing the head-to-head variability. This could be done by more careful set-up or by more careful control of the operation of the machine.

7-6   GAGE AND MEASUREMENT SYSTEM CAPABILITY STUDIES

7-6.1   Control Charts and Tabular Methods

An important aspect of many statistical process-control implementation efforts is ensuring adequate gage and inspection system capability. In any problem involving measurements, some of the observed variability will be due to variability in the product itself, and some will be due to measurement error or gage variability. Expressed mathematically,

$$\sigma^2_{\text{total}} = \sigma^2_{\text{product}} + \sigma^2_{\text{gage}} \qquad (7\text{-}20)$$

where σ^2_{total} is the total observed variance, $\sigma^2_{\text{product}}$ is the component of variance due to the product, and σ^2_{gage} is the component of variance due to measurement error. Control charts and other statistical methods can be used to separate these components of variance, as well as to give an assessment of gage capability.

······· **EXAMPLE 7-7** ···

Measuring Gage Capability

An instrument is to be used as part of a proposed SPC implementation. The quality-improvement team involved in designing the SPC system would like to get an assessment of gage capability. Twenty units of the product are obtained, and the process operator who will actually take the measurements for the control chart uses the instrument to measure each unit of product twice. The data are shown in Table 7-7.

Figure 7-13 shows the $\bar{x}$ and R charts for these data. Note that the $\bar{x}$ chart exhibits many out-of-control points. This is to be expected, because in this situation the $\bar{x}$ chart has an interpretation that is somewhat different from the usual interpretation. The $\bar{x}$ chart in this example shows the **discriminating power** of the instrument—literally, the ability of the gage to distinguish between units of product. The R chart directly shows the magnitude of measurement error, or the gage capability. The R values represent the difference between measurements made on the same unit using the same instrument. In this example, the R chart is in control. This indicates that the operator is having no difficulty in making consistent measurements. Out-of-control points on the R chart would indicate that the operator is having difficulty using the instrument.

Table 7-7   Parts Measurement Data

	Measurements			
Part Number	1	2	$\bar{x}$	R
1	21	20	20.5	1
2	24	23	23.5	1
3	20	21	20.5	1
4	27	27	27.0	0
5	19	18	18.5	1
6	23	21	22.0	2
7	22	21	21.5	1
8	19	17	18.0	2
9	24	23	23.5	1
10	25	23	24.0	2
11	21	20	20.5	1
12	18	19	18.5	1
13	23	25	24.0	2
14	24	24	24.0	0
15	29	30	29.5	1
16	26	26	26.0	0
17	20	20	20.0	0
18	19	21	20.0	2
19	25	26	25.5	1
20	19	19	19.0	0
			$\bar{\bar{x}} = 22.3$	$\bar{R} = 1.0$

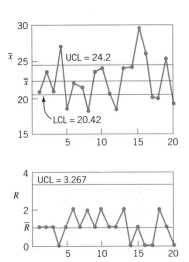

Figure 7-13   Control charts for the gage capability analysis in Example 7-7.

The standard deviation of measurement error, σ_{gage}, can be estimated as follows:

$$\sigma_{\text{gage}} = \frac{\bar{R}}{d_2} = \frac{1.0}{1.128} = 0.887$$

The distribution of measurement error is usually well approximated by the normal. Thus, $6\sigma_{\text{gage}}$ is a good estimate of gage capability. In our problem, this is

$$6\sigma_{\text{gage}} = 6(0.887) = 5.32$$

Individual measurements can be expected to vary as much as $\pm 3\sigma_{\text{gage}}$ (± 2.66) owing to gage error.

It is a fairly common (but not necessarily good) practice to compare the estimate of gage capability to the width of the specifications or the tolerance band (USL − LSL) for the part that is being measured. The ratio of $6\sigma_{\text{gage}}$ to the total tolerance band is often called the precision-to-tolerance or *P/T* ratio:

$$\frac{P}{T} = \frac{6\hat{\sigma}_{\text{gage}}}{\text{USL} - \text{LSL}} \qquad (7\text{-}21)$$

The part used in Example 7-7 has USL = 60 and LSL = 5. Therefore,

$$\frac{P}{T} = \frac{6(0.887)}{60 - 5} = \frac{5.32}{55} = 0.097$$

Values of *P/T* of 0.1 or less often are taken to imply adequate gage capability. This is based on the generally used rule that requires a measurement device to be calibrated in units one-tenth as large as the accuracy required in the final measurement. However, we should use **caution** in accepting this general rule-of-thumb in all cases. A gage must be sufficiently capable to measure product accurately enough and precisely enough so that the analyst can make the correct decision. This may not necessarily require that $P/T \leq$ 0.1. The best way to ensure that the gage can distinguish adequately between different grades of product is to include such different units in the gage capability experiment and construct the $\bar{x}$ and R control charts as we did in Example 7-7.

We can use the data from the gage capability experiment in Example 7-7 to estimate the variance components associated with total observed variability. From the actual sample measurements in Table 7-7, we can calculate $S = 3.17$. This is an estimate of the standard deviation of **total variability,** including both **product variability** and **gage variability.** Therefore,

$$\sigma_{\text{gage}}^2 = S^2 = (3.17)^2 = 10.05$$

Since

$$\sigma^2_{total} = \sigma^2_{product} + \sigma^2_{gage}$$

and since we have an estimate of $\sigma^2_{gage} = (0.887)^2 = 0.79$, we can obtain an estimate of $\sigma^2_{product}$ as

$$\hat{\sigma}^2_{product} = \hat{\sigma}^2_{total} - \hat{\sigma}^2_{gage}$$
$$= 10.05 - 0.79$$
$$= 9.26$$

Therefore, the standard deviation of the product characteristic is

$$\sigma_{product} = \sqrt{9.26} = 3.04$$

The measurement can also be expressed as a percentage of the product characteristic variability, or

$$\frac{\sigma_{gage}}{\sigma_{product}} \times 100 = \frac{0.887}{3.04} \times 100 = 29.2\%$$

This is often a much more meaningful expression of gage capability than the *P/T* ratio because, unlike the *P/T* ratio, it does not depend on the width of the specification limits.

It is also possible to design gage capability studies to investigate two components of measurement error, commonly called the **repeatability** and the **reproducibility** of the gage. We define reproducibility as the variability due to different operators using the gage (or different time periods, or different environments, or in general, different conditions) and repeatability as reflecting the basic inherent precision of the gage itself. That is,

$$\sigma^2_{measurement\ error} = \sigma^2_{gage} = \sigma^2_{repeatability} + \sigma^2_{reproducibility} \qquad (7\text{-}22)$$

We now illustrate a simple tabular procedure to estimate these components of variance.

········ **EXAMPLE 7-8** ···

Components of Measurement Error

We now extend the gage capability study of Example 7-7 to include the two components of measurement error—the repeatability and reproducibility of the gage. To measure these two components of measurement error, the original study was repeated using two additional operators who would also be involved in the work cell where the parts would be produced. The data from the completed measurement error study are shown in Table 7-8.

Table 7-8 Data for the Repeatability and Reproducibility Study in Example 7-8

Part Number	Operator 1				Operator 2				Operator 3			
	Measurements				Measurements				Measurements			
	1	2	$\bar{x}$	R	1	2	$\bar{x}$	R	1	2	$\bar{x}$	R
1	21	20	20.5	1	20	20	20.0	0	19	21	20.0	2
2	24	23	23.5	1	24	24	24.0	0	23	24	23.5	1
3	20	21	20.5	1	19	21	20.0	2	20	22	21.0	2
4	27	27	27.0	0	28	26	27.0	2	27	28	27.5	1
5	19	18	18.5	1	19	18	18.5	1	18	21	19.5	3
6	23	21	22.0	2	24	21	22.5	3	23	22	22.5	1
7	22	21	21.5	1	22	24	23.0	2	22	20	21.0	2
8	19	17	18.0	2	18	20	19.0	2	19	18	18.5	1
9	24	23	23.5	1	25	23	24.0	2	24	24	24.0	0
10	25	23	24.0	2	26	25	25.5	1	24	25	24.5	1
11	21	20	20.5	1	20	20	20.0	0	21	20	20.5	1
12	18	19	18.5	1	17	19	18.0	2	18	19	18.5	1
13	23	25	24.0	2	25	25	25.0	0	25	25	25.0	0
14	24	24	24.0	0	23	25	24.0	2	24	25	24.5	1
15	29	30	29.5	1	30	28	29.0	2	21	20	20.5	1
16	26	26	26.0	0	25	26	25.5	1	25	27	26.0	2
17	20	20	20.0	0	19	20	19.5	1	20	20	20.0	0
18	19	21	20.0	2	19	19	19.0	0	21	23	22.0	2
19	25	26	25.5	1	25	24	24.5	1	25	25	25.0	0
20	19	19	19.0	0	18	17	17.5	1	19	17	18.0	2
	$\bar{\bar{x}}_1 = 22.30$		$\bar{R}_1 = 1.00$		$\bar{\bar{x}}_2 = 22.28$		$\bar{R}_2 = 1.25$		$\bar{\bar{x}}_3 = 22.10$		$\bar{R}_3 = 1.20$	

The estimate of gage repeatability is obtained from the average of the three average ranges — say,

$$\bar{\bar{R}} = \tfrac{1}{3}(\bar{R}_1 + \bar{R}_2 + \bar{R}_3)$$
$$= \tfrac{1}{3}(1.00 + 1.25 + 1.20)$$
$$= 1.15$$

as follows:

$$\hat{\sigma}_{\text{repeatability}} = \frac{\bar{\bar{R}}}{d_2}$$
$$= \frac{1.15}{1.128}$$
$$= 1.02$$

Note that we have used the d_2 factor for samples of size two because each range was calculated from two repeat measurements on the same part made by the same operator.

Gage reproducibility is essentially variability that arises because of differences among the three operators. If the $\bar{x}_i$ values differ, the reason will be operator bias, since all three operators measure the same parts. Therefore, to estimate gage reproducibility let

$$\bar{\bar{x}}_{max} = \max(\bar{\bar{x}}_1, \bar{\bar{x}}_2, \bar{\bar{x}}_3)$$
$$\bar{\bar{x}}_{min} = \min(\bar{\bar{x}}_1, \bar{\bar{x}}_2, \bar{\bar{x}}_3)$$
$$R_{\bar{x}}^{=} = \bar{\bar{x}}_{max} - \bar{\bar{x}}_{min}$$

and

$$\sigma_{reproducibility} = \frac{R_{\bar{x}}^{=}}{d_2}$$

where we would use $d_2 = 1.693$, because $R_{\bar{x}}^{=}$ is the range of a sample of size three. Since for our example, $\bar{\bar{x}}_{max} = 22.30$, $\bar{\bar{x}}_{min} = 22.10$, $R_{\bar{x}}^{=} = 0.32$, and

$$\hat{\sigma}_{reproducibility} = \frac{0.20}{1.693}$$
$$= 0.12$$

We have now estimated both components of the measurement error standard deviation, σ_{gage}. Therefore,

$$\hat{\sigma}_{gage}^2 = \hat{\sigma}_{repeatability}^2 + \hat{\sigma}_{reproducibility}^2$$
$$= (1.02)^2 + (0.12)^2$$
$$= 1.0548$$

and $\hat{\sigma}_{gage} = \sqrt{1.0548} = 1.03$. Note that when both repeatability and reproducibility are taken into account, the standard deviation of measurement error increases. (Recall that the estimate of the standard deviation of measurement error based on a single operator was $\sigma_{gage} = 0.887$ from Example 7-7.) The P/T ratio for this gage would be

$$\frac{P}{T} = \frac{6\hat{\sigma}_{gage}}{USL - LSL}$$
$$= \frac{6(1.03)}{60 - 5}$$
$$= 0.11$$

Note that when both repeatability and reproducibility are taken into account, the gage capability is not as good as we would like. (Recall that as a rough guideline, P/T should be ≤ 0.10.) Training the operators to produce more uniform work methods in using the gage would help reduce $\sigma_{reproducibility}$, but since $\sigma_{repeatability}$ is the largest component of σ_{gage}, some effort should also be directed toward finding another inspection device.

....

The tabular procedure we have illustrated in Example 7-8 is often called a **gage R & R study.** Many individuals and organizations, such as the Automotive Industry Action Group (AIAG), have promoted its use. These advocates often do not emphasize the graphical control-charting aspect of a gage R & R study. Control charts should be constructed for the data from each operator, just as we illustrated for the data from operator 1 in Example 7-7. These control charts are often quite revealing. It is important to ensure that the gage or measurement system can discriminate between good and bad units of product. That information will not generally be directly obvious simply from inspecting the tabular results; it is necessary to examine the control charts.

There are also several disadvantages to the tabular approach. First, the components of gage variability are estimated by the range method, which is inefficient. Second, it is important in many cases to obtain confidence intervals on the components of gage variability, and that is not easily done from the tabular approach. Also, a gage R & R study is a **designed experiment,** and the principles of good experimental design and analysis should be brought to bear on the problem. For example, what *kinds* of parts should be used in the study? Routine production parts? Standards? Generally, a *variety* of parts, both conforming to specification and nonconforming, should be used. One of the objectives of the experiment is to determine whether the gage can distinguish between good and bad units of production. Montgomery and Runger (1993a, b) discuss this and many other practical aspects of designing a gage R & R experiment.

In this section we have dealt primarily with the **precision** of the gage, not its **accuracy.** These two concepts are illustrated in Fig. 7-14. In this figure, the "bull's-eye" of the target is considered to be the true value of the measured characteristic. Accuracy refers to the ability to measure the true value correctly on average, whereas precision is

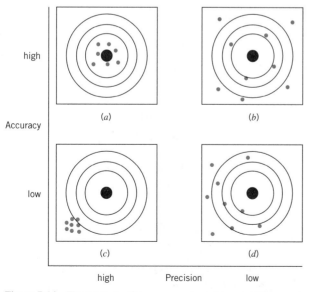

Figure 7-14 The concepts of accuracy and precision. (*a*) The gage is accurate and precise. (*b*) The gage is accurate but not precise. (*c*) The gage is not accurate but it is precise. (*d*) The gage is neither accurate nor precise.

a measure of the inherent variability in the measurements. Evaluating the accuracy of a gage or measurement system often requires use of a standard, for which the true value of the measured characteristic is known. Often the accuracy feature of a gage can be modified by making adjustments to the device or by the use of a properly constructed calibration curve.

7-6.2 Methods Based on Analysis of Variance

As noted in the previous section, a gage R & R study is a designed experiment. The problem considered in Example 7-8 is a factorial experiment, so called because each operator measures all of the parts. The analysis of variance introduced in Chapter 3 can be extended to analyze the data from such an experiment and to estimate the appropriate components of gage variability. We give only a brief overview of the procedure here; for more details, see Montgomery (1997), Montgomery and Runger (1993a, b), Borror, Montgomery, and Runger (1997), Burdick and Larson (1997), and the supplemental text material.

If there are a randomly selected parts and b randomly selected operators, and each operator measures every part n times, then the measurements (i = part, j = operator, k = measurement) could be represented by the model

$$y_{ijk} = \mu + \tau_i + \beta_j + (\tau\beta)_{ij} + \epsilon_{ijk} \qquad \begin{cases} i = 1, 2, \ldots, a \\ j = 1, 2, \ldots, b \\ k = 1, 2, \ldots, n \end{cases} \qquad (7\text{-}23)$$

where the model parameters τ_i, β_j, $(\tau\beta)_{ij}$, and ϵ_{ijk} are all independent random variables that represent the effects of parts, operators, the interaction or joint effects of parts and operators, and random error. This is a **random effects model** analysis of variance. We assume that the random variables τ_i, β_j, $(\tau\beta)_{ij}$, and ϵ_{ijk} are normally distributed with mean zero and variances given by $V(\tau_i) = \sigma_\tau^2$, $V(\beta_j) = \sigma_\beta^2$, $V[(\tau\beta)_{ij}] = \sigma_{\tau\beta}^2$, and $V(\epsilon_{ijk}) = \sigma^2$. Therefore, the variance of any observation is

$$V(y_{ijk}) = \sigma_\tau^2 + \sigma_\beta^2 + \sigma_{\tau\beta}^2 + \sigma^2 \qquad (7\text{-}24)$$

and σ_τ^2, σ_β^2, $\sigma_{\tau\beta}^2$, and σ^2 are the **variance components.** We want to estimate the variance components.

Analysis of variance methods can be used to estimate the variance components. The procedure involves partitioning the total variability in the measurements into the following component parts:

$$SS_{\text{Total}} = SS_{\text{Parts}} + SS_{\text{Operators}} + SS_{P\times0} + SS_{\text{Error}} \qquad (7\text{-}25)$$

where, as in Chapter 3, the notation SS represents a sum of squares. Although these sums of squares could be computed manually,[1] in practice we always use a computer software package to perform this task. Each sum of squares on the right-hand side of equation 7-25 is divided by its degrees of freedom to produce mean squares:

---

[1] The experimental structure here is that of a factorial design. See Chapter 12 and the supplemental text material for more details about the analysis of variance, including computing.

$$MS_{\text{Parts}} = \frac{SS_{\text{Parts}}}{a - 1}$$

$$MS_{\text{Operators}} = \frac{SS_{\text{Operators}}}{b - 1}$$

$$MS_{P \times 0} = \frac{SS_{P \times 0}}{(a - 1)(b - 1)}$$

$$MS_{\text{Error}} = \frac{SS_{\text{Error}}}{ab(n - 1)}$$

We can show that the expected values of the mean squares are as follows:

$$E(MS_{\text{Parts}}) = \sigma^2 + n\sigma_{\tau\beta}^2 + bn\sigma_\tau^2$$
$$E(MS_{\text{Operators}}) = \sigma^2 + n\sigma_{\tau\beta}^2 + an\sigma_\beta^2$$
$$E(MS_{P \times 0}) = \sigma^2 + n\sigma_{\tau\beta}^2$$

and

$$E(MS_{\text{Error}}) = \sigma^2$$

The variance components may be estimated by equating the calculated numerical values of the mean squares from an analysis of variance computer program to their expected values and solving for the variance components. This yields

$$\hat{\sigma}^2 = MS_{\text{Error}}$$
$$\hat{\sigma}_{\tau\beta}^2 = \frac{MS_{P \times 0} - MS_{\text{Error}}}{n}$$
$$\hat{\sigma}_\beta^2 = \frac{MS_{\text{Operators}} - MS_{P \times 0}}{an} \tag{7-26}$$
$$\hat{\sigma}_\tau^2 = \frac{MS_{\text{Parts}} - MS_{P \times 0}}{bn}$$

Table 7-9 shows the analysis of variance for this experiment. The computations were performed using the Balanced ANOVA routine in Minitab. Based on the P-values, we conclude that the effect of parts is large, operators may have a small effect, and there is no significant part–operator interaction. We may use equation 7-26 to estimate the variance components as follows:

$$\hat{\sigma}_\tau^2 = \frac{62.39 - 0.71}{(3)(2)} = 10.28$$

$$\hat{\sigma}_\beta^2 = \frac{1.31 - 0.71}{(20)(2)} = 0.015$$

$$\hat{\sigma}_{\tau\beta}^2 = \frac{0.71 - 0.99}{2} = -0.14$$

Table 7-9 Analysis of Variance (Minitab Balanced ANOVA) for the R & R Study of Example 7-8

Analysis of Variance (Balanced Designs)

Factor	Type	Levels	Values						
part	random	20	1	2	3	4	5	6	7
			8	9	10	11	12	13	14
			15	16	17	18	19	20	
operator	random	3	1	2	3				

Analysis of Variance for y

Source	DF	SS	MS	F	P
part	19	1185.425	62.391	87.65	0.000
operator	2	2.617	1.308	1.84	0.173
part*operator	38	27.050	0.712	0.72	0.861
Error	60	59.500	0.992		
Total	119	1274.592			

Source	Variance component	Error term	Expected Mean Square for Each Term (using unrestricted model)
1 part	10.2798	3	(4) + 2(3) + 6(1)
2 operator	0.0149	3	(4) + 2(3) + 40(2)
3 part*operator	−0.1399	4	(4) + 2(3)
4 Error	0.9917		(4)

and

$$\sigma^2 = 0.99$$

Note that these estimates also appear at the bottom of the Minitab output.

The estimate of one of the variance components, $\sigma^2_{\tau\beta}$, is negative. This is certainly not reasonable, since by definition variances are nonnegative. Unfortunately, negative estimates of variance components can result when we use the analysis of variance method of estimation (this is considered one of its drawbacks). There are a variety of ways to deal with this. One possibility is to assume that the negative estimate means that the variance component is really zero and just set it to zero, leaving the other nonnegative estimates unchanged. Another approach is to estimate the variance components with a method that assures nonnegative estimates. Finally, we could note that the P-value for the interaction term in Table 7-9 is very large, take this as evidence that $\sigma^2_{\tau\beta}$ really is zero, that there is no interaction effect, and fit a **reduced model** of the form

$$y_{ijk} = \mu + \tau_i + \beta_j + \epsilon_{ijk}$$

that does not include the interaction term. This is a relatively easy approach and one that often works nearly as well as more sophisticated methods.

Table 7-10 shows the analysis of variance for the reduced model. Since there is no interaction term in the model, both main effects are tested against the error term, and the

Table 7-10   Analysis of Variance for the Reduced Model, Example 7-8

Analysis of Variance (Balanced Designs)

```
Factor       Type Levels Values
part         random   20      1   2   3   4   5   6   7
                              8   9  10  11  12  13  14
                             15  16  17  18  19  20
operator random      3       1   2   3
```

Analysis of Variance for y

Source	DF	SS	MS	F	P
part	19	1185.425	62.391	70.64	0.000
operator	2	2.617	1.308	1.48	0.232
Error	98	86.550	0.883		
Total	119	1274.592			

Source	Variance component	Error term	Expected Mean Square for Each Term (using unrestricted model)
1 part	10.2513	3	(3) + 6(1)
2 operator	0.0106	3	(3) + 40(2)
3 Error	0.8832		(3)

estimates of the variance components are

$$\hat{\sigma}_\tau^2 = \frac{62.39 - 0.88}{(3)(2)} = 10.25$$

$$\hat{\sigma}_\beta^2 = \frac{1.31 - 0.88}{(20)(2)} = 0.0108$$

$$\hat{\sigma}^2 = 0.88$$

Typically we think of σ^2 as the **repeatability** variance component, and the gage **reproducibility** as the sum of the operator and the part $\times$ operator variance components,

$$\sigma_{\text{reproducibility}}^2 = \sigma_\beta^2 + \sigma_{\tau\beta}^2$$

In our example, $\sigma_{\tau\beta}^2 = 0$. Therefore,

$$\sigma_{\text{gage}}^2 = \sigma_{\text{reproducibility}}^2 + \sigma_{\text{repeatability}}^2$$

and the estimate is

$$\hat{\sigma}_{\text{gage}}^2 = \hat{\sigma}^2 + \hat{\sigma}_{\tau\beta}^2$$

$$= 0.88 + 0.0108$$

$$= 0.8908$$

The variability in the gage appears small relative to the variability in the product. This is generally a desirable situation, implying that the gage is capable of distinguishing among different grades of product.

In addition to the point estimates of these variance components, it is also possible to obtain confidence intervals. Refer to Montgomery (1997), Montgomery and Runger (1993a, b), Borror, Montgomery, and Runger (1997), and Burdick and Larson (1997) for the procedures to follow and several examples.

7-7   SETTING SPECIFICATION LIMITS ON DISCRETE COMPONENTS

It is often necessary to use information from a process capability study to set specifications on discrete parts or components that interact with other components to form the final product. This is particularly important in complex assemblies, or to prevent **tolerance stack-up** where there are many interacting dimensions. This section discusses some aspects of setting specifications on components to ensure that the final product meets specifications.

7-7.1   Linear Combinations

In many cases, the dimension of an item is a linear combination of the dimensions of the component parts. That is, if the dimensions of the components are $x_1, x_2, \ldots, x_n$, then the dimension of the final assembly is

$$y = a_1x_1 + a_2x_2 + \cdots + a_nx_n \qquad (7\text{-}27)$$

If the x_i are normally and independently distributed with mean μ_i and variance σ_i^2, then y is normally distributed with mean $\mu_y = \sum_{i=1}^{n} a_i\mu_i$ and variance $\sigma_y^2 = \sum_{i=1}^{n} a_i^2\sigma_i^2$. Therefore, if μ_i and σ_i^2 are known for each component, the fraction of assembled items falling outside the specifications can be determined.

······   EXAMPLE 7-9   ···

A linkage consists of four components as shown in Fig. 7-15. The lengths of x_1, x_2, x_3, and x_4 are known to be $x_1 \sim N(2.0, 0.0004)$, $x_2 \sim N(4.5, 0.0009)$, $x_3 \sim N(3.0, 0.0004)$, and $x_4 \sim N(2.5, 0.0001)$. The lengths of the components can be assumed independent, because they are produced on different machines. All lengths are in inches.

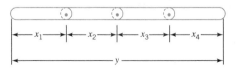

Figure 7-15   A linkage assembly with four components.

The design specifications on the length of the assembled linkage are 12.00 ± 0.10. To find the fraction of linkages that fall within these specification limits, note that y is normally distributed with mean

$$\mu_y = 2.0 + 4.5 + 3.0 + 2.5 = 12.0$$

and variance

$$\sigma_y^2 = 0.0004 + 0.0009 + 0.0004 + 0.0001 = 0.0018$$

To find the fraction of linkages that are within specification, we must evaluate

$$
\begin{aligned}
P\{11.90 \le y \le 12.10\} &= P\{y \le 12.10\} - P\{y \le 11.90\} \\
&= \Phi\left(\frac{12.10 - 12.00}{\sqrt{0.0018}}\right) - \Phi\left(\frac{11.90 - 12.00}{\sqrt{0.0018}}\right) \\
&= \Phi(2.36) - \Phi(-2.36) \\
&= 0.99086 - 0.00914 \\
&= 0.98172
\end{aligned}
$$

Therefore, we conclude that 98.172% of the assembled linkages will fall within the specification limits.

Sometimes it is necessary to determine specification limits on the individual components of an assembly so that specification limits on the final assembly will be satisfied. This is demonstrated in the following example.

EXAMPLE 7-10

Consider the assembly shown in Fig. 7-16. Suppose that the specifications on this assembly are 6.00 ± 0.06 in. Let each component x_1, x_2, and x_3 be normally and independently distributed with means $\mu_1 = 1.00$ in., $\mu_2 = 3.00$ in., and $\mu_3 = 2.00$ in., respectively. Suppose that we want the specification limits to fall inside the natural tolerance limits of the process for the final assembly so that $C_p = 1.50$, approximately, for the final assembly. From Table 7-3, this implies that about 7 ppm defective is allowable.

The length of the final assembly is normally distributed. Furthermore, if the allowable number of assemblies nonconforming to specifications is 7 ppm, this implies that

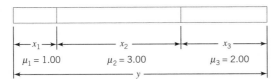

Figure 7-16 Assembly for Example 7-10.

the natural tolerance limits must be located at $\mu \pm 4.49\sigma_y$. Now $\mu_y = \mu_1 + \mu_2 + \mu_3 = 1.00 + 3.00 + 2.00 = 6.00$, so the process is centered at the nominal value. Therefore, the maximum possible value of σ_y that would yield the desired value of C_p is

$$\sigma_y = \frac{0.06}{4.49} = 0.0134$$

That is, if $\sigma_y \leq 0.0134$, then the number of nonconforming assemblies produced will be less than or equal to 7 ppm.

Now let us see how this affects the specifications on the individual components. The variance of the length of the final assembly is

$$\sigma_y^2 = \sigma_1^2 + \sigma_2^2 + \sigma_3^2 \leq (0.0134)^2 = 0.00018$$

Suppose that the variances of the component lengths are all equal; that is, $\sigma_1^2 = \sigma_2^2 = \sigma_3^2 = \sigma^2$ (say). Then

$$\sigma_y^2 = 3\sigma^2$$

and the maximum possible value for the variance of the length of any component is

$$\sigma^2 = \frac{\sigma_y^2}{3} = \frac{0.00018}{3} = 0.00006$$

Effectively, if $\sigma^2 \leq 0.00006$ for each component, then the natural tolerance limits for the final assembly will be inside the specification limits such that $C_p = 1.50$.

This can be translated into specification limits on the individual components. If we assume that the natural tolerance limits and the specification limits for the components are to coincide exactly, then the specification limits for each component are as follows:

$$x_1: \quad 1.00 \pm 3.00\sqrt{0.00006} = 1.00 \pm 0.0232$$
$$x_2: \quad 3.00 \pm 3.00\sqrt{0.00006} = 3.00 \pm 0.0232$$
$$x_3: \quad 2.00 \pm 3.00\sqrt{0.00006} = 2.00 \pm 0.0232$$

· ·

It is possible to give a general solution to the problem in Example 7-10. Let the assembly consist of n components having common variance σ^2. If the natural tolerances of the assembly are defined so that no more than $\alpha\%$ of the assemblies will fall outside these limits, and $2W$ is the width of the specification limits, then

$$\sigma_y^{2*} = \left(\frac{W}{Z_{\alpha/2}}\right)^2$$

is the maximum possible value for the variance of the final assembly that will permit the natural tolerance limits and the specification limits to coincide. Consequently, the maximum permissible value of the variance for the individual components is

$$\sigma^{2*} = \frac{\sigma_y^{2*}}{n}$$

······ **EXAMPLE 7-11** ··

A shaft is to be assembled into a bearing. The internal diameter of the bearing is a normal random variable—say, x_1—with mean $\mu_1 = 1.500$ in. and standard deviation $\sigma_1 = 0.0020$ in. The external diameter of the shaft—say, x_2—is normally distributed with mean $\mu_2 = 1.480$ in. and standard deviation $\sigma_2 = 0.0040$ in. The assembly is shown in Fig. 7-17.

When the two parts are assembled, interference will occur if the shaft diameter is larger than the bearing diameter—that is, if

$$y = x_1 - x_2 < 0$$

Note that the distribution of y is normal with mean

$$\mu_y = \mu_1 - \mu_2 = 1.500 - 1.480 = 0.020$$

and variance

$$\sigma_y^2 = \sigma_1^2 + \sigma_2^2 = (0.0020)^2 + (0.0040)^2 = 0.00002$$

Therefore, the probability of interference is

$$P\{\text{interference}\} = P\{y < 0\}$$

$$= \Phi\left(\frac{0 - 0.020}{\sqrt{0.00002}}\right)$$

$$= \Phi(-4.47)$$

$$= 0.000004 \ (4 \text{ ppm})$$

which indicates that very few assemblies will have interference.

In problems of this type, we occasionally define a minimum clearance—say, C—such that

$$P\{\text{clearance} < C\} = \alpha$$

Thus, C becomes the natural tolerance for the assembly and can be compared with the design specification. In our example, if we establish $\alpha = 0.0001$ (i.e., only 1 out of 10,000 assemblies or 100 ppm will have clearance less than or equal to C), then we have

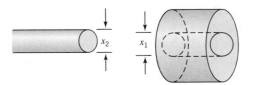

Figure 7-17 Assembly of a shaft and a bearing.

$$\frac{C - \mu_y}{\sigma_y} = -Z_{0.0001}$$

or

$$\frac{C - 0.020}{\sqrt{0.00002}} = -3.71$$

which implies that $C = 0.020 - (3.71)\sqrt{0.00002} = 0.0034$. That is, only 1 out of 10,000 assemblies will have clearance less than 0.0034 in.

7-7.2 Nonlinear Combinations

In some problems, the dimension of interest may be a nonlinear function of the n component dimensions $x_1, x_2, \ldots, x_n$—say,

$$y = g(x_1, x_2, \ldots, x_n) \tag{7-28}$$

In problems of this type, the usual approach is to approximate the nonlinear function g by a linear function of the x_i in the region of interest. If $\mu_1, \mu_2, \ldots, \mu_n$ are the nominal dimensions associated with the components $x_1, x_2, \ldots, x_n$, then by expanding the right-hand side of equation 7-28 in a Taylor series about $\mu_1, \mu_2, \ldots, \mu_n$, we obtain

$$y = g(x_1, x_2, \ldots, x_n)$$

$$= g(\mu_1, \mu_2, \ldots, \mu_n) + \sum_{i=1}^{n} (x_i - \mu_i) \frac{\partial g}{\partial x_i}\bigg|_{\mu_1, \mu_2, \ldots, \mu_n} + R \tag{7-29}$$

where R represents the higher-order terms. Neglecting the terms of higher order, we have

$$\mu_y \cong g(\mu_1, \mu_2, \ldots, \mu_n) \tag{7-30}$$

and

$$\sigma_y^2 \cong \sum_{i=1}^{n} \left(\frac{\partial g}{\partial x_i}\bigg|_{\mu_1, \mu_2, \ldots, \mu_n} \right)^2 \sigma_i^2 \tag{7-31}$$

Equation 7-31 is often called the **transmission of error formula.**

The following example illustrates how these results are useful in tolerance problems.

EXAMPLE 7-12

Consider the simple DC circuit components shown in Fig. 7-18. Suppose that the voltage across the points (a, b) is required to be 100 ± 2 V. The specifications on the current and the resistance in the circuit are shown in Fig. 7-18. We assume that the component random variables I and R are normally and independently distributed with means equal to their nominal values.

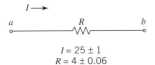

$I = 25 \pm 1$
$R = 4 \pm 0.06$

Figure 7-18 Electrical circuit for Example 7-12.

From Ohm's law, we know that the voltage is

$$V = IR$$

Since this involves a nonlinear combination, we expand V in a Taylor series about mean current μ_I and mean resistance μ_R, yielding

$$V \simeq \mu_I \mu_R + (I - \mu_I)\mu_R + (R - \mu_R)\mu_I$$

neglecting the terms of higher order. Now the mean and variance of voltage are

$$\mu_V \simeq \mu_I \mu_R$$

and

$$\sigma_V^2 \simeq \mu_R^2 \sigma_I^2 + \mu_I^2 \sigma_R^2$$

approximately, where σ_I^2 and σ_R^2 are the variances of I and R, respectively.

Now suppose that I and R are centered at their nominal values and that the natural tolerance limits are defined so that $\alpha = 0.0027$ is the fraction of values of each component falling outside these limits. Assume also that the specification limits are exactly equal to the natural tolerance limits. For the current I we have $I = 25 \pm 1$ A. That is, $24 \le I \le 26$ A correspond to the natural tolerance limits *and* the specifications. Since $I \sim N(25, \sigma_I^2)$, and since $Z_{\alpha/2} = Z_{0.00135} = 3.00$, we have

$$\frac{26 - 25}{\sigma_I} = 3.00$$

or $\sigma_I = 0.33$. For the resistance, we have $R = 4 \pm 0.06$ ohm as the specification limits *and* the natural tolerance limits. Thus,

$$\frac{4.06 - 4.00}{\sigma_R} = 3.00$$

and $\sigma_R = 0.02$. Note that σ_I and σ_R are the largest possible values of the component standard deviations consistent with the natural tolerance limits falling inside or equal to the specification limits.

Using these results, and if we assume that the voltage V is approximately normally distributed, then

$$\mu_V \simeq \mu_I \mu_R = (25)(4) = 100 \text{ V}$$

and

$$\sigma_V^2 \simeq \mu_R^2 \sigma_I^2 + \mu_I^2 \sigma_R^2$$

$$= (4)^2(0.33)^2 + (25)^2(0.02)^2 = 1.99$$

approximately. Thus $\sigma_V = \sqrt{1.99} = 1.41$. Therefore, the probability that the voltage will fall within the design specifications is

$$P\{98 \le V \le 102\} = P\{V \le 102\} - P\{V \le 98\}$$

$$= \Phi\left(\frac{102 - 100}{1.41}\right) - \Phi\left(\frac{98 - 100}{1.41}\right)$$

$$= \Phi(1.42) - \Phi(-1.42)$$

$$= 0.92219 - 0.07781$$

$$= 0.84438$$

That is, only 84% of the observed output voltages will fall within the design specifications. Note that the natural tolerance limits or process capability for the output voltage is

$$\mu_V \pm 3.00\sigma_V$$

or

$$100 \pm 4.23 \text{ V}$$

In this problem the process capability ratio is

$$C_p = \frac{\text{USL} - \text{LSL}}{6\sigma}$$

$$= \frac{102 - 98}{6(1.41)} = 0.47$$

Note that, although the individual current and resistance variations are not excessive relative to their specifications, because of "tolerance stack-up," they interact to produce a circuit whose performance relative to the voltage specifications is very poor.

··

7-8 ESTIMATING THE NATURAL TOLERANCE LIMITS OF A PROCESS

In many types of production processes, it is customary to think of the natural tolerance limits as those limits that contain a certain fraction—say, $1 - \alpha$—of the distribution. In this section we present some approaches to estimating the natural tolerance limits of a process.

   If the underlying distribution of the quality characteristic and its parameters are known—say, on the basis of long experience—then the tolerance limits may be readily established. For example, in Section 7-6, we studied several problems involving tolerances where the quality characteristic was normally distributed with known mean μ and known variance σ^2. If in this case we define the tolerance limits as those limits that contain $100(1 - \alpha)\%$ of the distribution of this quality characteristic, then these limits are simply $\mu \pm Z_{\alpha/2}\,\sigma$. If $\alpha = 0.05$ (say), then the tolerance limits are given by $\mu \pm 1.96\sigma$.

In most practical problems, both the form of the distribution and its parameters will be unknown. However, the parameters may usually be estimated from sample data. In certain cases, then, it is possible to estimate the tolerance limits of the process by use of these sample statistics. We will discuss two procedures for estimating natural tolerance limits, one for those situations in which the normality assumption is reasonable, and a nonparametric approach useful in cases where the normality assumption is inappropriate.

The estimation of the natural tolerance limits of a process is an important problem with many significant practical implications. As noted previously, unless the product specifications exactly coincide with or exceed the natural tolerance limits of the process (PCR ≥ 1), an extremely high percentage of the production will be outside specifications, resulting in a high loss or rework rate.

7-8.1 Tolerance Limits Based on the Normal Distribution

Suppose a random variable x is normally distributed with mean μ and variance σ^2, both unknown. From a random sample of n observations, the sample mean $\bar{x}$ and sample variance S^2 may be computed. A logical procedure for estimating the natural tolerance limits $\mu \pm Z_{\alpha/2}\sigma$ is to replace μ by $\bar{x}$ and σ by S, yielding

$$\bar{x} \pm Z_{\alpha/2}S \tag{7-32}$$

Since $\bar{x}$ and S are only *estimates* and not the *true* parameter values, we cannot say that the above interval always contains $100(1 - \alpha)\%$ of the distribution. However, one may determine a constant K, such that in a large number of samples a fraction γ of the intervals $\bar{x} \pm KS$ will include at least $100(1 - \alpha)\%$ of the distribution. Values of K for $2 \leq n \leq 1000$, $\gamma = 0.90, 0.95, 0.99$, and $\alpha = 0.10, 0.05$, and 0.01 are given in Appendix Table VII.

······· **EXAMPLE 7-13** ···
The manufacturer of a solid-fuel rocket propellant is interested in finding the tolerance limits of the process such that 95% of the burning rates will lie within these limits with probability 0.99. It is known from previous experience that the burning rate is normally distributed. A random sample of 25 observations shows that the sample mean and variance of burning rate are $\bar{x} = 40.75$ and $S^2 = 1.87$, respectively. Since $\alpha = 0.05$, $\gamma = 0.99$, and $n = 25$, we find $K = 2.972$ from Appendix Table VII. Therefore, the required tolerance limits are found as $\bar{x} \pm 2.972S = 40.75 \pm (2.972)(1.37) = 40.75 \pm 4.06 = [36.69, 44.81]$.

We note that there is a fundamental difference between confidence limits and tolerance limits. **Confidence limits** are used to provide an interval estimate of the parameter of a distribution, whereas **tolerance limits** are used to indicate the limits between which we can expect to find a specified proportion of a population. Note that as n approaches infinity, the length of a confidence interval approaches zero, while the tolerance limits approach the corresponding value for the population. Thus, in Appendix Table VII, as n approaches infinity for $\alpha = 0.05$, say, K approaches 1.96.

It is also possible to specify one-sided tolerance limits based on the normal distribution. That is, we may wish to state that with probability γ at least $100(1 - \alpha)\%$ of the distribution is greater than the lower tolerance limit $\bar{x} - KS$ or less than the upper tolerance limit $\bar{x} + KS$. Values of K for these one-sided tolerance limits for $2 \leq n \leq 1000$, $\gamma = 0.90, 0.95, 0.99$, and $\alpha = 0.10, 0.05$, and 0.01 are given in Appendix Table VIII.

7-8.2 Nonparametric Tolerance Limits

It is possible to construct **nonparametric** (or **distribution-free**) **tolerance limits** that are valid for any continuous probability distribution. These intervals are based on the distribution of the extreme values (largest and smallest sample observation) in a sample from an arbitrary continuous distribution. For two-sided tolerance limits, the number of observations that must be taken to ensure that with probability γ at least $100(1 - \alpha)\%$ of the distribution will lie between the largest and smallest observations obtained in the sample is

$$n \simeq \frac{1}{2} + \left(\frac{2 - \alpha}{\alpha}\right)\frac{\chi^2_{1-\gamma, 4}}{4}$$

approximately. Thus, to be 99% certain that at least 95% of the population will be included between the sample extreme values, we have $\alpha = 0.05$, $\gamma = 0.99$, and consequently,

$$n \simeq \frac{1}{2} + \left(\frac{1.95}{0.05}\right)\frac{13.28}{4} = 130$$

For one-sided nonparametric tolerance limits such that with probability γ at least $100(1 - \alpha)\%$ of the population exceeds the smallest sample value (or is less than the largest sample value), we must take a sample of

$$n = \frac{\log(1 - \gamma)}{\log(1 - \alpha)}$$

observations. Thus, the upper nonparametric tolerance limit that contains at least 90% of the population with probability at least 0.95 ($\alpha = 0.10$ and $\gamma = 0.95$) is the largest observation in a sample of

$$n = \frac{\log(1 - \gamma)}{\log(1 - \alpha)} = \frac{\log(0.05)}{\log(0.90)} = 28$$

observations.

In general, nonparametric tolerance limits have limited practical value, because to construct suitable intervals that contain a relatively large fraction of the distribution with high probability, large samples are required. In some cases, the sample sizes required may be so large as to prohibit their use. If one can specify the *form* of the distribution, it is possible for a given sample size to construct tolerance intervals that are narrower than those obtained from the nonparametric approach.

7-9 EXERCISES

7-1. Consider the piston-ring data in Table 5-1. Estimate the process capability assuming that specifications are 74.00 ± 0.035 mm.

7-2. Perform a process capability analysis using $\bar{x}$ and R charts for the data in Exercise 5-1.

7-3. Estimate process capability using $\bar{x}$ and R charts for the power supply voltage data in Exercise 5-2. If specifications are at 350 ± 5 V, calculate C_p, C_{pk}, and C_{pkm}. Interpret these capability ratios.

7-4. Consider the hole diameter data in Exercise 5-3. Estimate process capability using $\bar{x}$ and R charts. If specifications are at 0 ± 0.01, calculate C_p, C_{pk}, and C_{pkm}. Interpret these ratios.

7-5. A process is in control with $\bar{\bar{x}} = 100$, $\bar{S} = 1.05$, and $n = 5$. The process specifications are at 95 ± 10. The quality characteristic has a normal distribution.
(a) Estimate the potential capability.
(b) Estimate the actual capability.
(c) How much could the fallout in the process be reduced if the process were corrected to operate at the nominal specification?

7-6. A process is in control with $\bar{\bar{x}} = 75$ and $\bar{S} = 2$. The process specifications are at 80 ± 8. The sample size $n = 5$.
(a) Estimate the potential capability.
(b) Estimate the actual capability.
(c) How much could process fallout be reduced by shifting the mean to the nominal dimension? Assume that the quality characteristic is normally distributed.

7-7. Consider the two processes shown here (the sample size $n = 5$):

Process A	Process B
$\bar{\bar{x}}_A = 100$	$\bar{\bar{x}}_B = 105$
$\bar{S}_A = 3$	$\bar{S}_B = 1$

Specifications are at 100 ± 10. Calculate C_p, C_{pk}, and C_{pm} and interpret these ratios. Which process would you prefer to use?

7-8. Suppose that 20 of the parts manufactured by the processes in Exercise 7-7 were assembled so that their dimensions were additive; that is,

$$x = x_1 + x_2 + \cdots + x_{20}$$

Specifications on x are 2000 ± 200. Would you prefer to produce the parts using process A or process B? Why? Do the capability ratios computed in Exercise 7-7 provide any guidance for process selection?

7-9. The weights of nominal 1-1b containers of a concentrated chemical ingredient are shown here. Prepare a normal probability plot of the data and estimate process capability.

0.9475	0.9775	0.9965	1.0075	1.0180
0.9705	0.9860	0.9975	1.0100	1.0200
0.9770	0.9960	1.0050	1.0175	1.0250

7-10. Consider the hardness data in Exercise 5-47. Use a probability plot to assess normality. Estimate process capability.

7-11. The failure time in hours of 10 LSI memory devices is shown here. Plot the data on normal probability paper and, if appropriate, estimate process capability. Is it safe to estimate the proportion of circuits that fail below 1200 h?

1210	2105
1275	2230
1400	2250
1695	2500
1900	2625

7-12. A normally distributed process has specifications of LSL = 75 and USL = 85 on the output. A random sample of 25 parts indicates that the process is centered at the middle of the specification band and the standard deviation is $S = 1.5$.
(a) Find a point estimate of C_p.
(b) Find a 95% confidence interval on C_p. Comment on the width of this interval.

7-13. A company has been asked by an important customer to demonstrate that its process capability ratio C_p exceeds 1.33. It has taken a sample of 50 parts and obtained the point estimate $\hat{C}_p = 1.52$. Assume that the quality characteristic follows a normal distribution. Can the company demonstrate that C_p exceeds 1.33 at the 95% level of confidence? What level of confidence would give a one-sided lower confidence limit on C_p that exceeds 1.33?

7-14. Suppose that a quality characteristic has a normal distribution with specification limits at USL = 100 and LSL = 90. A random sample of 30 parts results in $\bar{x} = 97$ and $S = 1.6$.
(a) Calculate a point estimate of C_{pk}.
(b) Find a 95% confidence interval on C_{pk}.

7-15. The molecular weight of a particular polymer should fall between 2100 and 2350. Fifty samples of this material were analyzed with the results $\bar{x} = 2275$ and $S = 60$. Assume that molecular weight is normally distributed.
(a) Calculate a point estimate of C_{pk}.
(b) Find a 95% confidence interval on C_{pk}.

7-16. Consider a simplified version of equation 7-16:

$$\hat{C}_{pk}\left[1 - Z_{\alpha/2}\sqrt{\frac{1}{2(n-1)}}\right]$$
$$\le C_{pk}$$
$$\le \hat{C}_{pk}\left[1 + Z_{\alpha/2}\sqrt{\frac{1}{2(n-1)}}\right]$$

Note that this was obtained by assuming that the term $9n$ in equation 7-16 will probably be large. Rework Exercise 7-14 using this equation and compare your answer to the original answer obtained from equation 7-16. How good is the approximation suggested in this problem?

7-17. An operator–instrument combination is known to test parts with an average error of zero; however, the standard deviation of measurement error is estimated to be 3. Samples from a controlled process were analyzed, and the total variability was estimated to be $\hat{\sigma} = 5$. What is the true process standard deviation?

7-18. Consider the situation in Example 7-7. A new gage is being evaluated for this process. The same operator measures the same 20 parts twice using the new gage and obtains the data shown here.
(a) What can you say about the performance of the new gage relative to the old one?
(b) If specifications are at 25 ± 15, what is the P/T ratio for the new gage?

Part	Measurements	
Number	1	2
1	19	23
2	22	28
3	19	24
4	28	23
5	16	19
6	20	19
7	21	24
8	17	15
9	24	26
10	25	23
11	20	25
12	16	15
13	25	24
14	24	22
15	31	27
16	24	23
17	20	24
18	17	19
19	25	23
20	17	16

7-19. Ten parts are measured three times by the same operator in a gage capability study. The data are shown here.

Part	Measurements		
Number	1	2	3
1	100	101	100
2	95	93	97
3	101	103	100
4	96	95	97
5	98	98	96
6	99	98	98
7	95	97	98
8	100	99	98
9	100	100	97
10	100	98	99

(a) Describe the measurement error that results from the use of this gage.

(b) Estimate total variability and product variability.
(c) What percentage of total variability is due to the gage?
(d) If specifications on the part are at 100 ± 15, find the *P/T* ratio for this gage. Comment on the adequacy of the gage.

7-20. In a study to isolate both gage repeatability and gage reproducibility, two operators use the same gage to measure 10 parts three times each. The data are shown here.

Part	Operator 1 Measurements			Operator 2 Measurements		
Number	1	2	3	1	2	3
1	50	49	50	50	48	51
2	52	52	51	51	51	51
3	53	50	50	54	52	51
4	49	51	50	48	50	51
5	48	49	48	48	49	48
6	52	50	50	52	50	50
7	51	51	51	51	50	50
8	52	50	49	53	48	50
9	50	51	50	51	48	49
10	47	46	49	46	47	48

(a) Estimate gage repeatability and reproducibility.
(b) Estimate the standard deviation of measurement error.
(c) If the specifications are at 50 ± 10, what can you say about gage capability?

7-21. The following data were taken by one operator during a gage capability study.

Part	Measurements	
Number	1	2
1	20	20
2	19	20
3	21	21

Part	Measurements	
Number	1	2
4	24	20
5	21	21
6	25	26
7	18	17
8	16	15
9	20	20
10	23	22
11	28	22
12	19	25
13	21	20
14	20	21
15	18	18

(a) Estimate gage capability.
(b) Does the control chart analysis of these data indicate any potential problem in using the gage?

7-22. Three parts are assembled in series so that their critical dimensions x_1, x_2, and x_3 add. The dimensions of each part are normally distributed with the following parameters: $\mu_1 = 100$, $\sigma_1 = 4$, $\mu_2 = 75$, $\sigma_2 = 4$, $\mu_3 = 75$, and $\sigma_3 = 2$. What is the probability that an assembly chosen at random will have a combined dimension in excess of 262?

7-23. Two parts are assembled as shown in the figure. The distributions of x_1 and x_2 are normal, with $\mu_1 = 20$, $\sigma_1 = 0.3$, $\mu_2 = 19.6$, and $\sigma_2 = 0.4$. The specifications of the clearance between the mating parts are 0.5 ± 0.4. What fraction of assemblies will fail to meet specifications if assembly is at random?

7-24. A product is packaged by filling a container completely full. This container is shaped as shown in the figure. The process that produces these containers is examined, and the following information collected on the three critical dimensions:

Variable	Mean	Variance
L—Length	6.0	0.01
H—Height	3.0	0.01
W—Width	4.0	0.01

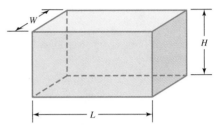

Assuming the variables to be independent, what are approximate values for the mean and variance of container volume?

7-25. A rectangular piece of metal of width W and length L is cut from a plate of thickness T. If W, L, and T are independent random variables with means and standard deviations as given here and the density of the metal is 0.08 g/cm³, what would be the estimated mean and standard deviation of the weights of pieces produced by this process?

Variable	Mean	Standard Deviation
W	10 cm	0.2 cm
L	20 cm	0.3 cm
T	3 cm	0.1 cm

7-26. The surface tension of a chemical product, measured on a coded scale, is given by the relationship

$$s = (3 + 0.05x)^2$$

where x is a component of the product with probability distribution

$$f(x) = \tfrac{1}{26}(5x - 2) \qquad 2 \le x \le 4$$

Find the mean and variance of s.

7-27. Two resistors are connected to a battery as shown in the figure. Find approximate expressions for the mean and variance of the resulting current (I). E, R_1, and R_2 are random variables with means μ_E, μ_{R_1}, μ_{R_2}, and variances σ_E^2, $\sigma_{R_1}^2$, and $\sigma_{R_2}^2$, respectively.

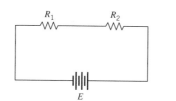

7-28. Two mating parts have critical dimensions x_1 and x_2 as shown in the figure. Assume that x_1 and x_2 are normally distributed with means μ_1 and μ_2 and standard deviations $\sigma_1 = 0.400$ and $\sigma_2 = 0.300$. If it is desired that the probability of a smaller clearance (i.e., $x_1 - x_2$) than 0.09 should be 0.006, what distance between the average dimension of the two parts (i.e., $\mu_1 - \mu_2$) should be specified by the designer?

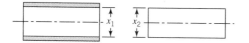

7-29. An assembly of two parts is formed by fitting a shaft into a bearing. It is known that the inside diameters of bearings are normally distributed with mean 2.010 cm and standard deviation 0.002 cm, and that the outside diameters of the shafts are normally distributed with mean 2.004 cm and standard deviation 0.001 cm. Determine the distribution of clearance between the parts if random assembly is used. What is the probability that the clearance is positive?

7-30. We wish to estimate a two-sided natural tolerance interval that will include 99% of the values of a random variable with probability 0.80. If nothing is known about the distribution of the random variable, how large should the sample be?

7-31. A sample of 10 items from a normal population had a mean of 300 and a standard deviation of 10. Using these data, estimate a value for the random variable such that the probability is 0.95 that 90% of the measurements on this random variable will lie below the value.

7-32. A sample of 25 measurements on a normally distributed quality characteristic has a mean of 85 and a standard deviation of 1. Using a confidence probability of 0.95, find a value such that 90% of the future measurements on this quality characteristic will lie above it.

7-33. A sample of 20 measurements on a normally distributed quality characteristic had $\bar{x} = 350$ and $S = 10$. Find an upper natural tolerance limit that has probability 0.90 of containing 95% of the distribution of this quality characteristic.

7-34. How large a sample is required to obtain a natural tolerance interval that has probability 0.90 of containing 95% of the distribution? After the data are collected, how would you construct the interval?

7-35. A random sample of $n = 40$ pipe sections resulted in a mean wall thickness of 0.1264 in. and a standard deviation of 0.0003 in. We assume that wall thickness is normally distributed.
(a) Between what limits can we say with 95% confidence that 95% of the wall thicknesses should fall?
(b) Construct a 95% confidence interval on the true mean thickness. Explain the difference between this interval and the one constructed in part (a).

7-36. Find the sample size required to construct an upper nonparametric tolerance limit that contains at least 95% of the population with probability at least 0.95. How would this limit actually be computed from sample data?

PART II

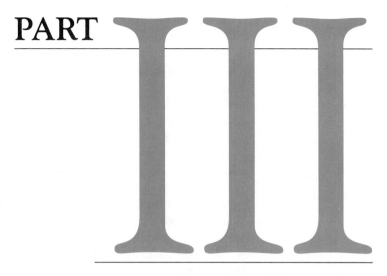

Other Statistical Process Monitoring and Control Techniques

Part II focused on the basic methods of statistical process control and capability analysis. Many of these techniques, such as the Shewhart control chart, have been in use for well over 50 years. However, the increasing emphasis on variability reduction, yield enhancement, and process improvement along with the success of the basic methods has led to the development of many new techniques for statistical process monitoring and control. This part contains four chapters describing some of these techniques. Chapter 8 presents the cumulative sum (cusum) control chart and the exponentially weighted moving average (EWMA) control chart. These procedures are not really new, since they date from the 1950s, but they are generally considered somewhat more advanced techniques than the Shewhart chart. As we will see, the cusum and the EWMA offer considerable performance improvement relative to Shewhart charts. Chapter 9 is a survey of several univariate process control techniques, including methods for short production runs, and monitoring techniques suitable for processes in which the data are autocorrelated. Chapter 10 is an introduction to multivariate process monitoring and control, techniques that are applicable when there are two or more related process variables that are of interest. Chapter 11 presents techniques for process control by feedback adjustment. In these systems the output quality characteristic of interest is influenced by a manipulatable process variable, and we use the deviation of the current output from its desired or target value to determine how much adjustment to make so that the next observation will be on target. These feedback control schemes are also called engineering process control, and they are widely used in the chemical and process industries.

Some of the topics presented in this part may require more statistical and mathematical background than the material in Part II. Two very useful references to accompany this section are the panel discussion on statistical process monitoring and control that appeared in the *Journal of Quality Technology* in 1997 [see Montgomery and Woodall (1997)] and the paper on research issues in SPC in the *Journal of Quality Technology* in 1999 [see Woodall and Montgomery (1999)].

Cumulative Sum and Exponentially Weighted Moving Average Control Charts

CHAPTER OUTLINE

8-1 THE CUMULATIVE SUM CONTROL CHART

    8-1.1 Basic Principles: The Cusum Control Chart for Monitoring the Process Mean

    8-1.2 The Tabular or Algorithmic Cusum for Monitoring the Process Mean

    8-1.3 Recommendations for Cusum Design

    8-1.4 The Standardized Cusum

    8-1.5 Rational Subgroups

    8-1.6 Improving Cusum Responsiveness for Large Shifts

    8-1.7 The Fast Initial Response or Headstart Feature

    8-1.8 One-Sided Cusums

    8-1.9 A Cusum for Monitoring Process Variability

    8-1.10 Cusums for Other Sample Statistics

    8-1.11 The V-Mask Procedure

8-2 THE EXPONENTIALLY WEIGHTED MOVING AVERAGE CONTROL CHART

    8-2.1 The Exponentially Weighted Moving Average Control Chart for Monitoring the Process Mean

    8-2.2 Design of an EWMA Control Chart

    8-2.3 Rational Subgroups

    8-2.4 Robustness of the EWMA to Nonnormality

    8-2.5 Extensions of the EWMA

8-3 THE MOVING AVERAGE CONTROL CHART

CHAPTER OVERVIEW

Chapters 4, 5, and 6 have concentrated on the basic methods of SPC. The control charts discussed in these chapters are usually called **Shewhart control charts,** as they are based on the principles of control charts developed by Dr. Walter A. Shewhart. A major disadvantage of any Shewhart control chart is that it only uses the information about the process contained in the last plotted point, and it ignores any information given by the entire sequence of points. This feature makes the Shewhart control chart relatively insensitive to small shifts in the process—say, on the order of about 1.5σ or less. Of course, other criteria can be applied to Shewhart charts, such as tests for runs and the use of warning limits, which attempt to incorporate information from the entire set of points into the decision procedure. However, the use of these supplemental sensitizing rules reduces the simplicity and ease of interpretation of the Shewhart control chart. Furthermore, as we have previously observed, the use of these sensitizing rules can dramatically reduce the average run length of the control chart when the process is in control, and this may be undesirable.

Two very effective alternatives to the Shewhart control chart may be used when small shifts are of interest: the **cumulative sum** (or **cusum**) **control chart,** and the **exponentially weighted moving average (EWMA) control chart.** These control charts will be presented in this chapter.

8-1   THE CUMULATIVE SUM CONTROL CHART

8-1.1   Basic Principles: The Cusum Control Chart for Monitoring the Process Mean

Consider the data in Table 8-1, column (a). The first 20 of these observations were drawn at random from a normal distribution with mean $\mu = 10$ and standard deviation $\sigma = 1$. These observations have been plotted on a Shewhart control chart in Fig. 8-1.

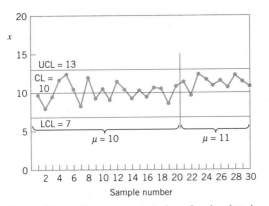

Figure 8-1   A Shewhart control chart for the data in Table 8-1.

Table 8-1   Data for the Cusum Example

Sample, i	(a) x_i	(b) $x_i - 10$	(c) $C_i = (x_i - 10) + C_{i-1}$
1	9.45	-0.55	-0.55
2	7.99	-2.01	-2.56
3	9.29	-0.71	-3.27
4	11.66	1.66	-1.61
5	12.16	2.16	0.55
6	10.18	0.18	0.73
7	8.04	-1.96	-1.23
8	11.46	1.46	0.23
9	9.20	-0.80	-0.57
10	10.34	0.34	-0.23
11	9.03	-0.97	-1.20
12	11.47	1.47	0.27
13	10.51	0.51	0.78
14	9.40	-0.60	0.18
15	10.08	0.08	0.26
16	9.37	-0.63	-0.37
17	10.62	0.62	0.25
18	10.31	0.31	0.56
19	8.52	-1.48	-0.92
20	10.84	0.84	-0.08
21	10.90	0.90	0.82
22	9.33	-0.67	0.15
23	12.29	2.29	2.44
24	11.50	1.50	3.94
25	10.60	0.60	4.54
26	11.08	1.08	5.62
27	10.38	0.38	6.00
28	11.62	1.62	7.62
29	11.31	1.31	8.93
30	10.52	0.52	9.45

The center line and three-sigma control limits on this chart are at

$$UCL = 13$$
$$\text{Center line} = 10$$
$$LCL = 7$$

Note that all 20 observations plot in control.

The last 10 observations in column (a) of Table 8-1 were drawn from a normal distribution with mean $\mu = 11$ and standard deviation $\sigma = 1$. Consequently, we can think of these last 10 observations as having been drawn from the process when it is out of control—that is, after the process has experienced a shift in the mean of 1σ.

These last 10 observations are also plotted on the control chart in Fig. 8-1. None of these points plots outside the control limits, so we have no strong evidence that the process is out of control. Note that there is an indication of a shift in process level for the last 10 points, because all but one of the points plot above the center line. However, if we rely on the traditional signal of an out-of-control process, one or more points beyond a three-sigma control limit, then the Shewhart control chart has failed to detect the shift.

The reason for this failure, of course, is the relatively small magnitude of the shift. The Shewhart chart for averages is very effective if the magnitude of the shift is 1.5σ to 2σ or larger. For smaller shifts, it is not as effective. The cumulative sum (or cusum) control chart is a good alternative when small shifts are important.

The cusum chart directly incorporates all the information in the sequence of sample values by plotting the cumulative sums of the deviations of the sample values from a target value. For example, suppose that samples of size $n \geq 1$ are collected, and $\bar{x}_j$ is the average of the jth sample. Then if μ_0 is the target for the process mean, the cumulative sum control chart is formed by plotting the quantity

$$C_i = \sum_{j=1}^{i} (\bar{x}_j - \mu_0) \tag{8-1}$$

against the sample i. C_i is called the cumulative sum up to and including the ith sample. Because they combine information from *several* samples, cumulative sum charts are more effective than Shewhart charts for detecting small process shifts. Furthermore, they are particularly effective with samples of size $n = 1$. This makes the cumulative sum control chart a good candidate for use in the chemical and process industries where rational subgroups are frequently of size one, and in discrete parts manufacturing with automatic measurement of each part and on-line control using a microcomputer directly at the work center.

Cumulative sum control charts were first proposed by Page (1954) and have been studied by many authors; in particular, see Ewan (1963), Page (1961), Gan (1991), Lucas (1976), Hawkins (1981) (1993a), and Woodall and Adams (1993). In this section, we concentrate on the cumulative sum chart for the process mean. It is possible to devise cumulative sum procedures for other variables, such as Poisson and binomial variables for modeling nonconformities and process fallout. We will show subsequently how the cusum can be used for monitoring process variability.

We note that if the process remains in control at the target value μ_0, the cumulative sum defined in equation 8-1 is a random walk with mean zero. However, if the mean shifts upward to some value $\mu_1 > \mu_0$, say, then an upward or positive drift will develop in the cumulative sum C_i. Conversely, if the mean shifts downward to some $\mu_1 < \mu_0$, then a downward or negative drift in C_i will develop. Therefore, if a trend develops in the plotted points either upward or downward, we should consider this as evidence that the process mean has shifted, and a search for some assignable cause should be performed.

This theory can be easily demonstrated by using the data in column (a) of Table 8-1 again. To apply the cusum in equation 8-1 to these observations, we would take $\bar{x}_i = x_i$ (since our sample size is $n = 1$) and let the target value $\mu_0 = 10$. Therefore, the cusum becomes

$$C_i = \sum_{j=1}^{i} (x_j - 10)$$

$$= (x_i - 10) + \sum_{j=1}^{i-1} (x_j - 10)$$

$$= (x_i - 10) + C_{i-1}$$

Column (b) of Table 8-1 contains the differences $x_i - 10$, and the cumulative sums are computed in column (c). The starting value for the cusum, C_0, is taken to be zero. Figure 8-2 plots the cusum from column (c) of Table 8-1. Note that for the first 20 observations where $\mu = 10$, the cusum tends to drift slowly, in this case maintaining values near zero. However, in the last 10 observations, where the mean has shifted to $\mu = 11$, a strong upward trend develops.

Of course, the cusum plot in Fig. 8-2 is not a control chart because it lacks statistical control limits. There are two ways to represent cusums, the **tabular** (or **algorithmic**) **cusum,** and the **V-mask** form of the cusum. Of the two representations, the tabular cusum is preferable. We now present the construction and use of the tabular cusum. We will also briefly discuss the V-mask procedure and indicate why it is not the best representation of a cusum.

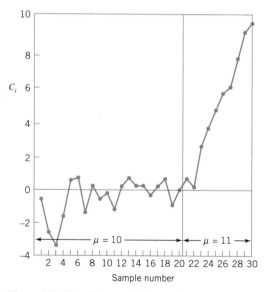

Figure 8-2 Plot of the cumulative sum from column (c) of Table 8-1.

8-1.2   The Tabular or Algorithmic Cusum for Monitoring the Process Mean

We now show how a tabular cusum may be constructed for monitoring the mean of a process. Cusums may be constructed both for individual observations and for the averages of rational subgroups. The case of individual observations occurs very often in practice, so that situation will be treated first. Later we will see how to modify these results for rational subgroups.

Let x_i be the ith observation on the process. When the process is in control, x_i has a normal distribution with mean μ_0 and standard deviation σ. We assume that either σ is known or that an estimate is available. Later we will discuss monitoring σ with a cusum.

Sometimes we think of μ_0 as a "target" value for the quality characteristic x. This viewpoint is often taken in the chemical and process industries when the objective is to control x (viscosity, say) to a particular target value (such as 2000 centistokes at 100°C). If the process drifts or shifts off this target value, the cusum will signal, and an adjustment is made to some manipulatable variable (such as the catalyst feed rate) to bring the process back on target. Also, in some cases a signal from a cusum indicates the presence of an assignable cause that must be investigated just as in the Shewhart chart case.

The tabular cusum works by accumulating derivations from μ_0 that are above target with one statistic C^+ and accumulating derivations from μ_0 that are below target with another statistic C^-. The statistics C^+ and C^- are called **one-sided upper and lower cusums,** respectively. They are computed as follows:

The Tabular Cusum

$$C_i^+ = \max[0, x_i - (\mu_0 + K) + C_{i-1}^+] \qquad (8\text{-}2)$$

$$C_i^- = \max[0, (\mu_0 - K) - x_i + C_{i-1}^-] \qquad (8\text{-}3)$$

where the starting values are $C_0^+ = C_0^- = 0$.

In equations 8-2 and 8-3, K is usually called the **reference value** (or the **allowance,** or the **slack value**), and it is often chosen about halfway between the target μ_0 and the out-of-control value of the mean μ_1 that we are interested in detecting quickly.

Thus, if the shift is expressed in standard deviation units as $\mu_1 = \mu_0 + \delta\sigma$ (or $\delta = |\mu_1 - \mu_0|/\sigma$), then K is one-half the magnitude of the shift or

$$K = \frac{\delta}{2}\sigma = \frac{|\mu_1 - \mu_0|}{2} \qquad (8\text{-}4)$$

Note that C_i^+ and C_i^- accumulate deviations from the target value μ_0 that are greater than K, with both quantities reset to zero on becoming negative. If either C_i^+ or C_i^- exceed the **decision interval** H, the process is considered to be out of control.

We have briefly mentioned how to choose K, but how does one choose H? Actually, the proper selection of these two parameters is quite important, as it has substantial impact on the performance of the cusum. We will talk more about this later, but a reasonable value for H is five times the process standard deviation σ.

••••••• **EXAMPLE 8-1** ••

A Tabular Cusum

We will demonstrate the calculations for the tabular cusum by using the data from Table 8-1. Recall that the target value is $\mu_0 = 10$, the subgroup size is $n = 1$, the process standard deviation is $\sigma = 1$, and suppose that the magnitude of the shift we are interested in detecting is $1.0\sigma = 1.0(1.0) = 1.0$. Therefore, the out-of-control value of the process mean is $\mu_1 = 10 + 1 = 11$. We will use a tabular cusum with $K = \frac{1}{2}$ (because the shift size is 1.0σ and $\sigma = 1$) and $H = 5$ (because the recommended value of the decision interval is $H = 5\sigma = 5(1) = 5$).

Table 8-2 presents the tabular cusum scheme. To illustrate the calculations, consider period 1. The equations for C_1^+ and C_1^- are

$$C_1^+ = \max[0, x_1 - 10.5 + C_0^+]$$

and

$$C_1^- = \max[0, 9.5 - x_1 + C_0^-]$$

since $K = 0.5$ and $\mu_0 = 10$. Now $x_1 = 9.45$, so since $C_0^+ = C_0^- = 0$,

$$C_1^+ = \max[0, 9.45 - 10.5 + 0] = 0$$

and

$$C_1^- = \max[0, 9.5 - 9.45 + 0] = 0.05$$

For period 2, we would use

$$C_2^+ = \max[0, x_2 - 10.5 + C_1^+]$$
$$= \max[0, x_2 - 10.5 + 0]$$

and

$$C_2^- = \max[0, 9.5 - x_2 + C_1^-]$$
$$= \max[0, 9.5 - x_2 + 0.05]$$

Since $x_2 = 7.99$, we obtain

$$C_2^+ = \max[0, 7.99 - 10.5 + 0] = 0$$

and

$$C_2^- = \max[0, 9.5 - 7.99 + 0.05] = 1.56$$

Panels (a) and (b) of Table 8-2 summarize the remaining calculations. The quantities N^+ and N^- in Table 8-2 indicate the number of consecutive periods that the cusums C_i^+ or C_i^- have been nonzero.

Table 8-2 The Tabular Cusum for Example 8-1

Period i	x_i	a			b		
		$x_i - 10.5$	C_i^+	N^+	$9.5 - x_i$	C_i^-	N^-
1	9.45	-1.05	0	0	0.05	0.05	1
2	7.99	-2.51	0	0	1.51	1.56	2
3	9.29	-1.21	0	0	0.21	1.77	3
4	11.66	1.16	1.16	1	-2.16	0	0
5	12.16	1.66	2.82	2	-2.66	0	0
6	10.18	-0.32	2.50	3	-0.68	0	0
7	8.04	-2.46	0.04	4	1.46	1.46	1
8	11.46	0.96	1.00	5	-1.96	0	0
9	9.20	-1.3	0	0	0.30	0.30	1
10	10.34	-0.16	0	0	-0.84	0	0
11	9.03	-1.47	0	0	0.47	0.47	1
12	11.47	0.97	0.97	1	-1.97	0	0
13	10.51	0.01	0.98	2	-1.01	0	0
14	9.40	-1.10	0	0	0.10	0.10	1
15	10.08	-0.42	0	0	-0.58	0	0
16	9.37	-1.13	0	0	0.13	0.13	1
17	10.62	0.12	0.12	1	-1.12	0	0
18	10.31	-0.19	0	0	-0.81	0	0
19	8.52	-1.98	0	0	0.98	0.98	1
20	10.84	0.34	0.34	1	-1.34	0	0
21	10.90	0.40	0.74	2	-1.40	0	0
22	9.33	-1.17	0	0	0.17	0.17	1
23	12.29	1.79	1.79	1	-2.79	0	0
24	11.50	1.00	2.79	2	-2.00	0	0
25	10.60	0.10	2.89	3	-1.10	0	0
26	11.08	0.58	3.47	4	-1.58	0	0
27	10.38	-0.12	3.35	5	-0.88	0	0
28	11.62	1.12	4.47	6	-2.12	0	0
29	11.31	0.81	5.28	7	-1.81	0	0
30	10.52	0.02	5.30	8	-1.02	0	0

The cusum calculations in Table 8-2 show that the upper-side cusum at period 29 is $C_{29}^+ = 5.28$. Since this is the first period at which $C_i^+ > H = 5$, we would conclude that the process is out of control at that point. The tabular cusum also indicates when the shift probably occurred. The counter N^+ records the number of consecutive periods since the upper-side cusum C_i^+ rose above the value of zero. Since $N^+ = 7$ at period 29, we would conclude that the process was last in control at period $29 - 7 = 22$, so the shift likely occurred between periods 22 and 23.

It is useful to present a graphical display for the tabular cusum. These charts are sometimes called **cusum status charts.** They are constructed by plotting C_i^+ and C_i^-

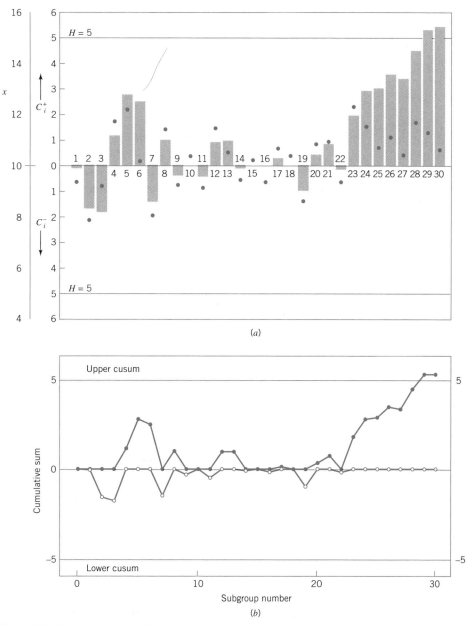

Figure 8-3   Cusum status charts for Example 8-1. (*a*) Manual chart. (*b*) Minitab chart.

versus the sample number. Figure 8-3*a* shows the cusum status chart for the data in Example 8-1. Each vertical bar represents the value of C_i^+ and C_i^- in period i. With the decision interval plotted on the chart, the cusum status chart resembles a Shewhart control chart. We have also plotted the observations x_i for each period on the cusum status chart as the solid dots. This frequently helps the user of the control chart visualize the actual process performance that has led to a particular value of the cusum. Some computer software packages have implemented the cusum status chart. Figure 8-3b shows the Minitab version.

....

The action taken following an out-of-control signal on a cusum control scheme is identical to that with any control chart; one should search for the assignable cause, take any corrective action required, and then reinitialize the cusum at zero. The cusum is particularly helpful in determining when the assignable cause has occurred; as we noted in the previous example, just count backward from the out-of-control signal to the time period when the cusum lifted above zero to find the first period following the process shift. The counters N^+ and N^- are used in this capacity.

In situations where an adjustment to some manipulatable variable is required in order to bring the process back to the target value μ_0, it may be helpful to have an estimate of the new process mean following the shift. This can be computed from

$$
\hat{\mu} = \begin{cases} \mu_0 + K + \dfrac{C_i^+}{N^+}, & \text{if } C_i^+ > H \\[2ex] \mu_0 - K - \dfrac{C_i^-}{N^-}, & \text{if } C_i^- > H \end{cases} \tag{8-5}
$$

To illustrate the use of equation 8-5, consider the cusum in period 29 with $C_{29}^+ = 5.28$. From equation 8-5, we would estimate the new process average as

$$
\begin{aligned}
\hat{\mu} &= \mu_0 + K + \frac{C_{29}^+}{N^+} \\[1.5ex]
&= 10.0 + 0.5 + \frac{5.28}{7} \\[1.5ex]
&= 11.25
\end{aligned}
$$

So, for example, if the process characteristic is viscosity, then we would conclude that mean viscosity has shifted from 10 to 11.25, and if the manipulatable variable that affects viscosity is catalyst feed rate, then we would need to make an adjustment in catalyst feed rate that would result in moving the viscosity down by 1.25 units.

Finally, we should note that runs tests, and other sensitizing rules such as the zone rules, cannot be safely applied to the cusum, because successive values of C_i^+ and C_i^- are not independent. In fact, the cusum can be thought of as a weighted average, where the weights are stochastic or random. For example, consider the cusum shown in Table 8-2. The cusum at period 30 is $C_{30}^+ = 5.30$. This can be thought of as a weighted average in which we give equal weight to the last $N^+ = 8$ observations and weight zero to all other observations.

8-1.3 Recommendations for Cusum Design

The tabular cusum is designed by choosing values for the reference value K and the decision interval H. It is usually recommended that these parameters be selected to provide good average run length performance. There have been many analytical studies of cusum ARL performance. Based on these studies, we may give some general recommendations for selecting H and K.

Define $H = h\sigma$ and $K = k\sigma$, where σ is the standard deviation of the sample variable used in forming the cusum. Using $h = 4$ or $h = 5$ and $k = \frac{1}{2}$ will generally provide a cusum that has good ARL properties against a shift of about 1σ in the process mean.

To illustrate how well the recommendations of $h = 4$ or $h = 5$ with $k = \frac{1}{2}$ work, consider the two-sided average run lengths shown in Table 8-3. Note that a 1σ shift would be detected in either 8.38 samples (with $k = \frac{1}{2}$ and $h = 4$) or 10.4 samples (with $k = \frac{1}{2}$ and $h = 5$). By comparison, a Shewhart control chart for individual measurements would require 43.96 samples, on the average, to detect this shift.

Note also from Table 8-3 that $h = 4$ results in an in-control $\text{ARL}_0 = 168$ samples, whereas $h = 5$ results in $\text{ARL}_0 = 465$ samples. If we choose $h = 4.77$, this will provide a cusum with $\text{ARL}_0 = 370$ samples, which matches the ARL_0 value for a Shewhart control chart with the usual 3σ limits.

Generally, we want to choose k relative to the size of the shift we want to detect; that is, $k = \frac{1}{2}\delta$, where δ is the size of the shift in standard deviation units. This approach comes very close to minimizing the ARL_1 value for detecting a shift of size δ for fixed ARL_0. As mentioned earlier, a widely used value in practice is $k = \frac{1}{2}$. Then, once k is selected, you should choose h to give the desired in-control ARL_0 performance.

Table 8-3   ARL Performance of the Tabular Cusum with $k = \frac{1}{2}$ and $h = 4$ or $h = 5$

Shift in Mean (multiple of σ)	$h = 4$	$h = 5$
0	168	465
0.25	74.2	139
0.50	26.6	38.0
0.75	13.3	17.0
1.00	8.38	10.4
1.50	4.75	5.75
2.00	3.34	4.01
2.50	2.62	3.11
3.00	2.19	2.57
4.00	1.71	2.01

Table 8-4   Values of k and the Corresponding Values of h That Give
$ARL_0 = 370$ for the Two-Sided Tabular Cusum [from
Hawkins (1993a)]

k	0.25	0.5	0.75	1.0	1.25	1.5
h	8.01	4.77	3.34	2.52	1.99	1.61

Hawkins (1993a) gives a table of k values and the corresponding h values that will achieve $ARL_0 = 370$. These are reproduced in Table 8-4.

Several techniques can be used to calculate the ARL of a cusum. Vance (1986) provides a very accurate computer program. A number of authors have used an approach to calculating ARLs due to Brook and Evans (1972) that is based on approximating transitions from the in-control to the out-of-control state with a Markov chain.[1] Hawkins (1992) has provided a simple but very accurate ARL calculation procedure based on an approximating equation. His approximation requires a table of constants to apply and is accurate to within 1–3% of the true ARL value. Woodall and Adams (1993) recommend the ARL approximation given by Siegmund (1985) because of its simplicity. For a **one-sided cusum** (that is, C_i^+ or C_i^-) with parameters h and k, Siegmund's approximation is

$$ARL = \frac{\exp(-2\Delta b) + 2\Delta b - 1}{2\Delta^2} \tag{8-6}$$

for $\Delta \neq 0$, where $\Delta = \delta^* - k$ for the upper one-sided cusum C_i^+, $\Delta = -\delta^* - k$ for the lower one-sided cusum C_i^-, $b = h + 1.166$, and $\delta^* = (\mu_1 - \mu_0)/\sigma$. If $\Delta = 0$, one can use $ARL = b^2$.

The quantity δ^* represents the shift in the mean, in the units of σ, for which the ARL is to be calculated. Therefore, if $\delta^* = 0$, we would calculate ARL_0 from equation 8-6, whereas if $\delta^* \neq 0$, we would calculate the value of ARL_1 corresponding to a shift of size δ^*. To obtain the ARL of the two-sided cusum from the ARLs of the two one-sided statistics — say, ARL^+ and ARL^- — use

$$\frac{1}{ARL} = \frac{1}{ARL^+} + \frac{1}{ARL^-} \tag{8-7}$$

---

[1] The Brook and Evans approach is discussed in the supplemental text material.

To illustrate, consider the two-sided cusum with $k = \frac{1}{2}$ and $h = 5$. To find ARL_0 we would first calculate the ARL_0 values for the two-sided schemes—say, ARL_0^+ and ARL_0^-. Set $\delta^* = 0$; then $\Delta = \delta^* - k = 0 - \frac{1}{2} = -\frac{1}{2}$, $b = h + 1.166 = 5 + 1.166 = 6.166$, and from equation 8-6

$$ARL_0^+ \simeq \frac{\exp[-2(-\frac{1}{2})(6.166)] + 2(-\frac{1}{2})(6.166) - 1}{2(-\frac{1}{2})^2} = 938.2$$

By symmetry, we have $ARL_0^- = ARL_0^+$, and so from equation 8-7, the in-control ARL for the two-sided cusum is

$$\frac{1}{ARL_0} = \frac{1}{938.2} + \frac{1}{938.2}$$

or

$$ARL_0 = 469.1$$

This is very close to the true ARL_0 value of 465 shown in Table 8-3. If the mean shifts by 2σ, then $\delta^* = 2$, $\Delta = 1.5$ for the upper one-sided cusum, $\Delta = -2.5$ for the lower one-sided cusum, and from equations 8-6 and 8-7 we can calculate the approximate ARL_1 of the two-sided cusum as $ARL_1 \simeq 3.89$. The exact value shown in Table 8-3 is 4.01.

One could use Siegmund's approximation and trial-and-error arithmetic to give a control limit that would have any desired ARL. Alternatively, numerical root-finding methods would also work well. Woodall and Adams (1993) give an excellent discussion of this.

8-1.4   The Standardized Cusum

Many users of the cusum prefer to standardize the variable x_i before performing the calculations. Let

$$y_i = \frac{x_i - \mu_0}{\sigma} \tag{8-8}$$

be the standardized value of x_i. Then the standardized cusums are defined as follows.

The Standardized Two-Sided Cusum

$$C_i^+ = \max[0, y_i - k + C_{i-1}^+] \tag{8-9}$$

$$C_i^- = \max[0, -k - y_i + C_{i-1}^-] \tag{8-10}$$

There are two advantages to standardizing the cusum. First, many cusum charts can now have the same values of k and h, and the choices of these parameters are not scale dependent (that is, they do not depend on σ). Second, a standardized cusum leads naturally to a cusum for controlling variability, as we will see in Section 8-1.9.

8-1.5   Rational Subgroups

Although we have given the development of the tabular cusum for the case of individual observations ($n = 1$), it is easily extended to the case of averages of rational subgroups ($n > 1$). Simply replace x_i by $\bar{x}_i$ (the sample or subgroup average) in the above formulas, and replace σ with $\sigma_{\bar{x}} = \sigma/\sqrt{n}$.

With Shewhart charts, the use of averages of rational subgroups substantially improves control chart performance. However, this does not always happen with the cusum. If, for example, you have a choice of taking a sample of size $n = 1$ every half-hour or a rational subgroup sample of size $n = 5$ every 2.5 hours (note that both choices have the same sampling intensity), the cusum will often work best with the choice $n = 1$ every half-hour. Only if there is some significant economy of scale or some other valid reason for taking samples of size greater than one should subgroups of size greater than one be used with the cusum.

8-1.6   Improving Cusum Responsiveness for Large Shifts

We have observed that the cusum control chart is very effective in detecting small shifts. However, the cusum control chart is not as effective as the Shewhart chart in detecting large shifts. An approach to improving the ability of the cusum control chart to detect large process shifts is to use a **combined cusum-Shewhart procedure** for on-

Table 8-5   ARL Values for Some Modifications of the Basic Cusum with $k = \frac{1}{2}$ and $h = 5$ (If subgroups of size $n > 1$ are used, then $\sigma = \sigma_{\bar{x}} = \sigma/\sqrt{n}$)

Shift in Mean (multiple of σ)	(a) Basic Cusum	(b) Cusum-Shewhart (Shewhart limits at 3.5σ)	(c) Cusum with FIR	(d) FIR Cusum-Shewhart (Shewhart limits at 3.5σ)
0	465	391	430	360
0.25	139	130.9	122	113.9
0.50	38.0	37.20	28.7	28.1
0.75	17.0	16.80	11.2	11.2
1.00	10.4	10.20	6.35	6.32
1.50	5.75	5.58	3.37	3.37
2.00	4.01	3.77	2.36	2.36
2.50	3.11	2.77	1.86	1.86
3.00	2.57	2.10	1.54	1.54
4.00	2.01	1.34	1.16	1.16

line control. Adding the Shewhart control chart is a very simple modification of the cumulative sum control procedure. The Shewhart control limits should be located approximately 3.5 standard deviations from the center line or target value μ_0. An out-of-control signal on either (or both) charts constitutes an action signal. Lucas (1982) gives a good discussion of this technique. Column (a) of Table 8-5 presents the ARLs of the basic cusum with $k = \frac{1}{2}$ and $h = 5$. Column (b) of Table 8-5 presents the ARLs of the cusum with Shewhart limit added to the individual measurements. As suggested above, the Shewhart limits are at 3.5σ. Note from examining these ARL values that the addition of the Shewhart limits has improved the ability of the procedure to detect larger shifts and has only slightly decreased the in-control ARL$_0$. We conclude that a combined cusum-Shewhart procedure is an effective way to improve cusum responsiveness to large shifts.

8-1.7   The Fast Initial Response or Headstart Feature

This procedure was devised by Lucas and Crosier (1982) to improve the sensitivity of a cusum at process start-up. Increased sensitivity at process start-up would be desirable if the corrective action did not reset the mean to the target value. The **fast initial response** (FIR) or headstart essentially just sets the starting values C_0^+ and C_0^- equal to some nonzero value, typically $H/2$. This is called a 50% headstart.

To illustrate the headstart procedure, consider the data in Table 8-6. These data have a target value of 100, $K = 3$, and $H = 12$. We will use a 50% headstart value of $C_0^+ = C_0^- = H/2 = 6$. The first 10 samples are in control with mean equal to the target value of 100. Since $x_1 = 102$, the cusums for the first period will be

$$C_1^+ = \max\,[0, x_1 - 103 + C_0^+]$$

$$= \max\,[0, 102 - 103 + 6] = 5$$

$$C_1^- = [0, 97 - x_1 + C_0^-]$$

$$= \max\,[0, 97 - 102 + 6] = 1$$

Note that the starting cusum value is the headstart $H/2 = 6$. In addition, we see from panels (a) and (b) of Table 8-6 that both cusums decline rapidly to zero from the starting value. In fact, from period 2 onward C_1^+ is unaffected by the headstart, and from period 3 onward C_1^- is unaffected by the headstart. This has occurred because the process is in control at the target value of 100, and several consecutive observations near the target value were observed.

Now suppose the process had been out of control at process start-up, with mean 105. Table 8-7 presents the data that would have been produced by this process and the resulting cusums. Note that the third sample causes C_3^+ to exceed the limit $H = 12$. If no headstart had been used, we would have started with $C_0^+ = 0$, and the cusum would not exceed H until sample number 6.

This example demonstrates the benefits of a headstart. If the process starts in control at the target value, the cusums will quickly drop to zero and the headstart will have

Table 8-6 A Cusum with a Headstart, Process Mean Equal to 100

Period		a			b		
i	x_i	$x_i - 103$	C_i^+	N^+	$97 - x_i$	C_i^-	N^-
1	102	-1	5	1	-5	1	1
2	97	-6	0	0	0	1	2
3	104	1	1	1	-7	0	0
4	93	-6	0	0	4	4	1
5	100	-3	0	0	-3	1	2
6	105	2	2	1	-8	0	0
7	96	-7	0	0	1	1	1
8	98	-5	0	0	-1	0	0
9	105	2	2	1	-8	0	0
10	99	-4	0	0	-2	0	0

little effect on the performance of the cusum procedure. Figure 8-4 illustrates this property of the headstart using the data from Table 8-1. The cusum chart was produced using Minitab. However, if the process starts at some level different from the target value, the headstart will allow the cusum to detect it more quickly, resulting in shorter out-of-control ARL values.

Column (c) of Table 8-5 presents the ARL performance of the basic cusum with the headstart or FIR feature. The ARLs were calculated using a 50% headstart. Note that the ARL values for the FIR cusum are valid for the case when the process is out of control at the time the cusums are reset. When the process is in control, the headstart value quickly drops to zero. Thus, if the process is in control when the cusum is reset but shifts out of control later, the more appropriate ARL for such a case should be read from column (a)—that is, the cusum without the FIR feature.

Table 8-7 A Cusum with a Headstart, Process Mean Equal to 105

Period		a			b		
i	x_i	$x_i - 103$	C_i^+	N^+	$97 - x_i$	C_i^-	N^-
1	107	4	10	1	-10	0	0
2	102	-1	9	2	-5	0	0
3	109	6	15	3	-12	0	0
4	98	-5	10	4	-1	0	0
5	105	2	12	5	-8	0	0
6	110	7	19	6	-13	0	0
7	101	-2	17	7	-4	0	0
8	103	0	17	8	-6	0	0·
9	110	7	24	9	-13	0	0
10	104	1	25	10	-7	0	0

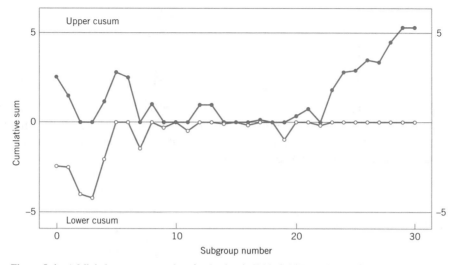

Figure 8-4   A Minitab cusum status chart for the data in Table 8-1 illustrating the fast initial response or headstart feature.

8-1.8   One-Sided Cusums

We have focused primarily on the two-sided cusum. Note that the tabular procedure is constructed by running two one-sided procedures, C_i^+ and C_i^-. There are situations in which only a single one-sided cusum procedure is useful.

For example, consider a chemical process for which the critical quality characteristic is the product viscosity. If viscosity drops below the target ($\mu_0 = 2000$ centistokes at 100°C, say), there is no significant problem, but any increase in viscosity should be detected quickly. A one-sided upper cusum would be an ideal process-monitoring scheme. Siegmund's procedure (equation 8-6) could be used to calculate the ARLs for the one-sided scheme.

It is also possible to design cusums that have different sensitivity on the upper and lower side. This could be useful in situations where shifts in either direction are of interest, but shifts above the target (say) are more critical than shifts below the target.

8-1.9   A Cusum for Monitoring Process Variability

It is also possible to construct cusum control charts for monitoring process variability. Since cusums are usually employed with individual observations, the procedure due to Hawkins (1981) is a very useful one. As before, let x_i be the normally distributed process measurement with mean or target value μ_0 and standard deviation σ. The standardized value of x_i is $y_i = (x_i - \mu_0)/\sigma$. Hawkins (1981) (1993a) suggests creating a new standardized quantity

$$v_i = \frac{\sqrt{|y_i|} - 0.822}{0.349} \tag{8-11}$$

He suggests that the v_i are sensitive to variance changes rather than mean changes. In fact, the statistic v_i is sensitive to both mean and variance changes. Since the in-control distribution of v_i is approximately $N(0, 1)$, two one-sided standardized scale (i.e., standard deviation) cusums can be established as follows.

The Scale Cusum

$$S_i^+ = \max[0, v_i - k + S_{i-1}^+] \tag{8-12}$$

$$S_i^- = \max[0, -k - v_i + S_{i-1}^-] \tag{8-13}$$

---

where $S_0^+ = S_0^- = 0$ (unless a FIR feature is used) and the values of k and h are selected as in the cusum for controlling the process mean.

The interpretation of the scale cusum is similar to the interpretation of the cusum for the mean. If the process standard deviation increases, the values of S_i^+ will increase and eventually exceed h, whereas if the standard deviation decreases, the values of S_i^- will increase and eventually exceed h.

Although one could maintain separate cusum status charts for the mean and standard deviation, Hawkins (1993) suggests plotting them on the same graph. He also provides several excellent examples and further discussion of this procedure. Study of his examples will be of value in improving your ability to detect changes in process variability from the scale cusum. If the scale cusum signals, one would suspect a change in variance, but if both cusums signal, one would suspect a shift in the mean.

8-1.10   Cusums for Other Sample Statistics

We have concentrated on cusums for sample averages. However, it is possible to develop cusums for other sample statistics such as the ranges and standard deviations of rational subgroups, fractions nonconforming, and defects. Some of these cusums are discussed in the papers by Lowery, Champ, and Woodall (1995), Gan (1993), Lucas (1985), and White, Keats, and Stanley (1997).

One variation of the cusum is extremely useful when working with count data and the count rate is very low. In this case, it is frequently more effective to form a cusum using the time between events. The most common situation encountered in practice is to use the time-between-events cusum to detect an *increase* in the count rate. This is equivalent to detecting a *decrease* in the time between events. An appropriate cusum scheme is

$$C_i^- = \max [0, K - T_i + C_{i-1}^-] \qquad (8\text{-}14)$$

where K is the reference value and T_i is the time that has elapsed since that last observed count. Lucas (1985) and Bourke (1991) discuss the choice of K and H for this procedure.

An alternative and very effective procedure would be to transform the time between observed counts to an approximately normally distributed random variable, as discussed in Section 6-3.5, and use the cusum for monitoring the mean of a normal distribution in Section 8-1.2 instead of equation 8-14.

8-1.11 The V-Mask Procedure

An alternative procedure to the use of a tabular cusum is the V-mask control scheme proposed by Barnard (1959). The V-mask is applied to successive values of the cusum statistic

$$C_i = \sum_{j=1}^{i} y_j$$

$$= y_i + C_{i-1}$$

where y_i is the standardized observation $y_i = (x_i - \mu_0)/\sigma$. A typical V-mask is shown in Fig. 8-5.

The decision procedure consists of placing the V-mask on the cumulative sum control chart with the point O on the last value of C_i and the line OP parallel to the horizontal axis. If all the previous cumulative sums, $C_1, C_2, \ldots, C_i$ lie within the two arms of the V-mask, the process is in control. However, if any of the cumulative sums lie outside the arms of the mask, the process is considered to be out of control. In actual use, the V-mask would be applied to each new point on the cusum chart as soon as it was plotted, and the arms are assumed to extend backward to the origin. The performance of the V-mask is determined by the lead distance d and the angle θ shown in Fig. 8-5.

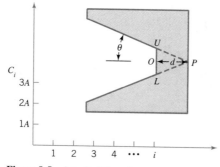

Figure 8-5 A typical V-mask.

The tabular cusum and the V-mask scheme are equivalent if

$$k = A \tan \theta \qquad (8\text{-}15)$$

and

$$h = A \, d \tan(\theta) = dk \qquad (8\text{-}16)$$

In these two equations, A is the horizontal distance on the V-mask plot between successive points in terms of unit distance on the vertical scale. Refer to Fig. 8-5. For example, to construct a V-mask equivalent to the tabular cusum scheme used in Example 8-1, where $k = \frac{1}{2}$ and $h = 5$, we would select $A = 1$ (say), and then equations 8-15 and 8-16 would be solved as follows:

$$k = A \tan \theta$$
$$\tfrac{1}{2} = (1) \tan \theta$$

or

$$\theta = 26.57°$$

and

$$h = dk$$
$$5 = d(\tfrac{1}{2})$$

or

$$d = 10$$

That is, the lead distance of the V-mask would be 10 horizontal plotting positions, and the angle opening on the V-mask would be 26.57°.

Johnson (1961) [also see Johnson and Leone (1962a, b, c)] has suggested a method for designing the V-mask; that is, selecting d and θ. He recommends the V-mask parameters

$$\theta = \tan^{-1}\left(\frac{\delta}{2A}\right) \qquad (8\text{-}17)$$

and

$$d = \left(\frac{2}{\delta^2}\right) \ln\left(\frac{1-\beta}{\alpha}\right) \qquad (8\text{-}18)$$

where 2α is the greatest allowable probability of a signal when the process mean is on target (a false alarm) and β is the probability of not detecting a shift of size δ. If β is small, which is usually the case, then

$$d \simeq -2\frac{\ln(\alpha)}{\delta} \qquad (8\text{-}19)$$

We strongly advise the quality engineer **not to use the V-mask procedure**. Some of the disadvantages and problems associated with this scheme are as follows:

Table 8-8   Actual Values of ARL_0 for a V-Mask Scheme Designed
Using Johnson's Method [Adapted from Table 2 in
Woodall and Adams (1993)]

Shift to Be Detected, δ	Values of α [Desired Value of $ARL_0 = 1/(2\alpha)$]	
	0.00135 (370)	0.001 (500)
1.0	2350.6	3184.5
2.0	1804.5	2435.8
3.0	2194.8	2975.4

1. The V-mask is a two-sided scheme; it is not very useful for one-sided process-monitoring problems.

2. The headstart feature, which is very useful in practice, cannot be implemented with the V-mask.

3. It is sometimes difficult to determine how far backwards the arms of the V-mask should extend, thereby making interpretation difficult for the practitioner.

4. Perhaps the biggest problem with the V-mask is the ambiguity associated with α and β.

Adams, Lowry, and Woodall (1992) point out that defining 2α as the probability of a false alarm is incorrect. Essentially, 2α cannot be the probability of a false alarm on any single sample, because this probability changes over time on the cusum, nor can 2α be the probability of eventually obtaining a false alarm (this probability is, of course, 1). In fact, 2α must be the long-run proportion of observations resulting in false alarms. If this is so, then the in-control ARL should be $ARL_0 = 1/(2\alpha)$. However, Johnson's design method produces values of ARL_0 that are substantially larger than $1/(2\alpha)$.

Table 8-8 shows values of ARL_0 for a V-mask scheme designed using Johnson's method. Note that the actual values of ARL_0 are about five times the desired value used in the design procedure. The schemes will also be much less sensitive to shifts in the process mean. Consequently, the use of the V-mask scheme is not a good idea. Unfortunately, it's still the default cusum in many SPC software packages. Hopefully, computer software developers will soon replace this procedure with the algorithmic or tabular cusum.

8-2 THE EXPONENTIALLY WEIGHTED MOVING AVERAGE CONTROL CHART

The exponentially weighted moving average (or EWMA) control chart is also a good alternative to the Shewhart control chart when we are interested in detecting small shifts. The performance of the EWMA control chart is approximately equivalent to that of the cumulative sum control chart, and in some ways it is easier to set up and operate. As with the cusum, the EWMA is typically used with individual observations, and so consequently, we will discuss that case first. We will also give the results for rational subgroups of size $n > 1$.

8-2.1 The Exponentially Weighted Moving Average Control Chart for Monitoring the Process Mean

The EWMA control chart was introduced by Roberts (1959). See also Crowder (1987a) (1989) and Lucas and Saccucci (1990) for good discussions of the EWMA. The exponentially weighted moving average is defined as

$$z_i = \lambda x_i + (1 - \lambda)z_{i-1} \tag{8-20}$$

where $0 < \lambda \le 1$ is a constant and the starting value (required with the first sample at $i = 1$) is the process target, so that

$$z_0 = \mu_0$$

Sometimes the average of preliminary data is used as the starting value of the EWMA, so that $z_0 = \bar{x}$.

To demonstrate that the EWMA z_i is a weighted average of all previous sample means, we may substitute for z_{i-1} on the right-hand side of equation 8-20 to obtain

$$z_i = \lambda x_i + (1 - \lambda)[\lambda x_{i-1} + (1 - \lambda)z_{i-2}]$$
$$= \lambda x_i + \lambda(1 - \lambda)x_{i-1} + (1 - \lambda)^2 z_{i-2}$$

Continuing to substitute recursively for z_{i-j}, $j = 2, 3, \ldots, t$, we obtain

$$z_i = \lambda \sum_{j=0}^{i-1} (1 - \lambda)^j x_{i-j} + (1 - \lambda)^i z_0 \tag{8-21}$$

The weights $\lambda(1 - \lambda)^j$ decrease geometrically with the age of the sample mean. Furthermore, the weights sum to unity, since

$$\lambda \sum_{j=0}^{i-1} (1 - \lambda)^j = \lambda \left[\frac{1 - (1 - \lambda)^i}{1 - (1 - \lambda)} \right] = 1 - (1 - \lambda)^i$$

If $\lambda = 0.2$, then the weight assigned to the current sample mean is 0.2 and the weights given to the preceding means are 0.16, 0.128, 0.1024, and so forth. A comparison of these weights with those of a five-period moving average is shown in Fig. 8-6. Because these weights decline geometrically when connected by a smooth curve, the EWMA is sometimes called a geometric moving average (GMA). The EWMA is used extensively in time series modeling and in forecasting [see Box, Jenkins, and Reinsel (1994) and Montgomery, Johnson, and Gardiner (1990)]. Since the EWMA can be viewed as a

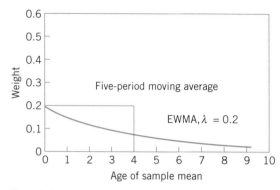

Figure 8-6   Weights of past sample means.

weighted average of all past and current observations, it is very insensitive to the normality assumption. It is therefore an ideal control chart to use with individual observations.

If the observations x_i are independent random variables with variance σ^2, then the variance of z_i is

$$\sigma_{z_i}^2 = \sigma^2 \left(\frac{\lambda}{2 - \lambda} \right) [1 - (1 - \lambda)^{2i}] \tag{8-22}$$

Therefore, the EWMA control chart would be constructed by plotting z_i versus the sample number i (or time). The center line and control limits for the EWMA control chart are as follows.

The EWMA Control Chart

$$UCL = \mu_0 + L\sigma \sqrt{\frac{\lambda}{(2 - \lambda)} [1 - (1 - \lambda)^{2i}]} \tag{8-23}$$

$$\text{Center line} = \mu_0$$

$$LCL = \mu_0 - L\sigma \sqrt{\frac{\lambda}{(2 - \lambda)} [1 - (1 - \lambda)^{2i}]} \tag{8-24}$$

In equations 8-23 and 8-24, the factor L is the width of the control limits. We will discuss the choice of the parameters L and λ shortly.

Note that the term $[1 - (1 - \lambda)^{2i}]$ in equations 8-23 and 8-24 approaches unity as i gets larger. This means that after the EWMA control chart has been running for several time periods, the control limits will approach steady-state values given by

$$\text{UCL} = \mu_0 + L\sigma\sqrt{\frac{\lambda}{(2-\lambda)}} \tag{8-25}$$

and

$$\text{LCL} = \mu_0 - L\sigma\sqrt{\frac{\lambda}{(2-\lambda)}} \tag{8-26}$$

However, we strongly recommend using the exact control limits in equations 8-23 and 8-24 for small values of i. This will greatly improve the performance of the control chart in detecting an off-target process immediately after the EWMA is started up.

······· **EXAMPLE 8-2** ··

We will apply the EWMA control chart with $\lambda = 0.10$ and $L = 2.7$ to the data in Table 8-1. Recall that the target value of the mean is $\mu_0 = 10$ and the standard deviation is $\sigma = 1$. The calculations for the EWMA control chart are summarized in Table 8-9 and the control chart is shown in Fig. 8-7.

To illustrate the calculations, consider the first observation, $x_1 = 9.45$. The first value of the EWMA is

$$z_1 = \lambda x_1 + (1 - \lambda)z_0$$
$$= 0.1(9.45) + 0.9(10)$$
$$= 9.945$$

Therefore, $z_1 = 9.945$ is the first value plotted on the control chart in Fig. 8-7. The second value of the EWMA is

$$z_2 = \lambda x_2 + (1 - \lambda)z_1$$
$$= 0.1(7.99) + 0.9(9.945)$$
$$= 9.7495$$

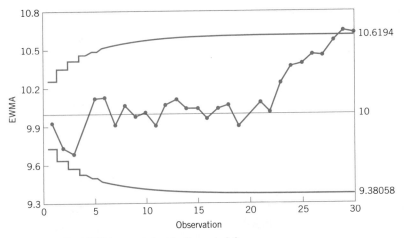

Figure 8-7 The EWMA control chart for Example 8-2.

Table 8-9   EWMA Calculations for Example 8-2

Subgroup, i	x_i	EWMA, z_i
	* = Beyond Limits	
1	9.45	9.945
2	7.99	9.7495
3	9.29	9.70355
4	11.66	9.8992
5	12.16	10.1253
6	10.18	10.1307
7	8.04	9.92167
8	11.46	10.0755
9	9.2	9.98796
10	10.34	10.0232
11	9.03	9.92384
12	11.47	10.0785
13	10.51	10.1216
14	9.4	10.0495
15	10.08	10.0525
16	9.37	9.98426
17	10.62	10.0478
18	10.31	10.074
19	8.52	9.91864
20	10.84	10.0108
21	10.9	10.0997
22	9.33	10.0227
23	12.29	10.2495
24	11.5	10.3745
25	10.6	10.3971
26	11.08	10.4654
27	10.38	10.4568
28	11.62	10.5731
29	11.31	10.6468*
30	10.52	10.6341*

The other values of the EWMA statistic are computed similarly.

The control limits in Fig. 8-7 are found using equations 8-23 and 8-24. For period $i = 1$,

$$\text{UCL} = \mu_0 + L\sigma \sqrt{\frac{\lambda}{(2-\lambda)}[1-(1-\lambda)^{2i}]}$$

$$= 10 + 2.7(1)\sqrt{\frac{0.1}{(2-0.1)}[1-(1-0.1)^{2(1)}]}$$

$$= 10.27$$

and

$$\text{LCL} = \mu_0 - L\sigma\sqrt{\frac{\lambda}{(2-\lambda)}[1-(1-\lambda)^{2i}]}$$

$$= 10 - 2.7(1)\sqrt{\frac{0.1}{(2-0.1)}[1-(1-0.1)^{2(1)}]}$$

$$= 9.73$$

For period 2, the limits are

$$\text{UCL} = \mu_0 + L\sigma\sqrt{\frac{\lambda}{(2-\lambda)}[1-(1-\lambda)^{2i}]}$$

$$= 10 + 2.7(1)\sqrt{\frac{0.1}{(2-0.1)}[1-(1-0.1)^{2(2)}]}$$

$$= 10.36$$

and

$$\text{LCL} = \mu_0 - L\sigma\sqrt{\frac{\lambda}{(2-\lambda)}[1-(1-\lambda)^{2i}]}$$

$$= 10 - 2.7(1)\sqrt{\frac{0.1}{(2-0.1)}[1-(1-0.1)^{2(2)}]}$$

$$= 9.64$$

Note from Fig. 8-7 that the control limits increase in width as i increases from $i = 1$, 2, . . . , until they stabilize at the steady-state values given by equations 8-25 and 8-26:

$$\text{LCL} = \mu_0 + L\sigma\sqrt{\frac{\lambda}{(2-\lambda)}}$$

$$= 10 + 2.7(1)\sqrt{\frac{0.1}{(2-0.1)}}$$

$$= 10.62$$

and

$$\text{LCL} = \mu_0 - L\sigma\sqrt{\frac{\lambda}{(2-\lambda)}}$$

$$= 10 - 2.7(1)\sqrt{\frac{0.1}{(2-0.1)}}$$

$$= 9.38$$

The EWMA control chart in Fig. 8-7 signals at observation 28, so we would conclude that the process is out of control.

....

8-2.2   Design of an EWMA Control Chart

The EWMA control chart is very effective against small process shifts. The design parameters of the chart are the multiple of sigma used in the control limits (L) and the value of λ. It is possible to choose these parameters to give ARL performance from the EWMA control chart that closely approximates cusum ARL performance for detecting small shifts.

There have been several theoretical studies of the average run length properties of the EWMA control chart. For example, see the papers by Crowder (1987a, 1989) and Lucas and Saccucci (1990). These studies provide average run-length tables or graphs for a range of values of λ and L. The average run-length performance for several EWMA control schemes is shown in Table 8-10. The optimal design procedure would consist of specifying the desired in-control and out-of-control average run lengths and the magnitude of the process shift that is anticipated, and then to select the combination of λ and L that provide the desired ARL performance.

In general, we have found that values of λ in the interval $0.05 \le \lambda \le 0.25$ work well in practice, with $\lambda = 0.05$, $\lambda = 0.10$, and $\lambda = 0.20$ being popular choices. A good rule of thumb is to use smaller values of λ to detect smaller shifts. We have also found that $L = 3$ (the usual three-sigma limits) works reasonably well, particularly with the larger value of λ, although when λ is small—say, $\lambda \le 0.1$—there is an advantage in reducing the width of the limits by using a value of L between about 2.6 and 2.8. Recall that in Example 8-2, we used $\lambda = 0.1$ and $L = 2.7$. We would expect this choice of parameters to result in an in-control ARL of $\mathrm{ARL}_0 \cong 500$ and an ARL for detecting a

Table 8-10   Average Run Lengths for Several EWMA Control Schemes [Adapted from Lucas and Saccucci (1990)]

Shift in Mean (multiple of σ)	$L = 3.054$ $\lambda = 0.40$	2.998 0.25	2.962 0.20	2.814 0.10	2.615 0.05
0	500	500	500	500	500
0.25	224	170	150	106	84.1
0.50	71.2	48.2	41.8	31.3	28.8
0.75	28.4	20.1	18.2	15.9	16.4
1.00	14.3	11.1	10.5	10.3	11.4
1.50	5.9	5.5	5.5	6.1	7.1
2.00	3.5	3.6	3.7	4.4	5.2
2.50	2.5	2.7	2.9	3.4	4.2
3.00	2.0	2.3	2.4	2.9	3.5
4.00	1.4	1.7	1.9	2.2	2.7

shift of one standard deviation in the mean of $ARL_1 \cong 10.3$. Thus this design is approximately equivalent to the cusum with $h = 5$ and $k = \frac{1}{2}$.

Hunter (1989) has also studied the EWMA and suggested choosing λ so that the weight given to current and previous observations matches as closely as possible the weights given to these observations by a Shewhart chart with the Western Electric rules. This results in a recommended value of $\lambda = 0.4$. If $L = 3.054$, then Table 8-10 indicates that this chart would have $ARL_0 \cong 500$ and for detecting a shift of one standard deviation in the process mean, the $ARL_1 \cong 14.3$.

Like the cusum, the EWMA performs well against small shifts but does not react to large shifts as quickly as the Shewhart chart. However, the EWMA is often superior to the cusum for large shifts, particularly if $\lambda > 0.10$. A good way to further improve the sensitivity of the control procedure to large shifts without sacrificing the ability to detect small shifts quickly is to combine a Shewhart chart with the EWMA. These combined Shewhart-EWMA control procedures are effective against both large and small shifts. When using such schemes, we have found it helpful to use slightly wider than usual limits on the Shewhart chart (say, 3.25 sigma, or even 3.5 sigma). It is also possible to plot *both* x_i (or $\bar{x}_i$) and the EWMA statistic z_i on the same control chart along with both the Shewhart and EWMA limits. This produces one chart for the combined control procedure that operators quickly become adept at interpreting. When the plots are computer generated, different colors or plotting symbols can be used for the two sets of control limits and statistics.

8-2.3 Rational Subgroups

The EWMA control chart is often used with individual measurements. However, if rational subgroups of size $n > 1$ are taken, then simply replace x_i with $\bar{x}_i$ and σ with $\sigma_{\bar{x}} = \sigma/\sqrt{n}$ in the previous equations.

8-2.4 Robustness of the EWMA to Nonnormality

When discussing the Shewhart control chart for individuals in Chapter 5, we observed that the individuals chart was very sensitive to nonnormality in the sense that the actual in-control ARL (ARL_0) would be considerably less than the "advertised" or expected value based on the assumption of a normal distribution. Borror, Montgomery, and Runger (1999) compared the ARL performance of the Shewhart individuals chart and the EWMA control chart for the case of nonnormal distributions. Specifically, they used the gamma distribution to represent the case of skewed distributions and the t distribution to represent symmetric distributions with heavier tails than the normal.

The ARL_0 of the Shewhart individuals chart and several EWMA control charts for these nonnormal distributions are given in Tables 8-11 and 8-12. Two aspects of the information in these tables are very striking. First, even moderately nonnormal distributions have the effect of greatly reducing the in-control ARL of the Shewhart individuals

Table 8-11 In-Control ARLs for the EWMA and the Individuals Control Charts for Various Gamma Distributions

	EWMA			Shewhart
λ	0.05	0.1	0.2	
L	2.492	2.703	2.86	3.00
Normal	370.4	370.8	370.5	370.4
Gam(4, 1)	372	341	259	97
Gam(3, 1)	372	332	238	85
Gam(2, 1)	372	315	208	71
Gam(1, 1)	369	274	163	55
Gam(0.5, 1)	357	229	131	45

chart. This will, of course, dramatically increase the rate of false alarms. Second, an EWMA with $\lambda = 0.05$ or $\lambda = 0.10$ and an appropriately chosen control limit will perform very well against both normal and nonnormal distributions. With $\lambda = 0.05$ and $L = 2.492$ the ARL_0 for the EWMA is within approximately 8% of the advertised normal theory in-control ARL_0 of 370, except in the most extreme cases. Furthermore, the shift detection properties of the EWMA are uniformly superior to the Shewhart chart for individuals.

Based on this information, we would recommend a properly designed EWMA as a control chart for individual measurements in a wide range of applications. It is almost a perfectly nonparametric (distribution-free) procedure.

8-2.5   Extensions of the EWMA

There have been numerous extensions and variations of the basic EWMA control chart. In this section, we describe a few of these procedures.

Table 8-12 In-Control ARLs for the EWMA and the Individuals Control Charts for Various t Distributions

	EWMA			Shewhart
λ	0.05	0.1	0.2	1
L	2.492	2.703	2.86	3.00
Normal	370.4	370.8	370.5	370.4
t_{50}	369	365	353	283
t_{40}	369	363	348	266
t_{30}	368	361	341	242
t_{20}	367	355	325	204
t_{15}	365	349	310	176
t_{10}	361	335	280	137
t_8	358	324	259	117
t_6	351	305	229	96
t_4	343	274	188	76

Fast Initial Response Feature

It is possible to add the headstart or fast initial response feature to the EWMA. As with the cusum, the advantage of the procedure would be to more quickly detect a process that is off target at start-up.

Two approaches have been suggested. Rhoads, Montgomery, and Mastrangelo (1996) set up two one-sided EWMA charts and start them off at values midway between the target and the control limit. Both one-sided charts are assumed to have the time-varying limits.

Steiner (1999) uses a single control chart but narrows the time-varying limits even further for the first few sample points. He uses an exponentially decreasing adjustment to further narrow the limits, so that the control limits are located a distance

$$\pm L\sigma \left\{ (1 - (1 - f)^{1 + a(t-1)}) \sqrt{\frac{\lambda}{2 - \lambda} [1 - (1 - \lambda)^{2t}]} \right\}$$

around the target. The constants f and a are to be determined. Steiner suggests choosing a so that the FIR has little effect after about 20 observations. This leads to choosing $a = [-2/\log(1 - f) - 1]/19$. For example, if $f = 0.5$, then $a = 0.3$. The choice of $f = 0.5$ is attractive because it mimics the 50% headstart often used with cusums.

Both of these procedures perform very well in reducing the ARL to detect an off-target process at start-up. The Steiner procedure is easier to implement in practice.

Monitoring Variability

MacGregor and Harris (1993) discuss the use of EWMA-based statistics for monitoring the process standard deviation. Let x_i be normally distributed with mean μ and standard deviation σ. The **exponentially weighted mean square error (EWMS)** is defined as

$$S_i^2 = \lambda(x_i - \mu)^2 + (1 - \lambda)S_{i-1}^2 \tag{8-27}$$

It can be shown that $E(S_i^2) = \sigma^2$ (for large i) and if the observations are independent, this estimate has an approximate chi-square distribution with $v = (2 - \lambda)/\lambda$ degrees of freedom. Therefore, if σ_0 represents the in-control or target value of the process standard deviation, we could plot $\sqrt{S_i^2}$ on an exponentially weighted root mean square or EWRMS control chart with control limits given by

$$\text{UCL} = \sigma_0 \sqrt{\frac{\chi_{v,\alpha/2}^2}{v}} \tag{8-28}$$

and

$$\text{LCL} = \sigma_0 \sqrt{\frac{\chi_{v,1-(\alpha/2)}^2}{v}} \tag{8-29}$$

MacGregor and Harris (1993) point out that the EWMS statistic can be sensitive to shifts in both the process mean and the standard deviation. They suggest replacing μ in equation 8-27 with an estimate $\hat{\mu}_i$ at each point in time. A logical estimate of μ turns out to be the ordinary EWMA z_i. They derive control limits for the resulting exponentially weighted moving variance (EWMV)

$$S_i^2 = \lambda(x_i - z_i)^2 + (1 - \lambda)S_{i-1}^2 \qquad (8\text{-}30)$$

Another approach to monitoring the process standard deviation with an EWMA is in Crowder and Hamilton (1992).

The EWMA for Poisson Data

Just as the cusum can be used as the basis of an effective control chart for Poisson counts, so can a suitably designed EWMA. Borror, Champ, and Rigdon (1998) describe the procedure, show how to design the control chart, and provide an example. If x_i is a count, then the basic EWMA recursion remains unchanged:

$$z_i = \lambda x_i + (1 - \lambda)z_{i-1}$$

with $z_0 = \mu_0$ the in-control or target count rate. The control chart parameters are as follows:

$$\text{UCL} = \mu_0 + A_U \sqrt{\frac{\lambda \mu_0}{2 - \lambda}[1 - (1 - \lambda)^{2i}]}$$

$$\text{Center line} = \mu_0$$

$$\text{LCL} = \mu_0 - A_L \sqrt{\frac{\lambda \mu_0}{2 - \lambda}[1 - (1 - \lambda)^{2i}]}$$

where A_U and A_L are the upper and lower control limit factors. In many applications we would choose $A_U = A_L = A$. Borror, Champ, and Rigdon (1998) give graphs of the ARL performance of the Poisson EWMA control chart as a function of λ and A and for various in-control or target count rates μ_0. Once μ_0 is determined and a value is specified for λ, these charts can be used to select the value of A that results in the desired in-control ARL_0. The authors also show that this control chart has considerably better ability to detect assignable causes than the Shewhart c chart. The Poisson EWMA should be used much more widely in practice.

The EWMA as a Predictor of Process Level

Although we have discussed the EWMA primarily as a statistical process-monitoring tool, it actually has a much broader interpretation. From an SPC viewpoint, the EWMA is roughly equivalent to the cusum in its ability to *monitor* a process and detect the presence of assignable causes that result in a process shift. However, the EWMA provides a *forecast* of where the process mean will be at the next time period. That is, z_i is actually a forecast of the value of the process mean μ at time $i + 1$. Thus, the EWMA could be used as the basis for a dynamic process-control algorithm.

In computer-integrated manufacturing where sensors are used to measure every unit manufactured, a forecast of the process mean based on previous behavior would be very useful. If the forecast of the mean is different from target by a critical amount, then either the operator or some electro-mechanical control system can make the necessary process adjustment. If the operator makes the adjustment, then he or she must exercise caution and not make adjustments too frequently because this will actually cause process variability to increase. The control limits on the EWMA chart can be used to signal *when* an adjustment is necessary, and the difference between the target and the forecast of the mean $\mu_{i + 1}$ can be used to determine *how much* adjustment is necessary.

The EWMA can be modified to enhance its ability to forecast the mean. Suppose that the process mean trends or drifts steadily away from the target. The forecasting performance of the EWMA can be improved in this case. First, note that the usual EWMA can be written as

$$z_i = \lambda x_i + (1 - \lambda)z_{i - 1}$$

$$= z_{i - 1} + \lambda(x_i - z_{i - 1})$$

and if we view $z_{i - 1}$ as a forecast of the process mean in period i, we can think of $x_i - z_{i - 1}$ as the forecast error e_i for period i. Therefore,

$$z_i = z_{i - 1} + \lambda e_i$$

Thus, the EWMA for period i is equal to the EWMA for period $i - 1$ plus a fraction λ of the forecast error for the mean in period i. Now add a second term to this last equation to give

$$z_i = z_{i-1} + \lambda_1 e_i + \lambda_2 \sum_{j=1}^{i} e_j$$

where λ_1 and λ_2 are constants that weight the error at time i and the *sum* of the errors accumulated to time i. If we let $\nabla e_i = e_i - e_{i - 1}$ be the first difference of the errors, then we can arrive at a final modification of the EWMA:

$$z_i = z_{i-1} + \lambda_1 e_i + \lambda_2 \sum_{j=1}^{i} e_j + \lambda_3 \nabla e_i$$

Note that in this empirical control equation the EWMA in period i (which is the forecast of the process mean in period $i + 1$) equals the current estimate of the mean ($z_{i - 1}$ estimates μ_i), plus a term proportional to the error, plus a term related to the sum of the errors, plus a term related to the first difference of the errors. These three terms can be thought of as **proportional, integral,** and **differential** adjustments. The parameters λ_1, λ_2, and λ_3 would be chosen to give the best forecasting performance.

Because the EWMA statistic z_i can be viewed as a **forecast** of the mean of the process at time $i + 1$, we often plot the EWMA statistic one time period ahead. That is, we actually plot z_i at time period $i + 1$ on the control chart. This allows the analyst to visually see how much difference there is between the current observation and the estimate of the current mean of the process. In statistical process-control applications where the mean may "wander" over time, this approach has considerable appeal.

8-3 THE MOVING AVERAGE CONTROL CHART

The EWMA chart uses a weighted average as the chart statistic. Occasionally, another type of time-weighted control chart based on a simple, unweighted moving average may be of interest.

Suppose that individual observations have been collected, and let x_1, x_2, . . . denote these observations. The moving average of span w at time i is defined as

$$M_i = \frac{x_i + x_{i-1} + \cdots + x_{i-w+1}}{w} \tag{8-31}$$

That is, at time period i, the oldest observation in the moving average set is dropped and the newest one added to the set. The variance of the moving average M_i is

$$V(M_i) = \frac{1}{w^2} \sum_{j=i-w+1}^{i} V(x_j) = \frac{1}{w^2} \sum_{j=i-w+1}^{i} \sigma^2 = \frac{\sigma^2}{w} \tag{8-32}$$

Therefore, if μ_0 denotes the target value of the mean used as the center line of the control chart, then the three-sigma control limits for M_i are

$$\text{UCL} = \mu_0 + \frac{3\sigma}{\sqrt{w}} \tag{8-33}$$

and

$$\text{LCL} = \mu_0 - \frac{3\sigma}{\sqrt{w}} \tag{8-34}$$

The control procedure would consist of calculating the new moving average M_i as each observation x_i becomes available, plotting M_i on a control chart with upper and lower control limits given by equations 8-33 and 8-34, and concluding that the process is out of control if M_i exceeds the control limits. In general, the magnitude of the shift of interest and w are inversely related; smaller shifts would be guarded against more effectively by longer-span moving averages, at the expense of quick response to large shifts.

· · · · · · · **EXAMPLE 8-3** ·

We will set up a moving average control chart for the data in Table 8-1, using $w = 5$. The observations x_i for periods $1 \le i \le 30$ are shown in Table 8-13. The statistic plotted

Table 8-13 Moving Average Chart for Example 8-3

Observation, i	x_i	M_i
1	9.45	9.45
2	7.99	8.72
3	9.29	8.91
4	11.66	9.5975
5	12.16	10.11
6	10.18	10.256
7	8.04	10.266
8	11.46	10.7
9	9.2	10.208
10	10.34	9.844
11	9.03	9.614
12	11.47	10.3
13	10.51	10.11
14	9.4	10.15
15	10.08	10.098
16	9.37	10.166
17	10.62	9.996
18	10.31	9.956
19	8.52	9.78
20	10.84	9.932
21	10.9	10.238
22	9.33	9.98
23	12.29	10.376
24	11.5	10.972
25	10.6	10.924
26	11.08	10.96
27	10.38	11.17
28	11.62	11.036
29	11.31	10.998
30	10.52	10.982

on the moving average control chart will be

$$M_i = \frac{x_i + x_{i-1} + \cdots + x_{i-4}}{5}$$

for periods $i \geq 5$. For time periods $i < 5$ the average of the observations for periods 1, 2, . . . , i is plotted. The values of these moving averages are shown in Table 8-13.

The control limits for the moving average control chart may be easily obtained from equations 8-33 and 8-34. Since we have $\sigma = 1.0$, then

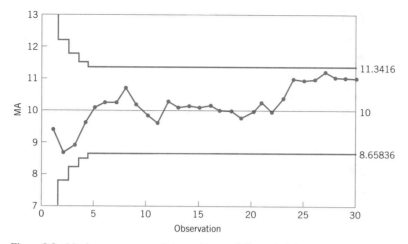

Figure 8-8   Moving average control chart with $w = 5$, Example 8-3.

$$\text{UCL} = \mu_0 + \frac{3\sigma}{\sqrt{w}} = 10 + \frac{3(1.0)}{\sqrt{5}} = 11.34$$

and

$$\text{LCL} = \mu_0 - \frac{3\sigma}{\sqrt{w}} = 10 - \frac{3(1.0)}{\sqrt{5}} = 8.66$$

The control limits for M_i apply for periods $i \geq 5$. For periods $0 < i < 5$, the control limits are given by $\mu_0 \pm 3\sigma/\sqrt{i}$. An alternative procedure that avoids using special control limits for periods $i < w$ is to use an ordinary Shewhart chart until at least w sample means have been obtained.

The moving average control chart is shown in Fig. 8-8. No points exceed the control limits. Note that for the initial periods $i < w$ the control limits are wider than their final steady-state value. Moving averages that are less than w periods apart are highly correlated, which often complicates interpreting patterns on the control chart. This is easily seen by examining Fig. 8-8.

The moving average control chart is more effective than the Shewhart chart in detecting small process shifts. However, it is generally not as effective against small shifts as either the cusum or the EWMA. The moving average control chart is considered by some to be simpler to implement than the cusum. This author prefers the EWMA to the moving average control chart.

8-4 EXERCISES

8-1. The following data represent individual observations on molecular weight taken hourly from a chemical process.

Observation Number	x	Observation Number	x
1	1045	11	1139
2	1055	12	1169
3	1037	13	1151
4	1064	14	1128
5	1095	15	1238
6	1008	16	1125
7	1050	17	1163
8	1087	18	1188
9	1125	19	1146
10	1146	20	1167

The target value of molecular weight is 1050 and the process standard deviation is thought to be about $\sigma = 25$.

(a) Set up a tabular cusum for the mean of this process. Design the cusum to quickly detect a shift of about 1.0σ in the process mean.

(b) Is the estimate of σ used in part (a) of this problem reasonable?

8-2. Rework Exercise 8-1 using a standardized cusum.

8-3. (a) Add a headstart feature to the cusum in Exercise 8-1.

(b) Use a combined Shewhart-cusum scheme on the data in Exercise 8-1. Interpret the results of both charts.

8-4. A machine is used to fill cans with motor oil additive. A single sample can is selected every hour and the weight of the can is obtained. Since the filling process is automated, it has very stable variability, and long experience indicates that $\sigma = 0.05$ oz. The individual observations for 24 hours of operation are shown here.

Sample Number	x	Sample Number	x
1	8.00	13	8.05
2	8.01	14	8.04
3	8.02	15	8.03
4	8.01	16	8.05
5	8.00	17	8.06
6	8.01	18	8.04
7	8.06	19	8.05
8	8.07	20	8.06
9	8.01	21	8.04
10	8.04	22	8.02
11	8.02	23	8.03
12	8.01	24	8.05

(a) Assuming that the process target is 8.02 oz, set up a tabular cusum for this process. Design the cusum using the standardized values $h = 4.77$ and $k = \frac{1}{2}$.

(b) Does the value of $\sigma = 0.05$ seem reasonable for this process?

8-5. Rework Exercise 8-4 using the standardized cusum parameters of $h = 8.01$ and $k = 0.25$. Compare the results with those obtained previously in Exercise 8-4. What can you say about the theoretical performance of those two cusum schemes?

8-6. Reconsider the data in Exercise 8-4. Suppose the data there represent observations taken immediately after a process adjustment that was intended to reset the process to a target of $\mu_0 = 8.00$. Set up and apply an FIR cusum to monitor this process.

8-7. The data that follow are temperature readings from a chemical process in °C, taken every 2 minutes. (Read the observations down, from left.)

953	985	949	937	959	948	958	952
945	973	941	946	939	937	955	931
972	955	966	954	948	955	947	928
945	950	966	935	958	927	941	937

975	948	934	941	963	940	938	950
970	957	937	933	973	962	945	970
959	940	946	960	949	963	963	933
973	933	952	968	942	943	967	960
940	965	935	959	965	950	969	934
936	973	941	956	962	938	981	927

The target value for the mean is $\mu_0 = 950$.

(a) Estimate the process standard deviation.

(b) Set up and apply a tabular cusum for this process, using standardized values $h = 5$ and $k = \frac{1}{2}$. Interpret this chart.

8-8. Bath concentrations are measured hourly in a chemical process. Data (in ppm) for the last 32 hours are shown here (read down from left).

160	186	190	206
158	195	189	210
150	179	185	216
151	184	182	212
153	175	181	211
154	192	180	202
158	186	183	205
162	197	186	197

The process target is $\mu_0 = 175$ ppm.

(a) Estimate the process standard deviation.

(b) Construct a tabular cusum for this process using standardized values of $h = 5$ and $k = \frac{1}{2}$.

8-9. Viscosity measurements on a polymer are made every 10 minutes by an on-line viscometer. Thirty-six observations are shown here (read down from left). The target viscosity for this process is $\mu_0 = 3200$.

3169	3205	3185	3188
3173	3203	3187	3183
3162	3209	3192	3175
3154	3208	3199	3174
3139	3211	3197	3171

3145	3214	3193	3180
3160	3215	3190	3179
3172	3209	3183	3175
3175	3203	3197	3174

(a) Estimate the process standard deviation.

(b) Construct a tabular cusum for this process using standardized values of $h = 8.01$ and $k = 0.25$.

(c) Discuss the choice of h and k in part (b) of this problem on cusum performance.

8-10. Set up a tabular cusum scheme for the piston-ring data used in Example 5-1 (see Tables 5-1 and 5-2). When the procedure is applied to all 40 samples, does the cusum react more quickly than the $\bar{x}$ chart to the shift in the process mean? Use $\sigma = 0.01$ in setting up the cusum, and design the procedure to quickly detect a shift of about 1σ.

8-11. Apply the scale cusum discussed in Section 8-1.9 to the data in Exercise 8-1.

8-12. Apply the scale cusum discussed in Section 8-1.9 to the concentration data in Exercise 8-8.

8-13. Consider a standardized two-sided cusum with $k = 0.2$ and $h = 8$. Use Siegmund's procedure to evaluate the in-control ARL performance of this scheme. Find ARL_1 for $\delta^* = 0.5$.

8-14. Consider the viscosity data in Exercise 8-9. Suppose that the target value of viscosity is $\mu_0 = 3150$ and that it is only important to detect disturbances in the process that result in increased viscosity. Set up and apply an appropriate one-sided cusum for this process.

8-15. Rework Exercise 8-1 using an EWMA control chart with $\lambda = 0.1$ and $L = 2.7$. Compare your results to those obtained with the cusum.

8-16. Consider a process with $\mu_0 = 10$ and $\sigma = 1$. Set up the following EWMA

control charts:
(a) $\lambda = 0.1, L = 3$
(b) $\lambda = 0.2, L = 3$
(c) $\lambda = 0.4, L = 3$
Discuss the effect of λ on the behavior of the control limits.

8-17. Reconsider the data in Exercise 8-4. Set up an EWMA control chart with $\lambda = 0.2$ and $L = 3$ for this process. Interpret the results.

8-18. Reconstruct the control chart in Exercise 8-17 using $\lambda = 0.1$ and $L = 2.7$. Compare this chart with the one constructed in Exercise 8-17.

8-19. Reconsider the data in Exercise 8-7. Apply an EWMA control chart to these data using $\lambda = 0.1$ and $L = 2.7$.

8-20. Reconstruct the control chart in Exercise 8-19 using $\lambda = 0.4$ and $L = 3$. Compare this chart to the one constructed in Exercise 8-19.

8-21. Reconsider the data in Exercise 8-8. Set up and apply an EWMA control chart to these data using $\lambda = 0.05$ and $L = 2.6$.

8-22. Reconsider the data in Exercise 8-9. Set up and apply an EWMA control chart to these data using $\lambda = 0.1$ and $L = 2.7$.

8-23. Analyze the data in Exercise 8-1 using a moving average control chart with $w = 6$. Compare the results obtained with the cumulative sum control chart in Exercise 8-1.

8-24. Analyze the data in Exercise 8-4 using a moving average control chart with $w = 5$. Compare the results obtained with the cumulative sum control chart in Exercise 8-4.

8-25. Show that if the process is in control at the level μ, the exponentially weighted moving average is an unbiased estimator of the process mean.

8-26. Derive the variance of the exponentially weighted moving average z_i.

8-27. **Equivalence of moving average and exponentially weighted moving average control charts.** Show that if $\lambda = 2/(w + 1)$ for the EWMA control chart, then this chart is equivalent to a w-period moving average control chart in the sense that the control limits are identical in the steady state.

8-28. **Continuation of Exercise 8-27.** Show that if $\lambda = 2/(w + 1)$, then the average "age" of the data used in computing the statistics z_i and M_i is identical.

8-29. Show how to modify the control limits for the moving average control chart if rational subgroups of size $n > 1$ are observed every period, and the objective of the control chart is to monitor the process mean.

8-30. A Shewhart $\bar{x}$ chart has center line at 10 with UCL = 16 and LCL = 4. Suppose you wish to supplement this chart with an EWMA control chart using $\lambda = 0.1$ and the same control limit width in σ-units as employed on the $\bar{x}$ chart. What are the values of the steady-state upper and lower control limits on the EWMA chart?

8-31. An EWMA control chart uses $\lambda = 0.4$. How wide will the limits be on the Shewhart control chart, expressed as a multiple of the width of the steady-state EWMA limits?

8-32. Consider the valve failure data in Example 6-6. Set up a cusum chart for monitoring the time between events using the transformed variable approach illustrated in that example. Use standardized values of $h = 5$ and $k = \frac{1}{2}$.

8-33. Consider the valve failure data in Example 6-6. Set up a one-sided cusum chart for monitoring and detecting an increase in failure rate of the valve. Assume that the target value of the mean time between failures is 700 hr.

8-34. Set up an appropriate EWMA control chart for the valve failure data in Example 6-6. Use the transformed variable approach illustrated in that example.

8-35. Discuss how you could set up one-sided EWMA control charts.

Other Univariate
Statistical Process
Monitoring and
Control Techniques

CHAPTER OUTLINE

9-1 STATISTICAL PROCESS CONTROL FOR SHORT PRODUCTION RUNS

    9-1.1 $\bar{x}$ and R Charts for Short Production Runs

    9-1.2 Attributes Control Charts for Short Production Runs

    9-1.3 Other Methods

9-2 MODIFIED AND ACCEPTANCE CONTROL CHARTS

    9-2.1 Modified Control Limits for the $\bar{x}$ Chart

    9-2.2 Acceptance Control Charts

9-3 CONTROL CHARTS FOR MULTIPLE-STREAM PROCESSES

    9-3.1 Multiple-Stream Processes

    9-3.2 Group Control Charts

    9-3.3 Other Approaches

9-4 SPC WITH AUTOCORRELATED PROCESS DATA

    9-4.1 Sources and Effects of Autocorrelation in Process Data

9-4.2 Model-Based Approaches

9-4.3 A Model-Free Approach

9-5 ADAPTIVE SAMPLING PROCEDURES

9-6 ECONOMIC DESIGN OF CONTROL CHARTS

    9-6.1 Designing a Control Chart

    9-6.2 Process Characteristics

    9-6.3 Cost Parameters

    9-6.4 Early Work and Semieconomic Designs

    9-6.5 An Economic Model of the $\bar{x}$ Control Chart

    9-6.6 Other Work

9-7 OVERVIEW OF OTHER PROCEDURES

    9-7.1 Tool Wear

    9-7.2 Control Charts Based on Other Sample Statistics

    9-7.3 Fill Control Problems

    9-7.4 Precontrol

CHAPTER OVERVIEW

The widespread successful use of the basic SPC methods described in Part II and the cusum and EWMA control charts in the previous chapter have led to the development of many new techniques and procedures over the last 20 years. This chapter is an overview of some of the more useful recent developments. We begin with a discussion of SPC methods for short production runs and concentrate on how conventional control charts can be modified for this situation. Although there are other techniques that can be applied to the short-run scenario, this approach seems to be most widely used in practice. We then discuss modified and acceptance control charts. These techniques find some application in situations where process capability is high, such as the "six-sigma" manufacturing environment. Multiple-stream processes are encountered in many industries. An example is container filling on a multiple-head machine. We present the group control chart (a classical method for the multiple-stream process) and another procedure based on control charts for monitoring the specific types of assignable causes associated with these systems. We also discuss techniques for monitoring processes with autocorrelated data, a topic of considerable importance in the chemical and process industries. Other chapter topics include a discussion of formal consideration of process economics in designing a monitoring scheme, adaptive control charts in which the sample size or time between samples (or both) may be modified based on the current value of the sample statistic, methods for tool wear, fill control problems, and control charts for sample statistics other than the conventional ones considered in previous chapters. In many cases we give only a brief summary of the topic and provide references to more complete descriptions.

9-1 STATISTICAL PROCESS CONTROL FOR SHORT PRODUCTION RUNS

Statistical process control methods have found wide application in almost every type of business. Some of the most interesting applications occur in job-shop manufacturing systems, or generally in any type of system characterized by **short production runs.** Some of the SPC methods for these situations are straightforward adaptations of the standard concepts and require no new methodology. In fact, Example 5-8 illustrated one of the basic techniques of control charting used in the short-run environment—using deviation from the nominal dimension as the variable on the control chart. In this section, we present a summary of several techniques that have proven successful in the short production run situation.

9-1.1 $\bar{x}$ and R Charts for Short Production Runs

The simplest technique for using $\bar{x}$ and R charts in the short production run situation was introduced previously in Example 5-8; namely, use deviation from nominal instead

Table 9-1 Data for Short-Run $\bar{x}$ and R Charts (deviation from nominal approach)

| Sample Number | Part Number | (a) Measurements | | | (b) Deviation from Nominal | | | | |
		M_1	M_2	M_3	x_1	x_2	x_3	$\bar{x}$	R
1	A	50	51	52	0	1	2	1.00	2
2	A	49	50	51	-1	0	1	0.00	2
3	A	48	49	52	-2	-1	2	-0.33	4
4	A	49	53	51	-1	3	1	1.00	4
5	B	24	27	26	-1	2	1	0.67	2
6	B	25	27	24	0	2	-1	0.33	2
7	B	27	26	23	2	1	-2	0.33	4
8	B	25	24	23	0	-1	-2	-1.00	2
9	B	24	25	25	-1	0	0	-0.33	1
10	B	26	24	25	1	-1	0	0.00	2
						$\bar{\bar{x}} = 0.17$		$\bar{R} = 2.7$	

of the measured variable on the control chart. This is sometimes called the **DNOM control chart.** To illustrate the procedure, consider the data in Table 9-1. The first four samples represent hole diameters in a particular part (say, part A). Panel (a) of this table shows the actual diameters in millimeters. For this part, the nominal diameter is $T_A = 50$ mm. Thus, if M_i represents the ith actual sample measurement in millimeters, then

$$x_i = M_i - T_A$$

would be the deviation from nominal. Panel (b) of Table 9-1 shows the deviations from nominal x_i, as well as the $\bar{x}$ and R values for each sample. Now consider the last six samples in Table 9-1. These hole diameters are from a different part number, B, for which the nominal dimension is $T_B = 25$ mm. Panel (b) of Table 9-1 presents the deviations from nominal and the averages and ranges of the deviations from nominal for the part B data.

The control charts for $\bar{x}$ and R using deviation from nominal are shown in Fig. 9-1. Note that control limits have been calculated using the data from all 10 samples. In practice, we would recommend waiting until approximately 20 samples are available before calculating control limits. However, for purposes of illustration we have calculated the limits based on 10 samples to show that, when using deviation from nominal as the variable on the chart, it is not necessary to have a long production run for each part number. It is also customary to use a dashed vertical line to separate different products or part numbers and to identify clearly which section of the chart pertains to each part number, as shown in Fig. 9-1.

Three important points should be made relative to the DNOM approach:

1. An assumption is that the process standard deviation is approximately the same for all parts. If this assumption is invalid, use a *standardized* $\bar{x}$ and R chart (see the next subsection).

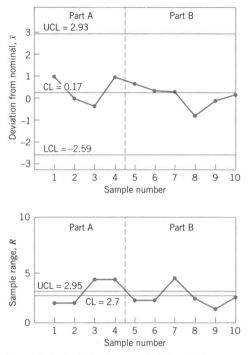

Figure 9-1   Deviation from nominal $\bar{x}$ and R charts.

2. This procedure works best when the sample size is constant for all part numbers.

3. Deviation from nominal control charts have intuitive appeal when the nominal specification is the desired target value for the process.

This last point deserves some additional discussion. In some situations the process should not (or cannot) be centered at the nominal dimension. For example, when the part has one-sided specifications, frequently a nominal dimension will not be specified. (For an example, see the data on bottle bursting strength in Chapter 7.) In cases where either no nominal is given or a nominal is not the desired process target, then in constructing the control chart, use the historical process average ($\bar{\bar{x}}$) instead of the nominal dimension. In some cases it will be necessary to compare the historical average to a desired process target to determine whether the true process mean is different from the target. Standard hypothesis testing procedures can be used to perform this task.

Standardized $\bar{x}$ and R Charts

If the process standard deviations are different for different part numbers, the deviation from nominal (or the deviation from process target) control charts described above will not work effectively. However, **standardized $\bar{x}$ and R charts** will handle this situation easily. Consider the jth part number. Let $\bar{R}_j$ and T_j be the average range and nominal value of x for this part number. Then for all the samples from this part number, plot

$$R_i^s = \frac{R_i}{\overline{R}_j} \tag{9-1}$$

on a standardized R chart with control limits at LCL $= D_3$ and UCL $= D_4$, and plot

$$\overline{x}_i^s = \frac{\overline{M}_i - T_j}{\overline{R}_j} \tag{9-2}$$

on a standardized $\overline{x}$ chart with control limits at LCL $= -A_2$ and UCL $= +A_2$. Note that the center line of the standardized $\overline{x}$ chart is zero because $\overline{M}_i$ is the average of the *original* measurements for subgroups of the jth part number. We point out that for this to be meaningful, there must be some logical justification for "pooling" parts on the same chart.

The target values $\overline{R}_j$ and T_j for each part number can be determined by using specifications for T_j and taking $\overline{R}_j$ from prior history (often in the form of a control chart, or by converting an estimate of σ into $\overline{R}_j$ by the relationship $\overline{R}_j \simeq Sd_2/c_4$). For new parts, it is a common practice to utilize prior experience on similar parts to set the targets.

Farnum (1992) has presented a generalized approach to the DNOM procedure that can incorporate a variety of assumptions about process variability. The standardized control chart approach discussed above is a special case of his method. His method would allow construction of DNOM charts in the case where the coefficient of variation (σ/μ) is approximately constant. This situation probably occurs fairly often in practice.

9-1.2   Attributes Control Charts for Short Production Runs

Dealing with attributes data in the short production run environment is extremely simple; the proper method is to use a standardized control chart for the attribute of interest. This method will allow different part numbers to be plotted on the same chart and will automatically compensate for variable sample size.

Standardized control charts for attributes have been discussed previously in Chapter 6. For convenience, the relevant formulas are presented in Table 9-2. All standardized attributes control charts have the center line at zero, and the upper and lower control limits are at $+3$ and -3, respectively.

Table 9-2   Standardized Attributes Control Charts Suitable for Short Production Runs

Attribute	Target Value	Standard Deviation	Statistic to Plot on the Control Chart
$\hat{p}_i$	$\bar{p}$	$\sqrt{\dfrac{\bar{p}(1-\bar{p})}{n}}$	$Z_i = \dfrac{\hat{p}_i - \bar{p}}{\sqrt{\bar{p}(1-\bar{p})/n}}$
$n\hat{p}_i$	$n\bar{p}$	$\sqrt{n\bar{p}(1-\bar{p})}$	$Z_i = \dfrac{n\hat{p}_i - n\bar{p}}{\sqrt{n\bar{p}(1-\bar{p})}}$
c_i	$\bar{c}$	$\sqrt{\bar{c}}$	$Z_i = \dfrac{c_i - \bar{c}}{\sqrt{\bar{c}}}$
u_i	$\bar{u}$	$\sqrt{\bar{u}/n}$	$Z_i = \dfrac{u_i - \bar{u}}{\sqrt{\bar{u}/n}}$

9-1.3   Other Methods

A variety of other approaches can be applied to the short-run production environment. For example, the cusum and EWMA control charts discussed in Chapter 8 have potential application to short production runs, because they have shorter average run-length performance than Shewhart-type charts, particularly in detecting small shifts. Since most production runs in the short-run environment will not, by definition, consist of many units, the rapid shift detection capability of those charts would be useful. Furthermore, cusum and EWMA control charts are very effective with subgroups of size one, another potential advantage in the short-run situation.

The number of subgroups used in calculating the trial control limits for Shewhart charts impacts the false alarm rate of the chart; in particular, when a small number of subgroups are used, the false alarm rate is inflated. Hillier (1969) studied this problem and presented a table of factors to use in setting limits for $\bar{x}$ and R charts based in a small number of subgroups for the case of $n = 5$ [see also Wang and Hillier (1970)]. Quesenberry (1993) has investigated a similar problem for both $\bar{x}$ and individuals control charts. Since control limits in the short-run environment will typically be calculated from a relatively small number of subgroups, these papers present techniques of some interest.

Quesenberry (1991a, b, c) has presented procedures for short-run SPC using a transformation that is different from the standardization approach discussed above. He refers to these as *Q-charts,* and notes that they can be used for both short or long production runs. Del Castillo and Montgomery (1994) have investigated the average run-length performance of the Q-chart for variables and show that in some cases the ARL performance is inadequate. They suggest some modifications to the Q-chart procedure and some alternate methods based on the EWMA and a related technique called the Kalman filter that have better ARL performance than the Q-chart. Crowder (1992) has also reported a short-run procedure based on the Kalman filter. In a subsequent series of papers,

Quesenberry (1995a, b, c, d) reports some refinements to the use of Q-charts that also enhance their performance in detecting process shifts. He also suggests that the probability that a shift is detected within a specified number of samples following its occurrence is a more appropriate measure of the performance of a short-run SPC procedure than its average run length. The interested reader should refer to the July and October 1995 issues of the *Journal of Quality Technology* that contain these papers and a discussion of Q-charts by several authorities. These papers and the discussion include a number of useful additional references.

9-2 MODIFIED AND ACCEPTANCE CONTROL CHARTS

In most situations in which control charts are used, the focus is on statistical monitoring or control of the process, reduction of variability, and continuous process improvement. When a high level of process capability has been achieved, it is sometimes useful to relax the level of surveillance provided by the standard control chart. One method for doing this with $\bar{x}$ charts uses **modified** (or **reject**) **control limits,** and the second uses the **acceptance control chart.**

9-2.1 Modified Control Limits for the $\bar{x}$ Chart

Modified control limits are generally used in situations where the natural variability or "spread" of the process is considerably smaller than the spread in the specification limits; that is, C_p or C_{pk} is much greater than 1. This situation occurs occasionally in practice. In fact, this should be the natural eventual result of a successful variability reduction effort—reduction of process variability with a corresponding increase in the process capability ratio. The "six-sigma" variability reduction program developed by Motorola focuses on improving processes until the minimum value of C_{pk} is 2.0.

Suppose, for example, that the specification limits on the fill volume of a carbonated beverage container are LSL = 10.00 oz and USL = 10.20 oz, but as the result of a program of engineering and operating refinements, the filling machine can operate with a standard deviation of fill volume of approximately $\sigma = 0.01$ oz. Therefore, the distance USL − LSL is approximately 20-sigma, or much greater than the 6-sigma natural tolerance limits on the process, and the process capability ratio is PCR = (USL − LSL)/6σ = $0.20/[6(0.01)] = 3.33$. This is clearly a "six-sigma" process, by the Motorola definition.

In situations where six-sigma is much smaller than the spread in the specifications (USL − LSL), the process mean can sometimes be allowed to vary over an interval without appreciably affecting the overall performance of the process.[1] For example, see Fig. 9-2.

---

[1] In the original Motorola definition of a "six-sigma" process, the process mean was assumed to drift about, wandering as far as 1.5σ from the desired target. If this is the actual behavior of the process and this type of behavior is acceptable, then modified control charts are a useful alternative to standard charts. There are also many cases where the mean should not be allowed to vary even if C_p or C_{pk} is large. The conventional control charts should be used in such situations.

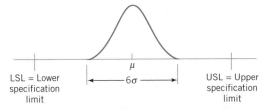

Figure 9-2 A process with the spread of the natural tolerance limits less than the spread of the specification limits, or $6\sigma < \text{USL} - \text{LSL}$.

When this situation occurs, we can use a **modified control chart** for $\bar{x}$ instead of the usual $\bar{x}$ chart. The modified $\bar{x}$ control chart is concerned only with detecting whether the true process mean μ is located such that the process is producing a fraction nonconforming in excess of some specified value δ. In effect, μ is allowed to vary over an interval—say, $\mu_L \le \mu \le \mu_U$—where μ_L and μ_U are chosen as the smallest and largest permissible values of μ, respectively, consistent with producing a fraction nonconforming of at most δ. We will assume that the process variability σ is in control. Good general discussions of the modified control chart are in Hill (1956) and Duncan (1986). As noted in Duncan (1986), the procedure is sometimes used when a process is subject to tool wear (see Section 9-7.1).

To specify the control limits for a modified $\bar{x}$ chart, we will assume that the process output is normally distributed. For the process fraction nonconforming to be less than δ, we must require that the true process mean is in the interval $\mu_L \le \mu \le \mu_U$. Consequently, we see from Fig. 9-3a that we must have

$$\mu_L = \text{LSL} + Z_\delta \sigma$$

and

$$\mu_U = \text{USL} - Z_\delta \sigma$$

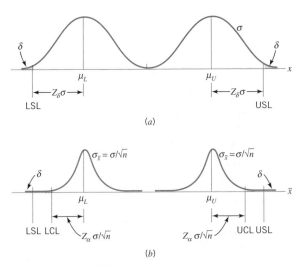

(a)

(b)

Figure 9-3 Control limits on a modified control chart. (*a*) Distribution of process output. (*b*) Distribution of the sample mean $\bar{x}$.

where Z_δ is the upper $100(1 - \delta)$ percentage point of the standard normal distribution. Now if we specify a type I error of α, the upper and lower control limits are

$$\text{UCL} = \mu_U + \frac{Z_\alpha \sigma}{\sqrt{n}}$$

$$= \text{USL} - Z_\delta \sigma + \frac{Z_\alpha \sigma}{\sqrt{n}}$$

$$= \text{USL} - \left(Z_\delta - \frac{Z_\alpha}{\sqrt{n}} \right) \sigma \qquad (9\text{-}3a)$$

and

$$\text{LCL} = \mu_L - \frac{Z_\alpha \sigma}{\sqrt{n}}$$

$$= \text{LSL} + Z_\delta \sigma - \frac{Z_\alpha \sigma}{\sqrt{n}}$$

$$= \text{LSL} + \left(Z_\delta - \frac{Z_\alpha}{\sqrt{n}} \right) \sigma \qquad (9\text{-}3b)$$

respectively. The control limits are shown on the distribution of $\bar{x}$ in Fig. 9-3b. Instead of specifying a type I error, one may use the following:

$$\text{UCL} = \text{USL} - \left(Z_\delta - \frac{3}{\sqrt{n}} \right) \sigma \qquad (9\text{-}4a)$$

and

$$\text{LCL} = \text{LSL} + \left(Z_\delta - \frac{3}{\sqrt{n}} \right) \sigma \qquad (9\text{-}4b)$$

Some quality engineers recommend the use of two-sigma limits on the modified control chart, arguing that the tighter control limits afford better protection (smaller β-risk) against critical shifts in the mean at little loss in the α-risk. A discussion of this subject is in Freund (1957). Note that the modified control chart is equivalent to testing the hypothesis that the process mean lies in the interval $\mu_L \leq \mu \leq \mu_U$.

To design a modified control chart, we must have a good estimate of σ available. If the process variability shifts, then the modified control limits are not appropriate. Consequently, an R or an S chart should always be used in conjunction with the modified control chart. Furthermore, the initial estimate of σ required to set up the modified control chart would usually be obtained from an R or an S chart.

······· **EXAMPLE 9-1** ···

A Control Chart for a "Six-Sigma" Process

Consider a normally distributed process with a target value of the mean of 20 and standard deviation $\sigma = 2$. The upper and lower process specifications are at LSL = 8 and USL = 32, so that if the process is centered at the target, $C_p = C_{pk} = 2.0$. This is a process with "six-sigma" capability. In a six-sigma process it is assumed that the mean may drift as much as 1.5 standard deviations off target without causing serious problems. Suppose that we want to set up a control chart for monitoring the mean of this six-sigma process with a sample size of $n = 4$.

Figure 9-4a shows this six-sigma process and Fig. 9-4b illustrates the control limit calculation. Notice that we can use equation 9-4 with Z_δ replaced by 4.5. Therefore, the upper and lower control limits become

$$
\begin{aligned}
\text{UCL} &= \text{USL} - \left(4.5 - \frac{3}{\sqrt{2}}\right)\sigma \\
&= 32 - (4.5 - 1.5)2 \\
&= 26
\end{aligned}
$$

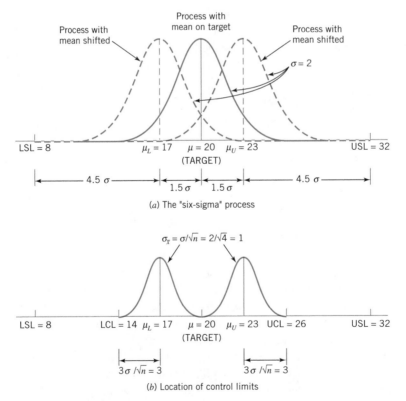

(a) The "six-sigma" process

(b) Location of control limits

Figure 9-4   Control limits for a six-sigma process.

and

$$\text{LCL} = \text{LSL} + \left(4.5 - \frac{3}{\sqrt{2}}\right)\sigma$$

$$= 8 + (4.5 - 1.5)2$$

$$= 14$$

9-2.2 Acceptance Control Charts

The second approach to using an $\bar{x}$ chart to monitor the fraction of nonconforming units, or the fraction of units exceeding specifications, is called the acceptance control chart. In the modified control chart design of Section 9-2.1, the chart was based on a specified sample size n, a process fraction nonconforming δ, and type I error probability α. Thus, we interpret δ as a process fraction nonconforming that we will accept with probability $1 - \alpha$. Freund (1957) developed the **acceptance control chart** to take into account both the risk of rejecting a process operating at a satisfactory level (type I error or α-risk) and the risk of accepting a process that is operating at an unsatisfactory level (type II error or β-risk).

There are two ways to design the control chart. In the first approach, we design the control chart based on a specified n and a process fraction nonconforming γ that we would like to reject with probability $1 - \beta$. In this case, the control limits for the chart are

$$\text{UCL} = \mu_U - \frac{Z_\beta \sigma}{\sqrt{n}}$$

$$= \text{USL} - Z_\gamma \sigma - \frac{Z_\beta \sigma}{\sqrt{n}}$$

$$= \text{USL} - \left(Z_\gamma + \frac{Z_\beta}{\sqrt{n}}\right)\sigma \qquad (9\text{-}5a)$$

and

$$\text{LCL} = \mu_L + \frac{Z_\beta \sigma}{\sqrt{n}}$$

$$= \text{LSL} + Z_\gamma \sigma + \frac{Z_\beta \sigma}{\sqrt{n}}$$

$$= \text{LSL} + \left(Z_\gamma + \frac{Z_\beta}{\sqrt{n}}\right)\sigma \qquad (9\text{-}5b)$$

Note that when n, γ, and $1 - \beta$ (or β) are specified, the control limits are inside the μ_L and μ_U values that produce the fraction nonconforming γ. In contrast, when n, δ, and α are specified, the lower control limit falls between μ_L and LSL and the upper control limit falls between μ_U and USL.

It is also possible to choose a sample size for an acceptance control chart so that specified values of δ, α, γ, and β are obtained. By equating the upper control limits (say) for a specified δ and α (equation 9-3a) and a specified γ and β (equation 9-5a), we obtain

$$\text{USL} - \left(Z_\delta - \frac{Z_\alpha}{\sqrt{n}} \right) \sigma = \text{USL} - \left(Z_\gamma + \frac{Z_\beta}{\sqrt{n}} \right) \sigma$$

Therefore, a sample size of

$$n = \left(\frac{Z_\alpha + Z_\beta}{Z_\delta - Z_\gamma} \right)^2$$

will yield the required values of δ, α, γ, and β. For example, if $\delta = 0.01$, $\alpha = 0.00135$, $\gamma = 0.05$, and $\beta = 0.20$, we must use a sample of size

$$n = \left(\frac{Z_{0.00135} + Z_{0.20}}{Z_{0.01} - Z_{0.05}} \right)^2 = \left(\frac{3.00 + 0.84}{2.33 - 1.645} \right)^2 = 31.43 \approx 32$$

on the acceptance control chart. Obviously, to use this approach, n must not be seriously restricted by cost or other factors such as rational subgrouping considerations.

9-3   CONTROL CHARTS FOR MULTIPLE-STREAM PROCESSES

9-3.1   Multiple-Stream Processes

A **multiple-stream process (MSP)** is a process with data at a point in time consisting of measurements from several individual sources or streams. When the process is in control, the sources or streams are assumed to be identical. Another characteristic of the MSP is that we can monitor and adjust each of the streams individually or in small groups.

The MSP occurs often in practice. For example, a machine may have several heads, with each head producing (we hope) identical units of product. In such situations several possible control procedures may be followed. One possibility is to use separate control charts on each stream. This approach usually results in a prohibitively large number of control charts. If the output streams are highly correlated—say, nearly perfectly correlated—then control charts on only one stream may be adequate. The most common situation is that the streams are only moderately correlated, so monitoring only one of the streams is not appropriate. There are at least two types of situations involving the occurrence of assignable causes in the MSP.

1. The output of one stream (or a few streams) has shifted off target.
2. The output of all streams has shifted off target.

In the first case, we are trying to detect an assignable cause that affects only one stream (or at most a few streams), whereas in the second, we are looking for an assignable cause that impacts *all* streams (such as would result from a change in raw materials).

The standard control chart for the MSP is the group control chart, introduced by Boyd (1950). We will also discuss other approaches to monitoring the MSP. Throughout this section we assume that the process has s streams and that the output quality characteristic from each stream is normally distributed.

9-3.2 Group Control Charts

The group control chart (GCC) was introduced by Boyd (1950) and remains the basic procedure for monitoring a MSP. To illustrate the methods of construction and use, suppose that the process has $s = 6$ streams and that each stream has the same target value and inherent variability. Variables measurement is made on the items produced, and the distribution of the measurement is well approximated by the normal. To establish a group control chart, the sampling is performed as if separate control charts were to be set up on each stream. Suppose, for purposes of illustration, that a sample size of $n = 4$ is used. This means that 4 units will be taken from each of the $s = 6$ streams over a short period of time. This will be repeated until about 20 such groups of samples have been taken. At this point we would have $20 \times 6 = 120$ averages of $n = 4$ observations each and 120 corresponding ranges. These averages and ranges would be averaged to produce a grand average $\bar{\bar{x}}$ and an average range $\bar{R}$. The limits on the group control charts would be at

$$\text{UCL} = \bar{\bar{x}} + A_2 \bar{R}$$
$$\text{LCL} = \bar{\bar{x}} - A_2 \bar{R}$$

for the $\bar{x}$ chart and at

$$\text{UCL} = D_4 \bar{R}$$
$$\text{LCL} = D_3 \bar{R}$$

for the R chart, with $A_2 = 0.729$, $D_3 = 0$, and $D_4 = 2.282$. Note that the sample size $n = 4$ determines the control chart constants.

When the group control chart is used to control the process, we plot only the largest and smallest of the $s = 6$ means observed at any time period on the $\bar{x}$ chart. If these means are inside the control limits, then all other means will also lie inside the limits. Similarly, only the largest range will be plotted on the range chart. Each plotted point is identified on the chart by the number of the stream that produced it. The process is out of control if a point exceeds a three-sigma limit. Runs tests cannot be applied to these charts, because the conventional runs tests were not developed to test averages or ranges that are the extremes of a group of averages or ranges.

It is useful to examine the stream numbers on the chart. In general, if a stream consistently gives the largest (or smallest) value several times in a row, that may constitute evidence that this stream is different from the others. If the process has s streams and if r is the number of consecutive times that a particular stream is the largest (or smallest) value, then the one-sided in-control average run length for this event is given by Nelson (1986) as

$$ARL(1)_0 = \frac{s^r - 1}{s - 1} \tag{9-6}$$

if all streams are identical. To illustrate the use of this equation, if $s = 6$ and $r = 4$, then

$$ARL(1)_0 = \frac{6^4 - 1}{6 - 1} = 259$$

That is, if the process is in control, we will expect to see the same stream producing an extreme value on the group control chart four times in a row only once every 259 samples.

One way to select the value of r to detect the presence of one stream that is different from the others is to use equation 9-6 to find an ARL that is roughly consistent with the ARL of a conventional control chart. The ARL for an in-control process for a single point beyond the upper control limit (say) is 740. Thus, using $r = 4$ for a six-stream process results in an ARL that is too short and that will give too many false alarms. A better choice is $r = 5$, since

$$ARL(1)_0 = \frac{6^5 - 1}{6 - 1} = 1555$$

Thus, if we have six streams and if the same stream produces an extreme value on the control chart in five consecutive samples, then we have strong evidence that this stream is different from the others.

Using equation 9-6, we can generate some general guidelines for choosing r given the number of streams s. Suitable pairs (s, r) would include (3, 7), (4, 6), (5–6, 5), and (7–10, 4). All of these combinations would give reasonable values of the one-sided ARL when the process is in control.

The two-sided in-control average run length $ARL(2)_0$ is defined as the expected number of trials until r consecutive largest or r consecutive smallest means come from the same stream while the MSP is in control. Mortell and Runger (1995) and Nelson and Stephenson (1996) used the Markov chain approach of Brook and Evans (1972) to compute $ARL(2)_0$ numerically. A closed form expression is not possible for $ARL(2)_0$, but Nelson and Stephenson (1996) give a lower bound on $ARL(2)_0$ as

$$ARL(2)_0 = \frac{s^r - 1}{2(r - 1)} \tag{9-7}$$

This approximation agrees closely with the numerical computations from the Markov chain approach.

There are some drawbacks connected with the GCC. These may be summarized as follows:

1. At each point in time where samples are taken, all s streams must be sampled. When s is small, this may be a satisfactory approach. However, some high-speed filling and packaging machines may have $100 \leq s \leq 200$ streams. The GCC is not practical for these situations.

2. There is no information about nonextreme streams at each trial, thus, we cannot utilize past values to form an EWMA or a cusum statistic to improve on the GCC performance.

3. The supplementary run tests fail if more than one stream shifts simultaneously (all up or all down) to the same off-target level, because the shifted streams will most likely alternate having the extreme value, and no single stream shift can be identified.

4. The selection of the run or pattern length r for a specific problem with s streams is a discrete process, and the difference in ARL_0 for $r = r*$ or $r = r* + 1$ choice can be substantial, as observed by Mortell and Runger (1995). This problem is evident in the original run rule devised by Nelson (1986), but can be alleviated somewhat by using the modified run rules in Nelson and Stephenson (1996), which are based on an $r - 1$ out of r in a row approach. In fact, it can be easily shown that the difference in the ARL_0 between $r = r*$ and $r = r* + 1$ choices equals s^{r*}, which becomes very large for the problems with large numbers of streams.

5. Although using an $r - 1$ out of r in a row run rule allows some flexibility in the choice of the critical value $r = r*$ and ARL_0, it adds some complexity since it is no longer a simple counting procedure.

9-3.3 Other Approaches

Mortell and Runger (1995) proposed the following model for the MSP to accommodate the practical case of dependent (cross-correlated) streams:

$$x_{tjk} = \mu + A_t + \epsilon_{tjk} \tag{9-8}$$

where x_{tjk} is the kth measurement on the jth stream at time t, $A_t \sim N(0, \sigma_a^2)$ represents the difference of the mean over all the streams at time t from the MSP target μ, and ϵ_{tjk} is a $N(0, \sigma^2)$ random variable that represents the difference of the kth measurements on the jth stream from the process mean over all streams at time t. In this representation of MSP data, the total variation is allocated into two sources: σ_a^2 accounting for the variation over time common to all streams, and σ^2 accounting for the variation between the streams at specific time t. From this model, the cross correlation between any pair of streams is given by $\rho = \sigma_a^2/(\sigma_a^2 + \sigma^2)$.

Mortell and Runger proposed monitoring the average at time t of the means across all streams with an individuals control chart to detect an overall assignable cause. Also, they proposed monitoring the range of the stream's means at time t denoted by $R_t = \max(\bar{x}_{tj}) - \min(\bar{x}_{tj})$ or the maximum **residual** at time t given by $\max(\bar{x}_{tj}) - \bar{x}_t$, with any individuals control chart to detect a relative assignable cause affecting one or a few streams. These proposed control charts are better than the GCC, especially when the variation in the process means over time is greater than the between-stream variability (i.e., when $\sigma_a > \sigma$). Mortell and Runger compared the performance of these proposed charts with the runs test of the GCC by Nelson (1986) using simulations of various MSPs and concluded that the cusum chart on the range outperforms the Shewhart and EWMA on either the range or the maximum residual. On the other hand, little difference in performance exists between the runs test of Nelson (1986) and the proposed cusum chart, but it is preferred because of the flexibility in the choice of in-control ARL.

Runger, Alt, and Montgomery (1996) related control charts for the MSP with correlated data to principal components analysis. (See Chapter 10 for information on principal components.) The first control chart monitors the major principal component variable, which equals the mean of the streams at any time t. The second control chart is based on the Hotelling T^2 statistic (see Chapter 10) of the last $p - 1$ principal components, which is shown to be equivalent to the S^2 chart; thus it is sensitive to relative assignable causes that affect the uniformity across the streams.

Lanning (1998) studied the use of "fractional" and "adaptive" sampling in the multiple-stream process with a large number of identical and independent streams, where simultaneous monitoring of all streams is impractical. In his procedures, only a fraction of the s streams are sampled at each time period, and this fraction is increased when there are indications that an out-of-control condition may be present. (Adaptive sampling schemes are introduced in Section 9-5.) He used the average time to signal (ATS) to compare performance of the competing charts, and focused on the scenarios where assignable causes tend to impact all or most of the process streams together. He concluded that the adaptive fraction approach gives better ATS results than the fixed fraction scheme, and often yields satisfactory results compared to a complete sampling situation.

9-4   SPC WITH AUTOCORRELATED PROCESS DATA

9-4.1   Sources and Effects of Autocorrelation in Process Data

The standard assumptions that are usually cited in justifying the use of control charts are that the data generated by the process when it is in control are normally and independently distributed with mean μ and standard deviation σ. Both μ and σ are considered fixed and unknown. An out-of-control condition is a change or shift in μ or σ (or both) to some different value. Therefore, we could say that when the process is in control the quality characteristic at time t, x_t, is represented by the model

$$x_t = \mu + \epsilon_t \qquad t = 1, 2, \ldots \qquad (9\text{-}9)$$

where ϵ_t is normally and independently distributed with mean zero and standard deviation σ. This is often called the **Shewhart model** of the process.

When these assumptions are satisfied, one may apply conventional control charts and draw conclusions about the state of statistical control of the process. Furthermore, the statistical properties of the control chart, such as the false-alarm rate with three-sigma control limits, or the average run length, can be easily determined and used to provide guidance for chart interpretation. Even in situations in which the normality assumption is violated to a slight or moderate degree, these control charts will still work reasonably well.

The most important of the assumptions made concerning control charts is that of independence of the observations, for conventional control charts do not work well if the quality characteristic exhibits even low levels of correlation over time. Specifically, these control charts will give misleading results in the form of too many false alarms if the data are correlated. This point has been made by numerous authors, including Berthouex, Hunter, and Pallesen (1978), Alwan and Roberts (1988), Montgomery and Friedman (1989), Alway (1992), Harris and Ross (1991), Montgomery and Mastrangelo (1991), and Maragah and Woodall (1992).

Unfortunately, the assumption of uncorrelated or independent observations is not even approximately satisfied in some manufacturing processes. Examples include chemical processes where consecutive measurements on process or product characteristics are often highly correlated, or automated test and inspection procedures, where every quality characteristic is measured on every unit in time order of production. An example of correlated process data was given in Fig. 4-6b. Basically, all manufacturing processes are driven by inertial elements, and when the interval between samples becomes small relative to these forces, the observations on the process will be correlated over time.

It is easy to give an analytical demonstration of this phenomena. Figure 9-5 shows a simple system consisting of a tank with volume V, with the input and output material streams having flow rate f. Let w_t be the concentration of a certain material in the input stream at time t and x_t be the corresponding concentration in the output stream at time t. Assuming homogeneity within the tank, the relationship between x_t and w_t is

$$x_t = w_t - T\frac{dx_t}{dt}$$

where $T = V/f$ is often called the **time constant** of the system. If the input stream experiences a step change of w_0 at time $t = 0$ (say), then the output concentration at time t is

$$x_t = w_0(1 - e^{-t/T})$$

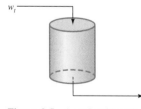

Figure 9-5   A tank with volume V and input and output material streams.

Now in practice, we do not observe x_t continuously, but only at small, equally spaced intervals of time Δt. In this case,

$$x_t = x_{t-1} + (w_t - x_{t-1})(1 - e^{-\Delta t/T})$$
$$= aw_t + (1 - a)x_{t-1}$$

where $a = 1 - e^{-\Delta t/T}$.

The properties of the output stream concentration x_t depend on those of the input stream concentration w_t and the sampling interval Δt. If we assume that the w_t are uncorrelated random variables, then the correlation between successive values of x_t (or the autocorrelation between x_t and x_{t-1}) is given by

$$\rho = 1 - a = e^{-\Delta t/T}$$

Note that if Δt is much greater than T, $\rho \approx 0$. That is, if the interval between samples Δt in the output stream is long—much longer than the time constant T—the observations on output concentration will be uncorrelated. However, if $\Delta t \leq T$, this will not be the case. For example, if

$\Delta t/T = 1$	$\rho = 0.37$
$\Delta t/T = 0.5$	$\rho = 0.61$
$\Delta t/T = 0.25$	$\rho = 0.78$
$\Delta t/T = 0.10$	$\rho = 0.90$

Clearly if we sample at least once per time constant there will be significant autocorrelation present in the observations. For instance, sampling four times per time constant ($\Delta t/T = 0.25$) results in autocorrelation between x_t and x_{t-1} of $\rho = 0.78$. Autocorrelation between successive observations as small as 0.25 can cause a substantial increase in the false alarm rate of a control chart, so clearly this is an important issue to consider in control chart implementation.

We can also give an empirical demonstration of this phenomena. Figure 9-6 is a plot of 1000 observations on a process quality characteristic x_t. Close examination of this plot will reveal that the behavior of the variable is nonrandom in the sense that a value of x_t that is above the long-term average (about 66) tends to be followed by other values above the average, whereas a value below the average tends to be followed by other similar values. This is also reflected in Fig. 9-7, a scatter plot of x_t (the observation at time t) versus x_{t-1} (the observation one period earlier). Note that the observations cluster around a straight line with a positive slope. That is, a relatively low observation on x at time $t - 1$ tends to be followed by another low value at time t, whereas a relatively large observation at time $t - 1$ tends to be followed by another large value at time t. This type of behavior is indicative of **positive autocorrelation** in the observations.

It is also possible to measure the level of autocorrelation analytically. The autocorrelation over a series of time-oriented observations (called a **time series**) is measured by the **autocorrelation function**

$$\rho_k = \frac{\text{Cov}(x_t, x_{t-k})}{V(x_t)} \qquad k = 0, 1, \ldots$$

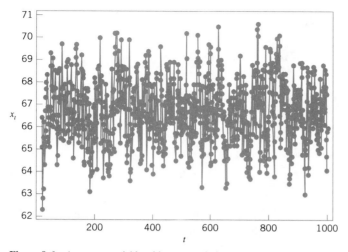

Figure 9-6 A process variable with autocorrelation.

where $\text{Cov}(x_t, x_{t-k})$ is the covariance of observations that are k time periods apart, and we have assumed that the observations have constant variance given by $V(x_t)$. We usually estimate the values of ρ_k with the **sample autocorrelation function:**

$$r_k = \frac{\displaystyle\sum_{t=1}^{n-k}(x_t - \bar{x})(x_{t-k} - \bar{x})}{\displaystyle\sum_{t=1}^{n}(x_t - \bar{x})^2} \qquad k = 0, 1, \ldots, K \qquad (9\text{-}10)$$

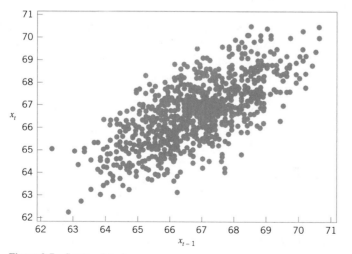

Figure 9-7 Scatter plot of x_t versus x_{t-1}.

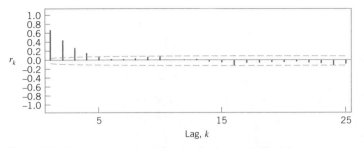

Figure 9-8 Sample autocorrelation function for the data in Fig. 9-6.

As a general rule, we usually need to compute values of r_k for a few values of k, $k \leq n/4$. Many software programs for statistical data analysis can perform these calculations.

The sample autocorrelation function for the data in Fig. 9-6 is shown in Fig. 9-8. This sample autocorrelation function was constructed using Minitab. The dashed lines on the graph are two standard deviation limits on the autocorrelation parameter ρ_k at lag k. They are useful in detecting nonzero autocorrelations; in effect, if a sample autocorrelation exceeds its two standard deviation limit, the corresponding autocorrelation parameter ρ_k is likely nonzero. Note that the autocorrelation at lag 1 is $r_1 \simeq 0.7$. This is certainly large enough to severely distort control chart performance.

Figure 9-9 presents control charts for the data in Fig. 9-6, constructed by Minitab. Note that both the individuals chart and the EWMA exhibit many out-of-control points.

Based on our previous discussion, we know that less frequent sampling can break up the autocorrelation in process data. To illustrate, consider Fig. 9-10a, which is a plot of every 10th observation from Fig. 9-6. The sample autocorrelation function, shown in Fig. 9-10b, indicates that there is very little autocorrelation at low lag. The control charts in Fig. 9-11 now indicate that the process is essentially stable. Clearly, then, one approach to dealing with autocorrelation is simply to sample from the process data stream less frequently. Although this seems to be an easy solution, on reconsideration it has some disadvantages. For example, we are making very inefficient use of the available data. Literally, in the above example, we are discarding 90% of the information! Also, since we are only using every 10th observation, it may take much longer to detect a real process shift than if we used all of the data.

Clearly, a better approach is needed. In the next two sections we present several approaches to monitoring autocorrelated process data.

9-4.2 Model-Based Approaches

Time Series Models
An approach that has proved useful in dealing with autocorrelated data is to directly model the correlative structure with an appropriate time series model, use that model to

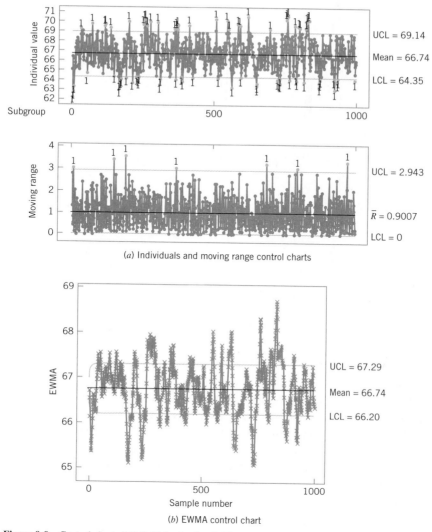

Figure 9-9 Control charts (Minitab) for the data in Fig. 9-6.

remove the autocorrelation from the data, and apply control charts to the **residuals.** For example, suppose that we could model the quality characteristic x_t as

$$x_t = \xi + \phi x_{t-1} + \epsilon_t \qquad (9\text{-}11)$$

where ξ and ϕ $(-1 < \phi < 1)$ are unknown constants, and ϵ_t is normally and independently distributed with mean zero and standard deviation σ. Note how intuitive this

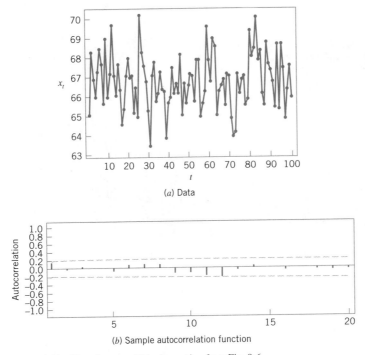

(a) Data

(b) Sample autocorrelation function

Figure 9-10   Plots for every 10th observation from Fig. 9-6.

model is from examining Figs. 9-6, 9-7, and 9-8. Equation 9-11 is called a **first-order autoregressive model;** the observations x_t from such a model have mean $\xi/(1 - \phi)$, standard deviation $\sigma/(1 - \phi^2)^{1/2}$, and the observations that are k periods apart (x_t and x_{t-k}) have correlation coefficient ϕ^k. That is, the autocorrelation function should decay exponentially just as the sample autocorrelation function did in Fig. 9-8. Suppose that $\hat{\phi}$ is an estimate of ϕ, obtained from analysis of sample data from the process, and $\hat{x}_t$ is the fitted value of x_t. Then the residuals

$$e_t = x_t - \hat{x}_t$$

are approximately normally and independently distributed with mean zero and constant variance. Conventional control charts could now be applied to the sequence of residuals. Points out of control or unusual patterns on such charts would indicate that the parameter ϕ had changed, implying that the original variable x_t was out of control. For details of identifying and fitting time series models such as this one, see Montgomery, Johnson, and Gardiner (1990) and Box, Jenkins, and Reinsel (1994).

······   **EXAMPLE 9-2**   ···
Figure 9-12 presents a control chart for individual measurements applied to viscosity measurements from a chemical process taken every hour. Note that many points are outside the control limits on this chart. Because of the nature of the production process, and the visual appearance of the viscosity measurements in Fig. 9-12, which

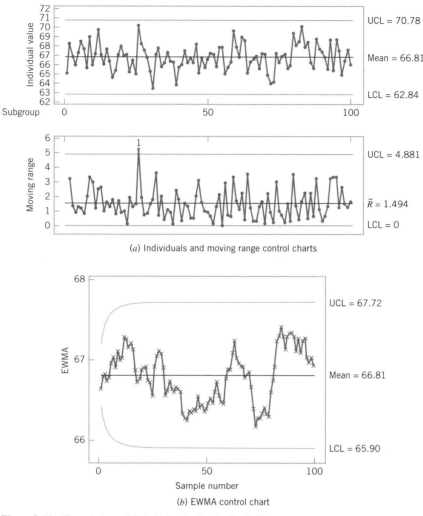

Figure 9-11   Control charts (Minitab) for the data in Fig. 9-10a.

appear to "drift" or "wander" slowly over time, we would probably suspect that viscosity is autocorrelated.

The sample autocorrelation function for the viscosity data is shown in Fig. 9-13. Note that there is a strong positive correlation at lag 1; that is, observations that are one period apart are positively correlated with $r_1 = 0.88$. This level of autocorrelation is sufficiently high to distort greatly the performance of a Shewhart control chart. In particular, because we know that positive correlation greatly increases the frequence of false alarms, we should be very suspicious about the out-of-control signals on the control chart in Fig. 9-12. Based on the behavior of the sample autocorrelation function, it seems logical to use the first-order autoregressive model to describe this process.

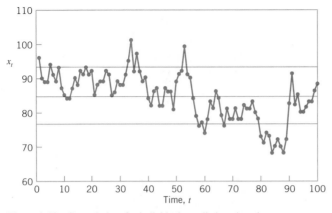

Figure 9-12   Control chart for individuals applied to viscosity.

The parameters in the autoregressive model equation 9-11 may be estimated by the method of least squares; that is, by choosing the values of ξ and ϕ that minimize the sum of squared errors ϵ_t. Many statistics software packages have routines for fitting these time series models. The fitted value of this model for the viscosity data is

$$x_t = 13.04 + 0.847 x_{t-1}$$

We may think of this as an alternative to the Shewhart model for this process.

Figure 9-14 is an individuals control chart of the residuals from the fitted first-order autoregressive model. Note that now no points are outside the control limits, although around period 90 two points are near the upper control limit. In contrast to the control chart on the individual measurements on Fig. 9-12, we would conclude that this process is in a reasonable state of statistical control.

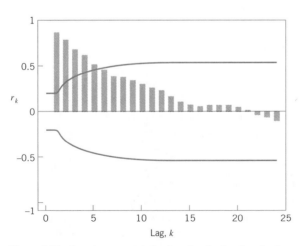

Figure 9-13   Sample autocorrelation function for the viscosity data.

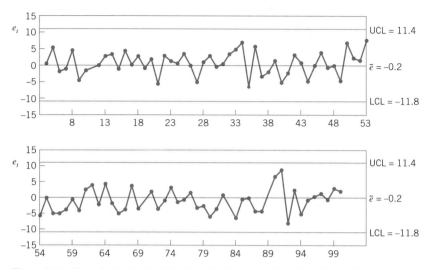

Figure 9-14 Control chart for individuals applied to the residuals from the model $x_t = 13.04 + 0.847x_{t-1}$.

Other Time Series Models

The first-order autoregressive model used in the viscosity example (equation 9-11) is not the only possible model for time-oriented data that exhibits correlative structure. An obvious extension to equation 9-11 is

$$x_t = \xi + \phi_1 x_{t-1} + \phi_2 x_{t-2} + \epsilon_t \tag{9-12}$$

which is a **second-order autoregressive model.** In general, in autoregressive-type models, the variable x_t is directly dependent on previous observations x_{t-1}, x_{t-2}, and so forth. Another possibility is to model the dependency through the random component ϵ_t. A simple way to do this is

$$x_t = \mu + \epsilon_t - \theta\epsilon_{t-1} \tag{9-13}$$

This is called a **first-order moving average model.** In this model, the correlation between x_t and x_{t-1} is $\rho_1 = -\theta/(1 + \theta^2)$ and is zero at all other lags. Thus, the correlative structure in x_t only extends backwards one time period.

Sometimes combinations of autoregressive and moving average terms are useful. A **first-order mixed model** is

$$x_t = \xi + \phi x_{t-1} + \epsilon_t - \theta\epsilon_{t-1} \tag{9-14}$$

This model often occurs in the chemical and process industries. The reason is that if the underlying process variable x_t is first-order autoregressive and a random error component is added to x_t, the result is the mixed model in equation 9-14. In the chemical and process industries first-order autoregressive process behavior is fairly common. Furthermore, the quality characteristic is often measured in a laboratory (or by an on-line instrument) that has measurement error, which we can usually think of as random or uncorrelated. The reported or observed measurement then consists of an autoregressive component plus random variation, so the mixed model in equation 9-14 is required as the process model.

We also encounter the **first-order integrated moving average** model

$$x_t = x_{t-1} + \epsilon_t - \theta\epsilon_{t-1} \tag{9-15}$$

in some applications. Whereas the previous models are used to describe stationary behavior (that is, x_t wanders around a "fixed" mean), the model in equation 9-15 describes nonstationary behavior (the variable x_t "drifts" as if there is no fixed value of the process mean). This model often arises in chemical and process plants when x_t is an "uncontrolled" process output; that is, when no control actions are taken to keep the variable close to a target value.

The models we have been discussing in equations 9-11 through 9-15 are members of a class of time series models called **autoregressive integrated moving average (ARIMA)** models. Montgomery, Johnson, and Gardiner (1990) and Box, Jenkins, and Reinsel (1994) discuss these models in detail. Although these models appear very different than the Shewhart model (equation 9-9), they are actually relatively similar and include the Shewhart model as a special case. Note that if we let $\phi = 0$ in equation 9-11, the Shewhart model results. Similarly, if we let $\theta = 0$ in equation 9-13, the Shewhart model results.

An Approximate EWMA Procedure for Correlated Data

The time series modeling approach illustrated in the viscosity example can be awkward in practice. Typically, we apply control charts to several process variables and developing an explicit time series model for each variable of interest is potentially time consuming. Some authors have developed automatic time series model building to partially alleviate this difficulty. [See Yourstone and Montgomery (1989) and the references therein.] However, unless the time series model itself is of intrinsic value in explaining process dynamics (as it sometimes is), this approach will frequently require more effort than may be justified in practice.

Montgomery and Mastrangelo (1991) have suggested an approximate procedure based on the EWMA. They utilize the fact that the EWMA can be used in certain situations where the data are autocorrelated. Suppose that the process can be modeled by the integrated moving average model in equation 9-15. It can be easily shown that the EWMA with $\lambda = 1 - \theta$ is the optimal one-step-ahead forecast for this process. That is, if $\hat{x}_{t+1}(t)$ is the forecast for the observation in period $t + 1$ made at the end of period t, then

$$\hat{x}_{t+1}(t) = z_t$$

where $z_t = \lambda x_t + (1 - \lambda)z_{t-1}$ is the EWMA. The sequence of one-step-ahead prediction errors

$$e_t = x_t - \hat{x}_t(t-1) \tag{9-16}$$

is independently and identically distributed with mean zero. Therefore, control charts could be applied to these one-step-ahead prediction errors. The parameter λ (or equivalently, θ) would be found by minimizing the sum of squares of the errors e_t.

Now suppose that the process is not modeled exactly by equation 9-15. In general, if the observations from the process are positively autocorrelated and the process mean does not drift too quickly, the EWMA with an appropriate value for λ will provide an excellent one-step-ahead predictor. The forecasting and time-series analysis field has used this result for many years; for examples, see Montgomery, Johnson, and Gardiner (1990). Consequently, we would expect many processes that obey first-order dynamics (that is, they follow a slow "drift") to be well represented by the EWMA.

Consequently, under the conditions just described, we may use the EWMA as the basis of a statistical process monitoring procedure that is an approximation of the exact time-series model approach. The procedure would consist of plotting one-step-ahead EWMA prediction errors (or model residuals) on a control chart. This chart could be accompanied by a run chart of the original observations on which the EWMA forecast is superimposed. Our experience indicates that both charts are usually necessary, as operational personnel feel that the control chart of residuals sometimes does not provide a direct frame of reference to the process. The run chart of original observations allows process dynamics to be visualized.

Figure 9-15 is a graph of the sum of squares of the EWMA prediction errors versus λ for the viscosity data. The minimum squared prediction error occurs at $\lambda = 0.825$. Figure 9-16 presents a control chart for individuals applied to the EWMA prediction errors. This chart is slightly different from the control chart of the exact autoregressive model residuals shown in Fig. 9-14, but not significantly so. Both indicate a process that is reasonably stable, with a period around $t = 90$ where an assignable cause may be present.

Montgomery and Mastrangelo (1991) point out that it is possible to combine information about the state of statistical control and process dynamics on a single control chart. Assume that the one-step-ahead prediction errors (or model residuals) e_t are normally distributed. Then the usual three-sigma control limits on the control chart on these errors satisfy the following probability statement

$$P[-3\sigma \le e_t \le 3\sigma] = 0.9973$$

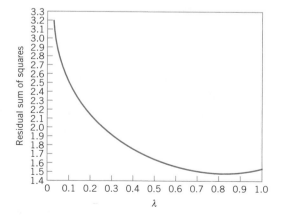

Figure 9-15   Residual sum of squares versus λ.

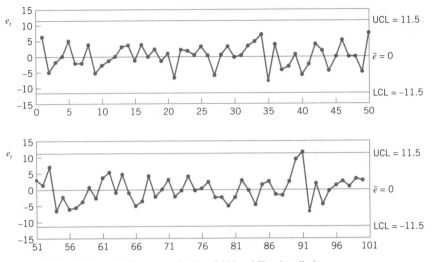

Figure 9-16   EWMA prediction errors with $\lambda = 0.825$ and Shewhart limits.

where σ is the standard deviation of the errors or residuals e_t. We may rewrite this as

$$P[-3\sigma \le x_t - \hat{x}_t(t-1) \le 3\sigma] = 0.9973$$

or

$$P[\hat{x}_t(t-1) - 3\sigma \le x_t \le \hat{x}_t(t-1) + 3\sigma] = 0.9973 \qquad (9\text{-}17)$$

Equation 9-17 suggests that if the EWMA is a suitable one-step-ahead predictor, then one could use z_t as the center line on a control chart for period $t+1$ with upper and lower control limits at

$$UCL_{t+1} = z_t + 3\sigma \qquad (9\text{-}18a)$$

and

$$LCL_{t+1} = z_t - 3\sigma \qquad (9\text{-}18b)$$

and the observation x_{t+1} would be compared to these limits to test for statistical control. We can think of this as a **moving center-line EWMA control chart.** As mentioned above, in many cases this would be preferable from an interpretation standpoint to a control chart of residuals and a separate chart of the EWMA as it combines information about process dynamics and statistical control on one chart.

Figure 9-17 is the moving center-line EWMA control chart for the viscosity data, with $\lambda = 0.825$. It conveys the same information about statistical control as the residual or EWMA prediction error control chart in Fig. 9-16, but operating personnel often feel more comfortable with this display.

Estimation and Monitoring of σ

The standard deviation of the one-step-ahead errors or model residuals σ may be estimated in several ways. If λ is chosen as suggested above over a record of n observations,

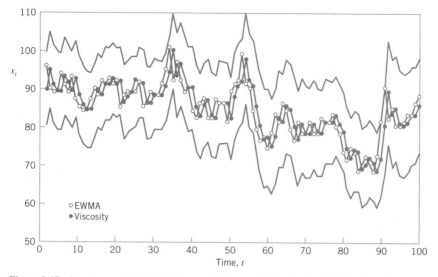

Figure 9-17   Moving center-line EWMA control chart applied to the viscosity data ($\lambda = 0.825$).

then dividing the sum of squared prediction errors for the optimal λ by n will produce an estimate of σ^2. This is the method used in many time-series analysis computer programs.

Another approach is to compute the estimate of σ as typically done in forecasting systems. The mean absolute deviation (MAD) could be used in this regard. The MAD is computed by applying an EWMA to the absolute value of the prediction error

$$\Delta_t = \alpha|e_t| + (1 - \alpha)\Delta_{t-1} \qquad 0 < \alpha \le 1$$

Since the MAD of a normal distribution is related to the standard deviation by $\sigma \cong 1.25\Delta$ [see Montgomery, Johnson, and Gardiner (1990)], we could estimate the standard deviation of the prediction errors at time t by

$$\hat{\sigma}_t \cong 1.25\Delta_t \tag{9-19}$$

Another approach is to directly calculate a smoothed variance

$$\hat{\sigma}_t^2 = \alpha e_t^2 + (1 - \alpha)\hat{\sigma}_{t-1}^2 \qquad 0 < \alpha \le 1 \tag{9-20}$$

MacGregor and Harris (1993) discuss the use of exponentially weighted moving variance estimates in monitoring the variability of a process. They show how to find control limits for these quantities for both correlated and uncorrelated data.

The Sensitivity of Residual Control Charts

Several authors have pointed out that residual control charts are not sensitive to small process shifts [see Wardell, Moskowitz, and Plante (1994)]. To improve sensitivity, we would recommend using cusum or EWMA control charts on residuals instead of Shewhart charts. Tseng and Adams (1994) note that because the EWMA is not an optimal forecasting scheme for most processes (except the model in equation 9-15), it will not completely account for the autocorrelation, and this can affect the statistical performance of control charts based on EWMA residuals or prediction errors. Montgomery and Mastrangelo (1991) suggest the use of supplementary procedures called **tracking signals** combined with the control chart for residuals. There is evidence that these supplementary procedures enhance considerably the performance of residual control charts. Furthermore, Mastrangelo and Montgomery (1995) show that if an appropriately designed tracking signal scheme is combined with the EWMA-based procedure we have described, good in-control performance and adequate shift detection can be achieved.

Other EWMA Control Charts for Autocorrelated Data

Lu and Reynolds (1999a) give a very thorough study of applying the EWMA control chart to monitoring the mean of an autocorrelated process. They consider the process to be modeled by a first-order autoregressive process with added white noise (uncorrelated error). This is equivalent to the first-order mixed model in equation 9-13. They provide charts for designing EWMA control charts for direct monitoring of the process variable that will give an in-control ARL value of $ARL_0 = 370$. They also present an extensive study of both the EWMA applied directly to the data and the EWMA of the residuals. Some of their findings may be summarized as follows:

1. When there is significant autocorrelation in process data and this autocorrelation is an inherent part of the process, traditional methods of estimating process parameters and constructing control charts should not be used. Instead, one should model the autocorrelation so that reliable control charts can be constructed.

2. A large data set should be used in the process of fitting a model for the process observations and estimating the parameters of this model. If a control chart must be constructed using a small data set, then signals from this chart should be interpreted with caution and the process of model fitting and parameter estimation should be repeated as soon as additional data become available. That is, the control limits for the chart are relatively sensitive to poor estimates of the process parameters.

3. For the low to moderate levels of correlation, a Shewhart chart of the observations will be much better at detecting a shift in the process mean than a Shewhart chart of the residuals. Unless interest is only in detecting large shifts, an EWMA chart will be better than a Shewhart chart. An EWMA chart of the residuals will be better than an EWMA chart of the observations for large shifts, and the EWMA of the observations will be a little better for small shifts.

In a subsequent paper Lu and Reynolds (1999b) present control charts for monitoring both the mean and variance of autocorrelated process data. Several types of control charts and combinations of control charts are studied. Some of these are control charts of the original observations with control limits that are adjusted to account for the autocorrelation and others are control charts of residuals from a time series model. Although there is no combination that emerges as best overall, an EWMA control chart of the observations and a Shewhart chart of residuals is a good combination for many practical situations.

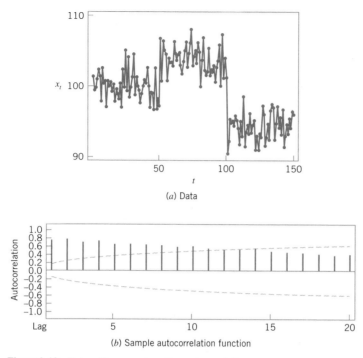

Figure 9-18 Data with apparent positive autocorrelation.

Know Your Process!

When autocorrelation is observed, we must be careful to ensure that the autocorrelation is really an inherent part of the process and not the result of some assignable cause. For example, consider the data in Fig. 9-18a for which the sample autocorrelation function is shown in Fig. 9-18b. The sample autocorrelation function gives a clear indication of positive autocorrelation in the data. Closer inspection of the data, however, reveals that there may have been an assignable cause around time $t = 50$ that resulted in a shift in the mean from 100 to about 105, and another shift may have occurred around time $t = 100$ resulting in a shift in the mean to about 95.

When these potential shifts are accounted for, the apparent autocorrelation may vanish. For example, Fig. 9-19 presents the sample autocorrelation functions for observations $x_1 - x_{50}$, $x_{51} - x_{100}$, and $x_{101} - x_{150}$. There is no evidence of autocorrelation in any of the three groups of data. Therefore, the autocorrelation in the original data is likely due to assignable causes and is not an inherent characteristic of the process.

9-4.3   A Model-Free Approach

The Batch Means Control Chart

Runger and Willemain (1996) proposed a control chart based on unweighted batch means (UBM) for monitoring autocorrelated process data. The batch means approach

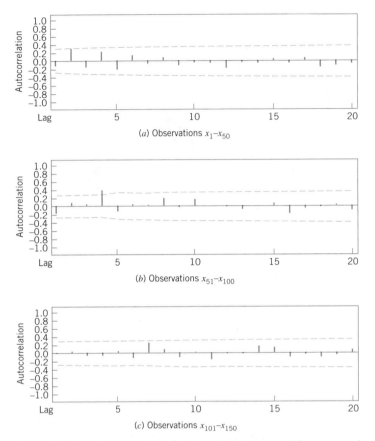

(a) Observations x_1-x_{50}

(b) Observations $x_{51}-x_{100}$

(c) Observations $x_{101}-x_{150}$

Figure 9-19   Sample autocorrelation functions after the process shifts are removed.

has been used extensively in the analysis of the output from computer simulation models, another area where highly autocorrelated data often occurs. The UBM chart breaks successive groups of sequential observations into batches, with equal weights assigned to every point in the batch. Let the jth unweighted batch mean be

$$\bar{x}_j = \frac{1}{b} \sum_{i=1}^{b} x_{(j-1)b+i} \qquad j = 1, 2, \ldots \tag{9-21}$$

The important implication of equation 9-21 is that although one has to determine an appropriate batch size b, it is not necessary to construct an ARMA model of the data. This approach is quite standard in simulation output analysis, which also focuses on inference for long time series with high autocorrelation.

Runger and Willemain (1996) showed that the batch means can be plotted and analyzed on a standard individuals control chart. Distinct from residuals plots, UBM charts retain the basic simplicity of averaging observations to form a point in a control chart. With UBMs, the control chart averaging is used to dilute the autocorrelation of the data.

Procedures for determining an appropriate batch size have been developed by researchers in the simulation area. These procedures are empirical and do not depend on identifying and estimating a time series model. Of course, a time series model can guide the process of selecting the batch size and also provide analytical insights.

Runger and Willemain (1996) provided a detailed analysis of batch sizes for AR(1) models. They recommend that the batch size be selected so as to reduce the lag 1 autocorrelation of the batch means to approximately 0.10. They suggest starting with $b = 1$ and doubling b until the lag 1 autocorrelation of the batch means is sufficiently small. This parallels the logic of the Shewhart chart in that larger batches are more effective for detecting smaller shifts; smaller batches respond more quickly to larger shifts.

EXAMPLE 9-3

To illustrate the utility of the batch means control chart, consider the data in Fig. 9-6. In Fig. 9-20a we give a plot of batch means computed using $b = 10$. The sample autocorrelation function in Fig. 9-20b indicates that the autocorrelation has been reduced dramatically by the batch means approach. The control charts for the batch means are shown in Fig. 9-21. The general indication is that the process is stable.

The batch means procedure is extremely useful when data become available very often. In many chemical and process plants some process data are observed every few

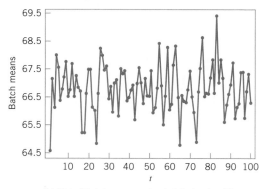

(a) Plot of batch means using batch size $b = 10$

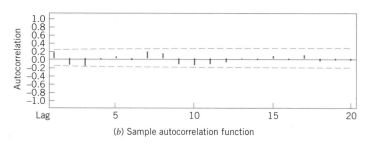

(b) Sample autocorrelation function

Figure 9-20   The batch means procedure applied to the data from Fig. 9-6.

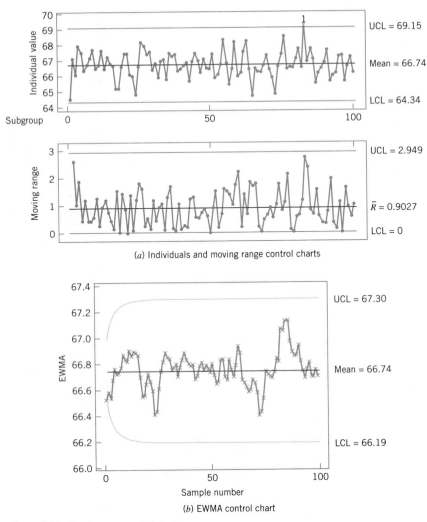

(a) Individuals and moving range control charts

(b) EWMA control chart

Figure 9-21   Batch means control charts.

seconds. Batch means clearly have great potential application in these situations. Also, note that batch means are not the same as sampling periodically from the process, because the averaging procedure uses information from all observations in the batch.

Summary

Figure 9-22 presents some guidelines for using univariate control charts to monitor processes with both correlated and uncorrelated data. The correlated data branch of the flow chart assumes that the sample size is $n = 1$. Note that one of the options in the autocorrelated data branch of the flow chart is a suggestion to eliminate the autocorrelation by using an engineering controller. This option exists frequently in the process

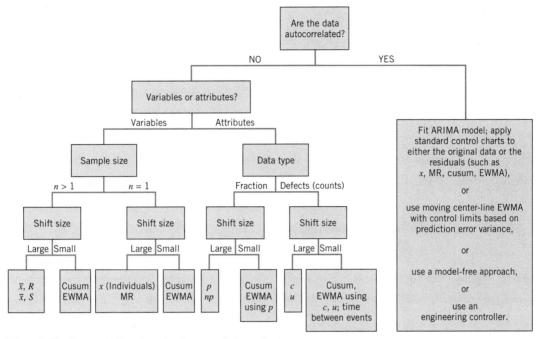

Figure 9-22   Some guidelines for univariate control chart selection.

industries, where the monitored output may be related to a manipulatable input variable, and by making a series of adjustments to this input variable, we may be able to consistently keep the output close to a desired target. These adjustments are usually made by some type of engineering process control system. We will briefly discuss these types of controllers in Chapter 11. In cases where they can be successfully employed, the effect of keeping the output at the desired target is often to eliminate (or greatly reduce) the autocorrelation in the output.

Figure 9-23 summarizes some situations in which various procedures are useful for process monitoring. On the left axis we find that as the interval between samples

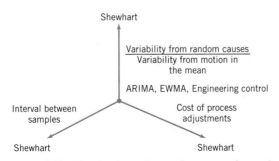

Figure 9-23   The situations where various types of control charts are useful.

increases, the Shewhart control chart becomes an appropriate choice because the larger sampling interval usually negates the effects of autocorrelation. As the interval between samples gets smaller, autocorrelation plays a much more important role, which leads to the ARIMA or EWMA approach. On the right axis we find that as the cost of process adjustment increases, we are driven to the Shewhart chart for process monitoring. On the other hand, if adjustment cost is low, we are driven to some type of engineering process control system, which we will discuss in the next section. Finally, on the vertical axis, we see that as the variability due to random causes or noise dominates any motion in the mean, the Shewhart chart becomes more appropriate. However, if the motion in the mean is large relative to random noise, we are driven once again to ARIMA- or EWMA-type procedures, or an engineering controller.

9-5   ADAPTIVE SAMPLING PROCEDURES

Traditional SPC techniques usually employ samples of fixed size taken at a fixed sampling interval. In practice, however, it is not uncommon to vary these design parameters on occasion. For example, if the sample average $\bar{x}_i$ falls sufficiently close to the upper control limit (say) on an $\bar{x}$ chart, the control chart analyst may decide to take the next sample sooner than he or she would ordinarily have, because the location of $\bar{x}_i$ on the chart could be an indication of potential trouble with the process. In fact, some practitioners use warning limits in this manner routinely.

A control chart in which either the sampling interval or the sample size (or both) can be changed depending on the value of the sample statistic is called an **adaptive SPC procedure.** The formal study of these procedures is fairly recent. For example, see Reynolds et al. (1988) and Runger and Pignatiello (1991), who studied the variable sampling interval strategy applied to the $\bar{x}$ chart, and Prabhu, Montgomery, and Runger (1994), who evaluated the performance of a combined adaptive procedure for the $\bar{x}$ chart in which both the sampling interval and the sample size depend on the current value of the sample average. These papers contain many other useful references on the subject.

The general approach used by these authors is to divide the region between the upper and lower control limits into zones, such that

$$\text{LCL} \leq -w \leq \text{CL} \leq w \leq \text{UCL}$$

If the sample statistic falls between $-w$ and w, then the standard sampling interval (and possibly sample size) are used for the next sample. However, if $w < \bar{x}_i < \text{UCL}$ or if $\text{LCL} < \bar{x}_i < -w_i$, then a shorter sampling interval (and possibly a larger sample size) is used for the next sample. It can be shown that these procedures greatly enhance control chart performance in that they reduce the average time to signal (ATS) particularly for small process shifts, when compared to an ordinary nonadaptive control chart that has a sample size and sampling interval equal to the average of those quantities for the adaptive chart when the process is in control. Prabhu, Montgomery, and Runger (1995) have given a FORTRAN program for evaluating the ATS for several different adaptive versions of the $\bar{x}$ chart.

······· **EXAMPLE 9-4** ···

An engineer is currently monitoring a process with an $\bar{x}$ chart using a sample size of five, with samples taken every hour. He is interested in detecting a shift of one standard deviation in the mean. It is easy to show that the average run length for detecting this shift is 4.5 samples. Therefore, if samples are taken every hour, it will take about 4.5 h to detect this shift on the average.

Suppose that we try to improve this performance with an adaptive control chart. We still want the sample size to be $n = 5$, but the time between samples could be variable. Suppose that the shortest time allowable between samples is 0.25 h to allow time for analysis and charting, but we want the average time between samples when the process is in control to still be one hour. Then it can be shown [see Prabhu, Montgomery, and Runger (1994), Table 7] that if we set the warning limit $w = 0.67\sigma_{\bar{x}}$ and choose the shortest and longest times between samples as $t_1 = 0.25$ h and $t_2 = 1.75$ h, respectively, the average time to signal can be reduced to 2.26 h. In this scheme, if $-0.67\sigma_{\bar{x}} \leq \bar{x} \leq 0.67\sigma_{\bar{x}}$, the time until the next sample is 1.75 h, whereas if the sample average $0.67\sigma_{\bar{x}} < \bar{x} < \text{UCL} = 3\sigma_{\bar{x}}$ or $\text{LCL} = -3\sigma_{\bar{x}} < \bar{x} < -0.67\sigma_{\bar{x}}$, the next sample is taken at 0.25 h. This adaptive scheme essentially reduces the ATS by 50%.

It is possible to do even better if one is willing to adapt both the sample size and the sampling interval. This chart will require that we plot a standardized value

$$z_i = \frac{\bar{x}_i - \mu_0}{\sigma_0 / \sqrt{n(i)}}$$

where μ_0 and σ_0 are the in-control values of μ and σ, and $n(i)$ is the sample size taken at the ith sample. Now if we want the average sample size to be five and the average interval between samples to be 1 h when the process is in control but both chart parameters can be variable, we have what Prabhu, Montgomery, and Runger (1994) called the combined adaptive $\bar{x}$ chart. If we use $n_1 = 3$, $n_2 = 8$, $t_1 = 0.25$, $t_2 = 1.50$ and $w = 0.84$, then the average time to signal for a one-sigma shift in the mean is 1.72 h [see Prabhu, Montgomery, and Runger (1994), Table 7]. In this scheme, if $-0.84 \leq z_i \leq 0.84$, use $n_1 = 3$ and $t_2 = 1.5$ h for the next sample, whereas if $0.84 < z_i < 3$ or if $-3 < z_i < -0.84$, use $n_2 = 8$ and $t_1 = 0.25$ h for the next sample.

···

Adaptive sampling procedures have considerable potential for application, if some flexibility exists with respect to the frequency and size of samples. If only one chart parameter can be adapted, we have found that more improvement in performance usually results from varying the sample size than from varying the sampling interval. The Supplemental Text Material contains more information about adaptive control charting methods.

9-6 ECONOMIC DESIGN OF CONTROL CHARTS

9-6.1 Designing a Control Chart

Control charts are widely used to establish and maintain statistical control of a process. They are also effective devices for estimating process parameters, particularly in process capability studies. The use of a control chart requires that the engineer or analyst select a sample size, a sampling frequency or interval between samples, and the control limits for the chart. Selection of these three parameters is usually called the **design** of the control chart.

Traditionally, control charts have been designed with respect to **statistical criteria** only. This usually involves selecting the sample size and control limits such that the average run length of the chart to detect a particular shift in the quality characteristic and the average run length of the procedure when the process is in control are equal to specified values. The frequency of sampling is rarely treated analytically, and usually the practitioner is advised to consider such factors as the production rate, the expected frequency of shifts to an out-of-control state, and the possible consequences of such process shifts in selecting the sampling interval. The use of statistical criteria and practical experience has led, in many cases, to general guidelines for the design of control charts. Many of these guidelines, as well as the approach used in developing them, have been discussed for specific types of control charts in previous chapters.

The design of a control chart has economic consequences in that the costs of sampling and testing, costs associated with investigating out-of-control signals and possibly correcting assignable causes, and costs of allowing nonconforming units to reach the consumer are all affected by the choice of the control chart parameters. Therefore, it is logical to consider the design of a control chart from an economic viewpoint. In recent years, considerable research has been devoted to this problem. This section will present several models for the optimal economic design of control charts. Some of the practical implications of these models will also be discussed.

9-6.2 Process Characteristics

To formulate an economic model for the design of a control chart, it is necessary to make certain assumptions about the behavior of the process. The assumptions summarized next are relatively standard in that most economic models incorporate them to some degree. In later sections we see how these assumptions are used in building specific models, and we discuss their relative importance.

The process is assumed to be characterized by a single in-control state. For example, if the process has one measurable quality characteristic, then the in-control state will correspond to the mean of this quality characteristic when no assignable causes are present. Similarly, when the quality characteristic is an attribute, the in-control state will be represented by the fraction nonconforming (say) produced by the process when no assignable causes are present. The process may have, in general, $s \geq 1$ out-of-control states. Each out-of-control state is usually associated with a particular type of assignable cause.

Determining the nature of the transitions between the in-control and out-of-control states requires certain assumptions. It is customary to assume that assignable causes occur during an interval of time according to a Poisson process. This implies that the length of time the process remains in the in-control state, given that it begins in control, is an exponential random variable. This assumption allows considerable simplification in the development of economic models, and in some situations results in a Markov chain model structure. The nature in which process shifts occur is sometimes called the **process-failure mechanism.** This can be a very critical assumption. We also observe that the assumption of discrete states and the nature of the failure mechanism imply that process transitions between states are instantaneous. Processes that "drift" slowly from an in-control state—as, for example, in the case of tool wear—have received little analytical attention.

It is also usually assumed that the process is not self-correcting. That is, once a transition to an out-of-control state has occurred, the process can be returned to the in-control condition only by management intervention following an appropriate out-of-control signal on the control chart. In some cases, however, transitions between different out-of-control states are allowed, provided the transitions are always consistent with further quality deterioration.

9-6.3   Cost Parameters

Three categories of costs are customarily considered in the economic design of control charts: the costs of sampling and testing, the costs associated with investigating an out-of-control signal and with the repair or correction of any assignable causes found, and the costs associated with the production of nonconforming items.

The costs of sampling and testing include the out-of-pocket expenses of inspectors' and technicians' salaries and wages, the costs of any necessary test equipment, and, in the case of destructive testing, the unit costs of the items sampled. Usually, the cost of sampling and testing is assumed to consist of both fixed and variable components—say, a_1 and a_2, respectively—such that the total cost of sampling and testing is

$$a_1 + a_2 n$$

Because of the difficulty of obtaining and evaluating cost information, use of more complex relationships is probably inappropriate.

The costs of investigating and possibly correcting the process following an out-of-control signal have been treated in several ways. Some authors have suggested that the costs of investigating false alarms will differ from the costs of correcting assignable causes; consequently, these two situations must be represented in the model by different cost coefficients. Furthermore, the cost of repairing or correcting the process could depend on the type of assignable cause present. Thus, in models having s out-of-control states, $s + 1$ cost coefficients might be necessary to model the search and adjustment procedures associated with out-of-control signals. Usually, these cost coefficients would be chosen so that larger process shifts incurred larger costs of repair or adjustment. Other authors have argued that this level of modeling detail is unnecessary because in many cases small shifts are difficult to find but easy to correct, whereas large shifts are easy to find but difficult to correct. Hence, one loses little accuracy by using a single cost coefficient to represent the *average* costs of investigating and possibly correcting out-of-control signals.

The costs associated with producing nonconforming items consist of typical *failure costs*—that is, the costs of rework or scrap for internal failures, or replacement or repair costs for units covered by warranties in the case of external failures. With external failures, there may also be secondary effects from the production of nonconforming items if the customer's dissatisfaction with the product causes an alteration in future purchases of the product or other products manufactured by the company. Finally, there may be losses resulting from product liability claims against the company. Most authors model these costs with a single, average cost coefficient, expressed on either a per unit time or per item basis.

Economic models are generally formulated using a total cost function, which expresses the relationships between the control chart design parameters and the three types of costs discussed above. The production, monitoring, and adjustment process may be thought of as a series of independent cycles over time. Each cycle begins with the production process in the in-control state and continues until process monitoring via the control chart results in an out-of-control signal. Following an adjustment in which the process is returned to the in-control state, a new cycle begins. Let $E(T)$ be the *expected length* (that is, the *long-term average length,* or *mean length*) of a cycle, and let $E(C)$ be the *expected total cost incurred* during a cycle. Then the expected cost per unit time is

$$E(A) = \frac{E(C)}{E(T)} \tag{9-22}$$

Optimization techniques are then applied to equation 9-22 to determine the economically optimal control chart design. Minor variations in equation 9-22 have appeared in the literature. For example, some authors have elected to replace $E(T)$ in equation 9-22 by the expected number of units produced during the cycle, resulting in the expected cost expressed on a per item rather than a per unit time basis. In other studies, a somewhat different definition of a cycle is used, depending on whether the process is shut down or allowed to continue operation while out-of-control signals are investigated.

The general model structure in equation 9-22 has a disturbing appearance. Note that C and T are dependent random variables, yet we have represented the expected value of their ratio $E(A)$ by the ratio of expectations $E(C)/E(T)$. Now it is well known that the expected value of a ratio is not equal to the ratio of expected values (even for independent random variables), so some further explanation of the structure of equation 9-22 seems warranted. The sequence of production–monitoring–adjustment, with accumulation of costs over the cycle, can be represented by a particular type of stochastic process called a *renewal reward process*. Stochastic processes of this type have the property that their average time cost is given by the ratio of the expected reward per cycle to the expected cycle length, as shown in equation 9-22.

9-6.4 Early Work and Semieconomic Designs

A fundamental paper in the area of cost modeling of quality control systems was published by Girshick and Rubin (1952). They consider a process model in which a machine producing items characterized by a measurable quality characteristic x can be in one of four states. States 1 and 2 are production states, and, in state i the output quality characteristic is described by the probability density function $f_i(x)$, $i = 1, 2$. State 1 is the "in-control" state. While in state 1, there is a constant probability of a shift state 2. The process is not self-correcting; repair is necessary to return the process to state 1. States $j = 3$ and $j = 4$ are repair states, if we assume that the machine was previously in state $j - 2$. In state $j = 3, 4$, n_j units of time are required for repair, where a time unit is defined as the time to produce one unit of product. Girshick and Rubin treat both 100% inspection and periodic inspection rules. The economic criterion is to maximize the expected income from the process. The optimal control rules are difficult to derive, as they depend on the solution to complex integral equations. Consequently, the model's use in practice has been very limited.

Although it has had little or no practical application, Girshick and Rubin's work is of significant theoretical value. They were the first researchers to propose the expected cost (or income) per unit time criterion (equation 9-22), and rigorously show its appropriateness for this problem. Later analysts' use of this criterion (equation 9-22) rests directly on its development by Girshick and Rubin. Other researchers have investigated generalized formulations of the Girshick–Rubin model, including Bather (1963), Ross (1971), Savage (1962), and White (1974). Again, their results are primarily of theoretical interest, as they do not lead to process control rules easily implemented by practitioners.

Economic design of conventional Shewhart control charts was investigated by several early researchers. Most of their work could be classified as semieconomic design procedures, in that either the proposed model did not consider all relevant costs or no formal optimization techniques were applied to the cost function. Weiler (1952) suggested that for an $\bar{x}$ chart, the optimum sample size should minimize the total amount of inspection required to detect a specified shift. If the shift is from an in-control state μ_0 to an out-of-control state $\mu_1 = \mu_0 + \delta\sigma$, then Weiler shows that the optimal sample size is

$$n = \frac{12.0}{\delta^2} \qquad \text{when } \pm 3.09\text{-sigma control limits are used}$$

$$n = \frac{11.1}{\delta^2} \qquad \text{when } \pm 3\text{-sigma control limits are used}$$

$$n = \frac{6.65}{\delta^2} \qquad \text{when } \pm 2.58\text{-sigma control limits are used}$$

$$n = \frac{4.4}{\delta^2} \qquad \text{when } \pm 2.33\text{-sigma control limits are used}$$

Note that Weiler did not formally consider costs; the implication is that minimizing total inspection will minimize total costs.

Taylor (1965) has shown that control procedures based on taking a sample of constant size at fixed intervals of time is nonoptimal. He suggests that sample size and sampling frequency should be determined at each point in time based on the posterior probability that the process is in an out-of-control state. Dynamic programming-type methods are utilized extensively in the development. In a subsequent paper, Taylor (1967) derives the optimal control rule for a two-state process with a normally distributed quality characteristic. Although Taylor's work has indicated their nonoptimality, fixed sample size–fixed sampling interval control rules are widely used in practice because of their administrative simplicity. Consequently, most researchers have concentrated on the optimal economic design of such procedures.

9-6.5   An Economic Model of the $\bar{x}$ Control Chart

Much of the research in the development of economic models of control charts has been devoted to the $\bar{x}$ chart. The interest of analysts in this control chart follows directly from its widespread use in practice. In this section, we discuss one of the economic models that has been widely studied.

In 1956, Duncan (1956) proposed an economic model for the optimum economic design of the $\bar{x}$ control chart. His paper was the first to deal with a fully economic model of a Shewhart-type control chart and to incorporate formal optimization methodology into determining the control chart parameters. Duncan's paper was the stimulus for much of the subsequent research in this area.

Duncan drew on the earlier work of Girshick and Rubin (1952) in that he utilized a design criterion that maximized the expected net income per unit of time from the process. In the development of the cost model, Duncan assumed that the process is characterized by an in-control state μ_0 and that a single assignable cause of magnitude δ, which occurs at random, results in a shift in the mean from μ_0 to either $\mu_0 + \delta\sigma$ or $\mu_0 - \delta\sigma$. The process is monitored by an $\bar{x}$ chart with center line μ_0 and upper and lower control limits $\mu_0 \pm k(\sigma/\sqrt{n})$. Samples are to be taken at intervals of h hours. When one point exceeds the control limits, a search for the assignable cause is initiated. During the search for the assignable cause, the process is allowed to continue in operation. Furthermore, it is assumed that the cost of adjustment or repairs (if necessary) is not charged against the net income from the process. The parameters μ_0, δ, and σ are assumed known, whereas n, k, and h are to be determined.

The assignable cause is assumed to occur according to a Poisson process with an intensity of λ occurrences per hour. That is, assuming that the process begins in the in-control state, the time interval that the process remains in control is an exponential random variable with mean $1/\lambda$ h. Therefore, given the occurrence of the assignable cause between the jth and $(j + 1)$st samples, the expected time of occurrence within this interval is

$$\tau = \frac{\displaystyle\int_{jh}^{(j+1)h} e^{-\lambda t}\lambda(t - jh)\, dt}{\displaystyle\int_{jh}^{(j+1)h} e^{-\lambda t}\lambda\, dt} = \frac{1 - (1 + \lambda h)e^{-\lambda h}}{\lambda(1 - e^{-\lambda h})} \tag{9-23}$$

When the assignable cause occurs, the probability that it will be detected on any subsequent sample is

$$1 - \beta = \int_{-\infty}^{-k - \delta\sqrt{n}} \phi(z) \, dz + \int_{k - \delta\sqrt{n}}^{\infty} \phi(z) \, dz \tag{9-24}$$

where $\phi(z) = (2\pi)^{-1/2} \exp(-z^2/2)$ is the standard normal density. The quantity $1 - \beta$ is the power of the test, and β is the type II error probability. The probability of a false alarm is

$$\alpha = 2 \int_{k}^{\infty} \phi(z) \, dz \tag{9-25}$$

A production cycle is defined as the interval of time from the start of production (the process is assumed to start in the in-control state) following an adjustment to the detection and elimination of the assignable cause. The cycle consists of four periods: (1) the in-control period, (2) the out-of-control period, (3) the time to take a sample and interpret the results, and (4) the time to find the assignable cause. The expected length of the in-control period is $1/\lambda$. Noting the number of samples required to produce an out-of-control signal given that the process is actually out of control is a geometric random variable with mean $1/(1 - \beta)$, we conclude that the expected length of the out-of-control period is $h/(1 - \beta) - \tau$. The time required to take a sample and interpret the results is a constant g proportional to the sample size, so that gn is the length of this segment of the cycle. The time required to find the assignable cause following an action signal is a constant D. Therefore, the expected length of a cycle is

$$E(T) = \frac{1}{\lambda} + \frac{h}{1 - \beta} - \tau + gn + D \tag{9-26}$$

The net income per hour of operation in the in-control state is V_0, and the net income per hour of operation in the out-of-control state is V_1. The cost of taking a sample of size n is assumed to be of the form $a_1 + a_2 n$; that is, a_1 and a_2 represent, respectively, the fixed and variable components of sampling cost. The expected number of samples taken within a cycle is the expected cycle length divided by the interval between samples, or $E(T)/h$. The cost of finding an assignable cause is a_3, and the cost of investigating a false alarm is a_3'. The expected number of false alarms generated during a cycle is α times the expected number of samples taken before the shift, or

$$\alpha \sum_{j=0}^{\infty} \int_{jh}^{(j+1)h} je^{-\lambda t} \, dt = \frac{\alpha e^{-\lambda h}}{1 - e^{-\lambda h}} \tag{9-27}$$

Therefore, the expected net income per cycle is

$$E(C) = V_0 \frac{1}{\lambda} + V_1 \left(\frac{h}{1 - \beta} - \tau + gn + D \right) - a_3 - \frac{a_3' e^{-\lambda h}}{1 - e^{-\lambda h}}$$

$$- (a_1 + a_2 n) \frac{E(T)}{h} \tag{9-28}$$

The expected net income per hour is found by dividing the expected net income per cycle (equation 9-28) by the expected cycle length (equation 9-26), resulting in

$$E(A) = \frac{E(C)}{E(T)}$$

$$= \frac{V_0(1/\lambda) + V_1[h/(1 - \beta) - \tau + gn + D] - a_3 - a_3{}'\alpha e^{-\lambda h}/(1 - e^{-\lambda h})}{1/\lambda + h/(1 - \beta) - \tau + gn + D}$$

$$- \frac{a_1 + a_2 n}{h} \tag{9-29}$$

Let $a_4 = V_0 - V_1$; that is, a_4 represents the hourly penalty cost associated with production in the out-of-control state. Then equation 9-29 may be rewritten as

$$E(A) = V_0 - \frac{a_1 + a_2 n}{h}$$

$$- \frac{a_4[h/(1 - \beta) - \tau + gn + D] + a_3 + a_3'\alpha e^{-\lambda h}/(1 - e^{-\lambda/h})}{1/\lambda + h/(1 - \beta) - \tau + gn + D} \tag{9-30}$$

or

$$E(A) = V_0 - E(L)$$

where

$$E(L) = \frac{a_1 + a_2 n}{h}$$

$$+ \frac{a_4[h/(1 - \beta) - \tau + gn + D] + a_3 + a_3'\alpha e^{-\lambda h}/(1 - e^{-\lambda h})}{1/\lambda + h/(1 - \beta) - \tau + gn + D} \tag{9-31}$$

The expression $E(L)$ represents the expected loss per hour incurred by the process. $E(L)$ is a function of the control chart parameters n, k, and h. Clearly, maximizing the expected net income per hour is equivalent to minimizing $E(L)$.

Duncan introduces several approximations to develop an optimization procedure for this model.[2] The optimization procedure suggested is based on solving numerical approximations to the system for first partial derivatives of $E(L)$ with respect to n, k, and h. An iterative procedure is required to solve for the optimal n and k. A closed-form solution for h is given using the optimal values of n and k.

Several authors have reported optimization methods for Duncan's model. Chiu and Wetherill (1974) have developed a simple, approximate procedure for optimizing Duncan's model. Their procedure utilizes a constraint on the power of the test $(1 - \beta)$. The recommended values are either $1 - \beta = 0.90$ or $1 - \beta = 0.95$. Tables are provided to generate the optimum design subject to this constraint. This procedure usually produces a design close to the true optimum. We also note that $E(L)$ could be easily minimized by

---

[2] Several numerical approximations are also introduced in the actual structure of the model. Approximations used are for $\tau \simeq h/2 - \lambda h^2/12$, and for the expected number of false alarms $\alpha e^{-\lambda h}/(1 - e^{-\lambda h}) \simeq \alpha/\lambda h$.

using an unconstrained optimization or search technique coupled with a digital computer program for repeated evaluations of the cost function. This is the approach to optimization most frequently used. Montgomery (1982) has given a FORTRAN program for the optimization of Duncan's model.

······· **EXAMPLE 9-5** ··

A manufacturer produces nonreturnable glass bottles for packaging a carbonated soft-drink beverage. The wall thickness of the bottles is an important quality characteristic. If the wall is too thin, internal pressure generated during filling will cause the bottle to burst. The manufacturer has used $\bar{x}$ and R charts for process surveillance for some time. These control charts have been designed with respect to statistical criteria. However, in an effort to reduce costs, the manufacturer wishes to design an economically optimum $\bar{x}$ chart for the process.

Based on an analysis of quality control technicians' salaries and the costs of test equipment, it is estimated that the fixed cost of taking a sample is $1. The variable cost of sampling is estimated to be $0.10 per bottle, and it takes approximately 1 min (0.0167 h) to measure and record the wall thickness of a bottle.

The process is subject to several different types of assignable causes. However, on the average, when the process goes out of control, the magnitude of the shift is approximately two standard deviations. Process shifts occur at random with a frequency of about one every 20 h of operation. Thus, the exponential distribution with parameter $\lambda = 0.05$ is a reasonable model of the run length in control. The average time required to investigate an out-of-control signal is 1 h. The cost of investigating an action signal that results in the elimination of an assignable cause is $25, whereas the cost of investigating a false alarm is $50.

The bottles are sold to a soft-drink bottler. If the walls are too thin, an excessive number of bottles will burst when they are filled. When this happens, the bottler's standard practice is to backcharge the manufacturer for the costs of cleanup and lost production. Based on this practice, the manufacturer estimates that the penalty cost of operating in the out-of-control state for one hour is $100.

The expected cost per hour associated with the use of an $\bar{x}$ chart for this process is given by equation 9-31, with $a_1 = \$1$, $a_2 = \$0.10$, $a_3 = \$25$, $a_3' = \$50$, $a_4 = \$100$, $\lambda = 0.05$, $\delta = 2.0$, $g = 0.0167$, and $D = 1.0$. Montgomery's computer program referenced earlier is used to optimize this problem. The output from this program, using the values of the model parameters given above, is shown in Fig. 9-24. The program calculates the optimal control limit width k and sampling frequency h for several values of n, and computes the value of the cost function equation 9-31. The corresponding α-risk and power for each combination of n, k, and h are also provided. The optimal control chart design may be found by inspecting the values of the cost function to find the minimum. From Fig. 9-24, we note that the minimum cost is $10.38 per hour, and the economically optimal $\bar{x}$ chart would use samples of size $n = 5$, the control limits would be located at $\pm k\sigma$, with $k = 2.99$, and samples would be taken at intervals of $h = 0.76$ h (approximately every 45 min). The α risk for this control chart is $\alpha = 0.0028$, and the power of the test is $1 - \beta = 0.9308$.

n	Optimum k	Optimum h	Alpha	Power	Cost
1	2.30	.45	.0214	.3821	14.71
2	2.51	.57	.0117	.6211	11.91
3	2.68	.66	.0074	.7835	10.90
4	2.84	.71	.0045	.8770	10.51
5	2.99	.76	.0028	.9308	10.38
6	3.13	.79	.0017	.9616	10.39
7	3.27	.82	.0011	.9784	10.48
8	3.40	.85	.0007	.9880	10.60
9	3.53	.87	.0004	.9932	10.75
10	3.66	.89	.0003	.9961	10.90
11	3.78	.92	.0002	.9978	11.06
12	3.90	.94	.0001	.9988	11.23
13	4.02	.96	.0001	.9993	11.39
14	4.14	.98	.0000	.9996	11.56
15	4.25	1.00	.0000	.9998	11.72

Figure 9-24   Optimum solution to Example 9-5.

After studying the optimal $\bar{x}$ chart design, the bottle manufacturer suspects that the penalty cost of operating out of control (a_4) may not have been precisely estimated. At worst, a_4 may have been underestimated by about 50%. Therefore, they decided to rerun the computer program with $a_4 = \$150$ to investigate the effect of misspecifying this parameter. The results of this additional run are shown in Fig. 9-25. We see that the optimal solution is now $n = 5$, $k = 2.99$, and $h = 0.62$, and the cost per hour is $13.88. Note that the optimal sample size and control limit width are unchanged. The primary effect of increasing a_4 by 50% is to reduce the optimal sampling frequency from 0.76 h to 0.62 h (i.e., from 45 min to 37 min). Based on this analysis, the manufacturer decides to adopt a sampling frequency of 45 min because of its administrative convenience.

From analysis of numerical problems such as those in Example 9-5, it is possible to draw several general conclusions about the optimum economic design of the $\bar{x}$ control chart. Some of these conclusions are illustrated next.

1. The optimum sample size is largely determined by the magnitude of the shift δ. For example, Fig. 9-26 illustrates the solution to the problem in Example 9-5 with $\delta = 1.0$. Note that the optimum sample size has increased considerably to $n = 14$. The optimum control limits are now slightly narrower, and the optimum sampling interval is slightly greater. In general, relatively large shifts— say, $\delta \geq 2$—often result in relatively small optimum sample size—say, $2 \leq$

n	Optimum k	Optimum h	Alpha	Power	Cost
1	2.31	.37	.0209	.3783	19.17
2	2.52	.46	.0117	.6211	15.71
3	2.68	.54	.0074	.7835	14.48
4	2.84	.58	.0045	.8770	14.01
5	2.99	.62	.0028	.9308	13.88
6	3.13	.65	.0017	.9616	13.91
7	3.27	.67	.0011	.9784	14.04
8	3.40	.69	.0007	.9880	14.21
9	3.53	.71	.0004	.9932	14.41
10	3.66	.73	.0003	.9961	14.62
11	3.78	.75	.0002	.9978	14.84
12	3.90	.77	.0001	.9988	15.06
13	4.02	.78	.0001	.9993	15.28
14	4.14	.80	.0000	.9996	15.50

Figure 9-25 Optimum $\bar{x}$ chart design for Example 9-5 with $a_4 = \$150$.

$n \leq 10$. Smaller shifts require much larger samples, with $1 \leq \delta \leq 2$ frequently producing optimum sample sizes in the range $10 \leq n \leq 20$. Very small shifts—say, $\delta \leq 0.5$—may require sample sizes as large as $n \geq 40$.

2. The hourly penalty cost for production in the out-of-control state a_4 mainly affects the interval between samples h. Larger values of a_4 imply smaller values of h (more frequent sampling), and smaller values of a_4 imply larger values of h (less frequent sampling). The effect of increasing a_4 is illustrated in Figs. 9-24 and 9-25 for the data in Example 9-5.

3. The costs associated with looking for assignable causes (a_3 and a_3') mainly affect the width of the control limits. They also have a slight effect on the sample size n. Figure 9-27 shows the solution to Example 9-5 with $a_3 = a_3' = \$100$. Note that the optimal $\bar{x}$ chart design is now $n = 6$, $k = 3.27$, and $h = 0.78$. The increased width of the control limits and the small change in n have decreased the α-risk considerably to 0.0011 and increased the power of the test slightly to 0.9483. This is intuitively appealing; as the cost of investigating action signals increases, we would want to reduce the incidence of false alarms (i.e., reduce α).

4. Variation in the costs of sampling affects all three design parameters. Increasing the fixed cost of sampling increases the interval between samples. It also usually results in slightly larger samples. Figure 9-28 presents the solution to Example 9-5 with $a_1 = \$2$. Note that in the optimum solution, the sampling frequency has increased to $h = 1.01$ h, and the sample size and control limit width have increased to $n = 6$ and $k = 3.03$, respectively. Large values of a_2,

n	Optimum k	Optimum h	Alpha	Power	Cost
4	2.22	.53	.0264	.4129	15.47
5	2.27	.58	.0232	.4865	14.57
6	2.31	.64	.0209	.5555	13.95
7	2.35	.69	.0188	.6163	13.50
8	2.39	.73	.0168	.6695	13.18
9	2.43	.77	.0151	.7157	12.95
10	2.47	.81	.0135	.7556	12.79
11	2.50	.85	.0124	.7929	12.69
12	2.54	.88	.0111	.8223	12.62
13	2.58	.91	.0099	.8474	12.59
14	2.61	.94	.0091	.8711	12.59
15	2.65	.96	.0080	.8893	12.61
16	2.69	.99	.0071	.9049	12.65
17	2.72	1.02	.0065	.9197	12.71
18	2.76	1.04	.0058	.9309	12.77
19	2.79	1.06	.0053	.9417	12.85
20	2.83	1.08	.0047	.9497	12.94
21	2.87	1.10	.0041	.9566	13.03
22	2.90	1.12	.0037	.9633	13.13
23	2.94	1.14	.0033	.9683	13.24
24	2.97	1.16	.0030	.9731	13.35

Figure 9-26   Optimum $\bar{x}$ chart design for Example 9-5 with $\delta = 1.0$.

the variable cost of sampling, imply small, relatively infrequent samples, and narrow control limits. Figure 9-29 gives the solution to Example 9-5 with $a_1 = \$0.5$. Note that the optimum sample size has decreased to 4, the optimum frequency of sampling has increased to $h = 1.07$, and the optimum control limit width has decreased to $k = 2.65$.

5. Changes in the mean number of occurrences of the assignable cause per hour primarily affect the interval between samples. Figure 9-30 presents the optimum solution to Example 9-5 with $\lambda = 0.01$ (i.e., the mean duration in control is 100 h). Note that the optimum sampling interval has increased considerably to 1.76 h. The optimum sample size and control limits have also increased slightly.

6. The optimum economic design is relatively insensitive to errors in estimating the cost coefficients. That is, the cost surface is relatively flat in the vicinity of the optimum. This may be seen in one dimension by examining Figs. 9-24 through 9-30. Note that as n varies, there is little change in the optimum cost.

(text continued on page 493)

n	Optimum k	Optimum h	Alpha	Power	Cost
1	2.53	.40	.0114	.2981	19.59
2	2.72	.53	.0065	.5432	16.03
3	2.87	.62	.0041	.7238	14.74
4	3.00	.70	.0027	.8413	14.20
5	3.13	.75	.0017	.9102	13.99
6	3.27	.78	.0011	.9483	13.95
7	3.40	.82	.0007	.9707	14.00
8	3.52	.84	.0004	.9837	14.10
9	3.65	.87	.0003	.9906	14.24
10	3.77	.89	.0002	.9947	14.38
11	3.89	.92	.0001	.9970	14.53
12	4.00	.94	.0001	.9983	14.69
13	4.12	.96	.0000	.9990	14.85
14	4.23	.98	.0000	.9994	15.01
15	4.34	1.00	.0000	.9997	15.17

Figure 9-27 Optimum $\bar{x}$ chart design for Example 9-5 with $a_3 = a_3' = \$100$.

n	Optimum k	Optimum h	Alpha	Power	Cost
1	2.07	.71	.0385	.4721	16.31
2	2.33	.82	.0198	.6909	13.25
3	2.53	.90	.0114	.8249	12.10
4	2.71	.95	.0067	.9015	11.63
5	2.87	.98	.0041	.9454	11.45
6	3.03	1.01	.0024	.9692	11.43
7	3.18	1.04	.0015	.9826	11.48
8	3.32	1.06	.0009	.9903	11.58
9	3.46	1.08	.0005	.9945	11.70
10	3.59	1.10	.0003	.9969	11.84
11	3.72	1.11	.0002	.9982	11.98
12	3.85	1.13	.0001	.9990	12.12
13	3.97	1.15	.0001	.9994	12.27
14	4.09	1.17	.0000	.9997	12.42
15	4.21	1.18	.0000	.9998	12.56

Figure 9-28 Optimum $\bar{x}$ chart design for Example 9-5 with $a_1 = \$2$.

n	Optimum k	Optimum h	Alpha	Power	Cost
1	2.19	.56	.0285	.4247	15.44
2	2.36	.78	.0183	.6803	13.02
3	2.51	.94	.0121	.8300	12.31
4	2.65	1.07	.0080	.9115	12.18
5	2.80	1.17	.0051	.9528	12.31
6	2.94	1.26	.0033	.9749	12.57
7	3.08	1.34	.0021	.9865	12.89
8	3.22	1.41	.0013	.9926	13.23
9	3.35	1.48	.0008	.9960	13.58
10	3.48	1.55	.0005	.9978	13.93
11	3.61	1.61	.0003	.9987	14.28
12	3.74	1.67	.0002	.9993	14.62
13	3.86	1.73	.0001	.9996	14.95
14	3.98	1.79	.0001	.9998	15.28

Figure 9-29 Optimum $\bar{x}$ chart design for Example 9-5 with $a_2 = \$0.5$.

n	Optimum k	Optimum h	Alpha	Power	Cost
1	2.33	.98	.0198	.3707	5.42
2	2.54	1.25	.0111	.6135	4.03
3	2.70	1.45	.0069	.7776	3.52
4	2.85	1.59	.0044	.8749	3.31
5	3.14	1.76	.0017	.9607	3.21
6	3.28	1.83	.0010	.9779	3.24
7	3.41	1.89	.0006	.9877	3.28
8	3.54	1.95	.0004	.9931	3.33
9	3.76	2.00	.0002	.9960	3.38
10	3.79	2.05	.0002	.9978	3.44
11	3.91	2.10	.0001	.9987	3.50
12	4.03	2.14	.0001	.9993	3.56
13	4.15	2.19	.0000	.9996	3.62
14	4.26	2.24	.0000	.9998	3.68

Figure 9-30 Optimum $\bar{x}$ chart design for Example 9-5 with $\lambda = 0.01$.

There is some indication that the cost surface is steeper near the origin, so that it would be preferable to overestimate the optimum n slightly rather than underestimate it. The optimum economic design is relatively sensitive to errors in estimating the magnitude of the shift (δ), the in-control state (μ_0), and the process standard deviation (σ).

7. One should exercise caution in using arbitrarily designed $\bar{x}$ control charts. Duncan (1956) has compared the optimum economic design with the arbitrary design $n = 5$, $k = 3.00$, and $h = 1$ for several sets of system parameters. Depending on the values of the system parameters, very large economic penalties may result from the use of the arbitrary design.

9-6.6 Other Work

The economic design of control charts is a rich area for research into the performance of control charts. Essentially, cost is simply another metric in which we can evaluate the performance of a control scheme. There is substantial literature in this field; see the review papers by Montgomery (1980), Svoboda (1991), Ho and Case (1994), and Keats et al. (1997) for discussion of most of the key work. A particularly useful paper by Lorenzen and Vance (1986) generalized Duncan's original model so that it was directly applicable to most types of control charts.

Woodall (1986, 1987) has criticized the economic design of control charts, noting that in many economic designs the type I error of the control chart is considerably higher than it usually would be in a statistical design, and that this will lead to more false alarms—an undesirable situation. The occurrence of excessive false alarms is always a problem, as managers will be reluctant to shut down a process if the control scheme has a history of many false alarms. Furthermore, if the type I error is high, then this could lead to excessive process adjustment, which often increases the variability of the quality characteristic. Woodall also notes that economic models assign a cost to passing defective items, which would include liability claims and customer dissatisfaction costs, among other components, and this is counter to Deming's philosophy that these costs cannot be measured and that customer satisfaction is necessary to staying in business.

These concerns can be easily overcome. An economic design should always be evaluated for statistical properties, such as type I and type II errors, average run lengths, and so forth. If any of these properties are at undesirable levels, this may indicate that inappropriate costs have been assigned, or that a constrained solution is necessary. It is recommended to optimize the cost function with suitable constraints on type I error, type II error, average run length, or other statistical properties. Saniga (1989) has reported such a study relating to the joint economic statistical design of $\bar{x}$ and R charts. Saniga uses constraints on type I error, power, and the average time to signal for the charts. His economic statistical designs have higher cost than the pure economic designs, but give superior protection over a wider range of process shifts and also have statistical properties that are as good as control charts designed entirely from statistical considerations. We strongly recommend that Saniga's approach be used in practice.

Saniga and Shirland (1977) and Chiu and Wetherill (1975) report that very few practitioners have implemented economic models for the design of control charts. This is somewhat surprising, as most quality engineers claim that a major objective in the use of statistical process control procedures is to reduce costs. There are at least two reasons for the lack of practical implementation of this methodology. First, the mathematical models and their associated optimization schemes are relatively complex, and are often presented in a manner that is difficult for the practitioner to understand and use. The availability of computer programs for these models and the development of simplified optimization procedures and methods for handling constraints is increasing. The availability of microcomputers and the ease with which these applications may be implemented on them should alleviate this problem. A second problem is the difficulty in estimating costs and other model parameters. Fortunately, costs do not have to be estimated with high precision, although other model components, such as the magnitude of the shift, require relatively accurate determination. Sensitivity analysis of the specific model could help the practitioner decide which parameters are critical in the problem.

As a guideline to implementation of these models, it is suggested that the cost or value of the items be evaluated and initial efforts in economic design of the control procedures be devoted to the products of greatest value. Frequently, we find that a relatively small percentage of the products produced account for most of the cost of production or sales revenue. In inventory control, the "*ABC* system" defines these as "*A* items," whereas "*B* items" and "*C* items" are more numerous but account for proportionately less value to the company. Usually, more sophisticated inventory-control techniques would be concentrated on the *A* items. Similarly, these items are probably appropriate candidates for the optimum economic design of their process-control procedure.

9-7 OVERVIEW OF OTHER PROCEDURES

There are many useful process control techniques in addition to those presented previously. This section gives a brief overview of some of these methods, along with some basic references. The selection of topics is far from exhaustive but does reflect a collection of ideas that are useful in practice.

9-7.1 Tool Wear

Many production processes are subject to tool wear. When tool wear occurs, we usually find that the process variability at any one point in time is considerably less than the allowable variability over the entire life of the tool. Furthermore, as the tool wears out, there will generally be an upward drift or trend in the mean caused by the worn tool

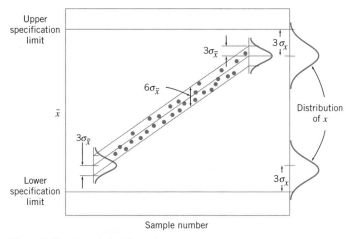

Figure 9-31   Control chart for tool wear.

producing larger dimensions. In such cases, the distance between specification limits is generally much greater than, say, 6σ. Consequently, the modified control chart concept can be applied to the tool-wear problem. The procedure is illustrated in Figure 9-31.

The initial setting for the tool is at some multiple of σ_x above the lower specification limit—say, $3\sigma_x$—and the maximum permissible process average is at the same multiple of σ_x below the upper specification limit. If the rate of wear is known or can be estimated from the data, we can construct a set of slanting control limits about the tool-wear trend line. If the sample values of $\bar{x}$ fall within these limits, the tool wear is in control. When the trend line exceeds the maximum permissible process average, the process should be reset or the tool replaced.

Control charts for tool wear are discussed in more detail by Duncan (1986) and Manuele (1945). The regression control chart [see Mandel (1969)] can also be adapted to the tool-wear problem. Quesenberry (1988) points out that these approaches essentially assume that resetting the process is expensive and that they attempt to minimize the number of adjustments made to keep the parts within specifications rather than reducing overall variability. Quesenberry develops a two-part tool-wear compensator that centers the process periodically and protects against assignable causes, as well as adjusting for the estimated mean tool wear since the previous adjustment.

9-7.2   Control Charts Based on Other Sample Statistics

Some authors have suggested the use of sample statistics other than the average and range (or standard deviation) for construction of control charts. For example, Ferrell (1953) proposed that subgroup midranges and ranges be used, with control limits determined by the median midrange and the median range. The author noted that ease of computation would be a feature of such control charts and that they would do a better job of detecting "outlier" points than conventional control charts. The median has been used frequently instead of $\bar{x}$ as a center line on charts of individuals when the underlying distribution is skewed. Similarly, medians of R and S have been proposed as the center lines of those charts so that the asymmetrical distribution of these statistics will not influence the number of runs above and below the center line.

The recent interest in robust statistical methods has generated some application of these ideas to control charts. Generally speaking, the presence of assignable causes produces "outlier" values that stretch or extend the control limits, thereby reducing the sensitivity of the control chart. One approach to this problem has been to develop control charts using statistics that are themselves outlier-resistant. Examples include the median and midrange control charts [see Clifford (1959)] and plotting subgroup boxplots [(see Inglewitz and Hoaglin (1987) and White and Schroeder (1987)]. These procedures are typically not as effective in assignable-cause or outlier detection as are conventional $\bar{x}$ and R (or S) charts.

A better approach is to plot a sample statistic that is sensitive to assignable causes ($\bar{x}$ and R or S), but to base the control limits on some outlier-resistant method. The paper by Ferrell (1953) mentioned above is an example of this approach, as is plotting $\bar{x}$ and R on charts with control limits determined by the trimmed mean of the sample means and the trimmed mean of the ranges, as suggested by Langenberg and Inglewitz (1986).

Rocke (1989) has reported that plotting an outlier-sensitive statistic on a control chart with control limits determined using an outlier-resistant method works well in practice. The suggested procedures in Ferrell (1953), Langenberg and Inglewitz (1986), and his own method are very effective in detecting assignable causes. Interestingly enough, Rocke also notes that the widely used two-stage method of setting control limits, wherein the initial limits are treated as trial control limits, the samples that plot outside these trial limits are then removed, and a final set of limits are then calculated, performs nearly as well as the more complex robust methods. In other words, the use of this two-stage method creates a robust control chart.

In addition to issues of robustness, other authors have suggested control charts for other sample statistics for process-specific reasons. For example, when pairs of measurements are made on each unit or when comparison with a standard unit is desired, one may plot the difference $x_{1j} - x_{2j}$ on a **difference control chart** [see Grubbs (1946) and the Supplemental Text Material]. In some cases, the largest and smallest sample values may be of interest. These charts have been developed by Howell (1949).

9-7.3   Fill Control Problems

Many products are filled on a high-speed, multiple-head circular filling machine that operates continuously. It is not unusual to find machines in the beverage industry that have from 40 to 72 heads and operate at speeds of from 800 to 1000 bottles per minute. In such cases, it is difficult to sample products from specific heads because there is no automatic method of identifying a filled container with its filling head. Furthermore, in addition to assignable causes that affect all filling heads simultaneously, some assignable causes affect only certain heads. Special sampling and control charting methods are needed for these types of fill control problems. Ott and Snee (1973) present some techniques useful for this problem, particularly for filling machines with a moderate number of heads. Many of the methods described in Section 9-3 for multiple-stream processes are also useful in fill control problems.

9-7.4 Precontrol

Precontrol is a technique that is used to detect shifts or upsets in the process that may result in the production of nonconforming units. The technique differs from statistical process control in that conventional control charts are designed to detect shifts in process parameters that are statistically significant, and precontrol requires no plotting of graphs and no computations. Precontrol uses the normal distribution in determining changes in the process mean or standard deviation that could result in increased production of nonconforming units. Only three units are required to give control information.

To demonstrate the procedure, suppose that the quality characteristic is normally distributed and that the natural tolerance limits ($\mu \pm 3\sigma$) exactly coincide with the specification limits. Furthermore, the process average μ is halfway between the specifications, so that the process is producing 0.27% fallout. Construct two precontrol limits (called upper and lower PC lines), each of which is one-fourth the distance in from the modified limit, as in Fig. 9-32. Since the distribution of the quality characteristic is normal, approximately 86% of the process output will lie inside the PC lines, and approximately 7% will lie in each of the regions between the PC line and specification limit. This means that only about one item in 14 will fall outside a PC line if the process mean and standard deviation are on the target values.

If the probability is one-fourteenth that one unit falls outside a PC line, the probability that two consecutive units fall outside a PC line is $(\frac{1}{14})(\frac{1}{14}) = \frac{1}{196} \approx \frac{1}{200}$. That is, if the process is operating correctly, the probability of finding two consecutive units outside of a given PC line is only about $\frac{1}{200}$. When two such consecutive units are found, it is likely that the process has shifted to an out-of-control state. Similarly, it is unlikely to find the first unit beyond one PC line and the next beyond the other PC line. In this case, we suspect that process variability has increased.

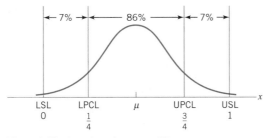

Figure 9-32   Location of precontrol lines.

A set of rules follow that describe the operation of precontrol. These rules assume that 1 to 3% nonconforming production is acceptable and that the process capability ratio is at least 1.15.

1. Start the process. If the first item is outside specifications, reset and start again.

2. If an item is inside specifications but outside a PC line, check the next item.

3. If the second item is outside the same PC line, reset the process.

4. If the second item is inside the PC line, continue. The process is reset only when two consecutive items are outside a given PC line.

5. If one item is outside a PC line and the next item is outside the other PC line, the process variability is out of control.

6. When five consecutive units are inside the PC lines, shift to frequency gaging.

7. When frequency gaging, do not adjust process until an item exceeds a PC line. Then examine the next consecutive item, and proceed as in step 4.

8. When the process is reset, five consecutive items must fall inside the PC line before frequency gaging can be resumed.

9. If the operator samples from the process more than 25 times without having to reset the process, reduce the gaging frequency so that more units are manufactured between samples. If we must rest before 25 samples are taken, increase the gaging frequency. An average of 25 samples to a reset indicates that the sampling frequency is satisfactory.

Precontrol is closely related to a technique called **narrow-limit gaging** (or **compressed-limit gaging**), in which inspection procedures are determined using tightened limits located so as to meet established risks of accepting nonconforming product. Narrow-limit gaging is discussed in more general terms by Ott (1975).

Although precontrol has the advantage of simplicity, it should not be used indiscriminately. The procedure has serious drawbacks. First, because no control chart is usually constructed, all the aspects of pattern recognition associated with the control chart cannot be used. Thus, the diagnostic information about the process contained in the pattern of points on the control chart, along with the logbook aspect of the chart, is lost. Second, the small sample sizes greatly reduce the ability of the procedure to detect even moderate-to-large shifts. Third, precontrol does not provide information that is helpful in bringing the process into control or that would be helpful in reducing variability (which is the goal of statistical process control). Finally, the assumption of an in-control process and adequate process capability is extremely important. Precontrol should only be considered in manufacturing processes where the process capability ratio is much greater than one (perhaps at least two or three), and where a near-zero defects environment has been achieved. Ledolter and Swersey (1997) in a comprehensive analysis of precontrol also observe that its use will likely lead to unnecessary tampering with the process; this can actually increase variability. This author believes that precontrol is a poor substitute for standard control charts and would never recommend it in practice.

9-8 EXERCISES

9-1. Use the following data to set up short-run $\bar{x}$ and R charts using the DNOM approach. The nominal dimensions for each part are $T_A = 100$, $T_B = 60$, $T_C = 75$, and $T_D = 50$.

Sample Number	Part Type	M_1	M_2	M_3
1	A	105	102	103
2	A	101	98	100
3	A	103	100	99
4	A	101	104	97
5	A	106	102	100
6	B	57	60	59
7	B	61	64	63
8	B	60	58	62
9	C	73	75	77
10	C	78	75	76
11	C	77	75	74
12	C	75	72	79
13	C	74	75	77
14	C	73	76	75
15	D	50	51	49
16	D	46	50	50
17	D	51	46	50
18	D	49	50	53
19	D	50	52	51
20	D	53	51	50

9-2. Use the following data to set up appropriate short-run $\bar{x}$ and R charts, assuming that the standard deviations of the measured characteristic for each part type are not the same. The normal dimensions for each part are $T_A = 100$, $T_B = 200$, and $T_C = 2000$.

Sample Number	Part Type	M_1	M_2	M_3	M_4
1	A	120	95	100	110
2	A	115	123	99	102
3	A	116	105	114	108
4	A	120	116	100	96
5	A	112	100	98	107
6	A	98	110	116	105
7	B	230	210	190	216
8	B	225	198	236	190
9	B	218	230	199	195
10	B	210	225	200	215
11	B	190	218	212	225
12	C	2150	2230	1900	1925
13	C	2200	2116	2000	1950
14	C	1900	2000	2115	1990
15	C	1968	2250	2160	2100
16	C	2500	2225	2475	2390
17	C	2000	1900	2230	1960
18	C	1960	1980	2100	2150
19	C	2320	2150	1900	1940
20	C	2162	1950	2050	2125

9-3. Discuss how you would use a cusum in the short production run situation. What advantages would it have relative to a Shewhart chart, such as a DNOM version of the $\bar{x}$ chart?

9-4. Printed circuit boards used in several different avionics devices are 100% tested for defects. The batch size for each board type is relatively small, and management wishes to establish SPC using a short-run version of the c chart. Defect data from the last 2 weeks of production are shown here. What chart would you recommend? Set up the chart and examine the process for control.

Production Day	Part Number	Total Number of Defects
245	1261	16
	1261	10
	1261	15

Production Day	Part Number	Total Number of Defects
246	1261	8
	1261	11
	1385	24
	1395	21
247	1385	28
	1385	35
	1261	10
248	1261	8
	8611	47
	8611	45
249	8611	53
	8611	41
	1385	21
250	1385	25
	1385	29
	1385	30
	4610	6
	4610	8
251	4610	10
	4610	0
	1261	20
	1261	21
252	1261	15
	1261	8
	1261	10
	1130	64
	1130	75
	1130	53
253	1055	16
	1055	15
	1055	10
254	1055	12
	8611	47
	8611	60
255	8611	51
	8611	57
	4610	0
	4610	4

9-5. A machine has four heads. Samples of $n = 3$ units are selected from each head, and the $\bar{x}$ and R values for an important quality characteristic are computed. The data are shown here. Set up group control charts for this process.

Sample Number	Head 1 $\bar{x}$	R	Head 2 $\bar{x}$	R	Head 3 $\bar{x}$	R	Head 4 $\bar{x}$	R
1	53	2	54	1	56	2	55	3
2	51	1	55	2	54	4	54	4
3	54	2	52	5	53	3	57	2
4	55	3	54	3	52	1	51	5
5	54	1	50	2	51	2	53	1
6	53	2	51	1	54	2	52	2
7	51	1	53	2	58	5	54	1
8	52	2	54	4	51	2	55	2
9	50	2	52	3	52	1	51	3
10	51	1	55	1	53	3	53	5
11	52	3	57	2	52	4	55	1
12	51	2	55	1	54	2	58	2
13	54	4	58	2	51	1	53	1
14	53	1	54	4	50	3	54	2
15	55	2	52	3	54	2	52	6
16	54	4	51	1	53	2	58	5
17	53	3	50	2	57	1	53	1
18	52	1	49	1	52	1	49	2
19	51	2	53	3	51	2	50	3
20	52	4	52	2	50	3	52	2

9-6. Consider the group control charts constructed in Exercise 9-5. Suppose the next 10 samples are as follows:

Sample Number	Head 1 $\bar{x}$	R	Head 2 $\bar{x}$	R	Head 3 $\bar{x}$	R	Head 4 $\bar{x}$	R
21	50	3	54	1	57	2	55	5
22	51	1	53	2	54	4	54	3
23	53	2	52	4	55	3	57	1

Sample Number	Head							
	1		2		3		4	
	$\bar{x}$	R	$\bar{x}$	R	$\bar{x}$	R	$\bar{x}$	R
24	54	4	54	3	53	1	56	2
25	50	2	51	1	52	2	58	4
26	51	2	55	5	54	5	54	3
27	53	1	50	2	51	4	60	1
28	54	3	51	4	54	3	61	4
29	52	2	52	1	53	2	62	3
30	52	1	53	3	50	4	60	1

Plot the new data on the control charts and discuss your findings.

9-7. Reconsider the data in Exercises 9-5 and 9-6. Suppose the process measurements are individual data values, not subgroup averages.

(a) Use observations 1–20 in Exercise 9-5 to construct appropriate group control charts.

(b) Plot observations 21–30 from Exercise 9-6 on the charts from part (a). Discuss your findings.

(c) Using observations 1–20, construct an individuals chart using the average of the readings on all four heads as an individual measurement and an S control chart using the individual measurements on each head. Discuss how these charts function relative to the group control chart.

(d) Plot observations 21–30 on the control charts from part (c). Discuss your findings.

9-8. The $\bar{x}$ and R values for 20 samples of size five are shown here. Specifications on this product have been established as 0.550 ± 0.02.

Sample Number	$\bar{x}$	R
1	0.549	0.0025
2	0.548	0.0021
3	0.548	0.0023
4	0.551	0.0029
5	0.553	0.0018
6	0.552	0.0017
7	0.550	0.0020
8	0.551	0.0024
9	0.553	0.0022
10	0.556	0.0028
11	0.547	0.0020
12	0.545	0.0030
13	0.549	0.0031
14	0.552	0.0022
15	0.550	0.0023
16	0.548	0.0021
17	0.556	0.0019
18	0.546	0.0018
19	0.550	0.0021
20	0.551	0.0022

(a) Construct a modified control chart with three-sigma limits, assuming that if the true process fraction nonconforming is as large as 1%, the process is unacceptable.

(b) Suppose that if the true process fraction nonconforming is as large as 1%, we would like an acceptance control chart to detect this out-of-control condition with probability 0.90. Construct this acceptance control chart, and compare it to the chart obtained in part (a).

9-9. A sample of five units is taken from a process every half hour. It is known that the process standard deviation is in control with $\sigma = 2.0$. The $\bar{x}$ values for the last 20 samples are:

Sample Number	$\bar{x}$	Sample Number	$\bar{x}$
1	41.5	11	40.6
2	42.7	12	39.4

Sample Number	$\bar{x}$	Sample Number	$\bar{x}$
3	40.5	13	38.6
4	39.8	14	42.5
5	41.6	15	41.8
6	44.7	16	40.7
7	39.6	17	42.8
8	40.2	18	43.4
9	41.4	19	42.0
10	43.9	20	41.9

Specifications on the product are 40 ± 8.

(a) Set up a modified control chart on this process. Use three-sigma limits on the chart and assume that the largest fraction nonconforming that is tolerable is 0.1%.

(b) Reconstruct the chart in part (a) using two-sigma limits. Is there any difference in the analysis of the data?

(c) Suppose that if the true process fraction nonconforming is 5%, we would like to detect this condition with probability 0.95. Construct the corresponding acceptance control chart.

9-10. A manufacturing process operates with an in-control fraction of nonconforming production of at most 0.1%, which management is willing to accept 95% of the time; however, if the fraction nonconforming increases to 2% or more, management wishes to detect this shift with probability 0.90. Design an appropriate acceptance control chart for this process.

9-11. Consider a modified control chart with center line at $\mu = 0$, and $\sigma = 1.0$ (known). If $n = 5$, the tolerable fraction nonconforming is $\delta = 0.00135$, and the control limits are at three-sigma, sketch the OC curve for the chart. On the same set of axes, sketch the OC curve corresponding to the chart with two-sigma limits.

9-12. Specifications on a bearing diameter are established at 8.0 ± 0.01 cm. Samples of size $n = 8$ are used, and a control chart for S shows statistical control, with the best current estimate of the population standard deviation $S = 0.001$. If the fraction of nonconforming product that is barely acceptable is 0.135%, find the three-sigma limits on the modified control chart for this process.

9-13. An $\bar{x}$ chart is to be designed for a quality characteristic assumed to be normal with a standard deviation of 4. Specifications on the product quality characteristics are 50 ± 20. The control chart is to be designed so that if the fraction nonconforming is 1%, the probability of a point falling inside the control limits will be 0.995. The sample size is $n = 4$. What are the control limits and center line for the chart?

9-14. An $\bar{x}$ chart is to be established to control a quality characteristic assumed to be normally distributed with a standard deviation of 4. Specifications on the quality characteristic are 800 ± 20. The control chart is to be designed so that if the fraction nonconforming is 1%, the probability of a point falling inside the control limits will be 0.90. The sample size is $n = 4$. What are the control limits and center line for the chart?

9-15. A normally distributed quality characteristic is controlled by $\bar{x}$ and R charts having the following parameters ($n = 4$, both charts are in control):

R Chart	$\bar{x}$ Chart
UCL = 18.795	UCL = 626.00
Center line = 8.236	Center line = 620.00
LCL = 0	LCL = 614.00

(a) What is the estimated standard deviation of the quality characteristic x?

(b) If specifications are 610 ± 15, what is your estimate of the fraction of nonconforming material produced by this process when it is in control at the given level?

(c) Suppose you wish to establish a modified $\bar{x}$ chart to substitute for the original $\bar{x}$ chart. The process mean is to be controlled so that the fraction nonconforming is less than 0.005. The probability of type I error is to be 0.01. What control limits do you recommend?

9-16. The data that follow are molecular weight measurements made every 2 hours on a polymer (read down, then across from left to right).

2048	2039	2051	2002	2029
2025	2015	2056	1967	2019
2017	2021	2018	1994	2016
1995	2010	2030	2001	2010
1983	2012	2023	2013	2006
1943	2003	2036	2016	2009
1940	1979	2019	2019	1990
1947	2006	2000	2036	1986
1972	2042	1986	2015	1947
1983	2000	1952	2032	1958
1935	2002	1988	2016	1983
1948	2010	2016	2000	2010
1966	1975	2002	1988	2000
1954	1983	2004	2010	2015
1970	2021	2018	2015	2032

(a) Calculate the sample autocorrelation function and provide an interpretation.

(b) Construct an individuals control chart with the standard deviation estimated using the moving range method. How would you interpret the chart? Are you comfortable with this interpretation?

(c) Fit a first-order autoregressive model $x_t = \xi + \phi x_{t-1} + \epsilon_t$ to the molecular weight data. Set up an individuals control chart on the residuals from this model. Interpret this control chart.

9-17. Consider the molecular weight data in Exercise 9-16. Construct a cusum control chart on the residuals from the model you fit to the data in part (c) of that exercise.

9-18. Consider the molecular weight data in Exercise 9-16. Construct an EWMA control chart on the residuals from the model you fit to the data in part (c) of that exercise.

9-19. Set up a moving center-line EWMA control chart for the molecular weight data in Exercise 9-16. Compare it to the residual control chart in Exercise 9-16, part (c).

9-20. The data shown here are concentration readings from a chemical process, made every 30 minutes (read down, then across from left to right).

204	190	208	207	200
202	196	209	204	202
201	199	209	201	202
202	203	206	197	207
197	199	200	189	206
201	207	203	189	211
198	204	202	196	205
188	207	195	193	210
195	209	196	193	210
189	205	203	198	198
195	202	196	194	194
192	200	197	198	192
196	208	197	199	189
194	214	203	204	188
196	205	205	200	189
199	211	194	203	194
197	212	199	200	194
197	214	201	197	198
192	210	198	196	196
195	208	202	202	200

(a) Calculate the sample autocorrelation function and provide an interpretation.

(b) Construct an individuals control chart with the standard deviation estimated using the moving range method. Provide an interpretation of this control chart.

(c) Fit a first-order autoregressive model $x_t = \xi + \phi x_{t-1} + \epsilon_t$ to the data. Set up an individuals control chart on the residuals from this model. Interpret this chart.

(d) Are the residuals from the model in part (c) uncorrelated? Does this have any impact on your interpretation of the control chart from part (c)?

9-21. Consider the concentration data in Exercise 9-20. Construct a cusum chart in the residuals from the model you fit in part (c) of that exercise.

9-22. Consider the concentration data in Exercise 9-20. Construct an EWMA control chart in the residuals from the model you fit in part (c) of that exercise.

9-23. Set up a moving center-line EWMA control chart for the concentration data in Exercise 9-20. Compare it to the residuals control chart in Exercise 9-20, part (c).

9-24. The data shown here are temperature measurements made every 2 minutes on an intermediate chemical product (read down, then across from left to right).

491	526	489	502	528
482	533	496	494	513
490	533	489	492	511
495	527	494	490	512
499	520	496	489	522
499	514	514	495	523
507	517	505	498	517
503	508	511	501	522
510	515	513	518	518
509	501	508	521	505
510	497	498	535	510
510	483	500	533	508

515	491	502	524	510
513	489	506	515	487
520	496	500	529	481
518	501	495	525	483
517	496	489	526	473
526	495	509	528	479
525	488	511	534	475
519	491	508	530	484

(a) Calculate the sample autocorrelation function. Interpret the results that you have obtained.

(b) Construct an individuals control chart, using the moving range method to estimate the standard deviation. Interpret the results you have obtained.

(c) Fit a first-order autoregressive model $x_t = \xi + \phi x_{t-1} + \epsilon_t$ to the temperature data. Set up an individuals control chart on the residuals from this model. Compare this chart to the individuals chart in the original data in part (a).

9-25. Consider the temperature data in Exercise 9-24. Set up a cusum control chart on the residuals from the model you fit to the data in part (c) of that exercise. Compare it to the individuals chart you constructed using the residuals.

9-26. Consider the temperature data in Exercise 9-24. Set up an EWMA control chart on the residuals from the model you fit to the data in part (c) of that exercise. Compare it to the individuals chart you constructed using the residuals.

9-27. Set up a moving center-line EWMA control chart for the temperature data in Exercise 9-24. Compare it to the residuals control chart in Exercise 9-24, part (c).

9-28. (a) Discuss the use of the moving range method to estimate the process standard deviation when the data are positively autocorrelated.

(b) Discuss the use of the sample variance S^2 with positively autocorrelated data. Specifically, if the observations at lag i have autocorrelation ρ_i, is S^2 still an unbiased estimation for σ^2?

(c) Does your answer in part (b) imply that S^2 would really be a good way (in practice) to estimate σ^2 in constructing a control chart for autocorrelated data?

9-29. The viscosity of a chemical product is read every 2 minutes. Some data from this process are shown here (read down, then across from left to right).

(a) Is there a serious problem with autocorrelation in these data?

(b) Set up a control chart for individuals with a moving range used to estimate process variability. What conclusion can you draw from this chart?

(c) Design a cusum control scheme for this process, assuming that the observations are uncorrelated. How does the cusum perform?

(d) Set up an EWMA control chart with $\lambda = 0.15$ for the process. How does this chart perform?

(e) Set up a moving center-line EWMA scheme for these data.

(f) Suppose that a reasonable model for the viscosity data is an AR (2) model. How could this model be used to assist in the development of a statistical process control procedure for viscosity? Set up an appropriate control chart and use it to assess the current state of statistical control.

29.330	33.220	27.990	24.280
19.980	30.150	24.130	22.690
25.760	27.080	29.200	26.600
29.000	33.660	34.300	28.860
31.030	36.580	26.410	28.270
32.680	29.040	28.780	28.170
33.560	28.080	21.280	28.580
27.500	30.280	21.710	30.760

26.750	29.350	21.470	30.620
30.550	33.600	24.710	20.840
28.940	30.290	33.610	16.560
28.500	20.110	36.540	25.230
28.190	17.510	35.700	31.790
26.130	23.710	33.680	32.520
27.790	24.220	29.290	30.280
27.630	32.430	25.120	26.140
29.890	32.440	27.230	19.030
28.180	29.390	30.610	24.340
26.650	23.450	29.060	31.530
30.010	23.620	28.480	31.950
30.800	28.120	32.010	31.680
30.450	29.940	31.890	29.100
36.610	30.560	31.720	23.150
31.400	32.300	29.090	26.740
30.830	31.580	31.920	32.440

9-30. An $\bar{x}$ chart is used to maintain current control of a process. A single assignable cause of magnitude 2σ occurs, and the time that the process remains in control is an exponential random variable with mean 100 h. Suppose that sampling costs are $0.50 per sample and $0.10 per unit, it costs $5 to investigate a false alarm, $2.50 to find the assignable cause, and $100 is the penalty cost per hour to operate in the out-of-control state. The time required to collect and evaluate a sample is 0.05 h, and it takes 2 h to locate the assignable cause. Assume that the process is allowed to continue operating during searches for the assignable cause.

(a) What is the cost associated with the arbitrary control chart design $n = 5, k = 3$, and $h = 1$?

(b) Find the control chart design that minimizes the cost function given by equation 9-31.

9-31. An $\bar{x}$ chart is used to maintain current control of a process. The cost parameters are $a_1 = \$0.50$, $a_2 = \$0.10$, $a_3 =$

$25, $a_3' = \$50$, and $a_4 = \$100$. A single assignable cause of magnitude $\delta = 2$ occurs, and the duration of the process in control is an exponential random variable with mean 100 h. Sampling and testing require 0.05 h, and it takes 2 h to locate the assignable cause. Assume that equation 9-31 is the appropriate process model.

(a) Evaluate the cost of the arbitrary control chart design $n = 5$, $k = 3$, $h = 1$.

(b) Evaluate the cost of the arbitrary control chart design $n = 5$, $k = 3$, $h = 0.5$.

(c) Determine the economically optimum design.

9-32. Consider the cost information given in Exercise 9-30. Suppose that the process model represented by equation 9-31 is appropriate. It requires 2 h to investigate a false alarm, the profit per hour of operating in the in-control state is $500, and it costs $25 to eliminate the assignable cause. Evaluate the cost of the arbitrary control chart design $n = 5$, $k = 3$, and $h = 1$.

9-33. An $\bar{x}$ chart is used to maintain current control of a process. The cost parameters are $a_1 = \$2$, $a_2 = \$0.50$, $a_3 = \$50$, $a_3' = \$75$, and $a_4 = \$200$. A single assignable cause occurs, with magnitude $\delta = 1$, and the run length of the process in control is exponentially distributed with mean 100 h. It requires 0.05 h to sample and test, and 1 h to locate the assignable cause.

(a) Evaluate the cost of the arbitrary $\bar{x}$ chart design $n = 5$, $k = 3$, and $h = 0.5$.

(b) Find the economically optimum design.

9-34. **A control chart for tool wear.** A sample of five units of product is taken from a production process every hour. The following results are obtained.

Sample Number	$\bar{x}$	R
1	1.0020	0.0008
2	1.0022	0.0009
3	1.0025	0.0006
4	1.0028	0.0007
5	1.0029	0.0005
6	1.0032	0.0006
Tool Reset		
7	1.0018	0.0005
8	1.0021	0.0006
9	1.0024	0.0005
10	1.0026	0.0008
11	1.0029	0.0005
12	1.0031	0.0007

Assume that the specifications on this quality characteristic are at 1.0015 and 1.0035. Set up the R chart on this process. Set up a control chart to monitor the tool wear.

10

Multivariate Process Monitoring and Control

CHAPTER OUTLINE

10-1 THE MULTIVARIATE QUALITY CONTROL PROBLEM

10-2 DESCRIPTION OF MULTIVARIATE DATA

    10-2.1 The Multivariate Normal Distribution

    10-2.2 The Sample Mean Vector and Covariance Matrix

10-3 THE HOTELLING T^2 CONTROL CHART

10-3.1 Subgrouped Data

10-3.2 Individual Observations

10-4 THE MULTIVARIATE EWMA CONTROL CHART

10-5 REGRESSION ADJUSTMENT

10-6 CONTROL CHARTS FOR MONITORING VARIABILITY

10-7 LATENT STRUCTURE METHODS

    10-7.1 Principal Components

    10-7.2 Partial Least Squares

CHAPTER OVERVIEW

In previous chapters we have addressed process monitoring and control primarily from the univariate perspective; that is, we have assumed that there is only one process output variable or quality characteristic of interest. In practice, however, many if not most process monitoring and control scenarios involve several related variables. Although applying univariate control charts to each individual variable is a possible solution, we will see that this is inefficient and can lead to erroneous conclusions. Multivariate methods that consider the variables jointly are required.

This chapter presents control charts that can be regarded as the multivariate extensions of some of the univariate charts of previous chapters. The Hotelling T^2 chart is the analog of the Shewhart $\bar{x}$ chart. We will also discuss a multivariate version of the EWMA control chart, and some methods for monitoring variability in the multivariate case. These multivariate control charts work well when the number of process variables is not too large—say, 10 or fewer. As the number of variables grows, however, traditional multivariate control charts lose efficiency with regard to shift detection. A popular approach in these situations is to reduce the dimensionality of the problem. We show how this can be done with principal components.

10-1   THE MULTIVARIATE QUALITY CONTROL PROBLEM

There are many situations in which the **simultaneous monitoring** or control of two or more related quality characteristics is necessary. For example, suppose that a bearing has both an inner diameter (x_1) and an outer diameter (x_2) that together determine the usefulness of the part. Suppose that x_1 and x_2 have independent normal distributions. Because both quality characteristics are measurements, they could be monitored by applying the usual $\bar{x}$ chart to each characteristic, as illustrated in Fig. 10-1. The process is considered to be in control only if the sample means $\bar{x}_1$ and $\bar{x}_2$ fall within their respective control limits. This is equivalent to the pair of means $(\bar{x}_1, \bar{x}_2)$ plotting within the shaded region in Fig. 10-2.

Monitoring these two quality characteristics independently can be very misleading. For example, note from Fig. 10-2 that one observation appears somewhat unusual with respect to the others. That point would be inside the control limits on both of the univariate $\bar{x}$ charts for x_1 and x_2, yet when we examine the two variables **simultaneously,** the unusual behavior of the point is fairly obvious. Furthermore, note that the probability that

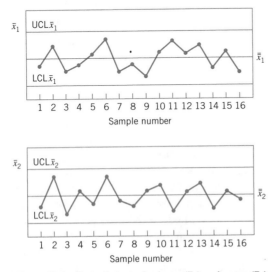

Figure 10-1   Control charts for inner $(\bar{x}_1)$ and outer $(\bar{x}_2)$ bearing diameters.

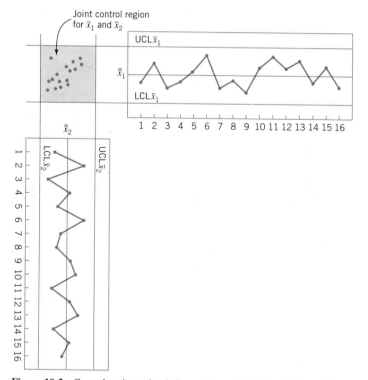

Figure 10-2 Control region using independent control limits for $\bar{x}_1$ and $\bar{x}_2$.

either $\bar{x}_1$ or $\bar{x}_2$ exceeds three-sigma control limits is 0.0027. However, the joint probability that both variables exceed their control limits simultaneously when they are both in control is $(0.0027)(0.0027) = 0.00000729$, which is considerably smaller than 0.0027. Furthermore, the probability that both $\bar{x}_1$ and $\bar{x}_2$ will simultaneously plot inside the control limits when the process is really in control is $(0.9973)(0.9973) = 0.99460729$. Therefore, the use of two independent $\bar{x}$ charts has distorted the simultaneous monitoring of $\bar{x}_1$ and $\bar{x}_2$, in that the type I error and the probability of a point correctly plotting in control are not equal to their advertised levels for the individual control charts.

This distortion in the process-monitoring procedure increases as the number of quality characteristics increases. In general, if there are p statistically independent quality characteristics for a particular product and if an $\bar{x}$ chart with $P\{\text{type I error}\} = \alpha$ is maintained on each, then the true probability of type I error for the joint control procedure is

$$\alpha' = 1 - (1 - \alpha)^p \tag{10-1}$$

and the probability that all p means will simultaneously plot inside their control limits when the process is in control is

$$P\{\text{all } p \text{ means plot in control}\} = (1 - \alpha)^p \tag{10-2}$$

Clearly, the distortion in the joint control procedure can be severe, even for moderate values of p. Furthermore, if the p quality characteristics are not independent, which usually would be the case if they relate to the same product, then equations 10-1 and 10-2 do not hold, and we have no easy way even to measure the distortion in the joint control procedure.

Process-monitoring problems in which several related variables are of interest are sometimes called **multivariate quality control** (or **process-monitoring**) problems. The original work in multivariate quality control was done by Hotelling (1947), who applied his procedures to bombsight data during World War II. Subsequent papers dealing with control procedures for several related variables include Hicks (1955); Jackson (1956) (1959) (1985); Crosier (1988); Hawkins (1991) (1993b); Lowry et al. (1992); Lowry and Montgomery (1995); Pignatiello and Runger (1990); Tracy, Young, and Mason (1992); Montgomery and Wadsworth (1972); and Alt (1985). This subject is particularly important today, as automatic inspection procedures make it relatively easy to measure many parameters on each unit of product manufactured. Many chemical and process plants and semiconductor manufacturers (for examples) routinely maintain manufacturing databases with process and quality data on hundreds of variables. Often the total size of these databases is measured in millions of individual records. Monitoring or analysis of these data with univariate SPC procedures is often ineffective. The use of multivariate methods has increased greatly in recent years for this reason.

10-2   DESCRIPTION OF MULTIVARIATE DATA

10-2.1   The Multivariate Normal Distribution

In univariate statistical quality control, we generally use the normal distribution to describe the behavior of a continuous quality characteristic. The univariate normal probability density function is

$$f(x) = \frac{1}{\sqrt{2\pi\sigma^2}} e^{-\frac{1}{2}\left(\frac{x-\mu}{\sigma}\right)^2} \qquad -\infty < x < \infty \tag{10-3}$$

The mean of the normal distribution is μ and the variance is σ^2. Note that the term in the exponent of the normal distribution can be written as follows:

$$(x - \mu)(\sigma^2)^{-1}(x - \mu) \tag{10-4}$$

This quantity measures the squared standardized distance from x to the mean μ, where by the term "standardized" we mean that the distance is expressed in standard deviation units.

This same approach can be used in the multivariate case. Suppose that we have p variables, given by $x_1, x_2, \ldots, x_p$. Arrange these variables in a p-component vector $\mathbf{x}' = [x_1, x_2, \ldots, x_p]$. Let $\boldsymbol{\mu}' = [\mu_1, \mu_2, \ldots, \mu_p]$ be the vector of the means of the x's, and let the variances and covariances of the random variables in $\mathbf{x}$ be contained in a $p \times p$ **covariance matrix** $\boldsymbol{\Sigma}$. The main diagonal elements of $\boldsymbol{\Sigma}$ are the variances of the x's and the off-diagonal elements are the covariances. Now the squared standardized

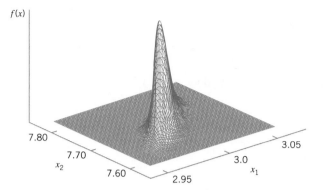

Figure 10-3 A multivariate normal distribution with $p = 2$ variables (bivariate normal).

(generalized) distance from $\mathbf{x}$ to $\boldsymbol{\mu}$ is

$$(\mathbf{x} - \boldsymbol{\mu})'\boldsymbol{\Sigma}^{-1}(\mathbf{x} - \boldsymbol{\mu}) \tag{10-5}$$

The multivariate normal density function is obtained simply by replacing the standardized distance in equation 10-4 by the multivariate generalized distance in equation 10-5 and changing the constant term $1/\sqrt{2\pi\sigma^2}$ to a more general form that makes the area under the probability density function unity regardless of the value of p. Therefore, the multivariate normal probability density function is

$$f(\mathbf{x}) = \frac{1}{(2\pi)^{p/2}|\boldsymbol{\Sigma}|^{1/2}} \, e^{-\frac{1}{2}(\mathbf{x} - \boldsymbol{\mu})'\boldsymbol{\Sigma}^{-1}(\mathbf{x} - \boldsymbol{\mu})} \tag{10-6}$$

where $-\infty < x_j < \infty, j = 1, 2, \ldots , p$.

A multivariate normal distribution for $p = 2$ variables (called a **bivariate normal**) is shown in Fig. 10-3. Note that the density function is a surface. The correlation coefficient between the two variables in this example is 0.8, and this causes the probability to concentrate closely along a line.

10-2.2   The Sample Mean Vector and Covariance Matrix

Suppose that we have a random sample from a multivariate normal distribution — say,

$$\mathbf{x}_1, \mathbf{x}_2, \ldots , \mathbf{x}_n$$

where the ith sample vector contains observations $x_{i1}, x_{i2}, \ldots , x_{ip}$. Then the sample mean vector is

$$\overline{\mathbf{x}} = \frac{1}{n} \sum_{i=1}^{n} \mathbf{x}_i \tag{10-7}$$

and the sample covariance matrix is

$$\mathbf{S} = \frac{1}{n-1} \sum_{i=1}^{n} (\mathbf{x}_i - \bar{\mathbf{x}})(\mathbf{x}_i - \bar{\mathbf{x}})' \tag{10-8}$$

That is, the sample variances on the main diagonal of the matrix $\mathbf{S}$ are computed as

$$S_j = \frac{1}{n-1} \sum_{i=1}^{n} (x_{ij} - \bar{x}_j)^2 \tag{10-9}$$

and the sample covariances are

$$S_{jk} = \frac{1}{n-1} \sum_{i=1}^{n} (x_{ij} - \bar{x}_j)(x_{ik} - \bar{x}_k) \tag{10-10}$$

We can show that the sample mean vector and sample covariance matrix are unbiased estimators of the corresponding population quantities; that is,

$$E(\bar{\mathbf{x}}) = \boldsymbol{\mu} \quad \text{and} \quad E(\mathbf{S}) = \boldsymbol{\Sigma}$$

10-3 THE HOTELLING T^2 CONTROL CHART

The most familiar multivariate process monitoring and control procedure is the Hotelling T^2 control chart for monitoring the mean vector of the process. It is a direct analog of the univariate Shewhart $\bar{x}$ chart. We present two versions of the Hotelling T^2 chart: one for subgrouped data, and another for individual observations.

10-3.1 Subgrouped Data

Suppose that two quality characteristics x_1 and x_2 are jointly distributed according to the bivariate normal distribution (see Fig. 10-3). Let μ_1 and μ_2 be the mean values of the quality characteristics, and let σ_1 and σ_2 be the standard deviations of x_1 and x_2, respectively. The covariance between x_1 and x_2 is denoted by σ_{12}. We assume that σ_1, σ_2, and σ_{12} are known. If $\bar{x}_1$ and $\bar{x}_2$ are the sample averages of the two quality characteristics computed from a sample of size n, then the statistic

$$\chi_0^2 = \frac{n}{\sigma_1^2 \sigma_2^2 - \sigma_{12}^2} [\sigma_2^2 (\bar{x}_1 - \mu_1)^2 + \sigma_1^2 (\bar{x}_2 - \mu_2)^2$$
$$- 2\sigma_{12}(\bar{x}_1 - \mu_1)(\bar{x}_2 - \mu_2)] \tag{10-11}$$

will have a chi-square distribution with 2 degrees of freedom. This equation can be used as the basis of a control chart for the process means μ_1 and μ_2. If the process means

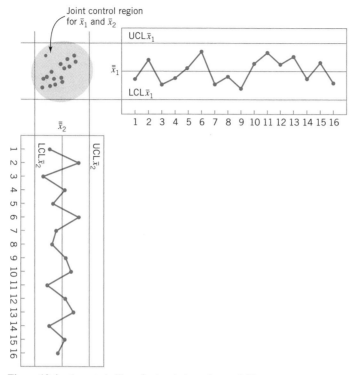

Figure 10-4   A control ellipse for two independent variables.

remain at the values μ_1 and μ_2, then values of χ_0^2 should be less than the upper control limit UCL $= \chi_{\alpha,2}^2$ where $\chi_{\alpha,2}^2$ is the upper α percentage point of the chi-square distribution with 2 degrees of freedom. If at least one of the means shifts to some new (out-of-control) value, then the probability that the statistic χ_0^2 exceeds the upper control limit increases.

The process monitoring procedure may be represented graphically. Consider the case in which the two random variables x_1 and x_2 are independent; that is, $\sigma_{12} = 0$. If $\sigma_{12} = 0$, then equation 10-11 defines an ellipse centered at (μ_1, μ_2) with principal axes parallel to the $\bar{x}_1, \bar{x}_2$ axes, as shown in Fig. 10-4. Taking χ_0^2 in equation 10-11 equal to $\chi_{\alpha,2}^2$ implies that a pair of sample averages $(\bar{x}_1, \bar{x}_2)$ yielding a value of χ_0^2 plotting inside the ellipse indicates that the process is in control, whereas if the corresponding value of χ_0^2 plots outside the ellipse the process is out of control. Figure 10-4 is often called a **control ellipse.**

In the case where the two quality characteristics are dependent, then $\sigma_{12} \neq 0$, and the corresponding control ellipse is shown in Fig. 10-5. When the two variables are dependent, the principal axes of the ellipse are no longer parallel to the $\bar{x}_1, \bar{x}_2$ axes. Also, note that sample point number 11 plots outside the control ellipse, indicating that an assignable cause is present, yet point 11 is inside the control limits on both of the individual control charts for $\bar{x}_1$ and $\bar{x}_2$. Thus there is nothing apparently unusual about point 11 when

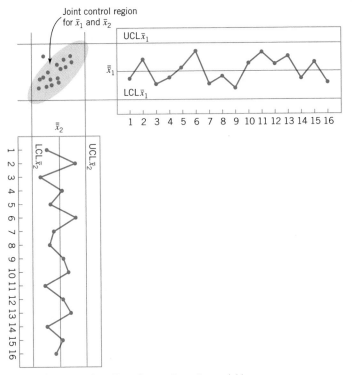

Figure 10-5   A control ellipse for two dependent variables.

viewed individually, yet the customer who received that shipment of material would quite likely observe very different performance in the product. It is nearly impossible to detect an assignable cause such as this one by maintaining individual control charts.

Two disadvantages are associated with the control ellipse. The first is that the time sequence of the plotted points is lost. This could be overcome by numbering the plotted points or by using special plotting symbols to represent the most recent observations. The second and more serious disadvantage is that it is difficult to construct the ellipse for more than two quality characteristics. To avoid these difficulties, it is customary to plot the values of χ_0^2 computed from equation 10-11 for each sample on a control chart with only an upper control limit at $\chi_{\alpha,2}^2$, as shown in Fig. 10-6. This control chart is usually called the **chi-square control chart.** Note that the time sequence of the data is preserved by this control chart, so that runs or other nonrandom patterns can be investigated. Furthermore, it has the additional advantage that the "state" of the process is characterized by a single number (the value of the statistic χ_0^2). This is particularly helpful when there are two or more quality characteristics of interest.

It is possible to extend these results to the case where p related quality characteristics are controlled jointly. It is assumed that the joint probability distribution of the p quality characteristics is the p-variate normal distribution. The procedure requires computing the sample mean for each of the p quality characteristics from a sample of

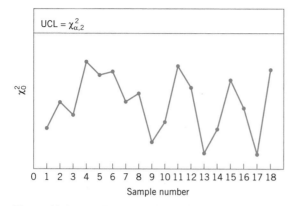

Figure 10-6 A chi-square control chart for $p = 2$ quality characteristics.

size n. This set of quality characteristic means is represented by the $p \times 1$ vector

$$\bar{\mathbf{x}} = \begin{bmatrix} \bar{x}_1 \\ \bar{x}_2 \\ \cdot \\ \cdot \\ \cdot \\ \bar{x}_p \end{bmatrix}$$

The test statistic plotted on the chi-square control chart for each sample is

$$\chi_0^2 = n(\bar{\mathbf{x}} - \boldsymbol{\mu})' \boldsymbol{\Sigma}^{-1} (\bar{\mathbf{x}} - \boldsymbol{\mu}) \qquad (10\text{-}12)$$

where $\boldsymbol{\mu}' = [\mu_1, \mu_2, \ldots, \mu_p]$ is the vector of in-control means for each quality characteristic and $\boldsymbol{\Sigma}$ is the covariance matrix. The upper limit on the control chart is

$$\text{UCL} = \chi_{\alpha,p}^2 \qquad (10\text{-}13)$$

Estimating μ and Σ
In practice, it is usually necessary to estimate $\boldsymbol{\mu}$ and $\boldsymbol{\Sigma}$ from the analysis of preliminary samples of size n, taken when the process is assumed to be in control. Suppose that m such samples are available. The sample means and variances are calculated from each

sample as usual; that is,

$$\bar{x}_{jk} = \frac{1}{n} \sum_{i=1}^{n} x_{ijk} \qquad \begin{cases} j = 1, 2, \ldots, p \\ k = 1, 2, \ldots, m \end{cases} \qquad (10\text{-}14)$$

$$S_{jk}^2 = \frac{1}{n-1} \sum_{i=1}^{n} (x_{ijk} - \bar{x}_{jk})^2 \qquad \begin{cases} j = 1, 2, \ldots, p \\ k = 1, 2, \ldots, m \end{cases} \qquad (10\text{-}15)$$

where x_{ijk} is the ith observation on the jth quality characteristic in the kth sample. The covariance between quality characteristic j and quality characteristic h in the kth sample is

$$S_{jhk} = \frac{1}{n-1} \sum_{i=1}^{n} (x_{ijk} - \bar{x}_{jk})(x_{ihk} - \bar{x}_{hk}) \qquad \begin{cases} k = 1, 2, \ldots, m \\ j \neq h \end{cases} \qquad (10\text{-}16)$$

The statistics $\bar{x}_{jk}$, S_{jk}^2, and S_{jhk} are then averaged over all m samples to obtain

$$\bar{\bar{x}}_j = \frac{1}{m} \sum_{k=1}^{m} \bar{x}_{jk} \qquad j = 1, 2, \ldots, p \qquad (10\text{-}17\text{a})$$

$$\bar{S}_j^2 = \frac{1}{m} \sum_{k=1}^{m} S_{jk}^2 \qquad j = 1, 2, \ldots, p \qquad (10\text{-}17\text{b})$$

and

$$\bar{S}_{jh} = \frac{1}{m} \sum_{k=1}^{m} S_{jhk} \qquad j \neq h \qquad (10\text{-}17\text{c})$$

The $\{\bar{\bar{x}}_j\}$ are the elements of the vector $\bar{\bar{\mathbf{x}}}$, and the $p \times p$ average of sample covariance matrices $\mathbf{S}$ is formed as

$$\mathbf{S} = \begin{bmatrix} \bar{S}_1^2 & \bar{S}_{12} & \bar{S}_{13} & \cdots & \bar{S}_{1p} \\ & \bar{S}_2^2 & \bar{S}_{23} & \cdots & \bar{S}_{2p} \\ & & \bar{S}_3^2 & & \vdots \\ & & & \ddots & \\ & & & & \bar{S}_p^2 \end{bmatrix} \qquad (10\text{-}18)$$

The average of the sample covariance matrices $\mathbf{S}$ is an unbiased estimate of $\mathbf{\Sigma}$ when the process is in control.

The T^2 Control Chart

Now suppose that $\mathbf{S}$ from equation 10-18 is used to estimate $\mathbf{\Sigma}$ and that the vector $\bar{\bar{\mathbf{x}}}$ is taken as the in-control value of the mean vector of the process. If we replace $\boldsymbol{\mu}$ with $\bar{\bar{\mathbf{x}}}$ and $\mathbf{\Sigma}$ with $\mathbf{S}$ in equation 10-12, the test statistic now becomes

$$T^2 = n(\bar{\mathbf{x}} - \bar{\bar{\mathbf{x}}})'\mathbf{S}^{-1}(\bar{\mathbf{x}} - \bar{\bar{\mathbf{x}}}) \qquad (10\text{-}19)$$

In this form, the procedure is usually called the **Hotelling T^2 control chart.**

Alt (1985) has pointed out that in multivariate quality control applications one must be careful to select the control limits for Hotelling's T^2 statistic (equation 10-19) based on how the chart is being used. There are two distinct phases of control chart usage. **Phase 1** is the use of the charts for establishing control; that is, testing whether the process was in control when the m preliminary subgroups were drawn and the sample statistics $\bar{\bar{x}}$ and S computed. The objective in phase 1 is to obtain an in-control set of observations so that control limits can be established for phase 2, which is the monitoring of future production. This is sometimes called a **retrospective analysis.**

The phase 1 control limits for the T^2 control chart are given by

$$\text{UCL} = \frac{p(m-1)(n-1)}{mn - m - p + 1} F_{\alpha,\, p,\, mn-m-p+1}$$

$$\text{LCL} = 0 \tag{10-20}$$

In **phase 2,** when the chart is used for monitoring future production, the control limits are as follows:

$$\text{UCL} = \frac{p(m+1)(n-1)}{mn - m - p + 1} F_{\alpha,\, p,\, mn-m-p+1}$$

$$\text{LCL} = 0 \tag{10-21}$$

When μ and Σ are estimated from a large number of preliminary samples, it is customary to use $\text{UCL} = \chi^2_{\alpha,p}$ as the upper control limit in both phase 1 and phase 2. Retrospective analysis of the preliminary samples to test for statistical control and establish control limits also occurs in the univariate control chart setting. For the $\bar{x}$ chart, it is well known that if we use $m \geq 20$ or 25 preliminary samples, the distinction between phase 1 and phase 2 limits is usually unnecessary, because the phase 1 and phase 2 limits will nearly coincide. However, with multivariate control charts, we must be careful.

Lowry and Montgomery (1995) show that in many situations a large number of preliminary samples would be required before the exact phase 2 control limits are well approximated by the chi-square limits. These authors present tables indicating the recommended minimum value of m for sample sizes of $n = 3, 5,$ and 10 and for $p = 2, 3, 4, 5, 10,$ and 20 quality characteristics. The recommended values of m are always greater than 20 preliminary samples, and often more than 50 samples.

······· **EXAMPLE 10-1** ··

The tensile strength and diameter of a textile fiber are two important quality characteristics that are to be jointly controlled. The quality engineer has decided to use $n = 10$ fiber specimens in each sample. He has taken 20 preliminary samples, and on the basis of these data he concludes that $\bar{\bar{x}}_1 = 115.59$ psi, $\bar{\bar{x}}_2 = 1.06(\times 10^{-2})$ inch, $\bar{S}_1^2 = 1.23$, $\bar{S}_2^2 = 0.83$, and $\bar{S}_{12} = 0.79$. Therefore, the statistic he will use for process control purposes is

$$T^2 = \frac{10}{(1.23)(0.83) - (0.79)^2} [0.83(\bar{x}_1 - 115.59)^2 + 1.23(\bar{x}_2 - 1.06)^2$$
$$- 2(0.79)(\bar{x}_1 - 115.59)(\bar{x}_2 - 1.06)]$$

The data used in this analysis and the summary statistics are in Table 10-1, panels (a) and (b).

Figure 10-7 presents the Hotelling T^2 control chart for this example. We will consider this to be phase 1, establishing statistical control in the preliminary samples, and calculate the upper control limit from equation 10-20. If $\alpha = 0.001$, then the UCL is

$$\text{UCL} = \frac{p(m-1)(n-1)}{mn - m - p + 1} F_{\alpha, p, mn-m-p+1}$$

$$= \frac{2(19)(9)}{20(10) - 20 - 2 + 1} F_{0.001,2,20(10)-20-2+1}$$

$$= \frac{342}{179} F_{0.001,2,179}$$

$$= (1.91)7.18$$

$$= 13.72$$

This control limit is shown on the chart in Fig. 10-7. Notice that no points exceed this limit, so we would conclude that the process is in control. Phase 2 control limits could be calculated from equation 10-21. If $\alpha = 0.001$, the upper control limit is UCL = 15.16. If

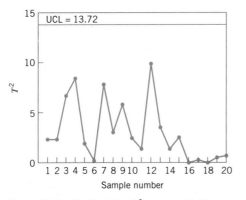

Figure 10-7 The Hotelling T^2 control chart for tensile strength and diameter, Example 10-1.

Table 10-1 Data for Example 10-1

Sample Number k	(a) Sample Means		(b) Variances and Covariances			(c) Control Chart Statistics	
	Tensile Strength ($\bar{x}_{1k}$)	Diameter ($\bar{x}_{2k}$)	S_{1k}^2	S_{2k}^2	S_{12k}	T_k^2	$\|S_k\|$
1	115.25	1.04	1.25	0.87	0.80	2.16	0.45
2	115.91	1.06	1.26	0.85	0.81	2.14	0.41
3	115.05	1.09	1.30	0.90	0.82	6.77	0.50
4	116.21	1.05	1.02	0.85	0.81	8.29	0.21
5	115.90	1.07	1.16	0.73	0.80	1.89	0.21
6	115.55	1.06	1.01	0.80	0.76	0.03	0.23
7	114.98	1.05	1.25	0.78	0.75	7.54	0.41
8	115.25	1.10	1.40	0.83	0.80	3.01	0.52
9	116.15	1.09	1.19	0.87	0.83	5.92	0.35
10	115.92	1.05	1.17	0.86	0.95	2.41	0.10
11	115.75	0.99	1.45	0.79	0.78	1.13	0.54
12	114.90	1.06	1.24	0.82	0.81	9.96	0.36
13	116.01	1.05	1.26	0.55	0.72	3.86	0.17
14	115.83	1.07	1.17	0.76	0.75	1.11	0.33
15	115.29	1.11	1.23	0.89	0.82	2.56	0.42
16	115.63	1.04	1.24	0.91	0.83	0.70	0.19
17	115.47	1.03	1.20	0.95	0.70	0.19	0.65
18	115.58	1.05	1.18	0.83	0.79	0.00	0.36
19	115.72	1.06	1.31	0.89	0.76	0.35	0.59
20	115.40	1.04	1.29	0.85	0.68	0.62	0.63
Averages	$\bar{\bar{x}}_1 = 115.59$	$\bar{\bar{x}}_2 = 1.06$	$\bar{S}_1^2 = 1.23$	$\bar{S}_2^2 = 0.83$	$\bar{S}_{12} = 0.79$		

we had used the approximate chi-square control limit, we would have obtained $\chi_{0.001,2}^2 = 13.816$, which is reasonably close to the correct limit for phase 1 but somewhat too small for phase 2.

. . . .

The widespread industrial interest in multivariate quality control has led several software developers to include the Hotelling T^2 control chart in their library of control chart programs. These programs must be used carefully, as they often display incorrect control limits. Figure 10-8 is the Hotelling control chart from Statgraphics for the data of Example 10-1. The control limit (at $\alpha = 0.001$) reported by Statgraphics is computed from

$$\text{UCL} = \frac{p(m-1)}{m-p} F_{\alpha, p, m-p}$$

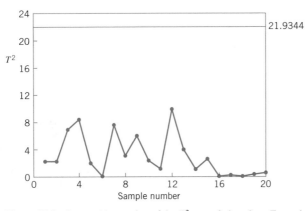

Figure 10-8   Statgraphics version of the T^2 control chart from Example 10-1.

This limit is obviously incorrect. This is the correct critical region to use in multivariate statistical hypothesis testing on the mean vector $\boldsymbol{\mu}$, where a sample of size m is taken at random from a p-dimensional normal distribution, but it is not directly applicable to the control chart.

Interpretation of Out-of-Control Signals

One difficulty encountered with any multivariate control chart is practical **interpretation** of an out-of-control signal. Specifically, which of the p variables (or which *subset* of them) is responsible for the signal? This question is not always easy to answer. The standard practice is to plot **univariate** $\bar{x}$ charts on the individual variables $x_1, x_2, \ldots, x_p$. However, this approach may not be successful, for reasons discussed previously. Alt (1985) suggests using $\bar{x}$ charts with Bonferroni-type control limits [i.e., replace $Z_{\alpha/2}$ in the $\bar{x}$ chart control limit calculation with $Z_{\alpha/(2p)}$]. This approach reduces the number of false alarms associated with using many simultaneous univariate control charts. Hayter and Tsui (1994) extend this idea by giving a procedure for exact simultaneous confidence intervals. Their procedure can also be used in situations where the normality assumption is not valid. Jackson (1980) recommends using control charts based on the p principal components (which are linear combinations of the original variables). Principal components are discussed in Section 10-7. The disadvantage of this approach is that the principal components do not always provide a clear interpretation of the situation with respect to the original variables. However, they are often effective in diagnosing an out-of-control signal, particularly in cases where the principal components do have an interpretation in terms of the original variables.

Another very useful approach to diagnosis of an out-of-control signal is to decompose the T^2 statistic into components that reflect the contribution of each individual variable. If T^2 is the current value of the statistic, and $T^2_{(i)}$ is the value of the statistic for all process variables except the ith one, then Runger, Alt, and

Montgomery (1996b) show that

$$d_i = T^2 - T^2_{(i)} \tag{10-22}$$

is an indicator of the relative contribution of the *i*th variable to the overall statistic. When an out-of-control signal is generated, we recommend computing the values of d_i ($i = 1, 2, \ldots, p$) and focusing attention on the variables for which d_i are relatively large. This procedure has an additional advantage in that the calculations can be performed using standard software packages.

To illustrate this procedure, consider the following example from Runger, Alt, and Montgomery (1996a). There are $p = 3$ quality characteristics and the covariance matrix is known. Assume that all three variables have been scaled as follows:

$$y_{ij} = \frac{x_{ij} - \mu_j}{\sqrt{(m - 1)\sigma_j^2}}$$

This scaling results in each process variable having mean zero and variance one. Therefore, the covariance matrix Σ is in **correlation form**; that is, the main diagonal elements are all one and the off-diagonal elements are the pairwise correlation between the process variables (the *x*'s). In our example,

$$\Sigma = \begin{bmatrix} 1 & 0.9 & 0.9 \\ 0.9 & 1 & 0.9 \\ 0.9 & 0.9 & 1 \end{bmatrix}$$

The in-control value of the process mean is $\mu' = [0, 0, 0]$. Consider the following display:

Observation Vector y'	Control Chart Statistic $T_0^2 (= \chi_0^2)$	$d_i = T^2 - T^2_{(i)}$		
		d_1	d_2	d_3
$(2, 0, 0)$	27.14	27.14	6.09	6.09
$(1, 1, -1)$	26.79	6.79	6.79	25.73
$(1, -1, 0)$	20.00	14.74	14.74	0
$(0.5, 0.5, 1)$	15.00	3.69	3.68	14.74

Since Σ is known, we can calculate the upper control limit for the chart from a chi-square distribution. We will choose $\chi^2_{0.005,3} = 12.84$ as the upper control limit. Clearly all four observation vectors in the above display would generate an out-of-control signal. Runger, Alt, and Montgomery (1996b) suggest that an approximate cutoff for the magnitude of an individual d_i is $\chi^2_{\alpha,1}$. Selecting $\alpha = 0.01$, we would find $\chi^2_{0.01,1} = 6.63$, so any d_i exceeding this value would be considered a large contributor. The decomposition statistics d_i computed above give clear guidance regarding *which* variables in the observation vector have shifted.

Other diagnostics have been suggested in the literature. For example, Murphy (1987) and Chua and Montgomery (1992) have developed procedures based on discrimination analysis, a statistical procedure for classifying observations into groups. Tracy, Mason, and Young (1996) also use decompositions of T^2 for diagnostic purposes, but their procedure requires more extensive computations and uses more elaborate decompositions than equation 10-22.

10-3.2   Individual Observations

In some industrial settings the subgroup size is naturally $n = 1$. This situation occurs frequently in the chemical and process industries. Since these industries frequently have multiple quality characteristics that must be monitored, multivariate control charts with $n = 1$ would be of interest there.

Suppose that m samples, each of size $n = 1$, are available and that p is the number of quality characteristics observed in each sample. Let $\mathbf{x}$ and $\mathbf{S}$ be the sample mean vector and covariance matrix, respectively, of these observations. The Hotelling T^2 statistic in equation 10-19 becomes

$$T^2 = (\mathbf{x} - \bar{\mathbf{x}})'\mathbf{S}^{-1}(\mathbf{x} - \bar{\mathbf{x}}) \qquad (10\text{-}23)$$

The phase 2 control limits for this statistic are

$$\text{UCL} = \frac{p(m + 1)(m - 1)}{m^2 - mp} F_{\alpha, p, m-p}$$

$$\text{LCL} = 0 \qquad (10\text{-}24)$$

When the number of preliminary samples m is large—say, $m > 100$—many practitioners use an approximate control limit, either

$$\text{UCL} = \frac{p(m - 1)}{m - p} F_{\alpha, p, m-p} \qquad (10\text{-}25)$$

or

$$\text{UCL} = \chi^2_{\alpha, p} \qquad (10\text{-}26)$$

For $m > 100$, equation 10-25 is a reasonable approximation. The chi-square limit in equation 10-26 is only appropriate if the covariance matrix is known, but it is widely used as an approximation. Lowry and Montgomery (1995) show that the chi-square limit should be used with caution. If p is large—say, $p \geq 10$—then at least 250 samples must be taken ($m \geq 250$) before the chi-square upper control limit is a reasonable approximation to the correct value.

Tracy, Young, and Mason (1992) point out that if $n = 1$, the phase 1 limits should be based on a beta distribution. This would lead to phase 1 limits defined as

$$\text{UCL} = \frac{(m - 1)^2}{m} \beta_{\alpha, p/2, (m-p-1)/2}$$

$$\text{LCL} = 0 \tag{10-27}$$

where $\beta_{\alpha, p/2, (m-p-1)/2}$ is the upper α percentage point of a beta distribution with parameters $p/2$ and $(m - p - 1)/2$. Approximations to the phase 1 limits based on the F and chi-square distributions are likely to be inaccurate.

A significant issue in the case of individual observations is estimating the covariance matrix $\boldsymbol{\Sigma}$. Sullivan and Woodall (1995) give an excellent discussion and analysis of this problem, and compare several estimators. One of these is the "usual" estimator obtained by simply pooling all m observations—say,

$$\mathbf{S}_1 = \frac{1}{m - 1} \sum_{i=1}^{m} (\mathbf{x}_i - \bar{\mathbf{x}})(\mathbf{x}_i - \bar{\mathbf{x}})'$$

Just as in the univariate case with $n = 1$, we would expect that $\mathbf{S}_1$ would be sensitive to outliers or out-of-control observations in the original sample of n observations. The second estimator [originally suggested by Holmes and Mergen (1993)], uses the difference between successive pairs of observations:

$$\mathbf{v}_i = \mathbf{x}_{i+1} - \mathbf{x}_i \qquad i = 1, 2, \ldots, m - 1 \tag{10-28}$$

Now arrange these vectors into a matrix $\mathbf{V}$, where

$$\mathbf{V} = \begin{bmatrix} \mathbf{v}'_1 \\ \mathbf{v}'_2 \\ \vdots \\ \mathbf{v}'_{m-1} \end{bmatrix}$$

The estimator for $\boldsymbol{\Sigma}$ is one-half the sample covariance matrix of these differences:

$$\mathbf{S}_2 = \frac{1}{2} \frac{\mathbf{V}'\mathbf{V}}{(m - 1)} \tag{10-29}$$

(Sullivan and Woodall originally denoted this estimator $\mathbf{S}_5$.)

Table 10-2 shows the example from Sullivan and Woodall (1996), in which they apply the T^2 chart procedure to the Holmes and Mergen (1993) data. There are 56 observations on the composition of "grit," where L, M, and S denote the percentages classified as large, medium, and small, respectively. Only the first two components were used because all those percentages add to 100%. The mean vector for these data is $\bar{\mathbf{x}}' = [5.682, 88.22]$. The two sample covariance matrices are

$$\mathbf{S}_1 = \begin{bmatrix} 3.770 & -5.495 \\ -5.495 & 13.53 \end{bmatrix}$$

$$\mathbf{S}_2 = \begin{bmatrix} 1.562 & -2.093 \\ -2.093 & 6.721 \end{bmatrix}$$

Figure 10-9 shows the T^2 control charts from this example. Sullivan and Woodall (1996) used simulation methods to find exact control limits for this data set (the false

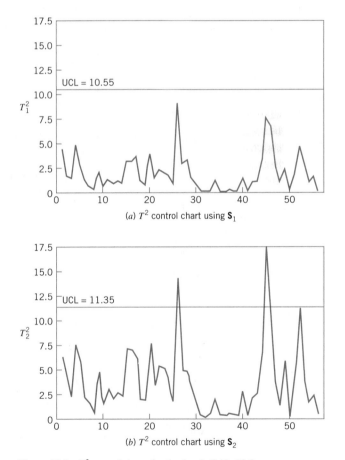

(a) T^2 control chart using $\mathbf{S}_1$

(b) T^2 control chart using $\mathbf{S}_2$

Figure 10-9 T^2 control charts for the data in Table 10-2.

Table 10-2   Example from Sullivan and Woodall (1996) Using the Data from Holmes and Mergen (1993) and the T^2 Statistics Using Estimators S_1 and S_2

i	$L = x_{i,1}$	$M = x_{i,2}$	$S = x_{i,3}$	$T^2_{1,i}$	$T^2_{2,i}$	i	$L = x_{i,1}$	$M = x_{i,2}$	$S = x_{i,3}$	$T^2_{1,i}$	$T^2_{2,i}$
1	5.4	93.6	1.0	4.496	6.439	29	7.4	83.6	9.0	1.594	3.261
2	3.2	92.6	4.2	1.739	4.227	30	6.8	84.8	8.4	0.912	1.743
3	5.2	91.7	3.1	1.460	2.200	31	6.3	87.1	6.6	0.110	0.266
4	3.5	86.9	9.6	4.933	7.643	32	6.1	87.2	6.7	0.077	0.166
5	2.9	90.4	6.7	2.690	5.565	33	6.6	87.3	6.1	0.255	0.564
6	4.6	92.1	3.3	1.272	2.258	34	6.2	84.8	9.0	1.358	2.069
7	4.4	91.5	4.1	0.797	1.676	35	6.5	87.4	6.1	0.203	0.448
8	5.0	90.3	4.7	0.337	0.645	36	6.0	86.8	7.2	0.193	0.317
9	8.4	85.1	6.5	2.088	4.797	37	4.8	88.8	6.4	0.297	0.590
10	4.2	89.7	6.1	0.666	1.471	38	4.9	89.8	5.3	0.197	0.464
11	3.8	92.5	3.7	1.368	3.057	39	5.8	86.9	7.3	0.242	0.353
12	4.3	91.8	3.9	0.951	1.986	40	7.2	83.8	9.0	1.494	2.928
13	3.7	91.7	4.6	1.105	2.688	41	5.6	89.2	5.2	0.136	0.198
14	3.8	90.3	5.9	1.019	2.317	42	6.9	84.5	8.6	1.079	2.062
15	2.6	94.5	2.9	3.099	7.262	43	7.4	84.4	8.2	1.096	2.477
16	2.7	94.5	2.8	3.036	7.025	44	8.9	84.3	6.8	2.854	6.666
17	7.9	88.7	3.4	3.803	6.189	45	10.9	82.2	6.9	7.677	17.666
18	6.6	84.6	8.8	1.167	1.997	46	8.2	89.8	2.0	6.677	10.321
19	4.0	90.7	5.3	0.751	1.824	47	6.7	90.4	2.9	2.708	3.869
20	2.5	90.2	7.3	3.966	7.811	48	5.9	90.1	4.0	0.888	1.235
21	3.8	92.7	3.5	1.486	3.247	49	8.7	83.6	7.7	2.424	5.914
22	2.8	91.5	5.7	2.357	5.403	50	6.4	88.0	5.6	0.261	0.470
23	2.9	91.8	5.3	2.094	4.959	51	8.4	84.7	6.9	1.995	4.731
24	3.3	90.6	6.1	1.721	3.800	52	9.6	80.6	9.8	4.732	11.259
25	7.2	87.3	5.5	0.914	1.791	53	5.1	93.0	1.9	2.891	4.303
26	7.3	79.0	13.7	9.226	14.372	54	5.0	91.4	3.6	0.989	1.609
27	7.0	82.6	10.4	2.940	4.904	55	5.0	86.2	8.8	1.770	2.495
28	6.0	83.5	10.5	3.310	4.771	56	5.9	87.2	6.9	0.102	0.166

alarm probability is 0.155). Note that only the control chart in Fig. 10-9*b* based on S_2 signals. It turns out that if we consider only samples 1–24, the sample mean vector is

$$\overline{\mathbf{x}}'_{1-24} = [4.23, 90.8]$$

and if we consider only the last 32 observations,

$$\overline{\mathbf{x}}'_{25-56} = [6.77, 86.3]$$

These are statistically significantly different, whereas the "within" covariance matrices are not significantly different. There is an apparent shift in the mean vector following sample 24, and this was correctly detected by the control chart based on S_2.

10-4   THE MULTIVARIATE EWMA CONTROL CHART

The chi-square and T^2 charts described in the previous section are Shewhart-type control charts. That is, they use information only from the current sample, so consequently, they are relatively insensitive to small and moderate shifts in the mean vector. Cumulative sum and EWMA control charts were developed to provide more sensitivity to small shifts in the univariate case, and they can be extended to multivariate quality control problems.

Crosier (1988) and Pignatiello and Runger (1990) have proposed several multivariate cusum procedures. Lowry et al. (1992) have developed a multivariate version of the EWMA (or an MEWMA) control chart. The MEWMA is a logical extension of the univariate EWMA and is defined as follows:

$$\mathbf{Z}_i = \lambda \mathbf{x}_i + (1 - \lambda)\mathbf{Z}_{i-1} \qquad\qquad (10\text{-}30)$$

where $0 < \lambda \leq 1$ and $\mathbf{Z}_0 = \mathbf{0}$. The quantity plotted on the control chart is

$$\mathbf{T}_i^2 = \mathbf{Z}_i' \mathbf{\Sigma}_{\mathbf{z}_i}^{-1} \mathbf{Z}_i \qquad\qquad (10\text{-}31)$$

where the covariance matrix is

$$\mathbf{\Sigma}_{\mathbf{z}_i} = \frac{\lambda}{2 - \lambda}[1 - (1 - \lambda)^{2i}]\mathbf{\Sigma} \qquad\qquad (10\text{-}32)$$

which is analogous to the variance of the univariate EWMA.

Prabhu and Runger (1997) have provided a thorough analysis of the average run-length performance of the MEWMA control chart, using a modification of the Brook and Evans (1972) Markov chain approach. They give tables and charts to guide selection of the upper control limit—say, UCL = H—for the MEWMA. Tables 10-3 and 10-4 contain this information. Table 10-3 contains ARL performance for MEWMA for various values of λ for $p = 2, 4, 6, 10,$ and 15 quality characteristics. The control limit H was chosen to give an in-control ARL$_0$ = 200. The ARLs in this table are all "zero-state" ARLs; that is, we assume that the process is in control when the chart is initiated. The shift size is reported in terms of a quantity

$$\delta = (\boldsymbol{\mu}' \mathbf{\Sigma}^{-1} \boldsymbol{\mu})^{1/2} \qquad\qquad (10\text{-}33)$$

usually called the **noncentrality parameter.** Basically, large values of δ correspond to bigger shifts in the mean. The value $\delta = 0$ is the in-control state (this is true because the

Table 10-3 Average Run Lengths (zero stats) for the MEWMA Control Chart [from Prabhu and Runger (1997)]

					λ				
p	δ	0.05	0.10	0.20	0.30	0.40	0.50	0.60	0.80
		$H = 7.35$	8.64	9.65	10.08	10.31	10.44	10.52	10.58
2	0.0	199.93	199.98	199.91	199.82	199.83	200.16	200.04	200.20
	0.5	26.61	28.07	35.17	44.10	53.82	64.07	74.50	95.88
	1.0	11.23	10.15	10.20	11.36	13.26	15.88	19.24	28.65
	1.5	7.14	6.11	5.49	5.48	5.78	6.36	7.25	10.28
	2.0	5.28	4.42	3.78	3.56	3.53	3.62	3.84	4.79
	3.0	3.56	2.93	2.42	2.20	2.05	1.95	1.90	1.91
		$H = 11.22$	12.73	13.87	14.34	14.58	14.71	14.78	14.85
4	0.0	199.84	200.12	199.94	199.91	199.96	200.05	199.99	200.05
	0.5	32.29	35.11	46.30	59.28	72.43	85.28	97.56	120.27
	1.0	13.48	12.17	12.67	14.81	18.12	22.54	28.06	42.58
	1.5	8.54	7.22	6.53	6.68	7.31	8.40	10.03	15.40
	2.0	6.31	5.19	4.41	4.20	4.24	4.48	4.93	6.75
	3.0	4.23	3.41	2.77	2.50	2.36	2.27	2.24	2.37
		$H = 14.60$	16.27	17.51	18.01	18.26	18.39	18.47	18.54
6	0.0	200.11	200.03	200.11	200.18	199.81	200.01	199.87	200.17
	0.5	36.39	40.38	54.71	70.30	85.10	99.01	111.65	133.91
	1.0	15.08	13.66	14.63	17.71	22.27	28.22	35.44	53.51
	1.5	9.54	8.01	7.32	7.65	8.60	10.20	12.53	20.05
	2.0	7.05	5.74	4.88	4.68	4.80	5.20	5.89	8.60
	3.0	4.72	3.76	3.03	2.72	2.58	2.51	2.51	2.77
		$H = 20.72$	22.67	24.07	24.62	24.89	25.03	25.11	25.17
10	0.0	199.91	199.95	200.08	200.01	199.98	199.84	200.12	200.00
	0.5	42.49	48.52	67.25	85.68	102.05	116.25	128.82	148.96
	1.0	17.48	15.98	17.92	22.72	29.47	37.81	47.54	69.71
	1.5	11.04	9.23	8.58	9.28	10.91	13.49	17.17	28.33
	2.0	8.15	6.57	5.60	5.47	5.77	6.48	7.68	12.15
	3.0	5.45	4.28	3.43	3.07	2.93	2.90	2.97	3.54
		$H = 27.82$	30.03	31.59	32.19	32.48	32.63	32.71	32.79
15	0.0	199.95	199.89	200.08	200.03	199.96	199.91	199.93	200.16
	0.5	48.20	56.19	78.41	98.54	115.36	129.36	141.10	159.55
	1.0	19.77	18.28	21.40	28.06	36.96	47.44	59.03	83.86
	1.5	12.46	10.41	9.89	11.08	13.53	17.26	22.38	37.07
	2.0	9.20	7.36	6.32	6.30	6.84	7.97	9.80	16.36
	3.0	6.16	4.78	3.80	3.43	3.29	3.31	3.49	4.49

control chart can be constructed using "standardized" data). Note that for a given shift size, ARLs generally tend to increase as λ increases, except for very large values of δ (or large shifts). Since the MEWMA with $\lambda = 1$ is equivalent to the T^2 (or chi-square) control chart, the MEWMA is more sensitive to smaller shifts. This is analogous to the univariate case.

Table 10-4 presents "optimum" MEWMA chart designs for various shifts (δ) and in-control target values of ARL_0 of either 500 or 1000. ARL_{min} is the minimum value of ARL_1 achieved for the value of λ specified.

To illustrate the design of a MEWMA control chart, suppose that $p = 6$ and the covariance matrix is

$$\Sigma = \begin{bmatrix} 1 & 0.7 & 0.9 & 0.3 & 0.2 & 0.3 \\ 0.7 & 1 & 0.8 & 0.1 & 0.4 & 0.2 \\ 0.9 & 0.8 & 1 & 0.1 & 0.2 & 0.1 \\ 0.3 & 0.1 & 0.1 & 1 & 0.2 & 0.1 \\ 0.2 & 0.4 & 0.2 & 0.2 & 1 & 0.1 \\ 0.3 & 0.2 & 0.1 & 0.1 & 0.1 & 1 \end{bmatrix}$$

Note that Σ is in **correlation** form. Suppose that we are interested in a process shift from $\mu' = 0$ to

$$\mu' = [1, 1, 1, 1, 1, 1]$$

Table 10-4   Optimal MEWMA Control Charts

δ	$ARL_0 =$	$p = 4$ 500	$p = 4$ 1000	$p = 10$ 500	$p = 10$ 1000	$p = 20$ 500	$p = 20$ 1000
0.5	λ	0.04	0.03	0.03	0.025	0.03	0.025
	H	13.37	14.68	22.69	24.70	37.09	39.63
	ARL_{min}	42.22	49.86	55.94	66.15	70.20	83.77
1.0	λ	0.105	0.09	0.085	0.075	0.075	0.065
	H	15.26	16.79	25.42	27.38	40.09	42.47
	ARL_{min}	14.60	16.52	19.29	21.74	24.51	27.65
1.5	λ	0.18	0.18	0.16	0.14	0.14	0.12
	H	16.03	17.71	26.58	28.46	41.54	43.80
	ARL_{min}	7.65	8.50	10.01	11.07	12.70	14.01
2.0	λ	0.28	0.26	0.24	0.22	0.20	0.18
	H	16.49	18.06	27.11	29.02	42.15	44.45
	ARL_{min}	4.82	5.30	6.25	6.84	7.88	8.60
3.0	λ	0.52	0.46	0.42	0.40	0.36	0.34
	H	16.84	18.37	27.55	29.45	42.80	45.08
	ARL_{min}	2.55	2.77	3.24	3.50	4.04	4.35

Note: ARL_0 and ARL_{min} are zero-state average run lengths.

This is essentially a one-sigma upward shift in all $p = 6$ variables. For this shift, $\delta = (\boldsymbol{\mu}'\boldsymbol{\Sigma}^{-1}\boldsymbol{\mu})^{1/2} = 1.75$. Table 10-3 suggests that $\lambda = 0.2$ and $H = 17.51$ would give an in-control $ARL_0 = 200$ and the ARL_1 would be between 4.88 and 7.32. It turns out that if the mean shifts by any constant multiple—say, k—of the original vector $\boldsymbol{\mu}$, then δ changes to $k\delta$. Therefore, ARL performance is easy to evaluate. For example, if $k = 1.5$, then the new δ is $\delta = 1.5(1.75) = 2.625$, and the ARL_1 would be between 3.03 and 4.88.

10-5 REGRESSION ADJUSTMENT

The Hotelling T^2 (and chi-square) control chart is based on the general idea of testing the hypothesis that the mean vector of a multivariate normal distribution is equal to a constant vector against the alternative hypothesis that the mean vector is not equal to that constant. In fact, it is an optimal test statistic for that hypothesis. However, it is not necessarily an optimal control-charting procedure for detecting mean shifts. The MEWMA can be designed to have faster detection capability (smaller values of the ARL_1). Furthermore, the Hotelling T^2 is not optimal for more structured shifts in the mean, such as shifts in only a few of the process variables. It also turns out that the Hotelling T^2, and any method that uses the *quadratic form* structure of the Hotelling T^2 test statistic (such as the MEWMA), will be sensitive to shifts in the variance as well as to shifts in the mean. Consequently, various researchers have developed methods to monitor multivariate processes that do not depend on the Hotelling T^2 statistic.

Hawkins (1991) has developed a procedure called **regression adjustment** that is potentially very useful. The scheme essentially consists of plotting univariate control charts of the **residuals** from each variable obtained when that variable is regressed on all the others. The procedure is very applicable to individual measurements, a case that occurs frequently in practice with multivariate data, and implementation is straightforward, since it requires only a least squares regression computer program to process the data prior to constructing the control charts. Hawkins shows that the ARL performance of this scheme is very competitive with other methods, but depends on the types of control charts applied to the residuals.

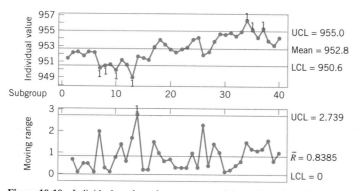

Figure 10-10   Individuals and moving range control charts for y_1 from Table 10-5.

Table 10-5 Cascade Process Data

Observation	x_1	x_2	x_3	x_4	x_5	x_6	x_7	x_8	x_9	y_1	Residuals	y_2
1	12.78	0.15	91	56	1.54	7.38	1.75	5.89	1.11	951.5	0.81498	87
2	14.97	0.1	90	49	1.54	7.14	1.71	5.91	1.109	952.2	− 0.31685	88
3	15.43	0.07	90	41	1.47	7.33	1.64	5.92	1.104	952.3	− 0.28369	86
4	14.95	0.12	89	43	1.54	7.21	1.93	5.71	1.103	951.8	− 0.45924	89
5	16.17	0.1	83	42	1.67	7.23	1.86	5.63	1.103	952.3	− 0.56512	86
6	17.25	0.07	84	54	1.49	7.15	1.68	5.8	1.099	952.2	− 0.22592	91
7	16.57	0.12	89	61	1.64	7.23	1.82	5.88	1.096	950.2	− 0.55431	99
8	19.31	0.08	99	60	1.46	7.74	1.69	6.13	1.092	950.5	− 0.18874	10
9	18.75	0.04	99	52	1.89	7.57	2.02	6.27	1.084	950.6	0.15245	10
10	16.99	0.09	98	57	1.66	7.51	1.82	6.38	1.086	949.8	− 0.33580	10
11	18.2	0.13	98	49	1.66	7.27	1.92	6.3	1.089	951.2	− 0.85525	98
12	16.2	0.16	97	52	2.16	7.21	2.34	6.07	1.089	950.6	0.47027	96
13	14.72	0.12	82	61	1.49	7.33	1.72	6.01	1.092	948.9	− 1.74107	93
14	14.42	0.13	81	63	1.16	7.5	1.5	6.11	1.094	951.7	0.62057	91
15	11.02	0.1	83	56	1.56	7.14	1.73	6.14	1.102	951.5	0.72583	91
16	9.82	0.1	86	53	1.26	7.32	1.54	6.15	1.112	951.3	− 0.03421	93
17	11.41	0.12	87	49	1.29	7.22	1.57	6.13	1.114	952.9	0.28093	91
18	14.74	0.1	81	42	1.55	7.17	1.77	6.28	1.114	953.9	− 1.87257	94
19	14.5	0.08	84	53	1.57	7.23	1.69	6.28	1.109	953.3	− 0.20805	96
20	14.71	0.09	89	46	1.45	7.23	1.67	6.12	1.108	952.6	− 0.66749	94
21	15.26	0.13	91	47	1.74	7.28	1.98	6.19	1.105	952.3	− 0.75390	99
22	17.3	0.12	95	47	1.57	7.18	1.86	6.06	1.098	952.6	− 0.03479	95
23	17.62	0.06	95	42	2.05	7.15	2.14	6.15	1.096	952.9	0.24439	92
24	18.21	0.06	93	41	1.46	7.28	1.61	6.11	1.096	953.9	0.67889	87
25	14.38	0.1	90	46	1.42	7.29	1.73	6.13	1.1	954.2	1.94313	89
26	12.13	0.14	87	50	1.76	7.21	1.9	6.31	1.112	951.9	− 0.92344	98
27	12.72	0.1	90	47	1.52	7.25	1.79	6.25	1.112	952.3	− 0.74707	95
28	17.42	0.1	89	51	1.33	7.38	1.51	6.01	1.111	953.7	− 0.21053	88
29	17.63	0.11	87	45	1.51	7.42	1.68	6.11	1.103	954.7	0.66802	86
30	16.17	0.05	83	57	1.41	7.35	1.62	6.14	1.105	954.6	1.35076	84
31	16.88	0.16	86	58	2.1	7.15	2.28	6.42	1.105	954.8	0.61137	91
32	13.87	0.16	85	46	2.1	7.11	2.16	6.44	1.106	954.4	0.56960	92
33	14.56	0.05	84	41	1.34	7.14	1.51	6.24	1.113	955	− 0.09131	88
34	15.35	0.12	83	40	1.52	7.08	1.81	6	1.114	956.5	1.03785	83
35	15.91	0.12	81	45	1.76	7.26	1.9	6.07	1.116	955.3	− 0.07282	83
36	14.32	0.11	85	47	1.58	7.15	1.72	6.02	1.113	954.2	0.53440	86
37	15.43	0.13	86	43	1.46	7.15	1.73	6.11	1.115	955.4	0.16379	85
38	14.47	0.08	85	54	1.62	7.1	1.78	6.15	1.118	953.8	− 0.37110	88
39	14.74	0.07	84	52	1.47	7.24	1.66	5.89	1.112	953.2	0.17177	83
40	16.28	0.13	86	49	1.72	7.05	1.89	5.91	1.109	954.2	0.47427	85

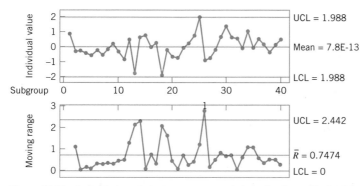

Figure 10-11 Individuals and moving range control charts for the residuals of the regression on y_1, Table 10-5.

A very important application of regression adjustment occurs when the process has a distinct hierarchy of variables, such as a set of *input* process variables (say, the x's) and a set of *output* variables (say, the y's). Sometimes we call this situation a **cascade process** [Hawkins (1993b)]. Table 10-5, on page 530, shows 40 observations from a cascade process, where there are nine input variables and two output variables. We will demonstrate the regression adjustment approach using only one of the output variables, y_1. Figure 10-10 is a control chart for individuals and a moving range control chart for the 40 observations on the output variable y_1. Note that there are seven out-of-control points on the individuals control chart. Using standard least squares regression techniques, we can fit the following regression model for y_1 to the process variables $x_1, x_2, \ldots, x_9$:

$$\hat{y}_1 = 826 + 0.474x_1 + 1.41x_2 - 0.117x_3 - 0.0824x_4 - 2.39x_5 - 1.30x_6 + 2.18x_7 + 2.98x_8 + 113x_9$$

The residuals are found simply by subtracting the fitted value from this equation from each corresponding observation on y_1. These residuals are shown in the next-to-last column of Table 10-5.

Figure 10-11 shows a control chart for individuals and a moving range control chart for the 40 residuals from this procedure. Note that there is now only one out-of-control point on the moving range chart, and the overall impression of process stability is rather different than was obtained from the control charts for y_1 alone, without the effects of the process variables taken into account.

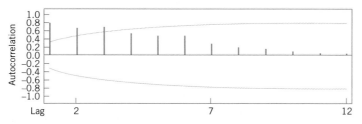

Figure 10-12 Sample autocorrelation function for y_1 from Table 10-5.

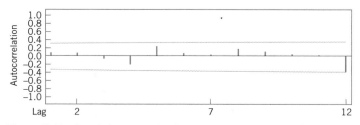

Figure 10-13   Sample autocorrelation function for the residuals from the regression on y_1, Table 10-5.

Regression adjustment has another nice feature. If the proper set of variables is included in the regression model, the residuals from the model will typically be uncorrelated, even though the original variable of interest y_1 exhibited correlation. To illustrate, Figure 10-12 is the sample autocorrelation function for y_1. Note that there is considerable autocorrelation at low lags in this variable. This is very typical behavior for data from a chemical or process plant. The sample autocorrelation function for the residuals is shown in Fig. 10-13. There is no evidence of autocorrelation in the residuals. Because of this nice feature, the regression adjustment procedure has many possible applications in chemical and process plants where there are often cascade processes with several inputs but only a few outputs, and where many of the variables are highly autocorrelated.

10-6   CONTROL CHARTS FOR MONITORING VARIABILITY

Just as it is important to monitor the process mean vector $\boldsymbol{\mu}$ in the multivariate case, it is also important to monitor process variability. Process variability is summarized by the $p \times p$ *covariance matrix* $\boldsymbol{\Sigma}$. The main diagonal elements of this matrix are the variances of the individual process variables, and the off-diagonal elements are the covariances. Alt (1985) gives a nice introduction to the problem and presents two useful procedures.

The first procedure is a direct extension of the univariate S^2 control chart. The procedure is equivalent to repeated tests of significance of the hypothesis that the process covariance matrix is equal to a particular matrix of constants $\boldsymbol{\Sigma}$. If this approach is used, the statistic plotted on the control chart for the ith sample is

$$W_i = -pn + pn \ln(n) - n \ln(|\mathbf{A}_i|/|\boldsymbol{\Sigma}|) + \mathrm{tr}(\boldsymbol{\Sigma}^{-1}\mathbf{A}_i) \qquad (10\text{-}34)$$

where $\mathbf{A}_i = (n-1)\mathbf{S}_i$, $\mathbf{S}_i$ is the sample covariance matrix for sample i, and tr is the trace operator. (The trace of a matrix is the sum of the main diagonal elements.) If the value of W_i plots above the upper control limit $\mathrm{UCL} = \chi^2_{\alpha, p(p+1)/2}$, the process is out of control.

The second approach is based on the sample *generalized* variance, $|\mathbf{S}|$. This statistic, which is the determinant of the sample covariance matrix, is a widely used measure of multivariate dispersion. Montgomery and Wadsworth (1972) used an asymptotic normal approximation to develop a control chart for $|\mathbf{S}|$. Another method would be to use the mean and variance of $|\mathbf{S}|$—that is, $E(|\mathbf{S}|)$ and $V(|\mathbf{S}|)$—and the property that most of the probability distribution of $|\mathbf{S}|$ is contained in the interval $E|\mathbf{S}| \pm 3\sqrt{V(|\mathbf{S}|)}$. It can be shown that

$$E(|\mathbf{S}|) = b_1|\mathbf{\Sigma}| \tag{10-35}$$

and

$$V(|\mathbf{S}|) = b_2|\mathbf{\Sigma}|^2$$

where

$$b_1 = \frac{1}{(n-1)^p} \prod_{i=1}^{p} (n - i)$$

and

$$b_2 = \frac{1}{(n-1)^{2p}} \prod_{i=1}^{p} (n - i) \left[\prod_{j=1}^{p} (n - j + 2) - \prod_{j=1}^{p} (n - j) \right]$$

Therefore, the parameters of the control chart for $|\mathbf{S}|$ would be

$$\begin{aligned}
\text{UCL} &= |\mathbf{\Sigma}|(b_1 + 3b_2^{1/2}) \\
\text{CL} &= b_1|\mathbf{\Sigma}| \\
\text{LCL} &= |\mathbf{\Sigma}|(b_1 - 3b_2^{1/2})
\end{aligned} \tag{10-36}$$

The lower control limit in equation 10-36 is replaced with zero if the calculated value is less than zero.

Usually, in practice $\mathbf{\Sigma}$ will be estimated by a sample covariance matrix $\mathbf{S}$, based on the analysis of preliminary samples. If this is the case, we should replace $|\mathbf{\Sigma}|$ in equation 10-36 by $|\mathbf{S}|/b_1$, since equation 10-35 has shown that $|\mathbf{S}|/b_1$ is an unbiased estimator of $|\mathbf{\Sigma}|$.

········ **EXAMPLE 10-2** ···

To illustrate controlling process variability in the multivariate case, we will return to Example 10-1 and construct a control chart for the generalized variance. Based on the 20 preliminary samples in Table 10-1, the sample covariance matrix is

$$\mathbf{S} = \begin{bmatrix} 1.23 & 0.79 \\ 0.79 & 0.83 \end{bmatrix}$$

so

$$|\mathbf{S}| = 0.3968$$

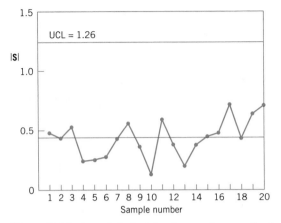

Figure 10-14 A control chart for the sample generalized variance, Example 10-2.

The constants b_1 and b_2 are (recall that $n = 10$)

$$b_1 = \frac{1}{81}(9)(8) = 0.8889$$

$$b_2 = \frac{1}{6561}(9)(8)[(11)(10) - (9)(8)] = 0.4170$$

Therefore, replacing $|\mathbf{\Sigma}|$ in equation 10-36 by $|\mathbf{S}|/b_1 = 0.3968/0.8889 = 0.4464$, we find that the control chart parameters are

$$\text{UCL} = (|\mathbf{S}|/b_1)(b_1 + 3b_2^{1/2}) = 0.4464[0.8889 + 3(0.4170)^{1/2}]$$

$$= 1.26$$

$$\text{CL} = |\mathbf{S}| = 0.3968$$

$$\text{LCL} = (|\mathbf{S}|/b_1)(b_1 - 3b_2^{1/2}) = 0.4464[0.8889 - 3(0.4170)^{1/2)}]$$

$$= -0.47 = 0$$

Figure 10-14 presents the control chart. The values of $|\mathbf{S}_i|$ for each sample are shown in the last column of panel (c) of Table 10-1.

Although the sample generalized variance is a widely used measure of multivariate dispersion, remember that it is a relatively simplistic scalar representation of a complex multivariable problem, and it is easy to be fooled if all we look at is $|\mathbf{S}|$. For example, consider the three covariance matrices:

$$\mathbf{S}_1 = \begin{bmatrix} 1 & 0 \\ 0 & 1 \end{bmatrix}$$

$$S_2 = \begin{bmatrix} 2.32 & 0.40 \\ 0.40 & 0.50 \end{bmatrix}$$

$$S_3 = \begin{bmatrix} 1.68 & -0.40 \\ -0.40 & 0.50 \end{bmatrix}$$

Now $|S_1| = |S_2| = |S_3| = 1$, yet the three matrices convey considerably different information about process variability and the correlation between the two variables. It is probably a good idea to use univariate control charts for variability in conjunction with the control chart for $|S|$.

10-7 LATENT STRUCTURE METHODS

Conventional multivariate control-charting procedures are reasonably effective as long as p (the number of process variables to be monitored) is not very large. However, as p increases, the average run-length performance to detect a specified shift in the mean of these variables for multivariate control charts also increases, because the shift is "diluted" in the p-dimensional space of the process variables. To illustrate this, consider the ARLs of the MEWMA control chart in Table 10-3. Suppose we choose $\lambda = 0.1$ and the magnitude of the shift is $\delta = 1.0$. Now in this table $ARL_0 = 200$ regardless of p, the number of parameters. However, note that as p increases, ARL_1 also increases. For $p = 2$, $ARL_1 = 10.15$, for $p = 6$, $ARL_1 = 13.66$, and for $p = 12$, $ARL_1 = 18.28$. Consequently, other methods are sometimes useful for process monitoring, particularly in situations where it is suspected that the variability in the process is not equally distributed among all p variables. That is, most of the "motion" of the process is in a relatively small subset of the original process variables.

Methods for discovering the subdimensions in which the process moves about are sometimes called **latent structure methods,** because of the analogy with photographic film on which a hidden or latent image is stored as a result of light interacting with the chemical surface of the film. We will discuss two of these methods, devoting most of our attention to the first one, called the method of **principal components.** We will also briefly discuss a second method called **partial least squares.**

10-7.1 Principal Components

The principal components of a set of process variables $x_1, x_2, \ldots, x_p$ are just a particular set of linear combinations of these variables—say,

$$
\begin{aligned}
z_1 &= c_{11}x_1 + c_{12}x_2 + \cdots + c_{1p}x_p \\
z_2 &= c_{21}x_1 + c_{22}x_2 + \cdots + c_{2p}x_p \\
&\qquad\qquad \vdots \\
z_p &= c_{p1}x_1 + c_{p2}x_2 + \cdots + c_{pp}x_p
\end{aligned}
\tag{10-37}
$$

where the c_{ij}'s are constants to be determined. Geometrically, the principal component variables $z_1, z_2, \ldots, z_p$ are the axes of a new coordinate system obtained by rotating the axes of the *original* system (the x's). The new axes represent the directions of maximum variability.

To illustrate, consider the two situations shown in Fig. 10-15. In Fig. 10-15a, there are two original variables x_1 and x_2, and two principal components z_1 and z_2. Note that the first principal component z_1 accounts for most of the variability in the two original variables. Figure 10-15b illustrates three original process variables. Most of the variability or "motion" in these two variables is in a plane, so only two principal components have been used to describe them. In this picture, once again z_1 accounts for most of the variability, but a nontrivial amount is also accounted for by the second principal component z_2. This is, in fact, the basic intent of principal components: Find the new set of orthogonal directions that define the maximum variability in the original data, and hopefully, this will lead to a description of the process requiring considerable fewer than the original p variables. The information contained in the complete set of all p principal components is exactly equivalent to the information in the complete set of all original process variables, but hopefully we can use far fewer than p principal components to obtain a satisfactory description.

It turns out that finding the c_{ij}'s that define the principal components is fairly easy. Let the random variables $x_1, x_2, \ldots, x_p$ be represented by a vector $\mathbf{x}$ with covariance matrix $\boldsymbol{\Sigma}$, and let the **eigenvalues** of $\boldsymbol{\Sigma}$ be $\lambda_1 \geq \lambda_2 \geq \cdots \geq \lambda_p \geq 0$. Then the constants c_{ij} are simply the elements of the ith **eigenvector** associated with the eigenvalue λ_i. Basically, if we let $\mathbf{C}$ be the matrix whose columns are the eigenvectors, then

$$\mathbf{C}'\boldsymbol{\Sigma}\mathbf{C} = \boldsymbol{\Lambda}$$

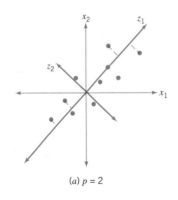

(a) $p = 2$

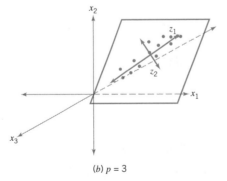

(b) $p = 3$

Figure 10-15 Principal components for $p = 2$ and $p = 3$ process variables.

where Λ is a $p \times p$ diagonal matrix with main diagonal elements equal to the eigenvalues $\lambda_1 \geq \lambda_2 \geq \cdots \geq \lambda_p \geq 0$. Many software packages will compute eigenvalues and eigenvectors and perform the principal components analysis.

The variance of the ith principal component is the ith eigenvalue λ_i. Therefore, the proportion of variability in the original data explained by the ith principal component is given by the ratio

$$\frac{\lambda_i}{\lambda_1 + \lambda_2 + \cdots + \lambda_p}$$

Therefore, one can easily see how much variability is explained by retaining just a few (say, r) of the p principal components simply by computing the sum of the eigenvalues for those r components and comparing that total to the sum of all p eigenvalues. It is a fairly typical practice to compute principal components using variables that have been standardized so that they have mean zero and unit variance. Then the covariance matrix Σ is in the form of a correlation matrix. The reason for this is that the original process variables are often expressed in different scales and as a result they can have very different magnitudes. Consequently, a variable may seem to contribute a lot to the total variability of the system just because its scale of measurement has larger magnitudes than the other variables. Standardization solves this problem nicely.

Once the principal components have been calculated and a subset of them selected, we can obtain new principal component observations z_{ij} simply by substituting the original observations x_{ij} into the set of retained principal components. This gives, for example,

$$z_{i1} = c_{11}x_{i1} + c_{12}x_{i2} + \cdots + c_{1p}x_{ip}$$
$$z_{i2} = c_{21}x \quad \cdots x_{i2} + \cdots + c_{2p}x_{ip}$$

$$(10\text{-}38)$$

$$z_{ir} = c_r \cdots$$

where we have retained the firs ... are
sometimes called the **principal**

We will illustrate this pro ... lysis
(or **PCA,** as it is often abbre ... $_2$, x_3,
and x_4 in Table 10-6, which ... ie first
20 observations in the uppe ... ther in
a pairwise manner in Fig. ... **scatter**
plots, and it indicates tha ... reas the
other two variables exhib ... 10-16 are
approximate 95% confid ... l distribution. The sample covari ... in correlation form, is

$$\Sigma = \begin{bmatrix} 0.9302 & \cdots & & \\ 0.2060 & 0.1669 & 1.000 & \\ 0.3595 & 0.4502 & 0.3439 & 1.0000 \end{bmatrix}$$

Table 10-6   Chemical Process Data

			Original Data			
Observation	x_1	x_2	x_3	x_4	z_1	z_2
1	10	20.7	13.6	15.5	0.291681	− 0.6034
2	10.5	19.9	18.1	14.8	0.294281	0.491533
3	9.7	20	16.1	16.5	0.197337	0.640937
4	9.8	20.2	19.1	17.1	0.839022	1.469579
5	11.7	21.5	19.8	18.3	3.204876	0.879172
6	11	20.9	10.3	13.8	0.203271	− 2.29514
7	8.7	18.8	16.9	16.8	− 0.99211	1.670464
8	9.5	19.3	15.3	12.2	− 1.70241	− 0.36089
9	10.1	19.4	16.2	15.8	− 0.14246	0.560808
10	9.5	19.6	13.6	14.5	− 0.99498	− 0.31493
11	10.5	20.3	17	16.5	0.944697	0.504711
12	9.2	19	11.5	16.3	− 1.2195	− 0.09129
13	11.3	21.6	14	18.7	2.608666	− 0.42176
14	10	19.8	14	15.9	− 0.12378	− 0.08767
15	8.5	19.2	17.4	15.8	− 1.10423	1.472593
16	9.7	20.1	10	16.6	− 0.27825	− 0.94763
17	8.3	18.4	12.5	14.2	− 2.65608	0.135288
18	11.9	21.8	14.1	16.2	2.36528	− 1.30494
19	10.3	20.5	15.6	15.1	0.411311	− 0.21893
20	8.9	19	8.5	14.7	− 2.14662	− 1.17849

			New Data			
Observation	x_1	x_2	x_3	x_4	z_1	z_2
21	9.9	20	15.4	15.9	0.074196	0.239359
22	8.7	19	9.9	16.8	− 1.51756	− 0.21121
23	11.5	21.8	19.3	12.1	1.408476	− 0.87591
24	15.9	24.6	14.7	15.3	6.298001	− 3.67398
25	12.6	23.9	17.1	14.2	3.802025	− 1.99584
26	14.9	25	16.3	16.6	6.490673	− 2.73143
27	9.9	23.7	11.9	18.1	2.738829	− 1.37617
28	12.8	26.3	13.5	13.7	4.958747	− 3.94851
29	13.1	26.1	10.9	16.8	5.678092	− 3.85838
30	9.8	25.8	14.8	15	3.369657	− 2.10878

Note that the correlation coefficient between x_1 and x_2 is 0.9302, which confirms the visual impression obtained from the matrix of scatter plots.

Table 10-7 presents the results of a PCA (Minitab was used to perform the calculations) on the first 20 observations, showing the eigenvalues and eigenvectors, as well as the percentage and the cumulative percentage of the variability explained by each principal component. By using only the first two principal components, we can account for over 83% of the variability in the original four variables. Generally, we will want to

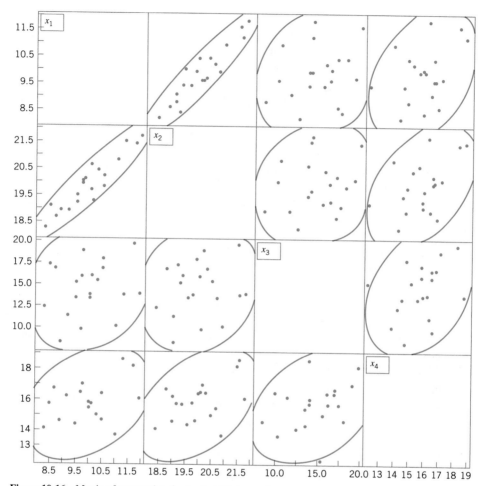

Figure 10-16   Matrix of scatter plots for the first 20 observations on x_1, x_2, x_3, and x_4 from Table 10-6.

Table 10-7   PCA for the First 20 Observations on x_1, x_2, x_3, and x_4 from Table 10-6

Eigenvalues:	2.3181	1.0118	0.6088	0.0613
Percent:	57.9516	25.2951	15.2206	1.5328
Cumulative Percent:	57.9516	83.2466	98.4672	100.0000

Eigenvectors				
x_1	0.59410	− 0.33393	0.25699	0.68519
x_2	0.60704	− 0.32960	0.08341	− 0.71826
x_3	0.28553	0.79369	0.53368	− 0.06092
x_4	0.44386	0.38717	− 0.80137	0.10440

retain enough components to explain a reasonable proportion of the total process variability, but there are no firm guidelines about how much variability needs to be explained in order to produce an effective process-monitoring procedure.

The last two columns in Table 10-6 contain the calculated values of the principal component scores z_{i1} and z_{i2} for the first 20 observations. Figure 10-17 is a scatter plot of these 20 principal component scores, along with the approximate 95% confidence contour. Note that all 20 scores for z_{i1} and z_{i2} are inside the ellipse. We typically regard this display as a monitoring device or control chart for the principal component variables, and the ellipse is an approximate control limit (obviously higher confidence level contours could be selected). Generally, we are using the scores as an **empirical reference distribution** to establish a control region for the process. When future values of the variables x_1, x_2, . . . , x_p are observed, the scores would be computed for the two principal components z_1 and z_2 and these scores plotted on the graph in Fig. 10-17. As long as the scores remain inside the ellipse, there is no evidence that the process mean has shifted. If subsequent scores plot outside the ellipse, then there is some evidence that the process is out of control.

The lower panel of Table 10-6 contains 10 new observations on the process variables x_1, x_2, . . . , x_p that were not used in computing the principal components. The principal component scores for these new observations are also shown in the table, and the scores are plotted on the control chart in Fig. 10-18. A different plotting symbol (x) has been used to assist in identifying the scores from the new points. Although the first few new scores are inside the ellipse, it is clear that beginning with observation 24 or 25 there has been a shift in the process. Control charts such as Fig. 10-18 based on principal component scores are often called principal component **trajectory plots.** Mastrangelo, Runger, and Montgomery (1996) also give an example of this procedure.

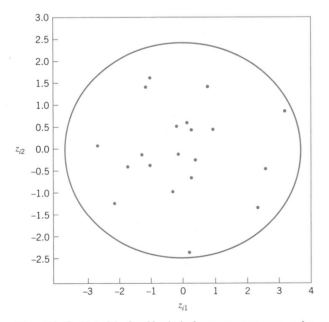

Figure 10-17   Plot of the first 20 principal component scores z_{i1} and z_{i2} from Table 10-6, with 95% confidence ellipse.

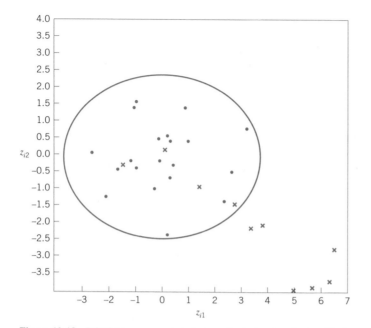

Figure 10-18   Principal components trajectory chart, showing the last 10 observations from Table 10-6.

If more than two principal components need to be retained, then pairwise scatter plots of the principal component scores would be used analogously to Fig. 10-18. However, if more than $r = 3$ or 4 components are retained, interpretation and use of the charts becomes cumbersome. Furthermore, interpretation of the principal components can be difficult, because they are not the original set of process variables but instead linear combinations of them. Sometimes principal components have a relatively simple interpretation and that can assist the analyst in using the trajectory chart. For instance, in our example the constants in the first principal component are all about the same size and have the same sign, so the first principal component can be thought of as an analog of the average of all $p = 4$ original variables. Similarly, the second component is roughly equivalent to the difference between the averages of the first two and the last two process variables. It's not always that easy.

A nice alternative to the trajectory plot is to collect the r retained principal component scores into a vector and apply the MEWMA control chart to them. Practical experience with this approach has been very promising, and the ARL of the MEWMA control chart to detect a shift will be much less using the set of retained principal components than it would have been if all p original process variables were used. Scranton et al. (1996) give more details of this technique.

Finally, note that control charts and trajectory plots based on PCA will be most effective in detecting shifts in the directions defined by the principal components. Shifts in other directions, particularly directions orthogonal to the retained principal component directions, may be very hard to detect. One possible solution to this would be to use a MEWMA control chart to monitor all the remaining principal components z_{r+1}, $z_{r+2}, \ldots, z_p$.

10-7.2 Partial Least Squares

The method of **partial least squares** (or **PLS**) is somewhat related to PCA, except that, like the regression adjustment procedure, it classifies the variables into x's (or inputs) and y's (or outputs). The goal is to create a set of weighted averages of the x's and y's that can be used for prediction of the y's or linear combinations of the y's. The procedure maximizes covariance in the same fashion that the principal component directions maximize variance.

The most common applications of partial least squares today are in the **chemometrics** field, where there are often many variables, both process and response. As of this writing, a major difficulty in practical application is the availability of good, easy-to-use computer software, such as exists for PCA and regression. This problem will be overcome in the near future, however. Another potential difficulty is that there has not been any extensive performance comparison of PLS reported, so there is minimal documentation and evidence concerning its performance and the ability to detect process upsets and shifts with control-charting approaches based on PLS.

10-8 EXERCISES

10-1. The data shown here come from a production process with two observable quality characteristics, x_1 and x_2. The data are sample means of each quality characteristic, based on samples of size $n = 25$. Assume that mean values of the quality characteristics and the covariance matrix were computed from 50 preliminary samples:

$$\bar{\bar{x}} = \begin{bmatrix} 55 \\ 30 \end{bmatrix} \qquad S = \begin{bmatrix} 200 & 130 \\ 130 & 120 \end{bmatrix}$$

Construct a T^2 control chart using these data. Use the phase 2 limits.

Sample Number	$\bar{x}_1$	$\bar{x}_2$
1	58	32
2	60	33
3	50	27
4	54	31
5	63	38
6	53	30
7	42	20
8	55	31
9	46	25
10	50	29

Sample Number	$\bar{x}_1$	$\bar{x}_2$
11	49	27
12	57	30
13	58	33
14	75	45
15	55	27

10-2. A product has three quality characteristics. The nominal values of these quality characteristics and their sample covariance matrix have been determined from the analysis of 30 preliminary samples of size $n = 10$ as follows:

$$\bar{\bar{x}} = \begin{bmatrix} 3.0 \\ 3.5 \\ 2.8 \end{bmatrix} \qquad S = \begin{bmatrix} 1.40 & 1.02 & 1.05 \\ 1.02 & 1.35 & 0.98 \\ 1.05 & 0.98 & 1.20 \end{bmatrix}$$

The sample means for each quality characteristic for 15 additional samples of size $n = 10$ are shown next. Is the process in statistical control?

Sample Number	$\bar{x}_1$	$\bar{x}_2$	$\bar{x}_3$
1	3.1	3.7	3.0
2	3.3	3.9	3.1
3	2.6	3.0	2.4
4	2.8	3.0	2.5

Sample Number	$\bar{x}_1$	$\bar{x}_2$	$\bar{x}_3$
5	3.0	3.3	2.8
6	4.0	4.6	3.5
7	3.8	4.2	3.0
8	3.0	3.3	2.7
9	2.4	3.0	2.2
10	2.0	2.6	1.8
11	3.2	3.9	3.0
12	3.7	4.0	3.0
13	4.1	4.7	3.2
14	3.8	4.0	2.9
15	3.2	3.6	2.8

10-3. Reconsider the situation in Exercise 10-1. Suppose that the sample mean vector and sample covariance matrix provided were the actual population parameters. What control limit would be appropriate for phase 2 of the control chart? Apply this limit to the data and discuss any differences in results that you find in comparison to the original choice of control limit.

10-4. Reconsider the situation in Exercise 10-2. Suppose that the sample mean vector and sample covariance matrix provided were the actual population parameters. What control limit would be appropriate for phase 2 of the control chart? Apply this limit to the data and discuss any differences in results that you find in comparison to the original choice of control limit.

10-5. Consider a T^2 control chart for monitoring $p = 6$ quality characteristics. Suppose that the subgroup size is $n = 3$ and there are 30 preliminary samples available to estimate the sample covariance matrix.
(a) Find the phase 2 control limits assuming that $\alpha = 0.005$.
(b) Compare the control limits from part (a) to the chi-square control limit. What is the magnitude of the difference in the two control limits?
(c) How many preliminary samples would have to be taken to ensure that the exact phase 2 control limit is within 1% of the chi-square control limit?

10-6. Rework Exercise 10-5, assuming that the subgroup size is $n = 5$.

10-7. Consider a T^2 control chart for monitoring $p = 10$ quality characteristics. Suppose that the subgroup size is $n = 3$ and there are 25 preliminary samples available to estimate the sample covariance matrix.
(a) Find the phase 2 control limits assuming that $\alpha = 0.005$.
(b) Compare the control limits from part (a) to the chi-square control limit. What is the magnitude of the difference in the two control limits?
(c) How many preliminary samples would have to be taken to ensure that the chi-square control limit is within 1% of the exact phase 2 control limit?

10-8. Rework Exercise 10-7, assuming that the subgroup size is $n = 5$.

10-9. Consider a T^2 control chart for monitoring $p = 10$ quality characteristics. Suppose that the subgroup size is $n = 3$ and there are 25 preliminary samples available to estimate the sample covariance matrix. Calculate both the phase 1 and the phase 2 control limits (use $\alpha = 0.01$).

10-10. Suppose that we have $p = 4$ quality characteristics, and in correlation form all four variables have variance unity and all pairwise correlation coefficients are 0.7. The in-control value of the process mean vector is $\mu' = [0, 0, 0, 0]$.
(a) Write out the covariance matrix Σ.
(b) What is the chi-square control limit for the chart, assuming that $\alpha = 0.01$?

(c) Suppose that a sample of observations results in the standardized observation vector $\mathbf{y}' = [3.5, 3.5, 3.5, 3.5]$. Calculate the value of the T^2 statistic. Is an out-of-control signal generated?

(d) Calculate the diagnostic quantities d_i, $i = 1, 2, 3, 4$ from equation 10-22. Does this information assist in identifying which process variables have shifted?

(e) Suppose that a sample of observations results in the standardized observation vector $\mathbf{y}' = [2.5, 2, -1, 0]$. Calculate the value of the T^2 statistic. Is an out-of-control signal generated?

(f) For the case in (e), calculate the diagnostic quantities d_i, $i = 1, 2, 3, 4$ from equation 10-22. Does this information assist in identifying which process variables have shifted?

10-11. Suppose that we have $p = 3$ quality characteristics, and in correlation form all three variables have variance unity and all pairwise correlation coefficients are 0.8. The in-control value of the process mean vector is $\boldsymbol{\mu}' = [0, 0, 0]$.

(a) Write out the covariance matrix $\boldsymbol{\Sigma}$.

(b) What is the chi-square control limit for the chart, assuming that $\alpha = 0.05$?

(c) Suppose that a sample of observations results in the standardized observation vector $\mathbf{y}' = [1, 2, 0]$. Calculate the value of the T^2 statistic. Is an out-of-control signal generated?

(d) Calculate the diagnostic quantities d_i, $i = 1, 2, 3$ from equation 10-22. Does this information assist in identifying which process variables have shifted?

(e) Suppose that a sample of observations results in the standardized observation vector $\mathbf{y}' = [2, 2, 1]$. Calculate the value of the T^2 sta-

tistic. Is an out-of-control signal generated?

(f) For the case in (e), calculate the diagnostic quantities d_i, $i = 1, 2, 3$ from equation 10-22. Does this information assist in identifying which process variables have shifted?

10-12. Consider the first two process variables in Table 10-5. Calculate an estimate of the sample covariance matrix using both estimators $\mathbf{S}_1$ and $\mathbf{S}_2$ discussed in Section 10-3.2.

10-13. Consider the first three process variables in Table 10-5. Calculate an estimate of the sample covariance matrix using both estimators $\mathbf{S}_1$ and $\mathbf{S}_2$ discussed in Section 10-3.2.

10-14. Consider all 30 observations on the first two process variables in Table 10-6. Calculate an estimate of the sample covariance matrix using both estimators $\mathbf{S}_1$ and $\mathbf{S}_2$ discussed in Section 10-3.2. Are the estimates very different? Discuss your findings.

10-15. Suppose that there are $p = 4$ quality characteristics, and in correlation form all four variables have variance unity and all pairwise correlation coefficients are 0.75. The in-control value of the process mean vector is $\boldsymbol{\mu}' = [0, 0, 0, 0]$, and we want to design an MEWMA control chart to provide good protection against a shift to a new mean vector of $\mathbf{y}' = [1, 1, 1, 1]$. If an in-control ARL_0 of 200 is satisfactory, what value of λ and what upper control limit should be used? Approximately, what is the ARL_1 for detecting the shift in the mean vector?

10-16. Suppose that there are $p = 4$ quality characteristics, and in correlation form all four variables have variance unity and that all pairwise correlation coefficients are 0.9. The in-control value of the process mean vector is $\boldsymbol{\mu}' = [0, 0, 0, 0]$, and we want to design an MEWMA control chart to provide good protection against a shift to a new mean vector of $y' = [1, 1, 1, 1]$. Suppose that an in-control ARL_0 of 500 is desired. What value of λ and what upper control limit would you recommend? Approximately, what is the ARL_1 for detecting the shift in the mean vector?

10-17.   Suppose that there are $p = 2$ quality characteristics, and in correlation form both variables have variance unity and the correlation coefficient is 0.8. The in-control value of the process mean vector is $\mu' = [0, 0]$, and we want to design an MEWMA control chart to provide good protection against a shift to a new mean vector of $y' = [1, 1]$. If an in-control ARL_0 of 200 is satisfactory, what value of λ and what upper control limit should be used? Approximately, what is the ARL_1 for detecting the shift in the mean vector?

10-18.   Consider the cascade process data in Table 10-5.
(a) Set up an individuals control chart on y_2.
(b) Fit a regression model to y_2, and set up an individuals control chart on the residuals. Comment on the differences between this chart and the one in part (a).
(c) Calculate the sample autocorrelation functions on y_2 and on the residuals from the regression model in part (b). Discuss your findings.

10-19.   Consider the cascade process data in Table 10-5. In fitting regression models to both y_1 and y_2 you will find that not all of the process variables are required to obtain a satisfactory regression model for the output variables. Remove the nonsignificant variables from these equations and obtain subset regression models for both y_1 and y_2. Then construct individuals control charts for both sets of residuals. Compare them to the residual control charts in the text (Fig. 10-11) and from Exercise 10-18. Are there any substantial differences between the charts from the two different approaches to fitting the regression models?

10-20.   **Continuation of Exercise 10-19.** Using the residuals from the regression models in Exercise 10-19, set up EWMA control charts. Compare these EWMA control charts to the Shewhart charts for individuals constructed previously. What are the potential advantages of the EWMA control chart in this situation?

10-21.   Consider the $p = 4$ process variables in Table 10-6. After applying the PCA procedure to the first 20 observations data (see Table 10-7), suppose that the first three principal components are retained.
(a) Obtain the principal component scores. (Hint: Remember that you must work in standardized variables).
(b) Construct an appropriate set of pairwise plots of the principal component scores.
(c) Calculate the principal component scores for the last 10 observations. Plot the scores on the charts from part (b) and interpret the results.

10-22.   Consider the $p = 9$ process variables in Table 10-5.
(a) Perform a PCA on the first 30 observations. Be sure to work with the standardized variables.
(b) How much variability is explained if only the first $r = 3$ principal components are retained?
(c) Construct an appropriate set of pairwise plots of the first $r = 3$ principal component scores.
(d) Now consider the last 10 observations. Obtain the principal component scores and plot them on the chart in part (c). Does the process seem to be in control?

11 Engineering Process Control and SPC

CHAPTER OUTLINE

11-1 PROCESS MONITORING AND PROCESS REGULATION

11-2 PROCESS CONTROL BY FEEDBACK ADJUSTMENT

11-2.1 A Simple Adjustment Scheme: Integral Control

11-2.2 The Adjustment Chart

11-2.3 Variations of the Adjustment Chart

11-2.4 Other Types of Feedback Controllers

11-3 COMBINING SPC AND EPC

CHAPTER OVERVIEW

Throughout this book we have stressed the importance of process control and variability reduction as essential ingredients of modern manufacturing strategy. There are two statistically based approaches for addressing this problem. The first of these is statistical process monitoring by control charts, or statistical process control (SPC). The focus of SPC is on identifying assignable causes so that they can be removed, thereby leading to permanent process improvement or reduction in variability. The second approach is based on adjusting the process using information about its current level or deviation from a desired target. This approach is often called feedback adjustment, and it is a form of engineering process control. Feedback adjustment regulates the process to account for sources of variability that cannot be removed by the SPC approach.

This chapter presents an introduction to simple methods of feedback adjustment and shows how these techniques can be easily implemented in processes where there is a manipulatable variable that affects the process output. We also show how simple SPC schemes can be combined or integrated with engineering process control.

11-1 PROCESS MONITORING AND PROCESS REGULATION

Reduction of variability is an important part of improving process performance in all industries. Statistical process control is an effective tool for reduction of variability through the ability of the control chart to detect assignable causes. When the assignable causes are removed, process variability is reduced and process performance is improved.

SPC has had a long history of successful use in **discrete parts manufacturing.** In continuous processes, such as those found in the chemical and process industries, another approach is often used to reduce variability. This approach is based on **process compensation and regulation,** in which some **manipulatible process variable** is **adjusted** with the objective of keeping the process output on target (or equivalently, minimizing the variability of the output around this target). These process compensation or regulation schemes are widely known as **engineering process control (EPC)**, stochastic control, or feedback or feedforward control depending on the nature of the adjustments.

SPC is always applied in a situation where we assume that it is possible to bring the process into a state of statistical control. By "statistical control," we mean that we observe only stable random variation around the process target. Furthermore, SPC also assumes that once the process is in this in-control state, it will tend to stay there for a relatively long period of time without continual ongoing adjustment. Now certainly if we eliminate assignable causes such as differences due to operators and variations in raw materials, it is often possible to obtain this in-control state. However, in some industrial settings, despite our best effort, the process may still have a tendency to **drift** or **wander** away from the target. This may occur because of phenomena such as continuous variation in input materials or effects of temperature, or it may be due entirely to unknown forces that impact the process. Process regulation through EPC assumes that there is another variable that can be adjusted to compensate for the drift in process output, and that a series of regular adjustments to this manipulatable variable will keep the process output close to the desired target.

There is considerable interest in combining or integrating these two strategies in an effort to provide an improved procedure—that is, an enhancement to EPC that would enable assignable-cause-type disturbances to be detected. For additional background and discussion, see Box and Kramer (1992); MacGregor (1987); Vander Weil, Tucker, Faltin, and Doganaksoy (1992); MacGregor and Harris (1990); Montgomery, Keats, Runger, and Messina (1994); Box, Jenkins, and Reinsel (1994); and Box and Luceño (1997).

It is natural to question the need for integrating EPC and SPC. Historically, these techniques have developed in somewhat different environments. SPC is often part of an organization's strategic thrust to improve quality, and it is usually a top-down, management driven, high-visibility activity, with emphasis on people, methods, and procedures. EPC on the other hand, is more tactical in nature with its origins in the process engineering organization, and its primary focus is on the process. The statistical framework of SPC is similar to that of **hypothesis testing,** whereas the statistical framework of EPC is **parameter estimation** — that is, estimating how much disturbance there is in the system forcing the process off target and then making an adjustment to cancel its effect. What these two procedures share is a common objective; **reduction of variability.** EPC assumes that there is a specific dynamic model that links the process input and output. If that model is correct, then the EPC process adjustment rules will minimize variation around the output target. However, when certain types of external disturbances or assignable causes occur that are outside the framework of this dynamic model, then the compensation rules will not completely account for this upset. As a result, variability will be increased. By applying SPC in a specific way, these assignable causes can be detected and the combined EPC/SPC procedure will be more effective than EPC alone.

11-2 PROCESS CONTROL BY FEEDBACK ADJUSTMENT

11-2.1 A Simple Adjustment Scheme: Integral Control

In this section we consider a simple situation involving a process in which feedback adjustment is appropriate and highly effective. The process output characteristic of interest at time period t is y_t, and we wish to keep y_t as close as possible to a target T. This process has a manipulatable variable x, and a change in x will produce all of its effect on y within one period; that is,

$$y_{t+1} - T = gx_t \tag{11-1}$$

where g is a constant usually called the **process gain.** The gain is like a regression coefficient, in that it relates the magnitude of a change in x_t to a change in y_t. Now, if no adjustment is made, the process drifts away from the target according to

$$y_{t+1} - T = N_{t+1} \tag{11-2}$$

where N_{t+1} is a **disturbance.** The disturbance in equation 11-2 is usually represented by an appropriate time-series model, often an autoregressive integrated moving average (ARIMA) model of the type discussed in Chapter 9, Section 9-4. Such a model is required because the uncontrolled output is usually autocorrelated (see the material in Section 9-4 about SPC with autocorrelated data).

Suppose that the disturbance can be predicted adequately using an EWMA:

$$\hat{N}_{t+1} = \hat{N}_t + \lambda(N_t - \hat{N}_t)$$
$$= \hat{N}_t + \lambda e_t \tag{11-3}$$

where $e_t = N_t - \hat{N}_t$ is the prediction error at time period t and $0 < \lambda \le 1$ is the weighting factor for the EWMA. This assumption is equivalent to assuming that the uncontrolled process is drifting according to the integrated moving average model in equation 9-15 with parameter $\theta = 1 - \lambda$. At time t, the adjusted process is

$$y_{t+1} - T = N_{t+1} + gx_t$$

This equation says that at time $t + 1$ the output deviation from target will depend on the disturbance in period $t + 1$ plus the level x_t to which we set the manipulatable variable in period t, or the **setpoint** in period t. Obviously, we should set x_t so as to exactly cancel out the disturbance. However, we can't do this, because N_{t+1} is unknown in period t. We can, however, forecast N_{t+1} by $\hat{N}_{t+1}$ using equation 11-3. Then we obtain

$$y_{t+1} - T = e_{t+1} + \hat{N}_{t+1} + gx_t \tag{11-4}$$

since $e_{t+1} = N_{t+1} - \hat{N}_{t+1}$.

From equation 11-4, it is clear that if we set $gx_t = -\hat{N}_{t+1}$ or the setpoint $x_t = -(1/g)\hat{N}_{t+1}$, then the adjustment should cancel out the disturbance, and in period $t + 1$ the output deviation from target should be $y_{t+1} - T = e_{t+1}$, where e_{t+1} is the prediction error in period t; that is, $e_{t+1} = N_{t+1} - \hat{N}_{t+1}$. The actual adjustment to the manipulatable variable made at time t is

$$x_t - x_{t-1} = -\frac{1}{g}(\hat{N}_{t+1} - \hat{N}_t) \tag{11-5}$$

Now the difference in the two EWMA predictions $\hat{N}_{t+1} - \hat{N}_t$ can be rewritten as

$$\begin{aligned} \hat{N}_{t+1} - \hat{N}_t &= \lambda N_t + (1 - \lambda)\hat{N}_t - \hat{N}_t \\ &= \lambda N_t - \lambda \hat{N}_t \\ &= \lambda(N_t - \hat{N}_t) \\ &= \lambda e_t \end{aligned}$$

and since the actual error at time t, e_t, is simply the difference between the output and the target, we can write

$$\hat{N}_{t+1} - \hat{N}_t = \lambda(y_t - T)$$

Therefore, the adjustment to be made to the manipulatable variable at time period t (equation 11-5) becomes

$$\begin{aligned} x_t - x_{t-1} &= -\frac{\lambda}{g}(y_t - T) \\ \\ &= -\frac{\lambda}{g}e_t \end{aligned} \tag{11-6}$$

The actual **setpoint** for the manipulatable variable at the end of period t is simply the sum of all the adjustments through time t, or

$$\begin{aligned} x_t &= \sum_{j=1}^{t}(x_j - x_{j-1}) \\ \\ &= -\frac{\lambda}{g}\sum_{j=1}^{t}e_j \end{aligned} \tag{11-7}$$

This type of process adjustment scheme is called **integral control.** It is a pure feedback control scheme that sets the level of the manipulatable variable equal to a weighted sum of all current and previous process deviations from target. It can be shown that if the deterministic part of the process model (equation 11-1) is correct, and if the disturbance N_t is predicted perfectly apart from random error by an EWMA, then this is an optimal control rule in the sense that it minimizes the mean-square error of the process output deviations from the target T. For an excellent discussion of this procedure, see Box (1991–1992) and Box and Luceño (1997).

EXAMPLE 11-1

An Example of Integral Control

Figure 11-1 shows 100 observations on the number average molecular weight of a polymer, taken every 4 hours. It is desired to maintain the molecular weight as close as possible to the target value $T = 2000$. Note that, despite our best efforts to bring the process into a state of statistical control, the molecular weight tends to wander away from the target. Individuals and moving range control charts are shown in Fig. 11-2, indicating the lack of statistical stability in the process. Note that the engineers have used the target value $T = 2000$ as the center line for the individuals chart. The actual sample average and standard deviation of molecular weight for these 100 observations is $\bar{x} = 2008$ and $s = 19.4$.

In this process, the drifting behavior of the molecular weight is likely caused by unknown and uncontrollable disturbances in the incoming material feedstock and other inertial forces, but it can be compensated for by making adjustments to the setpoint of the catalyst feed rate x. A change in the setpoint of the feed rate will have all of its effect on molecular weight within one period, so an integral control procedure such as the one discussed previously will be appropriate.

Suppose that the gain in the system is 1.2:1; that is, an increase in the feed rate of 1 unit increases the molecular weight by 1.2 units. Now for our example, the adjusted process would be

$$y_t - 2000 = N_{t+1} + 1.2x_t$$

We will forecast the disturbances with an EWMA having $\lambda = 0.2$. This is an arbitrary choice for λ. It is possible to use estimation techniques to obtain a precise value for λ,

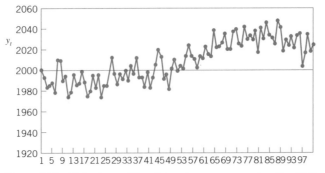

Figure 11-1 Molecular weight of a polymer, target value $T = 2000$ (uncontrolled process).

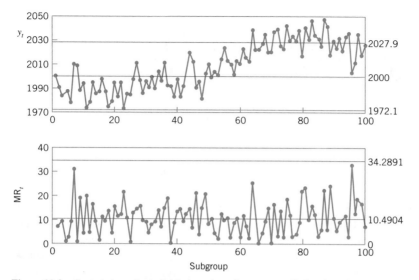

Figure 11-2   Control charts for individuals and moving range applied to the polymer molecular weight data.

but as we will see, often a value for λ between 0.2 and 0.4 works very well. Now the one-period-ahead forecast for the disturbance N_{t+1} is

$$
\begin{aligned}
\hat{N}_{t+1} &= \hat{N}_t + \lambda e_t \\
&= \hat{N}_t + 0.2(N_t - \hat{N}_t) \\
&= 0.2N_t + 0.8\hat{N}_t \\
&= 0.2(y_t - 2000) + 0.8\hat{N}_t
\end{aligned}
$$

Consequently, the setpoint for catalyst feed rate at the end of period t would be

$$
gx_t = -\hat{N}_{t+1}
$$

or

$$
1.2x_t = -[0.2(y_t - 2000) + 0.8\hat{N}_t]
$$

The adjustment made to the catalyst feed rate is

$$
\begin{aligned}
x_t - x_{t-1} &= -\frac{\lambda}{g}(y_t - 2000) \\
&= -\frac{0.2}{1.2}(y_t - 2000) \\
&= -\frac{1}{6}(y_t - 2000)
\end{aligned}
$$

Figure 11-3 plots the values of molecular weight after the adjustments are made to the catalyst feed rate. Note that the process is much closer to the target value of 2000. In

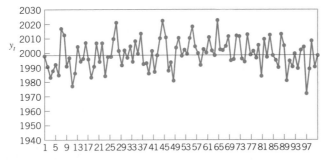

Figure 11-3   Values of molecular weight after adjustment.

fact, the sample average molecular weight for the 100 observations is now 2001, and the sample standard deviation is 10.35. Thus, the use of integral control has reduced process variability by nearly 50%. Figure 11-4 shows the setpoint for the catalyst feed rate used in each time period to keep the process close to the target value of $T = 2000$.

Figure 11-5 shows individuals and moving range control charts applied to the output deviation of the molecular weight from the target value of 2000. Note that now the process appears to be in a state of statistical control. Figure 11-6 is a set of similar control charts applied to the sequence of process adjustments (that is, the change in the setpoint value for feed rate).

In the foregoing example, the value of λ used in the EWMA was $\lambda = 0.2$. An "optimal" value for λ could be obtained by finding the value of λ that minimizes the sum of the squared forecast errors for the process disturbance. To perform this calculation, you will need a record of the process disturbances. Usually, you will have to construct this from past history. That is, you will typically have a history of the actual output and a history of whatever adjustments were made. The disturbance would be back-calculated from the historical deviation from target taken together with the adjustments. This will usually give you a disturbance series of sufficient accuracy to calculate the correct value for λ.

In some cases you may not be able to do this easily, and it may be necessary to choose λ arbitrarily. Figure 11-7 shows the effect of choosing λ arbitrarily when the true

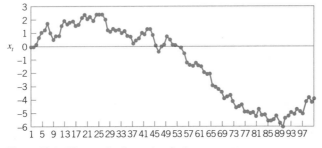

Figure 11-4   The setpoint for catalyst feed rate.

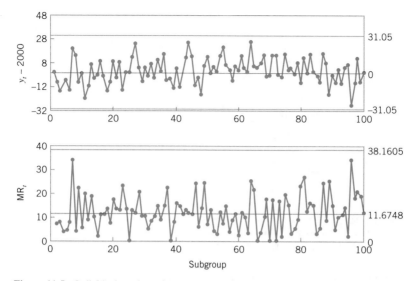

Figure 11-5 Individuals and moving range control charts applied to the output deviation of molecular weight from target, after integral control.

optimum value of λ is λ_0. The vertical scale ($\sigma_\lambda^2/\sigma_{\lambda_0}^2$) shows how much the variance of the output is inflated by choosing the arbitrary λ instead of λ_0.

Consider the case in Fig. 11-7 where $\lambda_0 = 0$. Now, since λ in the EWMA is equal to zero, this means that the process is in statistical control and it will not drift off target. Therefore, no adjustment to feed rate is necessary; see equation 11-6. Figure 11-7

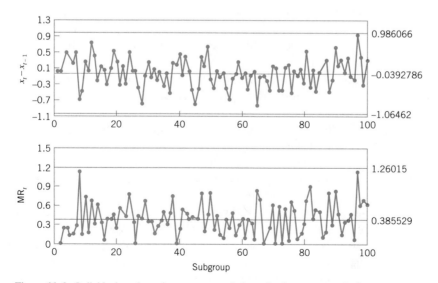

Figure 11-6 Individuals and moving range control charts for the sequence of adjustments to the catalyst feed rate.

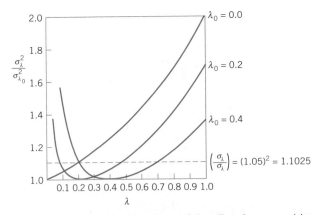

Figure 11-7 Inflation in the variance of the adjusted process arising from an arbitrary choice of λ when the true value of λ in the disturbance is λ_0 [adapted from Box (1991–1992), with permission].

shows very clearly that in this case *any* adjustments to the catalyst feed rate would increase the variance of the output. This is what Dr. Deming means by "tampering with the process." The worst case would occur with $\lambda = 1$, where the output variance would be doubled. Of course, $\lambda = 1$ implies that we are making an adjustment that (apart from the gain g) is exactly equal to the current deviation from target, something *no* rational control engineer would contemplate. Note, however, that a smaller value of λ ($\lambda \leq 0.2$, say) would not inflate the variance very much. Alternatively, if the true value of λ_0 driving the disturbance is *not* zero, meaning that the process drifts off target yet *no* adjustment is made, the output process variance will increase a lot. From the figure, we see that if you use a value of λ in the 0.2–0.4 range, that almost no matter what the *true* value of λ_0 is that drives the disturbances, the increase in output variance will be at most about 5% over what the true minimum variance would be if λ_0 were known exactly. Therefore, an "educated guess" about the value of λ in the 0.2–0.4 range will often work very well in practice.

We noted in Section 9-4 that one possible way to deal with autocorrelated data was to use an engineering controller to remove the autocorrelation. We can demonstrate that in the previous example.

Figure 11-8 is the sample autocorrelation function of the uncontrolled molecular weight measurements from Fig. 11-1. Obviously, the original unadjusted process observations exhibit strong autocorrelation. Figure 11-9 is the sample autocorrelation function of the output molecular weight deviations from target after the integral control adjustments. Note that the output deviations from target are now uncorrelated.

Engineering controllers cannot always be used to eliminate autocorrelation. For example, the process dynamics may not be understood well enough to implement an effective controller. Also note that any engineering controller essentially **transfers variability** from one part of the process to another. In our example, the integral controller transfers variability from molecular weight into the catalyst feed rate. To see this, examine Figs. 11-1, 11-3, and 11-4, and note that the reduction in variability in the out-

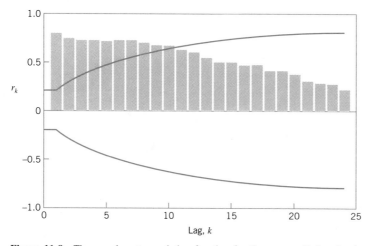

Figure 11-8   The sample autocorrelation function for the uncontrolled molecular weight observations from Figure 11-1.

put molecular weight was achieved by increasing the variability of the feed rate. There may be processes in which this is not always an acceptable alternative.

11-2.2   The Adjustment Chart

The feedback adjustment scheme based on integral control that we described in the previous section can be implemented so that the adjustments are made automatically. Usually this involves some combination of sensors or measuring devices, a logic device

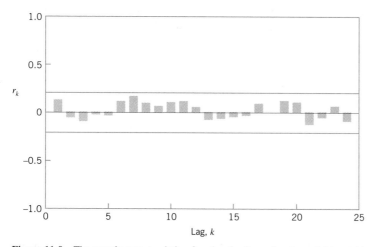

Figure 11-9   The sample autocorrelation function for the molecular weight variable after integral control.

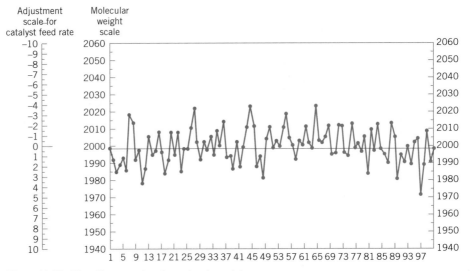

Figure 11-10   The adjustment chart for molecular weight.

or computer, and actuators to physically make the adjustments to the manipulatable variable x. When EPC or feedback adjustment is implemented in this manner, it is often called **automatic process control** or **APC.**

In many processes, feedback adjustments can be made manually. Operating person-nel routinely observe the current output deviation from target, compute the amount of adjustment to apply using equation 11-6, and then bring x_t to its new setpoint. When ad-justments are made manually by operating personnel, a variation of Fig. 11-3 called the **manual adjustment chart** is very useful.

Figure 11-10 is the manual adjustment chart corresponding to Fig. 11-3. Note that there is now a second scale, called the adjustment scale, on the vertical axis. Note also that the divisions on the adjustment scale are arranged so that one unit of adjustment exactly equals six units on the molecular weight scale. Furthermore, the units on the adjustment scale that correspond to molecular weight values above the target of 2000 are negative, whereas the units on the adjustment scale that correspond to molecular weight values below the target of 2000 are positive. The reason for this is that the spe-cific adjustment equation that is used for the molecular weight variable is

$$x_t - x_{t-1} = -\frac{1}{6}(y_t - 2000)$$

or

$$\text{adjustment to catalyst feed rate} = -\frac{1}{6}(\text{deviation of molecular weight from 2000})$$

That is, a **six-unit** change in molecular weight from its target of 2000 corresponds to a **one-unit** change in the catalyst feed rate. Furthermore, if the molecular weight is **above** the target, the catalyst feed rate must be **reduced** to drive molecular weight toward the target value, whereas if the molecular weight is **below** the target, the catalyst feed rate must be **increased** to drive molecular weight toward the target.

The adjustment chart is extremely easy for operating personnel to use. For example, consider Fig. 11-10 and, specifically, observation y_{13} as molecular weight. As soon as $y_{13} = 2006$ is observed and plotted on the chart, the operator simply reads off the corresponding value of -1 on the adjustment scale. This is the amount by which the operator should change the current setting of the catalyst feed rate. That is, the operator should **reduce** the catalyst feed rate by one unit. Now the next observation is $y_{14} = 1997$. The operator plots this point and observes that 1997 on the molecular weight scale corresponds to $+0.5$ on the adjustment scale. Thus, catalyst feed rate could now be **increased** by 0.5 units.

This is a very simple and highly effective procedure. Manual adjustment charts were first proposed by George Box and G. M. Jenkins [see Box, Jenkins, and Reinsel (1994); Box (1991); and Box and Luceño (1997) for more background]. They are often called **Box–Jenkins adjustment charts.**

11-2.3 Variations of the Adjustment Chart

The adjustment procedures in Sections 11-2.1 and 11-2.2 are very straightforward to implement, but they require that an adjustment be made to the process after each observation. In feedback adjustment applications in the chemical and process industries, this is not usually a serious issue because the major cost that must be managed is the cost of being off target, and the adjustments themselves are made with either no or very little cost. Indeed, they are often made automatically. However, situations can arise in which the cost or convenience of making an adjustment is a concern. For example, in discrete parts manufacturing it may be necessary to actually stop the process to make an adjustment. Consequently, it may be of interest to make some modification to the feedback adjustment procedure so that less frequent adjustments will be made.

There are several ways to do this. One of the simplest is the **bounded adjustment chart,** a variation of the procedure in Section 11-2.2 in which an adjustment will be made only in periods for which the EWMA forecast is outside one of the bounds given by $\pm L$. The boundary value L is usually determined from engineering judgment, taking the costs of being off target and the cost of making the adjustment into account. Box and Luceño (1997) discuss this situation in detail and, in particular, how costs can be used specifically for determining L.

We will use the data in Table 11-1 to illustrate the bounded adjustment chart. Column 1 of this table presents the unadjusted values of an important output characteristic from a chemical process. The values are reported as deviations from the actual target, so the target for this variable—say, y_t—is zero. Figure 11-11 plots these output data, along with an EWMA prediction made using $\lambda = 0.2$. Note that the variable does not stay very close to the desired target. The average of these 50 observations is 17.2 and the sum of the squared deviations from target is 21,468. The standard deviation of these observations is approximately 11.6.

Table 11-1   Chemical Process Data for the Bounded Adjustment Chart in Fig. 11-12

Observation	Original Process Output	Adjusted Process Output	EWMA	Adjustment	Cumulative Adjustment or Setpoint
1	0	0	0		0
2	16	16	3.200		0
3	24	24	7.360		0
4	29	29	11.688	− 7.250	− 7.250
5	34	26.750	5.350		− 7.250
6	24	16.750	7.630		− 7.250
7	31	23.750	10.854	− 5.938	− 13.188
8	26	12.813	2.563		− 13.188
9	38	24.813	7.013		− 13.188
10	29	15.813	8.773		− 13.188
11	25	11.813	9.381		− 13.188
12	26	12.813	10.067	− 3.213	− 16.391
13	23	6.609	1.322		− 16.391
14	34	17.609	4.579		− 16.391
15	24	7.609	5.185		− 16.391
16	14	− 2.391	3.670		− 16.391
17	41	24.609	7.858		− 16.391
18	36	19.609	10.208	− 4.904	− 21.293
19	29	7.707	1.541		− 21.293
20	13	− 8.293	− 0.425		− 21.293
21	26	4.707	0.601		− 21.293
22	12	− 9.293	− 1.378		− 21.293
23	15	− 6.293	− 2.361		− 21.293
24	34	12.707	0.653		− 21.293
25	7	− 14.293	− 2.336		− 21.293
26	20	− 1.293	− 2.128		− 21.293
27	16	− 5.293	− 2.761		− 21.293
28	7	− 14.293	− 5.067		− 21.293
29	0	− 21.293	− 8.312		− 21.293
30	8	− 13.293	− 9.308		− 21.293
31	23	1.707	− 7.105		− 21.293
32	10	− 11.293	− 7.943		− 21.293
33	12	− 9.293	− 8.213		− 21.293
34	− 2	− 23.293	− 11.229	5.823	− 15.470
35	10	− 5.970	− 1.094		− 15.470
36	28	12.530	1.631		− 15.470
37	12	− 3.470	0.611		− 15.470
38	8	− 7.470	− 1.005		− 15.470
39	11	− 4.470	− 1.698		− 15.470
40	4	− 11.470	− 3.653		− 15.470
41	9	− 6.470	− 4.216		− 15.470

Table 11-1   (*Continued*)

Observation	Original Process Output	Adjusted Process Output	EWMA	Adjustment	Cumulative Adjustment or Setpoint
42	15	− 0.470	− 3.467		− 15.470
43	5	− 10.470	− 4.867		− 15.470
44	13	− 2.470	− 4.388		− 15.470
45	22	6.530	− 2.204		− 15.470
46	− 9	− 24.470	− 6.657		− 15.470
47	3	− 12.470	− 7.820		− 15.470
48	12	− 3.470	− 6.950		− 15.470
49	3	− 12.470	− 8.054		− 15.470
50	12	− 3.470	− 7.137		− 15.470

There is a manipulatable variable in this process, and the relationship between the output and this variable is given by

$$y_t - T = 0.8x_t$$

That is, the process gain $g = 0.8$. The EWMA in Fig. 11-11 uses $\lambda = 0.2$. This value was chosen arbitrarily, but remember from our discussion in Section 11-2.1 that the procedure is relatively insensitive to this parameter.

Suppose that we decide to set $L = 10$. This means that we will only make an adjustment to the process when the EWMA exceeds $L = 10$ or $-L = -10$. Economics and the ease of making adjustments are typically factors in selecting L, but here we did it a slightly different way. Note that the standard deviation of the unadjusted process is approximately

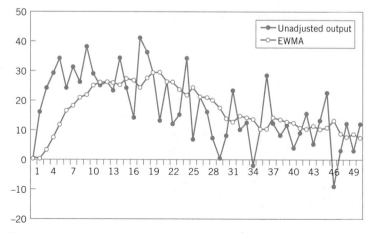

Figure 11-11   The unadjusted process data from Table 11-1 and an EWMA with $\lambda = 0.2$.

11.6, so the standard deviation of the EWMA in Fig. 11-11 is approximately

$$\hat{\sigma}_{\text{EWMA}} = \sqrt{\frac{\lambda}{2 - \lambda}}\ \hat{\sigma}_{\text{unadjusted process}}$$

$$= \sqrt{\frac{0.2}{2 - 0.2}}\ 11.6$$

$$= 3.87$$

Therefore, using $L = 10$ is roughly equivalent to using control limits on the EWMA that are about $2.6\sigma_{\text{EWMA}}$ in width. (Recall from Chapter 8 that we often use control limits on an EWMA that are slightly less than three-sigma.)

The computations for the EWMA are given in Table 11-1. Note that the EWMA is started off at zero, and the first period in which it exceeds $L = 10$ is period 4. The output deviation from target in period 4 is $+29$, so the adjustment would be calculated as usual in integral control as

$$x_t - x_{t-1} = -\frac{\lambda}{g}(y_t - T)$$

or

$$x_4 - x_3 = -\frac{0.2}{0.8}(y_4 - 0)$$

$$= -\frac{1}{4}(29)$$

$$= -7.250$$

That is, we would change the manipulatable variable from its previous setting in period 3 by -7.250 units. The full effect of this adjustment then would be felt in the next period, 5. The EWMA would be reset to zero at the end of period 4 and the forecasting procedure started afresh. The next adjustment occurs in period 7, where an additional -5.938 units of adjustment are made. The last column records the cumulative effect of all adjustments.

Note that only five adjustments are made over the 50 observations. Figure 11-12 is a plot of the original unadjusted output variable, the adjusted output, the EWMA forecasts, and the actual process adjustments. The variability in the adjusted output around the target has been reduced considerably; the sum of squared deviations from target is 9780 and the average deviation from target is 1.79. This is a reduction of over 50% in the output deviation from the target, achieved with only five adjustments to the process.

Bounded adjustment charts are often very good substitutes for making an adjustment every period. They usually result in slightly degraded performance when compared to the "always adjust" scheme, but usually the degradation is small.

Another variation of the adjustment chart encountered in practice is the **rounded adjustment chart.** This procedure is sometimes used to assist operating personnel in making simple adjustments. The adjustment scale is "rounded" to perhaps four or five zones on either side of the target, with each zone corresponding to adjustments that are easy to make (such as change the manipulatable variable by 1 unit, 2 units, and so

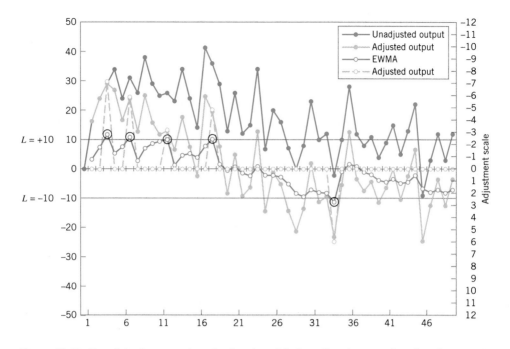

Figure 11-12 Bounded adjustment chart showing the original unadjusted output, the adjusted output, the EWMA, and the actual process adjustments. The circled EWMAs indicate points where adjustments are made.

forth). Often the central zone corresponds to making no adjustment. In this case, adjustments would not necessarily be made every period. See Box and Luceño (1997) for more discussion of these charts.

11-2.4   Other Types of Feedback Controllers

We have considered a feedback controller for which the process adjustment rule is

$$g(x_t - x_{t-1}) = -\lambda e_t \tag{11-8}$$

where e_t is the output deviation from target and λ is the EWMA parameter. By summing this equation we arrived at

$$x_t = -\frac{\lambda}{g} \sum_{i=1}^{t} e_i \tag{11-9}$$

where x_t is the level or setpoint of the manipulatable variable at time t. This is, of course, an integral control adjustment rule.

Now suppose that to make reasonable adjustments to the process we feel it is necessary to consider the last *two* errors, e_t and e_{t-1}. Suppose we write the adjustment equation in terms of two constants c_1 and c_2,

$$g(x_t - x_{t-1}) = c_1 e_t + c_2 e_{t-1} \tag{11-10}$$

If this expression is summed, the setpoint becomes

$$x_t = k_P e_t + k_I \sum_{i=1}^{t} e_i \qquad (11\text{-}11)$$

where $k_P = -(c_2/g)$ and $k_I = (c_1 + c_2)/g$. Note that the setpoint control equation contains a term calling for "proportional" control actions as well as the familiar integral action term. The two constants k_P and k_I are the proportional and integral action parameters, respectively. This is a discrete **proportional integral** (or **PI**) control equation.

Now suppose that the adjustment depends on the last *three* errors:

$$g(x_t - x_{t-1}) = c_1 e_t + c_2 e_{t-1} + c_3 e_{t-2} \qquad (11\text{-}12)$$

Summing this up leads to the discrete **proportional integral derivative** (or **PID**) control equation

$$x_t = k_P e_t + k_I \sum_{i=1}^{t} e_i + k_D(e_t - e_{t-1}) \qquad (11\text{-}13)$$

These models are widely used in practice, particularly in the chemical and process industries. Often two of the three terms will be used, such as **PI** or **PD** control. Choosing the constants (the k's or the c's) is usually called **tuning** the controller.

11-3   COMBINING SPC AND EPC

There is considerable confusion about process adjustment or regulation and the important role that it plays in reduction of variability. For example, the control chart is not always the best method for reducing variability around a target. In the chemical and process industries, techniques such as the simple integral control rule illustrated in Section 11-2.1 have been very effectively used for this purpose. In general, engineering control theory is based on the idea that if we can (1) predict the next observation on the process, (2) have some other variable that we can manipulate in order to affect the process output, and (3) know the effect of this manipulated variable so that we can determine how much control action to apply, then we can make the adjustment in the manipulated variable at time t that is most likely to produce an on-target value of the process output in period $t + 1$. Clearly, this requires good knowledge of the relationship between the output or controlled variable and the manipulated variable, as well as an understanding of process dynamics. We must also be able to easily change the manipulated variable. In fact, if the cost of taking control action is negligible, then the variability in the process output is minimized by taking control action every period. Note that this is in sharp contrast with SPC, where "control action" or a process adjustment is taken only when there is statistical evidence that the process is out of control. This statistical evidence is usually a point outside the limits of a control chart.

There are many processes where some type of feedback-control scheme would be preferable to a control chart. For example, consider the process of driving a car, with the objective of keeping it in the center of the right-hand lane (or equivalently, minimizing variation around the center of the right-hand lane). The driver can easily see the road ahead, and process adjustments (corrections to the steering wheel position) can be made at any time at negligible cost. Consequently, if the driver knew the relationship between the output variable (car position) and the manipulated variable (steering wheel adjustment), he would likely prefer to use a feedback-control scheme to control car position, rather than a statistical control chart. (Driving a car with a Shewhart control chart may be an interesting idea, but the author doesn't want to be in the car during the experiment.)

On the other hand, engineering process control makes no attempt to identify an assignable cause that may impact the process. The elimination of assignable causes can result in significant process improvement. All EPC schemes do is react to process upsets; they do not make any effort to remove the assignable causes. Consequently, in processes where feedback control is used there may be substantial improvement if control charts are also used for **statistical process monitoring** (as opposed to *control;* the control actions are based on the engineering scheme). Some authors refer to systems where both EPC and an SPC system for process monitoring have been implemented as **algorithmic SPC;** see Vander Weil et al. (1992).

The control chart should be applied to either the control error (the difference between the controlled variable and the target) or to the sequence of adjustments to the manipulated variable. Points that lie outside the control limits on these charts would identify periods where the control errors are large or where large changes to the manipulated variable are being made. These periods would likely be good opportunities to search for assignable causes. Montgomery et al. (1994) have demonstrated the effectiveness of such a system. Figure 11-13 illustrates how such a combination of engineering process control and statistical process monitoring might be employed.

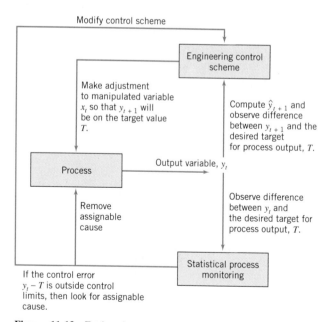

Figure 11-13 Engineering process control and statistical process monitoring.

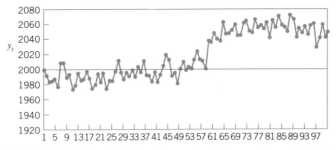

Figure 11-14   Molecular weight, with an assignable cause of magnitude 25 at
$t = 60$.

········   EXAMPLE 11-2   ···

Integrating SPC and EPC
Figure 11-14 shows the molecular weight measurements from Fig. 11-1 in Example 11-1,
except that now an assignable cause has impacted the process starting at period $t = 60$.
The effect of this assignable cause is to increase the molecular weight by 25 units, and this
results in adding variability to the process; the sample average and standard deviation of
molecular weight are $\bar{x} = 2019$ and $s = 30.4$ (compared to $\bar{x} = 2008$ and $s = 19.4$ when
there was no assignable cause present). Figure 11-15 shows the molecular weight after
adjustments to the catalyst feed rate by the integral control rule in Example 11-1 are
applied. Figure 11-16 shows the setpoints for feed rate. Process performance has improved, as the sample average and standard deviation are now $\bar{x} = 1992$ and $s = 15.4$.
Clearly the assignable cause is still adding to process variability, because when there was
no assignable cause in the system, $\bar{x} = 2001$ and $s = 10.35$ after the adjustments.

Figure 11-17 presents individuals and moving range control charts applied to the
output molecular weight deviation from the target value $T = 2000$, after the integral
control adjustments. An out-of-control signal is generated at period $t = 80$, indicating
that an assignable cause is present. An EWMA or cusum control chart on the output
deviation from target would generally detect the assignable cause more quickly.
Figure 11-18 is an EWMA with $\lambda = 0.1$, and it signals the assignable cause at period
$t = 70$.

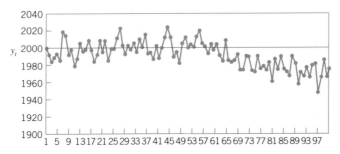

Figure 11-15   Molecular weight after integral control adjustments to catalyst
feed rate.

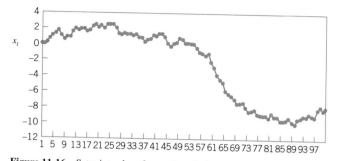

Figure 11-16   Setpoint values for catalyst feed rate, Example 11-2.

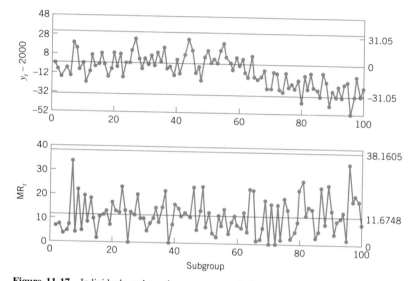

Figure 11-17   Individuals and moving range control charts applied to the output deviation from target, Example 11-2.

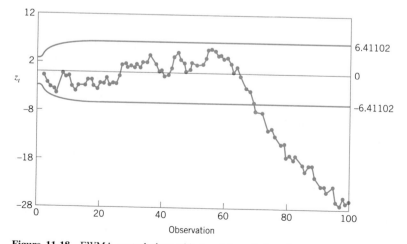

Figure 11-18   EWMA control chart with $\lambda = 0.1$ applied to the output deviation from target, Example 11-2.

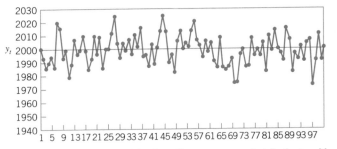

Figure 11-19 Molecular weight after adjustments to catalyst feed rate with the assignable cause removed following period $t = 70$.

Suppose that we detected and eliminated the assignable cause at period $t = 70$. Figure 11-19 is the resulting sequence of output molecular weight values. The sample average and standard deviation are now $\bar{x} = 1998$ and $s = 10.8$, so we see that coupling a control chart on the output deviation from target with the integral controller has improved the performance of the process when assignable causes are present.

11-4 EXERCISES

11-1. If y_t are the observations and z_t is the EWMA, show that the following relationships are true.
(a) $z_t - z_{t-1} = \lambda(y_t - z_{t-1})$
(b) $e_t - (1 - \lambda)e_{t-1} = y_t - y_{t-1}$

11-2. Consider the data in Table 11-1. Construct a bounded adjustment chart using $\lambda = 0.3$ and $L = 10$. Compare the performance of this chart to the one in Table 11-1 and Fig. 11-12.

11-3. Consider the data in Table 11-1. Construct a bounded adjustment chart using $\lambda = 0.4$ and $L = 10$. Compare the performance of this chart to the one in Table 11-1 and Fig. 11-12.

11-4. Consider the data in Table 11-1. Suppose that an adjustment is made to the output variable after every observation. Compare the performance of this chart to the one in Table 11-1 and Fig. 11-12.

11-5. **The Variogram.** Consider the variance of observations that are m periods apart; that is, $V_m = V(y_{t+m} - y_t)$. A graph of V_m/V_1 versus m is called a variogram. It is a nice way to check a data series for nonstationary (drifting

mean) behavior. If a data series is completely uncorrelated (white noise) the variogram will always produce a plot that stays near unity. If the data series is autocorrelated but stationary, the plot of the variogram will increase for a while, but as m increases the plot of V_m/V_1 will gradually stabilize and not increase any further. The plot of V_m/V_1 versus m will increase without bound for nonstationary data. Apply this technique to the data in Table 11-1. Is there an indication of nonstationary behavior? Calculate the sample autocorrelation function for the data. Compare the interpretation of both graphs.

11-6. Consider the observations shown in the following table. The target value for this process is 200.
(a) Set up an integral controller for this process. Assume that the gain for the adjustment variable is $g = 1.2$ and assume that $\lambda = 0.2$ in the EWMA forecasting procedure will provide adequate one-step-ahead predictions.

(b) How much reduction in variability around the target does the integral controller achieve?

(c) Rework parts (a) and (b) assuming that $\lambda = 0.4$. What change does this make in the variability around the target in comparison to that achieved with $\lambda = 0.2$?

Observation, t	y_t	Observation, t	y_t
1	215.8	26	171.9
2	195.8	27	170.4
3	191.3	28	169.4
4	185.3	29	170.9
5	216.0	30	157.2
6	176.9	31	172.4
7	176.0	32	160.7
8	162.6	33	145.6
9	187.5	34	159.9
10	180.5	35	148.6
11	174.5	36	151.1
12	151.6	37	162.1
13	174.3	38	160.0
14	166.5	39	132.9
15	157.3	40	152.8
16	166.6	41	143.7
17	160.6	42	152.3
18	155.6	43	111.3
19	152.5	44	143.6
20	164.9	45	129.9
21	159.0	46	122.9
22	174.2	47	126.2
23	143.6	48	133.2
24	163.1	49	145.0
25	189.7	50	129.5

11-7. Use the data in Exercise 11-6 to construct a bounded adjustment chart. Use $\lambda = 0.2$ and set $L = 12$. How

does the bounded adjustment chart perform relative to the integral control adjustment procedure in part (a) of Exercise 11-6?

11-8. Rework Exercise 11-7 using $\lambda = 0.4$ and $L = 15$. What differences in the results are obtained?

11-9. Consider the observations in the following table. The target value for this process is 50.

(a) Set up an integral controller for this process. Assume that the gain for the adjustment variable is $g = 1.6$ and assume that $\lambda = 0.2$ in the EWMA forecasting procedure will provide adequate one-step-ahead predictions.

(b) How much reduction in variability around the target does the integral controller achieve?

(c) Rework parts (a) and (b) assuming that $\lambda = 0.4$. What change does this make in the variability around the target in comparison to that achieved with $\lambda = 0.2$?

Observation, t	y_t	Observation, t	y_t
1	50	26	43
2	58	27	39
3	54	28	32
4	45	29	37
5	56	30	44
6	56	31	52
7	66	32	42
8	55	33	47
9	69	34	33
10	56	35	49
11	63	36	34
12	54	37	40
13	67	38	27
14	55	39	29
15	56	40	35

Observation, t	y_t	Observation, t	y_t
16	65	41	27
17	65	42	33
18	61	43	25
19	57	44	21
20	61	45	16
21	64	46	24
22	43	47	18
23	44	48	20
24	45	49	23
25	39	50	26

11-10. Use the data in Exercise 11-9 to construct a bounded adjustment chart. Use $\lambda = 0.2$ and set $L = 4$. How does the bounded adjustment chart perform relative to the integral control adjustment procedure in part (a) of Exercise 11-9?

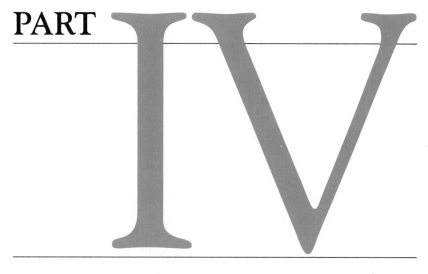

PART IV

Process Design and Improvement with Designed Experiments

Quality and productivity improvement are most effective when they are an integral part of the product and process development cycle. In particular, the formal introduction of **experimental design** methodology at the earliest stage of the development cycle, where new products are designed, existing product designs improved, and manufacturing processes optimized, is often the key to overall product success. This principle has been established in many different industries, including electronics and semiconductors, aerospace, automotive, medical devices, food and pharmaceuticals, and the chemical and process industries. The effective use of sound statistical experimental design methodology can lead to products that are easier to manufacture, have higher reliability, and have enhanced field performance. Experimental design can also greatly enhance process development and troubleshooting activities. This is the primary focus of this section.

Factorial and **fractional factorial designs** are introduced in Chapter 12 with particular emphasis on the two-level design system—that is, the 2^k factorial design and the fractions thereof. These designs are particularly useful for screening the variables in a process to determine those that are most important. Chapter 13 introduces **response surface methods,** a collection of techniques useful for process optimization. This chapter also discusses **process robustness studies,** an approach to reducing the variability in process output by minimizing the effects on the output transmitted by variables that are difficult to control during routine process operation. Finally, we present an overview of evolutionary operation, an experimental-design-based process monitoring scheme.

Throughout Part IV we use the analysis of variance as the basis for analyzing data from designed experiments. It is possible to introduce experimental design without using analysis of variance methods, but this author feels that it is a mistake to do so, primarily because students will encounter the analysis of variance in virtually every computer program they use, either in the classroom, or in professional practice. We also illustrate software packages supporting designed experiments.

The material in this section is not a substitute for a full course in experimental design. The reader who is interested in applying experimental design to process improvement will need additional background, but hopefully this presentation will serve to effectively illustrate some of the many applications of this powerful tool. In many industries, the effective use of statistical experimental design is the key to higher yields, reduced variability, reduced development lead times, better products, and satisfied customers.

12 Factorial and Fractional Factorial Experiments for Process Design and Improvement

CHAPTER OUTLINE

12-1 WHAT IS EXPERIMENTAL DESIGN?

12-2 EXAMPLES OF DESIGNED EXPERIMENTS IN PROCESS IMPROVEMENT

12-3 GUIDELINES FOR DESIGNING EXPERIMENTS

12-4 FACTORIAL EXPERIMENTS

    12-4.1 An Example

    12-4.2 Statistical Analysis

    12-4.3 Residual Analysis

12-5 THE 2^k FACTORIAL DESIGN

    12-5.1 The 2^2 Design

12-5.2 The 2^k Design for $k \geq 3$ Factors

12-5.3 A Single Replicate of the 2^k Design

12-5.4 Addition of Center Points to the 2^k Design

12-5.5 Blocking and Confounding in the 2^k Design

12-6 FRACTIONAL REPLICATION OF THE 2^k DESIGN

    12-6.1 The One-Half Fraction of the 2^k Design

    12-6.2 Smaller Fractions: The 2^{k-p} Fractional Factorial Design

CHAPTER OVERVIEW

Most experiments for process design and improvement involve several variables. Factorial experimental designs, and their variations, are used in such situations. This chapter gives an introduction to factorial designs, emphasizing their applications for process and quality improvement. We concentrate on designs where all the factors have two levels, and show how fractional versions of these designs can be used with great effectiveness in industrial experimentation. Important topics include the analysis of factorial experimental designs and the use of graphical methods in interpretation of the results. Both the interaction graph and a response surface plot are shown to be very useful in interpretation of results.

12-1 WHAT IS EXPERIMENTAL DESIGN?

As indicated in Chapter 1, a designed experiment is a test or series of tests in which purposeful changes are made to the input variables of a process so that we may observe and identify corresponding changes in the output response. The process, as shown in Fig. 12-1, can be visualized as some combination of machines, methods, and people that transforms an input material into an output product. This output product has one or more observable quality characteristics or responses. Some of the process variables $x_1, x_2, \ldots, x_p$ are **controllable,** whereas others $z_1, z_2, \ldots, z_q$ are **uncontrollable** (although they may be controllable for purposes of the test). Sometimes these uncontrollable factors are called **noise** factors. The objectives of the experiment may include

1. Determining which variables are most influential on the response, y.
2. Determining where to set the influential x's so that y is near the nominal requirement.
3. Determining where to set the influential x's so that variability in y is small.
4. Determining where to set the influential x's so that the effects of the uncontrollable variables z are minimized.

Thus, experimental design methods may be used either in process development or process troubleshooting to improve process performance or to obtain a process that is **robust** or **insensitive** to external sources of variability.

Statistical process control methods and experimental design, two very powerful tools for the improvement and optimization of processes, are closely interrelated. For example, if a process is in statistical control but still has poor capability, then to im-

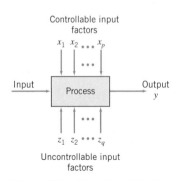

Figure 12-1 General model of a process.

prove process capability it will be necessary to reduce variability. Designed experiments may offer a more effective way to do this than SPC. Essentially, SPC is a **passive** statistical method: we watch the process and wait for some information that will lead to a useful change. However, if the process is in control, passive observation may not produce much useful information. On the other hand, experimental design is an **active** statistical method: We will actually perform a series of tests on the process making changes in the inputs and observing the corresponding changes in the outputs, and this will produce information that can lead to process improvement.

Experimental design methods can also be very useful in establishing statistical control of a process. For example, suppose that a control chart indicates that the process is out of control, and the process has many controllable input variables. Unless we know *which* input variables are the important ones, it may be very difficult to bring the process under control. Experimental design methods can be used to identify these influential process variables.

Experimental design is a critically important engineering tool for improving a manufacturing process. It also has extensive application in the development of new processes. Application of these techniques early in process development can result in

1. Improved yield
2. Reduced variability and closer conformance to nominal
3. Reduced development time
4. Reduced overall costs

Experimental design methods can also play a major role in **engineering design** activities, where new **products** are developed and existing ones improved. Some applications of statistical experimental design in engineering design include

1. Evaluation and comparison of basic design configurations
2. Evaluation of material alternatives
3. Determination of key product design parameters that impact performance

Use of experimental design in these areas can result in improved manufacturability of the product, enhanced field performance and reliability, lower product cost, and shorter product development time.

12-2   EXAMPLES OF DESIGNED EXPERIMENTS IN PROCESS IMPROVEMENT

In this section, we present several examples that illustrate the application of designed experiments in improving process and product quality. In subsequent sections, we will demonstrate the statistical methods used to analyze the data and draw conclusions from experiments such as these.

······   **EXAMPLE 12-1**   ··

Characterizing a Process

A manufacturing engineer has applied SPC to a process for soldering electronic components to printed circuit boards. Through the use of u charts and Pareto analysis he has established statistical control of the flow solder process and has reduced the average number of defective solder joints per board to around 1%. However, since the average board contains over 2000 solder joints, even 1% defective presents far too many solder joints requiring rework. The engineer would like to reduce defect levels even further; however, since the process is in statistical control, it is not obvious what machine adjustments will be necessary.

The flow solder machine has several variables that can be controlled. They include

1. Solder temperature
2. Preheat temperature
3. Conveyor speed
4. Flux type
5. Flux specific gravity
6. Solder wave depth
7. Conveyor angle

In addition to these controllable factors, several others cannot be easily controlled during routine manufacturing, although they could be controlled for purposes of a test. They are

1. Thickness of the printed circuit board
2. Types of components used on the board
3. Layout of the components on the board
4. Operator
5. Production rate

In this situation, the engineer is interested in **characterizing** the flow solder machine; that is, he wants to determine which factors (both controllable and uncontrollable) affect the occurrence of defects on the printed circuit boards. To accomplish this task he can design an experiment that will enable him to estimate the magnitude and direction of the factor effects. That is, how much does the response variable (defects per unit) change when each factor is changed, and does changing the factors *together* produce different results than are obtained from individual factor adjustments? A factorial experiment will be required to do this. Sometimes we call this kind of factorial experiment a **screening experiment.**

The information from this screening or characterization experiment will be used to identify the critical process factors and to determine the direction of adjustment for these factors to further reduce the number of defects per unit. The experiment may also provide information about which factors should be more carefully controlled during routine manufacturing to prevent high defect levels and erratic process performance.

Thus, one result of the experiment could be the application of control charts to one or more *process* variables (such as solder temperature) in addition to the *u* chart on process output. Over time, if the process is sufficiently improved, it may be possible to base most of the process control plan on controlling process input variables instead of control charting the output.

EXAMPLE 12-2

Optimizing a Process

In a characterization experiment, we are usually interested in determining which process variables affect the response. A logical next step is to **optimize**— that is, to determine the region in the important factors that lead to the best possible response. For example, if the response is yield, we will look for a region of maximum yield, and if the response is variability in a critical product dimension, we will look for a region of minimum variability.

Suppose we are interested in improving the yield of a chemical process. We know from the results of a characterization experiment (say) that the two most important process variables that influence yield are operating temperature and reaction time. The process currently runs at 155°F and 1.7 h of reaction time, producing yields around 75%. Figure 12-2 shows a view of the time–temperature region from above. In this graph the lines of constant yield are connected to form **response contours,** and we have shown the contour lines for 60, 70, 80, 90, and 95% yield.

To locate the optimum, it is necessary to perform an experiment that varies time and temperature together. This type of experiment is called a **factorial experiment;** an example of a factorial experiment with both time and temperature run at two levels is

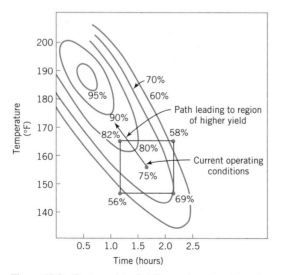

Figure 12-2 Contour plot of yield as a function of reaction time and reaction temperature, illustrating an optimization experiment.

shown in Fig. 12-2. The responses observed at the four corners of the square indicate that we should move in the general direction of increased temperature and decreased reaction time to increase yield. A few additional runs could be performed in this direction, which would be sufficient to locate the region of maximum yield. Once we are in the region of the optimum, a more elaborate experiment could be performed to give a very precise estimate of the optimum operating condition. This type of experiment, called a **response surface experiment,** is discussed in Chapter 13.

······· EXAMPLE 12-3 ··

A Product Design Example
Designed experiments can often be applied in the product design process. To illustrate, suppose that a group of engineers are designing a door hinge for an automobile. The quality characteristic of interest is the check effort, or the holding ability of the door latch that prevents the door from swinging closed when the vehicle is parked on a hill. The check mechanism consists of a spring and a roller. When the door is opened, the roller travels through an arc causing the leaf spring to be compressed. To close the door, the spring must be forced aside, which creates the check effort. The engineering team believes the check effort is a function of the following factors:

1. Roller travel distance
2. Spring height pivot to base
3. Horizontal distance from pivot to spring
4. Free height of the reinforcement spring
5. Free height of the main spring

The engineers build a prototype hinge mechanism in which all these factors can be varied over certain ranges. Once appropriate levels for these five factors are identified, an experiment can be designed consisting of various combinations of the factor levels, and the prototype hinge can be tested at these combinations. This will produce information concering which factors are most influential on latch check effort, and through use of this information the design can be improved.

······· EXAMPLE 12-4 ··

Determining System and Component Tolerances
The Wheatstone bridge shown in Fig. 10-3 is a device used for measuring an unknown resistance, Y. The adjustable resistor B is manipulated until a particular current flow is obtained through the ammeter (usually $X = 0$). Then the unknown resistance is calculated as

$$Y = \frac{BD}{C} - \frac{X^2}{C^2E}[A(D + C) + D(B + C)][B(C + D) + F(B + C)] \quad (12\text{-}1)$$

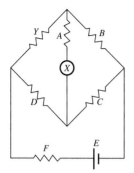

Figure 12-3 A Wheat-stone bridge.

The engineer wants to design the circuit so that overall gage capability is good; that is, he would like the standard deviation of measurement error to be small. He has decided that $A = 20$ Ω, $C = 2$ Ω, $D = 50$ Ω, $E = 1.5$ V, and $F = 2$ Ω is the best choice of the design parameters as far as gage capability is concerned, but the overall measurement error is still too high. This is likely due to the tolerances that have been specified on the circuit components. These tolerances are $\pm 1\%$ for each resistor A, B, C, D, and F, and $\pm 5\%$ for the power supply E. These tolerance bands can be used to define appropriate factor levels, and an experiment can be performed to determine which circuit components have the most critical tolerances and how much they must be tightened to produce adequate gage capability. The information from this experiment will result in a design specification that tightens only the most critical tolerances the minimum amount possible consistent with desired measurement capability. Consequently, a lower cost design that is easier to manufacture will be possible.

Notice that in this experiment is it unnecessary actually to build hardware, since the response from the circuit can be calculated via equation 12-1. The actual response variable for the experiment should be the standard deviation of Y. However, an equation for the transmitted variation in Y from the circuit can be found using the methods of Section 7-6.2. Therefore, the entire experiment can be performed using a computer model of the Wheatstone bridge.

12-3   GUIDELINES FOR DESIGNING EXPERIMENTS

Designed experiments are a powerful approach to improving a process. To use this approach, it is necessary that everyone involved in the experiment have a clear idea in advance of the objective of the experiment, exactly what factors are to be studied, how the experiment is to be conducted, and at least a qualitative understanding of how the data will be analyzed. Montgomery (1997) gives an outline of the

Pre-experimental planning
- **1.** Recognition of and statement of the problem
- **2.** Choice of factors and levels ⎤ often done simultane-
- **3.** Selection of the response variable ⎦ ously, or in reverse order
- **4.** Choice of experimental design
- **5.** Performing the experiment
- **6.** Data analysis
- **7.** Conclusions and recommendations

Figure 12-4 Procedure for designing an experiment.

recommended procedure, reproduced in Fig. 12-4. We now briefly amplify each point in this checklist.

1. Recognition of and statement of the problem. In practice, it is often difficult to realize that a problem requiring formal designed experiments exists, so it may not be easy to develop a clear and generally accepted statement of the problem. However, it is absolutely essential to fully develop all ideas about the problem and about the specific objectives of the experiment. Usually, it is important to solicit input from all concerned parties—engineering, quality, marketing, the customer, management, and the operators (who usually have much insight that is all too often ignored). A clear statement of the problem and the objectives of the experiment often contributes substantially to better process understanding and eventual solution of the problem.

2. Choice of factors and levels. The experimenter must choose the factors to be varied in the experiment, the ranges over which these factors will be varied, and the specific levels at which runs will be made. Process knowledge is required to do this. This process knowledge is usually a combination of practical experience and theoretical understanding. It is important to investigate all factors that may be of importance and to avoid being overly influenced by past experience, particularly when we are in the early stages of experimentation or when the process is not very mature. When the objective is factor screening or process characterization, it is usually best to keep the number of factor levels low. (Most often two levels are used.) As noted in Fig. 12-4, steps 2 and 3 are often carried out simultaneously, or step 3 may be done first in some applications.

3. Selection of the response variable. In selecting the response variable, the experimenter should be certain that the variable really provides useful information about the process under study. Most often the average or standard deviation (or both) of the measured characteristic will be the response variable. Multiple responses are not unusual. Gage capability is also an important factor. If gage capability is poor, then only relatively large factor effects will be detected by the experiment, or additional replication will be required.

4. Choice of experimental design. If the first three steps are done correctly, this step is relatively easy. Choice of design involves consideration of sample size (number of replicates), selection of a suitable run order for the experimental trials, and whether or not blocking or other randomization restrictions are involved. The next chapter illustrates some of the more important types of experimental designs.

5. Performing the experiment. When running the experiment, it is vital to carefully monitor the process to ensure that everything is being done according to plan. Errors in experimental procedure at this stage will usually destroy experimental validity. Up-front planning is crucial to success. It is easy to underestimate the logistical and planning aspects of running a designed experiment in a complex manufacturing environment.

6. Data analysis. Statistical methods should be used to analyze the data so that results and conclusions are objective rather than judgmental. If the experiment has been designed correctly and if it has been performed according to the design, then the type of statistical methods required is not elaborate. Many excellent software packages are available to assist in the data analysis, and simple graphical methods play an important role in data interpretation. Residual analysis and model validity checking are also important.

7. Conclusions and recommendations. Once the data have been analyzed, the experiment must draw *practical* conclusions about the results and recommend a course of action. Graphical methods are often useful in this stage, particularly in presenting the results to others. Follow-up runs and confirmation testing should also be performed to validate the conclusions from the experiment.

Steps 1 to 3 are usually called **pre-experimental planning.** It is vital that these steps be performed as well as possible if the experiment is to be successful. Coleman and Montgomery (1993) discuss this in detail and offer more guidance in pre-experimental planning, including worksheets to assist the experimenter in obtaining and documenting the required information.

Throughout this entire process, it is important to keep in mind that experimentation is an important part of the learning process, where we tentatively formulate hypotheses about a system, perform experiments to investigate these hypotheses, and on the basis of the results formulate new hypotheses, and so on. This suggests that experimentation is **iterative.** It is usually a major mistake to design a single, large comprehensive experiment at the start of a study. A successful experiment requires knowledge of the important factors, the ranges over which these factors should be varied, the appropriate number of levels to use, and the proper units of measurement for these variables. Generally, we do not know perfectly the answers to these questions, but we learn about them as we go along. As an experimental program progresses, we often drop some variables, add others, change the region of exploration for some factors, or add new response variables. Consequently, we usually experiment **sequentially,** and as a general rule, no more than about 25% of the available resources should be invested in the first experiment. This will ensure that sufficient resources are available to accomplish the final objective of the experiment.

12-4 FACTORIAL EXPERIMENTS

When there are several factors of interest in an experiment, a **factorial design** should be used. In such designs factors are varied together. Specifically, by a factorial experiment we mean that in each complete trial or replicate of the experiment all possible

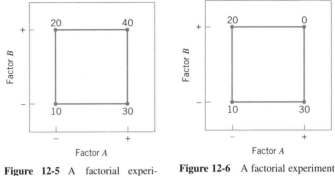

Figure 12-5 A factorial experiment with two factors.

Figure 12-6 A factorial experiment with interaction.

combinations of the levels of the factors are investigated. Thus, if there are two factors A and B with a levels of factor A and b levels of factor B, then each replicate contains all ab possible combinations.

The effect of a factor is defined as the change in response produced by a change in the level of the factor. This is called a **main effect** because it refers to the primary factors in the study. For example, consider the data in Fig. 12-5. In this factorial design, both the factors A and B have two levels, denoted by "−" and "+". These two levels are called "low" and "high," respectively. The main effect of factor A is the difference between the average response at the high level of A and the average response at the low level of A, or

$$A = \bar{y}_{A^+} - \bar{y}_{A^-} = \frac{30 + 40}{2} - \frac{10 + 20}{2} = 20$$

That is, changing factor A from the low level (−) to the high level (+) causes an average response increase of 20 units. Similarly, the main effect of B is

$$B = \bar{y}_{B^+} - \bar{y}_{B^-} = \frac{20 + 40}{2} - \frac{10 + 30}{2} = 10$$

In some experiments, the difference in response between the levels of one factor is not the same at all levels of the other factors. When this occurs, there is an interaction between the factors. For example, consider the data in Fig. 12-6. At the low level of factor B, the A effect is

$$A = 30 - 10 = 20$$

and at the high level of factor B, the A effect is

$$A = 0 - 20 = -20$$

Since the effect of A depends on the level chosen for factor B, there is interaction between A and B.

When an interaction is large, the corresponding main effects have little meaning. For example, by using the data in Fig. 12-6, we find the main effect of A as

$$A = \frac{30 + 0}{2} - \frac{10 + 20}{2} = 0$$

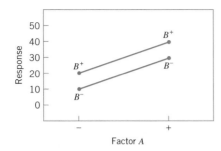

Figure 12-7 Factorial experiment, no interaction.

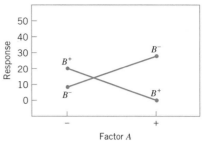

Figure 12-8 Factorial experiment with interaction.

and we would be tempted to conclude that there is no A effect. However, when we examine the main effect of A at *different levels of factor B*, we see that this is not the case. The effect of factor A depends on the levels of factor B. Thus, knowledge of the AB interaction is more useful than knowledge of the main effect. A significant interaction can mask the significance of main effects.

The concept of interaction can be illustrated graphically. Figure 12-7 plots the data in Fig. 12-5 against the levels of A for both levels of B. Note that the B^- and B^+ lines are roughly parallel, indicating that factors A and B do not interact. Figure 12-8 plots the data in Fig. 12-6. In Fig. 12-8, the B^- and B^+ lines are not parallel, indicating the interaction between factors A and B. Such graphical displays are often useful in presenting the results of experiments.

An alternative to the factorial design that is (unfortunately) used in practice is to change the factors one at a time rather than to vary them simultaneously. To illustrate the one-factor-at-a-time procedure, consider the optimization experiment described earlier in Example 12-2. The engineer is interested in finding the values of temperature and time that maximize yield. Suppose that we fix temperature at 155°F (the current operating level) and perform five runs at different levels of time—say, 0.5 h, 1.0 h, 1.5 h, 2.0 h, and 2.5 h. The results of this series of runs are shown in Fig. 12-9. This figure indicates that maximum yield is achieved at about 1.7 h of reaction time. To optimize temperature, the engineer fixes time at 1.7 h (the apparent optimum) and performs five runs

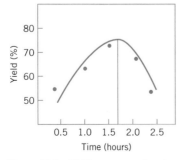

Figure 12-9 Yield versus reaction time with temperature constant at 155°F.

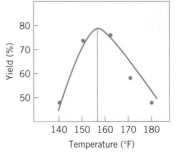

Figure 12-10 Yield versus temperature with reaction time constant at 1.7 h.

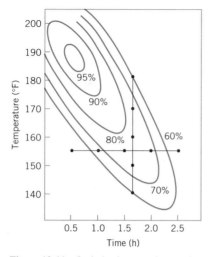

Figure 12-11 Optimization experiment using the one-factor-at-a-time method.

at different temperatures—say, 140°F, 150°F, 160°F, 170°F, and 180°F. The results of this set of runs are plotted in Figure 12-10. Maximum yield occurs at about 155°F. Therefore, we would conclude that running the process at 155°F and 1.7 h is the best set of operating conditions, resulting in yields around 75%.

Figure 12-11 displays the contour plot of yield as a function of temperature and time with the one-factor-at-a-time experiment shown on the contours. Clearly, the one-factor-at-a-time design has failed dramatically here, as the true optimum is at least 20 yield points higher and occurs at much lower reaction times and higher temperatures. The failure to discover the shorter reaction times is particularly important because it could have significant impact on production volume or capacity, production planning, manufacturing cost, and total productivity.

The one-factor-at-a-time method has failed here because it fails to detect the interaction between temperature and time. Factorial experiments are the only way to detect interactions. Furthermore, the one-factor-at-a-time method is inefficient; it will require more experimentation than a factorial, and as we have just seen, there is no assurance that it will produce the correct results. The factorial experiment shown in Fig. 12-2 that produced the information pointing to the region of the optimum is a simple example of a factorial experiment.

12-4.1 An Example

Aircraft primer paints are applied to aluminum surfaces by two methods—dipping and spraying. The purpose of the primer is to improve paint adhesion; some parts can be primed using either application method. An engineer interested in learning whether three different primers differ in their adhesion properties performed a factorial experi-

Table 12-1   Adhesion Force Data

Primer Type	Application Method				$y_{i..}$
	Dipping		Spraying		
1	4.0, 4.5, 4.3	(12.8)	5.4, 4.9, 5.6	(15.9)	28.7
2	5.6, 4.9, 5.4	(15.9)	5.8, 6.1, 6.3	(18.2)	34.1
3	3.8, 3.7, 4.0	(11.5)	5.5, 5.0, 5.0	(15.5)	27.0
$y_{.j.}$	40.2		49.6		$89.8 = y_{...}$

ment to investigate the effect of paint primer type and application method on paint adhesion. Three specimens were painted with each primer using each application method, a finish paint was applied, and the adhesion force was measured. The 18 runs from this experiment were run in random order. The resulting data are shown in Table 12-1. The circled numbers in the cells are the cell totals. The objective of the experiment was to determine which combination of primer paint and application method produced the highest adhesion force. It would be desirable if at least one of the primers produced high adhesion force *regardless* of application method, as this would add some flexibility to the manufacturing process.

12-4.2   Statistical Analysis

The analysis of variance described in Chapter 3 can be extended to handle the two-factor factorial experiment. Let the two factors be denoted A and B, with a levels of factor A and b levels of B. If the experiment is replicated n times, the data layout will look like Table 12-2. In general, the observation in the ijth cell in the kth replicate is y_{ijk}.

Table 12-2   Data for a Two-Factor Factorial Design

		Factor B			
		1	2	. . .	b
Factor A	1	$y_{111}, y_{112},$ $\ldots, y_{11n}$	$y_{121}, y_{122},$ $\ldots, y_{12n}$	. . .	$y_{1b1}, y_{1b2},$ $\ldots, y_{1bn}$
	2	$y_{211}, y_{212},$ $\ldots, y_{21n}$	$y_{221}, y_{222},$ $\ldots, y_{22n}$	. . .	$y_{2b1}, y_{2b2},$ $\ldots, y_{2bn}$
	$\vdots$	$\vdots$	$\vdots$	$\vdots$	$\vdots$
	a	$y_{a11}, y_{a12},$ $\ldots, y_{a1n}$	$y_{a21}, y_{a22},$ $\ldots, y_{a2n}$	. . .	$y_{ab1}, y_{ab2},$ $\ldots, y_{abn}$

In collecting the data, the abn observations would be run in *random* order. Thus, like the single-factor experiment studied in Chapter 3, the two-factor factorial is a **completely randomized design.** Both factors are assumed to be fixed effects.

The observations from a two-factor factorial experiment may be described by the model

$$y_{ijk} = \mu + \tau_i + \beta_j + (\tau\beta)_{ij} + \epsilon_{ijk} \qquad \begin{cases} i = 1, 2, \ldots, a \\ j = 1, 2, \ldots, b \\ k = 1, 2, \ldots, n \end{cases} \qquad (12\text{-}2)$$

where μ is the overall mean effect, τ_i is the effect of the ith level of factor A, β_j is the effect of the jth level of factor B, $(\tau\beta)_{ij}$ is the effect of the interaction between A and B, and ϵ_{ijk} is an NID$(0, \sigma^2)$ random error component. We are interested in testing the hypotheses of no significant factor A effect, no significant factor B effect, and no significant AB interaction.

Let $y_{i..}$ denote the total of the observations at the ith level of factor A, $y_{.j.}$ denote the total of the observations at the jth level of factor B, $y_{ij.}$ denote the total of the observations in the ijth cell of Table 12-2, and $y_{...}$ denote the grand total of all the observations. Define $\bar{y}_{i..}$, $\bar{y}_{.j.}$, $\bar{y}_{ij.}$, and $\bar{y}_{...}$ as the corresponding row, column, cell, and grand averages. That is,

$$y_{i..} = \sum_{j=1}^{b} \sum_{k=1}^{n} y_{ijk} \qquad \bar{y}_{i..} = \frac{y_{i..}}{bn} \qquad i = 1, 2, \ldots, a$$

$$y_{.j.} = \sum_{i=1}^{a} \sum_{k=1}^{n} y_{ijk} \qquad \bar{y}_{.j.} = \frac{y_{.j.}}{an} \qquad j = 1, 2, \ldots, b \qquad (12\text{-}3)$$

$$y_{ij.} = \sum_{k=1}^{n} y_{ijk} \qquad \bar{y}_{ij.} = \frac{y_{ij.}}{n} \qquad \begin{matrix} i = 1, 2, \ldots, a \\ j = 1, 2, \ldots, b \end{matrix}$$

$$y_{...} = \sum_{i=1}^{a} \sum_{j=1}^{b} \sum_{k=1}^{n} y_{ijk} \qquad \bar{y}_{...} = \frac{y_{...}}{abn}$$

The analysis of variance decomposes the total corrected sum of squares

$$SS_T = \sum_{i=1}^{a} \sum_{j=1}^{b} \sum_{k=1}^{n} (y_{ijk} - \bar{y}_{...})^2$$

as follows:

$$\sum_{i=1}^{a} \sum_{j=1}^{b} \sum_{k=1}^{n} (y_{ijk} - \bar{y}_{...})^2 = bn \sum_{i=1}^{a} (\bar{y}_{i..} - \bar{y}_{...})^2 + an \sum_{j=1}^{b} (\bar{y}_{.j.} - \bar{y}_{...})^2$$

$$+ n \sum_{i=1}^{a} \sum_{j=1}^{b} (\bar{y}_{ij.} - \bar{y}_{i..} - \bar{y}_{.j.} + \bar{y}_{...})^2$$

$$+ \sum_{i=1}^{a} \sum_{j=1}^{b} \sum_{k=1}^{n} (y_{ijk} - \bar{y}_{ij.})^2$$

or symbolically,

$$SS_T = SS_A + SS_B + SS_{AB} + SS_E \qquad (12\text{-}4)$$

The corresponding degree of freedom decomposition is

$$abn - 1 = (a - 1) + (b - 1) + (a - 1)(b - 1) + ab(n - 1) \qquad (12\text{-}5)$$

This decomposition is usually summarized in an analysis of variance table such as the one shown in Table 12-3.

To test for no row factor effects, no column factor effects, and no interaction effects, we would divide the corresponding mean square by mean square error. Each of these ra-

Table 12-3 The Analysis of Variance Table for a Two-Factor Factorial, Fixed Effects Model

Source of Variation	Sum of Squares	Degrees of Freedom	Mean Square	F_0
A	SS_A	$a - 1$	$MS_A = \dfrac{SS_A}{a - 1}$	$F_0 = \dfrac{MS_A}{MS_E}$
B	SS_B	$b - 1$	$MS_B = \dfrac{SS_B}{b - 1}$	$F_0 = \dfrac{MS_B}{MS_E}$
Interaction	SS_{AB}	$(a - 1)(b - 1)$	$MS_{AB} = \dfrac{SS_{AB}}{(a - 1)(b - 1)}$	$F_0 = \dfrac{MS_{AB}}{MS_E}$
Error	SS_E	$ab(n - 1)$	$MS_E = \dfrac{SS_E}{ab(n - 1)}$	
Total	SS_T	$abn - 1$		

tios will follow an F distribution, with numerator degrees of freedom equal to the number of degrees of freedom for the numerator mean square and $ab(n - 1)$ denominator degrees of freedom, when the null hypothesis of no factor effect is true. We would reject the corresponding hypothesis if the computed F exceeded the tabular value at an appropriate significance level, or alternatively, if the P-value were smaller than the specified significance level.

The analysis of variance is usually done using a computer, although simple computing formulas for the sums of squares may be obtained easily. The computing formulas for these sums of squares are given next.

$$SS_T = \sum_{i=1}^{a} \sum_{j=1}^{b} \sum_{k=1}^{n} y_{ijk}^2 - \frac{y_{...}^2}{abn} \tag{12-6}$$

Main effects

$$SS_A = \sum_{i=1}^{a} \frac{y_{i..}^2}{bn} - \frac{y_{...}^2}{abn} \tag{12-7}$$

$$SS_B = \sum_{j=1}^{b} \frac{y_{.j.}^2}{an} - \frac{y_{...}^2}{abn} \tag{12-8}$$

Interaction

$$SS_{AB} = \sum_{i=1}^{a} \sum_{j=1}^{b} \frac{y_{ij.}^2}{n} - \frac{y_{...}^2}{abn} - SS_A - SS_B \tag{12-9}$$

Error

$$SS_E = SS_T - SS_A - SS_B - SS_{AB} \tag{12-10}$$

······   **EXAMPLE 12-5**   ··

The Aircraft Primer Paint Problem

The analysis of variance described above may be applied to the aircraft primer paint experiment described in Section 12-4.1. The sums of squares required to perform the analysis of variance are

$$SS_T = \sum_{i=1}^{a} \sum_{j=1}^{b} \sum_{k=1}^{n} y_{ijk}^2 - \frac{y_{...}^2}{abn}$$

$$= (4.0)^2 + (4.5)^2 + \cdots + (5.0)^2 - \frac{(89.8)^2}{18} = 10.72$$

$$SS_{\text{primers}} = \sum_{i=1}^{a} \frac{y_{i..}^2}{bn} - \frac{y_{...}^2}{abn}$$

$$= \frac{(28.7)^2 + (34.1)^2 + (27.0)^2}{6} - \frac{(89.8)^2}{18} = 4.58$$

$$SS_{\text{methods}} = \sum_{j=1}^{b} \frac{y_{.j.}^2}{an} - \frac{y_{...}^2}{abn}$$

$$= \frac{(40.2)^2 + (49.6)^2}{9} - \frac{(89.8)^2}{18} = 4.91$$

$$SS_{\text{interaction}} = \sum_{i=1}^{a} \sum_{j=1}^{b} \frac{y_{ij.}^2}{n} - \frac{y_{...}^2}{abn} - SS_{\text{primers}} - SS_{\text{methods}}$$

$$= \frac{(12.8)^2 + (15.9)^2 + (11.5)^2 + (15.9)^2 + (18.2)^2 + (15.5)^2}{3}$$

$$- \frac{(89.8)^2}{18} - 4.58 - 4.91 = 0.24$$

and

$$SS_E = SS_T - SS_{\text{primers}} - SS_{\text{methods}} - SS_{\text{interaction}}$$

$$= 10.72 - 4.58 - 4.91 - 0.24 = 0.99$$

The analysis of variance is summarized in Table 12-4. Note that the P-values for both main effects are very small, indicating that the type of primer used and the application method significantly affect adhesion force. Since the P-value for the interaction effect F-ratio is relatively large, we would conclude that there is no interaction between primer type and application method. As an alternative to using the P-values, we could compare the computed F-ratios to a 5% (say) upper critical value of the F distribution. Since $F_{0.05,2,12} = 3.89$ and $F_{0.05,1,12} = 4.75$, we conclude that primer type and application method affect adhesion force. Furthermore, since $1.5 < F_{0.05,2,12}$, there is no indication of interaction between these factors.

In practice, the analysis of variance computations are performed on a computer using a statistics software package. Table 12-5 is the analysis of variance from Minitab.

Table 12-4 Analysis of Variance for Example 12-5

Source of Variation	Sum of Squares	Degrees of Freedom	Mean Square	F_0	P-value
Primer types	4.58	2	2.29	28.63	2.71×10^{-5}
Application methods	4.91	1	4.91	61.38	4.65×10^{-6}
Interaction	0.24	2	0.12	1.5	0.269
Error	0.99	12	0.08		
Total	10.72	17			

Table 12-5   Analysis of Variance Output from Minitab, Example 12-5

```
Two-way ANOVA: Force versus Primer, Method
Analysis of Variance for Force
Source          DF        SS        MS        F        P
Primer           2    4.5811    2.2906    27.86    0.000
Method           1    4.9089    4.9089    59.70    0.000
Interaction      2    0.2411    0.1206     1.47    0.269
Error           12    0.9867    0.0822
Total           17   10.7178

                          Individual 95% CI
Primer          Mean   ------+---------+---------+---------+-----
1               4.78          (----*----)
2               5.68                                (----*----)
3               4.50    (----*----)
                       ------+---------+---------+---------+-----
                          4.50      5.00      5.50      6.00

                          Individual 95% CI
Method          Mean   ---------+---------+---------+---------+--
Dip            4.467   (-----*-----)
Spray          5.511                               (----*-----)
                       ---------+---------+---------+---------+--
                          4.550     4.900     5.250     5.600
```

Confidence intervals on each mean calculated using MS_E as an estimate of σ^2 and applying the standard confidence interval procedure for the mean of a normal distribution with unknown variance.

Note the similarity of this display to Table 12-4. Because the computer carries more decimal places than we did in the manual calculations, the F-ratios in Tables 12-4 and 12-5 are slightly different. The P-value for each F-ratio is called "significance level" in Table 12-5, and when a P-value is less than 0.001, Minitab reports it as 0.000.

A graph of the adhesion force cell averages $\{\bar{y}_{ij.}\}$ versus the levels of primer type for each application method is shown in Fig. 12-12. This interaction graph was constructed by Minitab. The absence of interaction is evident in the parallelism of the two lines. Furthermore, since a large response indicates greater adhesion force, we conclude that spraying is a superior application method and that primer type 2 is most effective. Therefore, if we wish to operate the process so as to attain maximum adhesion force, we should use primer type 2 and spray all parts.

12-4.3   Residual Analysis

Just as in the single-factor experiments discussed in Chapter 3, the **residuals** from a factorial experiment play an important role in assessing model adequacy. The residuals

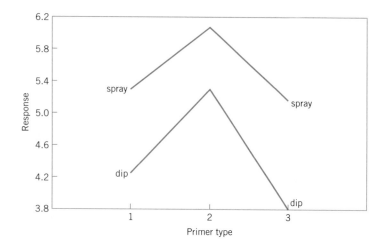

Figure 12-12 Graph of average adhesion force versus primer types for Example 12-5.

from a two-factor factorial are

$$e_{ijk} = y_{ijk} - \hat{\underline{y}}_{ijk}$$
$$= y_{ijk} - \bar{y}_{ij.}$$

That is, the residuals are simply the difference between the observations and the corresponding cell averages.

Table 12-6 presents the residuals for the aircraft primer paint data in Example 12-5. The normal probability plot of these residuals is shown in Fig. 12-13. This plot has tails that do not fall exactly along a straight line passing through the center of the plot, indicating some small problems with the normality assumption, but the departure from normality is not serious. Figures 12-14 and 12-15 plot the residuals versus the levels of primer types and application methods, respectively. There is some indication that primer type 3 results in slightly lower variability in adhesion force than the other two primers. The graph of residuals versus fitted values in Fig. 12-16 does not reveal any unusual or diagnostic pattern.

Table 12-6 Residuals for the Aircraft Primer Paint Experiment

Primer Type	Application Method	
	Dipping	Spraying
1	$-0.26,\quad 0.23, 0.03$	$0.10, -0.40,\quad 0.30$
2	$0.30, -0.40, 0.10$	$-0.26,\quad 0.03,\quad 0.23$
3	$-0.03, -0.13, 0.16$	$0.34, -0.17, -0.17$

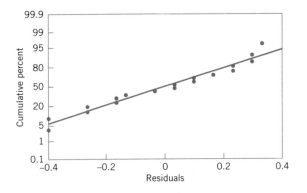

Figure 12-13 Normal probability plot of the residuals from Example 12-5.

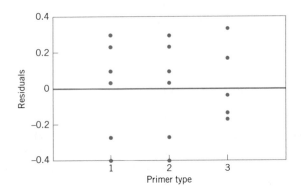

Figure 12-14 Plot of residuals versus primer type.

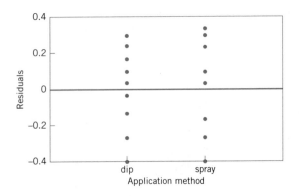

Figure 12-15 Plot of residuals versus application method.

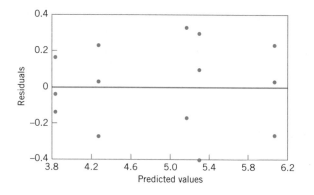

Figure 12-16   Plot of residuals versus predicted values y_{ijk}.

12-5   THE 2^k FACTORIAL DESIGN

Certain special types of factorial designs are very useful in process development and improvement. One of these is a factorial design with k factors, each at two levels. Because each complete replicate of the design has 2^k runs, the arrangement is called a 2^k factorial design. These designs have a greatly simplified analysis, and they also form the basis of many other useful designs.

12-5.1   The 2^2 Design

The simplest type of 2^k design is the 2^2—that is, two factors A and B, each at two levels. We usually think of these levels as the "low" or "−" and "high" or "+" levels of the factor. The 2^2 design is shown in Fig. 12-17. Note that the design can be represented geometrically as a square with the $2^2 = 4$ runs forming the corners of the square.

A special notation is used to represent the runs. In general, a run is represented by a series of lowercase letters. If a letter is present, then the corresponding factor is set at the high level in that run; if it is absent, the factor is run at its low level. For example, run a indicates that factor A is at the high level and factor B is at the low level. The run

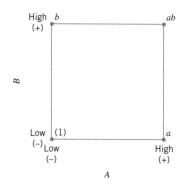

Figure 12-17   The 2^2 factorial design.

with both factors at the low level is represented by (1). This notation is used throughout the 2^k design series. For example, the run in a 2^4 with A and C at the high level and B and D at the low level is denoted by ac.

The effects of interest in the 2^2 design are the main effects A and B and the two-factor interaction AB. Let the letters (1), a, b, and ab also represent the totals of all n observations taken at these design points. It is easy to estimate the effects of these factors. To estimate the main effect of A, we would average the observations on the right side of the square when A is at the high level and subtract from this the average of the observations on the left side of the square where A is at the low level, or

$$
\begin{aligned}
A &= \bar{y}_{A^+} - \bar{y}_{A^-} \\
&= \frac{a + ab}{2n} - \frac{b + (1)}{2n} \\
&= \frac{1}{2n}[a + ab - b - (1)]
\end{aligned}
\tag{12-11}
$$

Similarly, the main effect of B is found by averaging the observations on the top of the square where B is at the high level and subtracting the average of the observations on the bottom of the square where B is at the low level:

$$
\begin{aligned}
B &= \bar{y}_{B^+} - \bar{y}_{B^-} \\
&= \frac{b + ab}{2n} - \frac{a + (1)}{2n} \\
&= \frac{1}{2n}[b + ab - a - (1)]
\end{aligned}
\tag{12-12}
$$

Finally, the AB interaction is estimated by taking the difference in the diagonal averages in Fig. 12-17, or

$$
\begin{aligned}
AB &= \frac{ab + (1)}{2n} - \frac{a + b}{2n} \\
&= \frac{1}{2n}[ab + (1) - a - b]
\end{aligned}
\tag{12-13}
$$

Table 12-7   Signs for Effects in the 2^2 Design

Run		Factorial Effect			
		I	A	B	AB
1	(1)	+	−	−	+
2	a	+	+	−	−
3	b	+	−	+	−
4	ab	+	+	+	+

The quantities in brackets in equations 12-11, 12-12, and 12-13 are called **contrasts.** For example, the A contrast is

$$\text{Contrast}_A = a + ab - b - (1)$$

In these equations, the contrast coefficients are always either $+1$ or -1. A table of plus and minus signs, such as Table 12-7, can be used to determine the sign on each run for a particular contrast. The column headings for the table are the main effects A and B, the AB interaction, and I, which represents the total. The row headings are the runs. Note that the signs in the AB column are the product of signs from columns A and B. To generate a contrast from this table, multiply the signs in the appropriate column of Table 12-7 by the runs listed in the rows and add.

To obtain the sums of squares for A, B, and AB, we use the following result.

$$SS = \frac{(\text{contrast})^2}{n\,\Sigma(\text{contrast coefficients})^2} \qquad (12\text{-}14)$$

Therefore, the sums of squares for A, B, and AB are

$$SS_A = \frac{[a + ab - b - (1)]^2}{4n}$$

$$SS_B = \frac{[b + ab - a - (1)]^2}{4n} \qquad (12\text{-}15)$$

$$SS_{AB} = \frac{[ab + (1) - a - b]^2}{4n}$$

The analysis of variance is completed by computing the total sum of squares SS_T (with $4n - 1$ degrees of freedom) as usual, and obtaining the error sum of squares SS_E [with $4(n - 1)$ degrees of freedom] by subtraction.

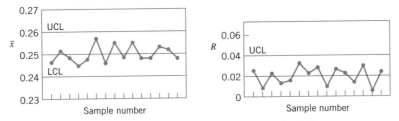

Figure 12-18 $\bar{x}$ and R control charts on notch dimension, Example 12-6.

······· **EXAMPLE 12-6** ··

The Router Experiment

A router is used to cut registration notches in printed circuit boards. The average notch dimension is satisfactory, and the process is in statistical control (see the $\bar{x}$ and R control charts in Fig. 12-18), but there is too much variability in the process. This excess variability leads to problems in board assembly. The components are inserted into the board using automatic equipment, and the variability in notch dimension causes improper board registration. As a result, the auto-insertion equipment does not work properly.

Since the process is in statistical control, the quality-improvement team assigned to this project decided to use a designed experiment to study the process. The team considered two factors: bit size (A) and speed (B). Two levels were chosen for each factor (bit size A at $\frac{1}{16}''$ and $\frac{1}{8}''$ and speed B at 40 rpm and 80 rpm) and a 2^2 design was set up. Since variation in notch dimension was difficult to measure directly, the team decided to measure it indirectly. Sixteen test boards were instrumented with accelerometers that allowed vibration on the (X, Y, Z) coordinate axes to be measured. The resultant vector of these three components was used as the response variable. Since vibration at the surface of the board when it is cut is directly related to variability in notch dimension, reducing vibration levels will also reduce the variability in notch dimension.

Four boards were tested at each of the four runs in the experiment, and the resulting data are shown in Table 12-8. Using equations 12-11, 12-12, and 12-13, we can compute

Table 12-8 Data from the Router Experiment

Run		Factors A	B	Vibration				Total
1	(1)	−	−	18.2	18.9	12.9	14.4	64.4
2	a	+	−	27.2	24.0	22.4	22.5	96.1
3	b	−	+	15.9	14.5	15.1	14.2	59.7
4	ab	+	+	41.0	43.9	36.3	39.9	161.1

the factor effect estimates as follows:

$$A = \frac{1}{2n}[a + ab - b - (1)]$$

$$= \frac{1}{2(4)}[96.1 + 161.1 - 59.7 - 64.4] = \frac{133.1}{8} = 16.64$$

$$B = \frac{1}{2n}[b + ab - a - (1)]$$

$$= \frac{1}{2(4)}[59.7 + 161.1 - 96.1 - 64.4] = \frac{60.3}{8} = 7.54$$

$$AB = \frac{1}{2n}[ab + (1) - a - b]$$

$$= \frac{1}{2(4)}[161.1 + 64.4 - 96.1 - 59.7] = \frac{69.7}{8} = 8.71$$

All the numerical effect estimates seem large. For example, when we change factor A from the low level to the high level (bit size from $\frac{1}{16}''$ to $\frac{1}{8}''$), the average vibration level increases by 16.64 cps.

The magnitude of these effects may be confirmed with the analysis of variance, which is summarized in Table 12-9. The sums of squares in this table for main effects and interaction were computed using equations 12-15. The analysis of variance confirms our conclusions that were obtained by initially examining the magnitude and direction of the factor effects; both bit size and speed are important, and there is interaction between two variables.

Regression Model and Residual Analysis

It is easy to obtain the residuals from a 2^k design by fitting a **regression model** to the data. For the router experiment, the regression model is

$$y = \beta_0 + \beta_1 x_1 + \beta_2 x_2 + \beta_{12} x_1 x_2 + \epsilon$$

where the factors A and B are represented by coded variables x_1 and x_2, and the AB interaction is represented by the cross-product term in the model, $x_1 x_2$. The low and high

Table 12-9 Analysis of Variance for the Router Experiment

Source of Variation	Sum of Squares	Degrees of Freedom	Mean Square	F_0	P-value
Bit size (A)	1107.226	1	1107.226	185.25	1.17×10^{-8}
Speed (B)	227.256	1	227.256	38.03	4.82×10^{-5}
AB	303.631	1	303.631	50.80	1.20×10^{-5}
Error	71.723	12	5.977		
Total	1709.836	15			

levels of each factor are assigned the values $x_j = -1$ and $x_j = +1$, respectively. The coefficients $\beta_0, \beta_1, \beta_2$, and β_{12} are called **regression coefficients,** and ϵ is a random error term, similar to the error term in an analysis of variance model.

The fitted regression model is

$$\hat{y} = 23.83 + \left(\frac{16.64}{2}\right)x_1 + \left(\frac{7.54}{2}\right)x_2 + \left(\frac{8.71}{2}\right)x_1 x_2$$

where the estimate of the intercept $\hat{\beta}_0$ is the grand average of all 16 observations $(\bar{y})$ and the estimates of the other regression coefficients $\hat{\beta}_j$ are one-half the effect estimate for the corresponding factor. [Each regression coefficient estimate is one-half the effect estimate because regression coefficients measure the effect of a unit change in x_i on the mean of y, and the effect estimate is based on a two-unit change (from -1 to $+1$).]

This model can be used to obtain the predicted values of vibration level at any point in the region of experimentation, including the four points in the design. For example, consider the point with the small bit $(x_1 = -1)$ and low speed $(x_2 = -1)$. The predicted vibration level is

$$\hat{y} = 23.83 + \left(\frac{16.64}{2}\right)(-1) + \left(\frac{7.54}{2}\right)(-1) + \left(\frac{8.71}{2}\right)(-1)(-1) = 16.1$$

The four residuals corresponding to the observations at this design point are found by taking the difference between the actual observation and the predicted value as follows:

$$e_1 = 18.2 - 16.1 = 2.1 \qquad e_3 = 12.9 - 16.1 = -3.2$$
$$e_2 = 18.9 - 16.1 = 2.8 \qquad e_4 = 14.4 - 16.1 = -1.7$$

The residuals at the other three runs would be computed similarly.

Figures 12-19 and 12-20 present the normal probability plot and the plot of residuals versus the fitted values, respectively. The normal probability plot is satisfactory, as is the plot of residuals versus $\hat{y}$, although this latter plot does give some indication that there may be less variability in the data at the point of lowest predicted vibration level.

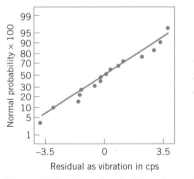

Figure 12-19 Normal probability plot, Example 12-6.

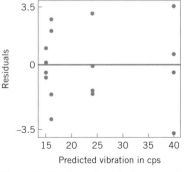

Figure 12-20 Plot of residuals versus $\hat{y}$, Example 12-6.

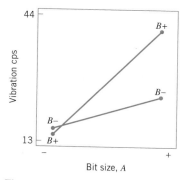

Figure 12-21 *AB* interaction plot.

Practical Interpretation

Since both factors *A* (bit size) and *B* (speed) have large, positive effects, we could reduce vibration levels by running both factors at the low level. However, with both bit size and speed at low level, the production rate could be unacceptably low. The *AB* interaction provides a solution to this potential dilemma. Figure 12-21 presents the two-factor *AB* interaction plot. Note that the large positive effect of speed occurs primarily when bit size is at the high level. If we use the small bit, then either speed level will provide lower vibration levels. If we run with speed high and use the small bit, the production rate will be satisfactory.

When manufacturing implemented this set of operating conditions, the result was a dramatic reduction in variability in the registration notch dimension. The process remained in statistical control, as the control charts in Fig. 12-22 imply, and the reduced variability dramatically improved the performance of the auto-insertion process.

Analysis Procedure for Factorial Experiments

Table 12-10 summarizes the sequence of steps that is usually employed to analyze factorial experiments. These steps were followed in the analysis of the router experiment in Example 12-6. Recall that our first activity, after the experiment was run, was to estimate the effect of the factors bit size, speed, and the two-factor interaction. The preliminary model that we used in the analysis was the two-factor factorial model with interaction. Generally, in any factorial experiment with replication, we will almost always

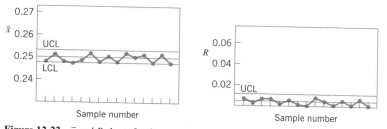

Figure 12-22 $\bar{x}$ and *R* charts for the router process after the experiment.

Table 12-10   Analysis Procedure for Factorial Designs

1.	Estimate the factor effects	4.	Analyze residuals
2.	Form preliminary model	5.	Refine model, if necessary
3.	Test for significance of factor effects	6.	Interpret results

use the full factorial model as the preliminary model. We tested for significance of factor effects by using the analysis of variance. Since the residual analysis was satisfactory, and both main effects and the interaction term were significant, there was no need to refine the model. Therefore, we were able to interpret the results in terms of the original full factorial model, using the two-factor interaction graph in Fig. 12-21. Sometimes refining the model includes deleting terms from the final model that are not significant, or taking other actions that may be indicated from the residual analysis.

Several statistics software packages include special routines for the analysis of two-level factorial designs. Many of these packages follow an analysis process similar to the one we have outlined. We will illustrate this analysis procedure again several times in this chapter.

12-5.2   The 2^k Design for $k \geq 3$ Factors

The methods presented in the previous section for factorial designs with $k = 2$ factors each at two levels can be easily extended to more than two factors. For example, consider $k = 3$ factors, each at two levels. This design is a 2^3 factorial design, and it has eight factor-level combinations. Geometrically, the design is a cube as shown in Fig. 12-23a, with the eight runs forming the corners of the cube. Figure 12-23b shows the eight runs in a tabular format, often called the **test matrix.** This design allows three main effects to be estimated (A, B, and C) along with three two-factor interactions (AB, AC, and BC) and a three-factor interaction (ABC). Thus, the full factorial model could be written symbolically as

$$y = \mu + A + B + C + AB + AC + BC + ABC + \epsilon$$

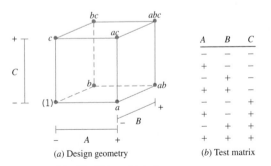

A	B	C
−	−	−
+	−	−
−	+	−
+	+	−
−	−	+
+	−	+
−	+	+
+	+	+

(a) Design geometry   (b) Test matrix

Figure 12-23   The 2^3 factorial design.

where μ is an overall mean, ϵ is a random error term assumed to be NID(0, σ^2), and the uppercase letters represent the main effects and interactions of the factors [note that we could have used Greek letters for the main effects and interactions, as in equation 12-2].

The main effects can be estimated easily. Remember that the lowercase letters (1), a, b, ab, c, ac, bc, and abc represent the total of all n replicates at each of the eight runs in the design. Referring to the cube in Fig. 12-23, we would estimate the main effect of A by averaging the four runs on the right side of the cube where A is at the high level and subtracting from that quantity the average of the four runs on the left side of the cube where A is at the low level. This gives

$$A = \bar{y}_{A^+} - \bar{y}_{A^-} = \frac{1}{4n}[a + ab + ac + abc - b - c - bc - (1)] \qquad (12\text{-}16)$$

In a similar manner, the effect of B is the average difference of the four runs in the back face of the cube and the four in the front, or

$$B = \bar{y}_{B^+} - \bar{y}_{B^-} = \frac{1}{4n}[b + ab + bc + abc - a - c - ac - (1)] \qquad (12\text{-}17)$$

and the effect of C is the average difference between the four runs in the top face of the cube and the four in the bottom, or

$$C = \bar{y}_{C^+} - \bar{y}_{C^-} = \frac{1}{4n}[c + ac + bc + abc - a - b - ab - (1)] \qquad (12\text{-}18)$$

The top row of Fig. 12-24 shows how the main effects of the three factors are computed.

Now consider the two-factor interaction AB. When C is at the low level, AB is simply the average difference in the A effect at the two levels of B, or

$$AB (C \text{ low}) = \frac{1}{2n}[ab - b] - \frac{1}{2n}[a - (1)]$$

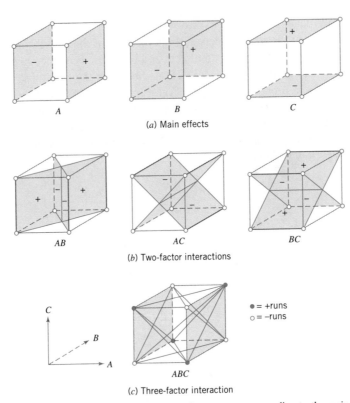

(a) Main effects

(b) Two-factor interactions

● = +runs
○ = −runs

(c) Three-factor interaction

Figure 12-24 Geometric presentation of contrasts corresponding to the main effects and interaction in the 2^3 design.

Similarly, when C is at the high level, the AB interaction is

$$AB \ (C \ \text{high}) = \frac{1}{2n} \ [abc - bc] - \frac{1}{2n} \ [ac - c]$$

The AB interaction is the average of these two components, or

$$AB = \frac{1}{4n} \ [ab + (1) + abc + c - b - a - bc - ac] \qquad (12\text{-}19)$$

Note that the AB interaction is simply the difference in averages on two diagonal planes in the cube (refer to the left-most cube in the middle row of Fig. 12-24).

Using a similar approach, we see from the middle row of Fig. 12-24 that the AC and BC interaction effect estimates are as follows:

$$AC = \frac{1}{4n} [ac + (1) + abc + b - a - c - ab - bc] \quad (12\text{-}20)$$

$$BC = \frac{1}{4n} [bc + (1) + abc + a - b - c - ab - ac] \quad (12\text{-}21)$$

The ABC interaction effect is the average difference between the AB interaction at the two levels of C. Thus

$$ABC = \frac{1}{4n} \{[abc - bc] - [ac - c] - [ab - b] + [a - (1)]\}$$

or

$$ABC = \frac{1}{4n} [abc - bc - ac + c - ab + b + a - (1)] \quad (12\text{-}22)$$

This effect estimate is illustrated in the bottom row of Fig. 12-24.

The quantities in brackets in equations 12-16 through 12-22 are contrasts in the eight factor-level combinations. These contrasts can be obtained from a table of plus and minus signs for the 2^3 design, shown in Table 12-11. Signs for the main effects (columns A, B, and C) are obtained by associating a plus with the high level and a minus with the low level. Once the signs for the main effects have been established, the signs for the remaining columns are found by multiplying the appropriate preceding columns, row by row. For example, the signs in column AB are the product of the signs in columns A and B.

Table 12-11 has several interesting properties:

1. Except for the identity column I, each column has an equal number of plus and minus signs.

2. The sum of products of signs in any two columns is zero; that is, the columns in the table are *orthogonal*.

3. Multiplying any column by column I leaves the column unchanged; that is, I is an *identity element*.

4. The product of any two columns yields a column in the table; for example, $A \times B = AB$, and $AB \times ABC = A^2B^2C = C$, since any column multiplied by itself is the identity column.

Table 12-11 Signs for Effects in the 2^3 Design

Treatment Combination	Factorial Effect							
	I	A	B	AB	C	AC	BC	ABC
(1)	+	−	−	+	−	+	+	−
a	+	+	−	−	−	−	+	+
b	+	−	+	−	−	+	−	+
ab	+	+	+	+	−	−	−	−
c	+	−	−	+	+	−	−	+
ac	+	+	−	−	+	+	−	−
bc	+	−	+	−	+	−	+	−
abc	+	+	+	+	+	+	+	+

The estimate of any main effect or interaction is determined by multiplying the factor-level combinations in the first column of the table by the signs in the corresponding main effect or interaction column, adding the result to produce a contrast, and then dividing the contrast by one-half the total number of runs in the experiment. Expressed mathematically,

$$\text{Effect} = \frac{\text{Contrast}}{n2^{k-1}} \qquad (12\text{-}23)$$

The sum of squares for any effect is

$$SS = \frac{(\text{Contrast})^2}{n2^k} \qquad (12\text{-}24)$$

EXAMPLE 12-7

An experiment was performed to investigate the surface finish of a metal part. The experiment is a 2^3 factorial design in the factors feed rate (A), depth of cut (B), and tool angle (C), with $n = 2$ replicates. Table 12-12 presents the observed surface-finish data for this experiment, and the design is shown graphically in Fig. 12-25.

Table 12-12   Surface-Finish Data for Example 12-7

Run		A	B	C	Surface Finish	Totals
1	(1)	−1	−1	−1	9, 7	16
2	a	1	−1	−1	10, 12	22
3	b	−1	1	−1	9, 11	20
4	ab	1	1	−1	12, 15	27
5	c	−1	−1	1	11, 10	21
6	ac	1	−1	1	10, 13	23
7	bc	−1	1	1	10, 8	18
8	abc	1	1	1	16, 14	30

The main effects may be estimated using equations 12-16 through 12-22. The effect of A, for example, is

$$A = \frac{1}{4n}[a + ab + ac + abc - b - c - bc - (1)]$$

$$= \frac{1}{4(2)}[22 + 27 + 23 + 30 - 20 - 21 - 18 - 16]$$

$$= \frac{1}{8}[27] = 3.375$$

and the sum of squares for A is found using equation 12-24:

$$SS_A = \frac{(\text{Contrast}_A)^2}{n2^k}$$

$$= \frac{(27)^2}{2(8)} = 45.5625$$

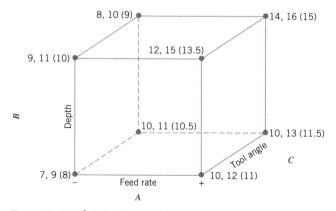

Figure 12-25 2^3 design for the surface finish experiment in Example 12-7 (the numbers in parentheses are the average responses at each design point).

It is easy to verify that the other effect estimates and sums of squares are

$$
\begin{array}{ll}
B = 1.625 & SS_B = 10.5625 \\
C = 0.875 & SS_C = 3.0625 \\
AB = 1.375 & SS_{AB} = 7.5625 \\
AC = 0.125 & SS_{AC} = 0.0625 \\
BC = -0.625 & SS_{BC} = 1.5625 \\
ABC = 1.125 & SS_{ABC} = 5.5625
\end{array}
$$

From examining the magnitude of the effects, feed rate (factor A) is clearly dominant, followed by depth of cut (B) and the AB interaction, although the interaction effect is relatively small. The analysis of variance for the full factorial model is summarized in Table 12-13. Based on the P-values, it is clear that the feed rate (A) is highly significant.

Many computer programs analyze the 2^k factorial design. Table 12-14 is the output from Minitab. Although at first glance the two tables seem somewhat different, they actually provide the same information. The analysis of variance displayed in the lower portion of Table 12-14 presents F-ratios computed on important groups of model terms; main effects, two-way interactions, and the three-way interaction. The mean square for each group of model terms was obtained by combining the sums of squares for each model component and dividing by the number of degrees of freedom associated with that group of model terms.

A t-test is used to test the significance of each individual term in the model. These t-tests are shown in the upper portion of Table 12-14. Note that a "coefficient estimate" is given for each variable in the full factorial model. These are actually the estimates of the coefficients in the regression model that would be used to predict surface finish in terms of the variables in the full factorial model. Each t-value is computed according to

$$
t_0 = \frac{\hat{\beta}}{s.e.\,(\hat{\beta})}
$$

Table 12-13 Analysis of Variance for the Surface-Finish Experiment

Source of Variation	Sum of Squares	Degrees of Freedom	Mean Square	F_0	P-value
A	45.5625	1	45.5625	18.69	2.54×10^{-3}
B	10.5625	1	10.5625	4.33	0.07
C	3.0625	1	3.0625	1.26	0.29
AB	7.5625	1	7.5625	3.10	0.12
AC	0.0625	1	0.0625	0.03	0.88
BC	1.5625	1	1.5625	0.64	0.45
ABC	5.5625	1	5.5625	2.08	0.19
Error	19.5000	8	2.4375		
Total	92.9375	15			

Table 12-14   Analysis of Variance from Minitab for the Surface-Finish Experiment

```
Factorial Design
Full Factorial Design
Factors:          3    Base Design:            3, 8
Runs:            16    Replicates:                2
Blocks:        none    Center pts (total):        0
All terms are free from aliasing

Fractional Factorial Fit: Finish versus A, B, C
Estimated Effects and Coefficients for Finish (coded units)
```

Term	Effect	Coef	SE Coef	T	P
Constant		11.0625	0.3903	28.34	0.000
A	3.3750	1.6875	0.3903	4.32	0.003
B	1.6250	0.8125	0.3903	2.08	0.071
C	0.8750	0.4375	0.3903	1.12	0.295
A*B	1.3750	0.6875	0.3903	1.76	0.116
A*C	0.1250	0.0625	0.3903	0.16	0.877
B*C	−0.6250	−0.3125	0.3903	−0.80	0.446
A*B*C	1.1250	0.5625	0.3903	1.44	0.188

```
Analysis of Variance for Finish (coded units)
```

Source	DF	Seq SS	Adj SS	Adj MS	F	P
Main Effects	3	59.187	59.187	19.729	8.09	0.008
2-Way Interactions	3	9.187	9.187	3.062	1.26	0.352
3-Way Interactions	1	5.062	5.062	5.062	2.08	0.188
Residual Error	8	19.500	19.500	2.437		
Pure Error	8	19.500	19.500	2.438		
Total	15	92.937				

where $\hat{\beta}$ is the coefficient estimate and $s.e.\ (\hat{\beta})$ is the estimated standard error of the coefficient. For a 2^k factorial design, the estimated standard error of the coefficient is

$$s.e.\ (\hat{\beta}) = \sqrt{\frac{\hat{\sigma}^2}{n2^k}}$$

We use the error or residual mean square from the analysis of variance as the estimate $\hat{\sigma}^2$. In our example,

$$s.e.\ (\hat{\beta}) = \sqrt{\frac{2.4375}{2(2^3)}} = 0.390312$$

as shown in Table 12-14. It is easy to verify that dividing any coefficient estimate by its estimated standard error produces the t-value for testing whether the corresponding regression coefficient is zero.

The t-tests in Table 12-14 are equivalent to the analysis of variance F-tests in Table 12-13. You may have suspected this already, since the P-values in the two tables are identical to two decimal places. Furthermore, note that the square of any t-value in Table 12-14 produces the corresponding F-ratio value in Table 12-13. In general, the square of a t random variable with ν degrees of freedom results in an F random variable with one numerator degree of freedom and ν denominator degrees of freedom. This explains the equivalence of the two procedures used to conduct the analysis of variance for the surface-finish experiment data.

Based on the analysis of variance, we conclude that the full factorial model in all these factors is unnecessary, and that a reduced model including fewer variables is more appropriate. The main effects of A and B both have relatively small P-values (<0.10), and this AB interaction is the next most important effect (P-value $\cong 0.12$). The regression model that we would use to represent this process is

$$y = \beta_0 + \beta_1 x_1 + \beta_2 x_2 + \beta_{12} x_1 x_2 + \varepsilon$$

where x_1 represents factor A, x_2 represents factor B, and $x_1 x_2$ represents the AB interaction. The regression coefficients $\hat{\beta}_1$, $\hat{\beta}_2$, and $\hat{\beta}_{12}$ are one-half the corresponding effect estimates and $\hat{\beta}_0$ is the grand average. Thus

$$\hat{y} = 11.0625 + \left(\frac{3.375}{2}\right)x_1 + \left(\frac{1.625}{2}\right)x_2 + \left(\frac{1.375}{2}\right)x_1 x_2$$

$$= 11.0625 + 1.6875x_1 + 0.8125x_2 + 0.6875x_1 x_2$$

Note that we can read the values of $\hat{\beta}_0$, $\hat{\beta}_1$, $\hat{\beta}_2$, and $\hat{\beta}_{12}$ directly from the "coefficient" column of Table 12-14.

This regression model can be used to predict surface finish at any point in the original experimental region. For example, consider the point where all three variables are at the low level. At this point, $x_1 = x_2 = -1$, and the predicted value is

$$\hat{y} = 11.0625 + 1.6875(-1) + 0.8125(-1) + 0.6875(-1)(-1) = 9.25$$

Figure 12-26 shows the predicted values at each point in the original experimental design.

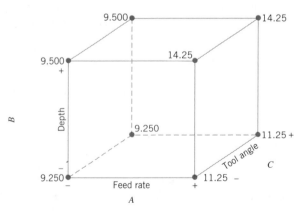

Figure 12-26 Predicted values of surface finish at each point in the original design, Example 12-7.

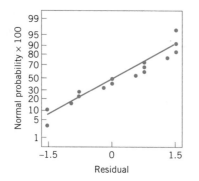

Figure 12-27 Normal probability plot of residuals, Example 12-7.

The residuals can be obtained as the difference between the observed and predicted values of surface finish at each design point. For the point where all three factors A, B, and C are at the low level, the observed values of surface finish are 9 and 7, so the residuals are $9 - 9.25 = -0.25$ and $7 - 9.25 = -2.25$.

A normal probability plot of the residuals is shown in Fig. 12-27. Since the residuals lie approximately along a straight line, we do not suspect any severe nonnormality in the data. There are also no indications of outliers. It would also be helpful to plot the residuals versus the predicted values and against each of the factors A, B, and C. These plots do not indicate any potential model problems.

Finally, we can provide a practical interpretation of the results of our experiment. Both main effects A and B are positive, and since small values of the surface finish response are desirable, this would suggest that both A (feed rate) and B (depth of cut) should be run at the low level. However, the model has an interaction term, and the effect of this interaction should be taken into account when drawing conclusions. We could do this by examining an interaction plot, as in Example 12-6. Alternatively, the cube plot of predicted responses in Fig. 12-26 can also be used for model interpretation. This figure indicates that the lowest values of predicted surface finish will be obtained when A and B are at the low level.

Some Comments on the Regression Model

In the two previous examples, we used a **regression model** to summarize the results of the experiment. In general, a regression model is an equation of the form

$$y = \beta_0 + \beta_1 x_1 + \beta_2 x_2 + \cdots + \beta_k x_k + \epsilon \qquad (12\text{-}25)$$

where y is the response variable, the x's are a set of regressor or predictor variables, the β's are the regression coefficients, and ϵ is an error term, assumed to be NID$(0, \sigma^2)$. In our examples, we had $k = 2$ factors and the models had an interaction term, so the specific form of the regression model that we fit was

$$y = \beta_0 + \beta_1 x_1 + \beta_2 x_2 + \beta_{12} x_1 x_2 + \epsilon$$

In general, the regression coefficients in these models are estimated using the **method of least squares;** that is, the $\hat{\beta}$'s are chosen so as to minimize the sum of the squares of the errors (the ϵ's). Most introductory statistics books give the details of the method of least squares [for example, see Montgomery and Runger (1999) and the Supplemental Text Material]. However, in the special case of a 2^k design, it is extremely easy to find the least squares estimates of the β's. The least squares estimate of any regression coefficient β is simply one-half of the corresponding factor effect estimate. Recall that we have used this result to obtain the regression models in Examples 12-6 and 12-7. Also, please remember that this result only works for a 2^k factorial design, and it assumes that the x's are coded variables over the range $-1 \le x \le +1$ that represent the design factors.

It is very useful to express the results of a designed experiment in terms of a **model,** which will be a valuable aid in interpreting the experiment. Recall that we used the cube plot of predicted values from the model in Fig. 12-27 to find appropriate settings for feed rate and depth of cut in Example 12-7. More general graphical displays can also be useful. For example, consider the model for surface finish in terms of feed rate (x_1) and depth of cut (x_2) without the interaction term

$$\hat{y} = 11.0625 + 1.6875x_1 + 0.8125x_2$$

Note that the model was obtained simply by deleting the interaction term from the original model. This can only be done if the variables in the experimental design are **orthogonal,** as they are in a 2^k design. Figure 12-28 plots the predicted value of surface finish ($\hat{y}$) in terms of the two process variables x_1 and x_2. Figure 12-28a is a three-dimensional plot showing the plane of predicted response values generated by the regression model. This type of display is called a **response surface plot,** and the regression model used to generate the graph is often called a **first-order response surface model.** The graph in Fig. 12-28b is a two-dimensional **contour plot** obtained by looking down on the three-dimensional response surface plot and connecting points of constant surface finish (response) in the x_1–x_2 plane. The lines of constant response are straight lines because the response surface is first-order; that is, it contains only the main effects x_1 and x_2.

In Example 12-7, we actually fit a first-order model with interaction:

$$\hat{y} = 11.0625 + 1.6875x_1 + 0.8125x_2 + 0.6875x_1x_2$$

Figure 12-29a is the three-dimensional response surface plot for this model and Fig. 12-29b is the contour plot. Note that the effect of adding the interaction term to the model is to introduce **curvature** into the response surface; in effect, the plane is "twisted" by the interaction effect.

Inspection of a response surface makes interpretation of the results of an experiment very simple. For example, note from Fig. 12-29 that if we wish to minimize the surface-finish response, we need to run x_1 and x_2 at (or near) their low levels. We reached the same conclusion by inspection of the cube plot in Fig. 12-26. However, suppose we needed to obtain a particular value of surface finish, say 10.25 (the surface might need to be this rough so that a coating will adhere properly). Figure 12-29b indicates that there are many combinations of x_1 and x_2 that will allow the process to operate on the contour line $\hat{y} = 10.25$. The experimenter might select a set of operating

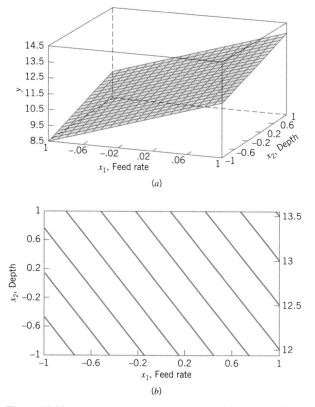

Figure 12-28 (a) Response surface for the model $\hat{y} = 11.0625 + 1.6875x_1 + 0.8125x_2$. (b) The contour plot.

conditions that maximized x_1 subject to x_1 and x_2 giving a predicted response on or near to the contour $\hat{y} = 10.25$, as this would satisfy the surface finish objective while simultaneously making the feed rate as large as possible, which would maximize the production rate.

Response surface models have many uses. In Chapter 13 we will give an overview of some aspects of response surfaces and how they can be used for process improvement and optimization. However, note how useful the response surface was, even in this simple example. This is why we tell experimenters that **the objective of every designed experiment is a quantitative model of the process.**

Projection of 2^k Designs

Any 2^k design will collapse or project into another two-level factorial design in fewer variables if one or more of the original factors are dropped. Usually this will provide additional insight into the remaining factors. For example, consider the surface-finish experiment. Since factor C and all its interactions are negligible, we could eliminate factor C from the design. The result is to collapse the cube in Fig. 12-25 into a square in the $A-B$ plane; however, each of the four runs in the new design has four replicates. In

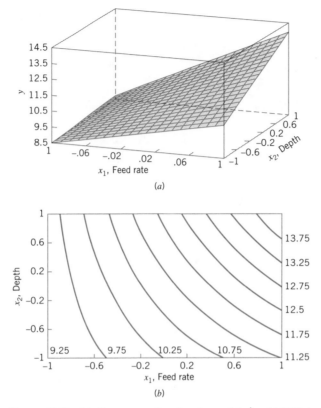

Figure 12-29 (a) Response surface for the model $\hat{y} = 11.0625 + 1.6875x_1 + 0.8125x_2 + 0.6875x_1x_2$. (b) The contour plot.

general, if we delete h factors so that $r = k - h$ factors remain, the original 2^k design with n replicates will project into a 2^r design with $n2^h$ replicates.

Other Methods for Judging the Significance of Effects

The analysis of variance is a formal way to determine which effects are nonzero. Two other methods are useful. In the first method, we can calculate the standard errors of the effects and compare the magnitude of the effects to their standard errors. The second method uses normal probability plots to assess the importance of the effects.

The standard error of any effect estimate in a 2^k design is given by

$$s.e. \text{ (Effect)} = \sqrt{\frac{\hat{\sigma}^2}{n2^{k-2}}} \tag{12-26}$$

where $\hat{\sigma}^2$ is an estimate of the experimental error variance σ^2. We usually take the error (or residual) mean square from the analysis of variance as the estimate $\hat{\sigma}^2$ of σ^2.

As an illustration for the surface-finish experiment, we find that $MS_E = \hat{\sigma}^2 = 2.4375$, and the standard error of each effect is

$$s.e. \text{ (Effect)} = \sqrt{\frac{\hat{\sigma}^2}{n2^{k-2}}}$$

$$= \sqrt{\frac{2.4375}{(2)2^{3-2}}}$$

$$= 0.78$$

Therefore, two standard deviation limits on the effect estimates are

$$
\begin{array}{rl}
A: & 3.375 \pm 1.56 \\
B: & 1.625 \pm 1.56 \\
C: & 0.875 \pm 1.56 \\
AB: & 1.375 \pm 1.56 \\
AC: & 0.125 \pm 1.56 \\
BC: & 0.625 \pm 1.56 \\
ABC: & 1.125 \pm 1.56
\end{array}
$$

These intervals are approximate 95% confidence intervals. They indicate that the two main effects A and B are important but that the other effects are not, since the intervals for all effects except A and B include zero. These conclusions are similar to those found in Example 12-7.

Normal probability plots can also be used to judge the significance of effects. We will illustrate that method in the next section.

12-5.3 A Single Replicate of the 2^k Design

As the number of factors in a factorial experiment grows, the number of effects that can be estimated also grows. For example, a 2^4 experiment has 4 main effect, 6 two-factor interactions, 4 three-factor interactions, and 1 four-factor interaction, whereas a 2^6 experiment has 6 main effects, 15 two-factor interactions, 20 three-factor interactions, 15 four-factor interactions, 6 five-factor interactions, and 1 six-factor interaction. In most situations the **sparsity of effects principle** applies; that is, the system is usually dominated by the main effects and low-order interactions. Three-factor and higher interactions are usually negligible. Therefore, when the number of factors is moderately large—say, $k \geq 4$ or 5—a common practice is to run only a single replicate of the 2^k design and then pool or combine the higher-order interactions as an estimate of error.

•••••••• **EXAMPLE 12-8** ••

An article in *Solid State Technology* ("Orthogonal Design for Process Optimization and Its Application in Plasma Etching," May 1987, pp. 127–132) describes the application of factorial designs in developing a nitride etch process on a single-wafer plasma etcher. The process uses C_2F_6 as the reactant gas. It is possible to vary the gas flow, the power applied to the cathode, the pressure in the reactor chamber, and the spacing between the anode and the cathode (gap). Several response variables would usually be of interest in this process, but in this example we will concentrate on etch rate for silicon nitride.

We will use a single replicate of a 2^4 design to investigate this process. Since it is unlikely that the three-factor and four-factor interactions are significant, we will tentatively plan to combine them as an estimate of error. The factor levels used in the design are shown here:

Design Factor

Level	Gap A (cm)	Pressure B (m Torr)	C_2F_6 Flow C (SCCM)	Power D (W)
Low (−)	0.80	450	125	275
High (+)	1.20	550	200	325

Table 12-15 presents the data from the 16 runs of the 2^4 design. The design is shown geometrically in Fig. 12-30. Table 12-16 is the table of plus and minus signs for the 2^4 design. The signs in the columns of this table can be used to estimate the factor effects. To illustrate, the estimate of A is

$$A = \tfrac{1}{8} [a + ab + ac + abc + ad + abd + acd + abcd - (1) - b$$
$$- c - d - bc - bd - cd - bcd]$$

$$= \tfrac{1}{8} [669 + 650 + 642 + 635 + 749 + 868 + 860 + 729 - 550$$
$$- 604 - 633 - 601 - 1037 - 1052 - 1075 - 1063]$$

$$= -101.625$$

Table 12-15   The 2^4 Design for the Plasma Etch Experiment

Run	A (Gap)	B (Pressure)	C (C_2F_6 flow)	D (Power)	Etch Rate (Å/min)
1	−1	−1	−1	−1	550
2	1	−1	−1	−1	669
3	−1	1	−1	−1	604
4	1	1	−1	−1	650
5	−1	−1	1	−1	633
6	1	−1	1	−1	642
7	−1	1	1	−1	601
8	1	1	1	−1	635
9	−1	−1	−1	1	1037
10	1	−1	−1	1	749
11	−1	1	−1	1	1052
12	1	1	−1	1	868
13	−1	−1	1	1	1075
14	1	−1	1	1	860
15	−1	1	1	1	1063
16	1	1	1	1	729

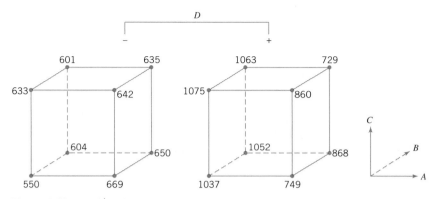

Figure 12-30 The 2^4 design for Example 12-8. The etch rate response is shown at the corners of the cubes.

Thus, the effect of increasing the gap between the anode and the cathode from 0.80 cm to 1.20 cm is to decrease the etch rate by 101.625 angstroms per minute. It is easy to verify that the complete set of effect estimates is

$$
\begin{aligned}
A &= -101.625 & AD &= -153.625 \\
B &= -1.625 & BD &= -0.625 \\
AB &= -7.875 & ABD &= 4.125
\end{aligned}
$$

Table 12-16 Contrast Constants for the 2^4 Design

Run		A	B	AB	C	AC	BC	ABC	D	AD	BD	ABD	CD	ACD	BCD	ABCD
1	(1)	−	−	+	−	+	+	−	−	+	+	−	+	−	−	+
2	a	+	−	−	−	−	+	+	−	−	+	+	+	+	−	−
3	b	−	+	−	−	+	−	+	−	+	−	+	+	−	+	−
4	ab	+	+	+	−	−	−	−	−	−	−	−	+	+	+	+
5	c	−	−	+	+	−	−	+	−	+	+	−	−	+	+	−
6	ac	+	−	−	+	+	−	−	−	−	+	+	−	−	+	+
7	bc	−	+	−	+	−	+	−	−	+	−	+	−	+	−	+
8	abc	+	+	+	+	+	+	+	−	−	−	−	−	−	−	−
9	d	−	−	+	−	+	+	−	+	−	−	+	−	+	+	−
10	ad	+	−	−	−	−	+	+	+	+	−	−	−	−	+	+
11	bd	−	+	−	−	+	−	+	+	−	+	−	−	+	−	+
12	abd	+	+	+	−	−	−	−	+	+	+	+	−	−	−	−
13	cd	−	−	+	+	−	−	+	+	−	−	+	+	−	−	+
14	acd	+	−	−	+	+	−	−	+	+	−	−	+	+	−	−
15	bcd	−	+	−	+	−	+	−	+	−	+	−	+	−	+	−
16	abcd	+	+	+	+	+	+	+	+	+	+	+	+	+	+	+

$$
\begin{array}{llll}
C = & 7.375 & CD = & -2.125 \\
AC = & -24.875 & ACD = & 5.625 \\
BC = & -43.875 & BCD = & -25.375 \\
ABC = & -15.625 & ABCD = & -40.125 \\
D = & 306.125 &
\end{array}
$$

A very helpful method in judging the significance of factors in a 2^k experiment is to construct a **normal probability plot** of the effect estimates. If none of the effects are significant, then the estimates will behave like a random sample drawn from a normal distribution with zero mean, and the plotted effects will lie approximately along a straight line. Those effects that do not plot on the line are significant factors.

The normal probability plot of effect estimates from the plasma etch experiment is shown in Fig. 12-31. Clearly, the main effects of A and D and the AD interaction are significant, as they fall far from the line passing through the other points. The analysis of variance summarized in Table 12-17 confirms these findings. Note that in the analysis of variance we have pooled the three- and four-factor interactions to form the error mean square. If the normal probability plot had indicated that any of these interactions were important, they would not be included in the error term.

Since $A = -101.625$, the effect of increasing the gap between the cathode and anode is to decrease the etch rate. However, $D = 306.125$, so applying higher power levels will increase the etch rate. Figure 12-32 is a plot of the AD interaction. This plot indicates that the effect of changing the gap width at low power settings is small, but that increasing the gap at high power settings dramatically reduces the etch rate. High etch rates are obtained at high power settings and narrow gap widths.

The regression model for this experiment is

$$
\hat{y} = 776.0625 - \left(\frac{101.625}{2}\right)x_1 + \left(\frac{306.125}{2}\right)x_4 - \left(\frac{153.625}{2}\right)x_1 x_4
$$

For example, when both A and D are at the low level, the predicted value from this

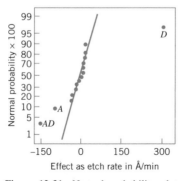

Figure 12-31 Normal probability plot of effects, Example 12-8.

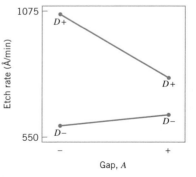

Figure 12-32 AD interaction in the plasma etch experiment.

Table 12-17 Analysis of Variance for the Plasma Etch Experiment

Source of Variation	Sum of Squares	Degrees of Freedom	Mean Square	F_0
A	41,310.563	1	41,310.563	20.28
B	10.563	1	10.563	< 1
C	217.563	1	217.563	< 1
D	374,850.063	1	374,850.063	183.99
AB	248.063	1	248.063	< 1
AC	2,475.063	1	2,475.063	1.21
AD	94,402.563	1	99,402.563	48.79
BC	7,700.063	1	7,700.063	3.78
BD	1.563	1	1.563	< 1
CD	18.063	1	18.063	< 1
Error	10,186.815	5	2,037.363	
Total	531,420.938	15		

model is

$$\hat{y} = 776.0625 - \left(\frac{101.625}{2}\right)(-1) + \left(\frac{306.125}{2}\right)(-1) - \left(\frac{153.625}{2}\right)(-1)(-1)$$

$$= 597$$

The four residuals at this run are

$$e_1 = 550 - 597 = -47$$
$$e_2 = 604 - 597 = 7$$
$$e_3 = 633 - 597 = 36$$
$$e_4 = 601 - 597 = 4$$

The residuals at the other three runs, (A high, D low), (A low, D high), and (A high, D high), are obtained similarly. A normal probability plot of the residuals is shown in Fig. 12-33. The plot is satisfactory.

12-5.4 Addition of Center Points to the 2^k Design

A potential concern in the use of two-level factorial designs is the assumption of linearity in the factor effects. Of course, perfect linearity is unnecessary, and the 2^k system will work quite well even when the linearity assumption holds only approximately. In fact, we have already observed that when an interaction term is added to a main-effects model, curvature is introduced into the response surface. Since a 2^k design will support a main effects plus interactions model, some protection against curvature is already inherent in the design.

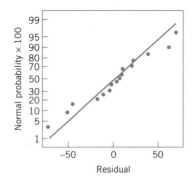

Figure 12-33 Normal probability plot of residuals, Example 12-8.

In some systems or processes, it will be necessary to incorporate **second-order effects** to obtain an adequate model. Consider the case of $k = 2$ factors. A model that includes second-order effects is

$$y = \beta_0 + \beta_1 x_1 + \beta_2 x_2 + \beta_{12} x_1 x_2 + \beta_{11} x_1^2 + \beta_{22} x_2^2 + \epsilon \quad (12\text{-}27)$$

where the coefficients β_{11} and β_{22} measure pure quadratic effects. Equation 12-27 is a **second-order response surface model.** This model cannot be fitted using a 2^2 design, because to fit a quadratic model, all factors must be run at at least three levels. It is important, however, to be able to determine whether the pure quadratic terms in equation 12-27 are needed.

There is a method of adding one point to a 2^k factorial design that will provide some protection against pure quadratic effects (in the sense that one can test to determine if the quadratic terms are necessary). Furthermore, if this point is replicated, then an independent estimate of experimental error can be obtained. The method consists of adding **center points** to the 2^k design. These center points consist of n_C replicates run at the point $x_i = 0$ ($i = 1, 2, \ldots , k$). One important reason for adding the replicate runs at the design center is that center points do not impact the usual effect estimates in a 2^k design. We assume that the k factors are quantitative; otherwise, a "middle" or center level of the factor would not exist.

To illustrate the approach, consider a 2^2 design with one observation at each of the factorial points $(-, -)$, $(+, -)$, $(-, +)$, and $(+, +)$ and n_C observations at the center points $(0, 0)$. Figure 12-34 illustrates the situation. Let $\bar{y}_F$ be the average of the four runs at the four factorial points, and let $\bar{y}_C$ be the average of the n_C run at the center point. If the difference $\bar{y}_F - \bar{y}_C$ is small, then the center points lie on or near the plane passing through the factorial points, and there is no curvature. On the other hand, if $\bar{y}_F - \bar{y}_C$ is large, then curvature is present. A single-degree-of-freedom sum of squares for pure

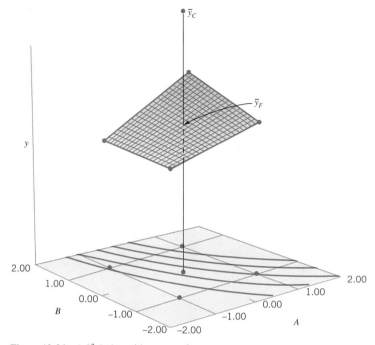

Figure 12-34 A 2^2 design with center points.

quadratic curvature is given by

$$SS_{\text{pure quadratic}} = \frac{n_F n_C (\bar{y}_F - \bar{y}_C)^2}{n_F + n_C} \tag{12-28}$$

where, in general, n_F is the number of factorial design points. This quantity may be compared to the error mean square to test for curvature.[1] More specifically, when points are added to the center of the 2^k design, then the model we may entertain is

$$y = \beta_0 + \sum_{j=1}^{k} \beta_j x_j + \sum_{i<j}\sum \beta_{ij} x_i x_j + \sum_{j=1}^{k} \beta_{jj} x_j^2 + \epsilon$$

where the β_{jj} are pure quadratic effects. The test for curvature actually tests the hypotheses

$$H_0: \quad \sum_{j=1}^{k} \beta_{jj} = 0$$

$$H_1: \quad \sum_{j=1}^{k} \beta_{jj} \neq 0$$

[1] A t-test can also be used; see the supplemental text material.

Furthermore, if the factorial points in the design are unreplicated, we may use the n_C center points to construct an estimate of error with $n_C - 1$ degrees of freedom.

······· **EXAMPLE 12-9** ···

We will illustrate the addition of center points to a 2^k design by reconsidering the plasma etch experiment in Example 12-8. Table 12-18 presents a modified version of the original unreplicated 2^4 design to which $n_C = 4$ center points have been added.

The average of the center points is $\bar{y}_C = 752.75$ and the average of the 16 factorial points is $\bar{y}_F = 776.0625$. The curvature sum of squares is computed from equation 12-28 as

$$SS_{\text{pure quadratic}} = \frac{n_F n_C (\bar{y}_F - \bar{y}_C)^2}{n_F + n_C}$$

$$= \frac{16(4)(776.0625 - 752.75)^2}{16 + 4}$$

$$= 1739.1$$

Table 12-18   The 2^4 Design for the Plasma Etch Experiment

Run	A (Gap)	B (Pressure)	C (C_2F_6 flow)	D (Power)	Etch Rate (Å/min)
1	−1	−1	−1	−1	550
2	1	−1	−1	−1	669
3	−1	1	−1	−1	604
4	1	1	−1	−1	650
5	−1	−1	1	−1	633
6	1	−1	1	−1	642
7	−1	1	1	−1	601
8	1	1	1	−1	635
9	−1	−1	−1	1	1037
10	1	−1	−1	1	749
11	−1	1	−1	1	1052
12	1	1	−1	1	868
13	−1	−1	1	1	1075
14	1	−1	1	1	860
15	−1	1	1	1	1063
16	1	1	1	1	729
17	0	0	0	0	706
18	0	0	0	0	764
19	0	0	0	0	780
20	0	0	0	0	761

Furthermore, an estimate of experimental error can be obtained by simply calculating the sample variance of the four center points as follows:

$$\hat{\sigma}^2 = \frac{\sum_{i=17}^{20} (y_i - 752.75)^2}{3} = 3122.7$$

This estimate of error has $n_C - 1 = 4 - 1 = 3$ degrees of freedom.

The pure quadratic sum of squares and the estimate of error may be incorporated into the analysis of variance for this experimental design. We would still use a normal probability plot of the effect estimates to preliminarily identify the important factors. The construction of this plot would not be affected by the addition of center points in the design, we would still identify A (Gap), D (Power), and the AD interaction is the most important effect.

Table 12-19 is the analysis of variance for this experiment obtained from Minitab. In the analysis, we included all four main effects and all six two-factor interactions in the model (just as we did in Example 12-8; see also Table 12-17). Note also that the

Table 12-19 Analysis of Variance Output from Minitab for Example 12-9

Estimated Effects and Coefficients for Etch Rate (coded units)

Term	Effect	Coef	SE Coef	T	P
Constant		776.06	10.20	76.11	0.000
A	-101.62	-50.81	10.20	-4.98	0.001
B	-1.63	-0.81	10.20	-0.08	0.938
C	7.37	3.69	10.20	0.36	0.727
D	306.12	153.06	10.20	15.01	0.000
A*B	-7.88	-3.94	10.20	-0.39	0.709
A*C	-24.88	-12.44	10.20	-1.22	0.257
A*D	-153.63	-76.81	10.20	-7.53	0.000
B*C	-43.87	-21.94	10.20	-2.15	0.064
B*D	-0.63	-0.31	10.20	-0.03	0.976
C*D	-2.13	-1.06	10.20	-0.10	0.920
Ct Pt		-23.31	22.80	-1.02	0.337

Analysis of Variance for Etch (coded units)

Source	DF	Seq SS	Adj SS	Adj MS	F	P
Main Effects	4	416389	416389	104097	62.57	0.000
2-Way Interactions	6	104845	104845	17474	10.50	0.002
Curvature	1	1739	1739	1739	1.05	0.337
Residual Error	8	13310	13310	1664		
Lack of Fit	5	10187	10187	2037	1.96	0.308
Pure Error	3	3123	3123	1041		
Total	19	536283				

pure quadratic sum of squares from equation 12-28 is called the "curvature" sum of squares, and the estimate of error calculated from the $n_C = 4$ center points is called the "pure error" sum of squares in Table 12-19. The "lack of fit" sum of squares in Table 12-19 is actually the total of the sums of squares for the three-factor and four-factor interactions. The F-test for lack of fit is computed as

$$F_0 = \frac{MS_{\text{lack of fit}}}{MS_{\text{pure error}}} = \frac{2037}{1041} = 1.96$$

and is not significant, indicating that none of the higher-order interaction terms are important. This computer program combines the pure error and lack-of-fit sums of squares to form a residual sum of squares with eight degrees of freedom. This residual sum of squares is used to test for pure quadratic curvature with

$$F_0 = \frac{MS_{\text{curvature}}}{MS_{\text{residual}}} = \frac{1739}{1664} = 1.05$$

The P-value in Table 12-19 associated with this F-ratio indicates that there is no evidence of pure quadratic curvature.

The upper portion of Table 12-19 shows the regression coefficient for each model effect, the corresponding t-value, and the P-value. Clearly the main effects of A and D and the AD interaction are the three largest effects.

12-5.5 Blocking and Confounding in the 2^k Design

It is often impossible to run all of the observations in a 2^k factorial design under constant or homogeneous conditions. For example, it might not be possible to conduct all the tests on one shift or use material from a single batch. When this problem occurs, **blocking** is an excellent technique for eliminating the unwanted variation that could be caused by the nonhomogeneous conditions. If the design is replicated, and if the block is of sufficient size, then one approach is to run each replicate in one **block** (set of homogeneous conditions). For example, consider a 2^3 design that is to be replicated twice. Suppose that it takes about 1 h to complete each run. Then by running the eight runs from replicate one on one day and the eight runs from the second replicate on another day, any time effect, or difference between how the process works on the two days, can be eliminated. Thus, the two days became the two blocks in the design. The average difference between the responses on the two days is the block effect.

Sometimes we cannot run a complete replicate of a factorial design under homogeneous experimental conditions. **Confounding** is a design technique for running a factorial experiment in blocks, where the block size is smaller than the number of runs in one complete replicate. The technique causes certain interactions to be indistinguishable from or **confounded with blocks.** We will illustrate confounding in the 2^k factorial design in 2^p blocks, where $p < k$.

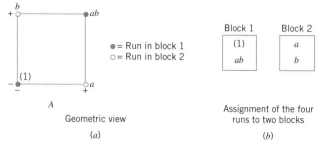

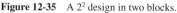

Figure 12-35   A 2^2 design in two blocks.

Consider a 2^2 design. Suppose that each of the $2^2 = 4$ runs require 4 h of laboratory analysis. Thus, 2 days are required to perform the experiment. If days are considered as blocks, then we must assign two of the four runs to each day.

Consider the design shown in Fig. 12-35. Note that block 1 contains the runs (1) and ab and that block 2 contains a and b. The contrasts for estimating the main effects A and B are

$$\text{Contrast}_A = ab + a - b - (1)$$
$$\text{Contrast}_B = ab + b - a - (1)$$

These contrasts are unaffected by blocking since in each contrast there is one plus and one minus run from each block. That is, any difference between block 1 and block 2 will cancel out. The contrast for the AB interaction is

$$\text{Contrast}_{AB} = ab + (1) - a - b$$

Since the two runs with the plus sign, ab and (1), are in block 1 and the two with the minus sign, a and b, are in block 2, the block effect and AB interaction are identical. That is, AB is confounded with blocks.

The reason for this is apparent from the table of plus and minus signs for the 2^2 design, shown in Table 12-7. From this table, we see that all runs that have a plus on AB are assigned to block 1, and all runs that have a minus sign on AB are assigned to block 2.

This scheme can be used to confound any 2^k design in two blocks. As a second example, consider a 2^3 design, run in two blocks. Suppose we wish to confound the three-factor interaction ABC with blocks. From the table of plus and minus signs, shown in Table 12-11, we assign the runs that are minus on ABC to block 1 and those that are plus on ABC to block 2. The resulting design is shown in Fig. 12-36.

For more information on confounding, refer to Montgomery (1997, Chapter 8). This book contains guidelines for selecting factors to confound with blocks so that main effects and low-order interactions are not confounded. In particular, the book contains a table of suggested confounding schemes for designs with up to seven factors and a range of block sizes, some as small as two runs.

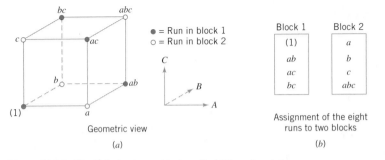

Figure 12-36   The 2^3 design in two blocks with *ABC* confounded.

12-6   FRACTIONAL REPLICATION OF THE 2^k DESIGN

As the number of factors in a 2^k design increases, the number of runs required increases rapidly. For example, a 2^5 requires 32 runs. In this design, only 5 degrees of freedom correspond to main effects and 10 degrees of freedom correspond to two-factor interactions. If we can assume that certain high-order interactions are negligible, then a fractional factorial design involving fewer than the complete set of 2^k runs can be used to obtain information on the main effects and low-order interactions. In this section, we will introduce fractional replication of the 2^k design. For a more complete treatment, see Montgomery (1997, Chapter 9).

12-6.1   The One-Half Fraction of the 2^k Design

A one-half fraction of the 2^k design contains 2^{k-1} runs and is often called a 2^{k-1} fractional factorial design. As an example, consider the 2^{3-1} design—that is, a one-half fraction of the 2^3. The table of plus and minus signs for the 2^3 design is shown in Table 12-20. Suppose we select the four runs a, b, c, and abc as our one-half fraction. These runs are shown in the top half of Table 12-20. The design is shown geometrically in Fig. 12-37a.

Note that the 2^{3-1} design is formed by selecting only those runs that yield a plus on the *ABC* effect. Thus, *ABC* is called the **generator** of this particular fraction. Furthermore, the identity element *I* is also plus for the four runs, so we call

$$I = ABC$$

the **defining relation** for the design.

The runs in the 2^{3-1} designs yield three degrees of freedom associated with the main effects. From Table 12-20, we obtain the estimates of the main effects as

$$A = \tfrac{1}{2}[a - b - c + abc]$$
$$B = \tfrac{1}{2}[-a + b - c + abc]$$
$$C = \tfrac{1}{2}[-a - b + c + abc]$$

Table 12-20 Plus and Minus Signs for the 2^3 Factorial Design

Run	Factorial Effect							
	I	A	B	C	AB	AC	BC	ABC
a	+	+	−	−	−	−	+	+
b	+	−	+	−	−	+	−	+
c	+	−	−	+	+	−	−	+
abc	+	+	+	+	+	+	+	+
ab	+	+	+	−	+	−	−	−
ac	+	+	−	+	−	+	−	−
bc	+	−	+	+	−	−	+	−
(1)	+	−	−	−	+	+	+	−

It is also easy to verify that the estimates of the two-factor interactions are

$$BC = \tfrac{1}{2}\,[a - b - c + abc]$$
$$AC = \tfrac{1}{2}\,[-a + b - c + abc]$$
$$AB = \tfrac{1}{2}\,[-a - b + c + abc]$$

Thus, the linear combination of observations in column A—say, ℓ_A—estimates $A + BC$. Similarly, ℓ_B estimates $B + AC$, and ℓ_C estimates $C + AB$. Two or more effects that have this property are called **aliases.** In our 2^{3-1} design, A and BC are aliases, B and AC are aliases, and C and AB are aliases. Aliasing is the direct result of fractional replication. In many practical situations, it will be possible to select the fraction so that the main effects and low-order interactions of interest will be aliased with high-order interactions (which are probably negligible).

The alias structure for this design is found by using the defining relation $I = ABC$. Multiplying any effect by the defining relation yields the aliases for that effect. In our example, the alias of A is

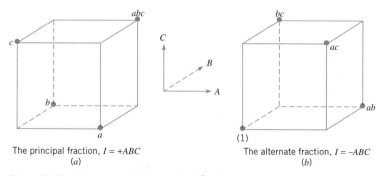

The principal fraction, $I = +ABC$
(a)

The alternate fraction, $I = -ABC$
(b)

Figure 12-37 The one-half fractions of the 2^3 design.

$$A = A \cdot ABC = A^2BC = BC$$

since $A \cdot I = A$ and $A^2 = I$. The aliases of B and C are

$$B = B \cdot ABC = AB^2C = AC$$

and

$$C = C \cdot ABC = ABC^2 = AB$$

Now suppose that we had chosen the other one-half fraction, that is, the runs in Table 12-20 associated with minus on ABC. This design is shown geometrically in Fig. 12-37b. The defining relation for this design is $I = -ABC$. The aliases are $A = -BC$, $B = -AC$, and $C = -AB$. Thus, the effects A, B, and C with this particular fraction really estimate $A - BC$, $B - AC$, and $C - AB$. In practice, it usually does not matter which one-half fraction we select. The fraction with the plus sign in the defining relation is usually called the **principal fraction;** the other fraction is usually called the **alternate fraction.**

Sometimes we use **sequences** of fractional factorial designs to estimate effects. For example, suppose we had run the principal fraction of the 2^{3-1} design. From this design we have the following effect estimates:

$$\ell_A = A + BC$$
$$\ell_B = B + AC$$
$$\ell_C = C + AB$$

Suppose we are willing to assume at this point that the two-factor interactions are negligible. If they are, then the 2^{3-1} design has produced estimates of the three main effects A, B, and C. However, if after running the principal fraction we are uncertain about the interactions, it is possible to estimate them by running the *alternate* fraction. The alternate fraction produces the following effect estimates:

$$\ell'_A = A - BC$$
$$\ell'_B = B - AC$$
$$\ell'_C = C - AC$$

If we combine the estimates from the two fractions, we obtain the following:

Effect, i	From $\frac{1}{2}(\ell_i + \ell'_i)$	From $\frac{1}{2}(\ell_i - \ell'_i)$
$i = A$	$\frac{1}{2}(A + BC + A - BC) = A$	$\frac{1}{2}[A + BC - (A - BC)] = BC$
$i = B$	$\frac{1}{2}(B + AC + B - AC) = B$	$\frac{1}{2}[B + AC - (B - AC)] = AC$
$i = C$	$\frac{1}{2}(C + AB + C - AB) = C$	$\frac{1}{2}[C + AB - (C - AB)] = AB$

Thus, by combining a sequence of two fractional factorial designs, we can isolate both the main effects and the two-factor interactions. This property makes the fractional factorial design highly useful in experimental problems because we can run sequences of small, efficient experiments, combine information across *several* experiments, and take advantage of learning about the process we are experimenting with as we go along.

A 2^{k-1} design may be constructed by writing down the treatment combinations for a full factorial in $k - 1$ factors and then adding the kth factor by identifying its plus and minus levels with the plus and minus signs of the highest-order interaction $\pm ABC \cdots$ $(K - 1)$. Therefore, a 2^{3-1} fractional factorial is obtained by writing down the full 2^2 factorial and then equating factor C to the $\pm AB$ interaction. Thus, to generate the principal fraction, we would use $C = +AB$ as follows:

Full 2^2		$2^{3-1}, I = ABC$		
A	B	A	B	$C = AB$
−	−	−	−	+
+	−	+	−	−
−	+	−	+	−
+	+	+	+	+

To generate the alternate fraction we would equate the last column to $C = -AB$.

······· **EXAMPLE 12-10** ···

To illustrate the use of a one-half fraction, consider the plasma etch experiment described in Example 12-8. Suppose we decide to use a 2^{4-1} design with $I = ABCD$ to investigate the four factors gap (A), pressure (B), C_2F_6 flow rate (C), and power setting (D). This design would be constructed by writing down a 2^3 in the factors A, B, and C and then setting $D = ABC$. The design and the resulting etch rates are shown in Table 12-21. The design is shown geometrically in Fig. 12-38.

In this design, the main effects are aliased with the three-factor interactions; note that the alias of A is

$$A \cdot I = A \cdot ABCD$$
$$A = A^2BCD$$
$$A = BCD$$

Similarly,

$$B = ACD$$
$$C = ABD$$
$$D = ABC$$

The two-factor interactions are aliased with each other. For example, the alias of AB is CD:

$$AB \cdot I = AB \cdot ABCD$$
$$AB = A^2B^2CD$$
$$AB = CD$$

Table 12-21 The 2^{4-1} Design with Defining Relation $I = ABCD$

Run		A	B	C	$D = ABC$	Etch Rate
1	(1)	−	−	−	−	550
2	ad	+	−	−	+	749
3	bd	−	+	−	+	1052
4	ab	+	+	−	−	650
5	cd	−	−	+	+	1075
6	ac	+	−	+	−	642
7	bc	−	+	+	−	601
8	$abcd$	+	+	+	+	729

The other aliases are

$$AC = BD$$
$$AD = BC$$

The estimates of the main effects (and their aliases) are found using the four columns of signs in Table 12-21. For example, from column A we obtain

$$\ell_A = A + BCD = \tfrac{1}{4}(-550 + 749 - 1052 + 650 - 1075 + 642 - 601 + 729)$$
$$= -127.00$$

The other columns produce

$$\ell_B = B + ACD = 4.00$$
$$\ell_C = C + ABD = 11.50$$

and

$$\ell_D = D + ABC = 290.51$$

Clearly, ℓ_A and ℓ_D are large, and if we believe that the three-factor interactions are negligible, then the main effects A (gap) and D (power setting) significantly affect the etch rate.

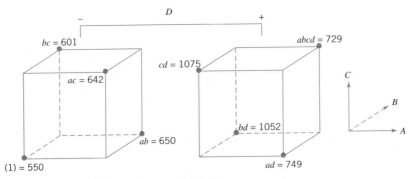

Figure 12-38 The 2^{4-1} design for Example 12-10.

The interactions are estimated by forming the AB, AC, and AD columns and adding them to the table. The signs in the AB column are $+, -, -, +, +, -, -, +$, and this column produces the estimate

$$\ell_{AB} = AB + CD = \tfrac{1}{4}(550 - 749 - 1052 + 650 + 1075 - 642 - 601 + 729)$$

$$= -10.00$$

From the AC and AD columns we find

$$\ell_{AC} = AC + BD = -25.50$$

$$\ell_{AD} = AD + BD = -197.50$$

The ℓ_{AD} estimate is large; the most straightforward interpretation of the results is that this is the A and D interaction. Thus, the results obtained from the 2^{4-1} design agree with the full factorial results in Example 12-8.

Normal Probability Plots and Residuals

The normal probability plot is very useful in assessing the significance of effects from a fractional factorial, especially when many effects are to be estimated. Residuals can be obtained from a fractional factorial by the regression model method shown previously. These residuals should be plotted against the predicted values, against the levels of the factors, and on normal probability paper as we have discussed before, both to assess the validity of the underlying model assumptions and to gain additional insight into the experimental situation.

Projection of the 2^{k-1} Design

If one or more factors from a one-half fraction of a 2^k can be dropped, the design will project into a full factorial design. For example, Fig. 12-39 presents a 2^{3-1} design. Note that this design will project into a full factorial in any two of the three original factors. Thus, if we think that at most two of the three factors are important, the 2^{3-1} design is an excellent design for identifying the significant factors. Sometimes experiments that seek to identify a relatively few significant factors from a larger number of factors are called **screening experiments.** This projection property is highly useful in factor screening because it allows negligible factors to be eliminated, resulting in a stronger experiment in the active factors that remain.

In the 2^{4-1} design used in the plasma etch experiment in Example 12-10, we found that two of the four factors (B and C) could be dropped. If we eliminate these two factors, the remaining columns in Table 12-21 form a 2^2 design in the factors A and D, with two replicates. This design is shown in Fig. 12-40.

Design Resolution

The concept of design resolution is a useful way to catalog fractional factorial designs according to the alias patterns they produce. Designs of resolution III, IV, and V are particularly important. The definitions of these terms and an example of each follow.

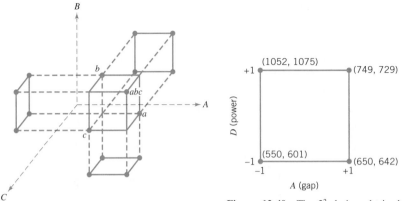

Figure 12-39   Projection of a 2^{3-1} design into three 2^2 designs.

Figure 12-40   The 2^2 design obtained by dropping factors B and C from the plasma etch experiment.

1. **Resolution III designs.** In these designs no main effects are aliased with any other main effect, but main effects are aliased with two-factor interactions and two-factor interactions may be aliased with each other. The 2^{3-1} design with $I = ABC$ is of resolution III. We usually employ a subscript roman numeral to indicate design resolution; thus, this one-half fraction is a 2_{III}^{3-1} design.

2. **Resolution IV designs.** In these designs no main effect is aliased with any other main effect or two-factor interaction, but two-factor interactions are aliased with each other. The 2^{4-1} design with $I = ABCD$ used in Example 12-10 is of resolution IV (2_{IV}^{4-1}).

3. **Resolution V designs.** In these designs no main effect or two-factor interaction is aliased with any other main effect or two-factor interaction, but two-factor interactions are aliased with three-factor interactions. A 2^{5-1} design with $I = ABCDE$ is of resolution V (2_V^{5-1}).

Resolution III and IV designs are particularly useful in factor screening experiments. The resolution IV design provides very good information about main effects and will provide some information about two-factor interactions.

12-6.2   Smaller Fractions: The 2^{k-p} Fractional Factorial Design

Although the 2^{k-1} design is valuable in reducing the number of runs required for an experiment, we frequently find that smaller fractions will provide almost as much useful information at even greater economy. In general, a 2^k design may be run in a $1/2^p$ fraction called a 2^{k-p} fractional factorial design. Thus, a $\frac{1}{4}$ fraction is called a 2^{k-2} fractional factorial design, a $\frac{1}{8}$ fraction is called a 2^{k-3} design, a $\frac{1}{16}$ fraction is called a 2^{k-4} design, and so on.

To illustrate a $\frac{1}{4}$ fraction, consider an experiment with six factors and suppose that the engineer is interested primarily in main effects but would also like to get some

Table 12-22 Construction of the 2^{6-2} Design with Generators $I = ABCE$ and $I = BCDF$

Run	A	B	C	D	$E = ABC$	$F = BCD$
1	−	−	−	−	−	−
2	+	−	−	−	+	−
3	−	+	−	−	+	+
4	+	+	−	−	−	+
5	−	−	+	−	+	+
6	+	−	+	−	−	+
7	−	+	+	−	−	−
8	+	+	+	−	+	−
9	−	−	−	+	−	+
10	+	−	−	+	+	+
11	−	+	−	+	+	−
12	+	+	−	+	−	−
13	−	−	+	+	+	−
14	+	−	+	+	−	−
15	−	+	+	+	−	+
16	+	+	+	+	+	+

information about the two-factor interactions. A 2^{6-1} design would require 32 runs and would have 31 degrees of freedom for estimation of effects. Since there are only 6 main effects and 15 two-factor interactions, the one-half fraction is inefficient—it requires too many runs. Suppose we consider a $\frac{1}{4}$ fraction, or a 2^{6-2} design. This design contains 16 runs and with 15 degrees of freedom will allow estimation of all six main effects, with some capability for examination of the two-factor interactions. To generate this design we would write down a 2^4 design in the factors A, B, C, and D, and then add two columns for E and F. Refer to Table 12-22. To find the new columns, we would select the two **design generators** $I = ABCE$ and $I = BCDF$. Thus, column E would be found from $E = ABC$ and column F would be $F = BCD$. Thus, columns $ABCE$ and $BCDF$ are equal to the identity column. However, we know that the product of any two columns in the table of plus and minus signs for a 2^k is just another column in the table; therefore, the product of $ABCE$ and $BCDF$ or $ABCE (ACDF) = AB^2C^2DEF = ADEF$ is also an identity column. Consequently, the **complete defining relation** for the 2^{6-2} design is

$$I = ABCE = BCDF = ADEF$$

To find the alias of any effect, simply multiply the effect by each word in the above defining relation. The complete alias structure is shown here.

$$A = BCE = DEF = ABCDF \qquad AB = CE = ACDF = BDEF$$
$$B = ACE = CDF = ABDEF \qquad AC = BE = ABDF = CDEF$$
$$C = ABE = BDF = ACDEF \qquad AD = EF = BCDE = ABCF$$
$$D = BCF = AEF = ABCDE \qquad AE = BC = DF = ABCDEF$$

$$E = ABC = ADF = BCDEF \quad AF = DE = BCEF = ABCD$$
$$F = BCD = ADE = ABCEF \quad BD = CF = ACDE = ABEF$$
$$ABD = CDE = ACF = BEF \quad BF = CD = ACEF = ABDE$$
$$ACD = BDE = ABF = CEF$$

Note that this is a resolution IV design; the main effects are aliased with three-factor and higher interactions, and two-factor interactions are aliased with each other. This design would provide very good information on the main effects and give some idea about the strength of the two-factor interactions.

Selection of Design Generators

In the foregoing example, we selected $I = ABCD$ and $I = BCDF$ as the generators to construct the 2^{6-2} fractional factorial design. This choice is not arbitrary; some generators will produce designs with more attractive alias structures than other generators. For a given number of factors and number of runs we wish to make, we want to select the generators so that the design has the highest possible resolution. Montgomery (1999) presents a set of designs of maximum resolution for 2^{k-p} designs with $p \leq 10$ factors. A portion of this table is reproduced in Table 12-23. In this table, each choice of generator is shown with a $\pm$ sign. If all generators are selected with a positive sign (as above), the principal fraction will result; selection of one or more negative signs for a set of generators will produce an alternate fraction.

Table 12-23 Selected 2^{k-p} Fractional Factorial Designs (from *Design and Analysis of Experiments*, 4th ed., by D. C. Montgomery, John Wiley, 1997)

Number of Factors, k	Fraction	Number of Runs	Design Generators
3	2_{III}^{3-1}	4	$C = \pm AB$
4	2_{IV}^{4-1}	8	$D = \pm ABC$
5	2_{V}^{5-1}	16	$E = \pm ABCD$
	2_{III}^{5-2}	8	$D = \pm AB$
			$E = \pm AC$
6	2_{VI}^{6-1}	32	$F = \pm ABCDE$
	2_{IV}^{6-2}	16	$E = \pm ABC$
			$F = \pm BCD$
	2_{III}^{6-3}	8	$D = \pm AB$
			$E = \pm AC$
			$F = \pm BC$
7	2_{VIII}^{7-1}	64	$G = \pm ABCDEF$
	2_{IV}^{7-2}	32	$F = \pm ABCD$
			$G = \pm ABDE$
	2_{IV}^{7-3}	16	$E = \pm ABC$
			$F = \pm BCD$
			$G = \pm ACD$
	2_{III}^{7-4}	8	$D = \pm AB$
			$E = \pm AC$
			$F = \pm BC$
			$G = \pm ABC$

Table 12-23   (*Continued*)

Number of Factors, k	Fraction	Number of Runs	Design Generators
8	2_V^{8-2}	64	$G = \pm ABCD$ $H = \pm ABEF$
	2_{IV}^{8-3}	32	$F = \pm ABC$ $G = \pm ABD$ $H = \pm BCDE$
	2_{IV}^{8-4}	16	$E = \pm BCD$ $F = \pm ACD$ $G = \pm ABC$ $H = \pm ABD$
9	2_{VI}^{9-2}	128	$H = \pm ACDFG$ $J = \pm BCEFG$
	2_{IV}^{9-3}	64	$G = \pm ABCD$ $H = \pm ACEF$ $J = \pm CDEF$
	2_{IV}^{9-4}	32	$F = \pm BCDE$ $G = \pm ACDE$ $H = \pm ABDE$ $J = \pm ABCE$
	2_{III}^{9-5}	16	$E = \pm ABC$ $F = \pm BCD$ $G = \pm ACD$ $H = \pm ABD$ $J = \pm ABCD$
10	2_V^{10-3}	128	$H = \pm ABCG$ $J = \pm BCDE$ $K = \pm ACDF$
	2_{IV}^{10-4}	64	$G = \pm BCDF$ $H = \pm ACDF$ $J = \pm ABDE$ $K = \pm ABCE$
	2_{IV}^{10-5}	32	$F = \pm ABCD$ $G = \pm ABCE$ $H = \pm ABDE$ $J = \pm ACDE$ $K = \pm BCDE$
	2_{III}^{10-6}	16	$E = \pm ABC$ $F = \pm BCD$ $G = \pm ACD$ $H = \pm ABD$ $J = \pm ABCD$ $K = \pm AB$

······   **EXAMPLE 12-11**   ··

Parts manufactured in an injection-molding process are experiencing excessive shrink-age, which is causing problems in assembly operations upstream from the injection-molding area. A quality-improvement team has decided to use a designed experiment to study the injection-molding process so that shrinkage can be reduced. The team decides to investigate seven factors: mold temperature (A), screw speed (B), holding time (C), cycle time (D), moisture content (E), gate size (F), and holding pressure (G). Each is examined at two levels, with the objective of learning how each factor affects shrinkage, as well as about how the factors interact.

The team decides to use a 16-run two-level fractional factorial design. Table 12-23 indicates that the appropriate design is a 2_{IV}^{7-3} design, with generators $I = ABCE$, $I = BCDF$, and $I = ACDG$. The design is shown in Table 12-24 and the alias structure for the design is shown in Table 12-25. The last column of Table 12-24 gives the observed shrinkage $\times$ 10 for the test part produced at each of the 16 runs in the design.

A normal probability plot of the effect estimates from this experiment is shown in Fig. 12-41. The only large effects are $A = 13.8750$ (mold temperature), $B = 35.6250$ (screw speed), and the AB interaction ($AB = 11.8750$). In light of the alias relationships in Table 12-25, it seems reasonable to tentatively adopt those conclusions. The AB inter-action plot in Fig. 12-42 shows that the process is very insensitive to temperature if screw speed is at the low level, but is very temperature-sensitive if screw speed is at the high level. With screw speed at the low level, the process should operate with average shrinkage around 10%, regardless of the temperature level chosen.

Table 12-24 2_{IV}^{7-3} Design for the Injection-Molding Experiment, Example 12-11

Run	A	B	C	D	$E\,(=ABC)$	$F\,(=BCD)$	$G\,(=ACD)$	Observed Shrinkage $(\times 10)$
1	−	−	−	−	−	−	−	6
2	+	−	−	−	+	−	+	10
3	−	+	−	−	+	+	−	32
4	+	+	−	−	−	+	+	60
5	−	−	+	−	+	+	+	4
6	+	−	+	−	−	+	−	15
7	−	+	+	−	−	−	+	26
8	+	+	+	−	+	−	−	60
9	−	−	−	+	−	+	+	8
10	+	−	−	+	+	+	−	12
11	−	+	−	+	+	−	+	34
12	+	+	−	+	−	−	−	60
13	−	−	+	+	+	−	−	16
14	+	−	+	+	−	−	+	5
15	−	+	+	+	−	+	−	37
16	+	+	+	+	+	+	+	52

Table 12-25   Aliases for the 2_{IV}^{7-3} Design Used in Example 12-11

$A = BCE = DEF = CDG = BFG$	$AB = CE = FG$
$B = ACE = CDF = DEG = AFG$	$AC = BE = DG$
$C = ABE = BDF = ADG = EFG$	$AD = EF = CF$
$D = BCF = AEF = ACG = BEG$	$AE = BC = DF$
$E = ABC = ADF = BDG = CFG$	$AF = DE = BG$
$F = BCD = ADE = ABG = CEG$	$AG = CD = BF$
$G = ACD = BDE = ABF = CEF$	$BD = CF = EG$

$$ABD = CDE = ACF = BEF = BCG = AEG = DFG$$

Based on this initial analysis, the team decided to set both mold temperature and screw speed at the low level. This set of conditions will reduce *mean* parts shrinkage to around 10%. However, the variability in shrinkage from part to part is still a potential problem. In effect, the mean shrinkage can be reduced effectively to nearly zero by appropriate modification of the tool; but the part-to-part variability in shrinkage over a production run could still cause problems in assembly, even if the average shrinkage over the run were nearly zero. One way to address this issue is to see whether any of the process variables affect *variability* in parts shrinkage.

Figure 12-43 presents the normal probability plot of the residuals. This plot appears satisfactory. The plots of residuals versus each variable were then constructed. One of these plots, that for residuals versus factor C (holding time), is shown in Fig. 12-44. The plot reveals that there is much less scatter in the residuals at low holding time than at high holding time. Now the residuals were obtained by first fitting a model for predicted shrinkage

$$\hat{y} = 27.3125 + 6.9375x_1 + 17.8125x_2 + 5.9375x_1x_2$$

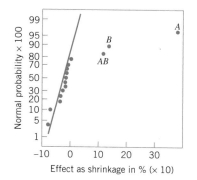

Figure 12-41 Normal probability plot of effects, Example 12-11.

Figure 12-42 *AB* or mold temperature–screw speed interaction plot, Example 12-11.

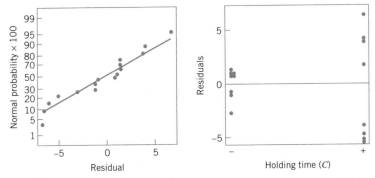

Figure 12-43 Normal probability plot of residuals, Example 12-11.

Figure 12-44 Residuals versus holding time (C), Example 12-11.

where x_1, x_2, and x_1x_2 are coded variables that correspond to the factors A, B, and the AB interaction, respectively. The residuals are then

$$e = y - \hat{y}$$

The regression model used to produce the residuals essentially removes the *location* effects of A, B, and AB from the data; the residuals therefore contain information about unexplained variability. Figure 12-44 indicates that there is a *pattern* in that variability and that variability in parts shrinkage may be smaller when holding time is at the low level. Note that this analysis assumes that the model for the location effects is a good one, and that it adequately models the mean of the process over the factor space.

Figure 12-45 shows the data from this experiment projected onto a cube in the factors A, B, and C. The average observed shrinkage and the range of observed shrinkage are shown at each corner of the cube. From inspection of this graph, we see that running

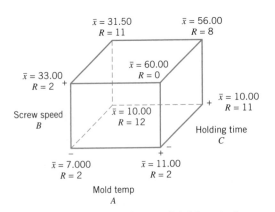

Figure 12-45 Average and range of shrinkage in factors A, B, C, Example 12-11.

the process with screw speed (*B*) at the low level is the key to reducing average parts shrinkage. If *B* is low, virtually any combination of temperature (*A*) and holding time (*C*) will result in low values of average parts shrinkage. However, from examining the ranges of the shrinkage values at each corner of the cube, it is immediately clear that holding time (*C*) at the low level is the only reasonable choice if we wish to keep the part-to-part variability in shrinkage low during a production run.

12-7 EXERCISES

12-1. An article in *Industrial Quality Control* (1956, pp. 5–8) describes an experiment to investigate the effect of glass type and phosphor type on the brightness of a television tube. The response measured is the current necessary (in microamps) to obtain a specified brightness level. The data are shown here. Analyze the data and draw conclusions.

	Phosphor Type		
Glass Type	1	2	3
1	280	300	290
	290	310	285
	285	295	290
2	230	260	220
	235	240	225
	240	235	230

12-2. A process engineer is trying to improve the life of a cutting tool. He has run a 2^3 experiment using cutting speed (*A*), metal hardness (*B*), and cutting angle (*C*) as the factors. The data from two replicates are shown here.

	Replicate	
Run	I	II
(1)	221	311
a	325	435
b	354	348
ab	552	472

	Replicate	
Run	I	II
c	440	453
ac	406	377
bc	605	500
abc	392	419

(a) Do any of the three factors affect tool life?

(b) What combination of factor levels produces the longest tool life?

(c) Is there a combination of cutting speed and cutting angle that always gives good results regardless of metal hardness?

12-3. Find the residuals from the tool life experiment in Exercise 12-2. Construct a normal probability plot of the residuals. Plot the residuals versus the predicted values. Comment on the plots.

12-4. Four factors are thought to possibly influence the taste of a soft-drink beverage: type of sweetener (*A*), ratio of syrup to water (*B*), carbonation level (*C*), and temperature (*D*). Each factor can be run at two levels, producing a 2^4 design. At each run in the design, samples of the beverage are given to a test panel consisting of 20 people. Each tester assigns point score from 1 to 10 to the beverage. Total score is the response variable, and the objective is to find a formulation that maximizes total score. Two replicates of this design are run, and the results shown here. Analyze the data and draw conclusions.

Treatment Combination	Replicate I	Replicate II	Treatment Combination	Replicate I	Replicate II
(1)	190	193	d	198	195
a	174	178	ad	172	176
b	181	185	bd	187	183
ab	183	180	abd	185	186
c	177	178	cd	199	190
ac	181	180	acd	179	175
bc	188	182	bcd	187	184
abc	173	170	abcd	180	180

12-5. Consider the experiment in Exercise 12-4. Plot the residuals against the levels of factors A, B, C, and D. Also construct a normal probability plot of the residuals. Comment on these plots.

12-6. Find the standard error of the effects for the experiment in Exercise 12-4. Using the standard errors as a guide, what factors appear significant?

12-7. Suppose that only the data from replicate I in Exercise 12-4 were available. Analyze the data and draw appropriate conclusions.

12-8. Suppose that only one replicate of the 2^4 design in Exercise 12-4 could be run, and we could only conduct eight tests each day. Set up a design that would block out the day effect. Show specifically which runs would be made on each day.

12-9. Show how a 2^5 experiment could be set up in two blocks of 16 runs each. Specifically, which runs would be made in each block?

12-10. R. D. Snee ("Experimenting with a Large Number of Variables," in *Ex-*

periments in Industry: Design, Analysis and Interpretation of Results, by R. D. Snee, L. B. Hare, and J. B. Trout, Editors, ASQC, 1985) describes an experiment in which a 2^{5-1} design with $I = ABCDE$ was used to investigate the effects of five factors on the color of a chemical product. The factors were $A =$ solvent/reactant, $B =$ catalyst/reactant, $C =$ temperature, $D =$ reactant purity, and $E =$ reactant pH. The results obtained are as follows:

$e =$	20.63	$d =$	6.79
$a =$	2.51	$ade =$	6.47
$b =$	−2.68	$bde =$	3.45
$abe =$	1.66	$abd =$	5.68
$c =$	2.06	$cde =$	5.22
$ace =$	1.22	$acd =$	4.38
$bce =$	−2.09	$bcd =$	4.30
$abc =$	1.93	$abcde =$	4.05

(a) Prepare a normal probability plot of the effects. Which effects seem active? Fit a model using these effects.

(b) Calculate the residuals for the model you fit in part (a). Construct a normal probability plot of the residuals and plot the residuals versus the fitted values. Comment on the plots.

(c) If any factors are negligible, collapse the 2^{5-1} design into a full factorial in the active factors. Comment on the resulting design and interpret the results.

12-11. An article in *Industrial and Engineering Chemistry* ("More on Planning Experiments to Increase Research Efficiency," 1970, pp. 60–65) uses a 2^{5-2} design to investigate the effect of $A =$ condensation temperature, $B =$ amount of material 1,

C = solvent volume, D = condensation time, and E = amount of material 2, on yield. The results obtained are as follows:

$e = 23.2$	$cd = 23.8$
$ab = 15.5$	$ace = 23.4$
$ad = 16.9$	$bde = 16.8$
$bc = 16.2$	$abcde = 18.1$

(a) Verify that the design generators used were $I = ACE$ and $I = BDE$.
(b) Write down the complete defining relation and the aliases from this design.
(c) Estimate the main effects.
(d) Prepare an analysis of variance table. Verify that the AB and AD interactions are available to use as error.
(e) Plot the residuals versus the fitted values. Also construct a normal probability plot of the residuals. Comment on the results.

12-12. A 2^4 factorial design has been run in a pilot plant to investigate the effect of four factors on the molecular weight of a polymer. The data from this experiment are as follows (values are coded by dividing by 10).

$(1) = 88$	$d = 86$
$a = 80$	$ad = 81$
$b = 89$	$bd = 85$
$ab = 87$	$abd = 86$
$c = 86$	$cd = 85$
$ac = 81$	$acd = 79$
$bc = 82$	$bcd = 84$
$abc = 80$	$abcd = 81$

(a) Construct a normal probability plot of the effects. Which effects are active?

(b) Construct an appropriate model. Fit this model and test for significant effects.
(c) Analyze the residuals from this model by constructing a normal probability plot of the residuals and plotting the residuals versus the predicted values of y.

12-13. Reconsider the data in Exercise 12-12. Suppose that four center points were added to this experiment. The molecular weights at the center point are 90, 87, 86, and 93.
(a) Analyze the data as you did in Exercise 12-12, but include a test for curvature.
(b) If curvature is significant in an experiment such as this one, describe what strategy you would pursue next to improve your model of the process.

12-14. Set up a 2^{8-4} fractional factorial design. Verify that this is a resolution IV design. Discuss the advantage of a resolution IV design relative to one of lower resolution.

12-15. A 2^{4-1} design has been used to investigate the effect of four factors on the resistivity of a silicon wafer. The data from this experiment are shown here.

Run	A	B	C	D	Resistivity
1	−	−	−	−	33.2
2	+	−	−	+	4.6
3	−	+	−	+	31.2
4	+	+	−	−	9.6
5	−	−	+	+	40.6
6	+	−	+	−	162.4
7	−	+	+	−	39.4
8	+	+	+	+	158.6
9	0	0	0	0	63.4
10	0	0	0	0	62.6
11	0	0	0	0	58.7
12	0	0	0	0	60.9

(a) Estimate the factor effects. Plot the effect estimates on a normal probability scale.

(b) Identify a tentative model for this process. Fit the model and test for curvature.

(c) Plot the residuals from the model in part (b) versus the predicted resistivity. Is there any indication on this plot of model inadequacy?

(d) Construct a normal probability plot of the residuals. Is there any reason to doubt the validity of the normality assumption?

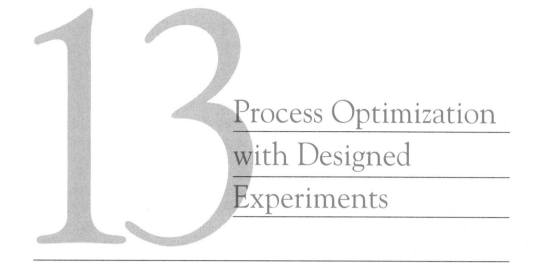

13 Process Optimization with Designed Experiments

CHAPTER OUTLINE

13-1 RESPONSE SURFACE METHODS AND DESIGNS

    13-1.1 The Method of Steepest Ascent

    13-1.2 Analysis of a Second-Order Response Surface

13-2 PROCESS ROBUSTNESS STUDIES

    13-2.1 Background

    13-2.2 The Response Surface Approach to Process Robustness Studies

13-3 EVOLUTIONARY OPERATION

CHAPTER OVERVIEW

In Chapter 12 we focused on factorial and fractional factorial designs. These designs are very useful for **factor screening**—that is, identifying the most important factors that affect the performance of a process. Sometimes this is called **process characterization.** Once the appropriate subset of process variables are identified, the next step is usually **process optimization,** or finding the set of operating conditions for the process variables that result in the best process performance. This chapter gives a brief account of how designed experiments can be used in process optimization.

We discuss and illustrate **response surface methodology,** an approach to optimization developed in the early 1950s and initially applied in the chemical and process industries. This is probably the most widely used and successful optimization technique based on designed experiments. Then we discuss how designed experiments can be used in **process robustness studies.** These are activities in which process engineering personnel try to reduce the variability in the output of a process by setting controllable factors to levels that minimize the variability transmitted into the responses of interest by other factors that are difficult to control during routine operation. We also present an example of **evolutionary operation,** an approach to maintaining optimum performance that is, in effect, an on-line or in-process application of the factorial design concepts of Chapter 12.

13-1 RESPONSE SURFACE METHODS AND DESIGNS

Response surface methodology (RSM) is a collection of mathematical and statistical techniques that are useful for modeling and analysis in applications where a response of interest is influenced by several variables and the objective is to **optimize this response.** The general RSM approach was developed in the early 1950s, and was initially applied in the chemical industry with considerable success. Over the last 20 years RSM has found extensive application in a wide variety of industrial settings, far beyond its origins in chemical processes, including semiconductor and electronics manufacturing, machining, metal cutting, and joining processes, among many others. Many statistics software packages have included the experimental designs and optimization techniques that make up the basics of RSM as standard features. For a recent comprehensive presentation of RSM, see Myers and Montgomery (1995).

To illustrate the general idea of RSM, suppose that a chemical engineer wishes to find the levels of reaction temperature (x_1) and reaction time (x_2) that maximize the yield (y) of a process. The process yield is a function of the levels of temperature and time—say,

$$y = f(x_1, x_2) + \epsilon$$

where ϵ represents the noise or error observed in the response Y. If we denote the expected value of the response by $E(y) = f(x_1, x_2)$, then the surface represented by

$$E(y) = f(x_1, x_2)$$

is called a **response surface.** Recall that we introduced the idea of a response surface in Chapter 12, where we presented an example of a response surface generated from a model that arose from a factorial design.

We may represent the response surface graphically as shown in Fig. 13-1, where $E(y)$ is plotted versus the levels of x_1 and x_2. Note that the response is represented as a

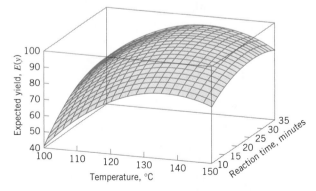

Figure 13-1 A three-dimensional response surface showing the expected yield as a function of reaction temperature and reaction time.

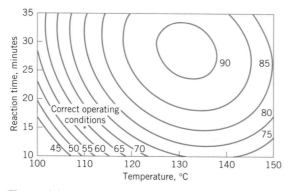

Figure 13-2 A contour plot of the yield response surface in Fig. 13-1.

surface plot in three-dimensional space. To help visualize the shape of a response surface, we often plot the contours of the response surface as shown in Fig. 13-2. In the contour plot, lines of constant response are drawn in the x_1, x_2 plane. Each contour corresponds to a particular height of the response surface. The contour plot is helpful in studying the levels of x_1, x_2 that result in changes in the shape or height of the response surface.

In most RSM problems, the form of the relationship between the response and the independent variables is unknown. Thus, the first step in RSM is to find a suitable approximation for the true relationship between y and the independent variables. Usually, a low-order polynomial in some region of the independent variables is employed. If the response is well modeled by a linear function of the independent variables, then the approximating function is the **first-order model**

$$y = \beta_0 + \beta_1 x_1 + \beta_2 x_2 + \cdots + \beta_k x_k + \epsilon \qquad (13\text{-}1)$$

If there is curvature in the system, then a polynomial of higher degree must be used, such as the **second-order model**

$$y = \beta_0 + \sum_{i=1}^{k} \beta_i x_i + \sum_{i=1}^{k} \beta_{ii} x_i^2 + \sum_{i<j=2}^{k} \beta_{ij} x_i x_j + \epsilon \qquad (13\text{-}2)$$

Many RSM problems utilize one or both of these approximating polynomials. Of course, it is unlikely that a polynomial model will be a reasonable approximation of the true functional relationship over the entire space of the independent variables, but for a relatively small region they usually work quite well.

The method of least squares is used to estimate the parameters in the approximating polynomials. That is, the estimates of the β's in equations 13-1 and 13-2 are those values of the parameters that minimize the sum of squares of the model errors. The response surface analysis is then done in terms of the fitted surface. If the fitted surface is an adequate approximation of the true response function, then analysis of the fitted surface will be approximately equivalent to analysis of the actual system.

RSM is a **sequential procedure.** Often, when we are at a point on the response surface that is remote from the optimum, such as the current operating conditions in Fig. 13-2, there is little curvature in the system and the first-order model will be appropriate. Our objective here is to lead the experimenter rapidly and efficiently to the general vicinity of the optimum. Once the region of the optimum has been found, a more elaborate model such as the second-order model may be employed, and an analysis may be performed to locate the optimum. From Fig. 13-2, we see that the analysis of a response surface can be thought of as "climbing a hill," where the top of the hill represents the point of maximum response. If the true optimum is a point of minimum response, then we may think of "descending into a valley."

The eventual objective of RSM is to determine the optimum operating conditions for the system or to determine a region of the factor space in which operating specifications are satisfied. Also, note that the word "optimum" in RSM is used in a special sense. The "hill climbing" procedures of RSM guarantee convergence to a local optimum only.

13-1.1   The Method of Steepest Ascent

Frequently, the initial estimate of the optimum operating conditions for the system will be far away from the actual optimum. In such circumstances, the objective of the experimenter is to move rapidly to the general vicinity of the optimum. We wish to use a simple and economically efficient experimental procedure. When we are remote from the optimum, we usually assume that a first-order model is an adequate approximation to the true surface in a small region of the x's.

The **method of steepest ascent** is a procedure for moving sequentially along the path of steepest ascent—that is, in the direction of the maximum increase in the response. Of course, if **minimization** is desired, then we would call this procedure the **method of steepest descent.** The fitted first-order model is

$$\hat{y} = \hat{\beta}_0 + \sum_{i=}^{k} \hat{\beta}_i x_i \qquad (13\text{-}3)$$

and the first-order response surface—that is, the contours of $\hat{y}$—is a series of parallel straight lines such as shown in Fig. 13-3. The direction of steepest ascent is the direction in which $\hat{y}$ increases most rapidly. This direction is normal to the fitted response surface contours. We usually take as the **path of steepest ascent** the line through the center of the region of interest and normal to the fitted surface contours. Thus, the steps along the

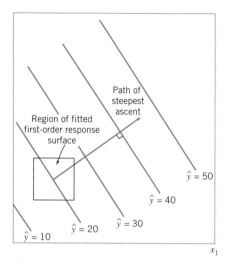

Figure 13-3 First-order response surface and path of steepest ascent.

path are proportional to the regression coefficients $\{\hat{\beta}_i\}$. The experimenter determines the actual amount of movement along this path based on process knowledge or other practical considerations.

Experiments are conducted along the path of steepest ascent until no further increase in response is observed or until the desired response region is reached. Then a new first-order model may be fitted, a new direction of steepest ascent determined, and if necessary, further experiments conducted in that direction until the experimenter feels that the process is near the optimum.

········ EXAMPLE 13-1 ··

In Example 12-8, we described an experiment on a plasma etching process in which four factors were investigated to study their effect on the etch rate in a semiconductor water-etching application. We found that two of the four factors, the gap (x_1) and the power (x_4), significantly affected etch rate. Recall from that example that if we fit a model using only these main effects we obtain

$$\hat{y} = 776.0625 + 50.8125x_1 + 153.0625x_4$$

as a prediction equation for the etch rate.

Figure 13-4 shows the contour plot from this model, over the original region of experimentation—that is, for gaps between 0.8 and 1.2 cm and power between 275 and 325 W. Note that within the original region of experimentation, the maximum etch rate that can be obtained is approximately 980 Å/m. The engineers would like to run this process at an etch rate of 1100–1150 Å/m. Therefore, it is necessary to move away from the original region of experimentation to increase the etch rate.

From examining the plot in Fig. 13-4 (or the fitted model) we see that to move away from the design center—the point $(x_1 = 0, x_2 = 0)$—along the path of steepest ascent,

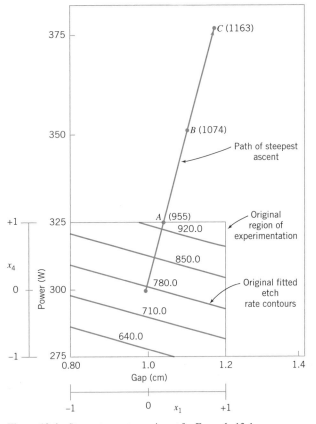

Figure 13-4   Steepest ascent experiment for Example 13-1.

we would move 50.8125 units in the x_1 direction for every 153.0625 units in the x_4 direction. Thus the path of steepest ascent passes through the point ($x_1 = 0$, $x_2 = 0$) and has slope $153.0625/50.8125 \cong 3$. The engineer decides to use 25 W of power as the basic step size. Now, 25 W of power is equivalent to a step in the coded variable x_4 of $\Delta x_4 = 1$. Therefore, the steps along the path of steepest ascent are $\Delta x_4 = 1$ and $\Delta x_1 = \Delta x_4/3 = 0.33$. A change of $\Delta x_1 = 0.33$ in the coded variable x_1 is equivalent to about 0.067 cm in the original variable gap. Therefore, the engineer will move along the path of steepest ascent by increasing power by 25 W and gap by 0.067 cm. An actual observation on etch rate will be obtained by running the process at each point.

Figure 13-4 shows three points along this path of steepest ascent and the etch rates actually observed from the process at those points. At points A, B, and C, the observed etch rates increase steadily. At point C, the observed etch rate is 1163 Å/m. Therefore, the steepest ascent procedure would terminate in the vicinity of power = 375 W and gap = 1.2 cm with an observed etch rate of 1163 Å/m. This region is very close to the desired operating region for the process.

13-1.2 Analysis of a Second-Order Response Surface

When the experimenter is relatively close to the optimum, a second-order model is usually required to approximate the response because of curvature in the true response surface. The fitted second-order response surface model is

$$\hat{y} = \hat{\beta}_0 + \sum_{i=1}^{k} \hat{\beta}_i x_i + \sum_{i=1}^{k} \hat{\beta}_{ii} x_i^2 + \sum_{\substack{i \ j \\ i<j}} \hat{\beta}_{ij} x_i x_j$$

where $\hat{\beta}$ denotes the least squares estimate of β. In the next example, we illustrate how a fitted second-order model can be used to find the optimum operating conditions for a process, and how to describe the behavior of the response function.

······ **EXAMPLE 13-2** ···

Continuation of Example 13-1

Recall that in applying the method of steepest ascent to the plasma etching process in Example 13-1 we had found a region near gap = 1.2 cm and power = 375 W, which would apparently give etch rates near the desired target of between 1100 and 1150 Å/m. The experimenters decided to explore this region more closely by running an experiment that would support a second-order response surface model. Table 13-1 and Fig. 13-5 show the experimental design, centered at gap = 1.2 cm and power = 375 W, which consists of a 2^2 factorial design with four center points and four runs located along the coordinate axes called axial runs. The resulting design is called a **central composite design,** and it is widely used in practice for fitting second-order response surfaces.

Two response variables were measured during this phase of the study: etch rate (in Å/m) and etch uniformity (this is the standard deviation of the thickness of the layer of material applied to the surface of the wafer after it has been etched to a particular

Table 13-1 Central Composite Design of Example 13-2

Observation	Gap (cm)	Power (W)	Coded x_1	Variables x_4	Etch Rate y_1 (Å/m)	Uniformity y_2 (Å/m)
1	1.000	350.0	−1.000	−1.000	1054.0	79.6
2	1.400	350.0	1.000	−1.000	936.0	81.3
3	1.000	400.0	−1.000	1.000	1179.0	78.5
4	1.400	400.0	1.000	1.000	1417.0	97.7
5	0.917	375.0	−1.414	0.000	1049.0	76.4
6	1.483	375.0	1.414	0.000	1287.0	88.1
7	1.200	339.6	0.000	−1.414	927.0	78.5
8	1.200	410.4	0.000	1.414	1345.0	92.3
9	1.200	375.0	0.000	0.000	1151.0	90.1
10	1.200	375.0	0.000	0.000	1150.0	88.3
11	1.200	375.0	0.000	0.000	1177.0	88.6
12	1.200	375.0	0.000	0.000	1196.0	90.1

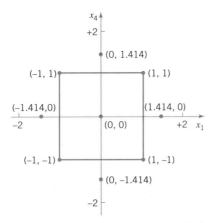

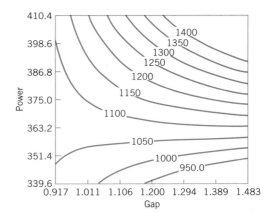

Figure 13-5 Central composite design in the coded variables for Example 13-2.

Figure 13-6 Contours of constant predicted etch rate, Example 13-2.

average thickness). Minitab can be used to analyze the data from this experiment. The Minitab output is in Table 13-2.

The second-order model fit to the etch rate response is

$$\hat{y}_1 = 1168.50 + 57.07x_1 + 149.64x_4 - 1.62x_1^2 - 17.63x_4^2 + 89.00x_1x_4$$

However, we note from the t-test statistics in Table 13-2 that the quadratic terms x_1^2 and x_4^2 are not statistically significant. Therefore, the experimenters decided to model etch rate with a first-order model with interaction:

$$\hat{y}_1 = 1155.7 + 57.1x_1 + 149.7x_4 + 89x_1x_4$$

Figure 13-6 shows the contours of constant etch rate from this model. There are obviously many combinations of x_1 (gap) and x_4 (power) that will give an etch rate in the desired range of 1100–1150 Å/m.

Table 13-2 Minitab Analysis of the Central Composite Design in Example 13-2

Response Surface Regression: Etch Rate versus A, B

The analysis was done using coded units.

Estimated Regression Coefficents for Etch Rate

Term	Coef	SE Coef	T	P
Constant	1168.50	17.59	66.417	0.000
A	57.07	12.44	4.588	0.004
B	149.64	12.44	12.029	0.000

```
A*A           -1.62      13.91     -0.117     0.911
B*B          -17.63      13.91     -1.267     0.252
A*B           89.00      17.59      5.059     0.002
```

S = 35.19   R-Sq = 97.0%   R-Sq(adj) = 94.5%

Analysis of Variance for Etch Rate

Source	DF	Seq SS	Adj SS	Adj MS	F	P
Regression	5	238898	238898	47780	38.59	0.000
Linear	2	205202	205202	102601	82.87	0.000
Square	2	2012	2012	1006	0.81	0.487
Interaction	1	31684	31684	31684	25.59	0.002
Residual Error	6	7429	7429	1238		
Lack-of-Fit	3	5952	5952	1984	4.03	0.141
Pure Error	3	1477	1477	492		
Total	11	246327				

Response Surface Regression: Uniformity versus A, B
The analysis was done using coded units.

Estimated Regression Coefficients for Uniformity

Term	Coef	SE Coef	T	P
Constant	89.275	0.5688	156.963	0.000
A	4.681	0.4022	11.639	0.000
B	4.352	0.4022	10.821	0.000
A*A	-3.400	0.4496	-7.561	0.000
B*B	-1.825	0.4496	-4.059	0.007
A*B	4.375	0.5688	7.692	0.000

S = 1.138   R-Sq = 98.4%   R-Sq(adj) = 97.1%

Analysis of Variance for Uniformity

Source	DF	Seq SS	Adj SS	Adj MS	F	P
Regression	5	486.085	486.085	97.217	75.13	0.000
Linear	2	326.799	326.799	163.399	126.28	0.000
Square	2	82.724	82.724	41.362	31.97	0.001
Interaction	1	76.563	76.563	76.563	59.17	0.000
Residual Error	6	7.764	7.764	1.294		
Lack-of-Fit	3	4.996	4.996	1.665	1.81	0.320
Pure Error	3	2.768	2.768	0.923		
Total	11	493.849				

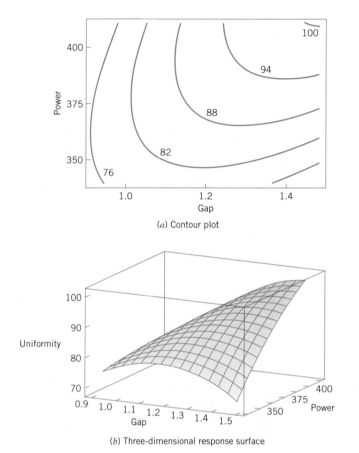

(a) Contour plot

(b) Three-dimensional response surface

Figure 13-7   Plots of the uniformity response, Example 13-2.

The second-order model for uniformity is

$$\hat{y}_2 = 89.275 + 4.681x_1 + 4.352x_4 - 3.400x_1^2 - 1.825x_4^2 + 4.375x_1x_4$$

Table 13-2 gives the t-statistics for each model term. Since all terms are significant, the experimenters decided to use the quadratic model for uniformity. Figure 13-7 gives the contour plot and response surface for uniformity.

As in most response surface problems, the experimenter in this example had conflicting objectives regarding the two responses. One objective was to keep the etch rate within the acceptable range of $1100 \le y_1 \le 1150$ but to simultaneously minimize the uniformity. Specifically, the uniformity must not exceed $y_2 = 80$, or many of the wafers will be defective in subsequent processing operations. When there are only a few independent variables, an easy way to solve this problem is to overlay the response surfaces to find the optimum. Figure 13-8 presents the overlay plot of both responses, with the contours of $\hat{y}_1 = 1100$, $\hat{y}_1 = 1150$, and $\hat{y}_2 = 80$ shown. The shaded areas on this plot

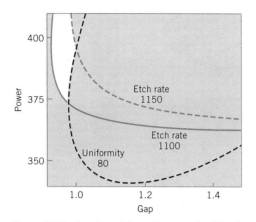

Figure 13-8 Overlay of the etch rate and uniformity response surfaces in Example 13-2 showing the region of the optimum (unshaded region).

identify infeasible combinations of gap and power. The graph indicates that several combinations of gap and power should result in acceptable process performance.

Example 13-2 illustrates the use of a **central composite design (CCD)** for fitting a second-order response surface model. These designs are widely used in practice because they are relatively efficient with respect to the number of runs required. In general, a CCD in k factors requires 2^k factorial runs, $2k$ axial runs, and at least one center point (3 to 5 center points are typically used). Designs for $k = 2$ and $k = 3$ factors are shown in Fig. 13-9.

The central composite design may be made **rotatable** by proper choice of the axial spacing α in Fig. 13-9. If the design is rotatable, the standard deviation of predicted response $\hat{y}$ is constant at all points that are the same distance from the center of the

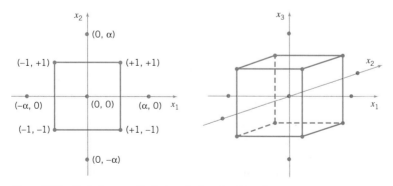

Figure 13-9 Central composite designs for $k = 2$ and $k = 3$.

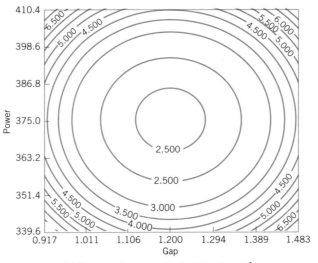

(a) Contours of constant standard deviation of $\hat{y}_2$

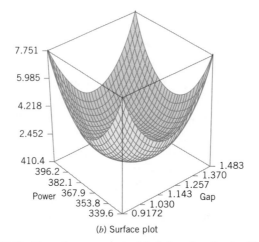

(b) Surface plot

Figure 13-10   Plots of constant standard deviation of predicted uniformity (y_2), Example 13-2.

design. For rotatability, choose $\alpha = (F)^{1/4}$, where F is the number of points in the factorial part of the design (usually $F = 2^k$). For the case of $k = 2$ factors, $\alpha = (2^2)^{1/4} = 1.414$, as was used in the design in Example 13-2. Figure 13-10 presents a contour plot and a surface plot of the standard deviation of prediction for the quadratic model used for the uniformity response. Note that the contours are concentric circles, implying that uniformity is predicted with equal precision for all points that are the same distance from the center of the design. Also, note that the precision of response estimation decreases with increasing distance from the design center.

13-2 PROCESS ROBUSTNESS STUDIES

13-2.1 Background

In Chapters 12 and 13 we have emphasized the importance of using statistically designed experiments for process design, development, and improvement. Since about 1980, engineers and scientists have become increasingly aware of the benefits of using designed experiments, and as a consequence, there have been many new application areas. One of the most important of these is in **process robustness studies,** where the focus is on the following:

1. Designing processes so that the manufactured product will be as close as possible to the desired target specifications even though some process variables (such as temperature), environmental factors (such as relative humidity), or raw material characteristics are impossible to control precisely.

2. Determining the operating conditions for a process so that critical product characteristics are as close as possible to the desired target value and the variability around this target is minimized. Examples of this type of problem occur frequently. For instance, in semiconductor manufacturing we would like the oxide thickness on a wafer to be as close as possible to the target mean thickness, and we would also like the variability in thickness across the wafer (a measure of uniformity) to be as small as possible.

In the early 1980s, a Japanese engineer, Dr. Genichi Taguchi, introduced an approach to solving these types of problems, which he referred to as the **robust parameter design (RPD)** problem [see Taguchi and Wu (1980), Taguchi (1986)]. His approach was based on classifying the variables in a process as either **control** (or **controllable) variables** and **noise** (or **uncontrollable) variables,** and then finding the settings for the controllable variables that minimized the variability transmitted to the response from the uncontrollable variables. We make the assumption that although the noise factors are uncontrollable in the full-scale system, they can be controlled for purposes of an experiment. Refer to Fig. 12-1 for a graphical view of controllable and uncontrollable variables in the general context of a designed experiment.

Taguchi introduced some novel statistical methods and some variations on established techniques as part of his RPD procedure. He made use of highly fractionated factorial designs and other types of fractional designs obtained from orthogonal arrays. His methodology generated considerable debate and controversy. Part of the controversy arose because Taguchi's methodology was advocated in the West initially (and primarily) by entrepreneurs, and the underlying statistical science had not been adequately peer-reviewed. By the late 1980s, the results of a very thorough and comprehensive peer review indicated that although Taguchi's engineering concepts and the overall objective of RPD were well founded, there were substantial problems with his experimental strategy and methods of data analysis. For specific details of these issues, see Box, Bisgaard, and Fung (1988); Hunter (1985, 1987); Montgomery (1999); Myers and Montgomery (1995); and Pignatiello and Ramberg (1991). Many of these concerns are

Table 13-3   Taguchi Parameter Design with Both Inner and Outer Arrays

					(b) Outer Array							
				E	1	1	1	1	2	2	2	2
				F	1	1	2	2	1	1	2	2
				G	1	2	1	2	1	2	1	2
	(a) Inner Array											
Run	A	B	C	D								
1	1	1	1	1	15.6	9.5	16.9	19.9	19.6	19.6	20.0	19.1
2	1	2	2	2	15.0	16.2	19.4	19.2	19.7	19.8	24.2	21.9
3	1	3	3	3	16.3	16.7	19.1	15.6	22.6	18.2	23.3	20.4
4	2	1	2	3	18.3	17.4	18.9	18.6	21.0	18.9	23.2	24.7
5	2	2	3	1	19.7	18.6	19.4	25.1	25.6	21.4	27.5	25.3
6	2	3	1	2	16.2	16.3	20.0	19.8	14.7	19.6	22.5	24.7
7	3	1	3	2	16.4	19.1	18.4	23.6	16.8	18.6	24.3	21.6
8	3	2	1	3	14.2	15.6	15.1	16.8	17.8	19.6	23.2	24.2
9	3	3	2	1	16.1	19.9	19.3	17.3	23.1	22.7	22.6	28.6

also summarized in the extensive panel discussion in the May 1992 issue of *Technometrics* [see Nair et al. (1992)]. The supplemental technical material for this chapter also discusses and illustrates many of the problems underlying Taguchi's technical methods.

Taguchi's methodology for the RPD problem revolves around the use of an orthogonal design for the controllable factors that is "crossed" with a separate orthogonal design for the noise factors. Table 13-3 presents an example from Byrne and Taguchi (1987) that involved the development of a method to assemble an elastometric connector to a nylon tube that would deliver the required pull-off force. There are four controllable factors, each at three levels (A = interference, B = connector wall thickness, C = insertion depth, D = percent adhesive), and three noise or uncontrollable factors, each at two levels (E = conditioning time, F = conditioning temperature, G = conditioning relative humidity). Panel (a) of Table 13-3 contains the design for the controllable factors. Note that the design is a three-level fractional factorial; specifically, it is a 3^{4-2} design. Taguchi calls this the **inner array design.** Panel (b) of Table 13-3 contains a 2^3 design for the noise factors, which Taguchi calls the **outer array design.** Each run in the inner array is performed for all treatment combinations in the outer array, producing the 72 observations on pull-off force shown in the table. This type of design is called a **crossed array design.**

Taguchi suggested that we summarize the data from a crossed array experiment with two statistics: the average of each observation in the inner array across all runs in the outer array, and a summary statistic that attempted to combine information about the mean and variance, called the **signal-to-noise ratio.** These signal-to-noise ratios are purportedly defined so that a maximum value of the ratio minimizes variability transmitted from the noise variables. Then an analysis is performed to determine which settings

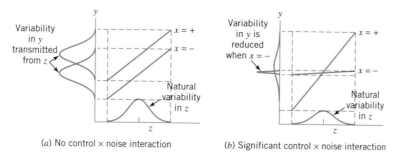

(a) No control × noise interaction

(b) Significant control × noise interaction

Figure 13-11   The role of the control × noise interaction in robust design.

of the controllable factors result in (1) the mean as close as possible to the desired target and (2) a maximum value of the signal-to-noise ratio.

Examination of Table 13-3 reveals a major problem with the Taguchi design strategy; namely, the crossed array approach will lead to a very large experiment. In our example, there are only seven factors, yet the design has 72 runs. Furthermore, the inner array design is a 3^{4-2} resolution III design [see Montgomery (1997), Chapter 9, for discussion of this design], so in spite of the large number of runs, we cannot obtain *any* information about interactions among the controllable variables. Indeed, even information about the main effects is potentially tainted, because the main effects are heavily aliased with the two-factor interactions. It also turns out that the Taguchi signal-to-noise ratios are problematic; maximizing the ratio does not necessarily minimize variability. Refer to the supplemental text material for more details.

An important point about the crossed array design is that it *does* provide information about controllable factor × noise factor interactions. These interactions are crucial to the solution of an RPD problem. For example, consider the two-factor interaction graphs in Fig. 13-11, where x is the controllable factor and z is the noise factor. In Fig. 13-11a, there is no $x \times z$ interaction; therefore, there is no setting for the controllable variable x that will affect the variability transmitted to the response by the variability in z. However, in Fig. 13-11b there is a strong $x \times z$ interaction. Note that when x is set to its low level there is much less variability in the response variable than when x is at the high level. Thus, unless there is at least one controllable factor × noise factor interaction, there is no robust design problem. As we will see in the next section, focusing on identifying and modeling these interactions is one of the keys to a more efficient and effective approach to investigating process robustness.

13-2.2   The Response Surface Approach to Process Robustness Studies

As noted in the previous section, interactions between controllable and noise factors are the key to a process robustness study. Therefore, it is logical to utilize a **model** for the response that includes both controllable and noise factors and their interactions. To illustrate, suppose that we have two controllable factors x_1 and x_2 and a single noise factor z_1.

We assume that both control and noise factors are expressed as the usual coded variables (that is, they are centered at zero and have lower and upper limits at $\pm a$). If we wish to consider a first-order model involving the controllable variables, then a logical model is

$$y = \beta_0 + \beta_1 x_1 + \beta_2 x_2 + \beta_{12} x_1 x_2 + \gamma_1 z_1 + \delta_{11} x_1 z_1 + \delta_{21} x_2 z_1 + \epsilon \quad (13\text{-}4)$$

Note that this model has the main effects of both controllable factors, the main effect of the noise variable, and both interactions between the controllable and noise variables. This type of model incorporating both controllable and noise variables is often called a **response model.** Unless at least one of the regression coefficients δ_{11} and δ_{21} is nonzero, there will be no robust design problem.

An important advantage of the response model approach is that both the controllable factors and the noise factors can be placed in a single experimental design; that is, the inner and outer array structure of the Taguchi approach can be avoided. We usually call the design containing both controllable and noise factors a **combined array design.**

As mentioned previously, we assume that noise variables are random variables, although they are controllable for purposes of an experiment. Specifically, we assume that the noise variables are expressed in coded units, that they have expected value zero, variance σ_z^2, and that if there are several noise variables, they have zero covariances. Under these assumptions, it is easy to find a model for the mean response just by taking the expected value of y in equation 13-4. This yields

$$E_z(y) = \beta_0 + \beta_1 x_1 + \beta_2 x_2 + \beta_{12} x_1 x_2$$

where the z subscript on the expectation operator is a reminder to take the expected value with respect to *both* random variables in equation 13-4, z_1 *and* ϵ. To find a model for the variance of the response y, first rewrite equation 13-4 as follows:

$$y = \beta_0 + \beta_1 x_1 + \beta_2 x_2 + \beta_{12} x_1 x_2 + (\gamma_1 + \delta_{11} x_1 + \delta_{21} x_2) z_1 + \epsilon$$

Now the variance of y can be obtained by applying the variance operator across this last expression. The resulting variance model is

$$V_z(y) = \sigma_z^2 (\gamma_1 + \delta_{11} x_1 + \delta_{21} x_2)^2 + \sigma^2$$

Once again, we have used the z subscript on the variance operator as a reminder that *both* z_1 *and* ϵ are random variables.

We have derived simple models for the mean and variance of the response variable of interest. Note the following:

1. The mean and variance models involve **only the controllable variables.** This means that we can potentially set the controllable variables to achieve a target value of the mean and minimize the variability transmitted by the noise variable.

2. Although the variance model involves only the controllable variables, it also involves the *interaction regression coefficients* between the controllable and noise variables. This is how the noise variable influences the response.

3. The variance model is a **quadratic function** of the controllable variables.

4. The variance model (apart from σ^2) is simply the square of the **slope** of the fitted response model in the direction of the noise variable.

To use these models operationally, we would:

1. Perform an experiment and fit an appropriate response model such as equation 13-4.

2. Replace the unknown regression coefficients in the mean and variance models with their least squares estimates from the response model and replace σ^2 in the variance model by the residual mean square found when fitting the response model.

3. Simultaneously optimize the mean and variance models. Often this can be done graphically. For more discussion of other optimization methods, refer to Myers and Montgomery (1995).

It is very easy to generalize these results. Suppose that there are k controllable variables $\mathbf{x'} = [x_1, x_2, \ldots, x_k]$, and r noise variables $\mathbf{z'} = [z_1, z_2, \ldots, z_r]$. We will write the general response model involving these variables as

$$y(\mathbf{x}, \mathbf{z}) = f(\mathbf{x}) + h(\mathbf{x}, \mathbf{z}) + \epsilon \tag{13-5}$$

where $f(\mathbf{x})$ is the portion of the model that involves only the controllable variables and $h(\mathbf{x}, \mathbf{z})$ are the terms involving the main effects of the noise factors and the interactions between the controllable and noise factors. Typically, the structure for $h(\mathbf{x}, \mathbf{z})$ is

$$h(\mathbf{x}, \mathbf{z}) = \sum_{i=1}^{r} \gamma_i z_i + \sum_{i=1}^{k} \sum_{j=1}^{r} \delta_{ij} x_i z_j$$

The structure for $f(\mathbf{x})$ will depend on what type of model for the controllable variables the experimenter thinks is appropriate. The logical choices are the first-order model with interaction and the second-order model. If we assume that the noise variables have mean zero, variance σ_z^2, and have zero covariances, and that the noise variables and the random errors ϵ have zero covariances, then the mean model for the response is simply

$$E_z[y(\mathbf{x}, \mathbf{z})] = f(\mathbf{x}) \tag{13-6}$$

To find the variance model, we will use the transmission of error approach from Section 7-7.2. This involves first expanding equation 13-5 around $\mathbf{z} = \mathbf{0}$ in a first-order Taylor series:

$$y(\mathbf{x}, \mathbf{z}) \simeq f(\mathbf{x}) + \sum_{i=1}^{r} \frac{\partial h(\mathbf{x}, \mathbf{z})}{\partial z_i} (z_i - 0) + R + \epsilon$$

where R is the remainder. If we ignore the remainder and apply the variance operator to this last expression, the variance model for the response is

$$V_z[y(\mathbf{x}, \mathbf{z})] = \sigma_z^2 \sum_{i=1}^{r} \left(\frac{\partial h(\mathbf{x}, \mathbf{z})}{\partial z_i} \right)^2 + \sigma^2 \tag{13-7}$$

Myers and Montgomery (1995) give a slightly more general form for equation 13-7 based on applying a conditional variance operator directly to the response model in equation 13-5.

······· **EXAMPLE 13-3** ···

To illustrate a process robustness study, consider an experiment [described in detail in Montgomery (1997)] in which four factors were studied in a 2^4 factorial design to investigate their effect on the filtration rate of a chemical product. We will assume that factor A, temperature, is hard to control in the full-scale process but it can be controlled during the experiment (which was performed in a pilot plant). The other three factors, pressure (B), concentration (C), and stirring rate (D) are easy to control. Thus the noise factor z_1 is temperature and the controllable variables x_1, x_2, and x_3 are pressure, concentration, and stirring rate, respectively. The experimenters conducted the (unreplicated) 2^4 design shown in Table 13-4. Since both the controllable factors and the noise factor are in the same design, the 2^4 factorial design used in this experiment is an example of a **combined array design.**

Using the methods for analysing a 2^k factorial design from Chapter 12, the response model is

$$\hat{y}(\mathbf{x}, z_1) = 70.06 + \left(\frac{21.625}{2}\right)z_1 + \left(\frac{9.875}{2}\right)x_2 + \left(\frac{14.625}{2}\right)x_3$$

$$- \left(\frac{18.125}{2}\right)x_2 z_1 + \left(\frac{16.625}{2}\right)x_3 z_1$$

$$= 70.06 + 10.81z_1 + 4.94x_2 + 7.31x_3 - 9.06x_2 z_1 + 8.31x_3 z_1$$

Using equations 13-6 and 13-7, we can find the mean and variance models as

$$E_z[y(\mathbf{x}, z_1)] = 70.06 + 4.94x_2 + 7.31x_3$$

and

$$V_z[y(\mathbf{x}, z_1)] = \sigma_z^2(10.81 - 9.06x_2 + 8.31x_3)^2 + \sigma^2$$

$$= \sigma_z^2(116.91 + 82.08x_2^2 + 69.06x_3^2 - 195.88x_2$$

$$+ 179.66x_3 - 150.58x_2 x_3) + \sigma^2$$

respectively. Now assume that the low and high levels of the noise variable temperature have been run at one standard deviation either side of its typical or average value, so that $\sigma_z^2 = 1$, and use $\hat{\sigma} = 19.51$ (this is the residual mean square obtained by fitting the response model). Therefore, the variance model becomes

$$V_z[y(\mathbf{x}, z_1)] = 136.42 - 195.88x_2 + 179.66x_3 - 150.58x_2 x_3 + 82.08x_2^2 + 69.06x_3^2$$

Figure 13-12 presents a contour plot of the response contours from the mean model. To construct this plot, we held the noise factor (temperature) at zero and the nonsignificant controllable factor (pressure) at zero. Note that mean filtration rate increases as both concentration and stirring rate increase. The square root of $V_z[y(\mathbf{x}, \mathbf{z})]$ is plotted in Fig. 13-13. Note that both a contour plot and a three-dimensional response surface plot

Table 13-4 Pilot Plant Filtration Rate Experiment

Run Number	Factor A	B	C	D	Run Label	Filtration Rate (gal/h)
1	−	−	−	−	(1)	45
2	+	−	−	−	a	71
3	−	+	−	−	b	48
4	+	+	−	−	ab	65
5	−	−	+	−	c	68
6	+	−	+	−	ac	60
7	−	+	+	−	bc	80
8	+	+	+	−	abc	65
9	−	−	−	+	d	43
10	+	−	−	+	ad	100
11	−	+	−	+	bd	45
12	+	+	−	+	abd	104
13	−	−	+	+	cd	75
14	+	−	+	+	acd	86
15	−	+	+	+	bcd	70
16	+	+	+	+	abcd	96

is given. This plot was also constructed by setting the noise factor temperature and the nonsignificant controllable factor to zero.

Suppose that the experimenter wants to maintain mean filtration rate above 75 and minimize the variability around this value. Figure 13-14 shows an overlay plot of the contours of mean filtration rate and the POE as a function of concentration and stirring

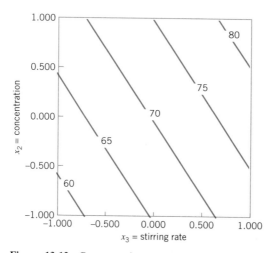

Figure 13-12 Contours of constant mean filtration rate, Example 13-3, with z_1 = temperature = 0.

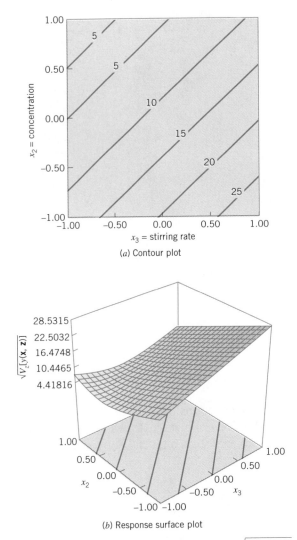

Figure 13-13 Contour plot and response surface of $\sqrt{V_z[y(\mathbf{x}, \mathbf{z})]}$ for Example 13-3, with z_1 = temperature = 0.

rate, the significant controllable variables. To achieve the desired objectives, it will be necessary to hold concentration at the high level and stirring rate very near the middle level.

Example 13-3 illustrates the use of a first-order model with interaction as the model for the controllable factors, $f(\mathbf{x})$. We now present an example that involves a second-order model.

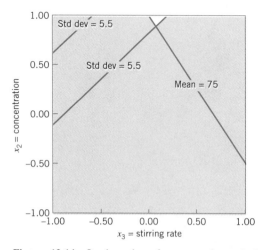

Figure 13-14 Overlay plot of mean and standard deviation for filtration rate, Example 13-3, with $Z_1 =$ temperature $= 0$.

EXAMPLE 13-4

A process robustness study was conducted in a semiconductor manufacturing plant involving two controllable variables x_1 and x_2 and a single noise factor z. Table 13-5 shows the experiment that was performed, and Fig. 13-15 gives a graphical view of the design. Note that the experimental design is a "modified" central composite

Table 13-5   The Modified Central Composite Design for the Process Robustness Study in Example 13-4

Run	x_1	x_2	z	y
1	-1.00	-1.00	-1.00	73.93
2	1.00	-1.00	-1.00	81.99
3	-1.00	1.00	-1.00	77.03
4	1.00	1.00	-1.00	99.29
5	-1.00	-1.00	1.00	70.21
6	1.00	-1.00	1.00	97.72
7	-1.00	1.00	1.00	83.20
8	1.00	1.00	1.00	125.50
9	-1.68	0.00	0.00	64.75
10	1.68	0.00	0.00	102.90
11	0.00	-1.68	0.00	70.20
12	0.00	1.68	0.00	100.30
13	0.00	0.00	0.00	100.50
14	0.00	0.00	0.00	100.00
15	0.00	0.00	0.00	98.86
16	0.00	0.00	0.00	103.90

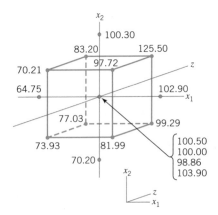

Figure 13-15   The modified central composite design in Example 13-4.

design in which the axial runs in the z direction have been eliminated. It is possible to delete these runs because no quadratic term (z^2) in the noise variable is included in the model.

Using equation 13-5, the response model for this process robustness study is

$$y(\mathbf{x}, \mathbf{z}) = f(\mathbf{x}) + h(\mathbf{x}, \mathbf{z}) + \epsilon$$
$$= \beta_0 + \beta_1 x_1 + \beta_2 x_2 + \beta_{11} x_1^2 + \beta_{22} x_2^2 + \beta_{12} x_1 x_2 + \gamma_1 z + \delta_{11} x_1 z$$
$$+ \delta_{21} x_2 z + \epsilon$$

The least squares fit is

$$\hat{y}(\mathbf{x}, \mathbf{z}) = 100.63 + 12.03x_1 + 8.19x_2 - 6.10x_1^2 - 5.60x_2^2 + 3.62x_1 x_2$$
$$+ 5.55z + 4.94x_1 z + 2.55x_2 z$$

Therefore, from equation 13-6, the mean model is

$$E_z[y(\mathbf{x}, \mathbf{z})] = 100.63 + 12.03x_1 + 8.19x_2 - 6.10x_1^2 - 5.60x_2^2 + 3.62x_1 x_2$$

Using equation 13-7 the variance model is

$$V_z[y(\mathbf{x}, \mathbf{z})] = \sigma_z^2 \left(\frac{\partial h(\mathbf{x}, \mathbf{z})}{\partial z} \right)^2 + \sigma^2$$
$$= \sigma_z^2 (5.55 + 4.94x_1 + 2.55x_2)^2 + \sigma^2$$

We will assume (as in the previous example) that $\sigma_z^2 = 1$ and since the residual mean square from fitting the response model is $MS_E = 3.73$, we will use $\hat{\sigma}^2 = MS_E = 3.73$. Therefore, the variance model is

$$V_z[y(\mathbf{x}, \mathbf{z})] = (5.55 + 4.94x_1 + 2.55x_2)^2 + 3.73$$

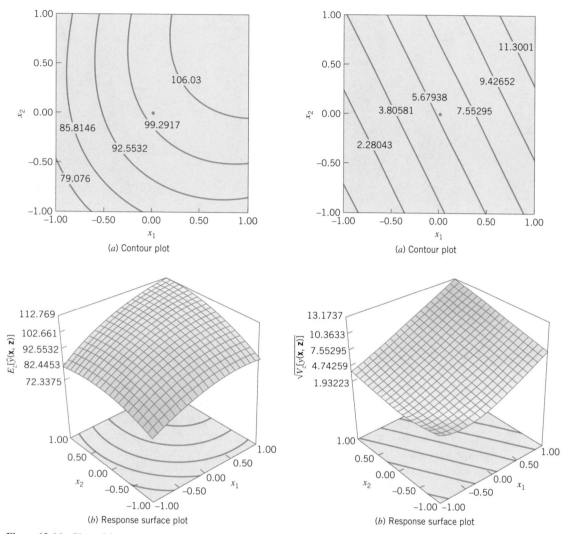

Figure 13-16 Plots of the mean model, Example 13-4.

Figure 13-17 Plots of the standard deviation of the response $[\sqrt{V_z[y(\mathbf{x}_1\mathbf{z})]}]$, Example 13-4.

Figures 13-16 and 13-17 show response surface contour plots and three-dimensional surface plots of the mean model and the standard deviation $\sqrt{V_z[y(\mathbf{x}, \mathbf{z})]}$, respectively.

The objective of the experimenters in this process robustness study was to find a set of operating conditions that would result in a mean response between 90 and 100 with low variability. Figure 13-18 is an overlay of the contours 90 and 100 from the mean model with the contour of constant standard deviation of 4. The unshaded region of this plot indicates operating conditions on x_1 and x_2, where the requirements for the mean response are satisfied and the response standard deviation does not exceed 4.

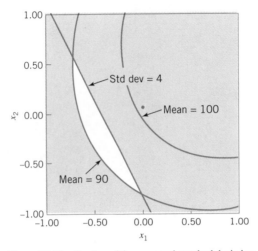

Figure 13-18   Overlay of the mean and standard deviation contours, Example 13-4.

13-3 EVOLUTIONARY OPERATION

Most process-monitoring techniques measure one or more output quality characteristics, and if these quality characteristics are satisfactory, no modification of the process is made. However, in some situations where there is a strong relationship between one or more **controllable process variables** and the observed quality characteristic or **response variable,** other process-monitoring methods can sometimes be employed. For example, suppose that a chemical engineer wishes to maximize the yield of the process. The yield is a function of two controllable process variables, temperature (x_1) and time (x_2)—say,

$$y = f(x_1, x_2) + \epsilon$$

where ϵ is a random error component. The chemical engineer has found a set of operating conditions or levels for x_1 and x_2 that maximizes yield and provides acceptable values for all other quality characteristics. This process optimization may have been done using RSM; however, even if the plant operates continuously at these levels, it will eventually "drift" away from the optimum as a result of variations in the incoming raw materials, environmental changes, operating personnel, and the like.

A method is needed for continuous operation and monitoring of a process with the goal of moving the operating conditions toward the optimum or following a "drift." The method should not require large or sudden changes in operating conditions that might disrupt production. **Evolutionary operation (EVOP)** was proposed by Box (1957) as such an operating procedure. It is designed as a method of routine plant operation that is carried out by operating personnel with minimum assistance from the quality or manufacturing engineering staff. EVOP makes use of principles of experimental design, which usually is considered to be an off-line quality engineering method. Thus, EVOP is an on-line application of designed experiments.

EVOP consists of systematically introducing small changes in the levels of the process operating variables. The procedure requires that each independent process variable be assigned a "high" and a "low" level. The changes in the variables are assumed to be small enough so that serious disturbances in product quality will not occur, yet large enough so that potential improvements in process performance will eventually be discovered. For two variables x_1 and x_2, the four possible combinations of high and low levels are shown in Fig. 13-19. This arrangement of points is the 2^2 factorial design introduced in Chapter 12. We have also included a point at the center of the design. Typically, the 2^2 design would be centered about the best current estimate of the optimum operating conditions.

The points in the 2^2 design are numbered 1, 2, 3, 4, and 5. Let y_1, y_2, y_3, y_4, and y_5 be the observed values of the dependent or response variable corresponding to these points. After one observation has been run at each point in the design, a **cycle** is said to have been completed. Recall that the main effect of a factor is defined as the average change in response produced by a change from the low level to the high level of the factor. Thus, the effect of x_1 is the average difference between the responses on the right-hand side of the design in Fig. 13-19 and the responses on the left-hand side, or

$$x_1 \text{ effect} = \tfrac{1}{2}[(y_3 + y_4) - (y_2 + y_5)]$$

$$= \tfrac{1}{2}[y_3 + y_4 - y_2 - y_5] \tag{13-8}$$

Similarly, the effect of x_2 is found by computing the average difference in the responses on the top of the design in Fig. 13-19 and the responses on the bottom; that is,

$$x_2 \text{ effect} = \tfrac{1}{2}[(y_3 + y_5) - (y_2 + y_4)]$$

$$= \tfrac{1}{2}[y_3 + y_5 - y_2 - y_4] \tag{13-9}$$

If the change from the low to the high level of x_1 produces an effect that is different at the two levels of x_2, then there is interaction between x_1 and x_2. The interaction effect is

$$x_1 \times x_2 \text{ interaction} = \tfrac{1}{2}[y_2 + y_3 - y_4 - y_5] \tag{13-10}$$

or simply the average difference between the diagonal totals in Fig. 13-19. After n cycles, there will be n observations at each of the five design points. The effects of x_1, x_2, and their interaction are then computed by replacing the individual observations y_i in equations 13-8, 13-9, and 13-10 by the averages $\bar{y}_i$ of the n observations at each point.

After several cycles have been completed, one or more process variables, or their interaction, may appear to have a significant effect on the response variable y. When this occurs, a decision may be made to change the basic operating conditions to improve the

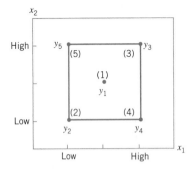

Figure 13-19 2^2 factorial design.

process output. When improved conditions are detected, a **phase** is said to have been completed.

In testing the significance of process variables and interactions, an estimate of experimental error is required. This is calculated from the cycle data. By comparing the response at the center point with the 2^k points in the factorial portion, we may check on the presence of curvature in the response function; that is, if the process is really centered at the maximum (say), then the response at the center should be significantly greater than the responses at the 2^k peripheral points.

In theory, EVOP can be applied to an arbitrary number of process variables. In practice, only two or three variables are usually considered at a time. We will give an example of the procedure for two variables. Box and Draper (1969) give a discussion of the three-variable case, including necessary forms and worksheets. Myers and Montgomery (1995) show how EVOP calculations can be easily performed in statistical software packages for factorial designs.

······ **EXAMPLE 13-5** ···

Consider a chemical process whose yield is a function of temperature (x_1) and pressure (x_2). The current operating conditions are $x_1 = 250°F$ and $x_2 = 145$ psi. The EVOP procedure uses the 2^2 design plus the center point shown in Fig. 13-20. The cycle is completed by running each design point in numerical order (1, 2, 3, 4, 5). The yields in the first cycle are shown in Fig. 13-20.

The yields from the first cycle are entered in the EVOP calculation sheet shown in Table 13-6. At the end of the first cycle, no estimate of the standard deviation can be made. The calculation of the main effects of temperature and pressure and their interaction are shown in the bottom half of Table 13-6.

A second cycle is then run, and the yield data are entered in another EVOP calculation sheet shown in Table 13-7. At the end of the second cycle, the experimental error can be estimated and the estimates of the effects compared to approximate 95% (two standard deviation) limits. Note that the range refers to the range of the differences in row (iv); thus, the range is $+1.0 - (-1.0) = 2.0$. Since none of the effects in Table

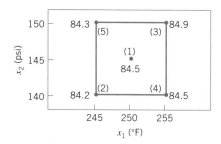

Figure 13-20 2^2 design for Example 13-5.

13-7 exceed their error limits, the true effect is probably zero, and no changes in operating conditions are contemplated.

The results of a third cycle are shown in Table 13-8. The effect of pressure now exceeds its error limit, and the temperature effect is equal to the error limit. A change in operating conditions is now probably justified.

Table 13-6   EVOP Calculation Sheet—Example 13-5, $n = 1$

| | | Cycle: $n = 1$ | | | | Phase 1 |
| | | Response: Yield | | | | Date: 6-14-00 |

		Calculation of Averages					Calculation of Standard Deviation
Operating Conditions	(1)	(2)	(3)	(4)	(5)	
(i) Previous cycle sum						Previous sum $S =$
(ii) Previous cycle average						Previous average $S =$
(iii) New observations	84.5	84.2	84.9	84.5	84.3	New $S =$ range $\times f_{5,n} =$
(iv) Differences [(ii) − (iii)]						Range of (iv) =
(v) New sums [(i) + (iii)]	84.5	84.2	84.9	84.5	84.3	New sum $S =$
(vi) New averages [$\bar{y}_i = $ (v)/n]	84.5	84.2	84.9	84.5	84.3	New average $S = \dfrac{\text{new sum } S}{n-1}$

Calculation of Effects	Calculation of Error Limits
Temperature effect $= \frac{1}{2}(\bar{y}_3 + \bar{y}_4 - \bar{y}_2 - \bar{y}_5) = 0.45$	For new average $\dfrac{2}{\sqrt{n}} S =$
Pressure effect $= \frac{1}{2}(\bar{y}_3 + \bar{y}_5 - \bar{y}_2 - \bar{y}_4) = 0.25$	For new effects $\dfrac{2}{\sqrt{n}} S =$
$T \times P$ interaction effect $= \frac{1}{2}(\bar{y}_2 + \bar{y}_3 - \bar{y}_4 - \bar{y}_5) = 0.15$	
Change-in-mean effect $= \frac{1}{5}(\bar{y}_2 + \bar{y}_3 + \bar{y}_4 + \bar{y}_5 - 4\bar{y}_1) = 0.02$	For change in mean $\dfrac{1.78}{\sqrt{n}} S =$

Table 13-7 EVOP Calculation Sheet—Example 13-5, $n = 2$

	5		3			Cycle: $n = 2$			Phase 1
	1					Response: Yield			Date: 6-14-00
	2		4						

			Calculation of Averages				**Calculation of Standard Deviation**
Operating Conditions		(1)	(2)	(3)	(4)	(5)	
(i) Previous cycle sum		84.5	84.2	84.9	94.5	84.3	Previous sum $S =$
(ii) Previous cycle average		84.5	84.2	84.9	84.5	84.3	Previous average $S =$
(iii) New observations		84.9	84.6	85.9	83.5	84.0	New $S = $ range $\times f_{5,n} = 0.60$
(iv) Differences [(ii) − (iii)]		− 0.4	− 0.4	− 1.0	+ 1.0	0.3	Range of (iv) = 2.0
(v) New sums [(i) + (iii)]		169.4	168.8	170.8	168.0	168.3	New sum $S = 0.60$
(vi) New averages		84.70	84.40	85.40	84.00	84.15	New average S
$[\bar{y}_i = (v)/n]$							$= \dfrac{\text{New sum } S}{n - 1} = 0.60$

Temperature effect $= \frac{1}{2}(\bar{y}_3 + \bar{y}_4 - \bar{y}_2 - \bar{y}_5) = 0.43$ For new average $\dfrac{2}{\sqrt{n}} S = 0.85$

Pressure effect $= \frac{1}{2}(\bar{y}_3 + \bar{y}_5 - \bar{y}_2 - \bar{y}_4) = 0.58$ For new effects $\dfrac{2}{\sqrt{n}} S = 0.85$

$T \times P$ interaction effect $= \frac{1}{2}(\bar{y}_2 + \bar{y}_3 - \bar{y}_4 - \bar{y}_5) = 0.83$

Change-in-mean effect $= \frac{1}{2}(\bar{y}_2 + \bar{y}_3 + \bar{y}_4 + \bar{y}_5 - 4\bar{y}_1) = -0.17$ For change in mean $\dfrac{1.78}{\sqrt{n}} S = 0.76$

In light of the results, it seems reasonable to begin a new EVOP phase about point (3). Thus, $x_1 = 225°F$ and $x_2 = 150$ psi would become the center of the 2^2 design in the second phase.

An important aspect of EVOP is feeding the information generated back to the process operators and supervisors. This is accomplished by a prominently displayed EVOP information board. The information board for this example at the end of cycle three is shown in Table 13-9.

• •

Most of the quantities on the EVOP calculation sheet follow directly from the analysis of the 2^k factorial design. For example, the variance of any effect such as $\frac{1}{2}(\bar{y}_3 + \bar{y}_5 - \bar{y}_2 - \bar{y}_4)$ is simply

$$V[\tfrac{1}{2}(\bar{y}_3 + \bar{y}_5 - \bar{y}_2 - \bar{y}_4)] = \tfrac{1}{4}(\sigma_{\bar{y}_3}^2 + \sigma_{\bar{y}_5}^2 + \sigma_{\bar{y}_2}^2 + \sigma_{\bar{y}_4}^2)$$

$$= \tfrac{1}{4}(4\sigma_{\bar{y}}^2) = \frac{\sigma^2}{n}$$

where σ^2 is the variance of the observations (y). Thus, two standard deviation (corre-

Table 13-8 EVOP Calculation Sheet—Example 13-5, $n = 3$

	Cycle: $n = 3$	Phase 1
(diagram: 5─3, 1, 2─4)	Response: Yield	Date: 6-14-00

		Calculation of Averages					Calculation of Standard Deviation
Operating Conditions	(1)	(2)	(3)	(4)	(5)		
(i) Previous cycle sum	169.4	168.8	170.8	168.0	168.3	Previous sum $S = 0.60$	
(ii) Previous cycle average	84.70	84.40	85.40	84.00	84.15	Previous average $S = 0.60$	
(iii) New observations	85.0	84.0	86.6	84.9	85.2	New $S = $ range $\times f_{5,n} = 0.56$	
(iv) Differences [(ii) − (iii)]	−0.30	+0.40	−1.20	−0.90	−1.05	Range of (iv) $= 1.60$	
(v) New sums [(i) + (iii)]	254.4	252.8	257.4	252.9	253.5	New sum $S = 1.16$	
(vi) New averages	84.80	84.27	85.80	84.30	84.50	New average S	
$[\bar{y} = (v)/n]$						$= \dfrac{\text{New sum } S}{n-1} = 0.58$	

Calculation of Effects	Calculation of Error Limits
Temperature effect $= \frac{1}{2}(\bar{y}_3 + \bar{y}_4 - \bar{y}_2 - \bar{y}_5) = 0.67$	For new average $\dfrac{2}{\sqrt{n}} S = 0.67$
Pressure effect $= \frac{1}{2}(\bar{y}_3 + \bar{y}_5 - \bar{y}_2 - \bar{y}_4) = 0.87$	For new effects $\dfrac{2}{\sqrt{n}} S = 0.67$
$T \times P$ interaction effect $= \frac{1}{2}(\bar{y}_2 + \bar{y}_3 - \bar{y}_4 - \bar{y}_5) = 0.64$	
Change-in-mean effect $= \frac{1}{5}(\bar{y}_2 + \bar{y}_3 + \bar{y}_4 + \bar{y}_5 - 4\bar{y}_1) = -0.07$	For change in mean $\dfrac{1.78}{\sqrt{n}} S = 0.60$

sponding to approximately 95%) error limits on any effect would be $\pm 2\sigma/\sqrt{n}$. The variance of the change in mean is

$$V(\text{CIM}) = V[\tfrac{1}{5}(\bar{y}_2 + \bar{y}_3 + \bar{y}_4 + \bar{y}_5 - 4\bar{y}_1)]$$

$$= \tfrac{1}{25}(4\sigma_{\bar{y}}^2 + 16\sigma_{\bar{y}}^2) = \frac{20}{25}\frac{\sigma^2}{n}$$

Thus, two standard deviation error limits on the CIM are $\pm 2\sigma\sqrt{(20/25)n} = \pm 1.78\sigma/\sqrt{n}$. For more information on the 2^k factorial design, see Chapter 12 and Montgomery (1997).

The standard deviation σ is estimated by the range method. Let $y_i(n)$ denote the observation at the ith design point in cycle n, and $\bar{y}_i(n)$ the corresponding average of $y_i(n)$ after n cycles. The quantities in row (iv) of the EVOP calculation sheet are the differences $y_i(n) - \bar{y}_i(n-1)$. The variance of these differences is $V[y_i(n) - \bar{y}_i(n-1)] \equiv \sigma^2[n/(n-1)]$. The range of the differences—say, R_D—is related to the estimate of the

Table 13-9   EVOP Information Board—Cycle Three

Response: Percent yield
Requirement: Maximize

Effects with 95% ErrorLimits		Error Limits for Averages: ± 0.67
Temperature	0.67	± 0.67
Pressure	0.87	± 0.67
$T \times P$	0.64	± 0.67
Change in mean	0.07	± 0.60

Standard deviation: 0.58

Table 13-10   Values of $f_{k,n}$

$n =$	2	3	4	5	6	7	8	9	10
$k = 5$	0.30	0.35	0.37	0.38	0.39	0.40	0.40	0.40	0.41
9	0.24	0.27	0.29	0.30	0.31	0.31	0.31	0.32	0.32
10	0.23	0.26	0.28	0.29	0.30	0.30	0.30	0.31	0.31

distribution of the differences by $\hat{\sigma}_D = R_D/d_2$. Now $R_D/d_2 = \hat{\sigma} \sqrt{n/(n-1)}$, so

$$\hat{\sigma} = \sqrt{\frac{(n-1)}{n}} \frac{R_D}{d_2} = (f_{k,n})R_D \equiv S$$

can be used to estimate the standard deviation of the observations, where k denotes the number of points used in the design. For a 2^2 with one center point we have $k = 5$, and for a 2^3 with one center point we have $k = 9$. Values of $f_{k,n}$ are given in Table 13-10.

13-4 EXERCISES

13-1.   Consider the first-order model

$$\hat{y} = 75 + 10x_1 + 6x_2$$

(a) Sketch the contours of constant predicted response over the range $-1 \le x_i \le +1$, $i = 1, 2$.
(b) Find the direction of steepest ascent.

13-2.   Consider the first-order model

$$\hat{y} = 50 + 2x_1 - 15x_2 + 3x_3$$

where $-1 \le x_i \le +1$, $i = 1, 2, 3$. Find the direction of steepest ascent.

13-3.   An experiment was run to study the effect of two factors, time and temperature, on the inorganic impurity levels in paper pulp. The results of this experiment are shown here:

x_1	x_2	y
-1	-1	210
1	-1	95
-1	1	218
1	1	100
-1.5	0	225
1.5	0	50
0	-1.5	175
0	1.5	180
0	0	145
0	0	175
0	0	158
0	0	166

(a) What type of experimental design has been used in this study? Is the design rotatable?

(b) Fit a quadratic model to the response, using the method of least squares.

(c) Construct the fitted impurity response surface. What values of x_1 and x_2 would you recommend if you wanted to minimize the impurity level?

(d) Suppose that

$$x_1 = \frac{\text{temp} - 750}{50} \qquad x_2 = \frac{\text{time} - 3}{15}$$

where temperature is in °C and time is in hours. Find the optimum operating conditions in terms of the natural variables temperature and time.

13-4.   A second-order response surface model in two variables is

$$\hat{y} = 69.0 + 1.6x_1 + 1.1x_2 - 1x_1^2$$
$$- 1.2x_2^2 + 0.3x_1x_2$$

(a) Generate a two-dimensional contour plot for this model over the region $-2 \le x_i \le +2$, $i = 1, 2$, and select the values of x_1 and x_2 that maximize $\hat{y}$.

(b) Find the two equations given by $\frac{\partial \hat{y}}{\partial x_1} = 0$ and $\frac{\partial \hat{y}}{\partial x_2} = 0$. Show that the solution to these equations for the optimum conditions x_1 and x_2 are the same as those found graphically in part (a).

13-5.   An article in *Rubber Chemistry and Technology* (Vol. 47, 1974, pp. 825–836) describes an experiment that studies the relationship of the Mooney viscosity of rubber to several variables, including silica filler (parts per hundred) and oil filler (parts per hundred). Some of the data from this experiment are shown here, where

$$x_1 = \frac{\text{silica} - 60}{15} \qquad x_2 = \frac{\text{oil} - 21}{1.5}$$

Coded Levels		
x_1	x_2	y
-1	-1	13.71
1	-1	14.15
-1	1	12.87
1	1	13.53
-1.4	0	12.99
1.4	0	13.89
0	-1.4	14.16
0	1.4	12.90
0	0	13.75
0	0	13.66
0	0	13.86
0	0	13.63
0	0	13.74

(a) What type of experimental design has been used? Is it rotatable?

(b) Fit a quadratic model to these data. What values of x_1 and x_2 will maximize the Mooney viscosity?

13-6.   In their book *Empirical Model Building and Response Surfaces* (John Wiley, 1987), G. E. P. Box and N. R. Draper describe an experiment with three factors. The data shown in the following table are a variation of the original experiment on p. 247 of their book. Suppose that these data were collected in a semiconductor manufacturing process.

x_1	x_2	x_3	y_1	y_2
-1	-1	-1	24.00	12.49
0	-1	-1	120.33	8.39
1	-1	-1	213.67	42.83
-1	0	-1	86.00	3.46
0	0	-1	136.63	80.41
1	0	-1	340.67	16.17
-1	1	-1	112.33	27.57
0	1	-1	256.33	4.62
1	1	-1	271.67	23.63
-1	-1	0	81.00	0.00
0	-1	0	101.67	17.67
1	-1	0	357.00	32.91
-1	0	0	171.33	15.01
0	0	0	372.00	0.00
1	0	0	501.67	92.50
-1	1	0	264.00	63.50
0	1	0	427.00	88.61
1	1	0	730.67	21.08
-1	-1	1	220.67	133.82
0	-1	1	239.67	23.46
1	-1	1	422.00	18.52
-1	0	1	199.00	29.44
0	0	1	485.33	44.67
1	0	1	673.67	158.21
-1	1	1	176.67	55.51
0	1	1	501.00	138.94
1	1	1	1010.00	142.45

(a) The response y_1 is the average of three readings on resistivity for a single wafer. Fit a quadratic model to this response.

(b) The response y_2 is the standard deviation of the three resistivity measurements. Fit a first-order model to this response.

(c) Where would you recommend that we set x_1, x_2, and x_3 if the objective is to hold mean resistiv-ity at 500 and minimize the standard deviation?

13-7.   An article by J. J. Pignatiello, Jr. and J. S. Ramberg in the *Journal of Quality Technology* (Vol. 17, 1985, pp. 198–206) describes the use of a replicated fractional factorial to investigate the effect of five factors on the free height of leaf springs used in an automotive application. The factors are A = furnace temperature, B = heating time, C = transfer time, D = hold down time, and E = quench oil temperature. The data are shown here.

A	B	C	D	E			
−	−	−	−	−	7.78,	7.78,	7.81
+	−	−	+	−	8.15,	8.18,	7.88
−	+	−	+	−	7.50,	7.56,	7.50
+	+	−	−	−	7.59,	7.56,	7.75
−	−	+	+	−	7.54,	8.00,	7.88
+	−	+	−	−	7.69,	8.09,	8.06
−	+	+	−	−	7.56,	7.52,	7.44
+	+	+	+	−	7.56,	7.81,	7.69
−	−	−	−	+	7.50,	7.25,	7.12
+	−	−	+	+	7.88,	7.88,	7.44
−	+	−	+	+	7.50,	7.56,	7.50
+	+	−	−	+	7.63,	7.75,	7.56
−	−	+	+	+	7.32,	7.44,	7.44
+	−	+	−	+	7.56,	7.69,	7.62
−	+	+	−	+	7.18,	7.18,	7.25
+	+	+	+	+	7.81,	7.50,	7.59

(a) Write out the alias structure for this design. What is the resolution of this design?

(b) Analyze the data. What factors influence mean free height?

(c) Calculate the range and standard deviation of free height for each run. Is there any indication that any of these factors affects variability in free height?

(d) Analyze the residuals from this experiment and comment on your findings.

(e) Is this the best possible design for five factors in 16 runs? Specifically, can you find a fractional design for five factors in 16 runs with higher resolution than this one?

13-8. Consider the leaf spring experiment in Exercise 13-7. Suppose that factor E (quench oil temperature) is very difficult to control during manufacturing. We want to have the mean spring height as close to 7.50 as possible with minimum variability. Where would you set factors A, B, C, and D to reduce variability in free height as much as possible regardless of the quench oil temperature used?

13-9. Consider the leaf spring experiment in Exercise 13-7. Rework this problem, assuming that factors A, B, and C are easy to control but factors D and E are hard to control.

13-10. The data shown in the following table were collected in an experiment to optimize crystal growth as a function

x_1	x_2	x_3	y
-1	-1	-1	66
-1	-1	1	70
-1	1	-1	78
-1	1	1	60
1	-1	-1	80
1	-1	1	70
1	1	-1	100
1	1	1	75
-1.682	0	0	100
1.682	0	0	80
0	-1.682	0	68
0	1.682	0	63
0	0	-1.682	65
0	0	1.682	82
0	0	0	113
0	0	0	100
0	0	0	118
0	0	0	88
0	0	0	100
0	0	0	85

of three variables x_1, x_2, and x_3. Large values of y (yield in grams) are desirable. Fit a second-order model and analyze the fitted surface. Under what set of conditions is maximum growth achieved?

13-11. The following data were collected by a chemical engineer. The response y is filtration time, x_1 is temperature, and x_2 is pressure. Fit a second-order model.

x_1	x_2	y
-1	-1	54
-1	1	45
1	-1	32
1	1	47
-1.414	0	50
1.414	0	53
0	-1.414	47
0	1.414	51
0	0	41
0	0	39
0	0	44
0	0	42
0	0	40

(a) What operating conditions would you recommend if the objective is to minimize the filtration time?

(b) What operating conditions would you recommend if the objective is to operate the process at a mean filtration rate very close to 46?

13-12. Reconsider the crystal growth experiment from Exercise 13-10. Suppose that $x_3 = z$ is now a noise variable, and that the modified experimental design shown here has been conducted. The experimenters want the growth rate to be as large as possible, but they also want the variability transmitted from z to be small. Under what set of conditions is growth greater than 90 with minimum variability achieved?

x_1	x_2	z	y
-1	-1	-1	66
-1	-1	1	70
-1	1	-1	78
-1	1	1	60
1	-1	-1	80
1	-1	1	70
1	1	-1	100
1	1	1	75
-1.682	0	0	100
1.682	0	0	80
0	-1.682	0	68
0	1.682	0	63
0	0	0	113
0	0	0	100
0	0	0	118
0	0	0	88
0	0	0	100
0	0	0	85

13-13. Consider the response model in equation 13-5 and the transmission of error approach to finding the variance model (equation 13-7). Suppose that in the response model we use

$$h(\mathbf{x}, \mathbf{z}) = \sum_{i=1}^{r} \gamma_i z_i + \sum_{i=1}^{k}\sum_{j=1}^{r} \delta_{ij} x_i z_j$$

$$+ \sum_{i<j=2}^{r} \lambda_{ij} z_i z_j$$

What effect does including the interaction terms between the noise variables have on the variance model?

13-14. Consider the response model in equation 13-5. Suppose that in the response model we allow for a complete second-order model in the noise factors so that

$$h(\mathbf{x}, \mathbf{z}) = \sum_{i=1}^{r} \gamma_i z_i + \sum_{i=1}^{k}\sum_{j=1}^{r} \delta_{ij} x_i z_j$$

$$+ \sum_{i<j=2}^{r} \lambda_{ij} z_i z_j + \sum_{i=1}^{r} \theta_i z_i^2$$

What effect does this have on the variance model?

PART

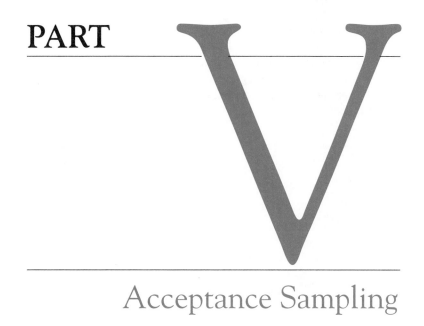

Acceptance Sampling

Inspection of raw materials, semifinished products, or finished products is one aspect of quality assurance. When inspection is for the purpose of acceptance or rejection of a product, based on adherence to a standard, the type of inspection procedure employed is usually called acceptance sampling. This section presents two chapters that deal with the design and use of sampling plans, schemes, and systems. The primary focus is on lot-by-lot acceptance sampling.

Chapter 14 presents lot-by-lot acceptance sampling plans for attributes. Included in this chapter is a discussion of MIL STD 105E and its civilian counterpart ANSI/ASQC Z1.4. Variables sampling plans are presented in Chapter 15, including MIL STD 414 and its civilian counterpart ANSI/ASQC Z1.9, along with a survey of several useful topics in acceptance sampling, including chain-sampling plans, sampling plans for continuous production, and skip-lot sampling plans.

The underlying philosophy here is that acceptance sampling is not a substitute for adequate process monitoring and control and use of other statistical methods to drive variability reduction. The successful use of these techniques at the early stages of manufacturing, including the vendor or supplier base, can greatly reduce and in some cases eliminate the need for extensive sampling inspection.

Lot-by-Lot Acceptance Sampling for Attributes

CHAPTER OUTLINE

14-1 THE ACCEPTANCE-SAMPLING PROBLEM

   14-1.1 Advantages and Disadvantages of Sampling

   14-1.2 Types of Sampling Plans

   14-1.3 Lot Formation

   14-1.4 Random Sampling

   14-1.5 Guidelines for Using Acceptance Sampling

14-2 SINGLE-SAMPLING PLANS FOR ATTRIBUTES

   14-2.1 Definition of a Single-Sampling Plan

   14-2.2 The OC Curve

   14-2.3 Designing a Single-Sampling Plan with a Specified OC Curve

   14-2.4 Rectifying Inspection

14-3 DOUBLE, MULTIPLE, AND SEQUENTIAL SAMPLING

   14-3.1 Double-Sampling Plans

   14-3.2 Multiple-Sampling Plans

   14-3.3 Sequential-Sampling Plans

14-4 MILITARY STANDARD 105E (ANSI/ASQC Z1.4, ISO 2859)

   14-4.1 Description of the Standard

   14-4.2 Procedure

   14-4.3 Discussion

14-5 THE DODGE–ROMIG SAMPLING PLANS

   14-5.1 AOQL Plans

   14-5.2 LTPD Plans

   14-5.3 Estimation of Process Average

CHAPTER OVERVIEW

This chapter presents lot-by-lot acceptance sampling plans for attributes. Key topics include the design and operation of single-sampling plans, the use of the operating characteristic curve, and the concepts of rectifying inspection, average outgoing quality, and average total inspection. Similar concepts are briefly introduced for types of sampling plans where more than one sample may be taken to determine the disposition of a lot (double, multiple, and sequential sampling). Two systems of standard sampling plans are also presented, the military standard plans known as MIL STD 105E and the Dodge–Romig plans. These plans are designed around different philosophies: MIL STD 105E has an acceptable quality level focus, whereas the Dodge–Romig plans are oriented around either the lot tolerance percent detective or the average outgoing quality limit perspective.

14-1   THE ACCEPTANCE-SAMPLING PROBLEM

As we observed in Chapter 1, acceptance sampling is concerned with inspection and decision making regarding products, one of the oldest aspects of quality assurance. In the 1930s and 1940s, acceptance sampling was one of the major components of the field of statistical quality control, and was used primarily for incoming or receiving inspection. In more recent years, it has become typical to work with suppliers to improve their process performance through the use of SPC and designed experiments, and not to rely as much on acceptance sampling as a primary quality assurance tool.

A typical application of acceptance sampling is as follows: A company receives a shipment of product from a vendor. This product is often a component or raw material used in the company's manufacturing process. A sample is taken from the lot, and some quality characteristic of the units in the sample is inspected. On the basis of the information in this sample, a decision is made regarding lot disposition. Usually, this decision is either to accept or to reject the lot. Sometimes we refer to this decision as **lot sentencing.** Accepted lots are put into production; rejected lots may be returned to the vendor or may be subjected to some other **lot-disposition action.**

Although it is customary to think of acceptance sampling as a receiving inspection activity, there are other uses of sampling methods. For example, frequently a manufacturer will sample and inspect its own product at various stages of production. Lots that are accepted are sent forward for further processing, and rejected lots may be reworked or scrapped.

Three aspects of sampling are important:

1.   It is the purpose of acceptance sampling to sentence lots, not to estimate the lot quality. Most acceptance-sampling plans are not designed for estimation purposes.

2.   Acceptance-sampling plans do not provide any *direct* form of quality control. Acceptance sampling simply accepts and rejects lots. Even if all lots are of the same quality, sampling will accept some lots and reject others, the accepted lots being no better than the rejected ones. Process controls are used to control and systematically improve quality, but acceptance sampling is not.

3. The most effective use of acceptance sampling is *not* to "inspect quality into the product," but rather as an audit tool to ensure that the output of a process conforms to requirements.

Generally, there are three approaches to lot sentencing: (1) accept with no inspection; (2) 100% inspection—that is, inspect every item in the lot, removing all defective[1] units found (defectives may be returned to the vendor, reworked, replaced with known good items, or discarded); and (3) acceptance sampling. The no-inspection alternative is useful in situations where either the vendor's process is so good that defective units are almost never encountered or where there is no economic justification to look for defective units. For example, if the vendor's process capability ratio is 3 or 4, acceptance sampling is unlikely to discover any defective units. We generally use 100% inspection in situations where the component is extremely critical and passing any defectives would result in an unacceptably high failure cost at subsequent stages, or where the vendor's process capability is inadequate to meet specifications. Acceptance sampling is most likely to be useful in the following situations:

1. When testing is destructive

2. When the cost of 100% inspection is extremely high

3. When 100% inspection is not technologically feasible or would require so much calendar time that production scheduling would be seriously impacted

4. When there are many items to be inspected and the inspection error rate is sufficiently high that 100% inspection might cause a higher percentage of defective units to be passed than would occur with the use of a sampling plan

5. When the vendor has an excellent quality history, and some reduction in inspection from 100% is desired, but the vendor's process capability is sufficiently low as to make no inspection an unsatisfactory alternative

6. When there are potentially serious product liability risks, and although the vendor's process is satisfactory, a program for continuously monitoring the product is necessary

14-1.1 Advantages and Disadvantages of Sampling

When acceptance sampling is contrasted with 100% inspection, it has the following advantages:

1. It is usually less expensive because there is less inspection.

2. There is less handling of the product, hence reduced damage.

3. It is applicable to destructive testing.

4. Fewer personnel are involved in inspection activities.

5. It often greatly reduces the amount of inspection error.

---

[1] In previous chapters, the terms "nonconforming" and "nonconformity" were used instead of defective and defect. This is because the popular meanings of defective and defect differ from their technical meanings and have caused considerable misunderstanding, particularly in product liability litigation. In the field of sampling inspection, however, "defective" and "defect" continue to be used in their technical sense—that is, nonconformance to requirements.

6. The rejection of entire lots as opposed to the simple return of defectives often provides a stronger motivation to the vendor for quality improvements.

Acceptance sampling also has several disadvantages, however. These include the following:

1. There are risks of accepting "bad" lots and rejecting "good" lots.

2. Less information is usually generated about the product or about the process that manufactured the product.

3. Acceptance sampling requires planning and documentation of the acceptance-sampling procedure whereas 100% inspection does not.

Although this last point is often mentioned as a disadvantage of acceptance sampling, proper design of an acceptance-sampling plan usually requires study of the actual level of quality required by the consumer. This resulting knowledge is often a useful input into the overall quality planning and engineering process. Thus, in many applications, it may not be a significant disadvantage.

We have pointed out that acceptance sampling is a "middle ground" between the extremes of 100% inspection and no inspection. It often provides a methodology for moving between these extremes as sufficient information is obtained on the control of the manufacturing process that produces the product. Although there is no direct control of quality in the application of an acceptance-sampling plan to an isolated lot, when that plan is applied to a stream of lots from a vendor, it becomes a means of providing protection for both the producer of the lot and the consumer. It also provides for an accumulation of quality history regarding the process that produces the lot, and it may provide feedback that is useful in process control, such as determining when process controls at the vendor's plant are not adequate. Finally, it may place economic or psychological pressure on the vendor to improve the production process.

14-1.2   Types of Sampling Plans

There are a number of different ways to classify acceptance-sampling plans. One major classification is by attributes and variables. **Variables,** of course, are quality characteristics that are measured on a numerical scale. **Attributes** are quality characteristics that are expressed on a "go, no-go" basis. This chapter deals with lot-by-lot acceptance-sampling plans for attributes. Variables sampling plans are the subject of Chapter 15, along with a brief discussion of several special acceptance-sampling procedures.

A **single-sampling plan** is a lot-sentencing procedure in which one sample of n units is selected at random from the lot, and the disposition of the lot is determined based on the information contained in that sample. For example, a single-sampling plan for attributes would consist of a sample size n and an acceptance number c. The procedure would operate as follows: Select n items at random from the lot. If there are c or fewer defectives in the sample, accept the lot, and if there are more than c defective items in the sample, reject the lot. We investigate this type of sampling plan extensively in Section 14-2.

Double-sampling plans are somewhat more complicated. Following an initial sample, a decision based on the information in that sample is made either to (1) accept the lot, (2) reject the lot, or (3) take a second sample. If the second sample is taken, the information from both the first and second sample is combined in order to reach a decision whether to accept or reject the lot. Double-sampling plans are discussed in Section 14-3.

A **multiple-sampling plan** is an extension of the double-sampling concept, in that more than two samples may be required in order to reach a decision regarding the disposition of the lot. Sample sizes in multiple sampling are usually smaller than they are in either single or double sampling. The ultimate extension of multiple sampling is **sequential sampling,** in which units are selected from the lot one at a time, and following inspection of each unit, a decision is made either to accept the lot, reject the lot, or select another unit. Multiple- and sequential-sampling plans are also discussed in Section 14-3.

Single-, double-, multiple-, and sequential-sampling plans can be designed so that they produce equivalent results. That is, these procedures can be designed so that a lot of specified quality has exactly the same probability of acceptance under all four types of sampling plans. Consequently, when selecting the type of sampling procedure, one must consider factors such as the administrative efficiency, the type of information produced by the plan, the average amount of inspection required by the procedure, and the impact that a given procedure may have on the material flow in the manufacturing organization. These issues are discussed in more detail in Section 14-3.

14-1.3   Lot Formation

How the lot is formed can influence the effectiveness of the acceptance-sampling plan. There are a number of important considerations in forming lots for inspection. Some of these are as follows:

1. **Lots should be homogeneous.** The units in the lot should be produced by the same machines, the same operators, and from common raw materials, at approximately the same time. When lots are nonhomogeneous, such as when the output of two different production lines is mixed, the acceptance-sampling scheme may not function as effectively as it could. Nonhomogeneous lots also make it more difficult to take corrective action to eliminate the source of defective products.

2. **Larger lots are preferred over smaller ones.** It is usually more economically efficient to inspect large lots than small ones.

3. **Lots should be conformable to the materials-handling systems used in both the vendor and consumer facilities.** In addition, the items in the lots should be packaged so as to minimize shipping and handling risks, and so as to make selection of the units in the sample relatively easy.

14-1.4   Random Sampling

The units selected for inspection from the lot should be chosen at random, and they should be representative of all the items in the lot. The random-sampling concept is extremely important in acceptance sampling. Unless random samples are used, bias will be introduced. For example, the vendor may ensure that the units packaged on the top of the lot are of extremely good quality, knowing that the inspector will select the sample from the top layer. "Salting" a lot in this manner is not a common practice, but if it occurs and nonrandom-sampling methods are used, the effectiveness of the inspection process is destroyed.

The technique often suggested for drawing a random sample is to first assign a number to each item in the lot. Then n random numbers are drawn, where the range of these numbers is from 1 to the maximum number of units in the lot. This sequence of random numbers determines which units in the lot will constitute the sample. If products have serial or other code numbers, these numbers can be used to avoid the process of actually assigning numbers to each unit. Another possibility would be to use a three-digit random number to represent the length, width, and depth in a container.

In situations where we cannot assign a number to each unit, utilize serial or code numbers, or randomly determine the location of the sample unit, some other technique must be employed to ensure that the sample is random or representative. Sometimes the inspector may "stratify" the lot. This consists of dividing the lot into strata or layers and then subdividing each strata into cubes, as shown in Fig. 14-1. Units are then selected from within each cube. Although this stratification of the lot is usually an imaginary activity performed by the inspector and does not necessarily ensure random samples, at least it ensures that units are selected from all locations in the lot.

We cannot overemphasize the importance of random sampling. If judgment methods are used to select the sample, the statistical basis of the acceptance-sampling procedure is lost.

14-1.5   Guidelines for Using Acceptance Sampling

An acceptance-sampling plan is a statement of the sample size to be used and the associated acceptance or rejection criteria for sentencing individual lots. A sampling scheme

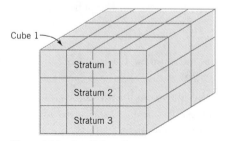

Figure 14-1   Stratifying a lot.

Table 14-1   Acceptance-Sampling Procedures

Objective	Attributes Procedure	Variables Procedure
Assure quality levels for consumer/producer	Select plan for specific OC curve	Select plan for specific OC curve
Maintain quality at a target	AQL system; MIL STD 105E, ANSI/ASQC Z1.4	AQL system; MIL STD 414, ANSI/ASQC Z1.9
Assure average outgoing quality level	AOQL system; Dodge–Romig plans	AOQL system
Reduce inspection, with small sample sizes, good-quality history	Chain sampling	Narrow-limit gaging
Reduce inspection after good-quality history	Skip-lot sampling; double sampling	Skip-lot sampling; double sampling
Assure quality no worse than target	LTPD plan; Dodge–Romig plans	LTPD plan; hypothesis testing

is defined as a set of procedures consisting of acceptance-sampling plans in which lot sizes, sample sizes, and acceptance or rejection criteria along with the amount of 100% inspection and sampling are related. Finally, a sampling system is a unified collection of one or more acceptance-sampling schemes. In this chapter, we see examples of sampling plans, sampling schemes, and sampling systems.

The major types of acceptance-sampling procedures and their applications are shown in Table 14-1. In general, the selection of an acceptance-sampling procedure depends on both the objective of the sampling organization and the history of the organization whose product is sampled. Furthermore, the application of sampling methodology is not static; that is, there is a natural evolution from one level of sampling effort to another. For example, if we are dealing with a vendor who enjoys an excellent quality history, we might begin with an attributes sampling plan. As our experience with the vendor grows, and its good-quality reputation is proved by the results of our sampling activities, we might transition to a sampling procedure that requires much less inspection, such as skip-lot sampling. Finally, after extensive experience with the vendor, and if its process capability is extremely good, we might stop all acceptance-sampling activities on the product. In another situation, where we have little knowledge of or experience with the vendor's quality-assurance efforts, we might begin with attributes sampling using a plan that assures us that the quality of accepted lots is no worse than a specified target value. If this plan proves successful, and if the vendor's performance is satisfactory, we might transition from attributes to variables inspection, particularly as we learn more about the nature of the vendor's process. Finally, we might use the information gathered in variables sampling plans in conjunction with efforts directly at the vendor's manufacturing facility to assist in the installation of process controls. A successful program of process controls at the vendor level might improve the vendor's process capability to the point where inspection could be discontinued.

These examples illustrate that there is a life cycle of application of acceptance-sampling techniques. This was also reflected in the phase diagram, Fig. 1-7, which presented the percentage of application of various quality-assurance techniques as a function of the maturity of the business organization. Typically, we find that organizations with relatively new quality-assurance efforts place a great deal of reliance on acceptance sampling. As their maturity grows and the quality organization develops, they begin to rely less on acceptance sampling and more on statistical process control and experimental design.

Manufacturers try to improve the quality of their products by reducing the number of vendors from whom they buy their components, and by working more closely with the ones they retain. Once again, the key tool in this effort to improve quality is statistical process control. Acceptance sampling can be an important ingredient of any quality-assurance program; however, remember that it is an activity that you try to avoid doing. It is much more cost effective to use statistically based process monitoring at the appropriate stage of the manufacturing process. Sampling methods can in some cases be a tool that you employ along the road to that ultimate goal.

14-2   SINGLE-SAMPLING PLANS FOR ATTRIBUTES

14-2.1   Definition of a Single-Sampling Plan

Suppose that a lot of size N has been submitted for inspection. A **single-sampling plan** is defined by the sample size n and the acceptance number c. Thus, if the lot size is $N = 10,000$, then the sampling plan

$$n = 89$$
$$c = 2$$

means that from a lot of size 10,000 a random sample of $n = 89$ units is inspected and the number of nonconforming or defective items d observed. If the number of observed defectives d is less than or equal to $c = 2$, the lot will be accepted. If the number of observed defectives d is greater than 2, the lot will be rejected. Since the quality characteristic inspected is an attribute, each unit in the sample is judged to be either conforming or nonconforming. One or several attributes can be inspected in the same sample; generally, a unit that is nonconforming to specifications on one or more attributes is said to be a defective unit. This procedure is called a single-sampling plan because the lot is sentenced based on the information contained in one sample of size n.

14-2.2   The OC Curve

An important measure of the performance of an acceptance-sampling plan is the **operating-characteristic (OC) curve.** This curve plots the probability of accepting the lot versus the lot fraction defective. Thus, the OC curve displays the discriminatory power of the sampling plan. That is, it shows the probability that a lot submitted with a certain fraction defective will be either accepted or rejected. The OC curve of the sampling plan

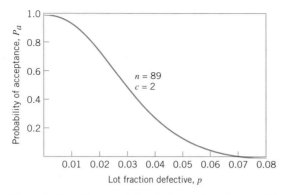

Figure 14-2 OC curve of the single-sampling plan $n = 89$, $c = 2$.

$n = 89$, $c = 2$ is shown in Fig. 14-2. It is easy to demonstrate how the points on this curve are obtained. Suppose that the lot size N is large (theoretically infinite). Under this condition, the distribution of the number of defectives d in a random sample of n items is binomial with parameters n and p, where p is the fraction of defective items in the lot. An equivalent way to conceptualize this is to draw lots of N items at random from a theoretically infinite process, and then to draw random samples of n from these lots. Sampling from the lot in this manner is the equivalent of sampling directly from the process. The probability of observing exactly d defectives is

$$P\{d \text{ defectives}\} = f(d) = \frac{n!}{d!(n-d)!} p^d (1-p)^{n-d} \qquad (14\text{-}1)$$

The probability of acceptance is simply the probability that d is less than or equal to c, or

$$P_a = P\{d \le c\} = \sum_{d=0}^{c} \frac{n!}{d!(n-d)!} p^d (1-p)^{n-d} \qquad (14\text{-}2)$$

For example, if the lot fraction defective is $p = 0.01$, $n = 89$, and $c = 2$, then

$$P_a = P\{d \le 2\} = \sum_{d=0}^{2} \frac{89!}{d!(89-d)!} (0.01)^d (0.99)^{89-d}$$

$$= \frac{89!}{0!\,89!} (0.01)^0 (0.99)^{89} + \frac{89!}{1!\,88!} (0.01)^1 (0.99)^{88} + \frac{89!}{2!(87)!} (0.01)^2 (0.99)^{87}$$

$$= 0.9397$$

The OC curve is developed by evaluating equation 14-2 for various values of p. Table 14-2 displays the calculated value of several points on the curve.

Table 14-2   Probabilities of Acceptance for the Single-Sampling Plan $n = 89, c = 2$

Fraction Defective, p	Probability of Acceptance, P_a
0.005	0.9897
0.010	0.9397
0.020	0.7366
0.030	0.4985
0.040	0.3042
0.050	0.1721
0.060	0.0919
0.070	0.0468
0.080	0.0230
0.090	0.0109

The OC curve shows the discriminatory power of the sampling plan. For example, in the sampling plan $n = 89$, $c = 2$, if the lots are 2% defective, the probability of acceptance is approximately 0.74. This means that if 100 lots from a process that manufactures 2% defective product are submitted to this sampling plan, we will expect to accept 74 of the lots and reject 26 of them.

Effect of n and c on OC Curves

A sampling plan that discriminated perfectly between good and bad lots would have an OC curve that looks like Fig. 14-3. The OC curve runs horizontally at a probability of acceptance $P_a = 1.00$ until a level of lot quality that is considered "bad" is reached, at which point the curve drops vertically to a probability of acceptance $P_a = 0.00$, and then the curve runs horizontally again for all lot fraction defectives greater than the undesirable level. If such a sampling plan could be employed, all lots of "bad" quality would be rejected, and all lots of "good" quality would be accepted.

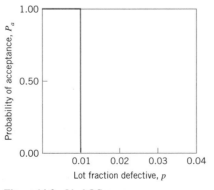

Figure 14-3   Ideal OC curve.

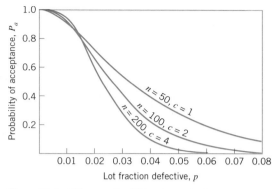

Figure 14-4   OC curves for different sample sizes.

Unfortunately, the **ideal OC curve** in Fig. 14-3 can almost never be obtained in practice. In theory, it could be realized by 100% inspection, if the inspection were error free. The ideal OC curve shape can be approached, however, by increasing the sample size. Figure 14-4 shows that the OC curve becomes more like the idealized OC curve shape as the sample size increases. (Note that the acceptance number c is kept proportional to n.) Thus, the precision with which a sampling plan differentiates between good and bad lots increases with the size of the sample. The greater is the slope of the OC curve, the greater is the discriminatory power.

Figure 14-5 shows how the OC curve changes as the acceptance number changes. Generally, changing the acceptance number does not dramatically change the slope of the OC curve. As the acceptance number is decreased, the OC curve is shifted to the left. Plans with smaller values of c provide discrimination at lower levels of lot fraction defective than do plans with larger values of c.

Specific Points on the OC Curve

Frequently, the quality engineer's interest focuses on certain points on the OC curve. The vendor or supplier is usually interested in knowing what level of lot or process quality

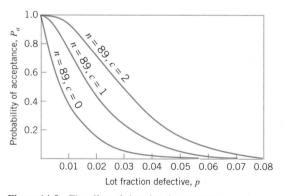

Figure 14-5   The effect of changing the acceptance number on the OC curve.

would yield a high probability of acceptance. For example, the vendor might be interested in the 0.95 probability of acceptance point. This would indicate the level of process fallout that could be experienced and still have a 95% chance that the lots would be accepted. Conversely, the consumer might be interested in the other end of the OC curve. That is, what level of lot or process quality will yield a low probability of acceptance?

A consumer often establishes a sampling plan for a continuing supply of components or raw material with reference to an **acceptable quality level** or **AQL.** The AQL represents the poorest level of quality for the vendor's process that the consumer would consider to be acceptable as a process average. Note that the AQL is a property of the vendor's manufacturing process; it is not a property of the sampling plan. The consumer will often design the sampling procedure so that the OC curve gives a high probability of acceptance at the AQL. Furthermore, the AQL is not usually intended to be a specification on the product, nor is it a target value for the vendor's production process. It is simply a standard against which to judge the lots. It is hoped that the vendor's process will operate at a fallout level that is considerably better than the AQL.

The consumer will also be interested in the other end of the OC curve—that is, in the protection that is obtained for individual lots of poor quality. In such a situation, the consumer may establish a **lot tolerance percent defective (LTPD).** The LTPD is the poorest level of quality that the consumer is willing to accept in an individual lot. Note that the lot tolerance percent defective is not a characteristic of the sampling plan, but is a level of lot quality specified by the consumer. Alternate names for the LTPD are the **rejectable quality level (RQL)** and the **limiting quality level (LQL).** It is possible to design acceptance-sampling plans that give specified probabilities of acceptance at the LTPD point. Subsequently, we will see how to design sampling plans that have specified performance at the AQL and LTPD points.

Type-A and Type-B OC Curves

The OC curves that were constructed in the previous examples are called type-B OC curves. In the construction of the OC curve it was assumed that the samples came from a large lot or that we were sampling from a stream of lots selected at random from a process. In this situation, the binomial distribution is the exact probability distribution for calculating the probability of lot acceptance. Such an OC curve is referred to as a type-B OC curve.

The type-A OC curve is used to calculate probabilities of acceptance for an isolated lot of finite size. Suppose that the lot size is N, the sample size is n, and the acceptance number is c. The exact sampling distribution of the number of defective items in the sample is the hypergeometric distribution.

Figure 14-6 shows the type-A OC curve for a single-sampling plan with $n = 50$, $c = 1$, where the lot size is $N = 500$. The probabilities of acceptance defining the OC curve were calculated using the hypergeometric distribution. Also shown on this graph is the type-A OC curve for $N = 2000$, $n = 50$, and $c = 1$. Note that the two OC curves are very similar. Generally, as the size of the lot increases, the lot size has a decreasing impact on the OC curve. In fact, if the lot size is at least 10 times the sample size ($n/N \leq 0.10$), the type-A and type-B OC curves are virtually indistinguishable. As an illustration, the type-B OC curve for the sampling plan $n = 50$, $c = 1$ is also shown in Fig. 14-6. Note that it is identical to the type-A OC curve based on a lot size of $N = 2000$.

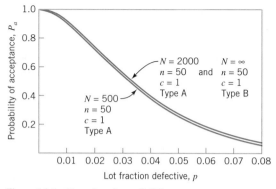

Figure 14-6   Type-A and type-B OC curves.

The type-A OC curve will always lie below the type-B OC curve. That is, if a type-B OC curve is used as an approximation for a type-A curve, the probabilities of acceptance calculated for the type-B curve will always be higher than they would have been if the type-A curve had been used instead. However, this difference is only significant when the lot size is small relative to the sample size. Unless otherwise stated, all discussion of OC curves in this text is in terms of the type-B OC curve.

Other Aspects of OC Curve Behavior

Two approaches to designing sampling plans that are encountered in practice have certain implications for the behavior of the OC curve. Since not all of these implications are positive, it is worthwhile to briefly mention these two approaches to sampling plan design. These approaches are the use of sampling plans with zero acceptance numbers ($c = 0$) and the use of sample sizes that are a fixed percentage of the lot size.

Figure 14-7 shows several OC curves for acceptance-sampling plans with $c = 0$. By comparing Fig. 14-7 with Fig. 14-5, it is easy to see that plans with zero acceptance

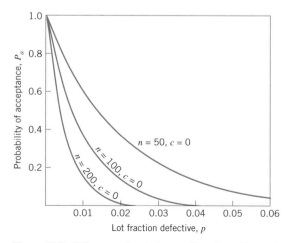

Figure 14-7   OC curves for single-sampling plan with $c = 0$.

numbers have OC curves that have a very different shape than the OC curves of sampling plans for which $c > 0$. Generally, sampling plans with $c = 0$ have OC curves that are convex throughout their range. As a result of this shape, the probability of acceptance begins to drop very rapidly, even for small values of the lot fraction defective. This is extremely hard on the vendor, and in some circumstances, it may be extremely uneconomical for the consumer. For example, consider the sampling plans in Fig. 14-5. Suppose the acceptable quality level is 1%. This implies that we would like to accept lots that are 1% defective or better. Note that if sampling plan $n = 89$, $c = 1$ is used, the probability of lot acceptance at the AQL is about 0.78. On the other hand, if the plan $n = 89$, $c = 0$ is used, the probability of acceptance at the AQL is approximately 0.41. That is, nearly 60% of the lots of AQL quality will be rejected if we use an acceptance number of zero. If rejected lots are returned to the vendor, then a large number of lots will be unnecessarily returned, perhaps creating production delays at the consumer's manufacturing site. If the consumer screens or 100% inspects all rejected lots, a large number of lots that are of acceptable quality will be screened. This is, at best, an inefficient use of sampling resources. In Chapter 15, we suggest an alternative approach to using zero acceptance numbers called **chain-sampling plans.** Under certain circumstances, chain sampling works considerably better than acceptance-sampling plans with $c = 0$. Also refer to the Supplemental Text Material for a discussion of lot-sensitive compliance sampling, another technique that utilizes zero acceptance numbers.

Figure 14-8 presents the OC curves for sampling plans in which the sample size is a fixed percentage of the lot size. The principal disadvantage of this approach is that the different sample sizes offer different levels of protection. It is illogical for the level of protection that the consumer enjoys for a critical part or component to vary as the size of the lot varies. Although sampling procedures such as this one were in wide use before the statistical principles of acceptance sampling were generally known, their use has (unfortunately) not entirely disappeared.

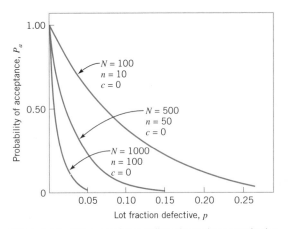

Figure 14-8   OC curves for sampling plans where sample size n is 10% of the lot size.

14-2.3   Designing a Single-Sampling Plan with a Specified OC Curve

A common approach to the design of an acceptance-sampling plan is to require that the OC curve pass through two designated points. Note that one point is not enough to fully specify the sampling plan; however, two points are sufficient. In general, it does not matter which two points are specified.

Suppose that we wish to construct a sampling plan such that the probability of acceptance is $1 - \alpha$ for lots with fraction defective p_1, and the probability of acceptance is β for lots with fraction defective p_2. Assuming that binomial sampling (with type-B OC curves) is appropriate, we see that the sample size n and acceptance number c are the solution to

$$
1 - \alpha = \sum_{d=0}^{c} \frac{n!}{d!(n-d)!} p_1^d (1 - p_1)^{n-d}
$$

$$
\beta = \sum_{d=0}^{c} \frac{n!}{d!(n-d)!} p_2^d (1 - p_2)^{n-d}
$$

(14-3)

Equation 14-3 was obtained by writing out the two points on the OC curve using the binomial distribution. The two simultaneous equations in equation 14-3 are nonlinear, and there is no simple, direct solution.

The nomograph in Fig. 14-9 can be used for solving these equations. The procedure for using the nomograph is very simple. Two lines are drawn on the nomograph, one connecting p_1 and $1 - \alpha$, and the other connecting p_2 and β. The intersection of these two lines gives the region of the nomograph in which the desired sampling plan is located. To illustrate the use of the nomograph, suppose we wish to construct a sampling plan for which $p_1 = 0.01$, $\alpha = 0.05$, $p_2 = 0.06$, and $\beta = 0.10$. Locating the intersection of the lines connecting ($p_1 = 0.01$, $1 - \alpha = 0.95$) and ($p_2 = 0.06$, $\beta = 0.10$) on the nomograph indicates that the plan $n = 89$, $c = 2$ is very close to passing through these two points on the OC curve. Obviously, since n and c must be integers, this procedure will actually produce several plans that have OC curves that pass close to the desired points. For instance, if the first line is followed either to the c-line just above the intersection point or to the c-line just below it, and the alternate sample sizes are read from the chart, this will produce two plans that pass almost exactly through the p_1, $1 - \alpha$ point, but may deviate somewhat from the p_2, β point. A similar procedure could be followed with the p_2, β-line. The result of following both of these lines would be four plans that pass approximately through the two points specified on the OC curve.

In addition to the graphical procedure that we have described for designing sampling plans with specified OC curves, tabular procedures are also available for the same purpose. Duncan (1986) gives a good description of these techniques.

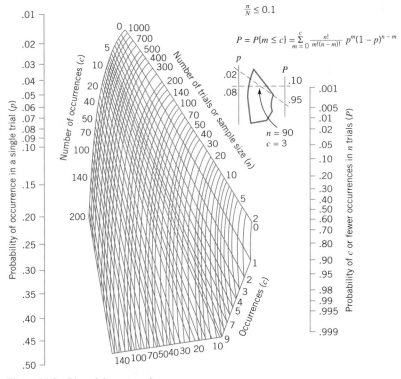

Figure 14-9 Binomial nomograph.

Although any two points on the OC curve could be used to define the sampling plan, it is customary in many industries to use the AQL and LTPD points for this purpose. When the levels of lot quality specified are $p_1 = $ AQL and $p_2 = $ LTPD, the corresponding points on the OC curve are usually referred to as the producer's risk point and the consumer's risk point, respectively. Thus, α would be called the producer's risk and β would be called the consumer's risk.

14-2.4 Rectifying Inspection

Acceptance-sampling programs usually require corrective action when lots are rejected. This generally takes the form of 100% inspection or **screening** of rejected lots, with all discovered defective items either removed for subsequent rework or return to the vendor, or replaced from a stock of known good items. Such sampling programs are called **rectifying inspection programs,** because the inspection activity affects the final quality of the outgoing product. This is illustrated in Fig. 14-10. Suppose that incoming lots to the inspection activity have fraction defective p_0. Some of these lots will be accepted, and others will be rejected. The rejected lots will be screened, and their final fraction defective will be zero. However, accepted lots have fraction defective p_0. Consequently, the outgoing lots from the inspection activity are a mixture of lots with fraction defective p_0 and fraction defective

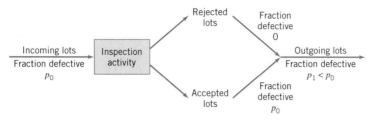

Figure 14-10 Rectifying inspection.

zero, so the average fraction defective in the stream of outgoing lots is p_1, which is less than p_0. Thus, a rectifying inspection program serves to "correct" lot quality.

Rectifying inspection programs are used in situations where the manufacturer wishes to know the average level of quality that is likely to result at a given stage of the manufacturing operations. Thus, rectifying inspection programs are used either at receiving inspection, in-process inspection of semifinished products, or at final inspection of finished goods. The objective of in-plant usage is to give assurance regarding the average quality of material used in the next stage of the manufacturing operations.

Rejected lots may be handled in a number of ways. The best approach is to return rejected lots to the vendor, and require it to perform the screening and rework activities. This has the psychological effect of making the vendor responsible for poor quality and may exert pressure on the vendor to improve its manufacturing processes or to install better process controls. However, in many situations, because the components or raw materials are required in order to meet production schedules, screening and rework take place at the consumer level. This is not the most desirable situation.

Average outgoing quality is widely used for the evaluation of a rectifying sampling plan. The average outgoing quality is the quality in the lot that results from the application of rectifying inspection. It is the average value of lot quality that would be obtained over a long sequence of lots from a process with fraction defective p. It is simple to develop a formula for average outgoing quality (AOQ). Assume that the lot size is N and that all defectives are replaced with good units. Then in lots of size N, we have

1. n items in the sample that, after inspection, contain no defectives, because all discovered defectives are replaced
2. $N - n$ items that, if the lot is rejected, also contain no defectives
3. $N - n$ items that, if the lot is accepted, contain $p(N - n)$ defectives

Thus, lots in the outgoing stage of inspection have an expected number of defective units equal to $P_a p(N - n)$, which we may express as an *average fraction defective,* called the **average outgoing quality** or

$$\text{AOQ} = \frac{P_a p (N - n)}{N} \qquad (14\text{-}4)$$

To illustrate the use of equation 14-4, suppose that $N = 10,000$, $n = 89$, and $c = 2$, and that the incoming lots are of quality $p = 0.01$. Now at $p = 0.01$, we have $P_a = 0.9397$, and the AOQ is

$$AOQ = \frac{P_a p(N - n)}{N}$$

$$= \frac{(0.9397)(0.01)(10,000 - 89)}{10,000}$$

$$= 0.0093$$

That is, the average outgoing quality is 0.93% defective. Note that as the lot size N becomes large relative to the sample size n, we may write equation 14-4 as

$$AOQ \simeq P_a p \qquad (14\text{-}5)$$

Average outgoing quality will vary as the fraction defective of the incoming lots varies. The curve that plots average outgoing quality against incoming lot quality is called an *AOQ curve*. The AOQ curve for the sampling plan $n = 89$, $c = 2$ is shown in Fig. 14-11. From examining this curve we note that when the incoming quality is very good, the average outgoing quality is also very good. In contrast, when the incoming lot

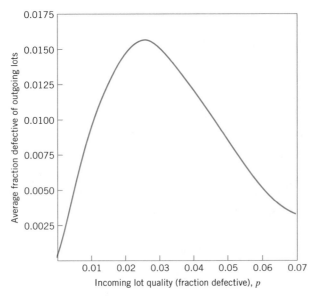

Figure 14-11   Average outgoing quality curve for $n = 89$, $c = 2$.

quality is very bad, most of the lots are rejected and screened, which leads to a very good level of quality in the outgoing lots. In between these extremes, the AOQ curve rises, passes through a maximum, and descends. The maximum ordinate on the AOQ curve represents the worst possible average quality that would result from the rectifying inspection program, and this point is called the **average outgoing quality limit (AOQL).** From examining Fig. 14-11, the AOQL is seen to be approximately 0.0155. That is, no matter how bad the fraction defective is in the incoming lots, the outgoing lots will never have a worse quality level on the average than 1.55% defective. Let us emphasize that this AOQL is an *average level of quality, across a large stream of lots.* It does not give assurance that an isolated lot will have quality no worse than 1.55% defective.

Another important measure relative to rectifying inspection is the total amount of inspection required by the sampling program. If the lots contain no defective items, no lots will be rejected, and the amount of inspection per lot will be the sample size n. If the items are all defective, every lot will be submitted to 100% inspection, and the amount of inspection per lot will be the lot size N. If the lot quality is $0 < p < 1$, the average amount of inspection per lot will vary between the sample size n and the lot size N. If the lot is of quality p and the probability of lot acceptance is P_a, then the **average total inspection** per lot will be

$$ATI = n + (1 - P_a)(N - n) \qquad (14\text{-}6)$$

To illustrate the use of equation 14-6, consider our previous example with $N = 10{,}000$, $n = 89$, $c = 2$, and $p = 0.01$. Then, since $P_a = 0.9397$, we have

$$ATI = n + (1 - P_a)(N - n)$$
$$= 89 + (1 - 0.9397)(10{,}000 - 89)$$
$$= 687$$

Remember that this is an average number of units inspected over *many* lots with fraction defective $p = 0.01$.

It is possible to draw a curve of average total inspection as a function of lot quality. Average total inspection curves for the sampling plan $n = 89$, $c = 2$, for lot sizes of 1000, 5000, and 10,000, are shown in Fig. 14-12.

The AOQL of a rectifying inspection plan is a very important characteristic. It is possible to design rectifying inspection programs that have specified values of AOQL. However, specification of the AOQL is not sufficient to determine a unique sampling plan. Therefore, it is relatively common practice to choose the sampling plan that has a specified AOQL and, in addition, yields a minimum ATI at a particular level of lot quality. The level of lot quality usually chosen is the most likely level of incoming

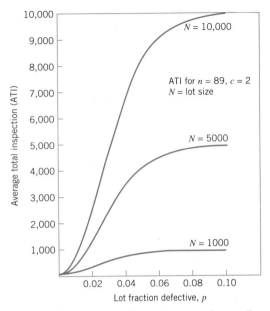

Figure 14-12   Average total inspection curves for sampling plan $n = 89$, $c = 2$, for lot sizes of 1000, 5000, and 10,000.

lot quality, which is generally called the *process average.* The procedure for generating these plans is relatively straightforward and is illustrated in Duncan (1986). Generally, it is unnecessary to go through this procedure, because tables of sampling plans that minimize ATI for a given AOQL and a specified process average p have been developed by Dodge and Romig. We describe the use of these tables in Section 14-5.

It is also possible to design a rectifying inspection program that gives a specified level of protection at the LTPD point and that minimizes the average total inspection for a specified process average p. The Dodge–Romig sampling inspection tables also provide these LTPD plans. Section 14-5 discusses the use of the Dodge–Romig tables to find plans that offer specified LTPD protection.

14-3   DOUBLE, MULTIPLE, AND SEQUENTIAL SAMPLING

A number of extensions of single-sampling plans for attributes are useful. These include **double-sampling plans, multiple-sampling plans,** and **sequential-sampling plans.** This section discusses the design and application of these sampling plans.

14-3.1   Double-Sampling Plans

A double-sampling plan is a procedure in which, under certain circumstances, a second sample is required before the lot can be sentenced. A double-sampling plan is defined by

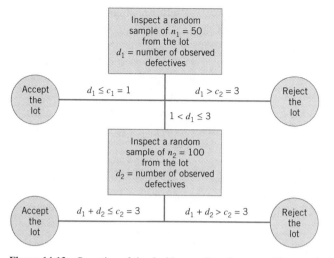

Figure 14-13   Operation of the double-sampling plan, $n_1 = 50$, $c_1 = 1$, $n_2 = 100$, $c_2 = 3$.

four parameters:[2]

$$n_1 = \text{sample size on the first sample}$$
$$c_1 = \text{acceptance number of the first sample}$$
$$n_2 = \text{sample size on the second sample}$$
$$c_2 = \text{acceptance number for both samples}$$

As an example, suppose $n_1 = 50$, $c_1 = 1$, $n_2 = 100$, and $c_2 = 3$. Thus, a random sample of $n_1 = 50$ items is selected from the lot, and the number of defectives in the sample, d_1, is observed. If $d_1 \leq c_1 = 1$, the lot is accepted on the first sample. If $d_1 > c_2 = 3$, the lot is rejected on the first sample. If $c_1 < d_1 \leq c_2$, a second random sample of size $n_2 = 100$ is drawn from the lot, and the number of defectives in this second sample, d_2, is observed. Now the combined number of observed defectives from both the first and second sample, $d_1 + d_2$, is used to determine the lot sentence. If $d_1 + d_2 \leq c_2 = 3$, the lot is accepted. However, if $d_1 + d_2 > c_2 = 3$, the lot is rejected. The operation of this double-sampling plan is illustrated graphically in Fig. 14-13.

The principal advantage of a double-sampling plan with respect to single sampling is that it may reduce the total amount of required inspection. Suppose that the first sample taken under a double-sampling plan is smaller than the sample that would be required using a single-sampling plan that offers the consumer the same protection. In all cases, then, in which a lot is accepted or rejected on the first sample, the cost of inspection will be lower for double sampling than it would be for single sampling. It is also possible to reject

---

[2] Some authors prefer the notation n_1, Ac_1, Re_1, n_2, Ac_2, $Re_2 = Ac_2 + 1$. Since the rejection number on the first sample Re_1 is not necessarily equal to Re_2, this gives some additional flexibility in designing double-sampling plans. MIL STD 105E and ANSI/ASQC Z1.4 currently use this notation. However, because assuming that $Re_1 = Re_2$ does not significantly affect the plans obtained, we have chosen to discuss this slightly simpler system.

a lot without complete inspection of the second sample. (This is called *curtailment* on the second sample.) Consequently, the use of double sampling can often result in lower total inspection costs. Furthermore, in some situations, a double-sampling plan has the psychological advantage of giving a lot a second chance. This may have some appeal to the vendor. However, there is no real advantage to double sampling in this regard, because single- and double-sampling plans can be chosen so that they have the same OC curves. Thus, both plans would offer the same risks of accepting or rejecting lots of specified quality.

Double sampling has two potential disadvantages. First, unless curtailment is used on the second sample, under some circumstances double sampling may require more total inspection than would be required in a single-sampling plan that offers the same protection. Thus, unless double sampling is used carefully, its potential economic advantage may be lost. The second disadvantage of double sampling is that it is administratively more complex, which may increase the opportunity for the occurrence of inspection errors. Furthermore, there may be problems in storing and handling raw materials or component parts for which one sample has been taken, but that are awaiting a second sample before a final lot dispositioning decision can be made.

The OC Curve

The performance of a double-sampling plan can be conveniently summarized by means of its operating-characteristic (OC) curve. The OC curve for a double-sampling plan is somewhat more involved than the OC curve for single sampling. In this section, we describe the construction of type-B OC curves for double sampling. A double-sampling plan has a primary OC curve that gives the probability of acceptance as a function of lot or process quality. It also has supplementary OC curves that show the probability of lot acceptance and rejection on the first sample. The OC curve for the probability of rejection on the first sample is simply the OC curve for the single-sampling plan $n = n_1$ and $c = c_2$. Primary and supplementary OC curves for the plan $n_1 = 50$, $c_1 = 1$, $n_2 = 100$, $c_2 = 3$ are shown in Fig. 14-14.

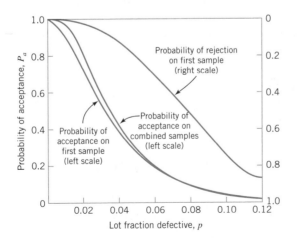

Figure 14-14 OC curves for the double-sampling plan, $n_1 = 50$, $c_1 = 1$, $n_2 = 100$, $c_2 = 3$.

We now illustrate the computation of the OC curve for the plan $n_1 = 50$, $c_1 = 1$, $n_2 = 100$, $c_2 = 3$. If P_a denotes the probability of acceptance on the combined samples, and P_a^{I} and P_a^{II} denote the probability of acceptance on the first and second samples, respectively, then

$$P_a = P_a^{\mathrm{I}} + P_a^{\mathrm{II}}$$

P_a^{I} is just the probability that we will observe $d_1 \le c_1 = 1$ defectives out of a random sample of $n_1 = 50$ items. Thus

$$P_a^{\mathrm{I}} = \sum_{d_1=0}^{1} \frac{50!}{d_1!(50 - d_1)!} p^{d_1}(1 - p)^{50 - d_1}$$

If $p = 0.05$ is the fraction defective in the incoming lot, then

$$P_a^{\mathrm{I}} = \sum_{d_1=0}^{1} \frac{50!}{d_1!(50 - d_1)!} (0.05)^{d_1}(0.95)^{50 - d_1} = 0.279$$

To obtain the probability of acceptance on the second sample, we must list the number of ways the second sample can be obtained. A second sample is drawn *only* if there are two or three defectives on the first sample — that is, if $c_1 < d_1 \le c_2$.

1. $d_1 = 2$ *and* $d_2 = 0$ or 1; that is, we find two defectives on the first sample and one or less defectives on the second sample. The probability of this is

$$P\{d_1 = 2, d_2 \le 1\} = P\{d_1 = 2\} \cdot P\{d_2 \le 1\}$$

$$= \frac{50!}{2!48!} (0.05)^2(0.95)^{48}$$

$$\times \sum_{d_2=0}^{1} \frac{100!}{d_2!(100 - d_2)!} (0.05)^{d_2}(0.95)^{100 - d_2}$$

$$= (0.261)(0.037)$$

$$= 0.009$$

2. $d_1 = 3$ *and* $d_2 = 0$; that is, we find three defectives on the first sample and no defectives on the second sample. The probability of this is

$$P\{d_1 = 3, d_2 = 0\} = P\{d_1 = 3\} \cdot P\{d_2 = 0\}$$

$$= \frac{50!}{3!(47)!} (0.05)^3(0.95)^{47} \frac{100!}{0!100!} (0.05)^0(0.95)^{100}$$

$$= (0.220)(0.0059)$$

$$= 0.001$$

Thus, the probability of acceptance on the second sample is

$$P_a^{\mathrm{II}} = P\{d_1 = 2, d_2 \le 1\} + P\{d_1 = 3, d_2 = 0\}$$

$$= 0.009 + 0.001$$

$$= 0.010$$

The probability of acceptance of a lot that has fraction defective $p = 0.05$ is therefore

$$P_a = P_a^{\mathrm{I}} + P_a^{\mathrm{II}}$$
$$= 0.279 + 0.010$$
$$= 0.289$$

Other points on the OC curve are calculated similarly.

The Average Sample Number Curve

The average sample number curve of a double-sampling plan is also usually of interest to the quality engineer. In single sampling, the size of the sample inspected from the lot is always constant, whereas in double sampling, the size of the sample selected depends on whether or not the second sample is necessary. The probability of drawing a second sample varies with the fraction defective in the incoming lot. With complete inspection of the second sample, the average sample size in double sampling is equal to the size of the first sample times the probability that there will only be one sample, plus the size of the combined samples times the probability that a second sample will be necessary. Therefore, a general formula for the average sample number in double sampling, if we assume complete inspection of the second sample, is

$$\mathrm{ASN} = n_1 P_{\mathrm{I}} + (n_1 + n_2)(1 - P_{\mathrm{I}})$$
$$= n_1 + n_2(1 - P_{\mathrm{I}}) \tag{14-7}$$

where P_{I} is the probability of making a lot-dispositioning decision on the *first* sample. This is

$P_{\mathrm{I}} = P\{\text{lot is accepted on the first sample}\} + P\{\text{lot is rejected on the first sample}\}$

If equation 14-7 is evaluated for various values of lot fraction defective p, the plot of ASN versus p is called an **average sample number curve.**

In practice, inspection of the second sample is usually terminated and the lot rejected as soon as the number of observed defective items in the combined sample exceeds the second acceptance number c_2. This is referred to as curtailment of the second sample. The use of curtailed inspection lowers the average sample number required in double sampling. It is not recommended that curtailment be used in single sampling, or in the first sample of double sampling, because it is usually desirable to have complete inspection of a fixed sample size in order to secure an unbiased estimate of the quality of the material supplied by the vendor. If curtailed inspection is used in single sampling or on the first sample of double sampling, the estimate of lot or process fallout obtained from these data is biased. For instance, suppose that the acceptance number is one. If the first two items in the sample are defective, and the inspection process is curtailed, the estimate of lot or process fraction defective is 100%. Based on this information, even nonstatistically trained managers or engineers will be very reluctant to believe that the lot is really 100% defective.

The ASN curve formula for a double-sampling plan with curtailment on the second sample is

$$
\text{ASN} = n_1 + \sum_{j=c_1+1}^{c_2} P(n_1, j) \left[n_2 P_L(n_2, c_2 - j) \right.
$$
$$
\left. + \frac{c_2 - j + 1}{p} P_M(n_2 + 1, c_2 - j + 2) \right] \qquad (14\text{-}8)
$$

In equation 14-8, $P(n_1, j)$ is the probability of observing exactly j defectives in a sample of size n_1, $P_L(n_2, c_2 - j)$ is the probability of observing $c_2 - j$ or fewer defectives in a sample of size n_2, and $P_M(n_2 + 1, c_2 - j + 2)$ is the probability of observing $c_2 - j + 2$ defectives in a sample of size $n_2 + 1$.

Figure 14-15 compares the average sample number curves for complete and curtailed inspection for the double-sampling plan $n_1 = 60$, $c_1 = 2$, $n_2 = 120$, $c_2 = 3$, and the average sample number that would be used in single sampling with $n = 89$, $c = 2$. Obviously, the sample size in the single-sampling plan is always constant. This double-sampling plan has been selected because it has an OC curve that is nearly identical to the OC curve for the single-sampling plan. That is, both plans offer equivalent protection to the producer and the

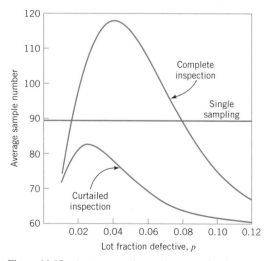

Figure 14-15   Average sample number curves for single and double sampling.

consumer. Note from inspection of Fig. 14-15 that the ASN curve for double sampling without curtailment on the second sample is not lower than the sample size used in single sampling throughout the entire range of lot fraction defective. If lots are of very good quality, they will usually be accepted on the first sample, whereas if lots are of very bad quality, they will usually be rejected on the first sample. This gives an ASN for double sampling that is smaller than the sample size used in single sampling for lots that are either very good or very bad. However, if lots are of intermediate quality, the second sample will be required in a large number of cases before a lot disposition decision can be made. In this range of lot quality, the ASN performance of double sampling is worse than single sampling.

This example points out that it is important to use double sampling very carefully. Unless care is exercised to ensure that lot or process quality is in the range where double sampling is most effective, then the economic advantages of double sampling relative to single sampling may be lost. It is a good idea to maintain a running estimate of the vendor's lot or process fallout, so that if it shifts into a range where double sampling is not economically effective, a change to single sampling (or some other appropriate strategy) can be made. Another way to do this would be to record the proportion of times that the second sample is required in order to make a decision.

Figure 14-15 also shows the ASN curve using curtailment on the second sample. Note that if curtailment is used, the average sample number curve for double sampling always lies below the sample size used in single sampling.

Designing Double-Sampling Plans with Specified p_1, $1 - \alpha$, p_2, and β

It is often necessary to be able to design a double-sampling plan that has a specified OC curve. Let $(p_1, 1 - \alpha)$ and (p_2, β) be the two points of interest on the OC curve. If, in addition, we impose another relationship on the parameters of the sampling plan, then a simple procedure can be used to obtain such plans. The most common constraint is to require that n_2 is a multiple of n_1. Refer to Duncan (1986) for a discussion of these techniques.

Rectifying Inspection

When rectifying inspection is performed with double sampling, the AOQ curve is given by

$$AOQ = \frac{[P_a^I(N - n_1) + P_a^{II}(N - n_1 - n_2)]p}{N} \qquad (14\text{-}9)$$

assuming that all defective items discovered, either in sampling or 100% inspection, are replaced with good ones. The average total inspection curve is given by

$$ATI = n_1 P_a^I + (n_1 + n_2)P_a^{II} + N(1 - P_a) \qquad (14\text{-}10)$$

Remember that $P_a = P_a^I + P_a^{II}$ is the probability of final lot acceptance and that the acceptance probabilities depend on the level of lot or process quality p.

14-3.2 Multiple-Sampling Plans

A multiple-sampling plan is an extension of double sampling in that more than two samples can be required to sentence a lot. An example of a multiple-sampling plan with five stages follows.

Cumulative-Sample Size	Acceptance Number	Rejection Number
20	0	3
40	1	4
60	3	5
80	5	7
100	8	9

This plan will operate as follows: If, at the completion of any stage of sampling, the number of defective items is less than or equal to the acceptance number, the lot is accepted. If, during any stage, the number of defective items equals or exceeds the rejection number, the lot is rejected; otherwise the next sample is taken. The multiple-sampling procedure continues until the fifth sample is taken, at which time a lot disposition decision must be made. The first sample is usually inspected 100%, although subsequent samples are usually subject to curtailment.

The construction of OC curves for multiple sampling is a straightforward extension of the approach used in double sampling. Similarly, it is also possible to compute the average sample number curve of multiple-sampling plans. One may also design a multiple-sampling plan for specified values of p_1, $1 - \alpha$, p_2, and β. For an extensive discussion of these techniques, see Duncan (1986).

The principal advantage of multiple-sampling plans is that the samples required at each stage are usually smaller than those in single or double sampling; thus, some economic efficiency is connected with the use of the procedure. However, multiple sampling is much more complex to administer.

14-3.3 Sequential-Sampling Plans

Sequential sampling is an extension of the double-sampling and multiple-sampling concept. In sequential sampling, we take a sequence of samples from the lot and allow the number of samples to be determined entirely by the results of the sampling process. In practice, sequential sampling can theoretically continue indefinitely, until the lot is inspected 100%. In practice, sequential-sampling plans are usually truncated after the number inspected is equal to three times the number that would have been inspected using a

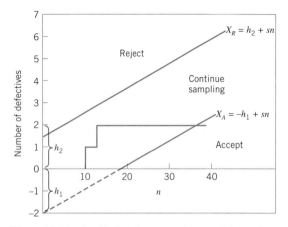

Figure 14-16   Graphical performance of sequential sampling.

corresponding single-sampling plan. If the sample size selected at each stage is greater than one, the process is usually called *group* sequential sampling. If the sample size inspected at each stage is one, the procedure is usually called **item-by-item sequential sampling.**

Item-by-item sequential sampling is based on the sequential probability ratio test (SPRT), developed by Wald (1947). The operation of an item-by-item sequential-sampling plan is illustrated in Fig. 14-16. The cumulative observed number of defectives is plotted on the chart. For each point, the abscissa is the total number of items selected up to that time, and the ordinate is the total number of observed defectives. If the plotted points stay within the boundaries of the acceptance and rejection lines, another sample must be drawn. As soon as a point falls on or above the upper line, the lot is rejected. When a sample point falls on or below the lower line, the lot is accepted. The equations for the two limit lines for specified values of $p_1, 1 - \alpha, p_2,$ and β are

$$X_A = -h_1 + sn \qquad \text{(acceptance line)} \qquad (14\text{-}11\text{a})$$

$$X_R = h_2 + sn \qquad \text{(rejection line)} \qquad (14\text{-}11\text{b})$$

where

$$h_1 = \left(\log \frac{1 - \alpha}{\beta}\right)k \qquad (14\text{-}12)$$

$$h_2 = \left(\log \frac{1 - \beta}{\alpha}\right)k \qquad (14\text{-}13)$$

$$k = \log \frac{p_2(1 - p_1)}{p_1(1 - p_2)} \qquad (14\text{-}14)$$

$$s = \log[(1 - p_1)/(1 - p_2)]/k \qquad (14\text{-}15)$$

To illustrate the use of these equations, suppose we wish to find a sequential-sampling plan for which $p_1 = 0.01$, $\alpha = 0.05$, $p_2 = 0.06$, and $\beta = 0.10$. Thus,

$$k = \log \frac{p_2(1 - p_1)}{p_1(1 - p_2)}$$

$$= \log \frac{(0.06)(0.99)}{(0.01)(0.94)}$$

$$= 0.80066$$

$$h_1 = \left(\log \frac{1 - \alpha}{\beta}\right) \Big/ k$$

$$= \left(\log \frac{0.95}{0.10}\right) \Big/ 0.80066$$

$$= 1.22$$

$$h_2 = \left(\log \frac{1 - \beta}{\alpha}\right) k$$

$$= \left(\log \frac{0.90}{0.05}\right) 0.80066$$

$$= 1.57$$

$$s = \log[(1 - p_1)/(1 - p_2)]/k$$

$$= (\log[0.99/0.94])/0.80066$$

$$= 0.028$$

Therefore, the limit lines are

$$X_A = -1.22 + 0.028n \qquad \text{(accept)}$$

and

$$X_R = 1.57 + 0.028n \qquad \text{(reject)}$$

Instead of using a graph to determine the lot disposition, the sequential-sampling plan can be displayed in a table such as Table 14-3. The entries in the table are found by substituting values of n into the equations for the acceptance and rejection lines and calculating acceptance and rejection numbers. For example, the calculations for $n = 45$ are

$$X_A = -1.22 + 0.028n$$
$$= -1.22 + 0.028(45) = 0.04 \qquad \text{(accept)}$$
$$X_R = 1.57 + 0.028n$$
$$= 1.57 + 0.028(45) = 2.83 \qquad \text{(reject)}$$

Acceptance and rejection numbers must be integers, so the acceptance number is the next integer less than or equal to X_A, and the rejection number is the next integer greater

Table 14-3 Item-by-Item Sequential-Sampling Plan $p_1 = 0.01$, $\alpha = 0.05$, $p_2 = 0.06$, $\beta = 0.10$ (first 46 units only)

Number of Items Inspected, n	Acceptance Number	Rejection Number	Number of Items Inspected, n	Acceptance Number	Rejection Number
1	a	b	24	a	3
2	a	2	25	a	3
3	a	2	26	a	3
4	a	2	27	a	3
5	a	2	28	a	3
6	a	2	29	a	3
7	a	2	30	a	3
8	a	2	31	a	3
9	a	2	32	a	3
10	a	2	33	a	3
11	a	2	34	a	3
12	a	2	35	a	3
13	a	2	36	a	3
14	a	2	37	a	3
15	a	2	38	a	3
16	a	3	39	a	3
17	a	3	40	a	3
18	a	3	41	a	3
19	a	3	42	a	3
20	a	3	43	a	3
21	a	3	44	0	3
22	a	3	45	0	3
23	a	3	46	0	3

"a" means acceptance not possible.

"b" means rejection not possible.

than or equal to X_R. Thus, for $n = 45$, the acceptance number is 0 and the rejection number is 3. Note that the lot cannot be accepted until at least 44 units have been tested. Table 14-3 shows only the first 46 units. Normally, the plan would be truncated after the inspection of 267 units, which is three times the sample size required for an equivalent single-sampling plan.

The OC Curve and ASN Curve for Sequential Sampling
The OC curve for sequential sampling can be easily obtained. Two points on the curve are $(p_1, 1 - \alpha)$ and (p_2, β). A third point, near the middle of the curve, is $p = s$ and $p_a = h_2/(h_1 + h_2)$.

The average sample number taken under sequential sampling is

$$\text{ASN} = P_a\left(\frac{A}{C}\right) + (1 - P_a)\frac{B}{C} \tag{14-16}$$

where

$$A = \log \frac{\beta}{1 - \alpha}$$

$$B = \log \frac{1 - \beta}{\alpha}$$

and

$$C = p \log\left(\frac{p_2}{p_1}\right) + (1 - p) \log\left(\frac{1 - p_2}{1 - p_1}\right)$$

Rectifying Inspection

The average outgoing quality (AOQ) for sequential sampling is given approximately by

$$\text{AOQ} \simeq P_a p \tag{14-17}$$

The average total inspection is also easily obtained. Note that the amount of sampling is A/C when a lot is accepted and N when it is rejected. Therefore, the average total inspection is

$$\text{ATI} = P_a\left(\frac{A}{C}\right) + (1 - P_a)N \tag{14-18}$$

14-4 MILITARY STANDARD 105E (ANSI/ASQC Z1.4, ISO 2859)

14-4.1 Description of the Standard

Standard sampling procedures for inspection by attributes were developed during World War II. MIL STD 105E is the most widely used acceptance-sampling system for attributes in the world today. The original version of the standard, MIL STD 105A, was issued in 1950. Since then, there have been four revisions; the latest version, MIL STD 105E, was issued in 1989.

The sampling plans discussed in previous sections of this chapter are individual sampling plans. A sampling scheme is an overall strategy specifying the way in which sampling plans are to be used. MIL STD 105E is a collection of sampling schemes; therefore, it is an acceptance-sampling system. Our discussion will focus primarily on MIL STD 105E; however, there is a derivative civilian standard, ANSI/ASQC Z1.4, which is quite similar to the military standard. The standard was also adopted by the International Organization for Standardization as ISO 2859.

The standard provides for three types of sampling: single sampling, double sampling, and multiple sampling. For each type of sampling plan, a provision is made for either normal inspection, tightened inspection, or reduced inspection. Normal inspection is used at the start of the inspection activity. Tightened inspection is instituted when the vendor's recent quality history has deteriorated. Acceptance requirements for lots under tightened inspection are more stringent than under normal inspection. Reduced inspection is instituted when the vendor's recent quality history has been exceptionally good. The sample size generally used under reduced inspection is less than that under normal inspection.

The primary focal point of MIL STD 105E is the acceptable quality level (AQL). The standard is indexed with respect to a series of AQLs. When the standard is used for percent defective plans, the AQLs range from 0.10% to 10%. For defects per units plans, there are an additional 10 AQLs running up to 1000 defects per 100 units. It should be noted that for the smaller AQL levels, the same sampling plan can be used to control either a fraction defective or a number of defects per unit. The AQLs are arranged in a progression, each AQL being approximately 1.585 times the preceding one.

The AQL is generally specified in the contract or by the authority responsible for sampling. Different AQLs may be designated for different types of defects. For example, the standard differentiates critical defects, major defects, and minor defects. It is relatively common practice to choose an AQL of 1% for major defects and an AQL of 2.5% for minor defects. No critical defects would be acceptable.

The sample size used in MIL STD 105E is determined by the lot size and by the choice of inspection level. Three general levels of inspection are provided. Level II is designated as normal. Level I requires about one-half the amount of inspection as Level II and may be used when less discrimination is needed. Level III requires about twice as much inspection as Level II and should be used when more discrimination is needed. There are also four special inspection levels, S-1, S-2, S-3, and S-4. The special inspection levels use very small samples, and should only be employed when the small sample sizes are necessary and when large sampling risks can or must be tolerated.

For a specified AQL and inspection level and a given lot size, MIL STD 105E provides a normal sampling plan that is to be used as long as the supplier is producing the product at AQL quality or better. It also provides a procedure for switching to tightened and reduced inspection whenever there is an indication that the vendor's quality has changed. The switching procedures between normal, tightened, and reduced inspection are illustrated graphically in Fig. 14-17 and are described next.

1. **Normal to tightened.** When normal inspection is in effect, tightened inspection is instituted when two out of five consecutive lots have been rejected on original submission.

2. **Tightened to normal.** When tightened inspection is in effect, normal inspection is instituted when five consecutive lots or batches are accepted on original inspection.

3. **Normal to reduced.** When normal inspection is in effect, reduced inspection is instituted provided all four of the following conditions are satisfied.

   a. The preceding 10 lots have been on normal inspection, and none of the lots has been rejected on original inspection.

   b. The total number of defectives in the samples from the preceding 10 lots is less than or equal to the applicable limit number specified in the standard.

   c. Production is at a steady rate; that is, no difficulty such as machine breakdowns, material shortages, or other problems have recently occurred.

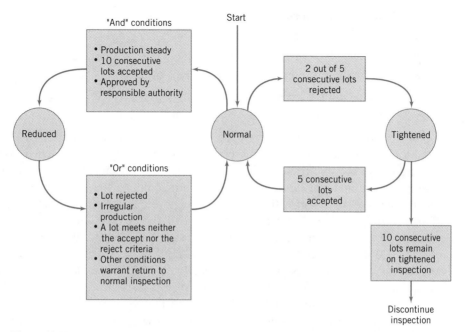

Figure 14-17   Switching rules for normal, tightened, and reduced inspection, MIL STD 105E.

    d. Reduced inspection is considered desirable by the authority responsible for sampling.

4. Reduced to normal.   When reduced inspection is in effect, normal inspection is instituted provided any of the following four conditions are satisfied.

    a. A lot or batch is rejected.

    b. When the sampling procedure terminates with neither acceptance nor rejection criteria having been met, the lot or batch is accepted, but normal inspection is reinstituted starting with the next lot.

    c. Production is irregular or delayed.

    d. Other conditions warrant that normal inspection be instituted.

5. Discontinuance of inspection.   In the event that 10 consecutive lots remain on tightened inspection, inspection under the provision of MIL STD 105E should be terminated, and action should be taken at the vendor level to improve the quality of submitted lots.

14-4.2   Procedure

A step-by-step procedure for using MIL STD 105E is as follows:

1. Choose the AQL.

2. Choose the inspection level.

3. Determine the lot size.
4. Find the appropriate sample size code letter from Table 14-4.
5. Determine the appropriate type of sampling plan to use (single, double, multiple).
6. Enter the appropriate table to find the type of plan to be used.
7. Determine the corresponding normal and reduced inspection plans to be used when required.

Table 14-4 presents the sample size code letters for MIL STD 105E. Tables 14-5, 14-6, and 14-7 present the single-sampling plans for normal inspection, tightened inspection, and reduced inspection, respectively. The standard also contains tables for double-sampling plans and multiple-sampling plans for normal, tightened, and reduced inspection.

To illustrate the use of MIL STD 105E, suppose that a product is submitted in lots of size $N = 2000$. The acceptable quality level is 0.65%. We will use the standard to generate normal, tightened, and reduced single-sampling plans for this situation. For lots of size 2000 under general inspection level II, Table 14-4 indicates that the appropriate sample size code letter is K. Therefore, from Table 14-5, for single-sampling plans under normal inspection, the normal inspection plan is $n = 125$, $c = 2$. Table 14-6 indicates that the corresponding tightened inspection plan is $n = 125$, $c = 1$. Note that in switching from normal to tightened inspection, the sample size remains the same, but the acceptance number is reduced by one. This general strategy is used throughout MIL STD 105E for a transition to tightened inspection. If the normal inspection acceptance

Table 14-4   Sample Size Code Letters (MIL STD 105E, Table 1)

Lot or Batch Size	Special Inspection Levels				General Inspection Levels		
	S-1	S-2	S-3	S-4	I	II	III
2 to 8	A	A	A	A	A	A	B
9 to 15	A	A	A	A	A	B	C
16 to 25	A	A	B	B	B	C	D
26 to 50	A	B	B	C	C	D	E
51 to 90	B	B	C	C	C	E	F
91 to 150	B	B	C	D	D	F	G
151 to 280	B	C	D	E	E	G	H
281 to 500	B	C	D	E	F	H	J
501 to 1200	C	C	E	F	G	J	K
1201 to 3200	C	D	E	G	H	K	L
3201 to 10000	C	D	F	G	J	L	M
10001 to 35000	C	D	F	H	K	M	N
35001 to 150000	D	E	G	J	L	N	P
150001 to 500000	D	E	G	J	M	P	Q
500001 and over	D	E	H	K	N	Q	R

Table 14-5 Master Table for Normal Inspection—Single Sampling (MIL STD 105E, Table II-A)

Acceptable Quality Levels (normal inspection)

Sample Size Code Letter	Sample Size	0.010		0.015		0.025		0.040		0.065		0.10		0.15		0.25		0.40		0.65		1.0		1.5		2.5		4.0		6.5		10		15		25		40		65		100		150		250		400		650		1000	
		Ac	Re	Ac	Re	Ac	Re	Ac	Re	Ac	Re	Ac	Re	Ac	Re	Ac	Re	Ac	Re	Ac	Re	Ac	Re	Ac	Re	Ac	Re	Ac	Re	Ac	Re	Ac	Re	Ac	Re	Ac	Re	Ac	Re	Ac	Re	Ac	Re	Ac	Re	Ac	Re	Ac	Re	Ac	Re	Ac	Re
A	2	↓		↓		↓		↓		↓		↓		↓		↓		↓		↓		↓		↓		↓		↓		↓		↓		0	1	1	2	2	3	3	4	5	6	7	8	10	11	14	15	21	22	30	31
B	3	↓		↓		↓		↓		↓		↓		↓		↓		↓		↓		↓		↓		↓		↓		↓		0	1	1	2	2	3	3	4	5	6	7	8	10	11	14	15	21	22	30	31	44	45
C	5	↓		↓		↓		↓		↓		↓		↓		↓		↓		↓		↓		↓		↓		↓		0	1	1	2	2	3	3	4	5	6	7	8	10	11	14	15	21	22	30	31	44	45	↑	
D	8	↓		↓		↓		↓		↓		↓		↓		↓		↓		↓		↓		↓		↓		0	1	1	2	2	3	3	4	5	6	7	8	10	11	14	15	21	22	30	31	44	45	↑		↑	
E	13	↓		↓		↓		↓		↓		↓		↓		↓		↓		↓		↓		↓		0	1	1	2	2	3	3	4	5	6	7	8	10	11	14	15	21	22	30	31	44	45	↑		↑		↑	
F	20	↓		↓		↓		↓		↓		↓		↓		↓		↓		↓		↓		0	1	1	2	2	3	3	4	5	6	7	8	10	11	14	15	21	22	30	31	44	45	↑		↑		↑		↑	
G	32	↓		↓		↓		↓		↓		↓		↓		↓		↓		↓		0	1	1	2	2	3	3	4	5	6	7	8	10	11	14	15	21	22	30	31	44	45	↑		↑		↑		↑		↑	
H	50	↓		↓		↓		↓		↓		↓		↓		↓		↓		0	1	1	2	2	3	3	4	5	6	7	8	10	11	14	15	21	22	30	31	44	45	↑		↑		↑		↑		↑		↑	
J	80	↓		↓		↓		↓		↓		↓		↓		↓		0	1	1	2	2	3	3	4	5	6	7	8	10	11	14	15	21	22	30	31	44	45	↑		↑		↑		↑		↑		↑		↑	
K	125	↓		↓		↓		↓		↓		↓		↓		0	1	1	2	2	3	3	4	5	6	7	8	10	11	14	15	21	22	30	31	44	45	↑		↑		↑		↑		↑		↑		↑		↑	
L	200	↓		↓		↓		↓		↓		↓		0	1	1	2	2	3	3	4	5	6	7	8	10	11	14	15	21	22	30	31	44	45	↑		↑		↑		↑		↑		↑		↑		↑		↑	
M	315	↓		↓		↓		↓		↓		0	1	1	2	2	3	3	4	5	6	7	8	10	11	14	15	21	22	30	31	44	45	↑		↑		↑		↑		↑		↑		↑		↑		↑		↑	
N	500	↓		↓		↓		↓		0	1	1	2	2	3	3	4	5	6	7	8	10	11	14	15	21	22	30	31	44	45	↑		↑		↑		↑		↑		↑		↑		↑		↑		↑		↑	
P	800	↓		↓		↓		0	1	1	2	2	3	3	4	5	6	7	8	10	11	14	15	21	22	30	31	44	45	↑		↑		↑		↑		↑		↑		↑		↑		↑		↑		↑		↑	
Q	1250	↓		↓		0	1	1	2	2	3	3	4	5	6	7	8	10	11	14	15	21	22	30	31	44	45	↑		↑		↑		↑		↑		↑		↑		↑		↑		↑		↑		↑		↑	
R	2000	↓		0	1	1	2	2	3	3	4	5	6	7	8	10	11	14	15	21	22	30	31	44	45	↑		↑		↑		↑		↑		↑		↑		↑		↑		↑		↑		↑		↑		↑	

↓ = Use first sampling plan below arrow. If sample size equals, or exceeds, lot or batch size, do 100% inspection.

↑ = Use first sampling plan above arrow.

Ac = Acceptance number.

Re = Rejection number.

709

Table 14-6   Master Table for Tightened Inspection—Single Sampling (MIL STD 105E, Table II-B)

Acceptable Quality Levels (tightened inspection)

Sample Size Code Letter	Sample Size	0.010		0.015		0.025		0.040		0.065		0.10		0.15		0.25		0.40		0.65		1.0		1.5		2.5		4.0		6.5		10		15		25		40		65		100		150		250		400		650		1000	
		Ac	Re	Ac	Re	Ac	Re	Ac	Re	Ac	Re	Ac	Re	Ac	Re	Ac	Re	Ac	Re	Ac	Re	Ac	Re	Ac	Re	Ac	Re	Ac	Re	Ac	Re	Ac	Re	Ac	Re	Ac	Re	Ac	Re	Ac	Re	Ac	Re	Ac	Re	Ac	Re	Ac	Re	Ac	Re	Ac	Re
A	2	↓		↓		↓		↓		↓		↓		↓		↓		↓		↓		↓		↓		↓		↓		↓		↓		↓		0	1	1	2	2	3	3	4	5	6	8	9	12	13	18	19	27	28
B	3	↓		↓		↓		↓		↓		↓		↓		↓		↓		↓		↓		↓		↓		↓		↓		↓		0	1	1	2	2	3	3	4	5	6	8	9	12	13	18	19	27	28	41	42
C	5	↓		↓		↓		↓		↓		↓		↓		↓		↓		↓		↓		↓		↓		↓		↓		0	1	1	2	2	3	3	4	5	6	8	9	12	13	18	19	27	28	41	42	↑	
D	8	↓		↓		↓		↓		↓		↓		↓		↓		↓		↓		↓		↓		↓		↓		0	1	1	2	2	3	3	4	5	6	8	9	12	13	18	19	27	28	41	42	↑		↑	
E	13	↓		↓		↓		↓		↓		↓		↓		↓		↓		↓		↓		↓		↓		0	1	1	2	2	3	3	4	5	6	8	9	12	13	18	19	27	28	41	42	↑		↑		↑	
F	20	↓		↓		↓		↓		↓		↓		↓		↓		↓		↓		↓		↓		0	1	1	2	2	3	3	4	5	6	8	9	12	13	18	19	27	28	41	42	↑		↑		↑		↑	
G	32	↓		↓		↓		↓		↓		↓		↓		↓		↓		↓		↓		0	1	1	2	2	3	3	4	5	6	8	9	12	13	18	19	27	28	41	42	↑		↑		↑		↑		↑	
H	50	↓		↓		↓		↓		↓		↓		↓		↓		↓		↓		0	1	1	2	2	3	3	4	5	6	8	9	12	13	18	19	27	28	41	42	↑		↑		↑		↑		↑		↑	
J	80	↓		↓		↓		↓		↓		↓		↓		↓		↓		0	1	1	2	2	3	3	4	5	6	8	9	12	13	18	19	27	28	41	42	↑		↑		↑		↑		↑		↑		↑	
K	125	↓		↓		↓		↓		↓		↓		↓		↓		0	1	1	2	2	3	3	4	5	6	8	9	12	13	18	19	27	28	41	42	↑		↑		↑		↑		↑		↑		↑		↑	
L	200	↓		↓		↓		↓		↓		↓		↓		0	1	1	2	2	3	3	4	5	6	8	9	12	13	18	19	27	28	41	42	↑		↑		↑		↑		↑		↑		↑		↑		↑	
M	315	↓		↓		↓		↓		↓		↓		0	1	1	2	2	3	3	4	5	6	8	9	12	13	18	19	27	28	41	42	↑		↑		↑		↑		↑		↑		↑		↑		↑		↑	
N	500	↓		↓		↓		↓		↓		0	1	1	2	2	3	3	4	5	6	8	9	12	13	18	19	27	28	41	42	↑		↑		↑		↑		↑		↑		↑		↑		↑		↑		↑	
P	800	↓		↓		↓		↓		0	1	1	2	2	3	3	4	5	6	8	9	12	13	18	19	27	28	41	42	↑		↑		↑		↑		↑		↑		↑		↑		↑		↑		↑		↑	
Q	1250	↓		↓		↓		0	1	1	2	2	3	3	4	5	6	8	9	12	13	18	19	27	28	41	42	↑		↑		↑		↑		↑		↑		↑		↑		↑		↑		↑		↑		↑	
R	2000	↓		↓		0	1	1	2	2	3	3	4	5	6	8	9	12	13	18	19	27	28	41	42	↑		↑		↑		↑		↑		↑		↑		↑		↑		↑		↑		↑		↑		↑	
S	3150	↓		0	1	1	2	2	3	3	4	5	6	8	9	12	13	18	19	27	28	41	42	↑		↑		↑		↑		↑		↑		↑		↑		↑		↑		↑		↑		↑		↑		↑	

⇩ = Use first sampling plan below arrow. If sample size equals, or exceeds, lot or batch size, do 100% inspection.

⇧ = Use first sampling plan above arrow.

Ac = Acceptance number.

Re = Rejection number.

710

Table 14-7 Master Table for Reduced Inspection—Single Sampling (MIL STD 105E, Table II-C)

Acceptable Quality Levels (reduced inspection)†

(Each data cell gives the pair "Ac Re"; ↓ = use first sampling plan below arrow, ↑ = use first sampling plan above arrow.)

Sample Size Code Letter	Sample Size	0.010	0.015	0.025	0.040	0.065	0.10	0.15	0.25	0.40	0.65	1.0	1.5	2.5	4.0	6.5	10	15	25	40	65	100	150	250	400	650	1000
A	2	↓	↓	↓	↓	↓	↓	↓	↓	↓	↓	↓	↓	↓	↓	↓	↓	0 1	1 2	2 3	3 4	5 6	7 8	10 11	14 15	21 22	30 31
B	2	↓	↓	↓	↓	↓	↓	↓	↓	↓	↓	↓	↓	↓	↓	↓	0 1	0 2	1 3	2 3	3 4	5 6	7 8	10 11	14 15	21 22	30 31
C	2	↓	↓	↓	↓	↓	↓	↓	↓	↓	↓	↓	↓	↓	↓	0 1	0 2	1 3	1 4	2 5	3 6	5 8	7 10	10 13	14 17	21 24	↑
D	3	↓	↓	↓	↓	↓	↓	↓	↓	↓	↓	↓	↓	↓	0 1	0 2	1 3	1 4	2 5	3 6	5 8	7 10	10 13	14 17	21 24	↑	↑
E	5	↓	↓	↓	↓	↓	↓	↓	↓	↓	↓	↓	↓	0 1	0 2	1 3	1 4	2 5	3 6	5 8	7 10	10 13	14 17	21 24	↑	↑	↑
F	8	↓	↓	↓	↓	↓	↓	↓	↓	↓	↓	↓	0 1	0 2	1 3	1 4	2 5	3 6	5 8	7 10	10 13	↑	↑	↑	↑	↑	↑
G	13	↓	↓	↓	↓	↓	↓	↓	↓	↓	↓	0 1	0 2	1 3	1 4	2 5	3 6	5 8	7 10	10 13	↑	↑	↑	↑	↑	↑	↑
H	20	↓	↓	↓	↓	↓	↓	↓	↓	↓	0 1	0 2	1 3	1 4	2 5	3 6	5 8	7 10	10 13	↑	↑	↑	↑	↑	↑	↑	↑
J	32	↓	↓	↓	↓	↓	↓	↓	↓	0 1	0 2	1 3	1 4	2 5	3 6	5 8	7 10	10 13	↑	↑	↑	↑	↑	↑	↑	↑	↑
K	50	↓	↓	↓	↓	↓	↓	↓	0 1	0 2	1 3	1 4	2 5	3 6	5 8	7 10	10 13	↑	↑	↑	↑	↑	↑	↑	↑	↑	↑
L	80	↓	↓	↓	↓	↓	↓	0 1	0 2	1 3	1 4	2 5	3 6	5 8	7 10	10 13	↑	↑	↑	↑	↑	↑	↑	↑	↑	↑	↑
M	125	↓	↓	↓	↓	↓	0 1	0 2	1 3	1 4	2 5	3 6	5 8	7 10	10 13	↑	↑	↑	↑	↑	↑	↑	↑	↑	↑	↑	↑
N	200	↓	↓	↓	↓	0 1	0 2	1 3	1 4	2 5	3 6	5 8	7 10	10 13	↑	↑	↑	↑	↑	↑	↑	↑	↑	↑	↑	↑	↑
P	315	↓	↓	↓	0 1	0 2	1 3	1 4	2 5	3 6	5 8	7 10	10 13	↑	↑	↑	↑	↑	↑	↑	↑	↑	↑	↑	↑	↑	↑
Q	500	↓	↓	0 1	0 2	1 3	1 4	2 5	3 6	5 8	7 10	10 13	↑	↑	↑	↑	↑	↑	↑	↑	↑	↑	↑	↑	↑	↑	↑
R	800	↓	0 1	0 2	1 3	1 4	2 5	3 6	5 8	7 10	10 13	↑	↑	↑	↑	↑	↑	↑	↑	↑	↑	↑	↑	↑	↑	↑	↑

⇩ = Use first sampling plan below arrow. If sample size equals, or exceeds, lot or batch size, do 100% inspection.

⇧ = Use first sampling plan above arrow.

Ac = Acceptance number.

Re = Rejection number.

† = If the acceptance number has been exceeded, but the rejection number has not been reached, accept the lot, but reinstate normal inspection.

number is 1, 2, or 3, the acceptance number for the corresponding tightened inspection plan is reduced by one. If the normal inspection acceptance number is 5, 7, 10, or 14, the reduction in acceptance number for tightened inspection is two. For a normal acceptance number of 21, the reduction is three. Table 14-7 indicates that under reduced inspection, the sample size for this example would be $n = 50$, the acceptance number would be $c = 1$, and the rejection number would be $r = 3$. Thus, if two defectives were encountered, the lot would be accepted, but the next lot would be inspected under normal inspection.

In examining the tables, note that if a vertical arrow is encountered, the first sampling plan above or below the arrow should be used. When this occurs, the sample size code letter and the sample size change. For example, if a single-sampling plan is indexed by an AQL of 1.5% and a sample size code letter of F, the code letter changes to G and the sample size changes from 20 to 32.

14-4.3 Discussion

MIL STD 105E presents the OC curves for single-sampling plans. These are all type-B OC curves. The OC curves for the matching double- and multiple-sampling plans are roughly comparable with those for the corresponding single-sampling plans. Figure 14-18 presents an example of these curves for code letter K. The OC curves presented in the standard are for the intitial sampling plan only. They are not the OC curves for the overall inspection program,[3] including shifts to and from tightened or reduced inspection.

Average sample number curves for double and multiple sampling are given, assuming that no curtailment is used. These curves are useful in evaluating the average sample sizes that may be expected to occur under the various sampling plans for a given lot or process quality.

There are several points about MIL STD 105E that should be emphasized. These include the following. One, MIL STD 105E is AQL-oriented. It focuses attention on the producer's risk end of the OC curve. The only control over the discriminatory power of the sampling plan (i.e., the steepness of the OC curve) is through the choice of inspection level.

Two, the sample sizes selected for use in MIL STD 105E are 2, 3, 5, 8, 13, 20, 32, 50, 80, 125, 200, 315, 500, 800, 1250, and 2000. Thus, not all sample sizes are possible. Note that there are some rather significant gaps, such as between 125 and 200, and between 200 and 315.

Three, the sample sizes in MIL STD 105E are related to the lot sizes. To see the nature of this relationship, calculate the midpoint of each lot size range, and plot the logarithm of the sample size for that lot size range against the logarithm of the lot size range midpoint. Such a plot will follow roughly a straight line up to $n = 80$, and

---

[3] ANSI/ASQC Z1.4 presents the scheme performance of the standard, giving scheme OC curves and the corresponding percentage points.

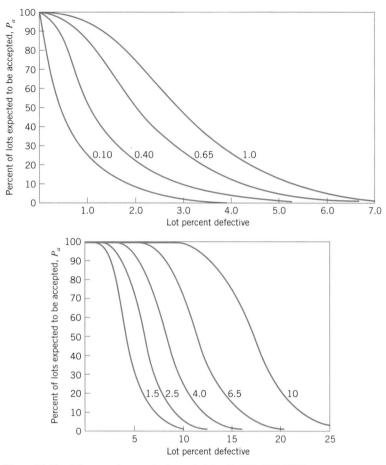

Figure 14-18 OC curves for sample size code letter K, MIL STD 105E.

thereafter another straight line with a shallower slope. Thus, the sample size will increase as the lot size increases. However, the ratio of sample size to lot size will decrease rapidly. This gives significant economy in inspection costs per unit when the vendor submits large lots. For a given AQL, the effect of this increase in sample size as the lot size increases is to increase the probability of acceptance for submitted lots of AQL quality. The probability of acceptance at a given AQL will vary with increasing sample size from about 0.91 to about 0.99. This feature of the standard was and still is subject to some controversy. The argument in favor of the approach in MIL STD 105E is that rejection of a large lot has more serious consequences for the vendor than rejection of a small lot, and if the probability of acceptance at the AQL increases with sample size, this reduces the risk of false rejection of a large lot. Furthermore, the large sample also gives a more discriminating OC curve, which means that the protection that the consumer receives against accepting an isolated bad lot will also be increased.

Four, the switching rules from normal to tightened inspection and from tightened to normal inspection are also subject to some criticism. In particular, some engineers dislike the switching rules because there is often a considerable amount of misswitching from normal to tightened or normal to reduced inspection when the process is actually producing lots of AQL quality. Also, there is a significant probability that production would even be discontinued, even though there has been no actual quality deterioration.

Five, a **flagrant** and **common abuse** of MIL STD 105E is failure to use the switching rules at all. When this is done, it results in ineffective and deceptive inspection and a substantial increase in the consumer's risk. It is not recommended that MIL STD 105E be implemented without use of the switching rules from normal to tightened and normal to reduced inspection.

As mentioned previously, a civilian standard, ANSI/ASQC Z1.4 or ISO 2859, is the counterpart of MIL STD 105E. It seems appropriate to conclude our discussion of MIL STD 105E with a comparison of the military and civilian standards. ANSI/ASQC Z1.4 was adopted in 1981. Aspects of ANSI/ASQC Z1.4 or ISO 2859 that differ from MIL STD 105E are presented next.

1. The terminology "nonconformity," "nonconformance," and "percent nonconforming" is used.

2. The switching rules were changed slightly to provide an option for reduced inspection without the use of limit numbers.

3. Several tables that show measures of scheme performance (*including* the switching rules) were introduced. Some of these performance measures include AOQL, limiting quality for which $P_a = 0.10$ and $P_a = 0.05$, ASN, and operating-characteristic curves.

4. A section was added describing proper use of individual sampling plans when extracted from the system.

5. A figure illustrating the switching rules was added.

These revisions modernize the terminology and emphasize the system concept of the standard. All tables, numbers, and procedures used in MIL STD 105E are retained in ANSI/ASQC Z1.4 and ISO 2859.

14-5   THE DODGE–ROMIG SAMPLING PLANS

H. F. Dodge and H. G. Romig have developed a set of sampling inspection tables for lot-by-lot inspection of product by attributes. Two types of sampling plans are presented in the tables—plans for lot tolerance percent defective (LTPD) protection and plans that provide a specified average outgoing quality limit (AOQL). For each of these approaches to sampling plan design, there are tables for single and double sampling.

Sampling plans that emphasize LTPD protection, such as the Dodge–Romig plans are often preferred to AQL-oriented sampling plans, such as those in MIL STD

105E, particularly for critical components and parts. Many manufacturers feel that they have relied too much on AQLs in the past, and they are now emphasizing other measures of performance, such as defective parts per million (ppm). Consider the following:

AQL	Defective Parts per Million
10%	100,000
1%	10,000
0.1%	1,000
0.01%	100
0.001%	10
0.0001%	1

Thus, even very small AQLs imply large numbers of defective ppm. In complex products, the effect of this can be devastating. For example, suppose that a printed circuit board contains 100 elements, each manufactured by a process operating at 0.5% defective. If the AQLs for these elements are 0.5% and if all elements on the printed circuit board must operate for the card to function properly, then the probability that a board works is

$$P(\text{function properly}) = (0.995)^{100} = 0.6058$$

Thus, there is an obvious need for sampling plans that emphasize LTPD protection, even when the process average fallout is low. The Dodge–Romig plans are often useful in these situations.

The Dodge–Romig AOQL plans are designed so that the average total inspection for a given AOQL and a specified process average p will be minimized. Similarly, the LTPD plans are designed so that the average total inspection is a minimum. This makes the Dodge–Romig plans very useful for in-plant inspection of semifinished product.

The Dodge–Romig plans apply only to programs that submit rejected lots to 100% inspection. Unless rectifying inspection is used, the AOQL concept is meaningless. Furthermore, to use the plans, we must know the process average—that is, the average fraction nonconforming of the incoming product. When a vendor is relatively new, we usually do not know its process fallout. Sometimes this may be estimated from a preliminary sample or from data provided by the vendor. Alternatively, the largest possible process average in the table can be used until enough information has been generated to provide a more accurate estimate of the vendor's process fallout. Obtaining a more accurate estimate of the incoming fraction nonconforming or process average will allow a more appropriate sampling plan to be adopted. It is not uncommon to find that sampling inspection begins with one plan, and after sufficient information is generated to reestimate the vendor's process fallout, a new plan is adopted. We discuss estimation of the process average in more detail in Section 14-5.3.

14-5.1 AOQL Plans

The Dodge–Romig (1959) tables give AOQL sampling plans for AOQL values of 0.1%, 0.25%, 0.5%, 0.75%, 1%, 1.5%, 2%, 2.5%, 3%, 4%, 5%, 7%, and 10%. For each of these AOQL values, six classes of values for the process average are specified. Tables are provided for both single and double sampling. These plans have been designed so that the average total inspection at the given AOQL and process average is approximately a minimum.

An example of the Dodge–Romig sampling plans is shown in Table 14-8.[4] To illustrate the use of the Dodge–Romig AOQL tables, suppose that we are inspecting LSI memory elements for a personal computer and that the elements are shipped in lots of size $N = 5000$. The vendor's process average fallout is 1% nonconforming. We wish to find a single-sampling plan with an AOQL $= 3\%$. From Table 14-8, we find that the plan is

$$n = 65 \qquad c = 3$$

Table 14-8 also indicates that the LTPD for this sampling plan is 10.3%. This is the point on the OC curve for which $P_a = 0.10$. Therefore, the sampling plan $n = 65, c = 3$ gives an AOQL of 3% nonconforming and provides assurance that 90% of incoming lots that are as bad as 10.3% defective will be rejected. Assuming that incoming quality is equal to the process average and that the probability of lot acceptance at this level of quality is $P_a = 0.9957$, we find that the average total inspection for this plan is

$$
\begin{aligned}
\text{ATI} &= n + (1 - P_a)(N - n) \\
&= 65 + (1 - 0.9957)(5000 - 65) \\
&= 86.22
\end{aligned}
$$

Thus, we will inspect approximately 86 units, on the average, in order to sentence a lot.

14-5.2 LTPD Plans

The Dodge–Romig LTPD tables are designed so that the probability of lot acceptance at the LTPD is 0.1. Tables are provided for LTPD values of 0.5%, 1%, 2%, 3%, 4%, 5%, 7%, and 10%. Table 14-9 for an LTPD of 1% is representative of these Dodge–Romig tables.

To illustrate the use of these tables, suppose that LSI memory elements for a personal computer are shipped from the vendor in lots of size $N = 5000$. The vendor's process average fallout is 0.25% nonconforming, and we wish to use a single-sampling plan with an LTPD of 1%. From inspection of Table 14-9, the sampling plan that should be used is

$$n = 770 \qquad c = 4$$

---

[4]Tables 14-8 and 14-9 are adapted from H. F. Dodge and H. G. Romig, *Sampling Inspection Tables, Single and Double Sampling*, 2nd ed., John Wiley, New York, 1959, with the permission of the publisher.

Table 14-8 Dodge–Romig Inspection Table—Single-Sampling Plans for AOQL = 3.0%

Lot Size	0–0.06%			0.07–0.60%			0.61–1.20%			1.21–1.80%			1.81–2.40%			2.41–3.00%		
	n	c	LTPD %	n	c	LTPD %	n	c	LTPD %	n	c	LTPD %	n	c	LTPD %	n	c	LTPD %
1–10	All	0	—	All	0	—	All	0	—	All	0	—	All	0	—	All	0	—
11–50	10	0	19.0	10	0	19.0	10	0	19.0	10	0	19.0	10	0	19.0	10	0	19.0
51–100	11	0	18.0	11	0	18.0	11	0	18.0	11	0	18.0	11	0	18.0	22	1	16.4
101–200	12	0	17.0	12	0	17.0	12	0	17.0	25	1	15.1	25	1	15.1	25	1	15.1
201–300	12	0	17.0	12	0	17.0	26	1	14.6	26	1	14.6	26	1	14.6	40	2	12.8
301–400	12	0	17.1	12	0	17.1	26	1	14.7	26	1	14.7	41	2	12.7	41	2	12.7
401–500	12	0	17.2	27	1	14.1	27	1	14.1	42	2	12.4	42	2	12.4	42	2	12.4
501–600	12	0	17.3	27	1	14.2	27	1	14.2	42	2	12.4	42	2	12.4	60	3	10.8
601–800	12	0	17.3	27	1	14.2	27	1	14.2	43	2	12.1	60	3	10.9	60	3	10.9
801–1000	12	0	17.4	27	1	14.2	44	2	11.8	44	2	11.8	60	3	11.0	80	4	9.8
1,001–2,000	12	0	17.5	28	1	13.8	45	2	11.7	65	3	10.2	80	4	9.8	100	5	9.1
2,001–3,000	12	0	17.5	28	1	13.8	45	2	11.7	65	3	10.2	100	5	9.1	140	7	8.2
3,001–4,000	12	0	17.5	28	1	13.8	65	3	10.3	85	4	9.5	125	6	8.4	165	8	7.8
4,001–5,000	28	1	13.8	28	1	13.8	65	3	10.3	85	4	9.5	125	6	8.4	210	10	7.4
5,001–7,000	28	1	13.8	45	2	11.8	65	3	10.3	105	5	8.8	145	7	8.1	235	11	7.1
7,001–10,000	28	1	13.9	46	2	11.6	65	3	10.3	105	5	8.8	170	8	7.6	280	13	6.8
10,001–20,000	28	1	13.9	46	2	11.7	85	4	9.5	125	6	8.4	215	10	7.2	380	17	6.2
20,001–50,000	28	1	13.9	65	3	10.3	105	5	8.8	170	8	7.6	310	14	6.5	560	24	5.7
50,001–100,000	28	1	13.9	65	3	10.3	125	6	8.4	215	10	7.2	385	17	6.2	690	29	5.4

Process Average

Table 14-9 Dodge–Romig Single-Sampling Table for Lot Tolerance Percent Defective (LTPD) = 1.0%

	Process Average																	
	0–0.01%			0.011%–0.10%			0.11–0.20%			0.21–0.30%			0.31–0.40%			0.41–0.50%		
Lot Size	n	c	AOQL %	n	c	AOQL %	n	c	AOQL %	n	c	AOQL %	n	c	AOQL %	n	c	AOQL %
1–120	All	0	0	All	0	0	All	0	0	All	0	0	All	0	0	All	0	0
121–150	120	0	0.06	120	0	0.06	120	0	0.06	120	0	0.06	120	0	0.06	120	0	0.06
151–200	140	0	0.08	140	0	0.08	140	0	0.08	140	0	0.08	140	0	0.08	140	0	0.08
201–300	165	0	0.10	165	0	0.10	165	0	0.10	165	0	0.10	165	0	0.10	165	0	0.10
301–400	175	0	0.12	175	0	0.12	175	0	0.12	175	0	0.12	175	0	0.12	175	0	0.12
401–500	180	0	0.13	180	0	0.13	180	0	0.13	180	0	0.13	180	0	0.13	180	0	0.13
501–600	190	0	0.13	190	0	0.13	190	0	0.13	190	0	0.13	190	0	0.13	305	1	0.14
601–800	200	0	0.14	200	0	0.14	200	0	0.14	330	1	0.15	330	1	0.15	330	1	0.15
801–1000	205	0	0.14	205	0	0.14	205	0	0.14	335	1	0.17	335	1	0.17	335	1	0.17
1,001–2,000	220	0	0.15	220	0	0.15	360	1	0.19	490	2	0.21	490	2	0.21	610	3	0.22
2,001–3,000	220	0	0.15	375	1	0.20	505	2	0.23	630	3	0.24	745	4	0.26	870	5	0.26
3,001–4,000	225	0	0.15	380	1	0.20	510	2	0.23	645	3	0.25	880	5	0.28	1,000	6	0.29
4,001–5,000	225	0	0.16	380	1	0.20	520	2	0.24	770	4	0.28	895	5	0.29	1,120	7	0.31
5,001–7,000	230	0	0.16	385	1	0.21	655	3	0.27	780	4	0.29	1,020	6	0.32	1,260	8	0.34
7,001–10,000	230	0	0.16	520	2	0.25	660	3	0.28	910	5	0.32	1,150	7	0.34	1,500	10	0.37
10,001–20,000	390	1	0.21	525	2	0.26	785	4	0.31	1,040	6	0.35	1,400	9	0.39	1,980	14	0.43
20,001–50,000	390	1	0.21	530	2	0.26	920	5	0.34	1,300	8	0.39	1,890	13	0.44	2,570	19	0.48
50,001–100,000	390	1	0.21	670	3	0.29	1,040	6	0.36	1,420	9	0.41	2,120	15	0.47	3,150	23	0.50

If we assume that rejected lots are screened 100% and that defective items are replaced with good ones, the AOQL for this plan is approximately 0.28%.

Note from inspection of the Dodge–Romig LTPD tables that values of the process average cover the interval from zero to one-half the LTPD. Provision for larger process averages is unnecessary, since 100% inspection is more economically efficient than inspection sampling when the process average exceeds one-half the desired LTPD.

14-5.3   Estimation of Process Average

As we have observed, selection of a Dodge–Romig plan depends on knowledge of the vendor's process average fallout or percent nonconforming. An estimate of the process average can be obtained using a fraction defective control chart. This chart is based on the first 25 lots submitted by the vendor. If double sampling is used, only the results from the first sample should be included in the computations. Any lot fraction defective that exceeds the upper control limit will be discarded, provided it has an assignable cause, and a new process average is calculated. Until results from 25 lots have been accumulated, the recommended procedure is to use the largest process average in the appropriate table.

14-6 EXERCISES

14-1.   Draw the type-B OC curve for the single-sampling plan $n = 50$, $c = 1$.

14-2.   Draw the type-B OC curve for the single-sampling plan $n = 100$, $c = 2$.

14-3.   Suppose that a product is shipped in lots of size $N = 5000$. The receiving inspection procedure used is single sampling with $n = 50$ and $c = 1$.
   (a) Draw the type-A OC curve for the plan.
   (b) Draw the type-B OC curve for this plan and compare it to the type-A OC curve found in part (a).
   (c) Which curve is appropriate for this situation?

14-4.   Find a single-sampling plan for which $p_1 = 0.01$, $\alpha = 0.05$, $p_2 = 0.10$, and $\beta = 0.10$.

14-5.   Find a single-sampling plan for which $p_1 = 0.05$, $\alpha = 0.05$, $p_2 = 0.15$, and $\beta = 0.10$.

14-6.   Find a single-sampling plan for which $p_1 = 0.02$, $\alpha = 0.01$, $p_2 = 0.06$, and $\beta = 0.10$.

14-7.   A company uses the following acceptance-sampling procedure. A sample equal to 10% of the lot is taken. If 2% or less of the items in the sample are defective, the lot is accepted; otherwise, it is rejected. If submitted lots vary in size from 5000 to 10,000 units, what can you say about the protection by this plan? If 0.05 is the desired LTPD, does this scheme offer reasonable protection to the consumer?

14-8.   A company uses a sample size equal to the square root of the lot size. If 1% or less of the items in the sample are defective, the lot is accepted; otherwise, it is rejected. Submitted lots vary in size from 1000 to 5000 units. Comment on the effectiveness of this procedure.

14-9.   Consider the single-sampling plan found in Exercise 14-4. Suppose that lots of $N = 2000$ are submitted. Draw the ATI curve for this plan. Draw the AOQ curve and find the AOQL.

14-10. Suppose that a single-sampling plan with $n = 150$ and $c = 2$ is being used for receiving inspection where the vendor ships the product in lots of size $N = 3000$.

(a) Draw the OC curve for this plan.

(b) Draw the AOQ curve and find the AOQL.

(c) Draw the ATI curve for this plan.

14-11.   Suppose that a vendor ships components in lots of size 5000. A single-sampling plan with $n = 50$ and $c = 2$ is being used for receiving inspection. Rejected lots are screened, and all defective items are reworked and returned to the lot.

(a) Draw the OC curve for this plan.

(b) Find the level of lot quality that will be rejected 90% of the time.

(c) Management has objected to the use of the above sampling procedure and wants to use a plan with an acceptance number $c = 0$, arguing that this is more consistent with their zero-defects program. What do you think of this?

(d) Design a single-sampling plan with $c = 0$ that will give a 0.90 probability of rejection of lots having the quality level found in part (b). Note that the two plans are now matched at the LTPD point. Draw the OC curve for this plan and compare it to the one for $n = 50$, $c = 2$ in part (a).

(e) Suppose that incoming lots are 0.5% nonconforming. What is the probability of rejecting these lots under both plans? Calculate the ATI at this point for both plans. Which plan do you prefer? Why?

14-12.   Draw the primary and supplementary OC curves for a double-sampling plan with $n_1 = 50$, $c_1 = 2$, $n_2 = 100$, $c_2 = 6$. If the incoming lots have fraction nonconforming $p = 0.05$, what is the probability of acceptance on the first sample? What is the probability of final acceptance? Calculate the probability of rejection on the first sample.

14-13.   (a) Derive an item-by-item sequential-sampling plan for which

$p_1 = 0.01$, $\alpha = 0.05$, $p_2 = 0.10$, and $\beta = 0.10$.

(b) Draw the OC curve for this plan.

14-14.   (a) Derive an item-by-item sequential-sampling plan for which $p_1 = 0.01$, $\alpha = 0.05$, $p_2 = 0.10$, and $\beta = 0.10$.

(b) Draw the OC curve for this plan.

14-15.   Consider rectifying inspection for single sampling. Develop an AOQ equation assuming that all defective items are removed but *not* replaced with good ones.

14-16.   A vendor ships a component in lots of size $N = 3000$. The AQL has been established for this product at 1%. Find the normal, tightened, and reduced single-sampling plans for this situation from MIL STD 105E, assuming that general inspection level II is appropriate.

14-17.   Repeat Exercise 14-16, using general inspection level I. Discuss the differences in the various sampling plans.

14-18.   A product is supplied in lots of size $N = 10,000$. The AQL has been specified at 0.10%. Find the normal, tightened, and reduced single-sampling plans from MIL STD 105E, assuming general inspection level II.

14-19.   MIL STD 105E is being used to inspect incoming lots of size $N = 5000$. Single sampling, general inspection level II, and an AQL of 0.65% are being used.

(a) Find the normal, tightened, and reduced inspection plans.

(b) Draw the OC curves of the normal, tightened, and reduced inspection plans on the same graph.

14-20.   A product is shipped in lots of size $N = 2000$. Find a Dodge–Romig single-sampling plan for which the LTPD = 1%, assuming that the process average is 0.25% defective. Draw the OC curve and the ATI curve for this plan. What is the AOQL for this sampling plan?

14-21. We wish to find a single-sampling plan for a situation where lots are shipped from a vendor. The vendor's process operates at a fallout level of 0.50% defective. We want the AOQL from the inspection activity to be 3%.
   (a) Find the appropriate Dodge–Romig plan.
   (b) Draw the OC curve and the ATI curve for this plan. How much inspection will be necessary, on the average, if the vendor's process operates close to the average fallout level?
   (c) What is the LTPD protection for this plan?

14-22. A vendor ships a product in lots of size $N = 8000$. We wish to have an AOQL of 3%, and we are going to use single sampling. We do not know the vendor's process fallout but suspect that it is at most 1% defective.
   (a) Find the appropriate Dodge–Romig plan.
   (b) Find the ATI for this plan, assuming that incoming lots are 1% defective.
   (c) Suppose that our estimate of the vendor's process average is incorrect and that it is really 0.25% defective. What sampling plan should we have used? What reduction in ATI would have been realized if we had used the correct plan?

15

Other Acceptance-Sampling Techniques

CHAPTER OUTLINE

15-1   ACCEPTANCE SAMPLING BY VARIABLES

    15-1.1   Advantages and Disadvantages of Variables Sampling

    15-1.2   Types of Sampling Plans Available

    15-1.3   Caution in the Use of Variables Sampling

15-2   DESIGNING A VARIABLES SAMPLING PLAN WITH A SPECIFIED OC CURVE

15-3   MIL STD 414 (ANSI/ASQC Z1.9)

    15-3.1   General Description of the Standard

    15-3.2   Use of the Tables

15-3.3   Discussion of MIL STD 414 and ANSI/ASQC Z1.9

15-4   OTHER VARIABLES SAMPLING PROCEDURES

    15-4.1   Sampling by Variables to Give Assurance Regarding the Lot or Process Mean

    15-4.2   Sequential Sampling by Variables

15-5   CHAIN SAMPLING

15-6   CONTINUOUS SAMPLING

    15-6.1   CSP-1

    15-6.2   Other Continuous-Sampling Plans

15-7   SKIP-LOT SAMPLING PLANS

CHAPTER OVERVIEW

This chapter summarizes several useful acceptance-sampling techniques, including variables sampling plans, which can be used as alternatives to attribute plans when measurement data are available. We also briefly discuss chain sampling, continuous sampling, and skip-lot sampling.

15-1 ACCEPTANCE SAMPLING BY VARIABLES

15-1.1 Advantages and Disadvantages of Variables Sampling

The primary advantage of variables sampling plans is that the same operating-characteristic curve can be obtained with a smaller sample size than would be required by an attributes sampling plan. Thus, a variables acceptance-sampling plan that has the same protection as an attributes acceptance-sampling plan would require less sampling. The measurements data required by a variables sampling plan would probably cost more per observation than the collection of attributes data. However, the reduction in sample size obtained may more than offset this increased cost. For example, suppose that an attributes sampling plan requires a sample of size 100 items, but the equivalent variables sampling plan requires a sample size of only 65. If the cost of measurement data is less than 1.61 times the cost of measuring the observations on an attributes scale, the variables sampling plan will be more economically efficient, considering sampling costs only. When destructive testing is employed, variables sampling is particularly useful in reducing the costs of inspection.

A second advantage is that measurement data usually provide more information about the manufacturing process or the lot than do attributes data. Generally, numerical measurements of quality characteristics are more useful than simple classification of the item as defective or nondefective.

A final point to be emphasized is that when acceptable quality levels are very small, the sample sizes required by attributes sampling plans are very large. Under these circumstances, there may be significant advantages in switching to variables measurement. Thus, as many manufacturers begin to emphasize allowable numbers of defective parts per million, variables sampling becomes very attractive.

Variables sampling plans have several disadvantages. Perhaps the primary disadvantage is that the distribution of the quality characteristic must be known. Furthermore, most standard variables acceptance-sampling plans assume that the distribution of the quality characteristic is normal. If the distribution of the quality characteristic is not normal, and a plan based on the normal assumption is employed, serious departures from the advertised risks of accepting or rejecting lots of given quality may be experienced. We discuss this point more completely in Section 15-1.3. The second disadvantage of variables sampling is that a separate sampling plan must be employed for each quality characteristic that is being inspected. For example, if an item is inspected for four quality characteristics, it is necessary to have four separate variables inspection sampling plans. If this same product were being inspected under attributes sampling, one attributes sampling plan could be employed. Finally, it is possible that the use of a variables sampling plan will lead to rejection of a lot even though the actual sample inspected does not contain any defective items. Although this does not happen very often, when it does occur it usually causes considerable unhappiness in both the vendors' and the consumers' organizations, particularly if rejection of the lot has caused a manufacturing facility to shut down or operate on a reduced production schedule.

15-1.2   Types of Sampling Plans Available

There are two general types of variables sampling procedures: plans that control the lot or process fraction defective (or nonconforming) and plans that control a lot or process parameter (usually the mean). Sections 15-2 and 15-3 present variables sampling plans to control the process fraction defective. Variables sampling plans for the process mean are presented in Section 15-4.

Consider a variables sampling plan to control the lot or process fraction nonconforming. Since the quality characteristic is a variable, there will exist either a lower specification limit (LSL), an upper specification limit (USL), or both, that define the acceptable values of this parameter. Figure 15-1 illustrates the situation in which the quality characteristic x is normally distributed and there is a lower specification limit on this parameter. The symbol p represents the fraction defective in the lot. Note that the fraction defective is a function of the lot or process mean μ and the lot or process standard deviation σ.

Suppose that the standard deviation σ is known. Under this condition, we may wish to sample from the lot to determine whether or not the value of the mean is such that the fraction defective p is acceptable. As described next, we may organize the calculations in the variables sampling plan in two ways.

Procedure 1.   Take a random sample of n items from the lot and compute the statistic

$$Z_{\text{LSL}} = \frac{\bar{x} - \text{LSL}}{\sigma} \tag{15-1}$$

Note that Z_{LSL} in equation 15-1 simply expresses the distance between the sample average $\bar{x}$ and the lower specification limit in standard deviation units. The larger is the value of Z_{LSL}, the farther the sample average $\bar{x}$ is from the lower specification limit, and consequently, the smaller is the lot fraction defective p. If there is a critical value of p of interest that should not be exceeded with stated probability, we can translate this value of p into a *critical distance* — say, k — for Z_{LSL}. Thus, if $Z_{\text{LSL}} \geq k$, we would accept the lot because the sample data imply that the lot mean is sufficiently far above the LSL to ensure that the lot fraction nonconforming is satisfactory. However, if $Z_{\text{LSL}} < k$, the mean is too close to the LSL, and the lot should be rejected.

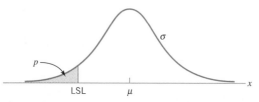

Figure 15-1   Relationship of the lot or process fraction defective p to the mean and standard deviation of a normal distribution.

Procedure 2. Take a random sample of n items from the lot and compute Z_{LSL} using equation 15-1. Use Z_{LSL} to estimate the fraction defective of the lot or process as the area under the standard normal curve below Z_{LSL}. (Actually, using $Q_{LSL} = Z_{LSL}\sqrt{n/(n-1)}$ as a standard normal variable is slightly better, because it gives a better estimate of p.) Let $\hat{p}$ be the estimate of p so obtained. If the estimate $\hat{p}$ exceeds a specified maximum value M, reject the lot; otherwise, accept it.

The two procedures can be designed so that they give equivalent results. When there is only a single specification limit (LSL or USL), either procedure may be used. Obviously, in the case of an upper specification limit, we would compute

$$Z_{USL} = \frac{USL - \bar{x}}{\sigma} \tag{15-2}$$

instead of using equation 15-1. When there are both lower and upper specifications, the M method, Procedure 2, should be used.

When the standard deviation σ is unknown, it is estimated by the sample standard deviation S, and σ in equations 15-1 and 15-2 is replaced by S. It is also possible to design plans based on the sample range R instead of S. However, these plans are not discussed in this chapter because using the sample standard deviation will lead to smaller sample sizes. Plans based on R were once in wide use because R is easier to compute by hand than is S, but computation is not a problem today.

15-1.3 Caution in the Use of Variables Sampling

We have remarked that the distribution of the quality characteristic must be of known form to use variables sampling. Furthermore, the usual assumption is that the parameter of interest follows the normal distribution. This assumption is critical because all variables sampling plans require that there be some method of converting a sample mean and standard deviation into a lot or process fraction defective. If the parameter of interest is not normally distributed, estimates of the fraction defective based on the sample mean and sample standard deviation will not be the same as if the parameter were normally distributed. The difference between these estimated fraction defectives may be large when we are dealing with very small fractions defective. For example, if the mean of a normal distribution lies three standard deviations below a single upper specification limit, the lot will contain no more than 0.135% defective. On the other hand, if the quality characteristic in the lot or process is very nonnormal, and the mean lies three standard deviations below the specification limit, it is entirely possible that 1% or more of the items in the lot might be defective.

It is possible to use variables sampling plans when the parameter of interest does not have a normal distribution. Provided that the form of the distribution is known, or that there is a method of determining the fraction defective from the sample average and sample standard deviation (or other appropriate sample statistics), it is possible to devise a procedure for applying a variables sampling plan. For example, Duncan (1986) presents a procedure for using a variables sampling plan when the distribution of the quality characteristic can be described by a Pearson type III distribution. A general discussion of variables sampling in the nonnormal case is, however, beyond the scope of this book.

15-2 DESIGNING A VARIABLES SAMPLING PLAN WITH A SPECIFIED OC CURVE

It is easy to design a variables sampling plan using Procedure 1, the k-method, that has a specified OC curve. Let $(p_1, 1 - \alpha)$, (p_2, β) be the two points on the OC curve of interest. Note that p_1 and p_2 may be the levels of lot or process fraction nonconforming that correspond to acceptable and rejectable levels of quality, respectively.

The nomograph shown in Fig. 15-2 enables the quality engineer to find the required sample size n and the critical value k to meet a set of given conditions p_1, $1 - \alpha$, p_2, β for both the σ known and the σ unknown cases. The nomograph contains separate scales for sample size for these two cases. The greater uncertainty in the case where the standard deviation is unknown requires a larger sample size than does the σ known case, but the same value of k is used. In addition, for a given sampling plan, the probability of acceptance for

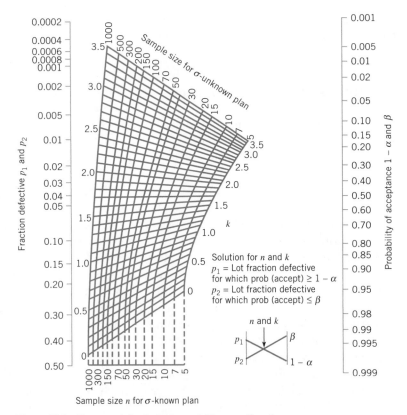

Figure 15-2   Nomograph for designing variables sampling plans.

any value of fraction defective can be found from the nomograph. By plotting several of these points, the quality engineer may construct an operating-characteristic curve of the sampling plan. The use of this nomograph is illustrated in the following example.

······· **EXAMPLE 15-1** ···

A soft-drink bottler buys nonreturnable bottles from a vendor. The bottler has established a lower specification on the bursting strength of the bottles at 225 psi. If 1% or less of the bottles burst below this limit, the bottler wishes to accept the lot with probability 0.95 ($p_1 = 0.01$, $1 - \alpha = 0.95$), whereas if 6% or more of the bottles burst below this limit, the bottler would like to reject the lot with probability 0.90 ($p_2 = 0.06$, $\beta = 0.10$). To find the sampling plan, draw a line connecting the point 0.01 on the fraction defective scale to the point 0.95 on the probability of acceptance scale. Then draw a similar line connecting the points $p_2 = 0.06$ and $P_a = 0.10$. At the intersection of these lines, we read $k = 1.9$. Suppose that σ is unknown. Following the curved line from the intersection point to the upper sample size scale gives $n = 40$. Therefore, the procedure is to take a random sample of $n = 40$ bottles, observe the bursting strengths, compute $\bar{x}$ and S, then calculate

$$Z_{\text{LSL}} = \frac{\bar{x} - \text{LSL}}{S}$$

and to accept the lot if

$$Z_{\text{LSL}} \geq k = 1.9$$

If σ is known, drop vertically from the intersection point to the σ-known scale. This would indicate a sample size of $n = 15$. Thus, if the standard deviation is known, a considerable reduction in sample size is possible.

···

It is also possible to design a variables acceptance-sampling plan from the nomograph using Procedure 2 (the M-method). To do so, an additional step is necessary. Figure 15-3 presents a chart for determining the maximum allowable fraction defective M. Once the values of n and k have been determined for the appropriate sampling plan from Fig. 15-2, a value of M can be read directly from Fig. 15-3. To use Procedure 2, it is necessary to convert the value of Z_{LSL} or Z_{USL} into an estimated fraction defective. Figure 15-4 can be used for this purpose. The following example illustrates how a single-sampling plan for variables with a one-sided specification limit using Procedure 2 can be designed.

······· **EXAMPLE 15-2** ···

Consider the situation described in Example 15-1. Since we know that $n = 40$ and $k = 1.9$, we enter Fig. 15-3 with $n = 40$ and abscissa value

$$\frac{1 - \dfrac{k\sqrt{n}}{(n-1)}}{2} = \frac{1 - \dfrac{1.9\sqrt{40}}{39}}{2} = 0.35$$

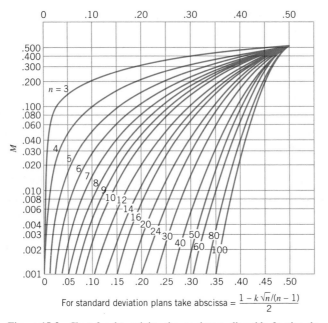

For standard deviation plans take abscissa $= \dfrac{1 - k\sqrt{n}/(n-1)}{2}$

Figure 15-3 Chart for determining the maximum allowable fraction defective M. (From A. J. Duncan, *Quality Control and Industrial Statistics*, 5th ed., Irwin, Homewood, Ill., 1986, with the permission of the publisher.)

This indicates that $M = 0.030$. Now suppose that a sample of $n = 40$ is taken, and we observe $\bar{x} = 255$ and $S = 15$. The value of Z_{LSL} is

$$Z_{\text{LSL}} = \frac{\bar{x} - \text{LSL}}{s} = \frac{255 - 225}{15} = 2$$

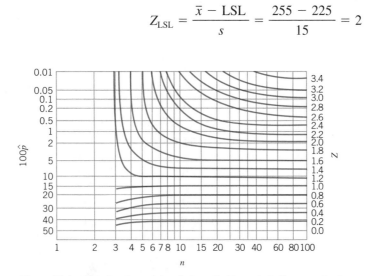

Figure 15-4 Chart for determining $\hat{p}$ from Z. (From A. J. Duncan, *Quality Control and Industrial Statistics*, 5th ed., Irwin, Homewood, Ill., 1986, with the permission of the publisher.)

From Fig. 15-4 we read $\hat{p} = 0.020$. Since $\hat{p} = 0.020$ is less than $M = 0.030$, we will accept the lot.

⋯⋯⋯

When there are double-specification limits, Procedure 2 can be used directly. We begin by first obtaining the sample size n and the critical value k for a single-limit plan that has the same values of p_1, p_2, α, and β as the desired double-specification-limit plan. Then the value of M is obtained directly from Fig. 15-3. Now in the operation of the acceptance-sampling plan, we compute Z_{LSL} and Z_{USL} and, from Fig. 15-4, find the corresponding fraction defective estimates—say, $\hat{p}_{LSL}$ and $\hat{p}_{USL}$. Then, if $\hat{p}_{LSL} + \hat{p}_{USL} \leq M$, the lot will be accepted; otherwise, it will be rejected.

It is also possible to use Procedure 1 for double-sided specification limits. However, the procedure must be modified extensively. Details of the modifications are in Duncan (1986).

15-3   MIL STD 414 (ANSI/ASQC Z1.9)

15-3.1   General Description of the Standard

MIL STD 414 is a lot-by-lot acceptance-sampling plan for variables. The standard was introduced in 1957. The focal point of this standard is the acceptable quality level (AQL), which ranges from 0.04% to 15%. There are five general levels of inspection, and level IV is designated as "normal." Inspection level V gives a steeper OC curve than level IV. When reduced sampling costs are necessary and when greater risks can or must be tolerated, lower inspection levels can be used. As with the attributes standard, MIL STD 105E, sample size code letters are used, but the same code letter does not imply the same sample size in both standards. In addition, the lot size classes are different in both standards. Sample sizes are a function of the lot size and the inspection level. Provision is made for normal, tightened, and reduced inspection. All the sampling plans and procedures in the standard assume that the quality characteristic of interest is normally distributed.

Figure 15-5 presents the organization of the standard. Note that acceptance-sampling plans can be designed for cases where the lot or process variability is either known or unknown, and where there are either single-specification limits or double-specification limits on the quality characteristic. In the case of single-specification limits, either Procedure 1 or Procedure 2 may be used. If there are double-specification limits, then Procedure 2 must be used. If the process or lot variability is known and stable, the variability known plans are the most economically efficient. When lot or process variability is unknown, either the standard deviation or the range of the sample may be used in operating the sampling plan. The range method requires a larger sample size, and we do not generally recommend its use.

MIL STD 414 is divided into four sections. Section A is a general description of the sampling plans, including definitions, sample size code letters, and OC curves for the various sampling plans. Section B of the standard gives variables sampling plans based

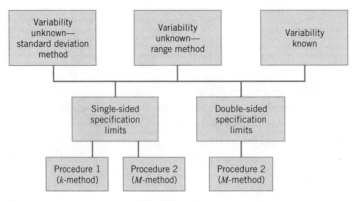

Figure 15-5 Organization of MIL STD 414.

on the sample standard deviation for the case in which the process or lot variability is unknown. Section C presents variables sampling plans based on the sample range method. Section D gives variables sampling plans for the case where the process standard deviation is known.

15-3.2 Use of the Tables

Two typical tables from MIL STD 414 are reproduced as Tables 15-1 and 15-2. The following example illustrates the use of these tables.

Table 15-1 (Table A-2. MIL STD 414) Sample Size Code Letters

	Inspection Levels				
Lot Size	I	II	III	IV	V
3 to 8	B	B	B	B	C
9 to 15	B	B	B	B	D
16 to 25	B	B	B	C	E
26 to 40	B	B	B	D	F
41 to 65	B	B	C	E	G
66 to 110	B	B	D	F	H
111 to 180	B	C	E	G	I
181 to 300	B	D	F	H	J
301 to 500	C	E	G	I	K
501 to 800	D	F	H	J	L
801 to 1,300	E	G	I	K	L
1,301 to 3,200	F	H	J	L	M
3,201 to 8,000	G	I	L	M	N
8,001 to 22,000	H	J	M	N	O
22,001 to 110,000	I	K	N	O	P
110,001 to 550,000	I	K	O	P	Q
550,001 and over	I	K	P	Q	Q

Table 15-2 Master Table for Normal and Tightened Inspection for Plans Based on Variability Unknown (Standard Deviation Method) (Single-Specification Limit—Form 1)(Table B-1, MIL STD 414)

Acceptable Quality Levels (normal inspection)

Sample Size Code Letter	Sample Size	.04	.065	.10	.15	.25	.40	.65	1.00	1.50	2.50	4.00	6.50	10.00	15.00
		k	k	k	k	k	k	k	k	k	k	k	k	k	k
B	3	→	→	→	→	→	→	→	▼	▼	1.12	.958	.765	.566	.341
C	4	→	→	→	→	→	→	→	1.45	1.34	1.17	1.01	.814	.617	.393
D	5	→	→	→	→	→	→	1.65	1.53	1.40	1.24	1.07	.874	.675	.455
E	7	→	→	→	→	2.00	1.88	1.75	1.62	1.50	1.33	1.15	.955	.755	.536
F	10	→	→	→	2.24	2.11	1.98	1.84	1.72	1.58	1.41	1.23	1.03	.828	.611
G	15	2.64	2.53	2.42	2.32	2.20	2.06	1.91	1.79	1.65	1.47	1.30	1.09	.886	.664
H	20	2.69	2.58	2.47	2.36	2.24	2.11	1.96	1.82	1.69	1.51	1.33	1.12	.917	.695
I	25	2.72	2.61	2.50	2.40	2.26	2.14	1.98	1.85	1.72	1.53	1.35	1.14	.936	.712
J	30	2.73	2.61	2.51	2.41	2.28	2.15	2.00	1.86	1.73	1.55	1.36	1.15	.946	.723
K	35	2.77	2.65	2.54	2.45	2.31	2.18	2.03	1.89	1.76	1.57	1.39	1.18	.969	.745
L	40	2.77	2.66	2.55	2.44	2.31	2.18	2.03	1.89	1.76	1.58	1.39	1.18	.971	.746
M	50	2.83	2.71	2.60	2.50	2.35	2.22	2.08	1.93	1.80	1.61	1.42	1.21	1.00	.774
N	75	2.90	2.77	2.66	2.55	2.41	2.27	2.12	1.98	1.84	1.65	1.46	1.24	1.03	.804
O	100	2.92	2.80	2.69	2.58	2.43	2.29	2.14	2.00	1.86	1.67	1.48	1.26	1.05	.819
P	150	2.96	2.84	2.73	2.61	2.47	2.33	2.18	2.03	1.89	1.70	1.51	1.29	1.07	.841
Q	200	2.97	2.85	2.73	2.62	2.47	2.33	2.18	2.04	1.89	1.70	1.51	1.29	1.07	.845
		.065	.10	.15	.25	.40	.65	1.00	1.50	2.50	4.00	6.50	10.00	15.00	

Acceptable Quality Levels (tightened inspection)

All AQL values are in percent defective.

→ Use first sampling plan below arrow, that is, both sample size as well as k value. When sample size equals or exceeds lot size, every item in the lot must be inspected.

731

········   **EXAMPLE 15-3**   ··· ·

Consider the soft-drink bottler in the previous two examples who is purchasing bottles from a vendor. The lower specification limit on bursting strength is 225 psi. Suppose that the AQL at this specification limit is 1%. Let us suppose that bottles are shipped in lots of size 100,000. We will obtain a variables sampling plan that uses Procedure 1 from MIL STD 414. We assume that the lot standard deviation is unknown.

From Table 15-1, if we use inspection level IV, the sample size code letter is O. From Table 15-2 we find that sample size code letter O implies a sample size of $n = 100$. For an acceptable quality level of 1%, on normal inspection, the value of k is 2.00. If tightened inspection is employed, the appropriate value of k is 2.14. Note that normal and tightened inspection use the same tables. The AQL values for normal inspection are indexed at the top of the table, and the AQL values for tightened inspection are indexed from the bottom of the table.

MIL STD 414 contains a provision for a shift to tightened or reduced inspection when this is warranted. The process average is used as the basis for determining when such a shift is made. The process average is taken as the average of the sample estimates of percent defective computed for lots submitted on original inspection. Usually, the process average is computed using information from the preceding 10 lots. Full details of the switching procedures are described in the standard and in a technical memorandum on MIL STD 414, published by the United States Department of the Navy, Bureau of Ordnance.

Estimation of the fraction defective is required in using Procedure 2 of MIL STD 414. It is also required in implementing the switching rules between normal, tightened, and reduced inspection. In the standard, three tables are provided for estimating the fraction defective.

When starting to use MIL STD 414, one can choose between the known standard deviation and unknown standard deviation procedures. When there is no basis for knowledge of σ, obviously the unknown standard deviation plan must be used. However, it is a good idea to maintain either an R or S chart on the results of each lot so that some information on the state of statistical control of the scatter in the manufacturing process can be collected. If this control chart indicates statistical control, it will be possible to switch to a known σ plan. Such a switch will reduce the required sample size. Even if the process were not perfectly controlled, the control chart could provide information leading to a conservative estimate of σ for use in a known σ plan. When a known σ plan is used, it is also necessary to maintain a control chart on either R or S as a continuous check on the assumption of stable and known process variability.

MIL STD 414 contains a special procedure for application of mixed variables/attributes acceptance-sampling plans. If the lot does not meet the acceptability criterion of the variables plan, an attributes single-sampling plan, using tightened inspection and the same AQL, is obtained from MIL STD 105E. A lot can be accepted by either of the plans in sequence but must be rejected by both the variables and attributes plan.

15-3.3   Discussion of MIL STD 414 and ANSI/ASQC Z1.9

In 1980, the American National Standards Institute and the American Society for Quality Control released an updated civilian version of MIL STD 414 known as ANSI/ASQC Z1.9. MIL STD 414 was originally structured to give protection essentially equivalent to that provided by MIL STD 105A (1950). When MIL STD 105D was adopted in 1963, this new standard contained substantially revised tables and procedures that led to differences in protection between it and MIL STD 414. Consequently, it is not possible to move directly from an attributes sampling plan in the current MIL STD 105E to a corresponding variables plan in MIL STD 414 if the assurance of continued protection is desired for certain lot sizes and AQLs.

The civilian counterpart of MIL STD 414, ANSI/ASQC Z1.9, restores this original match. That is, ANSI/ASQC Z1.9 now is directly compatible with MIL STD 105E (and its equivalent civilian counterpart ANSI/ASQC Z1.4). This equivalence was obtained by incorporating the following revisions in ANSI/ASQC Z1.9:

1. Lot size ranges were adjusted to correspond to MIL STD 105D.
2. The code letters assigned to the various lot size ranges were arranged to make protection equal to that of MIL STD 105E.
3. AQLs of 0.04, 0.065, and 15 were deleted.
4. The original inspection levels I, II, III, IV, and V were relabeled S3, S4, I, II, III, respectively.
5. The original switching rules were replaced by those of MIL STD 105E, with slight revisions.

In addition, to modernize terminology, the word *nonconformity* was substituted for defect, *nonconformance* was substituted for defective, and *percent nonconforming* was substituted for percent defective. The operating-characteristic curves were recomputed and replotted, and a number of editorial changes were made to the descriptive material of the standard to match MIL STD 105E as closely as possible. Finally, an appendix was included showing the match between ANSI/ASQC Z1.9, MIL STD 105E, and the corresponding civilian version ANSI Z1.4. This appendix also provided selected percentage points from the OC curves of these standards and their differences.

As of this writing, the Department of Defense has not officially adopted ANSI/ASQC Z1.9 and continues to use MIL STD 414. Both standards will probably be used for the immediate future. The principal advantage of the ANSI/ASQC Z1.9 standard is that it is possible to start inspection by using an attributes sampling scheme from MIL STD 105E or ANSI/ASQC Z1.4, collect sufficient information to use variables inspection, and then switch to the variables scheme, while maintaining the same AQL-code letter combination. It would then be possible to switch back to the attributes scheme if the assumption of the variables scheme appeared not to be satisfied. It is also possible to take advantage of the information gained in coordinated attributes and variables inspection to move in a logical manner from inspection sampling to statistical process control.

As in MIL STD 414, ANSI/ASQC Z1.9 assumes that the quality characteristic is normally distributed. This is an important assumption that we have commented on previously. We have suggested that a test for normality should be incorporated as part of the standard. One way this can be done is to plot a control chart for $\bar{x}$ and S (or $\bar{x}$ and R) from the variables data from each lot. After a sufficient number of observations have been obtained, a test for normality can be employed by plotting the individual measurements on normal probability paper or by conducting one of the specialized statistical tests for normality. It is recommended that a relatively large sample size be used in this statistical test. At least 100 observations should be collected before the test for normality is made, and it is our belief that the sample size should increase inversely with AQL. If the assumption of normality is badly violated, either a special variables sampling procedure must be developed, or we must return to attributes inspection.

An additional advantage of applying a control chart to the result of each lot is that if the process variability has been in control for at least 30 samples, it will be possible to switch to a known standard deviation plan, thereby allowing a substantial reduction in sample size. Although this can be instituted in any combined program of attributes and variables inspection, it is easy to do so using the ANSI/ASQC standards, because of the design equivalence between the attributes and variables procedures.

15-4   OTHER VARIABLES SAMPLING PROCEDURES

15-4.1   Sampling by Variables to Give Assurance Regarding the Lot or Process Mean

Variables sampling plans can also be used to give assurance regarding the **average** quality of a material, instead of the fraction defective. Sampling plans such as this are most likely to be employed in the sampling of bulk materials that come in bags, drums, or other containers. However, they can also be applied to discrete parts and to other variables, such as energy loss in power transformers. The general approach employed in this type of variables sampling is statistical hypothesis testing. We now present an example of the procedure.

······   **EXAMPLE 15-4** ··

A manufacturer of wood paneling samples the substrate blanks bought from an offshore vendor to determine their formaldehyde emission level. As long as the mean emission level is less than 0.3 ppm, the lot is satisfactory. We will design a variables sampling procedure that will give lots that have a mean emission level of 0.3 ppm a 0.95 probability of acceptance, and lots that have a mean emission level of 0.4 ppm a 0.10 probability of acceptance. The largest probable value of the standard deviation of emission level is known from past experience to be $\sigma = 0.10$ ppm.

Let $\bar{x}_A$ be the value of the sample average above which the lot will be accepted. Then, we know that

$$\frac{\bar{x}_A - 0.30}{\sigma/\sqrt{n}} = \frac{\bar{x}_A - 0.30}{0.10/\sqrt{n}}$$

is distributed as a standard normal variable. If lots of this type have a 0.95 probability of acceptance, then

$$\frac{\bar{x}_A - 0.30}{0.10/\sqrt{n}} = +1.645$$

Similarly, if lots that have a mean emission level of 0.40 ppm are to have a 0.10 probability of acceptance, then

$$\frac{\bar{x}_A - 0.40}{0.10/\sqrt{n}} = -1.282$$

These two equations may be solved for n and $\bar{x}_A$, giving $n = 9$ and $\bar{x}_A = 0.356$. It is also possible to design the sampling plan using an OC curve method.

We may also design variables acceptance-sampling procedures such as this one for the case where the standard deviation is unknown. Similarly, we may derive lot-by-lot variables acceptance-sampling plans to give assurance regarding the standard deviation of a lot or process. The standard techniques of statistical hypothesis testing on means and variances can be used to obtain sampling procedures that have specified OC curves. For an extensive discussion of designing these procedures, refer to Montgomery and Runger (1999).

15-4.2   Sequential Sampling by Variables

Just as sequential sampling proves useful in attributes inspection, it can also be applied in variables inspection. The usual assumptions are that the quality characteristic is normally distributed and that the standard deviation of the lot or process is known. The item-by-item sequential sampling plan by variables plots the cumulative sum of the measurements on the quality characteristic. Limit lines for accepting the lot, rejecting the lot, and continuing sampling are constructed much as they are in the case of attributes inspection. Duncan (1986) provides a good discussion of the design of these plans.

15-5   CHAIN SAMPLING

For situations in which testing is destructive or very expensive, sampling plans with small sample sizes are usually selected. These small sample size plans often have acceptance numbers of zero. Plans with zero acceptance numbers are often undesirable, however, in that their OC curves are convex throughout. This means that the probability of lot acceptance begins to drop very rapidly as the lot fraction defective becomes greater than zero. This is often unfair to the producer, and in situations where rectifying inspection is used, it can require the consumer to screen a large number of lots that are essentially of acceptable quality. Figures 14-5 and 14-7 in Chapter 14 present OC curves of sampling plans that have acceptance numbers of zero and acceptance numbers that are greater than zero.

Dodge (1955) suggested an alternate procedure, known as chain sampling, that might be a substitute for ordinary single-sampling plans with zero acceptance numbers in certain circumstances. Chain-sampling plans make use of the cumulative results of several preceding lots. The general procedure is as follows:

1. For each lot, select the sample of size n and observe the number of defectives.

2. If the sample has zero defectives, accept the lot; if the sample has two or more defectives, reject the lot; and if the sample has one defective, accept the lot provided there have been no defectives in the previous i lots.

Thus, for a chain-sampling plan given by $n = 5$, $i = 3$, a lot would be accepted if there were no defectives in the sample of five, or if there were one defective in the sample of five and no defectives had been observed in the samples from the previous three lots. This type of plan is known as a ChSP-1 plan.

The effect of chain sampling is to alter the shape of the OC curve near the origin so that it has a more desirable shape. That is, it is more difficult to reject lots with very small fraction defectives with a ChSP-1 plan than it is with ordinary single sampling. Figure 15-6 shows OC curves for ChSP-1 plans with $n = 5$, $c = 0$, and $i = 1, 2, 3,$ and 5. The curve for $i = 1$ is dotted, and it is not a preferred choice. In practice, values of i usually vary between three and five, since the OC curves of such plans approximate the

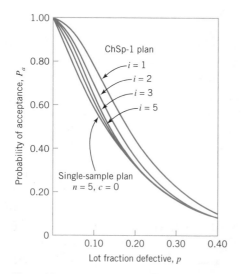

Figure 15-6   OC curves for ChSP-1 plan with $n = 5$, $c = 0$, and $i = 1, 2, 3, 5$. (Reproduced with permission from H. F. Dodge. "Chain Sampling Inspection Plans," *Industrial Quality Control*, Vol. 11, No. 4, 1955.)

single-sampling plan OC curve. The points on the OC curve of a ChSP-1 plan are given by the equation

$$P_a = P(0, n) + P(1, n)[P(0, n)]^i \qquad (15\text{-}3)$$

where $P(0, n)$ and $P(1, n)$ are the probabilities of obtaining 0 and 1 defectives, respectively, out of a random sample of size n. To illustrate the computations, consider the ChSP-1 plan with $n = 5$, $c = 0$, and $i = 3$. For $p = 0.10$, we have

$$P(0, n) = \frac{n!}{d!(n-d)!} p^d (1-p)^{n-d} = \frac{5!}{0!5!} (0.10)^0 (0.90)^5 = 0.590$$

$$P(1, n) = \frac{n!}{d!(n-d)!} p^d (1-p)^{n-d} = \frac{5!}{1!(5-1)!} (0.10)^1 (0.90)^4 = 0.328$$

and

$$\begin{aligned} P_a &= P(0, n) + P(1, n)[P(0, n)]^i \\ &= 0.590 + (0.328)(0.590)^3 \\ &= 0.657 \end{aligned}$$

The proper use of chain sampling requires that the following conditions be met:

1. The lot should be one of a series in a continuing stream of lots, from a process in which there is repetitive production under the same conditions, and in which the lots of products are offered for acceptance in substantially the order of production.
2. Lots should usually be expected to be of essentially the same quality.
3. The sampling agency should have no reason to believe that the current lot is of poorer quality than those immediately preceding.
4. There should be a good record of quality performance on the part of the vendor.
5. The sampling agency must have confidence in the supplier, in that the supplier will not take advantage of its good record and occasionally send a bad lot when such a lot would have the best chance of acceptance.

15-6 CONTINUOUS SAMPLING

All the sampling plans discussed previously are lot-by-lot plans. With these plans, there is an explicit assumption that the product is formed into lots, and the purpose of the sampling plan is to sentence the individual lots. However, many manufacturing operations, particularly complex assembly processes, do not result in the natural formation of lots. For example, manufacturing of many electronics products, such as personal computers, is performed on a conveyorized assembly line.

When production is continuous, two approaches may be used to form lots. The first procedure allows the accumulation of production at given points in the assembly process. This has the disadvantage of creating in-process inventory at various points, which requires additional space, may constitute a safety hazard, and is a generally inefficient approach to managing an assembly line. The second procedure arbitrarily marks off a given segment of production as a "lot." The disadvantage of this approach is that if a lot is ultimately rejected and 100% inspection of the lot is subsequently required, it may be necessary to recall products from manufacturing operations that are farther downstream. This may require disassembly or at least partial destruction of semifinished items.

For these reasons, special sampling plans for continuous production have been developed. Continuous-sampling plans consist of alternating sequences of sampling inspection and screening (100% inspection). The plans usually begin with 100% inspection, and when a stated number of units is found to be free of defects (the number of units i is usually called the **clearance number**), sampling inspection is instituted. Sampling inspection continues until a specified number of defective units is found, at which time 100% inspection is resumed. Continuous-sampling plans are rectifying inspection plans, in that the quality of the product is improved by the partial screening.

15-6.1   CSP-1

Continuous-sampling plans were first proposed by Harold F. Dodge (1943). Dodge's initial plan is called CSP-1. At the start of the plan, all units are inspected 100%. As soon as the clearance number has been reached—that is, as soon as i consecutive units of product are found to be free of defects—100% inspection is discontinued, and only a fraction (f) of the units are inspected. These sample units are selected one at a time at random from the flow of production. If a sample unit is found to be defective, 100% inspection is resumed. All defective units found are either reworked or replaced with good ones. The procedure for CSP-1 is shown in Fig. 15-7.

A CSP-1 plan has an overall AOQL. The value of the AOQL depends on the values of the clearance number i and the sampling fraction f. The same AOQL can be obtained by different combinations of i and f. Table 15-3 presents various values of i and f for CSP-1 that will lead to a stipulated AOQL. Note in the table that an AOQL of 0.79% could be obtained using a sampling plan with $i = 59$ and $f = \frac{1}{3}$, or with $i = 113$ and $f = \frac{1}{7}$.

The choice of i and f is usually based on practical considerations in the manufacturing process. For example, i and f may be influenced by the workload of the inspectors and operators in the system. It is a fairly common practice to use quality-assurance inspectors to do the sampling inspection, and place the burden of 100% inspection on manufacturing. As a general rule, however, it is not a good idea to choose values of f smaller than $\frac{1}{200}$ because the protection against bad quality in a continuous run of production then becomes very poor.

The average number of units inspected in a 100% screening sequence following the occurrence of a defect is equal to

$$u = \frac{1 - q^i}{pq^i} \qquad (15\text{-}4)$$

where $q = 1 - p$, and p is the fraction defective produced when the process is operating

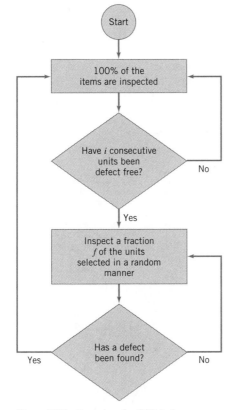

Figure 15-7 Procedure for CSP-1 plans.

Table 15-3 Values of i for CSP-1 Plans

	AOQL (%)															
f	0.018	0.033	0.046	0.074	0.113	0.143	0.198	0.33	0.53	0.79	1.22	1.90	2.90	4.94	7.12	11.46
$\frac{1}{2}$	1,540	840	600	375	245	194	140	84	53	36	23	15	10	6	5	3
$\frac{1}{3}$	2,550	1,390	1,000	620	405	321	232	140	87	59	38	25	16	10	7	5
$\frac{1}{4}$	3,340	1,820	1,310	810	530	420	303	182	113	76	49	32	21	13	9	6
$\frac{1}{5}$	3,960	2,160	1,550	965	630	498	360	217	135	91	58	38	25	15	11	7
$\frac{1}{7}$	4,950	2,700	1,940	1,205	790	623	450	270	168	113	73	47	31	18	13	8
$\frac{1}{10}$	6,050	3,300	2,370	1,470	965	762	550	335	207	138	89	57	38	22	16	10
$\frac{1}{15}$	7,390	4,030	2,890	1,800	1,180	930	672	410	255	170	108	70	46	27	19	12
$\frac{1}{25}$	9,110	4,970	3,570	2,215	1,450	1,147	828	500	315	210	134	86	57	33	23	14
$\frac{1}{50}$	11,730	6,400	4,590	2,855	1,870	1,477	1,067	640	400	270	175	110	72	42	29	18
$\frac{1}{100}$	14,320	7,810	5,600	3,485	2,305	1,820	1,302	790	500	330	215	135	89	52	36	22
$\frac{1}{200}$	17,420	9,500	6,810	4,235	2,760	2,178	1,583	950	590	400	255	165	106	62	43	26

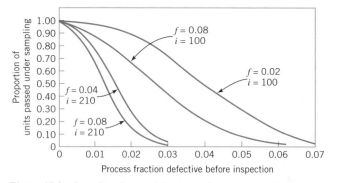

Figure 15-8   Operating-characteristic curves for various continuous sampling plans, CSP-1. (Adapted with permission from A. J. Duncan, *Quality Control and Industrial Statistics*, 5th ed., Irwin, Homewood, Ill., 1986.)

in control. The average number of units passed under the sampling inspection procedure before a defective unit is found is

$$v = \frac{1}{fp} \tag{15-5}$$

The average fraction of total manufactured units inspected in the long run is

$$\text{AFI} = \frac{u + fv}{u + v} \tag{15-6}$$

The average fraction of manufactured units passed under the sampling procedure is

$$P_a = \frac{v}{u + v} \tag{15-7}$$

When P_a is plotted as a function of p, we obtain an operating-characteristic curve for a continuous-sampling plan. Note that whereas an OC curve for a lot-by-lot acceptance-sampling plan gives the percentage of lots that would be passed under sampling inspection, the OC curve for a continuous-sampling plan gives the percentage of units passed under sampling inspection. Graphs of operating-characteristic curves for several values of f and i for CSP-1 plans are shown in Fig. 15-8. Note that for moderate-to-small values of f, i has much more effect on the shape of the curve than does f.

15-6.2   Other Continuous-Sampling Plans

There have been a number of variations in the original Dodge CSP-1 plan. One variation was designed to meet the objection that the occurrence of a single isolated defective unit sometimes does not warrant return to 100% inspection. This is particularly true when dealing with minor defects. To meet this objection, Dodge and Torrey (1951) proposed CSP-2 and CSP-3. Under CSP-2, 100% inspection will not be reinstated when production is under sampling inspection until two defective sample units have been found within a space of K sample units of each other. It is common practice to choose K equal to the clearance number i. CSP-2 plans are indexed by specific AOQLs that may be obtained by different combinations of i and f. CSP-3 is very similar to CSP-2, but is designed to give additional protection against spotty production. It requires that after a defective unit has been found in sampling inspection, the immediately following four units should be inspected. If any of these four units is defective, 100% inspection is immediately reinstituted. If no defectives are found, the plan continues as under CSP-2.

Another common objection to continuous-sampling plans is the abrupt transition between sampling inspection and 100% inspection. Lieberman and Solomon (1955) have designed multilevel continuous-sampling plans to overcome this objection. Multilevel continuous-sampling plans begin with 100% inspection, as does CSP-1, and then switch to inspecting a fraction f of the production as soon as the clearance number i has been reached. However, when under sampling inspection at rate f, a run of i consecutive sample units is found free of defects, then sampling continues at the rate f^2. If a further run of i consecutive units is found to be free of defects, then sampling may continue at the rate f^3. This reduction in sampling frequency may be continued as far as the sampling agency wishes. If at any time sampling inspection reveals a defective unit, return is immediately made to the next lower level of sampling. This type of multilevel continuous-sampling plan greatly reduces the inspection effort when the manufacturing process is operating very well, and increases it during periods of poor production. This transition in inspection intensity is also accomplished without abrupt changes in the inspection load.

Much of the work on continuous-sampling plans has been incorporated into MIL STD 1235C. The standard provides for five different types of continuous-sampling plans. Tables to assist the analyst in designing sampling plans are presented in the standard. CSP-1 and CSP-2 are a part of MIL STD 1235C. In addition, there are two other single-level continuous-sampling procedures, CSP-F and CSP-V. The fifth plan in the standard is CSP-T, a multilevel continuous-sampling plan.

The sampling plans in MIL STD 1235C are indexed by sampling frequency code letter and AOQL. They are also indexed by the AQLs of MIL STD 105E. This aspect of MIL STD 1235C has sparked considerable controversy. CSP plans are not AQL plans and do not have AQLs naturally associated with them. MIL STD 105E, which does focus on the AQL, is designed for manufacturing situations in which lotting is a natural aspect of production, and provides a set of decision rules for sentencing lots so that certain AQL protection is obtained. CSP plans are designed for situations in which production is continuous and lotting is not a natural aspect of the manufacturing situations. In MIL STD 1235C, the sampling plan tables are footnoted and indicate that the AQLs have no meaning relative to the plan, and are only an index.

15-7 SKIP-LOT SAMPLING PLANS

This section describes the development and evaluation of a system of lot-by-lot inspection plans in which a provision is made for inspecting only some fraction of the submitted lots. These plans are known as skip-lot sampling plans. Generally speaking, skip-lot sampling plans should be used only when the quality of the submitted product is good as demonstrated by the vendor's quality history.

Dodge (1956) initially presented skip-lot sampling plans as an extension of CSP-type continuous-sampling plans. In effect, a skip-lot sampling plan is the application of continuous sampling to lots rather than to individual units of production on an assembly line. The version of skip-lot sampling initially proposed by Dodge required a single determination or analysis to ascertain the lot's acceptability or unacceptability. These plans are called SkSP-1. Skip-lot sampling plans designated SkSP-2 follow the next logical step; that is, each lot to be sentenced is sampled according to a particular attribute lot inspection plan. Perry (1973) gives a good discussion of these plans.

A skip-lot sampling plan of type SkSP-2 uses a specified lot inspection plan called the "reference-sampling plan," together with the following rules:

1. Begin with normal inspection, using the reference plan. At this stage of operation, every lot is inspected.

2. When i consecutive lots are accepted on normal inspection, switch to skipping inspection. In skipping inspection, a fraction f of the lots is inspected.

3. When a lot is rejected on skipping inspection, return to normal inspection.

The parameters f and i are the parameters of the skip-lot sampling plan SkSP-2. In general, the clearance number i is a positive integer, and the sampling fraction f lies in the interval $0 < f < 1$. When the sampling fraction $f = 1$, the skip-lot sampling plan reduces to the original reference-sampling plan. Let P denote the probability of acceptance of a lot from the reference-sampling plan. Then $P_a(f, i)$ is the probability of acceptance for the skip-lot sampling plan SkSP-2, where

$$P_a(f, i) = \frac{fP + (1 - f)P^i}{f + (1 - f)P^i} \tag{15-8}$$

It can be shown that for $f_2 < f_1$, a given value of the clearance number i, and a specified reference-sampling plan,

$$P_a(f_1, i) \le P_a(f_2, i) \tag{15-9}$$

Furthermore, for integer clearance numbers $i < j$, a fixed value of f, and a given reference-sampling plan,

$$P_a(f, j) \le P_a(f, i) \tag{15-10}$$

These properties of a skip-lot sampling plan are shown in Figs. 15-9 and 15-10 for the reference-sampling plan $n = 20$, $c = 1$. The OC curve of the reference-sampling plan is also shown on these graphs.

A very important property of a skip-lot sampling plan is the average amount of inspection required. In general, skip-lot sampling plans are used where it is necessary to reduce the average amount of inspection required. The average sample number of a skip-lot sampling plan is

$$\text{ASN}(SkSP) = \text{ASN}(R)F \tag{15-11}$$

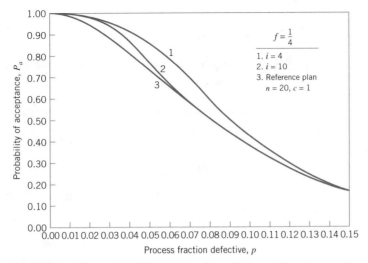

Figure 15-9 OC curves for SkSP-2 skip-lot plans: single-sampling reference plan, same *f*, different *i*. (From R. L. Perry, "Skip-Lot Sampling Plans," *Journal of Quality Technology*, Vol. 5, 1973, with permission of the American Society for Quality Control.)

where F is the average fraction of submitted lots that are sampled and ASN(R) is the average sample number of the reference-sampling plan. It can be shown that

$$F = \frac{f}{(1-f)P^i + f} \tag{15-12}$$

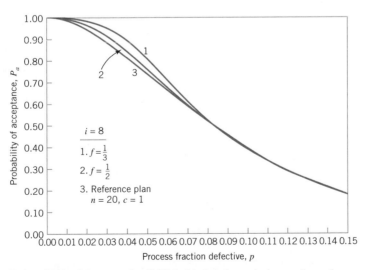

Figure 15-10 OC curves for SkSP-2 skip-lot plans: single-sampling reference plan, same *i*, different *f*. (From R. L. Perry, "Skip-Lot Sampling Plans," *Journal of Quality Technology*, Vol. 5, 1973, with permission of the American Society for Quality Control.)

Thus, since $0 < F < 1$, it follows that

$$\text{ASN}(SkSP) < \text{ASN}(R) \qquad (15\text{-}13)$$

Therefore, skip-lot sampling yields a reduction in the average sample number. For situations in which the quality of incoming lots is very high, this reduction in inspection effort can be significant.

To illustrate the average sample number behavior of a skip-lot sampling plan, consider a reference-sampling plan of $n = 20$ and $c = 1$. Since the average sample number for a single-sampling plan is $\text{ASN} = n$, we have

$$\text{ASN}(SkSP) = n(F)$$

Figure 15-11 presents the ASN curve for the reference-sampling plan $n = 20$, $c = 1$ and the following skip-lot sampling plans:

1. $f = \frac{1}{5}, i = 4$
2. $f = \frac{1}{5}, i = 14$
3. $f = \frac{2}{3}, i = 4$
4. $f = \frac{2}{3}, i = 14$

From examining Fig. 15-11, we note that for small values of incoming lot fraction defective, the reductions in average sample number are very substantial for the skip-lot sampling plans evaluated. If the incoming lot quality is very good, consistently close to

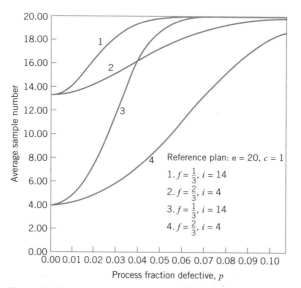

Figure 15-11   Average sample number (ASN) curves for SkSP-2 skip-lot plans with single-sampling reference plan. (From R. L. Perry, "Skip-Lot Sampling Plans," *Journal of Quality Technology*, Vol. 5, 1973, with permission of the American Society for Quality Control.)

zero fraction nonconforming, say, then a small value of f, perhaps $\frac{1}{4}$ or $\frac{1}{5}$, could be used. If incoming quality is slightly worse, then an appropriate value of f might be $\frac{1}{2}$.

Skip-lot sampling plans are an effective acceptance-sampling procedure and may be useful as a system of reduced inspection. Their effectiveness is partially good when the quality of submitted lots is very good. However, one should be careful to use skip-lot sampling plans only for situations in which there is a sufficient history of vendor quality to ensure that the quality of submitted lots is very good. Furthermore, if the vendor's process is highly erratic and there is a great deal of variability from lot to lot, skip-lot sampling plans are inappropriate. They seem to work best when the vendor's processes are in a state of statistical control and when the process capability is adequate to ensure virtually defect-free production.

15-8 EXERCISES

15-1. The density of a plastic part used in a cellular telephone is required to be at least 0.70 g/cm³. The parts are supplied in large lots, and a variables sampling plan is to be used to sentence the lots. It is desired to have $p_1 = 0.02$, $p_2 = 0.10$, $\alpha = 0.10$, and $\beta = 0.05$. The variability of the manufacturing process is unknown but will be estimated by the sample standard deviation.
   (a) Find an appropriate variables sampling plan, using Procedure 1.
   (b) Suppose that a sample of the appropriate size was taken, and $\bar{x} = 0.73$, $S = 1.05 \times 10^{-2}$. Should the lot be accepted or rejected?
   (c) Sketch the OC curve for this sampling plan. Find the probability of accepting lots that are 5% defective.

15-2. A belt that is used in a drive mechanism in a copier machine is required to have a minimum tensile strength of LSL = 150 lb. It is known from long experience that $\sigma = 5$ lb for this particular belt. Find a variables sampling plan so that $p_1 = 0.005$, $p_2 = 0.02$, $\alpha = 0.05$, and $\beta = 0.10$. Assume that Procedure 1 is to be used.

15-3. Describe how rectifying inspection can be used with variables sampling. What are the appropriate equations for the AOQ and the ATI, assuming single sampling and requiring that all defective items found in either sampling or 100% inspection are replaced by good ones?

15-4. An inspector for a military agency desires a variables sampling plan for use with an AQL of 1.5%, assuming that lots are of size 7000. If the standard deviation of the lot or process is unknown, derive a sampling plan using Procedure 1 from MIL STD 414.

15-5. How does the sample size found in Exercise 15-4 compare with what would have been used under MIL STD 105E?

15-6. A lot of 500 items is submitted for inspection. Suppose that we wish to find a plan from MIL STD 414, using inspection level II. If the AQL is 4%, find the Procedure 1 sampling plan from the standard.

15-7. A soft-drink bottler purchases nonreturnable glass bottles from a vendor. The lower specification on bursting strength in the bottles is 225 psi. The bottler wishes to use variables sampling to sentence the lots, and has decided to use an AQL of 1%. Find an appropriate set of normal and tightened sampling plans from the standard. Suppose that a lot is submitted, and the sample results yield

$$\bar{x} = 255 \qquad S = 10$$

Determine the disposition of the lot using Procedure 1. The lot size is $N = 100,000$.

15-8. A chemical ingredient is packed in metal containers. A large shipment of these containers has been delivered to a manufacturing facility. The mean bulk density of this ingredient should not be less than 0.15 g/cm^3. Suppose that lots of this quality are to have a 0.95 probability of acceptance. If the mean bulk density is as low as 0.1450, the probability of acceptance of the lot should be 0.10. Suppose we know that the standard deviation of bulk density is approximately 0.005 g/cm^3. Obtain a variables sampling plan that could be used to sentence the lots.

15-9. A standard of 0.3 ppm has been established for formaldehyde emission levels in wood products. Suppose that the standard deviation of emissions in an individual board is $\sigma = 0.10$ ppm. Any lot that contains 1% of its items above 0.3 ppm is considered acceptable. Any lot that has 8% or more of its items above 0.3 ppm is considered unacceptable. Good lots are to be accepted with probability 0.95, and bad lots are to be rejected with probability 0.90.
   (a) Derive a variables sampling plan for this situation.
   (b) Using the 1% nonconformance level as an AQL, and assuming that lots consist of 5000 panels, find an appropriate set of sampling plans from MIL STD 414, assuming σ unknown. Compare the sample sizes and the protection that both producer and consumer obtain from this plan with the plan derived in part (a).
   (c) Find an attributes sampling plan that has the same OC curve as the variables sampling plan derived in part (a). Compare the sample sizes required for equivalent protection. Under what circumstances would variables sampling be more economically efficient?
   (d) Using the 1% nonconforming as an AQL, find an attributes sampling plan from MIL STD 105E. Compare the sample sizes and the protection obtained from this plan with the plans derived in parts (a), (b), and (c).

15-10. Consider a single-sampling plan with $n = 25$, $c = 0$. Draw the OC curve for this plan. Now consider chain-sampling plans with $n = 25$, $c = 0$, and $i = 1, 2, 5, 7$. Sketch the OC curves for these chain-sampling plans on the same axis. Discuss the behavior of chain sampling in this situation compared to the conventional single-sampling plan with $c = 0$.

15-11. An electronics manufacturer buys memory devices in lots of 30,000 from a vendor. The vendor has a long record of good quality performance, with an average fraction defective of approximately 0.10%. The quality engineering department has suggested using a conventional acceptance-sampling plan with $n = 32$, $c = 0$.
   (a) Draw the OC curve of this sampling plan.
   (b) If lots are of a quality that is near the vendor's long-term process average, what is the average total inspection at that level of quality?
   (c) Consider a chain-sampling plan with $n = 32$, $c = 0$, and $i = 3$. Contrast the performance of this plan with the conventional sampling plan $n = 32$, $c = 0$.
   (d) How would the performance of this chain-sampling plan change if we substituted $i = 4$ in part (c)?

15-12. A ChSP-1 plan has $n = 4$, $c = 0$, and

$i = 3$. Draw the OC curve for this plan.

15-13.   A chain-sampling plan is used for the inspection of lots of size $N = 500$. The sample size is $n = 6$. If the sample contains no defectives, the lot is accepted. If one defective is found, the lot is accepted provided that the samples from the four previous lots are free of defectives. Determine the probability of acceptance of a lot that is 2% defective.

15-14.   Suppose that a manufacturing process operates in continuous production, such that continuous sampling plans could be applied. Determine three different CSP-1 sampling plans that could be used for an AOQL of 0.198%.

15-15.   For the sampling plans developed in Exercise 15-14, compare the plans' performance in terms of average fraction inspected, given that the process is in control at an average fallout level of 0.15%. Compare the plans in terms of their operating-characteristic curves.

15-16.   Suppose that CSP-1 is used for a manufacturing process where it is desired to maintain an AOQL of 1.90%. Specify two CSP-1 plans that would meet this AOQL target.

15-17.   Compare the plans developed in Exercise 15-16 in terms of average fraction inspected and their operating-characteristic curves. Which plan would you prefer if $p = 0.0375$?

Appendix

I.	Summary of Common Probability Distributions Often Used in Statistical Quality Control	751
II.	Cumulative Standard Normal Distribution	752
III.	Percentage Points of the χ^2 Distribution	754
IV.	Percentage Points of the t Distribution	755
V.	Percentage Points of the F Distribution	756
VI.	Factors for Constructing Variables Control Charts	761
VII.	Factors for Two-Sided Normal Tolerance Limits	762
VIII.	Factors for One-Sided Normal Tolerance Limits	763
IX.	Random Numbers	764

Appendix I Summary of Common Probability Distributions Often Used in Statistical Quality Control

Name	Probability Distribution	Mean	Variance
Discrete			
Uniform	$\dfrac{1}{b-a}, a \leq b$	$\dfrac{(b+a)}{2}$	$\dfrac{(b-a+1)^2-1}{12}$
Binomial	$\dbinom{n}{x} p^x(1-p)^{n-x},$ $x = 0, 1, \ldots, n, 0 \leq p \leq 1$	np	$np(1-p)$
Pascal (negative binomial)	$\dbinom{x-1}{r-1} p^r(1-p)^{x-r},$ $x = r, r+1, r+2, \ldots$	$\dfrac{r}{p}$	$\dfrac{r(1-p)}{p^2}$
Geometric	$(1-p)^{x-1}p,$ $x = 1, 2, \ldots, 0 \leq p \leq 1$	$\dfrac{1}{p}$	$\dfrac{(1-p)}{p^2}$
Hypergeometric	$\dfrac{\dbinom{D}{x}\dbinom{N-D}{n-x}}{\dbinom{N}{n}},$ $x = 0, 1, \ldots, \min(D,n), D \leq N,$ $n \leq N$	$np,$ where $p = \dfrac{D}{N}$	$np(1-p)\left(\dfrac{N-n}{N-1}\right)$
Poisson	$\dfrac{e^{-\lambda}\lambda^x}{x!}, x = 0, 1, 2, \ldots, \lambda > 0$	λ	λ
Continuous			
Uniform	$\dfrac{1}{b-a}, a \leq x \leq b$	$\dfrac{(b+a)}{2}$	$\dfrac{(b-a)^2}{12}$
Normal	$\dfrac{1}{\sqrt{2\pi}\sigma} e^{-\frac{1}{2}\left(\frac{x-\mu}{\sigma}\right)^2},$ $-\infty < x < \infty, -\infty < \mu < \infty, \sigma > 0$	μ	σ^2
Exponential	$\lambda e^{-\lambda x}, x \geq 0 \quad \lambda > 0$	$1/\lambda$	$1/\lambda^2$
Gamma	$\dfrac{\lambda x^{r-1} e^{-\lambda x}}{\Gamma(r)}, x > 0, r > 0, \lambda > 0$	r/λ	r/λ^2
Erlang	$\dfrac{\lambda^r x^{r-1} e^{-\lambda x}}{(r-1)!}, x > 0, r = 1, 2, \ldots$	r/λ	r/λ^2
Weibull	$\dfrac{\beta}{\theta}\left(\dfrac{x}{\theta}\right)^{\beta-1} e^{-(x/\theta)^\beta},$ $x > 0, \beta > 0, \theta > 0$	$\theta\Gamma\left(1+\dfrac{1}{\beta}\right)$	$\theta^2\Gamma\left(1+\dfrac{2}{\beta}\right)$ $-\theta^2\left[\Gamma\left(1+\dfrac{1}{\beta}\right)\right]^2$

(handwritten note next to Geometric: "Memorylessness")

(handwritten note next to Exponential: "memoryless")

Appendix II   Cumulative Standard Normal Distribution

$$\Phi(z) = \int_{-\infty}^{z} \frac{1}{\sqrt{2\pi}} \, e^{-u^2/2} \, du$$

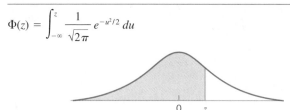

z	0.00	0.01	0.02	0.03	0.04	z
0.0	0.50000	0.50399	0.50798	0.51197	0.51595	0.0
0.1	0.53983	0.54379	0.54776	0.55172	0.55567	0.1
0.2	0.57926	0.58317	0.58706	0.59095	0.59483	0.2
0.3	0.61791	0.62172	0.62551	0.62930	0.63307	0.3
0.4	0.65542	0.65910	0.62276	0.66640	0.67003	0.4
0.5	0.69146	0.69497	0.69847	0.70194	0.70540	0.5
0.6	0.72575	0.72907	0.73237	0.73565	0.73891	0.6
0.7	0.75803	0.76115	0.76424	0.76730	0.77035	0.7
0.8	0.78814	0.79103	0.79389	0.79673	0.79954	0.8
0.9	0.81594	0.81859	0.82121	0.82381	0.82639	0.9
1.0	0.84134	0.84375	0.84613	0.84849	0.85083	1.0
1.1	0.86433	0.86650	0.86864	0.87076	0.87285	1.1
1.2	0.88493	0.88686	0.88877	0.89065	0.89251	1.2
1.3	0.90320	0.90490	0.90658	0.90824	0.90988	1.3
1.4	0.91924	0.92073	0.92219	0.92364	0.92506	1.4
1.5	0.93319	0.93448	0.93574	0.93699	0.93822	1.5
1.6	0.94520	0.94630	0.94738	0.94845	0.94950	1.6
1.7	0.95543	0.95637	0.95728	0.95818	0.95907	1.7
1.8	0.96407	0.96485	0.96562	0.96637	0.96711	1.8
1.9	0.97128	0.97193	0.97257	0.97320	0.97381	1.9
2.0	0.97725	0.97778	0.97831	0.97882	0.97932	2.0
2.1	0.98214	0.98257	0.98300	0.98341	0.98382	2.1
2.2	0.98610	0.98645	0.98679	0.98713	0.98745	2.2
2.3	0.98928	0.98956	0.98983	0.99010	0.99036	2.3
2.4	0.99180	0.99202	0.99224	0.99245	0.99266	2.4
2.5	0.99379	0.99396	0.99413	0.99430	0.99446	2.5
2.6	0.99534	0.99547	0.99560	0.99573	0.99585	2.6
2.7	0.99653	0.99664	0.99674	0.99683	0.99693	2.7
2.8	0.99744	0.99752	0.99760	0.99767	0.99774	2.8
2.9	0.99813	0.99819	0.99825	0.99831	0.99836	2.9
3.0	0.99865	0.99869	0.99874	0.99878	0.99882	3.0
3.1	0.99903	0.99906	0.99910	0.99913	0.99916	3.1
3.2	0.99931	0.99934	0.99936	0.99938	0.99940	3.2
3.3	0.99952	0.99953	0.99955	0.99957	0.99958	3.3
3.4	0.99966	0.99968	0.99969	0.99970	0.99971	3.4
3.5	0.99977	0.99978	0.99978	0.99979	0.99980	3.5
3.6	0.99984	0.99985	0.99985	0.99986	0.99986	3.6
3.7	0.99989	0.99990	0.99990	0.99990	0.99991	3.7
3.8	0.99993	0.99993	0.99993	0.99994	0.99994	3.8
3.9	0.99995	0.99995	0.99996	0.99996	0.99996	3.9

Appendix II (*Continued*)

$$\Phi(z) = \int_{-\infty}^{z} \frac{1}{\sqrt{2\pi}} e^{-u^2/2} \, du$$

z	0.05	0.06	0.07	0.08	0.09	z
0.0	0.51994	0.52392	0.52790	0.53188	0.53586	0.0
0.1	0.55962	0.56356	0.56749	0.57142	0.57534	0.1
0.2	0.59871	0.60257	0.60642	0.61026	0.61409	0.2
0.3	0.63683	0.64058	0.64431	0.64803	0.65173	0.3
0.4	0.67364	0.67724	0.68082	0.68438	0.68793	0.4
0.5	0.70884	0.71226	0.71566	0.71904	0.72240	0.5
0.6	0.74215	0.74537	0.74857	0.75175	0.75490	0.6
0.7	0.77337	0.77637	0.77935	0.78230	0.78523	0.7
0.8	0.80234	0.80510	0.80785	0.81057	0.81327	0.8
0.9	0.82894	0.83147	0.83397	0.83646	0.83891	0.9
1.0	0.85314	0.85543	0.85769	0.85993	0.86214	1.0
1.1	0.87493	0.87697	0.87900	0.88100	0.88297	1.1
1.2	0.89435	0.89616	0.89796	0.89973	0.90147	1.2
1.3	0.91149	0.91308	0.91465	0.91621	0.91773	1.3
1.4	0.92647	0.92785	0.92922	0.93056	0.93189	1.4
1.5	0.93943	0.94062	0.94179	0.94295	0.94408	1.5
1.6	0.95053	0.95154	0.95254	0.95352	0.95448	1.6
1.7	0.95994	0.96080	0.96164	0.96246	0.96327	1.7
1.8	0.96784	0.96856	0.96926	0.96995	0.97062	1.8
1.9	0.97441	0.97500	0.97558	0.97615	0.97670	1.9
2.0	0.97982	0.98030	0.98077	0.98124	0.98169	2.0
2.1	0.98422	0.98461	0.98500	0.98537	0.98574	2.1
2.2	0.98778	0.98809	0.98840	0.98870	0.98899	2.2
2.3	0.99061	0.99086	0.99111	0.99134	0.99158	2.3
2.4	0.99286	0.99305	0.99324	0.99343	0.99361	2.4
2.5	0.99461	0.99477	0.99492	0.99506	0.99520	2.5
2.6	0.99598	0.99609	0.99621	0.99632	0.99643	2.6
2.7	0.99702	0.99711	0.99720	0.99728	0.99736	2.7
2.8	0.99781	0.99788	0.99795	0.99801	0.99807	2.8
2.9	0.99841	0.99846	0.99851	0.99856	0.99861	2.9
3.0	0.99886	0.99889	0.99893	0.99897	0.99900	3.0
3.1	0.99918	0.99921	0.99924	0.99926	0.99929	3.1
3.2	0.99942	0.99944	0.99946	0.99948	0.99950	3.2
3.3	0.99960	0.99961	0.99962	0.99964	0.99965	3.3
3.4	0.99972	0.99973	0.99974	0.99975	0.99976	3.4
3.5	0.99981	0.99981	0.99982	0.99983	0.99983	3.5
3.6	0.99987	0.99987	0.99988	0.99988	0.99989	3.6
3.7	0.99991	0.99992	0.99992	0.99992	0.99992	3.7
3.8	0.99994	0.99994	0.99995	0.99995	0.99995	3.8
3.9	0.99996	0.99996	0.99996	0.99997	0.99997	3.9

Appendix III   Percentage Points of the χ^2 Distribution[a]

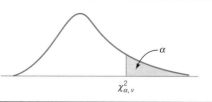

ν	0.995	0.990	0.975	0.950	α 0.500	0.050	0.025	0.010	0.005
1	0.00 +	0.00 +	0.00 +	0.00 +	0.45	3.84	5.02	6.63	7.88
2	0.01	0.02	0.05	0.10	1.39	5.99	7.38	9.21	10.60
3	0.07	0.11	0.22	0.35	2.37	7.81	9.35	11.34	12.84
4	0.21	0.30	0.48	0.71	3.36	9.49	11.14	13.28	14.86
5	0.41	0.55	0.83	1.15	4.35	11.07	12.38	15.09	16.75
6	0.68	0.87	1.24	1.64	5.35	12.59	14.45	16.81	18.55
7	0.99	1.24	1.69	2.17	6.35	14.07	16.01	18.48	20.28
8	1.34	1.65	2.18	2.73	7.34	15.51	17.53	20.09	21.96
9	1.73	2.09	2.70	3.33	8.34	16.92	19.02	21.67	23.59
10	2.16	2.56	3.25	3.94	9.34	18.31	20.48	23.21	25.19
11	2.60	3.05	3.82	4.57	10.34	19.68	21.92	24.72	26.76
12	3.07	3.57	4.40	5.23	11.34	21.03	23.34	26.22	28.30
13	3.57	4.11	5.01	5.89	12.34	22.36	24.74	27.69	29.82
14	4.07	4.66	5.63	6.57	13.34	23.68	26.12	29.14	31.32
15	4.60	5.23	6.27	7.26	14.34	25.00	27.49	30.58	32.80
16	5.14	5.81	6.91	7.96	15.34	26.30	28.85	32.00	34.27
17	5.70	6.41	7.56	8.67	16.34	27.59	30.19	33.41	35.72
18	6.26	7.01	8.23	9.39	17.34	28.87	31.53	34.81	37.16
19	6.84	7.63	8.91	10.12	18.34	30.14	32.85	36.19	38.58
20	7.43	8.26	9.59	10.85	19.34	31.41	34.17	37.57	40.00
25	10.52	11.52	13.12	14.61	24.34	37.65	40.65	44.31	46.93
30	13.79	14.95	16.79	18.49	29.34	43.77	46.98	50.89	53.67
40	20.71	22.16	24.43	26.51	39.34	55.76	59.34	63.69	66.77
50	27.99	29.71	32.36	34.76	49.33	67.50	71.42	76.15	79.49
60	35.53	37.48	40.48	43.19	59.33	79.08	83.30	88.38	91.95
70	43.28	45.44	48.76	51.74	69.33	90.53	95.02	100.42	104.22
80	51.17	53.54	57.15	60.39	79.33	101.88	106.63	112.33	116.32
90	59.20	61.75	65.65	69.13	89.33	113.14	118.14	124.12	128.30
100	67.33	70.06	74.22	77.93	99.33	124.34	129.56	135.81	140.17

ν = degrees of freedom.

[a] Adapted with permission from *Biometrika Tables for Statisticians*, Vol. 1, 3rd ed., by E. S. Pearson and H. O. Hartley, Cambridge University Press, Cambridge, 1966.

Appendix IV   Percentage Points of the t Distribution[a]

v	\multicolumn{10}{c}{α}
	0.40	0.25	0.10	0.05	0.025	0.01	0.005	0.0025	0.001	0.0005
1	0.325	1.000	3.078	6.314	12.706	31.821	63.657	127.32	318.31	636.62
2	0.289	0.816	1.886	2.920	4.303	6.965	9.925	14.089	23.326	31.598
3	0.277	0.765	1.638	2.353	3.182	4.541	5.841	7.453	10.213	12.924
4	0.271	0.741	1.533	2.132	2.776	3.747	4.604	5.598	7.173	8.610
5	0.267	0.727	1.476	2.015	2.571	3.365	4.032	4.773	5.893	6.869
6	0.265	0.727	1.440	1.943	2.447	3.143	3.707	4.317	5.208	5.959
7	0.263	0.711	1.415	1.895	2.365	2.998	3.49	4.019	4.785	5.408
8	0.262	0.706	1.397	1.860	2.306	2.896	3.355	3.833	4.501	5.041
9	0.261	0.703	1.383	1.833	2.262	2.821	3.250	3.690	4.297	4.781
10	0.260	0.700	1.372	1.812	2.228	2.764	3.169	3.581	4.144	4.587
11	0.260	0.697	1.363	1.796	2.201	2.718	3.106	3.497	4.025	4.437
12	0.259	0.695	1.356	1.782	2.179	2.681	3.055	3.428	3.930	4.318
13	0.259	0.694	1.350	1.771	2.160	2.650	3.012	3.372	3.852	4.221
14	0.258	0.692	1.345	1.761	2.145	2.624	2.977	3.326	3.787	4.140
15	0.258	0.691	1.341	1.753	2.131	2.602	2.947	3.286	3.733	4.073
16	0.258	0.690	1.337	1.746	2.120	2.583	2.921	3.252	3.686	4.015
17	0.257	0.689	1.333	1.740	2.110	2.567	2.898	3.222	3.646	3.965
18	0.257	0.688	1.330	1.734	2.101	2.552	2.878	3.197	3.610	3.992
19	0.257	0.688	1.328	1.729	2.093	2.539	2.861	3.174	3.579	3.883
20	0.257	0.687	1.325	1.725	2.086	2.528	2.845	3.153	3.552	3.850
21	0.257	0.686	1.323	1.721	2.080	2.518	2.831	3.135	3.527	3.819
22	0.256	0.686	1.321	1.717	2.074	2.508	2.819	3.119	3.505	3.792
23	0.256	0.685	1.319	1.714	2.069	2.500	2.807	3.104	3.485	3.767
24	0.256	0.685	1.318	1.711	2.064	2.492	2.797	3.091	3.467	3.745
25	0.256	0.684	1.316	1.708	2.060	2.485	2.787	3.078	3.450	3.725
26	0.256	0.684	1.315	1.706	2.056	2.479	2.779	3.067	3.435	3.707
27	0.256	0.684	1.314	1.703	2.052	2.473	2.771	3.057	3.421	3.690
28	0.256	0.683	1.313	1.701	2.048	2.467	2.763	3.047	3.408	3.674
29	0.256	0.683	1.311	1.699	2.045	2.462	2.756	3.038	3.396	3.659
30	0.256	0.683	1.310	1.697	2.042	2.457	2.750	3.030	3.385	3.646
40	0.255	0.681	1.303	1.684	2.021	2.423	2.704	2.971	3.307	3.551
60	0.254	0.679	1.296	1.671	2.000	2.390	2.660	2.915	3.232	3.460
120	0.254	0.677	1.289	1.658	1.980	2.358	2.617	2.860	3.160	3.373
∞	0.253	0.674	1.282	1.645	1.960	2.326	2.576	2.807	3.090	3.291

v = degrees of freedom.

[a] Adapted with permission from *Biometrika Tables for Statisticians*, Vol. 1, 3rd ed., by E. S. Pearson and H. O. Hartley, Cambridge University Press, Cambridge, 1966.

Appendix V Percentage Points of the F Distribution

$$F_{0.25,\nu_1,\nu_2}$$

ν_2 \ ν_1	1	2	3	4	5	6	7	8	9	10	12	15	20	24	30	40	60	120	∞
1	5.83	7.50	8.20	8.58	8.82	8.98	9.10	9.19	9.26	9.32	9.41	9.49	9.58	9.63	9.67	9.71	9.76	9.80	9.85
2	2.57	3.00	3.15	3.23	3.28	3.31	3.34	3.35	3.37	3.38	3.39	3.41	3.43	3.43	3.44	3.45	3.46	3.47	3.48
3	2.02	2.28	2.36	2.39	2.41	2.42	2.43	2.44	2.44	2.44	2.45	2.46	2.46	2.46	2.47	2.47	2.47	2.47	2.47
4	1.81	2.00	2.05	2.06	2.07	2.08	2.08	2.08	2.08	2.08	2.08	2.08	2.08	2.08	2.08	2.08	2.08	2.08	2.08
5	1.69	1.85	1.88	1.89	1.89	1.89	1.89	1.89	1.89	1.89	1.89	1.89	1.88	1.88	1.88	1.88	1.87	1.87	1.87
6	1.62	1.76	1.78	1.79	1.79	1.78	1.78	1.78	1.77	1.77	1.77	1.76	1.76	1.75	1.75	1.75	1.74	1.74	1.74
7	1.57	1.70	1.72	1.72	1.71	1.71	1.70	1.70	1.70	1.69	1.68	1.68	1.67	1.67	1.66	1.66	1.65	1.65	1.65
8	1.54	1.66	1.67	1.66	1.66	1.65	1.64	1.64	1.63	1.63	1.62	1.62	1.61	1.60	1.60	1.59	1.59	1.58	1.58
9	1.51	1.62	1.63	1.63	1.62	1.61	1.60	1.60	1.59	1.59	1.58	1.57	1.56	1.56	1.55	1.54	1.54	1.53	1.53
10	1.49	1.60	1.60	1.59	1.59	1.58	1.57	1.56	1.56	1.55	1.54	1.53	1.52	1.52	1.51	1.51	1.50	1.49	1.48
11	1.47	1.58	1.58	1.57	1.56	1.55	1.54	1.53	1.53	1.52	1.51	1.50	1.49	1.49	1.48	1.47	1.47	1.46	1.45
12	1.46	1.56	1.56	1.55	1.54	1.53	1.52	1.51	1.51	1.50	1.49	1.48	1.47	1.46	1.45	1.45	1.44	1.43	1.42
13	1.45	1.55	1.55	1.53	1.52	1.51	1.50	1.49	1.49	1.48	1.47	1.46	1.45	1.44	1.43	1.42	1.42	1.41	1.40
14	1.44	1.53	1.53	1.52	1.51	1.50	1.49	1.48	1.47	1.46	1.45	1.44	1.43	1.42	1.41	1.41	1.40	1.39	1.38
15	1.43	1.52	1.52	1.51	1.49	1.48	1.47	1.46	1.46	1.45	1.44	1.43	1.41	1.41	1.40	1.39	1.38	1.37	1.36
16	1.42	1.51	1.51	1.50	1.48	1.47	1.46	1.45	1.44	1.44	1.43	1.41	1.40	1.39	1.38	1.37	1.36	1.35	1.34
17	1.42	1.51	1.50	1.49	1.47	1.46	1.45	1.44	1.43	1.43	1.41	1.40	1.39	1.38	1.37	1.36	1.35	1.34	1.33
18	1.41	1.50	1.49	1.48	1.46	1.45	1.44	1.43	1.42	1.42	1.40	1.39	1.38	1.37	1.36	1.35	1.34	1.33	1.32
19	1.41	1.49	1.49	1.47	1.46	1.44	1.43	1.42	1.41	1.41	1.40	1.38	1.37	1.36	1.35	1.34	1.33	1.32	1.30
20	1.40	1.49	1.48	1.47	1.45	1.44	1.43	1.42	1.41	1.40	1.39	1.37	1.36	1.35	1.34	1.33	1.32	1.31	1.29
21	1.40	1.48	1.48	1.46	1.44	1.43	1.42	1.41	1.40	1.39	1.38	1.37	1.35	1.34	1.33	1.32	1.31	1.30	1.28
22	1.40	1.48	1.47	1.45	1.44	1.42	1.41	1.40	1.39	1.39	1.37	1.36	1.34	1.33	1.32	1.31	1.30	1.29	1.28
23	1.39	1.47	1.47	1.45	1.43	1.42	1.41	1.40	1.39	1.38	1.37	1.35	1.34	1.33	1.32	1.31	1.30	1.28	1.27
24	1.39	1.47	1.46	1.44	1.43	1.41	1.40	1.39	1.38	1.38	1.36	1.35	1.33	1.32	1.31	1.30	1.29	1.28	1.26
25	1.39	1.47	1.46	1.44	1.42	1.41	1.40	1.39	1.38	1.37	1.36	1.34	1.33	1.32	1.31	1.29	1.28	1.28	1.25
26	1.38	1.46	1.45	1.44	1.42	1.41	1.39	1.38	1.37	1.37	1.35	1.34	1.32	1.31	1.30	1.29	1.28	1.27	1.25
27	1.38	1.46	1.45	1.43	1.42	1.40	1.39	1.38	1.37	1.36	1.35	1.33	1.32	1.31	1.30	1.28	1.28	1.26	1.24
28	1.38	1.46	1.45	1.43	1.41	1.40	1.39	1.38	1.37	1.36	1.34	1.33	1.31	1.30	1.29	1.28	1.27	1.25	1.24
29	1.38	1.45	1.45	1.43	1.41	1.40	1.38	1.37	1.36	1.35	1.34	1.32	1.31	1.30	1.29	1.27	1.26	1.25	1.23
30	1.38	1.45	1.44	1.42	1.41	1.39	1.38	1.37	1.36	1.35	1.34	1.32	1.30	1.29	1.28	1.27	1.26	1.24	1.23
40	1.36	1.44	1.42	1.40	1.39	1.37	1.36	1.35	1.34	1.33	1.31	1.30	1.28	1.26	1.25	1.24	1.22	1.21	1.19
60	1.35	1.42	1.41	1.38	1.37	1.35	1.33	1.32	1.31	1.30	1.29	1.27	1.25	1.24	1.22	1.21	1.19	1.17	1.15
120	1.34	1.40	1.39	1.37	1.35	1.33	1.31	1.30	1.29	1.28	1.26	1.24	1.22	1.21	1.19	1.18	1.16	1.13	1.10
∞	1.32	1.39	1.37	1.35	1.33	1.31	1.29	1.28	1.27	1.25	1.24	1.22	1.19	1.18	1.16	1.14	1.12	1.08	1.00

Degrees of freedom for the numerator (ν_1)

Degrees of freedom for the denominator (ν_2)

Note: $F_{0.75,\nu_1,\nu_2} = 1/F_{0.25,\nu_2,\nu_1}$.

Source: Adapted with permission from *Biometrika Tables for Statisticians*, Vol. 1, 3rd ed., by E. S. Pearson and H. O. Hartley, Cambridge University Press, Cambridge, 1966.

$F_{0.10, v_1, v_2}$

v_2 \ v_1	Degrees of freedom for the numerator (n_1)																		
	1	2	3	4	5	6	7	8	9	10	12	15	20	24	30	40	60	120	∞
1	39.86	49.50	53.59	55.83	57.24	58.20	58.91	59.44	59.86	60.19	60.71	61.22	61.74	62.00	62.26	62.53	62.79	63.06	63.33
2	8.53	9.00	9.16	9.24	9.29	9.33	9.35	9.37	9.38	9.39	9.41	9.42	9.44	9.45	9.46	9.47	9.47	9.48	9.49
3	5.54	5.46	5.39	5.34	5.31	5.28	5.27	5.25	5.24	5.23	5.22	5.20	5.18	5.18	5.17	5.16	5.15	5.14	5.13
4	4.54	4.32	4.19	4.11	4.05	4.01	3.98	3.95	3.94	3.92	3.90	3.87	3.84	3.83	3.82	3.80	3.79	3.78	3.76
5	4.06	3.78	3.62	3.52	3.45	3.40	3.37	3.34	3.32	3.30	3.27	3.24	3.21	3.19	3.17	3.16	3.14	3.12	3.10
6	3.78	3.46	3.29	3.18	3.11	3.05	3.01	2.98	2.96	2.94	2.90	2.87	2.84	2.82	2.80	2.78	2.76	2.74	2.72
7	3.59	3.26	3.07	2.96	2.88	2.83	2.78	2.75	2.72	2.70	2.67	2.63	2.59	2.58	2.56	2.54	2.51	2.49	2.47
8	3.46	3.11	2.92	2.81	2.73	2.67	2.62	2.59	2.56	2.54	2.50	2.46	2.42	2.40	2.38	2.36	2.34	2.32	2.29
9	3.36	3.01	2.81	2.69	2.61	2.55	2.51	2.47	2.44	2.42	2.38	2.34	2.30	2.28	2.25	2.23	2.21	2.18	2.16
10	3.29	2.92	2.73	2.61	2.52	2.46	2.41	2.38	2.35	2.32	2.28	2.24	2.20	2.18	2.16	2.13	2.11	2.08	2.06
11	3.23	2.86	2.66	2.54	2.45	2.39	2.34	2.30	2.27	2.25	2.21	2.17	2.12	2.10	2.08	2.05	2.03	2.00	1.97
12	3.18	2.81	2.61	2.48	2.39	2.33	2.28	2.24	2.21	2.19	2.15	2.10	2.06	2.04	2.01	1.99	1.96	1.93	1.90
13	3.14	2.76	2.56	2.43	2.35	2.28	2.23	2.20	2.16	2.14	2.10	2.05	2.01	1.98	1.96	1.93	1.90	1.88	1.85
14	3.10	2.73	2.52	2.39	2.31	2.24	2.19	2.15	2.12	2.10	2.05	2.01	1.96	1.94	1.91	1.89	1.86	1.83	1.80
15	3.07	2.70	2.49	2.36	2.27	2.21	2.16	2.12	2.09	2.06	2.02	1.97	1.92	1.90	1.87	1.85	1.82	1.79	1.76
16	3.05	2.67	2.46	2.33	2.24	2.18	2.13	2.09	2.06	2.03	1.99	1.94	1.89	1.87	1.84	1.81	1.78	1.75	1.72
17	3.03	2.64	2.44	2.31	2.22	2.15	2.10	2.06	2.03	2.00	1.96	1.91	1.86	1.84	1.81	1.78	1.75	1.72	1.69
18	3.01	2.62	2.42	2.29	2.20	2.13	2.08	2.04	2.00	1.98	1.93	1.89	1.84	1.81	1.78	1.75	1.72	1.69	1.66
19	2.99	2.61	2.40	2.27	2.18	2.11	2.06	2.02	1.98	1.96	1.91	1.86	1.81	1.79	1.76	1.73	1.70	1.67	1.63
20	2.97	2.59	2.38	2.25	2.16	2.09	2.04	2.00	1.96	1.94	1.89	1.84	1.79	1.77	1.74	1.71	1.68	1.64	1.61
21	2.96	2.57	2.36	2.23	2.14	2.08	2.02	1.98	1.95	1.92	1.87	1.83	1.78	1.75	1.72	1.69	1.66	1.62	1.59
22	2.95	2.56	2.35	2.22	2.13	2.06	2.01	1.97	1.93	1.90	1.86	1.81	1.76	1.73	1.70	1.67	1.64	1.60	1.57
23	2.94	2.55	2.34	2.21	2.11	2.05	1.99	1.95	1.92	1.89	1.84	1.80	1.74	1.72	1.69	1.66	1.62	1.59	1.55
24	2.93	2.54	2.33	2.19	2.10	2.04	1.98	1.94	1.91	1.88	1.83	1.78	1.73	1.70	1.67	1.64	1.61	1.57	1.53
25	2.92	2.53	2.32	2.18	2.09	2.02	1.97	1.93	1.89	1.87	1.82	1.77	1.72	1.69	1.66	1.63	1.59	1.56	1.52
26	2.91	2.52	2.31	2.17	2.08	2.01	1.96	1.92	1.88	1.86	1.81	1.76	1.71	1.68	1.65	1.61	1.58	1.54	1.50
27	2.90	2.51	2.30	2.17	2.07	2.00	1.95	1.91	1.87	1.85	1.80	1.75	1.70	1.67	1.64	1.60	1.57	1.53	1.49
28	2.89	2.50	2.29	2.16	2.06	2.00	1.94	1.90	1.87	1.84	1.79	1.74	1.69	1.66	1.63	1.59	1.56	1.52	1.48
29	2.89	2.50	2.28	2.15	2.06	1.99	1.93	1.89	1.86	1.83	1.78	1.73	1.68	1.65	1.62	1.58	1.55	1.51	1.47
30	2.88	2.49	2.28	2.14	2.03	1.98	1.93	1.88	1.85	1.82	1.77	1.72	1.67	1.64	1.61	1.57	1.54	1.50	1.46
40	2.84	2.44	2.23	2.09	2.00	1.93	1.87	1.83	1.79	1.76	1.71	1.66	1.61	1.57	1.54	1.51	1.47	1.42	1.38
60	2.79	2.39	2.18	2.04	1.95	1.87	1.82	1.77	1.74	1.71	1.66	1.60	1.54	1.51	1.48	1.44	1.40	1.35	1.29
120	2.75	2.35	2.13	1.99	1.90	1.82	1.77	1.72	1.68	1.65	1.60	1.55	1.48	1.45	1.41	1.37	1.32	1.26	1.19
∞	2.71	2.30	2.08	1.94	1.85	1.77	1.72	1.67	1.63	1.60	1.55	1.49	1.42	1.38	1.34	1.30	1.24	1.17	1.00

Degrees of freedom for the denominator (v_2)

Note: $F_{0.90, v_1, v_2} = 1/F_{0.10, v_2, v_1}$.

Appendix V (Continued)

$$F_{0.05,v_1,v_2}$$

Degrees of freedom for the denominator (v_2)

v_2 \ v_1	1	2	3	4	5	6	7	8	9	10	12	15	20	24	30	40	60	120	∞
1	161.4	199.5	215.7	224.6	230.2	234.0	236.8	238.9	240.5	241.9	243.9	245.9	248.0	249.1	250.1	251.1	252.2	253.3	254.3
2	18.51	19.00	19.16	19.25	19.30	19.33	19.35	19.37	19.38	19.40	19.41	19.43	19.45	19.45	19.46	19.47	19.48	19.49	19.50
3	10.13	9.55	9.28	9.12	9.01	8.94	8.89	8.85	8.81	8.79	8.74	8.70	8.66	8.64	8.62	8.59	8.57	8.55	8.53
4	7.71	6.94	6.59	6.39	6.26	6.16	6.09	6.04	6.00	5.96	5.91	5.86	5.80	5.77	5.75	5.72	5.69	5.66	5.63
5	6.61	5.79	5.41	5.19	5.05	4.95	4.88	4.82	4.77	4.74	4.68	4.62	4.56	4.53	4.50	4.46	4.43	4.40	4.36
6	5.99	5.14	4.76	4.53	4.39	4.28	4.21	4.15	4.10	4.06	4.00	3.94	3.87	3.84	3.81	3.77	3.74	3.70	3.67
7	5.59	4.74	4.35	4.12	3.97	3.87	3.79	3.73	3.68	3.64	3.57	3.51	3.44	3.41	3.38	3.34	3.30	3.27	3.23
8	5.32	4.46	4.07	3.84	3.69	3.58	3.50	3.44	3.39	3.35	3.28	3.22	3.15	3.12	3.08	3.04	3.01	2.97	2.93
9	5.12	4.26	3.86	3.63	3.48	3.37	3.29	3.23	3.18	3.14	3.07	3.01	2.94	2.90	2.86	2.83	2.79	2.75	2.71
10	4.96	4.10	3.71	3.48	3.33	3.22	3.14	3.07	3.02	2.98	2.91	2.85	2.77	2.74	2.70	2.66	2.62	2.58	2.54
11	4.84	3.98	3.59	3.36	3.20	3.09	3.01	2.95	2.90	2.85	2.79	2.72	2.65	2.61	2.57	2.53	2.49	2.45	2.40
12	4.75	3.89	3.49	3.26	3.11	3.00	2.91	2.85	2.80	2.75	2.69	2.62	2.54	2.51	2.47	2.43	2.38	2.34	2.30
13	4.67	3.81	3.41	3.18	3.03	2.92	2.83	2.77	2.71	2.67	2.60	2.53	2.46	2.42	2.38	2.34	2.30	2.25	2.21
14	4.60	3.74	3.34	3.11	2.96	2.85	2.76	2.70	2.65	2.60	2.53	2.46	2.39	2.35	2.31	2.27	2.22	2.18	2.13
15	4.54	3.68	3.29	3.06	2.90	2.79	2.71	2.64	2.59	2.54	2.48	2.40	2.33	2.29	2.25	2.20	2.16	2.11	2.07
16	4.49	3.63	3.24	3.01	2.85	2.74	2.66	2.59	2.54	2.49	2.42	2.35	2.28	2.24	2.19	2.15	2.11	2.06	2.01
17	4.45	3.59	3.20	2.96	2.81	2.70	2.61	2.55	2.49	2.45	2.38	2.31	2.23	2.19	2.15	2.10	2.06	2.01	1.96
18	4.41	3.55	3.16	2.93	2.77	2.66	2.58	2.51	2.46	2.41	2.34	2.27	2.19	2.15	2.11	2.06	2.02	1.97	1.92
19	4.38	3.52	3.13	2.90	2.74	2.63	2.54	2.48	2.42	2.38	2.31	2.23	2.16	2.11	2.07	2.03	1.98	1.93	1.88
20	4.35	3.49	3.10	2.87	2.71	2.60	2.51	2.45	2.39	2.35	2.28	2.20	2.12	2.08	2.04	1.99	1.95	1.90	1.84
21	4.32	3.47	3.07	2.84	2.68	2.57	2.49	2.42	2.37	2.32	2.25	2.18	2.10	2.05	2.01	1.96	1.92	1.87	1.81
22	4.30	3.44	3.05	2.82	2.66	2.55	2.46	2.40	2.34	2.30	2.23	2.15	2.07	2.03	1.98	1.94	1.89	1.84	1.78
23	4.28	3.42	3.03	2.80	2.64	2.53	2.44	2.37	2.32	2.27	2.20	2.13	2.05	2.01	1.96	1.91	1.86	1.81	1.76
24	4.26	3.40	3.01	2.78	2.62	2.51	2.42	2.36	2.30	2.25	2.18	2.11	2.03	1.98	1.94	1.89	1.84	1.79	1.73
25	4.24	3.39	2.99	2.76	2.60	2.49	2.40	2.34	2.28	2.24	2.16	2.09	2.01	1.96	1.92	1.87	1.82	1.77	1.71
26	4.23	3.37	2.98	2.74	2.59	2.47	2.39	2.32	2.27	2.22	2.15	2.07	1.99	1.95	1.90	1.85	1.80	1.75	1.69
27	4.21	3.35	2.96	2.73	2.57	2.46	2.37	2.31	2.25	2.20	2.13	2.06	1.97	1.93	1.88	1.84	1.79	1.73	1.67
28	4.20	3.34	2.95	2.71	2.56	2.45	2.36	2.29	2.24	2.19	2.12	2.04	1.96	1.91	1.87	1.82	1.77	1.71	1.65
29	4.18	3.33	2.93	2.70	2.55	2.43	2.35	2.28	2.22	2.18	2.10	2.03	1.94	1.90	1.85	1.81	1.75	1.70	1.64
30	4.17	3.32	2.92	2.69	2.53	2.42	2.33	2.27	2.21	2.16	2.09	2.01	1.93	1.89	1.84	1.79	1.74	1.68	1.62
40	4.08	3.23	2.84	2.61	2.45	2.34	2.25	2.18	2.12	2.08	2.00	1.92	1.84	1.79	1.74	1.69	1.64	1.58	1.51
60	4.00	3.15	2.76	2.53	2.37	2.25	2.17	2.10	2.04	1.99	1.92	1.84	1.75	1.70	1.65	1.59	1.53	1.47	1.39
120	3.92	3.07	2.68	2.45	2.29	2.17	2.09	2.02	1.96	1.91	1.83	1.75	1.66	1.61	1.55	1.50	1.43	1.35	1.25
∞	3.84	3.00	2.60	2.37	2.21	2.10	2.01	1.94	1.88	1.83	1.75	1.67	1.57	1.52	1.46	1.39	1.32	1.22	1.00

Degrees of freedom for the numerator (v_1)

Note: $F_{0.95,v_1,v_2} = 1/F_{0.05,v_2,v_1}$.

$$F_{0.025,\,v_1,\,v_2}$$

$v_2 \backslash v_1$	1	2	3	4	5	6	7	8	9	10	12	15	20	24	30	40	60	120	∞
1	647.8	799.5	864.2	899.6	921.8	937.1	948.2	956.7	963.3	968.6	976.7	984.9	993.1	997.2	1001.0	1006.0	1010.0	1014.0	1018.0
2	38.51	39.00	39.17	39.25	39.30	39.33	39.36	39.37	39.39	39.40	39.41	39.43	39.45	39.46	39.46	39.47	39.48	39.49	39.50
3	17.44	16.04	15.44	15.10	14.88	14.73	14.62	14.54	14.47	14.42	14.34	14.25	14.17	14.12	14.08	14.04	13.99	13.95	13.90
4	12.22	10.65	9.98	9.60	9.36	9.20	9.07	8.98	8.90	8.84	8.75	8.66	8.56	8.51	8.46	8.41	8.36	8.31	8.26
5	10.01	8.43	7.76	7.39	7.15	6.98	6.85	6.76	6.68	6.62	6.52	6.43	6.33	6.28	6.23	6.18	6.12	6.07	6.02
6	8.81	7.26	6.60	6.23	5.99	5.82	5.70	5.60	5.52	5.46	5.37	5.27	5.17	5.12	5.07	5.01	4.96	4.90	4.85
7	8.07	6.54	5.89	5.52	5.29	5.12	4.99	4.90	4.82	4.76	4.67	4.57	4.47	4.42	4.36	4.31	4.25	4.20	4.14
8	7.57	6.06	5.42	5.05	4.82	4.65	4.53	4.43	4.36	4.30	4.20	4.10	4.00	3.95	3.89	3.84	3.78	3.73	3.67
9	7.21	5.71	5.08	4.72	4.48	4.32	4.20	4.10	4.03	3.96	3.87	3.77	3.67	3.61	3.56	3.51	3.45	3.39	3.33
10	6.94	5.46	4.83	4.47	4.24	4.07	3.95	3.85	3.78	3.72	3.62	3.52	3.42	3.37	3.31	3.26	3.20	3.14	3.08
11	6.72	5.26	4.63	4.28	4.04	3.88	3.76	3.66	3.59	3.53	3.43	3.33	3.23	3.17	3.12	3.06	3.00	2.94	2.88
12	6.55	5.10	4.47	4.12	3.89	3.73	3.61	3.51	3.44	3.37	3.28	3.18	3.07	3.02	2.96	2.91	2.85	2.79	2.72
13	6.41	4.97	4.35	4.00	3.77	3.60	3.48	3.39	3.31	3.25	3.15	3.05	2.95	2.89	2.84	2.78	2.72	2.66	2.60
14	6.30	4.86	4.24	3.89	3.66	3.50	3.38	3.29	3.21	3.15	3.05	2.95	2.84	2.79	2.73	2.67	2.61	2.55	2.49
15	6.20	4.77	4.15	3.80	3.58	3.41	3.29	3.20	3.12	3.06	2.96	2.86	2.76	2.70	2.64	2.59	2.52	2.46	2.40
16	6.12	4.69	4.08	3.73	3.50	3.34	3.22	3.12	3.05	2.99	2.89	2.79	2.68	2.63	2.57	2.51	2.45	2.38	2.32
17	6.04	4.62	4.01	3.66	3.44	3.28	3.16	3.06	2.98	2.92	2.82	2.72	2.62	2.56	2.50	2.44	2.38	2.32	2.25
18	5.98	4.56	3.95	3.61	3.38	3.22	3.10	3.01	2.93	2.87	2.77	2.67	2.56	2.50	2.44	2.38	2.32	2.26	2.19
19	5.92	4.51	3.90	3.56	3.33	3.17	3.05	2.96	2.88	2.82	2.72	2.62	2.51	2.45	2.39	2.33	2.27	2.20	2.13
20	5.87	4.46	3.86	3.51	3.29	3.13	3.01	2.91	2.84	2.77	2.68	2.57	2.46	2.41	2.35	2.29	2.22	2.16	2.09
21	5.83	4.42	3.82	3.48	3.25	3.09	2.97	2.87	2.80	2.73	2.64	2.53	2.42	2.37	2.31	2.25	2.18	2.11	2.04
22	5.79	4.38	3.78	3.44	3.22	3.05	2.93	2.84	2.76	2.70	2.60	2.50	2.39	2.33	2.27	2.21	2.14	2.08	2.00
23	5.75	4.35	3.75	3.41	3.18	3.02	2.90	2.81	2.73	2.67	2.57	2.47	2.36	2.30	2.24	2.18	2.11	2.04	1.97
24	5.72	4.32	3.72	3.38	3.15	2.99	2.87	2.78	2.70	2.64	2.54	2.44	2.33	2.27	2.21	2.15	2.08	2.01	1.94
25	5.69	4.29	3.69	3.35	3.13	2.97	2.85	2.75	2.68	2.61	2.51	2.41	2.30	2.24	2.18	2.12	2.05	1.98	1.91
26	5.66	4.27	3.67	3.33	3.10	2.94	2.82	2.73	2.65	2.59	2.49	2.39	2.28	2.22	2.16	2.09	2.03	1.95	1.88
27	5.63	4.24	3.65	3.31	3.08	2.92	2.80	2.71	2.63	2.57	2.47	2.36	2.25	2.19	2.13	2.07	2.00	1.93	1.85
28	5.61	4.22	3.63	3.29	3.06	2.90	2.78	2.69	2.61	2.55	2.45	2.34	2.23	2.17	2.11	2.05	1.98	1.91	1.83
29	5.59	4.20	3.61	3.27	3.04	2.88	2.76	2.67	2.59	2.53	2.43	2.32	2.21	2.15	2.09	2.03	1.96	1.89	1.81
30	5.57	4.18	3.59	3.25	3.03	2.87	2.75	2.65	2.57	2.51	2.41	2.31	2.20	2.14	2.07	2.01	1.94	1.87	1.79
40	5.42	4.05	3.46	3.13	2.90	2.74	2.62	2.53	2.45	2.39	2.29	2.18	2.07	2.01	1.94	1.88	1.80	1.72	1.64
60	5.29	3.93	3.34	3.01	2.79	2.63	2.51	2.41	2.33	2.27	2.17	2.06	1.94	1.88	1.82	1.74	1.67	1.58	1.48
120	5.15	3.80	3.23	2.89	2.67	2.52	2.39	2.30	2.22	2.16	2.05	1.94	1.82	1.76	1.69	1.61	1.53	1.43	1.31
∞	5.02	3.69	3.12	2.79	2.57	2.41	2.29	2.19	2.11	2.05	1.94	1.83	1.71	1.64	1.57	1.48	1.39	1.27	1.00

Degrees of freedom for the numerator (v_1)

Degrees of freedom for the denominator (v_2)

Note: $F_{0.95,\,v_1,\,v_2} = 1/F_{0.05,\,v_2,\,v_1}$

Appendix V (Continued)

$$F_{0.01, \nu_1, \nu_2}$$

Degrees of freedom for the numerator (ν_1)

ν_2	1	2	3	4	5	6	7	8	9	10	12	15	20	24	30	40	60	120	∞
1	4052.0	4999.5	5403.0	5625.0	5764.0	5859.0	5928.0	5982.0	6022.0	6056.0	6106.0	6157.0	6209.0	6235.0	6261.0	6287.0	6313.0	6339.0	6366.0
2	98.50	99.00	99.17	99.25	99.30	99.33	99.36	99.37	99.39	99.40	99.42	99.43	99.45	99.46	99.47	99.47	99.48	99.49	99.50
3	34.12	30.82	29.46	28.71	28.24	27.91	27.67	27.49	27.35	27.23	27.05	26.87	26.69	26.60	26.50	26.41	26.32	26.22	26.13
4	21.20	18.00	16.69	15.98	15.52	15.21	14.98	14.80	14.66	14.55	14.37	14.20	14.02	13.93	13.84	13.75	13.65	13.56	13.46
5	16.26	13.27	12.06	11.39	10.97	10.67	10.46	10.29	10.16	10.05	9.89	9.72	9.55	9.47	9.38	9.29	9.20	9.11	9.02
6	13.75	10.92	9.78	9.15	8.75	8.47	8.26	8.10	7.98	7.87	7.72	7.56	7.40	7.31	7.23	7.14	7.06	6.97	6.88
7	12.25	9.55	8.45	7.85	7.46	7.19	6.99	6.84	6.72	6.62	6.47	6.31	6.16	6.07	5.99	5.91	5.82	5.74	5.65
8	11.26	8.65	7.59	7.01	6.63	6.37	6.18	6.03	5.91	5.81	5.67	5.52	5.36	5.28	5.20	5.12	5.03	4.95	4.86
9	10.56	8.02	6.99	6.42	6.06	5.80	5.61	5.47	5.35	5.26	5.11	4.96	4.81	4.73	4.65	4.57	4.48	4.40	4.31
10	10.04	7.56	6.55	5.99	5.64	5.39	5.20	5.06	4.94	4.85	4.71	4.56	4.41	4.33	4.25	4.17	4.08	4.00	3.91
11	9.65	7.21	6.22	5.67	5.32	5.07	4.89	4.74	4.63	4.54	4.40	4.25	4.10	4.02	3.94	3.86	3.78	3.69	3.60
12	9.33	6.93	5.95	5.41	5.06	4.82	4.64	4.50	4.39	4.30	4.16	4.01	3.86	3.78	3.70	3.62	3.54	3.45	3.36
13	9.07	6.70	5.74	5.21	4.86	4.62	4.44	4.30	4.19	4.10	3.96	3.82	3.66	3.59	3.51	3.43	3.34	3.25	3.17
14	8.86	6.51	5.56	5.04	4.69	4.46	4.28	4.14	4.03	3.94	3.80	3.66	3.51	3.43	3.35	3.27	3.18	3.09	3.00
15	8.68	6.36	5.42	4.89	4.56	4.32	4.14	4.00	3.89	3.80	3.67	3.52	3.37	3.29	3.21	3.13	3.05	2.96	2.87
16	8.53	6.23	5.29	4.77	4.44	4.20	4.03	3.89	3.78	3.69	3.55	3.41	3.26	3.18	3.10	3.02	2.93	2.84	2.75
17	8.40	6.11	5.18	4.67	4.34	4.10	3.93	3.79	3.68	3.59	3.46	3.31	3.16	3.08	3.00	2.92	2.83	2.75	2.65
18	8.29	6.01	5.09	4.58	4.25	4.01	3.84	3.71	3.60	3.51	3.37	3.23	3.08	3.00	2.92	2.84	2.75	2.66	2.57
19	8.18	5.93	5.01	4.50	4.17	3.94	3.77	3.63	3.52	3.43	3.30	3.15	3.00	2.92	2.84	2.76	2.67	2.58	2.49
20	8.10	5.85	4.94	4.43	4.10	3.87	3.70	3.56	3.46	3.37	3.23	3.09	2.94	2.86	2.78	2.69	2.61	2.52	2.42
21	8.02	5.78	4.87	4.37	4.04	3.81	3.64	3.51	3.40	3.31	3.17	3.03	2.88	2.80	2.72	2.64	2.55	2.46	2.36
22	7.95	5.72	4.82	4.31	3.99	3.76	3.59	3.45	3.35	3.26	3.12	2.98	2.83	2.75	2.67	2.58	2.50	2.40	2.31
23	7.88	5.66	4.76	4.26	3.94	3.71	3.54	3.41	3.30	3.21	3.07	2.93	2.78	2.70	2.62	2.54	2.45	2.35	2.26
24	7.82	5.61	4.72	4.22	3.90	3.67	3.50	3.36	3.26	3.17	3.03	2.89	2.74	2.66	2.58	2.49	2.40	2.31	2.21
25	7.77	5.57	4.68	4.18	3.85	3.63	3.46	3.32	3.22	3.13	2.99	2.85	2.70	2.62	2.54	2.45	2.36	2.27	2.17
26	7.72	5.53	4.64	4.14	3.82	3.59	3.42	3.29	3.18	3.09	2.96	2.81	2.66	2.58	2.50	2.42	2.33	2.23	2.13
27	7.68	5.49	4.60	4.11	3.78	3.56	3.39	3.26	3.15	3.06	2.93	2.78	2.63	2.55	2.47	2.38	2.29	2.20	2.10
28	7.64	5.45	4.57	4.07	3.75	3.53	3.36	3.23	3.12	3.03	2.90	2.75	2.60	2.52	2.44	2.35	2.26	2.17	2.06
29	7.60	5.42	4.54	4.04	3.73	3.50	3.33	3.20	3.09	3.00	2.87	2.73	2.57	2.49	2.41	2.33	2.23	2.14	2.03
30	7.56	5.39	4.51	4.02	3.70	3.47	3.30	3.17	3.07	2.98	2.84	2.70	2.55	2.47	2.39	2.30	2.21	2.11	2.01
40	7.31	5.18	4.31	3.83	3.51	3.29	3.12	2.99	2.89	2.80	2.66	2.52	2.37	2.29	2.20	2.11	2.02	1.92	1.80
60	7.08	4.98	4.13	3.65	3.34	3.12	2.95	2.82	2.72	2.63	2.50	2.35	2.20	2.12	2.03	1.94	1.84	1.73	1.60
120	6.85	4.79	3.95	3.48	3.17	2.96	2.79	2.66	2.56	2.47	2.34	2.19	2.03	1.95	1.86	1.76	1.66	1.53	1.38
∞	6.63	4.61	3.78	3.32	3.02	2.80	2.64	2.51	2.41	2.32	2.18	2.04	1.88	1.79	1.70	1.59	1.47	1.32	1.00

Degrees of freedom for the denominator (ν_2)

Note: $F_{0.99, \nu_1, \nu_2} = 1/F_{0.01, \nu_2, \nu_1}$.

Appendix VI Factors for Constructing Variables Control Charts

Observations in Sample, n	Chart for Averages			Chart for Standard Deviations						Chart for Ranges						
	Factors for Control Limits			Factors for Center Line		Factors for Control Limits				Factors for Center Line			Factors for Control Limits			
	A	A_2	A_3	c_4	$1/c_4$	B_3	B_4	B_5	B_6	d_2	$1/d_2$	d_3	D_1	D_2	D_3	D_4
2	2.121	1.880	2.659	0.7979	1.2533	0	3.267	0	2.606	1.128	0.8865	0.853	0	3.686	0	3.267
3	1.732	1.023	1.954	0.8862	1.1284	0	2.568	0	2.276	1.693	0.5907	0.888	0	4.358	0	2.575
4	1.500	0.729	1.628	0.9213	1.0854	0	2.266	0	2.088	2.059	0.4857	0.880	0	4.698	0	2.282
5	1.342	0.577	1.427	0.9400	1.0638	0	2.089	0	1.964	2.326	0.4299	0.864	0	4.918	0	2.115
6	1.225	0.483	1.287	0.9515	1.0510	0.030	1.970	0.029	1.874	2.534	0.3946	0.848	0	5.078	0	2.004
7	1.134	0.419	1.182	0.9594	1.0423	0.118	1.882	0.113	1.806	2.704	0.3698	0.833	0.204	5.204	0.076	1.924
8	1.061	0.373	1.099	0.9650	1.0363	0.185	1.815	0.179	1.751	2.847	0.3512	0.820	0.388	5.306	0.136	1.864
9	1.000	0.337	1.032	0.9693	1.0317	0.239	1.761	0.232	1.707	2.970	0.3367	0.808	0.547	5.393	0.184	1.816
10	0.949	0.308	0.975	0.9727	1.0281	0.284	1.716	0.276	1.669	3.078	0.3249	0.797	0.687	5.469	0.223	1.777
11	0.905	0.285	0.927	0.9754	1.0252	0.321	1.679	0.313	1.637	3.173	0.3152	0.787	0.811	5.535	0.256	1.744
12	0.866	0.266	0.886	0.9776	1.0229	0.354	1.646	0.346	1.610	3.258	0.3069	0.778	0.922	5.594	0.283	1.717
13	0.832	0.249	0.850	0.9794	1.0210	0.382	1.618	0.374	1.585	3.336	0.2998	0.770	1.025	5.647	0.307	1.693
14	0.802	0.235	0.817	0.9810	1.0194	0.406	1.594	0.399	1.563	3.407	0.2935	0.763	1.118	5.696	0.328	1.672
15	0.775	0.223	0.789	0.9823	1.0180	0.428	1.572	0.421	1.544	3.472	0.2880	0.756	1.203	5.741	0.347	1.653
16	0.750	0.212	0.763	0.9835	1.0168	0.448	1.552	0.440	1.526	3.532	0.2831	0.750	1.282	5.782	0.363	1.637
17	0.728	0.203	0.739	0.9845	1.0157	0.466	1.534	0.458	1.511	3.588	0.2787	0.744	1.356	5.820	0.378	1.622
18	0.707	0.194	0.718	0.9854	1.0148	0.482	1.518	0.475	1.496	3.640	0.2747	0.739	1.424	5.856	0.391	1.608
19	0.688	0.187	0.698	0.9862	1.0140	0.497	1.503	0.490	1.483	3.689	0.2711	0.734	1.487	5.891	0.403	1.597
20	0.671	0.180	0.680	0.9869	1.0133	0.510	1.490	0.504	1.470	3.735	0.2677	0.729	1.549	5.921	0.415	1.585
21	0.655	0.173	0.663	0.9876	1.0126	0.523	1.477	0.516	1.459	3.778	0.2647	0.724	1.605	5.951	0.425	1.575
22	0.640	0.167	0.647	0.9882	1.0119	0.534	1.466	0.528	1.448	3.819	0.2618	0.720	1.659	5.979	0.434	1.566
23	0.626	0.162	0.633	0.9887	1.0114	0.545	1.455	0.539	1.438	3.858	0.2592	0.716	1.710	6.006	0.443	1.557
24	0.612	0.157	0.619	0.9892	1.0109	0.555	1.445	0.549	1.429	3.895	0.2567	0.712	1.759	6.031	0.451	1.548
25	0.600	0.153	0.606	0.9896	1.0105	0.565	1.435	0.559	1.420	3.931	0.2544	0.708	1.806	6.056	0.459	1.541

For $n > 25$.

$$A = \frac{3}{\sqrt{n}} \qquad A_3 = \frac{3}{c_4\sqrt{n}} \qquad c_4 \cong \frac{4(n-1)}{4n-3}$$

$$B_3 = 1 - \frac{3}{c_4\sqrt{2(n-1)}} \qquad B_4 = 1 + \frac{3}{c_4\sqrt{2(n-1)}}$$

$$B_5 = c_4 - \frac{3}{\sqrt{2(n-1)}} \qquad B_6 = c_4 + \frac{3}{\sqrt{2(n-1)}}$$

Appendix VII   Factors for Two-Sided Normal Tolerance Limits

	90% Confidence That Percentage of Population Between Limits Is			95% Confidence That Percentage of Population Between Limits Is			99% Confidence That Percentage of Population Between Limits Is		
n	90%	95%	99%	90%	95%	99%	90%	95%	99%
2	15.98	18.80	24.17	32.02	37.67	48.43	160.2	188.5	242.3
3	5.847	6.919	8.974	8.380	9.916	12.86	18.93	22.40	29.06
4	4.166	4.943	6.440	5.369	6.370	8.299	9.398	11.15	14.53
5	3.494	4.152	5.423	4.275	5.079	6.634	6.612	7.855	10.26
6	3.131	3.723	4.870	3.712	4.414	5.775	5.337	6.345	8.301
7	2.902	3.452	4.521	3.369	4.007	5.248	4.613	5.448	7.187
8	2.743	3.264	4.278	3.136	3.732	4.891	4.147	4.936	6.468
9	2.626	3.125	4.098	2.967	3.532	4.631	3.822	4.550	5.966
10	2.535	3.018	3.959	2.829	3.379	4.433	3.582	4.265	5.594
11	2.463	2.933	3.849	2.737	3.259	4.277	3.397	4.045	5.308
12	2.404	2.863	3.758	2.655	3.162	4.150	3.250	3.870	5.079
13	2.355	2.805	3.682	2.587	3.081	4.044	3.130	3.727	4.893
14	2.314	2.756	3.618	2.529	3.012	3.955	3.029	3.608	4.737
15	2.278	2.713	3.562	2.480	2.954	3.878	2.945	3.507	4.605
16	2.246	2.676	3.514	2.437	2.903	3.812	2.872	3.421	4.492
17	2.219	2.643	3.471	2.400	2.858	3.754	2.808	3.345	4.393
18	2.194	2.614	3.433	2.366	2.819	3.702	2.753	3.279	4.307
19	2.172	2.588	3.399	2.337	2.784	3.656	2.703	3.221	4.230
20	2.152	2.564	3.368	2.310	2.752	3.615	2.659	3.168	4.161
21	2.135	2.543	3.340	2.286	2.723	3.577	2.620	3.121	4.100
22	2.118	2.524	3.315	2.264	2.697	3.543	2.584	3.078	4.044
23	2.103	2.506	3.292	2.244	2.673	3.512	2.551	3.040	3.993
24	2.089	2.489	3.270	2.225	2.651	3.483	2.522	3.004	3.947
25	2.077	2.474	3.251	2.208	2.631	3.457	2.494	2.972	3.904
26	2.065	2.460	3.232	2.193	2.612	3.432	2.469	2.941	3.865
27	2.054	2.447	3.215	2.178	2.595	3.409	2.446	2.914	3.828
28	2.044	2.435	3.199	2.164	2.579	3.388	2.424	2.888	3.794
29	2.034	2.424	3.184	2.152	2.554	3.368	2.404	2.864	3.763
30	2.025	2.413	3.170	2.140	2.549	3.350	2.385	2.841	3.733
35	1.988	2.368	3.112	2.090	2.490	3.272	2.306	2.748	3.611
40	1.959	2.334	3.066	2.052	2.445	3.213	2.247	2.677	3.518
50	1.916	2.284	3.001	1.996	2.379	3.126	2.162	2.576	3.385
60	1.887	2.248	2.955	1.958	2.333	3.066	2.103	2.506	3.293
80	1.848	2.202	2.894	1.907	2.272	2.986	2.026	2.414	3.173
100	1.822	2.172	2.854	1.874	2.233	2.934	1.977	2.355	3.096
200	1.764	2.102	2.762	1.798	2.143	2.816	1.865	2.222	2.921
500	1.717	2.046	2.689	1.737	2.070	2.721	1.777	2.117	2.783
1000	1.695	2.019	2.654	1.709	2.036	2.676	1.736	2.068	2.718
∞	1.645	1.960	2.576	1.645	1.960	2.576	1.645	1.960	2.576

Appendix VIII Factors for One-Sided Normal Tolerance Limits

	90% Confidence That Percentage of Population Below (Above) Limits Is			95% Confidence That Percentage of Population Below (Above) Limits Is			99% Confidence That Percentage of Population Below (Above) Limits Is		
n	90%	95%	99%	90%	95%	99%	90%	95%	99%
3	4.258	5.310	7.340	6.158	7.655	10.552			
4	3.187	3.957	5.437	4.163	5.145	7.042			
5	2.742	3.400	4.666	3.407	4.202	5.741			
6	2.494	3.091	4.242	3.006	3.707	5.062	4.408	5.409	7.334
7	2.333	2.894	3.972	2.755	3.399	4.641	3.856	4.730	6.411
8	2.219	2.755	3.783	2.582	3.188	4.353	3.496	4.287	5.811
9	2.133	2.649	3.641	2.454	3.031	4.143	3.242	3.971	5.389
10	2.065	2.568	3.532	2.355	2.911	3.981	3.048	3.739	5.075
11	2.012	2.503	3.444	2.275	2.815	3.852	2.897	3.557	4.828
12	1.966	2.448	3.371	2.210	2.736	3.747	2.773	3.410	4.633
13	1.928	2.403	3.310	2.155	2.670	3.659	2.677	3.290	4.472
14	1.895	2.363	3.257	2.108	2.614	3.585	2.592	3.189	4.336
15	1.866	2.329	3.212	2.068	2.566	3.520	2.521	3.102	4.224
16	1.842	2.299	3.172	2.032	2.523	3.463	2.458	3.028	4.124
17	1.820	2.272	3.136	2.001	2.486	3.415	2.405	2.962	4.038
18	1.800	2.249	3.106	1.974	2.453	3.370	2.357	2.906	3.961
19	1.781	2.228	3.078	1.949	2.423	3.331	2.315	2.855	3.893
20	1.765	2.208	3.052	1.926	2.396	3.295	2.275	2.807	3.832
21	1.750	2.190	3.028	1.905	2.371	3.262	2.241	2.768	3.776
22	1.736	2.174	3.007	1.887	2.350	3.233	2.208	2.729	3.727
23	1.724	2.159	2.987	1.869	2.329	3.206	2.179	2.693	3.680
24	1.712	2.145	2.969	1.853	2.309	3.181	2.154	2.663	3.638
25	1.702	2.132	2.952	1.838	2.292	3.158	2.129	2.632	3.601
30	1.657	2.080	2.884	1.778	2.220	3.064	2.029	2.516	3.446
35	1.623	2.041	2.833	1.732	2.166	2.994	1.957	2.431	3.334
40	1.598	2.010	2.793	1.697	2.126	2.941	1.902	2.365	3.250
45	1.577	1.986	2.762	1.669	2.092	2.897	1.857	2.313	3.181
50	1.560	1.965	2.735	1.646	2.065	2.863	1.821	2.296	3.124

Appendix IX   Random Numbers

10480	15011	01536	02011	81647	91646	69179	14194	62590
22368	46573	25595	85393	30995	89198	27982	53402	93965
24130	48360	22527	97265	76393	64809	15179	24830	49340
42167	93093	06243	61680	07856	16376	39440	53537	71341
37570	39975	81837	16656	06121	91782	60468	81305	49684
77921	06907	11008	42751	27756	53498	18602	70659	90655
99562	72905	56420	69994	98872	31016	71194	18738	44013
96301	91977	05463	07972	18876	20922	94595	56869	69014
89579	14342	63661	10281	17453	18103	57740	84378	25331
85475	36857	53342	53988	53060	59533	38867	62300	08158
28918	69578	88231	33276	70997	79936	56865	05859	90106
63553	40961	48235	03427	49626	69445	18663	72695	52180
09429	93969	52636	92737	88974	33488	36320	17617	30015
10365	61129	87529	85689	48237	52267	67689	93394	01511
07119	97336	71048	08178	77233	13916	47564	81056	97735
51085	12765	51821	51259	77452	16308	60756	92144	49442
02368	21382	52404	60268	89368	19885	55322	44819	01188
01011	54092	33362	94904	31273	04146	18594	29852	71585
52162	53916	46369	58586	23216	14513	83149	98736	23495
07056	97628	33787	09998	42698	06691	76988	13602	51851
48663	91245	85828	14346	09172	30168	90229	04734	59193
54164	58492	22421	74103	47070	25306	76468	26384	58151
32639	32363	05597	24200	13363	38005	94342	28728	35806
29334	27001	87637	87308	58731	00256	45834	15398	46557
02488	33062	28834	07351	19731	92420	60952	61280	50001
81525	72295	04839	96423	24878	82651	66566	14778	76797
29676	20591	68086	26432	46901	20849	89768	81536	86645
00742	57392	39064	66432	84673	40027	32832	61362	98947
05366	04213	25669	26422	44407	44048	37937	63904	45766
91921	26418	64117	94305	26766	25940	39972	22209	71500
00582	04711	87917	77341	42206	35126	74087	99547	81817
00725	69884	62797	56170	86324	88072	76222	36086	84637
69011	65795	95876	55293	18988	27354	26575	08625	40801
25976	57948	29888	88604	67917	48708	18912	82271	65424
09763	83473	73577	12908	30883	18317	28290	35797	05998
91567	42595	27958	30134	04024	86385	29880	99730	55536
17955	56349	90999	49127	20044	59931	06115	20542	18059
46503	18584	18845	49618	02304	51038	20655	58727	28168
92157	89634	94824	78171	84610	82834	09922	25417	44137
14577	62765	35605	81263	39667	47358	56873	56307	61607
98427	07523	33362	64270	01638	92477	66969	98420	04880
34914	63976	88720	82765	34476	17032	87589	40836	32427
70060	28277	39475	46473	23219	53416	94970	25832	69975
53976	54914	06990	67245	68350	82948	11398	42878	80287
76072	29515	40980	07391	58745	25774	22987	80059	39911
90725	52210	83974	29992	65831	38857	50490	83765	55657
64364	67412	33339	31926	14883	24413	59744	92351	97473
08962	00358	31662	25388	61642	34072	81249	35648	56891
95012	68379	93526	70765	10592	04542	76463	54328	02349
15664	10493	20492	38391	91132	21999	59516	81652	27195

Bibliography

Adams, B. M., C. Lowry, and W. H. Woodall (1992). "The Use (and Misuse) of False Alarm Probabilities in Control Chart Design," in *Frontiers in Statistical Quality Control 4*, edited by H. J. Lenz, G. B. Wetherill, and P.-Th. Wilrich, Physica-Verlag, Heidelberg.

Alt, F. B. (1985). "Multivariate Quality Control," in *Encyclopedia of Statistical Sciences*, Vol. 6, edited by N. L. Johnson and S. Kotz, Wiley, New York.

Alwan, L. C. (1992). "Effects of Autocorrelation on Control Charts," *Communications in Statistics — Theory and Methods*, Vol. 21.

Alwan, L. C., and H. V. Roberts (1988). "Time Series Modeling for Statistical Process Control," *Journal of Business and Economic Statistics*, Vol. 6.

Barnard, G. A. (1959). "Control Charts and Stochastic Processes," *Journal of the Royal Statistical Society*, (B), Vol. 21.

Bather, J. A. (1963). "Control Charts and the Minimization of Costs," *Journal of the Royal Statistical Society*, (B), Vol. 25.

Berthouex, P. M., W. G. Hunter, and L. Pallesen (1978). "Monitoring Sewage Treatment Plants: Some Quality Control Aspects," *Journal of Quality Technology*, Vol. 10.

Bisgaard, S., W. G. Hunter, and L. Pallesen (1984). "Economic Selection of Quality of Manufactured Product," *Technometrics*, Vol. 26.

Bissell, A. F. (1990). "How Reliable Is Your Capability Index?" *Applied Statistics*, Vol. 39.

Borror, C. M., C. W. Champ, and S. E. Rigdon (1998). "Poisson EWMA Control Charts," *Journal of Quality Technology*, Vol. 30, No. 4.

Borror, C. M., D. C. Montgomery, and G. C. Runger (1997). "Confidence Intervals for Variance Components from Gauge Capability Studies," *Quality and Reliability Engineering International*, Vol. 13.

Borror, C. M., D. C. Montgomery, and G. C. Runger (1999). "Robustness of the EWMA Control Chart to Nonnormality," *Journal of Quality Technology*, Vol. 31, No. 3.

Bourke, P. O. (1991). "Detecting a Shift in the Fraction Nonconforming Using Run-Length Control Chart with 100% Inspection," *Journal of Quality Technology*, Vol. 23.

Bowker, A. H., and G. J. Lieberman (1972). *Engineering Statistics*, 2nd ed., Prentice-Hall, Englewood Cliffs, NJ.

Box, G. E. P. (1957). "Evolutionary Operation: A Method for Increasing Industrial Productivity," *Applied Statistics*, Vol. 6.

Box, G. E. P. (1991). "The Bounded Adjustment Chart," *Quality Engineering*, Vol. 4.

Box, G. E. P. (1991–1992). "Feedback Control by Manual Adjustment," *Quality Engineering*, Vol. 4.

Box, G. E. P., S. Bisgaard, and C. Fung (1988). "An Explanation and Critique of Taguchi's Contributions to Quality Engineering." *Quality and Reliability Engineering International*, Vol. 4.

Box, G. E. P., and N. R. Draper (1969). *Evolutionary Operation*, Wiley, New York.

Box, G. E. P., and N. R. Draper (1986). *Empirical Model Building and Response Surfaces*, Wiley, New York.

Box, G. E. P., G. M. Jenkins, and G. C. Reinsel (1994). *Time Series Analysis, Forecasting, and Control*, 3rd edition, Prentice-Hall, Englewood Cliffs, NJ.

Box, G. E. P., and T. Kramer (1992). "Statistical Process Monitoring and Feedback Adjustment—A Discussion," *Technometrics*, Vol. 34.

Box, G. E. P., and A. Luceño (1997). *Statistical Control by Monitoring and Feedback Adjustment*, Wiley, New York.

Boyd, D. F. (1950). "Applying the Group Control Chart for $\bar{x}$ and R," *Industrial Quality Control*, Vol. 6.

Boyles, R. A. (1991). "The Taguchi Capability Index," *Journal of Quality Technology*, Vol. 27.

Brook, D., and D. A. Evans (1972). "An Approach to the Probability Distribution of CUSUM Run Length," *Biometrika*, Vol. 59.

Bryce, G. R., M. A. Gaudard, and B. L. Joiner (1997–1998). "Estimating the Standard Deviation for Individuals Charts," *Quality Engineering*, Vol. 10, No. 2.

Burdick, R. K., and G. A. Larsen (1997). "Confidence Intervals on Measures of Variability in Gauge R & R Studies," *Journal of Quality Technology*, Vol. 29, No. 2.

Burr, I. J. (1967). "The Effect of Nonnormality on Constants for $\bar{x}$ and R Charts," *Industrial Quality Control*, Vol. 23.

Byrne, D. M., and S. Taguchi (1987). "The Taguchi Approach to Robust Parameter Design," *Quality Progress*, December.

Champ, C. W., and W. H. Woodall (1987). "Exact Results for Shewhart Control Charts with Supplementary Runs Rules," *Technometrics*, Vol. 29.

Chan, L. K., S. W. Cheng, and F. A. Spiring (1988). "A New Measure of Process Capability: C_{pm}," *Journal of Quality Technology*, Vol. 20.

Chan, L. K., K. P. Hapuarachchi, and B. D. Macpherson (1988). "Robustness of x and R Charts," *IEEE Transactions on Reliability*, Vol. 37, No. 3.

Chiu, W. K., and G. B. Wetherill (1974). "A Simplified Scheme for the Economic Design of $\bar{x}$-Charts," *Journal of Quality Technology*, Vol. 6.

Chiu, W. K., and G. B. Wetherill (1975). "Quality Control Practices," *International Journal of Production Research*, Vol. 13.

Chrysler, Ford, and GM (1995). *Measurement Systems Analysis Reference Manual*, AIAG, Detroit, MI.

Chua, M., and D. C. Montgomery (1992). "Investigation and Characterization of a Control Scheme for Multivariate Quality Control," *Quality and Reliability Engineering International*, Vol. 8.

Clements, J. A. (1989). Process Capability Calculations for Non-Normal Distributions," *Quality Progress*.

Clifford, P. C. (1959). "Control Charts Without Calculations," *Industrial Quality Control*, Vol. 15.

Coleman, D. E., and D. C. Montgomery (1993). "A Systematic Method for Planning for a Designed Industrial Experiment" (with discussion), *Technometrics,* Vol. 35.

Cornell, J. A., and A. I. Khuri (1996). *Response Surfaces*, 2nd ed., Dekker, New York.

Cowden, D. J. (1957). *Statistical Methods in Quality Control*, Prentice-Hall, Englewood Cliffs, NJ.

Crosier, R. B. (1988). "Multivariate Generalizations of Cumulative Sum Quality Control Schemes," *Technometrics*, Vol. 30.

Crowder, S. V. (1987a). "A Simple Method for Studying Run-Length Distributions of Exponentially Weighted Moving Average Charts," *Technometrics*, Vol. 29.

Crowder, S. V. (1987b). "Computation of ARL for Combined Individual Measurement and Moving Range Charts," *Journal of Quality Technology*, Vol. 19.

Crowder, S. V. (1989). "Design of Exponentially Weighted Moving Average Schemes," *Journal of Quality Technology*, Vol. 21.

Crowder, S. V. (1992). "An SPC Model for Short Production Runs: Minimizing Expected Costs," *Technometrics*, Vol. 34.

Crowder, S. V., and M. Hamilton (1992). "An EWMA for Monitoring a Process Standard Deviation," *Journal of Quality Technology*, Vol. 24.

Cruthis, E. N., and S. E. Rigdon (1992–1993). Comparing Two Estimates of the Variance to Determine the Stability of a Process," *Quality Engineering*, Vol. 5.

Del Castillo, E., and D. C. Montgomery (1994). "Short-Run Statistical Process Control: Q-Chart Enhancements and Alternative Methods," *Quality and Reliability Engineering International*, Vol. 10.

Dodge, H. F. (1943). "A Sampling Plan for Continuous Production," *Annals of Mathematical Statistics*, Vol. 14.

Dodge, H. F. (1955). "Chain Sampling Inspection Plans," *Industrial Quality Control*, Vol. 11.

Dodge, H. F. (1956). "Skip-Lot Sampling Plan," *Industrial Quality Control*, Vol. 11.

Dodge, H. F., and H. G. Romig (1959). *Sampling Inspection Tables, Single and Double Sampling*, 2nd ed., Wiley, New York.

Dodge, H. F., and M. N. Torrey (1951). "Additional Continuous Sampling Inspection Plans," *Industrial Quality Control*, Vol. 7.

Duncan, A. J. (1956). "The Economic Design of $\overline{X}$-Charts Used to Maintain Current Control of a Process," *Journal of the American Statistical Association*, Vol. 51.

Duncan, A. J. (1986). *Quality Control and Industrial Statistics*, 5th ed., Irwin, Homewood, IL.

Duncan, A. J. (1978). "The Economic Design of *p*-Charts to Maintain Current Control of a Process: Some Numerical Results," *Technometrics*, Vol. 20.

English, J. R., and G. D. Taylor (1993). "Process Capability Analysis—A Robustness Study," *International Journal of Production Research*, Vol. 31, No. 7.

Ewan, W. D. (1963). "When and How to Use Cu-Sum Charts," *Technometrics*, Vol. 5.

Farnum, N. R. (1992). "Control Charts for Short Runs: Nonconstant Process and Measurement Error," *Journal of Quality Technology*, Vol. 24.

Ferrell, E. B. (1953). "Control Charts Using Midranges and Medians," *Industrial Quality Control*, Vol. 9.

Freund, R. A. (1957). "Acceptance Control Charts," *Industrial Quality Control*, Vol. 12.

Gan, F. F. (1991). "An Optimal Design of CUSUM Quality Control Charts," *Journal of Quality Technology*, Vol. 23.

Gan, F. F. (1993). "An Optimal Design of CUSUM Control Charts for Binomial Counts," *Journal of Applied Statistics*, Vol. 20.

Gardiner, J. S. (1987). *Detecting Small Shifts in Quality Levels in a Near-Zero Defect Environment for Integrated Circuits*, Ph.D. Dissertation, Department of Mechanical Engineering, University of Washington, Seattle, WA.

Gardiner, J. S., and D. C. Montgomery (1987). "Using Statistical Control Charts for Software Quality Control," *Quality and Reliability Engineering International*, Vol. 3.

Garvin, D. A. (1987). "Competing in the Eight Dimensions of Quality," *Harvard Business Review*, Sept.–Oct.

Girshick, M. A., and H. Rubin (1952). "A Bayes' Approach to a Quality Control Model," *Annals of Mathematical Statistics*, Vol. 23.

Grant, E. L., and R. S. Leavenworth (1980). *Statistical Quality Control*, 5th ed., McGraw-Hill, New York.

Grubbs, F. E. (1946). "The Difference Control Chart with an Example of Its Use," *Industrial Quality Control*, Vol. 2.

Guenther, W. C. (1972). "Tolerance Intervals for Univariate Distributions," *Naval Research Logistics Quarterly*, Vol. 19.

Hahn, G. J., and S. S. Shapiro (1967). *Statistical Models in Engineering*, Wiley, New York.

Harris, T. J., and W. H. Ross (1991). "Statistical Process Control Procedures for Correlated Observations," *Canadian Journal of Chemical Engineering*, Vol. 69.

Hawkins, D. M. (1981). "A CUSUM for a Scale Parameter," *Journal of Quality Technology*, Vol. 13.

Hawkins, D. M. (1991). "Multivariate Quality Control Based on Regression Adjusted Variables," *Technometrics*, Vol. 33.

Hawkins, D. M. (1992). "A Fast, Accurate Approximation of Average Run Lengths of CUSUM Control Charts," *Journal of Quality Technology*, Vol. 24.

Hawkins, D. M. (1993a). "Cumulative Sum Control Charting: An Underutilized SPC Tool," *Quality Engineering*, Vol. 5.

Hawkins, D. M. (1993b). "Regression Adjustment for Variables in Multivariate Quality Control," *Journal of Quality Technology*, Vol. 25.

Hayter, A. J., and K.-L. Tsui (1994). "Identification and Quantification in Multivariate Quality Control Problems," *Journal of Quality Technology*, Vol. 26.

Hicks, C. R. (1955). "Some Applications of Hotelling's T^2," *Industrial Quality Control*, Vol. 11.

Hill, D. (1956). "Modified Control Limits," *Applied Statistics*, Vol. 5.

Hillier, F. S. (1969). "$\bar{x}$ and R Chart Control Limits Based on a Small Number of Subgroups," *Journal of Quality Technology*, Vol. 1.

Hines, W. W., and D. C. Montgomery (1990). *Probability and Statistics in Engineering and Management Science*, 3rd ed., Wiley, New York.

Ho, C., and K. E. Case (1994). "Economic Design of Control Charts: A Literature Review for 1981–1991," *Journal of Quality Technology*, Vol. 26.

Holmes, D. S., and A. E. Mergen (1993). "Improving the Performance of the T^2 Control Chart," *Quality Engineering*, Vol. 5.

Hotelling, H. (1947). "Multivariate Quality Control," *Techniques of Statistical Analysis*, edited by Eisenhart, Hastay, and Wallis, McGraw-Hill, New York.

Howell, J. M. (1949). "Control Charting Largest and Smallest Value," *Annals of Mathematical Statistics*, Vol. 20.

Hunter, J. S. (1985). "Statistical Design Applied to Product Design," *Journal of Quality Technology*, Vol. 17.

Hunter, J. S. (1986). "The Exponentially Weighted Moving Average," *Journal of Quality Technology*, Vol. 18.

Hunter, J. S. (1987). Letter to the Editor, *Quality Progress*, May 1987.

Hunter, J. S. (1989). "A One-Point Plot Equivalent to the Shewhart Chart with Western Electric Rules," *Quality Engineering*, Vol. 2.

Hunter, J. S. (1998). "The Box–Jenkins Manual Bounded Adjustment Chart," *Quality Progress*, December.

Hunter, W. G., and C. P. Kartha (1977). "Determining the Most Profitable Target Value for a Production Process," *Journal of Quality Technology*, Vol. 9.

Iglewitz, B., and D. Hoaglin (1987). "Use of Boxplots for Process Evaluation," *Journal of Quality Technology*, Vol. 19.

Jackson, J. E. (1956). "Quality Control Methods for Two Related Variables," *Industrial Quality Control*, Vol. 12.

Jackson, J. E. (1959). "Quality Control Methods for Several Related Variables," *Technometrics*, Vol. 1.

Jackson, J. E. (1972). "All Count Distributions Are Not Alike," *Journal of Quality Technology*, Vol. 4.

Jackson, J. E. (1980). "Principal Components and Factor Analysis: Part I—Principal Components," *Journal of Quality Technology*, Vol. 12.

Jackson, J. E. (1985). "Multivariate Quality Control," *Communications in Statistics—Theory and Methods*, Vol. 14.

Johnson, N. L. (1961). "A Simple Theoretical Approach to Cumulative Sum Control Charts," *Journal of the American Statistical Association*, Vol. 54.

Johnson, N. L., and S. Kotz (1969). *Discrete Distributions*, Houghton Mifflin, Boston.

Johnson, N. L., and F. C. Leone (1962a). "Cumulative Sum Control Charts—Mathematical Principles Applied to Their Construction and Use," Part I, *Industrial Quality Control*, Vol. 18.

Johnson, N. L., and F. C. Leone (1962b). "Cumulative Sum Control Charts—Mathematical Principles Applied to Their Construction and Use," Part II, *Industrial Quality Control*, Vol. 18.

Johnson, N. L., and F. C. Leone (1962c). "Cumulative Sum Control Charts—Mathematical Principles Applied to Their Construction and Use," Part III, *Industrial Quality Control*, Vol. 18.

Jones, L. A., W. H. Woodall, and M. D. Conerly (1999). "Exact Properties of Demerit Control Charts," *Journal of Quality Technology*, Vol. 31, No. 2.

Juran, J. M., and F. M. Gryna, Jr. (1980). *Quality Planning and Analysis*, 2nd ed., McGraw-Hill, New York.

Kane, V. E. (1986). "Process Capability Indices," *Journal of Quality Technology*, Vol. 18.

Keats, J. B., E. Del Castillo, E. von Collani, and E. M. Saniga (1997). Economic Modeling for Statistical Process Control," *Journal of Quality Technology*, Vol. 29, No. 2.

Kotz, S., and C. R. Lovelace (1998). *Process Capability Indices in Theory and Practice*, Arnold, London.

Kushler, R. H., and P. Hurley (1992). "Confidence Bounds for Capability Indices," *Journal of Quality Technology*, Vol. 24.

Langenberg, P., and B. Inglewitz (1986). "Trimmed Mean $\bar{x}$ and R Charts," *Journal of Quality Technology*, Vol. 18.

Lanning, J. W. (1998). *Methods for Monitoring Fractionally Sampled Multiple Stream Processes*, Ph.D. Dissertation, Department of Industrial Engineering, Arizona State University, Tempe.

Ledolter, J., and A. Swersey (1997). "An Evaluation of Pre-Control," *Journal of Quality Technology*, Vol. 29, No. 2.

Lieberman, G. J., and G. J. Resnikoff (1955). "Sampling Plans for Inspection by Variables," *Journal of the American Statistical Association*, Vol. 50.

Lieberman, G. J., and H. Solomon (1955). "Multi-Level Continuous Sampling Plans," *Annals of Mathematical Statistics*, Vol. 26.

Lorenzen, T. J., and L. C. Vance (1986). "The Economic Design of Control Charts: A Unified Approach," *Technometrics*, Vol. 28.

Lowry, C. A., C. W. Champ, and W. H. Woodall (1995). "The Performance of Control Charts for Monitoring Process Variation," *Communications in Statistics—Computation and Simulation*, Vol. 21.

Lowry, C. A., and D. C. Montgomery (1995). "A Review of Multivariate Control Charts," *IIE Transactions*, Vol. 26.

Lowry, C. A., W. H. Woodall, C. W. Champ, and S. E. Rigdon (1992). "A Multivariate Exponentially Weighted Moving Average Control Chart," *Technometrics*, Vol. 34.

Lu, C.-W., and M. R. Reynolds, Jr. (1999a). "EWMA Control Charts for Monitoring the Mean of Autocorrelated Processes," *Journal of Quality Technology*, Vol. 31, No. 2.

Lu, C.-W., and M. R. Reynolds, Jr. (1999b). "Control Charts for Monitoring the Mean and Variance of Autocorrelated Processes," *Journal of Quality Technology*, Vol. 31, No. 3.

Lucas, J. M. (1973). "A Modified V-Mask Control Scheme," *Technometrics*, Vol. 15.

Lucas, J. M. (1976). "The Design and Use of Cumulative Sum Quality Control Schemes," *Journal of Quality Technology*, Vol. 8.

Lucas, J. M. (1982). "Combined Shewhart–CUSUM Quality Control Schemes," *Journal of Quality Technology*, Vol. 14.

Lucas, J. M. (1985). "Counted Data CUSUM's," *Technometrics*, Vol. 27.

Lucas, J. M., and R. B. Crosier (1982). "Fast Initial Response for CUSUM Quality Control Schemes," *Technometrics*, Vol. 24.

Lucas, J. M., and M. S. Saccucci (1990), "Exponentially Weighted Moving Average Control Schemes: Properties and Enhancements," *Technometrics*, Vol. 32.

Luceño, A. (1996). "A Process Capability Ratio with Reliable Confidence Intervals," *Communications in Statistics—Simulation and Computation*, Vol. 25, No. 1.

MacGregor, J. F. (1987). "Interfaces Between Process Control and On-Line Statistical Process Control," *A.I.Ch.E. Cast Newsletter*, 9–19.

MacGregor, J. F., and T. J. Harris (1993). "The Exponentially Weighted Moving Variance," *Journal of Quality Technology*, Vol. 25.

MacGregor, J. F., and T. J. Harris (1990). "Discussion of: "EWMA Control Schemes: Properties and Enhancement," by Lucas and Sacucci," *Technometrics*, Vol. 32.

Mandel, J. (1969). "The Regression Control Chart," *Journal of Quality Technology*, Vol. 1.

Manuele, J. (1945). "Control Chart for Determining Tool Wear," *Industrial Quality Control*, Vol. 1.

Maragah, H. O., and W. H. Woodall (1992). "The Effect of Autocorrelation on the Retrospective X-chart," *Journal of Statistical Computation and Simulation*, Vol. 40.

Mason, R. L., N. D. Tracy, and J. C. Young (1995). "Decomposition of T^2 for Multivariate Control Chart Interpretation," *Journal of Quality Technology*, Vol. 27.

Mastrangelo, C. M., and D. C. Montgomery (1995). "SPC with Correlated Observations for the Chemical and Process Industries," *Quality and Reliability Engineering International,* Vol. 11.

Mastrangelo, C. M., G. C. Runger, and D. C. Montgomery (1996). "Statistical Process Monitoring with Principal Components," *Quality and Reliability Engineering International*, Vol. 12.

Molina, E. C. (1942). *Poisson's Exponential Binomial Limit*, Van Nostrand Reinhold, New York.

Montgomery, D. C. (1980). "The Economic Design of Control Charts: A Review and Literature Survey," *Journal of Quality Technology*, Vol. 14.

Montgomery, D. C. (1982). "Economic Design of an $\bar{x}$ Control Chart," *Journal of Quality Technology*, Vol. 14.

Montgomery, D. C. (1997). *Design and Analysis of Experiments,* 4th ed., Wiley, New York.

Montgomery, D. C. (1999). "Experimental Design for Product and Process Design and Development" (with commentary), *Journal of the Royal Statistical Society Series D (The Statistician)*, Vol. 38, part 2.

Montgomery, D. C., and J. J. Friedman (1989). "Statistical Process Control in a Computer-Integrated Manufacturing Environment," *Statistical Process Control in Automated Manufacturing,* edited by J. B. Keats and N. F. Hubele, Dekker, Series in Quality and Reliability, New York.

Montgomery, D. C., L. A. Johnson, and J. S. Gardiner (1990). *Forecasting and Time Series Analysis*, 2nd ed., McGraw-Hill, New York.

Montgomery, D. C., J. B. Keats, G. C. Runger, and W. S. Messina (1994). "Integrating Statistical Process Control and Engineering Process Control," *Journal of Quality Technology*, Vol. 26.

Montgomery, D. C., and C. M. Mastrangelo (1991). "Some Statistical Process Control Methods for Aurocorrelated Data" (with discussion**),** *Journal of Quality Technology*, Vol. 23.

Montgomery, D. C., and E. A. Peck (1992). *Introduction to Linear Regression Analysis*, 2nd ed., Wiley, New York.

Montgomery, D. C., and G. C. Runger (1993a). "Gauge Capability and Designed Experiments: Part I: Basic Methods," *Quality Engineering*, Vol. 6.

Montgomery, D. C., and G. C. Runger (1993b). "Gauge Capability Analysis and Designed Experiments: Part II: Experimental Design Models and Variance Component Estimation," *Quality Engineering*, Vol. 6.

Montgomery, D. C., and G. C. Runger (1999). *Applied Statistics and Probability for Engineers*, 2nd ed., Wiley, New York.

Montgomery, D. C., and H. M. Wadsworth, Jr. (1972). "Some Techniques for Multivariate Quality Control Applications," *ASQC Technical Conference Transactions*, Washington, DC.

Montgomery, D. C., and W. H. Woodall, editors (1997). "A Discussion of Statistically-Based Process Monitoring and Control," *Journal of Quality Technology*, Vol. 29, No. 2.

Mortell, R. R., and G. C. Runger (1995). "Statistical Process Control of Multiple Stream Processes," *Journal of Quality Technology*, Vol. 27.

Murphy, B. J. (1987). "Screening Out-of-Control Variables with T^2 Multivariate Quality Control Procedures," *The Statistician*, Vol. 36.

Myers, R. H., and D. C. Montgomery (1995). *Response Surface Methodology: Process and Product Optimization Using Designed Experiments*, Wiley, New York.

Nair, V. N., editor (1992). "Taguchi's Parameter Design: A Panel Discussion," *Technometrics*, Vol. 34.

Nelson, L. S. (1978). "Best Target Value for a Production Process," *Journal of Quality Technology*, Vol. 10.

Nelson, L. S. (1984). "The Shewhart Control Chart—Tests for Special Causes," *Journal of Quality Technology*, Vol. 16.

Nelson, L. S. (1986). "Control Chart for Multiple Stream Processes," *Journal of Quality Technology*, Vol. 18.

Nelson, L. S. (1994). "A Control Chart for Parts-per-Million Nonconforming Items," *Journal of Quality Technology*, Vol. 26.

Nelson, P. R., and P. L. Stephenson (1996). "Runs Tests for Group Control Charts," *Communications in Statistics—Theory and Methods*, Vol. 25, No. 10.

Ott, E. R. (1975). *Process Quality Control*, McGraw-Hill, New York.

Ott, E. R., and R. D. Snee (1973). "Identifying Useful Differences in a Multiple-Head Machine," *Journal of Quality Technology*, Vol. 5.

Page, E. S. (1954). "Continuous Inspection Schemes," *Biometrics*, Vol. 41.

Page, E. S. (1961). "Cumulative Sum Control Charts," *Technometrics*, Vol. 3.

Page, E. S. (1963). "Controlling the Standard Deviation by Cusums and Warning Lines," *Technometrics*, Vol. 5.

Pearn, W. L., S. Kotz, and N. L. Johnson (1992). "Distributional and Inferential Properties of Process Capability Indices," *Journal of Quality Technology*, Vol. 24.

Perry, R. L. (1973). "Skip-Lot Sampling Plans," *Journal of Quality Technology*, Vol. 5.

Pignatiello, J. J., Jr., and J. S. Ramberg (1991). "Top Ten Triumphs and Tragedies of Genichi Taguchi," *Quality Engineering*, Vol. 4.

Pignatiello, J. J., Jr., and G. C. Runger (1990). "Comparison of Multivariate CUSUM Charts," *Journal of Quality Technology*, Vol. 22.

Pignatiello, J. J., Jr., and G. C. Runger (1991). "Adaptive Sampling for Process Control," *Journal of Quality Technology*, Vol. 23.

Prabhu, S. S., D. C. Montgomery, and G. C. Runger (1994). "A Combined Adaptive Sample Size and Sampling Interval $\bar{x}$ Control Scheme," *Journal of Quality Technology*, Vol. 26.

Prabhu, S. S., D. C. Montgomery, and G. C. Runger (1995). "A Design Tool to Evaluate Average Time to Signal Properties of Adaptive $\bar{x}$ Charts," *Journal of Quality Technology*, Vol. 27.

Prabhu, S. S., and G. C. Runger (1997). "Designing a Multivariate EWMA Control Chart," *Journal of Quality Technology*, Vol. 29, No. 1.

Quesenberry, C. P. (1988). "An SPC Approach to Compensating a Tool-Wear Process," *Journal of Quality Technology*, Vol. 20.

Quesenberry, C. P. (1991a). "SPC Q Charts for Start-Up Processes and Short or Long Runs," *Journal of Quality Technology*, Vol. 23.

Quesenberry, C. P. (1991b). "SPC Q Charts for a Binomial Parameter p: Short or Long Runs," *Journal of Quality Technology*, Vol. 23.

Quesenberry, C. P. (1991c). "SPC Q Charts for a Poisson Parameter λ: Short or Long Runs," *Journal of Quality Technology*, Vol. 23.

Quesenberry, C. P. (1993). "The Effect of Sample Size on Estimated Limits for $\bar{x}$ and x Control Charts," *Journal of Quality Technology*, Vol. 25.

Quesenberry, C. P. (1995a). "On Properties of Q Charts for Variables," *Journal of Quality Technology*, Vol. 27.

Quesenberry, C. P. (1995b). "On Properties of Binomial Q-charts for Attributes," *Journal of Quality Technology*, Vol. 27.

Quesenberry, C. P. (1995c). "Geometric Q Charts for High Quality Processes," *Journal of Quality Technology*, Vol. 27.

Quesenberry, C. P. (1995d). "On Properties of Poisson Q Charts for Attributes" (with discussion), *Journal of Quality Technology*, Vol. 27.

Reynolds, M. R., Jr., R. W. Amin, J. C. Arnold, and J. A. Nachlas (1988). "$\bar{x}$ Charts with Variable Sampling Intervals," *Technometrics*, Vol. 30.

Rhoads, T. R., D. C. Montgomery, and C. M. Mastrangelo (1996). "Fast Initial Response Scheme for the EWMA Control Chart," *Quality Engineering*, Vol. 9.

Roberts, S. W. (1958). "Properties of Control Chart Zone Tests," *Bell System Technical Journal*, Vol. 37.

Roberts, S. W. (1959). "Control Chart Tests Based on Geometric Moving Averages," *Technometrics*, Vol. 1.

Rocke, D. M. (1989). "Robust Control Charts," *Technometrics*, Vol. 31.

Rodriguez, R. N. (1992). "Recent Developments in Process Capability Analysis," *Journal of Quality Technology*, Vol. 24.

Ross, S. M. (1971). "Quality Control Under Markovian Deterioration," *Management Science*, Vol. 17.

Runger, G. C., and T. R. Willemain (1996). "Batch Means Control Charts for Autocorrelated Data," *IIE Transactions*, Vol. 28.

Runger, G. C., F. B. Alt, and D. C. Montgomery (1996a). "Controlling Multiple Stream Processes with Principal Components," *International Journal of Production Research*, Vol. 34.

Runger, G. C., F. B. Alt, and D. C. Montgomery (1996b). "Contributors to a Multivariate Statistical Process Control Signal," *Communications in Statistics—Theory and Methods*, Vol. 25, No. 10.

Saniga, E. M. (1989). "Economic Statistical Control Chart Design with an Application to $\bar{x}$ and R Charts," *Technometrics*, Vol. 31.

Saniga, E. M., and L. E. Shirland (1977). "Quality Control in Practice—A Survey," *Quality Progress*, Vol. 10.

Savage, I. R. (1962). "Surveillance Problems," *Naval Research Logistics Quarterly*, Vol. 9.

Schilling, E. G., and P. R. Nelson (1976). "The Effect of Nonnormality on the Control Limits of $\bar{x}$ Charts," *Journal of Quality Technology*, Vol. 8.

Schmidt, S. R., and J. R. Boudot (1989). "A Monte Carlo Simulation Study Comparing Effectiveness of Signal-to-Noise Ratios and Other Methods for Identifying Dispersion Effects," presented at the 1989 Rocky Mountain Quality Conference.

Scranton, R., G. C. Runger, J. B. Keats, and D. C. Montgomery (1996). "Efficient Shift Detection Using Exponentially Weighted Moving Average Control Charts and Principal Components," *Quality and Reliability Engineering International*, Vol. 12.

Shapiro, S. S. (1980). *How to Test Normality and Other Distributional Assumptions*, Vol. 3, *The ASQC Basic References in Quality Control: Statistical Techniques*, ASQC, Milwaukee, WI.

Sheaffer, R. L., and R. S. Leavenworth (1976). "The Negative Binomial Model for Counts in Units of Varying Size," *Journal of Quality Technology*, Vol. 8.

Siegmund, D. (1985). *Sequential Analysis: Tests and Confidence Intervals*, Springer-Verlag, New York.

Somerville, S. E., and D. C. Montgomery (1996). "Process Capability Indices and Nonnormal Distributions," *Quality Engineering*, Vol. 9, No. 2.

Steiner, S. H. (1999). "EWMA Control Charts with Time-Varying Control Limits and Fast Initial Response," *Journal of Quality Technology*, Vol. 31, No. 1.

Stephens, K. S. (1979). *How to Perform Continuous Sampling (CSP)*, Vol. 2, *The ASQC Basic References in Quality Control: Statistical Techniques*, ASQC, Milwaukee, WI.

Sullivan, J. H., and W. H. Woodall (1995). "A Comparison of Multivariate Quality Control Charts for Individual Observations," *Journal of Quality Technology*, Vol. 27.

Svoboda, L. (1991). "Economic Design of Control Charts: A Review and Literature Survey (1979–1989)," in *Statistical Process Control in Manufacturing*, edited by J. B. Keats and D. C. Montgomery, Dekker, New York.

Taguchi, G. (1986). *Introduction to Quality Engineering*, Asian Productivity Organization, UNIPUB, White Plains, NY.

Taguchi, G., and Y. Wu (1980). *Introduction to Off-Line Quality Control*, Japan Quality Control Organization, Nagoya, Japan.

Taylor, H. M. (1965). "Markovian Sequential Replacement Processes," *Annals of Mathematical Statistics*, Vol. 36.

Taylor, H. M. (1967). "Statistical Control of a Gaussian Process," *Technometrics*, Vol. 9.

Taylor, H. M. (1968). "The Economic Design of Cumulative Sum Control Charts," *Technometrics*, Vol. 10.

Tracy, N. D., J. C. Young, and R. L. Mason (1992). "Multivariate Control Charts for Individual Observations," *Journal of Quality Technology*, Vol. 24.

Tseng, S., and B. M. Adams (1994), "Monitoring Autocorrelated Processes with an Exponentially Weighted Moving Average Forecast," *Journal of Statistical Computation and Simulation*, Vol. 50.

United States Department of Defense (1957). *Sampling Procedures and Tables for Inspection by Variables for Percent Defective*, MIL STD 414, U.S. Government Printing Office, Washington, DC.

United States Department of Defense (1989). *Sampling Procedures and Tables for Inspection by Attributes*, MIL STD 105E, U.S. Government Printing Office, Washington, DC.

Vance, L. C. (1986). "Average Run Lengths of Cumulative Sum Control Charts for Controlling Normal Means," *Journal of Quality Technology*, Vol. 18.

Vander Weil, S., W. T. Tucker, F. W. Faltin, and N. Doganaksoy (1992). "Algorithmic Statistical Process Control: Concepts and an Application," *Technometrics*, Vol. 34.

Wald, A. (1947). *Sequential Analysis*, Wiley, New York.

Wang, C.-H., and F. S. Hillier (1970). "Mean and Variance Control Chart Limits Based on a Small Number of Subgroups," *Journal of Quality Technology*, Vol. 2.

Wardell, D. G., H. Moskowitz, and R. D. Plante (1994). "Run Length Distributions of Special-Cause Control Charts for Correlated Processes," *Technometrics*, Vol. 36.

Weiler, H. (1952). "On the Most Economical Sample Size for Controlling the Mean of a Population," *Annals of Mathematical Statistics*, Vol. 23.

Western Electric (1956). *Statistical Quality Control Handbook*, Western Electric Corporation, Indianapolis, IN.

Wetherill, G. B., and D. W. Brown (1991). *Statistical Process Control: Theory and Practice*, Chapman and Hall, New York.

White, C. C. (1974). "A Markov Quality Control Process Subject to Partial Observation," *Management Science*, Vol. 23.

White, C. H., J. B. Keats, and J. Stanley (1997). "Poison Cusum Versus c Chart for Defect Data," *Quality Engineering*, Vol. 9, No. 4.

White, E. M., and R. Schroeder (1987). "A Simultaneous Control Chart," *Journal of Quality Technology*, Vol. 19.

Willemain, T. R., and G. C. Runger (1996). "Designing Control Charts Based on an Empirical Reference Distribution," *Journal of Quality Technology*, Vol. 28, No. 1.

Woodall, W. H. (1986). "Weakness of the Economic Design of Control Charts," Letter to the Editor, *Technometrics*, Vol. 28.

Woodall, W. H. (1987). "Conflicts Between Deming's Philosophy and the Economic Design of Control Charts," in *Frontiers in Statistical Quality Control,* 3, edited by H. J. Lenz, G. B. Wetherill, and P.-T. Wilrich, Physica-Verlag, Vienna.

Woodall, W. H., and B. M. Adams (1993). "The Statistical Design of CUSUM Charts," *Quality Engineering*, Vol. 5.

Woodall, W. H., and D. C. Montgomery (1999). "Research Issues and Ideas in Statistical Process Control," *Journal of Quality Technology*, Vol. 31, No. 2.

Woodall, W. H., and D. C. Montgomery (2000). "Using Ranges to Estimate Variability," *Quality Engineering*, to appear.

Yourstone, S. A., and D. C. Montgomery (1989). "Development of a Real-Time Statistical Process-Control Algorithm," *Quality and Reliability Engineering International*, Vol. 5.

Yourstone, S., and W. Zimmer (1992). "Non-normality and the Design of Control Charts for Averages," *Decision Sciences*, Vol. 32.

Zhang, N. F., G. A. Stenback, and D. M. Wardrop (1990). "Interval Estimation of Process Capability Index C_{pk}," *Communications in Statistics—Theory and Methods*, Vol. 19.

Answers to Selected Exercises

CHAPTER 2

2-1. (a) $\bar{x} = 10.028$
(b) $S = 0.015$

2-3. (a) $\bar{x} = 952.9$
(b) $S = 3.7$

2-5. (a) $\bar{x} = 121.25$
(b) $S = 22.63$

2-11. (a) $\bar{x} = 89.476$
(b) $S = 4.158$

2-15. *sample space*: $\{2, 3, 4, 5, 6, 7, 8, 9, 10, 11, 12\}$

$$p(x) = \begin{cases} 1/36; x = 2 \\ 2/36; x = 3 \\ 3/36; x = 4 \\ 4/36; x = 5 \\ 5/36; x = 6 \\ 6/36; x = 7 \\ 5/36; x = 8 \\ 4/36; x = 9 \\ 3/36; x = 10 \\ 2/36; x = 11 \\ 1/36; x = 12 \\ 0; \text{otherwise} \end{cases}$$

2-17. (a) 0.0196
(b) 0.0198
(c) Cutting occurrence rate reduces probability from 0.0198 to 0.0100.

2-19. (a) $k = 0.05$
(b) $\mu = 1.867$, $\sigma^2 = 0.615$

$$(c) \; F(x) = \begin{cases} \dfrac{1.15}{3} = 0.383; x = 1 \\[3mm] \dfrac{1.15 + 1.1}{3} = 0.750; x = 2 \\[3mm] \dfrac{1.15 + 1.1 + 0.75}{3} = 1.000; x = 3 \end{cases}$$

2-21. (a) Approximately 11.8%.
       (b) Decrease profit by $5.90/calculator.

2-23. Decision rule means 63.6% of samples will have one or more nonconforming units.

2-25. 0.818

2-27. (a) 0.496
       (b) 0.528. Approximation is not satisfactory.
       (c) $n/N = 0.033$. Approximation is satisfactory.
       (d) $n = 11$

2-29. $\Pr\{x = 0\} = 0.364$
       $\Pr\{x \geq 2\} = 0.264$

2-31. $\Pr\{x \geq 1\} = 0.00001$

2-33. $\mu = 1/p$

2-35. $\Pr\{x \leq 34\} = 0.2266$. Number failing minimum spec is 11,330.
       $\Pr\{x > 48\} = 0.1587$. Number failing maximum spec is 7935.

2-37. Process is centered at target, so shifting process mean in either direction increases nonconformities. Process variance must be reduced to 0.03^2 to have at least 999 of 1000 conform to specification.

2-39. $\Pr\{x > 1000\} = 0.0021$

2-41. If $c_2 > c_1 + 0.0620$, then choose process 1.

CHAPTER 3

3-1.   (a) $Z_0 = 6.78$. Reject H_0.
       (b) $P = 0$
       (c) $8.2525 \leq \mu \leq 8.2545$

3-3.   (a) $t_0 = 1.833$. Reject H_0.
       (b) $25.06 \leq \mu \leq 26.94$

3-5.   (a) $t_0 = -3.089$. Reject H_0.
       (b) $13.39216 \leq \mu \leq 13.40020$

3-7. $n = 246$

3-9.   (a) $t_0 = 2.131$. Reject H_0.
       (b) $9.727 \leq \mu \leq 10.791$
       (c) $\chi_0^2 = 14.970$. Do not reject H_0.
       (d) $0.738 \leq \sigma \leq 1.546$
       (e) $\sigma \leq 1.436$

3-11. (a) $t_0 = 0.11$. Do not reject H_0.
       (c) $-0.127 \leq (\mu_1 - \mu_2) \leq 0.141$
       (d) $F_0 = 0.8464$. Do not reject H_0.
       (e) $0.165 \leq \sigma_1^2/\sigma_2^2 \leq 4.821$
       (f) $0.007 \leq \sigma^2 \leq 0.065$

3-13. (a) $t_0 = -0.77$. Do not reject H_0.
       (b) $-6.7 \leq (\mu_1 - \mu_2) \leq 3.1$
       (c) $0.21 \leq \sigma_1^2/\sigma_2^2 \leq 3.34$

3-15. (a) $Z_0 = 4.0387$. Reject H_0.
(b) $P = 0.00006$
(c) $p \leq 0.155$

3-17. (a) $F_0 = 1.0417$. Do not reject H_0.
(b) $t_0 = 0.4174$. Do not reject H_0.

3-19. $t_0 = -0.7537$. There is no difference between mean measurements.

3-21. (a) $\chi_0^2 = 42.75$. Do not reject H_0.
(b) $1.1909 \leq \sigma \leq 2.0556$

3-23. $n = [(Z_{\alpha/2} + Z_\beta)\sigma/\delta]^2$

3-27. $Z_0 = 0.3162$. Do not reject H_0.

3-29. (a) $F_0 = 3.59$, $P = 0.053$.

3-31. (a) $F_0 = 1.87$, $P = 0.214$.

3-33. (a) $F_0 = 1.45$, $P = 0.258$.

3-35. (a) $F_0 = 30.85$, $P = 0.000$.

CHAPTER 4

4-17. Pattern is random.

4-19. There is a nonrandom, cyclic pattern.

4-21. Points 17, 18, 19, and 20 are outside lower 1-sigma area.

4-23. Points 16, 17, and 18 are 2 of 3 beyond 2 sigma of centerline. Points 5, 6, 7, 8, and 9 are of 5 at 1 sigma or beyond of centerline.

4-25. (a) $\sigma_{\bar{x}} = 0.0045$, $UCL2_S = 74.0090$, $LCL2_S = 73.9910$
(c) $ARL_{0\text{-}2S} = 1/0.046\_2$, $ARL_{0\text{-}3S} = 1/0.0027\_370$

4-27. Rule 1: $\_1 = 0.84939$. Rule 2: $\_2 = 0.75370$. Both rules: $\_ = 0.64019$.

4-29. (a) $\_(1\_\_)$
(b) $\_^{m\_1}(1\_\_)$
(c) $ARL_0 = 1/(1\_\_)$
(d) $\_$
(e) $\_$
(f) $(1\_\_)^2$
(g) $1 - \_^m$

CHAPTER 5

5-1. (a) Samples 12 and 15 exceed $\bar{x}$ UCL.
(b) $\hat{p} = 0.00050$

5-3. (a) $\bar{x}$ chart: CL = 10.9, UCL = 47.53, LCL = -25.73
R chart: CL = 63.5, UCL = 134.3, LCL = 0
Process is in statistical control.
(b) $\hat{\sigma}_x = 27.3$
(c) $\hat{C}_P = 1.22$

5-5. (a) $\bar{x}$ chart: CL = _0.003, UCL = 1.037, LCL = _1.043
       S chart: CL = 1.066, UCL = 1.830, LCL = 0.3025
       (b) R chart: CL = 3.2, UCL = 5.686, LCL = 0.713
       (c) S^2 chart: CL = 1.2057, UCL = 3.6297, LCL = 0.1663

5-7. $\bar{x}$ chart: CL = 10.38, UCL = 14.78, LCL = 5.974
      S chart: CL = 2.703, UCL = 6.125, LCL = 0

5-9. $\bar{x}$ chart: CL = 80, UCL = 89.49, LCL = 70.51
      S chart: CL = 9.727, UCL = 16.69, LCL = 2.76

5-11. (a) $\bar{x}$ chart: CL = 20, UCL = 21.93, LCL = 18.07
       S chart: CL = 1.50, UCL = 2.955, LCL = 0.045
       (b) LNTL = 15.27, UNTL = 24.73
       (c) $\hat{C}_P = 0.84$
       (d) $\hat{p}_{rework} = 0.02880$, $\hat{p}_{scrap} = 0.00078$, Total = 2.958%
       (e) $\hat{p}_{rework} = 0.00568$, $\hat{p}_{scrap} = 0.00568$, Total = 1.136%

5-13. (a) $\bar{x}$ chart: CL = 79.53, UCL = 84.58, LCL = 74.49
       R chart: CL = 8.745, UCL = 18.49, LCL = 0
       Process is in statistical control.
       (b) Several subgroups exceed UCL on R chart.

5-15. (a) $\bar{x}$ chart: CL = 34.00, UCL = 38.81, LCL = 29.19
       R chart: CL = 4.7, UCL = 12.1, LCL = 0
       (b) Detect shift more quickly.
       (c) $\bar{x}$ chart: CL = 34.00, UCL = 35.75, LCL = 32.25
       R chart: CL = 4.7, UCL = 8.76, LCL = 0.64

5-17. (a) $\bar{x}$ chart: CL = 26.50, UCL = 26.71, LCL = 26.29
       R chart: CL = 0.36, UCL = 0.76, LCL = 0
       (b) $\hat{p} = 0.00499$
       (c) $\hat{p} = 0.00124$

5-19. Pr{detect} = 0.84134

5-21. (a) $\bar{x}$ chart: CL = 10, UCL = 12.373, LCL = 7.628
       (b) R chart: CL = 7.695, UCL = 13.67, LCL = 1.72
       (c) S chart: CL = 2.432, UCL = 4.17, LCL = 0.69

5-23. (a) $\bar{x}$ chart: CL = 200, UCL = 202.42, LCL = 197.59
       R chart: CL = 5, UCL = 10.02, LCL = 0
       (b) $\hat{C}_P = 0.85$
       (c) Pr{not detect} = 0.95994

5-25. (a) $\bar{x}$ chart with all data: CL = 104.05, UCL = 106.329, LCL = 101.771
       R chart with all data: CL = 3.95, UCL = 8.354, LCL = 0
       $\bar{x}$ chart without sample 4: CL = 104, UCL = 106.064, LCL = 101.936
       R chart without sample 4: CL = 3.579, UCL = 7.567, LCL = 0
       (b) Without sample 4: $\hat{\sigma}_x = 1.539$
       (c) LNTL = 99.38, UNTL = 108.62
       (d) $\hat{p} = 0.0262$

5-27. (a) Sample 12 exceeds UCL on R chart.
       (b) R chart without sample 12: CL = 5.643, UCL = 11.93, LCL = 0
       (c) $\hat{\sigma}_{\bar{x}} = 2.426$
       (d) Without sample 12, $\hat{C}_P = 1.374$.

5-29. (a) $\hat{\sigma}_x = 0.4000$, $\hat{\sigma}_y = 0.300$
        (b) $z = 1.346$

5-31. (a) $\bar{S} = 13.667$, $UCL_S = 30.969$, $LCL_S = 0$
        (b) $\hat{\mu} = 429.0$, $\hat{\sigma}_x = 14.834$

5-33. Pr{out-of-control signal by at least 3rd plot point} = 0.7058

5-35. $\hat{C}_P = 0.7678$

5-37. (a) $\hat{C}_P = 1.111$

5-39. $UCL_{\bar{x}} = 610.3$, $LCL_{\bar{x}} = 589.7$

5-41. $ARL_1 = 6.30$

5-43. (a) $\hat{\sigma}_x = 4.000$
        (b) $\bar{S} = 3.865$, $UCL_S = 8.351$, $LCL_S = 0$
        (c) $\hat{p} = 0.1056$
        (e) Pr{detect on 1st sample} = 0.9772
        (f) Pr{detect by 3rd sample} = 1.0000

5-45. (a) $\hat{\mu} = 700$, $\hat{\sigma}_x = 8.661$
        (b) $\hat{p} = 0.1355$
        (c) $\alpha = 0.0208$
        (d) Pr{detect on 1st sample} = 0.3108
        (e) $ARL_1 = 3.22$

5-47. x chart: $\bar{x} = 53.27$, $UCL_x = 61.82$, $LCL_x = 44.72$
        MR chart: $\overline{MR2} = 3.214$, $UCL_{MR2} = 10.50$, $LCL_{MR2} = 0$

5-51. (b) x chart: $\bar{x} = 73.73$, $UCL_x = 104.9$, $LCL_x = 42.59$
        MR chart: $\overline{MR2} = 11.71$, $UCL_{MR2} = 38.26$, $LCL_{MR2} = 0$

5-53. x chart: $\bar{x} = 16.11$, $UCL_x = 16.17$, $LCL_x = 16.04$
        MR chart: $\overline{MR2} = 0.02365$, $UCL_{MR2} = .07726$, $LCL_{MR2} = 0$

5-55. x chart: $\bar{x} = 2929$, $UCLx = 3338$, $LCL_x = 2520$
        MR chart: $\overline{MR2} = 153.7$, $UCL_{MR2} = 502.2$, $LCL_{MR2} = 0$

5-57. (a) $\hat{\sigma}_x = 1.157$
        (b) $\hat{\sigma}_x = 1.682$
        (c) $\hat{\sigma}_x = 1.137$
        (d) $\hat{\sigma}_{x,\,span\,3} = 1.210$, $\hat{\sigma}_{x,\,span\,4} = 1.262$, . . . , $\hat{\sigma}_{x,\,span\,19} = 1.406$, $\hat{\sigma}_{x,\,span\,20} = 1.435$

5-59. (a) $\bar{x}$ chart: CL = 11.76, UCL = 11.79, LCL = 11.72
        R chart (within): CL = 0.06109, UCL = 0.1292, LCL = 0
        (c) I chart: CL = 11.76, UCL = 11.87, LCL = 11.65
        MR2 chart (between): CL = 0.04161, UCL = 0.1360, LCL = 0

5-61. (b) R chart (within): CL = 0.06725, UCL = 0.1422, LCL = 0
        (c) I chart: CL = 2.074, UCL = 2.159, LCL = 1.989
        MR2 chart (between): CL = 0.03210, UCL = 0.1049, LCL = 0
        (d) Need lot average, moving range between lot averages, and range within a lot.
        I chart: CL = 2.074, UCL = 2.133, LCL = 2.015
        MR2 chart (between): CL = 0.02205, UCL = 0.07205, LCL = 0
        R chart (within): CL = 0.1335, UCL = 0.2372, LCL = 0.02978

CHAPTER 6

6-1. $\bar{p} = 0.0585$, UCL $= 0.1289$, LCL $= 0$. Sample 12 exceeds UCL.
      Without sample 12: $\bar{p} = 0.0537$, UCL $= 0.1213$, LCL $= 0$

6-3. For $n = 80$, UCL$_i = 0.1397$, LCL$_i = 0$.
      Process is in statistical control.

6-5. (a) $\bar{p} = 0.1228$, UCL $= 0.1425$, LCL $= 0.1031$
      (b) Data should not be used since many subgroups are out of control.

6-7. Pr{detect shift on 1st sample} $= 0.278$, Pr{detect shift by 3rd sample} $= 0.624$

6-9. $\bar{p} = 0.10$, UCL $= 0.2125$, LCL $= 0$
      $p = 0.212$ to make _ $= 0.50$
      $n\_ 82$ to give positive LCL

6-11. $n = 81$

6-13. (a) $\bar{p} = 0.07$, UCL $= 0.108$, LCL $= 0.032$
      (b) Pr{detect shift on 1st sample} $= 0.297$
      (c) Pr{detect shift on 1st or 2nd sample} $= 0.506$

6-15. (a) $n\bar{p} = 16.4$, UCL $= 27.51$, LCL $= 5.292$
      (b) Pr{detect shift on 1st sample} $= 0.813$

6-17. (a) $n\bar{p} = 40$, UCL $= 58$, LCL $= 22$
      (b) Pr{detect shift on 1st sample} $= 0.583$

6-19. ARL$_1 = 1.715$ _2

6-21. (a) CL $= \bar{p} = 0.0221$
      for $n = 100$: UCL $= 0.0622$, LCL $= 0$
      for $n = 150$: UCL $= 0.0581$, LCL $= 0$
      for $n = 200$: UCL $= 0.0533$, LCL $= 0$
      for $n = 250$: UCL $= 0.0500$, LCL $= 0$
      (b) $Z_i = (\hat{p}_i - 0.0221)/\sqrt{0.0221(1 - 0.0221)/n_i}$

6-23. $Z_i = (\hat{p}_i - 0.0221)/\sqrt{0.0216/n_i}$

6-25. $n\_ 892$

6-27. (a) $k = 2.83$
      (b) $n\bar{p} = 20$, UCL $= 32.36$, LCL $= 7.64$
      (c) Pr{detect shift on 1st sample} $= 0.140$

6-29. (a) $n\_ 397$
      (b) $n = 44$

6-31. (a) $\bar{p} = 0.02$, UCL $= 0.062$, LCL $= 0$
      (b) Process has shifted to $\bar{p} = 0.038$.

6-33. $n\bar{p} = 2.505$, UCL $= 7.213$, LCL $= 0$

6-35. $Z_i = (\hat{p}_i - 0.06)/\sqrt{0.0564/n_i}$

6-37. Variable u:
      CL $= 0.7007$; UCL$_i = 0.7007 + 3\sqrt{0.7007/n_i}$; LCL$_i = 0.7007 - 3\sqrt{0.7007/n_i}$
      Averaged u: CL $= 0.7007$, UCL $= 1.249$, LCL $= 0.1527$

6-39. $Z_i = (u_i - 0.7007)/\sqrt{0.7007/n_i}$

6-41. $\bar{c} = 8.59$, UCL $= 17.384$, LCL $= 0$. Process is not in statistical control.

6-43. (a) c chart: CL $= 21.48$, UCL $= 35.38$, LCL $= 7.58$
       (b) u chart: CL $= 0.0086$, UCL $= 0.2868$, LCL $= 0$

6-45. (a) c chart: CL $= 4$, UCL $= 10$, LCL $= 0$
       (b) u chart: CL $= 1$, UCL $= 2.5$, LCL $= 0$

6-47. (a) c chart: CL $= 9$, UCL $= 18$, LCL $= 0$
       (b) u chart: CL $= 4$, UCL $= 7$, LCL $= 1$

6-49. $\bar{c} = 7.6$, UCL $= 13.00$, LCL $= 2.20$

6-51. $\bar{u} = 7$; UCL; $= 7 + 3\sqrt{7/n_i}$; LCL; $= 7 - 3\sqrt{7/n_i}$

6-53. $\bar{c} = 8.5$, UCL $= 17.25$, LCL $= 0$

6-55. $\bar{u} = 4$, UCL $= 9$

6-57. (a) $\bar{c} = 0.533$, UCL $= 1.993$, LCL $= 0$
       (b) _ $= 0.6872$
       (c) _ $= 0.271$
       (d) $\text{ARL}_1 = 3.69 \_ 4$

6-59. (a) $\bar{c} = 4$, UCL $= 10$, LCL $= 0$
       (b) _ $= 0.0264$

6-61. $n > L^2/\bar{c}$

CHAPTER 7

7-1. $\hat{C}_p = 1.17$, $\hat{C}_{pk} = 1.13$

7-3. $\hat{C}_p = 5.48$, $\hat{C}_{pk} = 4.34$, $\hat{C}_{pkm} = 0.43$

7-5.   (a) $\hat{C}_p = 2.98$
       (b) $\hat{C}_{pk} = 1.49$
       (c) $\hat{p}_{\text{actual}} = 0.000004$, $\hat{p}_{\text{potential}} = 0.000000$

7-7.   Process A: $\hat{C}_p = \hat{C}_{pk} = \hat{C}_{pkm} = 1.045$, $\hat{p} = 0.001726$
       Process B: $\hat{C}_p = 3.133$, $\hat{C}_{pk} = 1.566$, $\hat{C}_{pkm} = 0.652$, $\hat{p} = 0.000001$

7-9. $\hat{\sigma} = x_{84} - x_{50} = 1.0200 - 0.9975 = 0.0225$

7-13. $1.26 \_ C_p, \_ = 0.12$

7-15. (a) $\hat{C}_{pk} = 0.42$
       (b) $0.2957 \leq C_{pk} \leq 0.5443$

7-17. $\hat{\sigma}_{\text{process}} = 4$

7-19. (a) R chart indicates operator has no difficulty making consistent measurements.
       (b) $\hat{\sigma}^2_{\text{total}} = 4.717$, $\hat{\sigma}^2_{\text{product}} = 2.872$
       (c) 62.5%
       (d) $P/T = 0.272$

7-21. (a) $6\hat{\sigma}_{\text{gage}} = 8.154$
       (b) R chart indicates operator has difficulty using gage.

7-23. $\hat{p} = 0.4330$

7-25. $\hat{\mu}_{\text{Weight}} = 48$, $\hat{\sigma}_{\text{Weight}} = 0.04252$

7-27. $\mu_I \cong \mu_E/(\mu_{R_1} + \mu_{R_2})$
$\sigma_I^2 \cong \sigma_E^2/(\mu_{R_1} + \mu_{R_2})^2 + (\mu_E^2/(\mu_{R_1} + \mu_{R_2})^2)(\sigma_{R_1}^2 + \sigma_{R_2}^2)$

7-29. $C \sim N(0.006, 0.000005)$
Pr{positive clearance} $= 0.9964$

7-31. UTL $= 323.55$

7-33. UTL $= 372.08$

7-35. (a) $0.1257 \le x \le 0.1271$
(b) $0.1263 \le \bar{x} \le 0.1265$

CHAPTER 8

8-1. (a) $K = 12.5$, $H = 125$. Process is out of control on upper side after observation 7.
(b) $\hat{\sigma} = 34.43$

8-3. (a) $K = 12.5$, $H = 62.5$. Process is out of control on upper side after observation 7.
(b) Process is out of control on lower side at sample 6 and upper at sample 15.

8-5. Process is in control. $\text{ARL}_0 = 370.84$

8-7. (a) $\hat{\sigma} = 12.16$
(b) Process is out of control on upper side after reading 2.

8-9. (a) $\hat{\sigma} = 5.95$
(b) Process is out of control on lower side at start, then upper after observation 9.

8-11. Process is out of control on upper side after observation 7.

8-13. $\text{ARL}_0 = 215.23$, $\text{ARL}_1 = 25.02$

8-15. EWMA chart: CL $= 1050$, UCL $= 1065.49$, LCL $= 1034.51$.
Process exceeds upper control limit at sample 10.

8-17. EWMA chart: CL $= 8.02$, UCL $= 8.07$, LCL $= 7.97$.
Process is in control.

8-19. EWMA chart: CL $= 950$, UCL $= 957.535$, LCL $= 942.465$.
Process is out of control at samples 8, 12, and 13.

8-21. EWMA chart: CL $= 175$, UCL $= 177.346$, LCL $= 172.654$.
Process is out of control.

8-23. MA chart: CL $= 1050$, UCL $= 1080.62$, LCL $= 1019.38$.
Process is out of control at sample 10.

8-31. $k = 0.5L$

8-33. Transform data with $X = T^{0.2777}$. $X \sim N(5.386, 2.044^2)$.
Use $h = 0.5$, $k = 5$. Process is in control.

CHAPTER 9

9-1. $\bar{x}$ chart: CL $= 0.55$, UCL $= 4.438$, LCL $= -3.338$
R chart: CL $= 3.8$, UCL $= 9.782$, LCL $= 0$

9-5. $\bar{x}$ chart: CL = 52.988, UCL = 55.379, LCL = 50.596
       R chart: CL = 2.338, UCL = 6.017, LCL = 0

9-7.   (a) $\bar{x}$ chart: CL = 52.988, UCL = 58.727, LCL = 47.248
       R chart: CL = 2.158, UCL = 7.050, LCL = 0
       (c) $\bar{x}$ chart: CL = 52.99, UCL = 55.65, LCL = 50.33
       R chart: CL = 1.948, UCL = 4.415, LCL = 0

9-9.   (a) UCL = 44.503, LCL = 35.497
       (b) UCL = 43.609, LCL = 36.391
       (c) UCL = 46.181, LCL = 33.819

9-13. $\bar{x}$ chart: CL = 50, UCL = 65.304, LCL = 34.696

9-15. (a) $\hat{\sigma} = 4.000$
       (b) $\hat{p} = 0.1056$
       (c) UCL = 618.86, LCL = 601.14

9-17. $\mu_0 = 0.7072$, $\sigma = 20.41$, $\delta = 1\sigma$, $k = 0.5$, $h = 5$, $K = 10.20$, $H = 102.03$, no FIR.
       No observations exceed the control limit.

9-19. $\alpha = 0.1$, $\lambda = 0.150$, $\hat{\sigma} = 20.41$. Observations 6, 16, and 40 exceed control limits.

9-21. $\mu_0 = 0.050705$, $\sigma = 4.15$, $\delta = 1\sigma$, $k = 0.5$, $h = 5$, $K = 2.08$, $H = 20.77$, no FIR.
       No observations exceed the control limit.

9-23. $\alpha = 0.1$, $\lambda = 0.150$, $\hat{\sigma} = 4.15$. No observation exceeds control limits.

9-25. $\mu_0 = 0.2159$, $\sigma = 7.13$, $\delta = 1\sigma$, $k = 0.5$, $h = 5$, $K = 3.57$, $H = 35.65$, no FIR.
       No observations exceed the control limit.

9-27. $\alpha = 0.1$, $\lambda = 0.150$, $\hat{\sigma} = 7.349$. Several observations exceed the control limits.

9-29. (b) I chart: CL = 28.57, UCL = 37.11, LCL = 20.03
       MR2 chart: CL = 3.212, UCL = 10.49, LCL = 0
       (c) $\mu_0 = 28.57$, $\sigma = 2.85$, $\delta = 1\sigma$, $k = 0.5$, $h = 5$, $K = 1.42$, $H = 14.24$, no FIR.
       Several observations are out of control on both lower and upper sides.
       (d) EWMA chart: CL = 28.57, UCL = 31.00, LCL = 26.14.
       (e) Moving CL EWMA chart: $\alpha = 0.1$, $\lambda = 0.150$, $\hat{\sigma} = 2.85$.
       A few observations are beyond the lower control limit.
       (f) $\xi = 20.5017$, $\phi_1 = 0.7193$, $\phi_2 = -0.4349$.
       Set up an I and MR chart for residuals.
       I chart: CL = -0.03970, UCL = 9.598, LCL = -9.677
       MR2 chart: CL = 3.624, UCL = 11.84, LCL = 0

9-31. (a) $E(L) = \$4.12/\text{hr}$
       (b) $E(L) = \$4.98/\text{hr}$
       (c) $n = 5$, $k_{\text{opt}} = 3.080$, $h_{\text{opt}} = 1.368$, $\alpha = 0.00207$, $1 - \beta = 0.918$, $E(L) = \$4.01392/\text{hr}$

9-33. (a) $E(L) = \$16.17/\text{hr}$
       (b) $E(L) = \$10.39762/\text{hr}$

CHAPTER 10

10-1. $\text{UCL}_{\text{Phase 2}} = 14.186$, $\text{LCL}_{\text{Phase 2}} = 0$

10-3. $\text{UCL}_{\text{Phase 2}} = 13.186$

10-5. (a) $UCL_{Phase\ 2} = 23.882$, $LCL_{Phase\ 2} = 0$
   (b) $UCL_{chi\text{-}square} = 18.548$

10-7. (a) $UCL_{Phase\ 2} = 39.326$
   (b) $UCL_{chi\text{-}square} = 25.188$
   (c) $m = 988$

10-9. Assume $\alpha = 0.01$. $UCL_{Phase\ 1} = 32.638$, $UCL_{Phase\ 2} = 35.360$

10-11.

   (a) $\Sigma = \begin{bmatrix} 1 & 0.8 & 0.8 \\ 0.8 & 1 & 0.8 \\ 0.8 & 0.8 & 1 \end{bmatrix}$

   (b) $UCL_{chi\text{-}square} = 7.815$
   (c) $T^2 = 11.154$. Yes, an out-of-control signal is generated.
   (d) $d_1 = 0.043$, $d_2 = 8.376$, $d_3 = 6.154$
   (e) $T^2 = 21.800$
   (f) $d_1 = 16.800$, $d_2 = 16.800$, $d_3 = 17.356$

10-13.

   $\Sigma = \begin{bmatrix} 4.4397 & -0.0163 & 5.3947 \\ -0.0163 & 0.0010 & -0.0141 \\ 5.3947 & -0.0141 & 27.5994 \end{bmatrix}$

10-15. $\lambda = 0.1$, ARL_1 is between 7.22 and 12.17.

10-17. $\lambda = 0.2$ with $UCL = H = 9.65$. ARL_1 is between 5.49 and 10.20.

10-19. Significant variables for y_1 are x_1, x_3, x_4, x_8, and x_9.
   Control limits for y_1 model I chart: $CL = 0$, $UCL = 2.105$, $LCL = -2.105$
   Control limits for y_1 model MR chart: $CL = 0.7914$, $UCL = 2.586$, $LCL = 0$

   Significant variables for y_2 are x_3, x_8, and x_9.
   Control limits for y_2 model I chart: $CL = 0$, $UCL = 7.581$, $LCL = -7.581$
   Control limits for y_2 model MR chart: $CL = 2.850$, $UCL = 9.313$, $LCL = 0$

10-21. (a) $z_1 = \{0.29168, 0.29428, 0.19734, 0.83902, 3.20488, 0.20327, -0.99211,$
   $-1.70241, -0.14246, -0.99498, 0.94470, -1.21950, 2.60867, -0.12378, -1.10423,$
   $-0.27825, -2.65608, 2.36528, 0.41131, -2.14662\}$

CHAPTER 11

11-3. Process is adjusted at observations 3, 4, 7, and 29.

11-5. $m = 1$, $Var_1 = 147.11$, $Var_1/Var_1 = 1.000$
   $m = 2$, $Var_2 = 175.72$, $Var_2/Var_1 = 1.195$
   $m = 3$, $Var_3 = 147.47$, $Var_3/Var_1 = 1.002$
   $m = 4$, $Var_4 = 179.02$, $Var_4/Var_1 = 1.217$
   $m = 5$, $Var_5 = 136.60$, $Var_5/Var_1 = 0.929$
   . . . Variogram stabilizes near 1.5
   $r_1 = 0.44$, $r_2 = 0.33$, $r_3 = 0.44$, $r_4 = 0.32$, $r_5 = 0.30$, . . . Sample ACF slowly decays.

11-7. In each control scheme, adjustments are made after each observation following observation 2.
   There is no difference in results; variance for each procedure is the same.

11-9. (b) Average is closer to target (44.4 vs. 46.262), and variance is smaller (223.51 vs. 78.32).
       (c) Average is closer to target (47.833) and variance is smaller (56.40).

CHAPTER 12

12-1. Glass effect: $F_0 = 273.789$, P value $= 0.0000$
       Phosphor effect: $F_0 = 8.842$, P value $= 0.0044$
       Glass_Phosphor interaction: $F_0 = 1.263$, P value $= 0.3178$

12-3. Normality assumption is reasonable. Constant variance assumption is reasonable.

12-5. Plots of residuals versus factors A and C show unequal scatter. Residuals versus predicted shows variance not constant. Residuals are approximately normal.

12-7. Largest effect is factor A. Other large effects are ABC, D, ABD, AB, AD, ABCD.

12-9. Block 1: (1), ab, ac, bc, ad, bd, cd, ae, be, ce, de, $abcd$, $abce$, $abde$, $acde$, $bcde$
       Block 2: a, b, c, d, e, abc, abd, acd, bcd, abe, ace, bce, ade, bde, cde, $abcde$

12-11. (b) $I = ACE = BDE = ABCD$, $A = CE = BCD = ABDE$, $B = DE = ACD = ABCE$, $C = AE = ABD = BCDE$, $D = BE = ABC = ACDE$, $E = AC = BD = BCDE$, $AB = CD = ADE = BCE$, $AD = BC = ABE = CDE$
       (c) $l_A = -1.525$, $l_B = 5.175$, $l_C = 2.275$, $l_D = -0.675$, $l_E = 2.275$, $l_{AB} = 1.825$, $l_{AD} = -1.275$
       (d) With only main effect B: $F_0 = 8.88$, P value $= 0.0246$
       (e) Residuals plots are satisfactory.

12-13. (a) Model: $F_0 = 7.440$, P value $= 0.0028$
       Curvature: $F_0 = 18.19$, P value $= 0.0007$
       Lack of fit: $F_0 = 0.3562$, P value $= 0.915$

12-15. (a) $l_A = 23.85$, $l_B = -0.25$, $l_C = 40.30$, $l_D = -1.20$, $l_{AB} = 0.55$, $l_{AC} = 36.40$, $l_{AD} = -1.00$
       (b) Model with C, AC, A: $F_0 = 1854$, P value $= 0.0001$
       Curvature: $F_0 = 1.108$, P value $= 0.3275$
       Lack of fit: $F_0 = 1.297$, P value $= 0.4327$

CHAPTER 13

13-1. (b) $\Delta x = 1$, $\Delta x_2 = 0.6$

13-3. (a) CCD with $k = 2$ and $\alpha = 1.5$. The design is not rotatable.
       (b) $y = 160.9 - 58.294x_1 + 2.412x_2 - 10.855x_1^2 + 6.923x_2^2 - 0.750x_1x_2$
       (c) $x_1 = +1.5$, $x_2 = -0.1112$, $\hat{y} = 49.24$
       (d) Temp $= 842.90$, Time $= 7.05$

13-5. (a) CCD with $k = 2$ and $\alpha = 1.4$. The design is rotatable.
       (b) $y = 13.727 + 0.298x_1 - 0.407x_2 - 0.125x_1^2 - 0.079x_2^2 + 0.055x_1x_2$
       $x_1 = 0.575$, $x_2 = -1.321$ gives $\hat{y} = 14.22$

13-7. (a) The design is resolution IV with $A = BCD$, $B = ACD$, $C = ABD$, $D = ABC$, $E = ABCDE$, $AB = CD$, $AC = BD$, $AD = BC$, $AE = BCDE$, $BE = ACDE$, $CE = ABDE$, $DE = ABCE$, $ABE = CDE$, $ACE = BDE$, $ADE = BCE$.

(b) Factors A, B, E and interaction BE affect mean free height.

(c) Factors A, B and interaction CE affect variability in free height.

(e) A 2^{5-1}, resolution V design can be generated with $E = \pm ABCD$.

13-9. (b) $\bar{y}$ appears to be affected by factors A and B, while SN_N appears to be affected by factors B and C.

(c) $\log(S^2)$ appears to be affected by factors B and C in about the same proportion as for SN_N.

CHAPTER 14

14-1. Two points on OC curve are $P_a\{p = 0.007\} = 0.95190$ and $P_a\{p = 0.080\} = 0.08271$.

14-3. (a) Two points on OC curve are $P_a\{d = 35\} = 0.95271$ and $P_a\{d = 375\} = 0.10133$.

(b) Two points on OC curve are $P_a\{p = 0.0070\} = 0.9519$ and $P_a\{p = 0.0750\} = 0.1025$.

(c) Difference in curves is small. Either is appropriate.

14-5. $n = 80, c = 7$

14-7. Different sample sizes offer different levels of protection. Consumer is protected from an LTPD = 0.05 by $P_a\{N = 5000\} = 0.00046$ or $P_a\{N = 10,000\} = 0.00000$, but pays for high probability of rejecting acceptable lots (i.e., for $p = 0.025$, $P_a\{N = 5000\} = 0.294$ while $P_a\{N = 10,000\} = 0.182$).

14-9. AOQL = 0.0234

14-11. (a) Two points on OC curve are $P_a\{p = 0.016\} = 0.95397$ and $P_a\{p = 0.105\} = 0.09255$.

(b) $p = 0.103$

(d) $n = 20, c = 0$. This OC curve is much steeper.

(e) For $c = 2$, Pr{reject} = 0.00206, ATI = 60. For $c = 0$, Pr{reject} = 0.09539, ATI = 495

14-13. (a) Constants for limit lines are: $k = 1.0414$, $h_1 = 0.9389$, $h_2 = 1.2054$, and $s = 0.0397$.

(b) Three points on OC curve are $P_a\{p_1 = 0.01\} = 1 - \_ = 0.95$, $P_a\{p_2 = 0.10\} = \beta = 0.10$, and $P_a\{p = s = 0.0397\} = 0.5621$.

14-15. AOQ = $[P_a \_ p \_ (N - n)]/[N - P_a\_(np) - (1 - P_a)\_(Np)]$

14-17. Normal: sample size code letter = H, $n = 50$, Ac = 1, Re = 2
Tightened: sample size code letter = J, $n = 80$, Ac = 1, Re = 2
Reduced: sample size code letter = H, $n = 20$, Ac = 0, Re = 2

14-19. (a) Sample size code letter = L
Normal: $n = 200$, Ac = 3, Re = 4
Tightened: $n = 200$, Ac = 2, Re = 3
Reduced: $n = 80$, Ac = 1, Re = 4

14-21. (a) Minimum cost sampling effort that meets quality requirements is 50,001 $\_ N \leq 100,000$, $n = 65, c = 3$.

(b) ATI = 82

CHAPTER 15

15-1. (a) $n = 35, k = 1.7$

(b) $Z_{LSL} = 2.857 > 1.7$, so accept lot

(c) From nomograph, $P_a\{p = 0.05\} \_ 0.38$

15-3. $\text{AOQ} = P_a \_ p \_ (N - n)/N$
$\text{ATI} = n + (1 - P_a) \_ (N - n)$

15-5. From MIL-STD-105E, $n = 200$ for normal and tightened and $n = 80$ for reduced. Sample size required by MIL-STD-414 are considerably smaller than those for MIL-STD-105E.

15-7. Assume inspection level IV. Sample size code letter $= O$, $n = 100$, $k_{normal} = 2.00$, $k_{tightened} = 2.14$. $Z_{LSL} = 3.000 > 2.00$, so accept lot.

15-9. (a) From nomograph for variables: $n = 30$, $k = 1.8$
(b) Assume inspection level IV. Sample size code letter $= M$
Normal: $n = 50$, $M = 1.00$
Tightened: $n = 50$, $M = 1.71$
Reduced: $n = 20$, $M = 4.09$

σ known permits smaller sample sizes than $\_$ unknown.

(c) From nomograph for attributes: $n = 60$, $c = 2$
Variables sampling is more economic when $\_$ is known.

(d) Assume inspection level II. Sample size code letter $= L$
Normal: $n = 200$, Ac $= 5$, Re $= 6$
Tightened: $n = 200$, Ac $= 3$, Re $= 4$
Reduced: $n = 80$, Ac $= 2$, Re $= 5$
Much larger samples are required for this plan than others.

15-11. (a) Three points on OC curve are $P_a\{p = 0.001\} = 0.9685$, $P_a\{p = 0.015\} = 0.9531$, and $P_a\{p = 0.070\} = 0.0981$.
(b) ATI $= 976$
(c) $P_a\{p = 0.001\} = 0.9967$, ATI $= 131$
(d) $P_a\{p = 0.001\} = 0.9958$, ATI $= 158$

15-13. $i = 4$, $P_a\{p = 0.02\} = 0.9526$

15-15. For $f = 1/2$, $i = 140$: $u = 155.915$, $v = 1333.3$, AFI $= 0.5523$, $P_a\{p = 0.0015\} = 0.8953$
For $f = 1/10$, $i = 550$: $u = 855.530$, $v = 6666.7$, AFI $= 0.2024$, $P_a\{p = 0.0015\} = 0.8863$
For $f = 1/100$, $i = 1302$: $u = 4040.00$, $v = 66666.7$, AFI $= 0.0666$, $P_a\{p = 0.0015\} = 0.9429$

15-17. For $f = 1/5$, $i = 38$: AFI $= 0.5165$, $P_a\{p = 0.0375\} = 0.6043$
For $f = 1/25$, $i = 86$: AFI $= 0.5272$, $P_a\{p = 0.0375\} = 0.4925$

Index

100 percent inspection, 11, 677
2^2 factorial design, 592
2^k factorial design, 592, 598
2^{k-1} fractional design, 622, 625
2^{k-p} fractional factorial design, 628

A

acceptable quality level (AQL), 686, 706, 731
acceptance control chart, 449, 453
acceptance sampling, 11, 15, 673, 676, 724
acceptance sampling plans, 673, 706
accuracy of a confidence interval, 96
accuracy versus precision of a gauge, 383
action limits, 165
actual capability of a process, 363
adaptive sampling in process control, 166, 175, 458, 478
adjustment charts, 555, 557, 560
aesthetics and quality, 3, 34
AIAG, see Automotive Industry Action Group
algorithmic cusum, see tabular cusum
algorithmic SPC, see engineering process control and SPC
alias, 623
alternate fraction, 624, 630
alternative hypothesis, 93
analysis of variance, 131, 384, 570, 583, 593, 595
analysis procedure for factorials, 597
ANSI/ASQ Z1.4, 675, 707, 735
ANSI/ASQ Z1.9, 675, 731, 735
AOQL, see average outgoing quality limit
appraisal costs, 28
approximations to distributions, 74
AQL, see acceptable quality level
ARIMA models, 468, 478, 548
ARL, 167, 169, 235, 236–239, 253, 306–308, 416, 431, 456, 526
assignable cause, 155

ATS, 168, 169, 238
attribute control charts, see control charts for attributes
attribute sampling plans, 678, 680
attributes data, 7, 152, 283–284
autocorrelated process data, 163, 458, 548, 552, 554
autocorrelation, 163, 458, 460
autocorrelation function, 460, 461
automatic control, 15, 555, 556, see also engineering process control, feedback control
Automotive Industry Action Group, 372, 382
autoregressive integrated moving average models, see ARIMA models
autoregressive models, 464, 467
average outgoing quality, 691, 702
average outgoing quality limit, 693, 718, 740
average run length, see ARL
average sample number (ASN), 698, 706, 742
average time to signal, see ATS
average total inspection (ATI), 693, 702

B

batch means control chart, 473
Bernoulli distribution, 89
Bernoulli trials, 58, 89
between/within control charts, 264
binomial approximation to the hypergeometric, 75
binomial distribution, 53, 58, 89, 285
blocking in designed experiments, 620
bounded adjustment charts, 557
Box-Jenkins adjustment charts, 557
box plot, 48, 49

C

c chart, 284, see also control chart for nonconformities
cascade process, 531

cause and effect diagram, 154, 181, 186, 313
center points in factorial designs, 615, 618
central composite design, 645, 649
central limit theorem, 68
chain sampling, 688, 724, 735
chance cause of variation, 154
change-in-mean effect (in EVOP), 665, 667
changing sample size on control charts, 224
check sheet, 154, 177, 188, 198
chemometrics, 542
chi-square control chart, 514
chi-square distribution, 85
ChSP-1, 738, see also chain sampling
clearance number, 738
combined array design, 654
common cause, 155, see also chance cause of variation
company-wide quality control, 17
comparison of means, 95, 99, 115, 121, 125
completely randomized experiment, 133
compound Poisson distribution, 318
concurrent engineering, 8
confidence interval, 96, on population means, 96, 101, 117, 121, on population proportions, 108, 130, on population variances, 104, 127
confidence intervals on process capability ratios, 361, 367–370
conformance to standards and quality, 3, 4, 34
confounding in factorial designs, 620
consumer's risk, 94
continuous probability distributions, 62, exponential, 69, gamma, 71, normal, 63, uniform, 53, Weibull, 73
continuous sampling, 724, 737, 740
contour plot, 575, 608
contrast, 593, 621
control and capability, 361, 373, 375

control chart, 11, 13, 154, 352, statistical basis of, 156

control chart design, see guidelines for designing control charts

control chart for fraction nonconforming, 284, 286, 287, 295

control chart for individuals, 249, 253

control chart for nonconformities, 284, 308, 309

control chart for nonconformities per unit, 284, 313, 319

control chart performance, 157, 233–239

control charts and gage capability analysis, 377

control charts and hypothesis testing, 157, 159

control charts and process capability analysis, 161, 216, 352, 373, 375

control charts for attributes, 152, 161, 283

control charts for low count rates, 325

control charts for variables, 152, 161, 206

control charts as estimating devices, 161, 216, 243

control ellipse, 513

control limits, relationships to specifications limits and natural tolerance limits, 220

controllable factor, 376, 572, 651, 654, 662

correlated data, see autocorrelated process data

correlation, 184

correlation and causality, 184

covariance matrix, 510, 532

C_p, 216, 357, 360, 362, 368

C_{pc}, 364, 370

$C_p(q)$, 365

C_{pm}, 366

C_{pk}, 363, 369

critical region for a statistical test, 94

crossed array design, 652

CSP-1, 738, see also continuous sampling

cube plot, 607, 608

cumulative sum control chart or cusum, 254, 404, 406, 413, 415, 417, 421, 422, 448

curvature, 616, 617, 665

cycle (in EVOP), 664

cycle pattern on control chart, 173, 230

D

data transformation, 140

decision interval for the cusum, 410

defect concentration diagram, 154, 182

defective, 8, 283

defects, 8

defining relation for a fractional factorial, 622, 629

definition of quality improvement, 6

definitions of quality, 2, 4

demerit systems and control charts, 322–324

Deming philosophy, 18

Deming's 14 points, 18–20

demonstration of process capability, 370

derivative control actions, 562

descriptive statistics, 37

design generator, 622, 629, 630

design projection, 609, 627

design resolution, 627

designed experiments, 11, 12, 184, 191, 352, 376, 383, 569

destructive testing, 679

deviation from nominal control chart, 443

difference control chart, 496

digidot plot, 42

dimensions of quality, 2, 34

discrete probability distributions, 56, binomial, 52, 58, geometric, 62, hypergeometric, 56, negative binomial, 62, Pascal, 61, Poisson, 60

distributions: binomial, 52, 58, exponential, 69, gamma, 71, geometric, 62 hypergeometric, 56, negative binomial, 62, normal, 63, Pascal, 61, Poisson, 60, uniform, 53, Weibull, 73

Dodge-Romig sampling plans, 678, 694, 714

double sampling plans for attributes, 679, 694

drifting process mean, 547

durability, 3, 34

E

economic design of a control chart, 479, 484, 493

economic-statistical control chart design, 493

efficiency of the range in estimating variability, 93

engineering process control (EPC), 15, 163, 477, 546, 562, 563, see also feedback control, integral control

engineering process control and SPC, 477, 546

estimate of a parameter, 91

evolutionary operation (EVOP), 640, 662, 663

EWMA as a predictor of process level, 435, 468, 548, 550, 552

EWMA control chart, 254, 404, 406, 425, 433, 448, 457, 458, 468, 472

exponential distribution, 69, 325–326

exponentially weighted mean square (EWMS), 434

exponentially weight moving average control chart, see EWMA control chart

exponentially weighted moving variance, 435

external failure costs, 29

F

factorial design, 13, 14, 192, 384, 570, 575, 579

failure costs, 29

failure rate, 70

false alarm on a control chart, see type I error

fast initial response on the cusum, 419

fast initial response on the EWMA, 434

F-distribution, 87

features, 3, 34

feedback control, 15, 163, 546, 547, 548, 561, 562

Feigenbaum philosophy, 20

fill control, 496

first-order autoregressive model, 464

first-order integrated moving average model, 468

first-order mixed model, 467

first-order moving average model, 467

first-order response surface model, 608, 641

fitness for use and quality, 4

fixed effects analysis of variance, 133

flow charts, 194, 197

fraction defective control charts, see fraction nonconforming control charts

fraction nonconforming control charts, 284, 286, 287, 295

fractional factorial design, 192, 570, 622

frequency distribution, 43, 45

G

gage capability, 349, 377, 383, see also measurement systems capability analysis

gage R & R studies, 382, see also repeatability, reproducibility, and measurement systems capability analysis

gamma distribution, 71

generalized variance, 533

geometric distribution, 62, 168, 237

geometric moving average, see EWMA control chart

goodness-of-fit testing, 110

graduated response to out-of-control signals, 175

group control charts, 455

guidelines for designing control charts, 222, 295–297, 415–417, 431–432, 526

guidelines for designing experiments, 577

guidelines for implementing control charts, 333–339

guidelines for using acceptance sampling, 680–683, 728, 731

H

headstart procedure, see fast initial response for cusum and EWMA

histogram, 42, 45, 154, 352

Hotelling T^2 control chart, see T^2 control chart

hypergeometric distribution, 56

hypothesis testing framework for control charts, 157, 547

hypothesis testing, 94, on population means, 95, 99, 115, 121, 125, on population proportions, 105, 129, on population variances, 103, 127, on process capability ratios, 370

I

ideal OC curve, 685

implementation of SPC, 184

implementing TQM, 34

in-control process, definition, 154, 162

incoming inspection, 15

inertial forces in manufacturing processes, 459

inner array design, 652

inspection error, 677

inspection unit, 308, 313, 316

integral control, 548, 550

interaction, 580, 592, 600, 601

internal failure costs, 29

interquartile range, 42

ISO 2856, see MIL STD 105E and ANSI/ASQC Z1.4

ISO 9000, 22

J

joint monitoring of variables, see multivariate quality control

Juran philosophy, 20

K

kurtosis, 356

L

latent structure methods for process monitoring, 535, see also principal components and partial least squares

lean manufacturing, 25

least squares, 608

legal liability, 32

level of significance of statistical test, 98

limiting quality level, see lot tolerance percent defective

lot disposition, 676, 690

lot formation in acceptance sampling, 679

lot screening, 692

lot sentencing, 676

lot tolerance percent defective, 686, 714, 716

low count rates of defects, 325

lower confidence limit, 96

lower specification limit, 7

LTDP, see lot tolerance percent defective

M

magnificent seven, the, 154, 177

main effect of a factor, 580, 592, 600

Malcolm Baldrige National Quality Award, 10

management controllable defects, 291

managerial breakthrough, 20

managing quality costs, 30

manipulatable variable, 547, 548, 549

manual adjustment chart, 556

matrix of scatter plots, 537

mean absolute deviation, 471

mean of a distribution, 54

mean squares, 136, 384, 385

measurement systems capability, 349, 377, 379, 380, 382, 383, 384, 387

median control chart, 496

median of a distribution, 54

method of steepest ascent, 642, 643

MIL STD 105E, 675, 678, 705, 743

MIL STD 1235C, 741

MIL STD 414, 675, 722, 729, 735

military standard sampling plans, 675, 678, 707, 724, 731, 735

minimum variance estimate, 91

mixed variables-attributes acceptance sampling, 732

mixture patterns on a control chart, 230

mode of a distribution, 54

model adequacy checking, 588, 596, 627

modern definition of quality, 4

modified control charts, 231, 449

moving average control chart, 437

moving centerline EWMA control chart, 468, 470

moving range, 250, 258–260

moving range control chart, 250–253

multilevel continuous sampling, 743

multiple sampling plans for attributes, 679, 696, 701

multiple stream process, 454

multivariate EWMA control chart, 526–530

multivariate normal distribution, 510

multivariate quality control, 507, 508, 510, 512, 516, 520, 522, 526, 530, 532, 535, 542

N

narrow limit gaging, 498

natural tolerance limits of a process, 220, 350, 394–397

negative binomial distribution, 62, 319

Neyman Type A distribution for effects, 319

noise factors, 572

nominal value, 7

nonconforming product, 8

nonconformity, 8, see also defects

non-manufacturing applications of quality improvement methods, 180, 193–201, 262, 303–305, 328

nonnormality and process capability ratios, 364

nonnormality and control charts, 232, 254–255, 432

nonstationary process data, 163

nonvalue-add activities, 194

normal approximation to the binomial, 75

normal approximation to the Poisson, 76
normal distribution, 63, 85
normal inspection, 708
normal probability plot of effects, 611, 614, 627
normal probability plots, 111, 255, 355
normality assumption in variables control charts, see nonnormality and control charts
normality assumption in variables sampling, 725, 727
np control chart, 297
null hypothesis, 93

O
OC curve, see operating characteristic curve
off-line quality control, 14
one-factor-at-a-time experiments, 581
one-sided alternative hypothesis, 93
one-sided confidence interval, 96
one-sided cusum control chart, 421
one-sided EWMA control chart, 442
on-line quality control, 14
operating characteristic curves, 109, 233–236, 305–306, 324, 682, 684, 686, 689, 696, 700, 704, 712, 713, 726, 736, 740, 742
operation process chart, 194, 195
operator controllable defects, 291, 375
ordered stem-and-leaf display, 41
orthogonal design, 608
out-of-control action plan (OCAP), 160, 220, 312
out-of-control process, definition, 155
outer array design, 652
outgoing inspection, 15
over-the-wall approach to engineering design, 8

P
paired *t*-test, 125
parameters of a distribution, 90
Pareto analysis, 179–181, 188, 197, 312
Pareto chart, 154, 178, 188
partial least squares, 542
Pascal distribution, 61
path of steepest ascent, 642, 643
pattern recognition, 174
patterns on control charts, 172, 229–232
p chart, 284
perceived quality, 3, 34

percentiles, 42
performance (as a dimension of quality), 2, 34
phase (in EVOP), 665
PI and PID controller, 562
point estimation of parameters, 90, 91
point estimator, 91
Poisson approximation to the binomial, 75
Poisson distribution, 60, 89, 308–309
pooled estimator of variance, 118
pooled *t*-test, 119
population, 51
population mean, 54
population variance, 54
potential capability of a process, 363
power of a statistical test, 94
precision versus accuracy of a gage, 383
precision-to-tolerance (P/T) ratios, 379, 382
precontrol, 497
pre-experimental planning, 579
prevention costs, 27
principal components, 458, 520, 535–541
principal fraction, 624, 640
probability, 51
probability distribution, 40, 51, continuous, 52, discrete, 52, see also distributions
probability limits on control charts, 165, 226, 323
probability models for count data, 308–309, 318
probability plotting, 110, 352, 355
process adjustment, 547
process capability, 152, 216, 336, 349, 350
process capability analysis, 152, 216, 350, 351, 373, 375
process capability ratios (PCRs), 216, 357, 362, 364, see also C_p, C_{pk}, C_{pm}, and C_{pc}
process characterization, 574, 639
process charts, see operation process charts
process failure mechanism, 481
process gain, 548
process mapping, 194
process monitoring, see statistical process monitoring
process optimization, 575, 639
process performance indices, 372

process robustness study, 570, 639, 649, 653
process target, see target values for process
producer's risk, 94
product characterization, 351
product liability, 33
proportional control actions, 562
P-values in hypothesis testing, 98

Q
Q-charts, 448
quality (definitions), 2, 4
quality and productivity, 25
quality and strategic management, 34
quality and variability, 4, 5
quality characteristics, 6
quality costs, 26, 30
quality engineering, 7
quality improvement (definition), 6
"quality is free", 21
quality of conformance, 4
quality of design, 4
quality standards and registration, 22
quartiles, 42

R
random effects analysis of variance, 384
random sample, 83, 680
random variable, 52
randomization, 583
range, 92
range-based estimation of the standard deviation, 92, 209, 379, 381, 669
rational subgroups, 170, 209, 221, 263, 418, 432, 512
R-chart, 170, 207. 210
rectifying inspection, 15, 690, 700, 705
reduced inspection, 708
reference value for the cusum, 410
regression adjustment, 530
regression analysis, 15
regression control chart, 231
regression model, 595, 606, 607, 614
rejectable quality level, see lot tolerance percent defective
rejection region for a statistical test, 94
relative range, 92, 210
reliability engineering, 70
reliability, 2, 34
renewal reward process, 482
repeatability, 379, 387
replication, 578, 583, 611

reproducibility, 379, 387

residual analysis, 140, 588, 595

residual control charts, 457, 463, 470, 471, 530

residuals, 140, 457, 588, 627

response surface methods, 570, 576, 608, 639, 640, 653

revision of control limits, 218

rework, 5

robust parameter design, 651

robust process, 570, 639, 651

rotatable second-order design, 649

rounded adjustment charts, 560

run chart, 198–200

runs on control charts, 173

S

S control chart, 242

S^2 control chart, 248

sample autocorrelation function, 461

sample average, 46

sample fraction defective (or nonconforming), 59

sample median, 42

sample size code letters for MIL STD 105E, 708

sample size code letters for MIL STD 414, 730

sample size on control charts, 166

sample standard deviation, 47

sample variance, 45

sample, 40, 51, 83

sampling distribution, 85

sampling interval for control charts, 166, 169

sampling with replacement, 83

sampling without replacement, 83

scatter diagram, 154, 183

screening experiment, 574, 627, 639

second-order autoregressive model, 467

second-order response surface model, 616, 641, 645

sensitivity of control charts, 166, 175

sensitivity of residual control charts, 471

sensitizing rules for control charts, 166, 175–177

sequential experimentation, 579, 624, 627, 642

sequential sampling, 679, 696, 701, 704, 735

serviceability, 3, 34

setpoint, 549

setting specifications on components, 349, 388–394

Shewhart control chart, 151, 159, alternatives to, 254

Shewhart process model, 458

short-run manufacturing and SPC, 261, 444–449

signal-to-noise ratios, 652

single replicate of a factorial design, 611

single-sampling plans for attributes, 678, 682, 691

six-sigma, 17, 23–25, 452

six-sigma control chart, 452

six-sigma quality performance, 23

skewness, 356

skip-lot sampling, 724, 741

SkSP-1, 742

SkSP-2, 742

sources of variability, 222, 376

sparsity of effects principle, 611

special cause, 155

specifications, 7, 220, 349, see also upper and lower specification limits, and setting specifications on components

standard deviation of a distribution, 56

standard error of a regression coefficient, 605

standard error of an effect, 610

standard normal distribution, 64

standard values of process parameters on control charts, 228

standardized control charts, 303, 321, 446, 447

standardized cusum, 417

stationary process, 162

statistic, 84

statistical control, 154, 162

statistical hypothesis, 93

statistical inference, 37, 83, 84, 93, 114

statistical model, 133

statistical process control (SPC), 12, 13, 154, see also statistical process monitoring

statistical process monitoring, 13, 510, 546, 547, 548, 563, 564

statistical terrorism, 373

statistics, 37, 40

status chart for the cusum, 413

steepest ascent, see method of steepest ascent

stem-and-leaf display, 40, 41, 154

stratification pattern on control chart, 232

strict product liability, 33

supplier quality, 35, 261

switching rules in MIL STD 105E, 706

switching rules in MIL STD 414, 734

T

T^2 control chart, 512, 516

tabular cusum, 409, 410

tampering with the process, 554

target value for process, 7, 218, 408

t-distribution, 86

test statistic, 94

testing for curvature, 616, see also center points in factorial designs

three-sigma control limits, 158–159

three-sigma quality performance, 23

tier diagram, see tolerance diagram

tightened inspection, 708

time constant of system, 459

time-between events control charts, 325

time series analysis, 15, 462, 468

time series graph, 42

time series model, 462, 467

tolerance determination, 388, 390, 392

tolerance diagram, 190, 219

tolerance stack, 394

tool wear, 450, 494

TQM (Total Quality Management), 17, 21

tracking signals, 472

transformations and control charts, 255–258, 326

transformations and process capability ratios, 364

transmission of error, 392, 655

treatment effects, 133

trend pattern on a control chart, 231

trial control limits, 211

t-test, 99, 117

two-sided alternative hypothesis, 93

two-sided confidence interval, 96

type A OC curves, 686

type B OC curves, 686

type I error, 94, 166, 168

type II error, 94, 107

U

u-chart, 284, 313, 319

unbiased estimator, 91

uncontrollable variable, 572, 651

uncorrelated process data, 162

uniform distribution, 53

upper and lower specification limits, 7
upper confidence limit, 96

V
value engineering, 21
variability, 4, 5, 7, 17, 40
variable sample size on control charts, 166, 244, 298–303, 319–322, 478
variable sampling interval on control charts, 166, 478
variable width control limits, 248, 297, 319
variables data, 7, 152, 206, 207

variables sampling plans, 678, 722, 723, 724, 726, 729, 734, 735
variance components, 384
variance of a distribution, 54
variance model for process robustness study, 655, 656, 660
V-mask version of the cusum, 409, 423–425

W
Weibull distribution, 73, 326
warning limits, 165
Western Electric rules for control charts, 174
white noise, 163

X
x control chart, see individuals control chart
$\bar{x}$ control chart, 158, 170
$\bar{x}$ and R control charts, 207
$\bar{x}$ and S control charts, 239, 244

Y
yield control charts, 285

Z
zero acceptance number acceptance sampling plans, problems with, 689–690
zero defects, 21